B.Sc.
PRACTICAL
PHYSICS

For B.Sc. (Pass & Honours) and B. Tech Students of all Indian Universities

C.L. ARORA
Formerly Principal
D.A.V. College, Jalandhar
and
Dean of Colleges
Guru Nanak Dev University, Amritsar
Punjab

S Chand And Company Limited
(ISO 9001 Certified Company)

S Chand And Company Limited

(ISO 9001 Certified Company)

Head Office: D-92, Sector–2, Noida – 201301, U.P. (India), Ph. 91-120-4682700

Registered Office: A-27, 2nd Floor, Mohan Co-operative Industrial Estate, New Delhi – 110 044, Phone: 011-49731800

www.**schandpublishing.com**; e-mail: **info@schandpublishing.com**

Marketing Offices:

Chennai : Ph: 23632120; chennai@schandpublishing.com
Guwahati : Ph: 2738811, 2735640; guwahati@schandpublishing.com
Hyderabad : Ph: 40186018; hyderabad@schandpublishing.com
Jalandhar : Ph: 4645630; jalandhar@schandpublishing.com
Kolkata : Ph: 23357458, 23353914; kolkata@schandpublishing.com
Lucknow : Ph: 4003633; lucknow@schandpublishing.com
Mumbai : Ph: 25000297; mumbai@schandpublishing.com
Patna : Ph: 2260011; patna@schandpublishing.com

First Edition 1995
Subsequent Editions and Reprints 1997, 98, 2000, 2001, 2002, 2004, 2006, 2007, 2008, 2009, 2010, 2011, 2012, 2013, 2014, 2015, 2018 (Thrice), 2019, 2020 (Twice), 2021, 2022 (Thrice), 2023, 2024

Reprint 2025 (Twice)

ISBN: 978-81-219-0909-9 **Product Code:** H6PHP68PHYS10ENAU0XO

PRINTED IN INDIA

By Vikas Publishing House Private Limited, Plot 20/4, Site-IV, Industrial Area Sahibabad, Ghaziabad – 201 010 and Published by S Chand And Company Limited, A-27, 2nd Floor, Mohan Co-operative Industrial Estate, New Delhi – 110 044.

PREFACE TO THE TWENTY FIRST EDITION

This book has been written to meet the requirements of students of the New Three Year B.Sc. Degree Course in Indian Universities. It contains experiments on Mechanics and Properties of Matter, Heat and Statistical Physics, Optics, Sound, Magnetism, Electricity, Atomic and Molecular Physics, Nuclear Physics, Electronics and Solid State Physics according to the latest syllabi for the new three year degree course laid down by the Panjab University, Chandigarh; Guru Nanak Dev University, Amritsar; Panjabi University, Patiala; Kurukshetra University, Kurukshetra; Maharishi Dayanand University, Rohtak and Himachal Pradesh University, Simla in particular and other Indian Universities in general. The text has been thoroughly revised, rewritten and brought up-to-date by adding all the new experiments included in the latest syllabi. All experiments prescribed by Punjab Technical university have been included to make the book useful for B. Tech students. Many new drawings have been included for greater clarity. A very large number of new experiments included in the latest syllabi of various universities have been incorporated. The book is useful for engineering as well as honours students.

Some special features of this book are:-

1. According to the latest system of examination adopted by most of the Indian Universities, the candidates are required to write the theory of the experiment before starting the experimental work. With this aim in view a brief and relevant theoretical discussion of the experiment has been included in the body of the experiment proper, so that a student actually reads and understands it before he starts working. In addition, detailed theory has also been given separately wherever necessary.

2. A brief description of the apparatus required during the experiment has been given and the range of various measuring instruments has been specified so that the observations can be taken with a fair degree of accuracy.

3. The procedure has been divided into steps so that a student may be able to proceed systematically even without the direct guidance of a teacher.

4. The system of recording the observations in a tabular form has been encouraged throughout the text so that various readings can be compared at a glance.

5. The precautions have not only been given at the end of the experiment but also have been included in the instructions so that a student actually observes them even when he is performing the experiment.

6. As there is a growing tendency on the part of examiners to ask oral questions about the experiments done by a student in the examination, a number of such questions have been given at the end of each chapter for practice.

It is hoped that with all these and other features this book will prove very useful for the students.

Any suggestion towards further improvement will be thankfully acknowledged and incorporated in the next edition.

C.L. ARORA

CONTENTS

PART IX—ADDITIONAL EXPERIMENTS

PART X—LIST OF EXPERIMENTS FOR PUNJAB TECHNICAL UNIVERSITY

B.Sc.

Practical Physics

INTRODUCTION

Ours is the scientific age. Material prosperity, the conveniences of life and even life itself rests upon the development of science. Science may be defined as the systematic investigation of the physical world and its claim to be regarded as such is based largely on the method by which it searches for knowledge. The first and the fundamental step to ascertain the facts in connection with a problem under investigation is by an appeal to experience *i.e.*, by experimental investigations. Foundation of science, particularly of Physics is, therefore, based on experiment. When an experiment is to be performed we cannot rely too much on our senses of touch, sight, hearing etc. to make accurate observations. It is true that many scientific discoveries have been made in the past with what we now call the crudest of apparatus and equipment; it is the development of precision instruments and apparatus during the last several decades, that has led to discoveries of extreme practical importance and far reaching theoretical implications. Scientific work, therefore, requires making of measurements and observations with the greatest possible accuracy and reliability. In this lies the importance of the study of **Experimental Physics.**

Importance of an experiment. The experiments in a course of Practical Physics are designed to achieve definite quantitative results. The object of an experiment, in general, is

(*i*) *Determination of a physical constant*, such as specific resistance of a material.

(*ii*) *Measurement of a quantity connected with a particular body*, such as a focal length of a lens.

(*iii*) *Comparison of two quantities*, such as the electromotive forces of two cells.

(*iv*) *Testing of quantitative laws*, such as the laws of vibration of stretched strings etc.

In fact, the laws of Physics have already been verified and the physical constants have also been determined much more accurately than what can be expected of a new experimentalist in the laboratory. The reader might then ask "*what is the utility of doing such experiments* ?"

A laboratory experiment affords an opportunity to the students to familiarise themselves with various instruments which they read in theory and which find application in commerce, industry, engineering and everyday life. It further develops in them a habit of doing things *efficiently* and *regularly.* In the laboratory a closer contact is established between the teacher and the taught, and under the expert supervision of the teacher the students get a training of using their physical senses and create in them a scientific outlook. The work in the laboratory trains a student in the habit of *honesty, patience* and *calm reasoning*, which will enable him to think and act with the precision and accuracy of a scientist in after life.

Record of an experiment. The record of each experiment should be made in ink in a special laboratory note-book according to a definite scheme as given below :

(*a*) Date on which the experiment is performed.

(*b*) The name of experiment.

(*c*) The apparatus with diagram wherever necessary.
(*d*) A brief theory of the experiment with formulae.
(*e*) A concise account of the method and procedure.
(*f*) Observations.
(*g*) Calculations and result.
(*h*) Percentage error.
(*i*) Precautions.
(*j*) A brief discussion of the result with sources of error, if any.

Choice of the apparatus. The student should read the experiment carefully and make a detailed list of the apparatus required after careful thinking and should pay due regard to the purpose for which an instrument is required. For example, in an experiment '*To verify Ohm's law*' the most important point is the selection of a battery of a suitable *E.M.F.* and a voltmeter and an ammeter of a suitable range. The working table must be clean and the apparatus should be placed in an orderly manner.

Theory and procedure. Students should read the experiment before coming to the laboratory and make a record of the theory and the procedure step wise in their practical note-books.

Observations. Before starting to take the observations the necessary apparatus required in the experiment should be collected and should be arranged on the working table in such a manner that the parts which are to be read or adjusted lie at a convenient distance and it is easy to make the observations.

Physical instruments and apparatus should be used with great care. A great damage may be done by careless handling or through want of experience. Electrical instruments should be used with even greater care because they are very delicate.

While making observations the student should keep his mind free from prejudice. All the precautions should be observed and all the readings whether correct or wrong should be recorded. Even if the observation appears to be wrong it should not be set right by manipulation as it is a serious mistake and **all credit gained is lost if detected.**

Since all measurements are affected by errors of observation the student *should not* in general rely upon a single measurement for the value of a quantity such as length or time which can easily be re-measured. Several measurements should be made for such quantities and the arithmetic mean, which will give the most probable value of the quantity deducible from the observations, should be used in the calculations of the results of the experiment. Only the mean value **should not** be recorded but all observations taken should be recorded on the note-book.

In many experiments divergences from the mean value are not due to errors of observation only, but represent real differences in the quantities measured. As for example, when the diameter of a wire is measured at different places the difference may be due to the non-uniformity of the wire.

The observations should be recorded directly in the practical note-book, in the same order in which they have been taken during the course of the experiment. **Writing of observations on loose sheets should be strictly avoided.** Think well before recording an observation. Once the observation is recorded it should not be changed unless it is extremely necessary. There should be no **overwriting.** *Alterations, if any, should be shown to the professor incharge or the examiner as the case may be.*

Calculations and result. Calculations must be made neatly and legibly with the help of four figure logarithm tables so that the checking is possible. In certain experiments a large number of observations are taken to reduce error in measuring a quantity. The mean value usually gives the probable value. It is no use calculating this value to a greater degree of accuracy than that of the actual observation.

The percentage error in the result must be calculated.

The units must be specified along with the result. For example, in the determinations of density of brass the result must be recorded as 7.8 gm per c.c. and not merely as 7.8.

Note. *Intelligent work is more important than good results. Students who try to change their observations simply for getting an accurate result are not working with the true spirit of a scientist. The apparatus may not be good, but what is desired is honest work even without good results.*

SPECIMEN GRAPH

Verification of Boyles' law

Graph between P and 1/V

P	88.75	86.25	82.65	78.75	72.75	69.25	67.00	64.0
1/V	.062	060	.058	.055	.051	.049	.047	.045

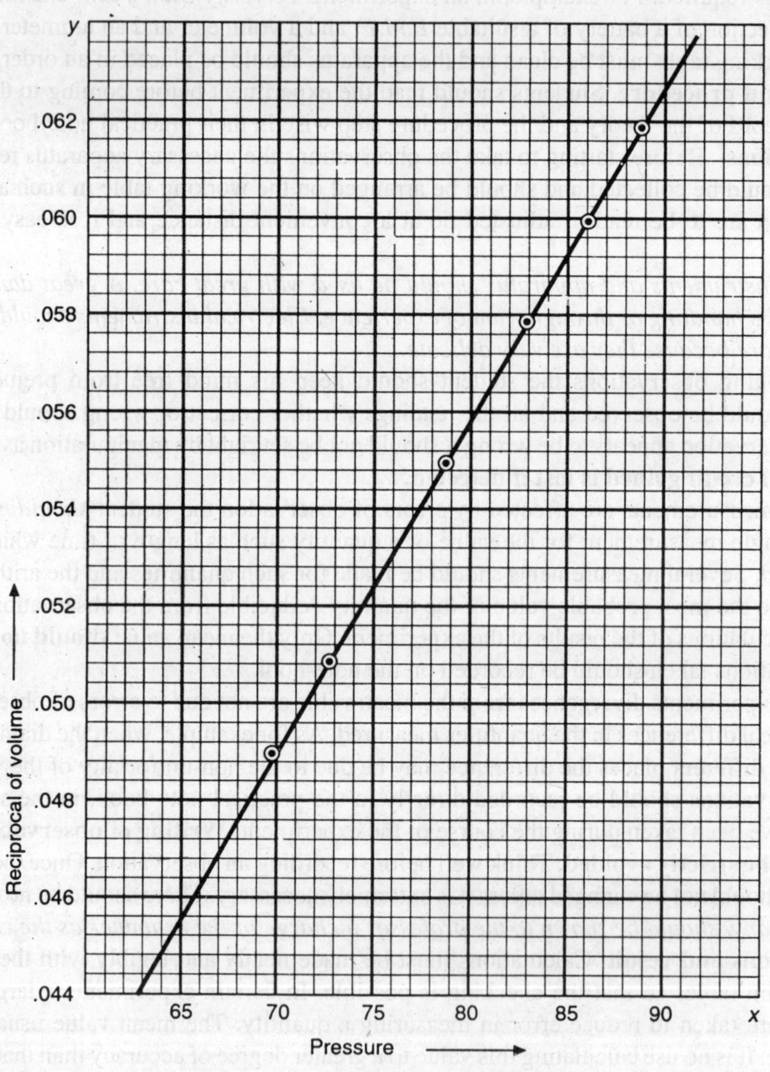

Fig. 0.1

1

Units, Errors and Graphs

1.1. Units. Every physical experiment requires the *measurement* of one or more quantities. To express a quantity completely we must give the unit in which it is measured and the *number* of times this unit is contained in it. There are three *fundamental* units, namely, those of **length, mass and time.** The systems of fundamental units in common use are:

(*i*) **The M.K.S. system.** The units of length, mass and time on this system are *metre, kilogram* and *second* respectively.

The **metre** is the distance between the centres of two transverse lines engraved upon the polished surface of a platinum-iridium bar at the temperature of melting ice, kept at the International Bureau of Weights and Measures at Sevres near Paris.

The General Conference on Weights and Measures (1960) defines the metre as the length equal to 1,650,673.73 wavelengths in vacuum of the radiation corresponding to the transition between the levels $^2p_{10}$ and $^5d_5$ of krypton –86. The wavelength of radiation being 6057.802×10^{-10} metre.

The international **kilogram** is the mass of a cylinder of platinum-iridium kept at the International Bureau of Weights and Measures at Sevres near Paris.

The mean solar **second** is $\frac{1}{24 \times 60 \times 60} = \frac{1}{86400}$ th part of the mean solar day.

The General Conference on Weights and Measures (1960) defines a second as the duration of 9,192,631,770 periods of the radiation corresponding to the transition between two hyperfine levels of the ground state of caesium-133 atom.

(*ii*) **The C.G.S. system.** The units of length, mass and time on this system are *centimetre, gram* and *second* respectively.

A **centimetre** is one hundredth $\left(\frac{1}{100}\right)$ part of a standard metre.

A **gram** is one thousandth $\left(\frac{1}{1000}\right)$ part of the international kilogram.

For all practical purposes one gram is the mass of 1 cc of pure water at 4°C.

(*iii*) **S.I. units.** The International system of units briefly written as **S.I.** has six fundamental units

(*i*)	Unit of length	Metre (m)
(*ii*)	Unit of mass	Kilogram (kg)
(*iii*)	Unit of time	Second (s)
(*iv*)	Unit of current	Ampere (A)
(*v*)	Unit of thermodynamic temperature	Kelvin (K)
(*vi*)	Unit of luminous intensity	Candela (Cd)

The metre, kilogram and second have already been defined.

The **ampere** is the strength of that constant current which flowing through each of the two parallel and straight conductors of infinite length and negligible cross-section placed exactly one metre apart in vacuum exert a mutual force of 2×10^{-7} newton per metre of their length.

The **kelvin** is $\frac{1}{273.16}$ of thermodynamic temperature of the triple point of water.

The **candela** is the luminous intensity in a perpendicular direction of a surface 1/600,000 square metre of a black body at a temperature of freezing platinum under a pressure of 101.325 newton per sq. metre.

Most of the quantities are measured in units, the magnitude of which depends upon those of fundamental units of length, mass and time. These units are called *derived units*. As an example, the S.I. unit of area is a square metre which is the area of a square of side one metre in length. Some of the derived units are given special names. For example, the *S.I. unit* of force is called a newton and the *S.I.* unit of work a *Joule*. When no special name has been given to a unit it is expressed in terms of the fundamental units, *e.g.*, acceleration is stated as *'metre per second per second'*.

1.2. Errors. There is some error or the other in every measurement we make. The errors are of two kinds:

(*i*) Errors due to known causes (systematic errors).

(*ii*) Errors due to unknown causes (random errors).

(*i*) **Errors due to known causes.** Some important causes of such errors are given below :—

(*a*) Error due to the temperature of the measuring scale being different from that at which it was graduated.

(*b*) Error due to buoyancy of air which arises when a large body is weighed.

(*c*) Error in measuring time with a stop watch, the watch running either too slow or too fast.

(*d*) Error due to radiation loss or gain in calorimetric experiments.

(*e*) Zero error in various measuring instruments.

All such errors can be eliminated by suitable methods since we know the cause *e.g.*, the error due to buoyancy can be calculated from the density of the weights, density of the substance and the density of air; the error in a stop watch can be checked by comparison with a standard chronometer; the radiation error can be minimised by suitable means or calculated by the preliminary experiment and the zero error can also be measured.

(*ii*) **Errors due to unknown causes.** If an observation is repeated a number of times by the same person under similar conditions, it is found that every time a different reading is obtained, even though the instrument used is very sensitive and accurate and the observer is an experienced one. These errors are not due to any definite cause. We, therefore, cannot depend upon a single observation.

When a large number of observations are taken it is likely that some of them may have a value slightly greater than the correct value and an equal number may have a value slightly less than the correct value. This is why it is recommended that each observation must be repeated at least three times.

The effect of random error may, therefore be minimised by taking a number of measurements of the quantity to be determined and using the *arithmetic mean* of the measured values as the best estimate of the true value of the quantity.

Thus if *n* measurements of the quantity are made, all equally reliable and the measured values are $X_1, X_2, \ldots\ldots X_n$, then

$$\text{Arithmetic mean} \quad \overline{X} = \frac{1}{n}(X_1 + X_2 + \ldots\ldots X_n) = \frac{1}{n}\sum_{i=1}^{n} X_i$$

The *true value* is defined as the mean value of an *infinite* set of measurements made under constant conditions and is denoted by μ. In practice, it is not possible to make an infinite number of observations. Hence when the number of observations is *sufficiently large*,

True value $\mu \cong$ Mean value $\overline{X}$

The *precision* with which a physical quantity is measured depends *inversely* upon the deviation or dispersion of the set of measured values X_i about their mean value $\overline{X}$. If the values are widely dispersed or the observed values have a large deviation from the true value, the precision is said to be *low*.

The deviation $\delta_i = X_i - \overline{X}$

Average deviation. The average value of the deviation of all the individual measurements from the arithmetic mean is known as *average deviation* and is denoted by $\overline{d}$

$$\therefore \quad \text{Average deviation} \quad \overline{d} = \frac{(X_1 - \overline{X}) + (X_2 - \overline{X}) \; ..... \; + (X_n - \overline{X})}{n}$$

$$= \frac{\delta_1 + \delta_2 + ... \delta_n}{n}$$

$$= \frac{\Sigma \, \delta_i}{n}$$

Standard deviation. The square root of the mean square deviation for an infinite set of measurements is known as *standard deviation* (or root mean square deviation) and is denoted by σ.

When the number of observations n is sufficiently large

$$\sigma \cong \sqrt{\frac{(X_1 - \overline{X})^2 + (X_2 - \overline{X})^2 \; .... \; (X_n - \overline{X})^2}{n - 1}}$$

$$= \sqrt{\frac{\delta_1^2 + \delta_2^2 + ... \delta_n^2}{n - 1}} = \sqrt{\frac{S}{n - 1}}$$

where $S = (X_1 - \overline{X})^2 + (X_2 - \overline{X})^2 \; .... \; (X_n - \overline{X})^2$

$$= \delta_1^2 + \delta_2^2 + ... \delta_n^2 = \sum_{i=1}^{n} \delta_i^2$$

For a fairly large sample of measurements which have a reasonably normal distribution about the mean value $\overline{X}$

$$\frac{\text{Average deviation}}{\text{Standard deviation}} = \frac{\overline{d}}{\sigma} \cong 0.80$$

Standard error. The quantity $\frac{\sigma}{\sqrt{n}}$ is known as *standard error* and is denoted by σ_m

$$\therefore \text{ Standard error} \quad \sigma_m = \sqrt{\frac{(X_1 - X)^2 + (X_2 - X)^2 \; ...... \; (X_n - X)^2}{n\,(n-1)}}$$

$$= \sqrt{\frac{S}{n\,(n-1)}}$$

The normal law of errors predicts that the probability that the mean value obtained from a finite number of observations n may be in the interval $\mu \pm \sigma_m$ is 0.68, that it may lie between $\mu \pm 2\sigma_m$ is 0.95 and for it to lie between $\mu \pm 3\sigma_m$ is 0.99.

Probable error. The probable error is a quantity e such that it is an even chance whether true value of the quantity measured differs from the mean value by an amount greater or less than e. For example, if 5.45 is the mean of all the determinations of the density of the earth and 0.25 is the probable error, then it means that there is an even chance that the density of earth lies between 5.20 (5.45 – 0.25) and 5.70 (5.45 + 0.25). It also implies that the chance that the value of the mean density of the earth may differ from the mean value by more than 4 or 5 times the probable error is very remote.

On the basis of the theory of probability it can be shown that

$$\text{Probable error} = \pm\, 0.6745 \sqrt{\frac{S}{n\,(n-1)}}$$

$$= \pm\, 0.6745 \text{ standard error}$$

where 0.6745 is a natural constant.

Let us consider the following values for the radius of curvature of the surface of a convex lens with a spherometer.

No.	*Radius of curvature*	δ	δ^2
1	15.25	–.07	.0049
2	15.42	+.10	.0100
3	15.30	–.02	.0004
4	15.20	–.12	.0144
5	15.35	+.03	.0009
6	15.40	+.08	.0064

Mean radius of curvature = 15.32

$$S = \delta_1^2 + \delta_2^2 \,....\, + \delta_n^2$$

$$= 0.0370$$

$$\text{Standard deviation } \sigma = \sqrt{\frac{S}{n-1}}$$

$$= \sqrt{\frac{0.0370}{5}} = 0.0860$$

$$\text{Standard error} \quad \sigma_m = \sqrt{\frac{S}{n\,(n-1)}}$$

$$= \sqrt{\frac{.0370}{6 \times 5}} = 0.03512$$

$$\text{Hence probable error} \pm 0.6745 \times \sigma_m = \pm\, .6745 \sqrt{\frac{0.0370}{6 \times 5}} = \pm\, 0.0237$$

$\therefore$ Radius of curvature = 15.32 ± 0.0237

1.3. Degree of accuracy. In an experiment all observations must be taken to the same "*degree of accuracy*". It is no use taking some observations to a much higher degree of accuracy than the rest of the observations because it will not make the result more accurate. The accuracy of the result is the same as that of the '*least accurate observation*'. It does not mean that the observations should not be taken accurately, but a *careful sense of proportion must be used*. For example, in the determination of specific heat by the method of mixtures it is useless to find the weight of the liquid in the calorimeter correct up to a fraction of a milligram. It is because apart from the errors due to the defects in the balance and the weights, it is not possible to measure the temperature to the desired accuracy. Great care, therefore, should be taken in reading the thermometer. To increase the accuracy practice should be made in estimating the temperature to at least 1/10th of a degree.

Percentage error. The students should have an idea of percentage error. For example, in the above experiment if the weight of the calorimeter is 50 gm an error of 5 mgm in the weight will mean only an error of .01%. The rise in the temperature is, however, only about 10°C and even if measured accurately up to one-tenth of a degree will cause an error of 1% which is 100 times as large as the error in the measurement of the weight.

1.4. Effect of combining errors. All observed values are subjected to mathematical treatment to find the results. *It is not possible to produce greater accuracy by mathematical manipulation*. The following are some of the important points:

(*i*) *Addition and subtraction.* Suppose the quantities having their *true values a* and *b* have *measured values* $a \pm \delta a$ and $b \pm \delta b$ respectively *where* δa *and* δb are their *absolute errors.* To find the error δQ in the *sum* $Q = a + b$, we have

$$Q \pm \delta Q = (a \pm \delta a) + (b \pm \delta b)$$

$$= (a + b) + (\pm \delta a \pm \delta b)$$

$$\therefore \quad \delta Q = \delta a + \delta b$$

To find the error in the difference $Q = a - b$, we have

$$Q \pm \delta Q = (a \pm \delta a) - (b \pm \delta b)$$

$$= (a - b) + (\pm \delta a \pm \delta b)$$

The maximum error in Q is again given by

$$\delta Q = \delta a + \delta b$$

Thus we find *that when two quantities are added or subtracted the absolute error in the final result is the sum of the absolute errors in the quantities.*

Example 1. Resistance $R_1 = 100 \pm 2$ ohm

Resistance $R_2 = 200 \pm 3$ ohm

Equivalent resistance when connected in series

$$R = (100 \pm 2) + (200 \pm 3)$$

$$= 300 \pm 5 \text{ ohm}$$

Example 2. Mass of a bulb with air $= 66.928 \pm .001$ gm

Mass of empty bulb $= 66.682 \pm .001$ gm

$\therefore$ Mass of air $= 0.246 \pm .002$ gm

Though the mass of air has been found by subtraction the errors are added.

(*ii*) *Multiplication and division.* Suppose the quantity $Q = ab$ and measured values of a and b are $a \pm \delta a$ and $b \pm \delta b$ respectively, then

$$Q \pm \delta Q = (a \pm \delta a)\ (b \pm \delta b) = ab \pm b\delta a \pm a\delta b \pm \delta a \delta b.$$

Dividing *L.H.S.* by Q and the *R.H.S.* by ab, as $Q = ab$, we have

$$1 \pm \frac{\delta Q}{Q} = 1 \pm \frac{\delta a}{a} \pm \frac{\delta b}{b}$$

neglecting $\pm \dfrac{\delta a}{a} \cdot \dfrac{\delta b}{b}$ which is the product of two very small quantities. Hence the *maximum* error in Q is given by

$$\frac{\delta Q}{Q} = \frac{\delta a}{a} + \frac{\delta b}{b}$$

and expressed as percentage error

$$\frac{\delta Q}{Q} \times 100 = \frac{\delta a}{a} \times 100 + \frac{\delta b}{b} \times 100$$

If the quantity $Q = a/b$, then also

$$Q \pm \delta Q = (a \pm \delta a)\ (b \pm \delta b)^{-1} = ab^{-1} + b^{-1}\ \delta a + ab^{-2}\ \delta b$$

Dividing L.H.S. by Q and R.H.S by ab^{-1} as

$$Q = a/b = ab^{-1} \quad \text{we have } 1 \pm \frac{\delta Q}{Q} = 1 \pm \frac{\delta a}{a} \pm \frac{\delta b}{b}$$

Hence the maximum error in Q

$$\frac{\delta Q}{Q} = \frac{\delta a}{a} + \frac{\delta b}{b}$$

and expressed as percentage

$$\frac{\delta Q}{Q} \times 100 = \frac{\delta a}{a} \times 100 + \frac{\delta b}{b} \times 100.$$

Thus *if the result Q is the product or quotient of two measured quantities a and b the fractional or percentage error in Q is the sum of the fractional or percentage errors in a and b.*

Example 1. Capacity of a capacitor $C = 2 \pm 0.1$ Farad

Applied voltage $V = 25 \pm 0.5$ Volt

$\therefore$ Charge on the capacitor $Q = CV = 2 \times 25 = 50$ coulomb

Percentage error in $C = \frac{0.1}{2} \times 100 = 5\%$

Percentage error in $V = \frac{0.5}{25} \times 100 = 2\%$

$\therefore$ Percentage error in $Q = 5 + 2 = 7\%$

or Error in $Q = 50 \times \frac{7}{100} = 3.5$ coulomb

Hence charge on the capacitor $Q = 50 \pm 3.5$ coulomb.

Example 2. Mass of an object $M = 345.1 \pm 0.1$ gm

Volume of the object $V = 41.55 \pm 0.05\ \text{cm}^3$

Density of the object $D = \frac{M}{V} = \frac{345.1}{41.55}$

$= 8.31\ \text{gm/cm}^3$

Percentage error in $M = \frac{0.1}{345.1} \times 100 = 0.03\%$

Percentage error in $V = \frac{0.05}{41.55} \times 100 = 0.12\%$

$\therefore$ Percentage error in $D = 0.03 + 0.12 = 0.15\%$

or Error in density $D = 8.31 \times \frac{0.15}{100} = 0.012\ \text{gm/cm}^3$

Hence density of the object $= 8.31 \pm .012\ \text{gm/cm}^3$

As a corollary, it follows that if the *result Q is some power n of a measured quantity a, then the fractional or percentage error in Q is n times the fractional or percentage error in a.*

(*iii*) *General case.* The final result of an experiment is generally calculated from a set of observations taken with a number of measuring instruments and connected by means of a formula, the process requiring the use of multiplication and division. It can be shown that all the quantities do not affect the result equally. Some affect more than others do. For example, consider a quantity Q which depends upon other quantities a, b and c connected by the relation

$$Q = ka^x b^y c^z$$

where k is a constant and x, y, z are numerical values, then

$$\frac{\delta Q}{Q} = x\frac{\delta a}{a} + y\frac{\delta b}{b} + z\frac{\delta c}{c}$$

or Percentage error in $Q = x$ times the % error in a plus y times the % error in b plus z times the % error in c.

Thus the effect on the final result is greater for the factor having higher power than that having smaller power. If in the above case x is the higher power the quantity a must be measured with the greatest possible accuracy to make δa as small as possible.

The area of a square is given by

$$A = l^2$$

$$\therefore \quad \frac{\delta A}{A} = 2\frac{\delta l}{l}$$

In other words, if the length of the side of a square is measured and found to be 10 cm and this measurement has a possible percentage error of $\pm$ 0.5, then the value of the area will have a possible error of $2 \times 0.5 = 1\%$.

To find the value of g by a simple pendulum, we have

$$g = 4\pi^2 \frac{l}{t^2}$$

$$\therefore \quad \frac{\delta g}{g} = \frac{\delta l}{l} - 2\frac{\delta t}{t}$$

Hence a slight error in the measurement of the time period t will produce double the error in the result. The time period, therefore, should be measured very accurately.

To find the viscosity of a liquid by Poiseuille's formula, we have

$$\eta = \frac{\pi p r^4}{8lv}$$

$$\therefore \quad \frac{\delta \eta}{\eta} = \frac{\delta p}{p} + 4\frac{\delta r}{r} - \frac{\delta l}{l} - \frac{\delta v}{v}$$

In the case of random errors *i.e.* errors due to unknown causes the sign should be chosen in such a manner as to give maximum error.

$$\therefore \quad \frac{\delta \eta}{\eta} = \frac{\delta p}{p} + 4\frac{\delta r}{r} + \frac{\delta l}{l} + \frac{\delta v}{v}.$$

It is to be noted that an error in the measurement of the radius of the capillary tube is magnified 4 times in the final result.

This is why in experiments where the quantities to be measured are raised to some power, *the quantity with the highest power should be measured with greatest possible accuracy.*

Significant figures. The number of significant figures to which the final result of an experiment should be stated depends upon the nature of the experiment and the accuracy with which the various measurements have been made. For example, in the measurements of the value of g at a place it is not desirable to write it as 980.432 cm/sec^2 because the length and the time period are not measured with the desired accuracy. Moreover, the various other sources of error, *e.g.*, friction due to air, amplitude not in one plane etc., cannot be removed.

In cases where the numbers dealt with are very large as 1053000000000 it may be written as 10.5×10^{11} if the accuracy is 1 in 100, and as 10.53×10^{11} if the accuracy is 1 in 1000.

Thus the number of significant figures automatically gives the accuracy of the result.

Result. Students often attach a great importance to the result of an experiment. They are usually alarmed when they find a large error in the result. But it is the *percentage error* that matters and not the *absolute error* as is clear from examples given below :

(*i*) In an experiment on the determination of '*g*' the acceleration due to gravity, the result is 9.6 ms^{-2}. This result has an error of $9.8 - 9.6 = 0.2\ ms^{-2}$.

(*ii*) In an experiment on the determination of specific gravity of common salt the result is 2.37 instead of 2.17. This result has an error of $2.37 - 2.17 = 0.20$.

It is clear that the absolute error in both the experiments is the same, but it does not mean that percentage accuracy is the same.

The percentage accuracy in the first case is $\frac{0.2}{9.8} \times 100 = 2\%$ (nearly)

and in the second case it is $\frac{0.20}{2.17} \times 100 = 9\%$ (nearly)

1.5. Graph. Physical laws express relationship between various physical quantities. These relationships can easily be expressed by means of graphs.

A graph is a pictorial representation of experimental data in the form of a curve which makes visible, at a glance, the main features of the relationship between two variables.

One of the two variable quantities is varied at will, generally in convenient equal steps and the corresponding values of the second variable quantity are observed experimentally. The corresponding values of the two quantities obtained are plotted on a squared paper and a graph is obtained. The quantity which is varied at will is called **independent variable** and the other which varies as a result of variation in the first is called **dependent variable.** As a general rule independent variable is plotted along the *X*-axis while the dependent variable is plotted along the *Y*-axis.

Plotting of a graph. The following rules must be observed to plot a good graph.

1. *Find the independent and the dependent variables. Represent the independent variable along the X-axis and the dependent variable along the Y-axis.*

Examples. (*i*) In an experiment with a simple pendulum the time period is measured for different lengths by changing the length. The *length* is the independent variable and *the time period* the dependent variable.

(*ii*) In the verification of Boyle's law where the volume is measured for various values of pressure, *the pressure* is the independent variable and *volume* (or $1/V$) the dependent variable.

(*iii*) In the experiment on the determination of Young's modulus where the extension of the wire is measured for various loads, *load* is the independent variable and extension the dependent variable.

2. *Determine the range of each of the variables* and count the number of big squares available for each along the two axes.

3. **Scale.** Choose a convenient scale for both the variables. It is not essential to have *the same scale for both.* The scale should neither be too narrow nor too wide. If the scales are too wide the irregularities due merely to experimental errors will become very much magnified and the shape of the graph will not be proper. If the scales are too narrow the points corresponding to accurate observations cannot be plotted with the same degree of accuracy. For convenience one big square should represent 1, 2, 10 or their multiples by any positive or negative power of 10.

A study of the specimen graph between P and $1/V$ for air in Fig. 0.1 shows that the pressure varies from 64.00 to 88.75 cm. Since pressure is the independent variable it is represented along the *X*-axis. The range of pressure lies roughly between 60 cm and 90 cm. There are seven big squares and the most convenient scale is to represent *one* big square by 5 cm thus utilising 6 squares.

Reciprocal of volume $1/V$ is the dependent variable and is represented along the *Y*-axis. The value of $1/V$ varies from .045 to .062. There are 10 big squares and the most convenient scale is to represent *one* big square by .002 thus utilising 9 squares.

Students sometimes unnecessarily trouble themselves in plotting and also make it less easy to read the values from the curves by selecting an inconvenient scale, *e.g.*, in the above case one big square equal to 4 will utilise more squares but will be an unwise choice.

4. **Origin.** If the relation between the two variables begins from zero or if it is desired to find the zero position on one of the variables where actual determination is not possible, then *zero must be taken at the origin on both the scales.*

In all other cases it is not necessary to take zero at the origin. When it is desired simply to test the proportionality of one quantity to the other the origin on both the axes should represent a **quantity a little less than the smallest value of the corresponding variable.**

In the specimen graph of Boyle's law the origin should be 60 along the X-axis and .044 along the Y-axis which represent pressure P and the reciprocal of volume $1/V$ respectively.

Of course, it will be more complicated to take the actual least value of the two variables at the origin.

5. **The quantities to be plotted along X-axis and Y-axis should be given on the top of the graph in a tabular form.**

6. Do not write all the values along the respective axes but only mark the ends of the thicker lines to indicate the values of the variables in round numbers. For example, in the graph Fig. 0.1 round values of pressure 60, 65, 70 etc. and those of $1/V$ .044, .046, .048 etc., are indicated along the respective axes.

7. Draw a small circle round each of the plotted points or put a cross mark neatly. Join the plotted points by a regular free hand curve and not by zig-zag lines from one point to the other. Maximum number of points should lie on the curve and the remaining points should be situated symmetrically on either side of it.

8. Take at least six observations (preferably more) whenever a graph is to be plotted. The range of observations should be as wide as possible.

Uses. 1. The graphs are used to extrapolate certain quantities beyond the limits of observations of the experiment, *e.g.* absolute zero can be determined by plotting a graph between pressure and temperature at constant volume. By extrapolating, the temperature at which pressure becomes zero is found and that gives the value of the absolute zero. In this case, it is evident that origin must be zero on both the axes.

2. Verification of certain laws can best be represented by graphs. There is always a mathematical relationship between the two variables. It may be simple or a complex relation. For the purpose of verification of a law the graphs may be divided into two groups.

(*i*) **Straight line graphs.** A straight line graph is very easy to interpret. The equation of a straight line is

$$y = mx + c$$

where y is the dependent variable, x the independent variable, m the slope or the gradient and c the intercept on the Y-axis.

If the graph plotted is a straight line, we can evaluate the constants m and c and thus we can find the relation that exists. For example, if we plot a graph between E the potential difference across the ends of a conductor and I the current through it the graph is a straight line passing through the origin. If the slope of the straight line is represented by a constant R (intercept c being zero) the relation can be put in the form

$$\frac{E}{I} = R \text{ (a constant)}$$

This is the statement of Ohm's law.

(*ii*) **Graphs which are not straight lines.** If the graph is not a straight line, there still may be a simple equation that gives the relation between the independent and the dependent variables. The relation can be converted into linear one by changing the variable, thus giving a straight line graph.

It should, however, be clearly understood that we do not at all change the *quantity* which is variable in the experiment. For example, in the experiment on verification of Boyle's law the independent variable is the *pressure P* and the dependent variable is the *volume V*. According to Boyle's law

$$PV = k \text{ (constant)}$$

The graph between P and V will not be a straight line. To get a straight line the relation is put in the form

$$P = k \cdot \frac{1}{V}$$

which shows that a graph between P and $1/V$ will be a straight line. It is for this reason that the specimen graph has been plotted between P and $1/V$.

Similarly in the verification of Newton's law of cooling a straight line graph is obtained by plotting *log of excess of temperature* and *time* and thus the law is verified.

In practice the matters often are not as simple as given above particularly when we do not know the relation that exists and we are required to find it out. If y varies as some power of x, say

$$y \propto x^m$$

then the best method is to plot log y (dependent variable) against log x (independent variable). Slope of the straight line gives the power of x in the original relation. To understand the theory suppose the relation is

$$y \propto x^m$$

or $$y = kx^m$$

then $$\log y = m \log x + \log k$$

or $$\log y = m \log x + c \qquad ...(i)$$

where c is another constant.

Hence a graph between log y *as dependent variable and log x* as an independent variable is a straight line.

Also consider the relation

$$y = b\, a^{kx}$$

where a, b and k are constants

$\therefore$ $$\log y = \log b + kx \log a$$

or $$\log y = \log b + (k \log a)\, x$$

Substituting $\log b = c$

and $k \log a = m$

where c and m are also constants, we get

$$\log y = mx + c \qquad ...(ii)$$

Hence a graph between log y as dependent variable and x as independent variable is a straight line.

3. Graphs are used to calibrate or graduate a given instrument for ready calculations. For example, a graph is plotted between galvanometer deflection and temperature for a thermocouple to calibrate it and find any temperature from the graph.

4. A graph is used to find the mean value from a large number of observations. For example, in the experiment with a simple pendulum the mean value of l/t^2 is found from the graph and the value of g calculated from it. The method has the advantage that we can see at a glance whether the determination is likely to be a reliable one by noting how closely the plotted points lie with respect to the mean line. Any point which lies at a large distance may be ignored as it represents an observation which has some abnormal error. If time and conditions of the experiment allow, the observations may be repeated.

5. A graph is used to find the maximum and minimum values of the dependent variable. For example, we find the value of minimum deviation by plotting a graph between i the angle of incidence and D the corresponding angle of deviation.

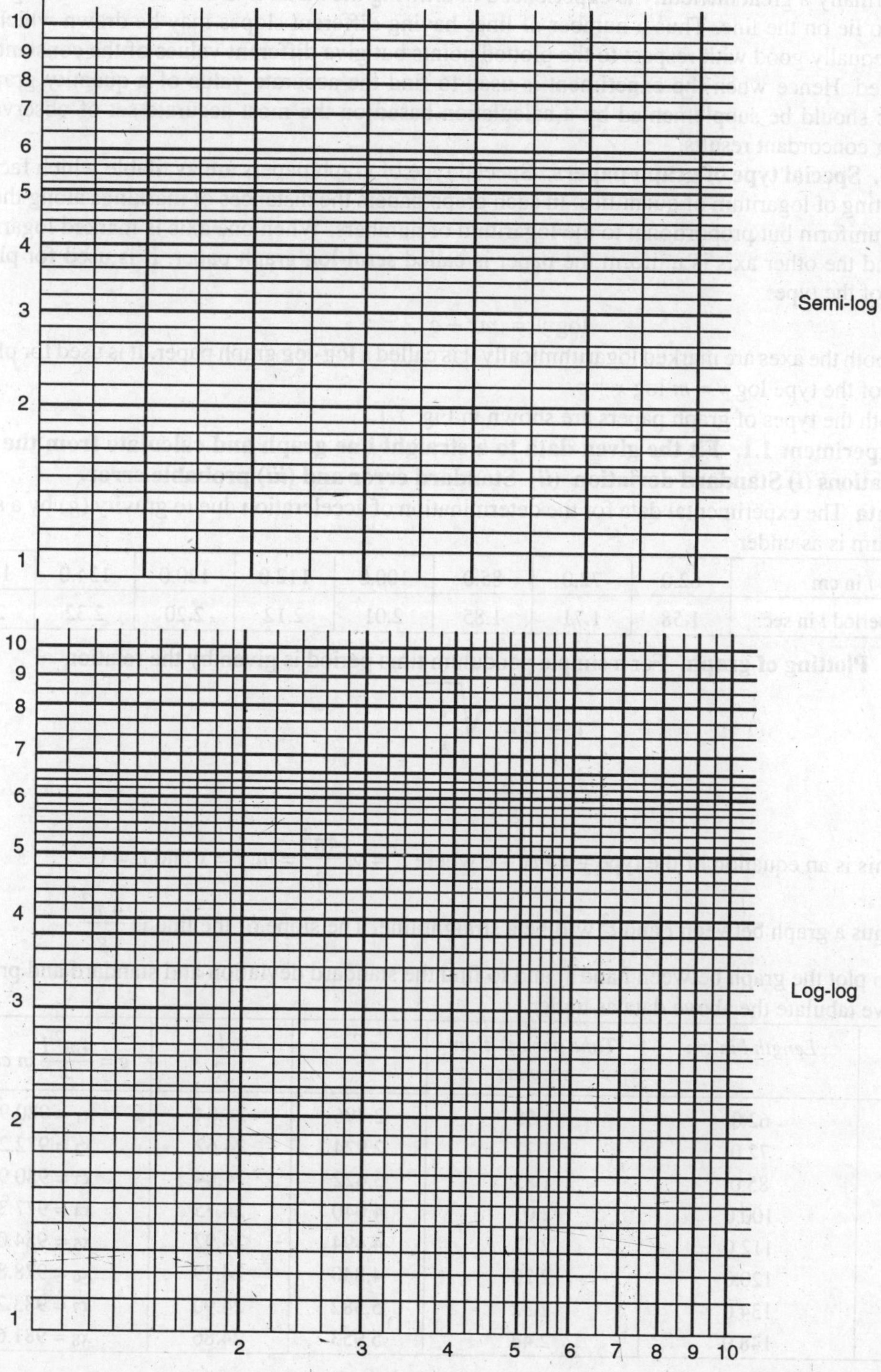

Fig. 1.1

Limitations of a graph. Although a graph is a good method of *visualising* the results of an experiment yet it is not the best method of finding the value of a constant. A graph enables us to see how closely the plotted points lie on a straight line and helps us to reject a point which lies a long way from the mean line.

Normally a great difficulty is experienced in drawing the *mean line* and also in deciding which points to lie on the line. Thus a number of lines having different slopes may be drawn which may appear equally good with respect to the plotted points but give different values of the constant to be calculated. Hence when the experiment is used to find the accurate value of a quantity graphical analysis should be supplemented by a calculation based on the most accurate set of observations yielding concordant results.

1.6. Special type of graph papers. Special type of graph papers are available which facilitate the plotting of logarithm of quantities. In such graph papers the distances of markings along the axes are not uniform but proportional to the logarithm of numbers. When one axis is marked logarithmically and the other axis is uniform the paper is called **semi-log** graph paper. It is used for plotting graphs of the type

$$\log y = mx + c$$

If both the axes are marked logarithmically it is called a **log-log** graph paper. It is used for plotting graphs of the type $\log y = m \log x + c$.

Both the types of graph papers are shown in Fig. 1.1.

Experiment 1.1. Fit the given data to a straight line graph and calculate from the given observations (*i*) Standard deviation (*ii*) Standard error and (*iii*) probable error.

Data The experimental data for the determination of acceleration due to gravity (g) by a simple pendulum is as under

Length l in cm	62.0	72.0	85.0	100.0	112.0	120.0	134.0	148.0
Time period t in sec.	1.58	1.71	1.85	2.01	2.12	2.20	2.32	2.44

(*i*) **Plotting of graph.** For a simple pendulum time period is given by the relation

$$t = 2\pi \sqrt{\frac{l}{g}}$$

$$\therefore \quad t^2 = \frac{4\pi^2}{g} l$$

This is an equation of the type $y = mx + c$ where $t^2 = y$, $\frac{4\pi^2}{g} = m$, $l = x$ and $c = 0$.

Thus a graph between l and t^2 will be a straight line. The slope of the line is $\frac{4\pi^2}{g}$.

To plot the graph between l and t^2 and to find the standard deviation and standard and probable error we tabulate the above data as under.

No.	*Length l in cm*	*Time period 't' in seconds*	t^2	$\frac{l}{t^2}$	$g = \frac{4\pi^2 l}{t^2}$ *in cm* s^{-2}
1.	62.0	1.58	2.496	24.84	$x_1 = 980.90$
2.	72.0	1.71	2.924	24.62	$x_2 = 972.21$
3.	85.0	1.85	3.422	24.84	$x_3 = 980.90$
4.	100.0	2.01	4.040	24.75	$x_4 = 977.34$
5.	112.0	2.12	4.494	24.92	$x_5 = 984.06$
6.	120.0	2.20	4.840	24.79	$x_6 = 978.82$
7.	134.0	2.32	5.382	24.90	$x_7 = 983.27$
8.	148.0	2.44	5.953	24.86	$x_8 = 981.69$

Mean value of $g = 979.91$ cm s^{-2}

To plot the graph:–

(*i*) As l is the independent variable it is taken along the X-axis. t^2 is the dependent variable it is taken along the Y-axis.

(*ii*) The scale along the X-axis is 1 small division = 2 cm and along Y-axis 1 small division is $t^2 = 0.1$ sec^2. The origin is taken as $l = 50$ and $t^2 = 2.0$.

(*iii*) Plotting various points and joining them we get a straight line as shown in Fig. 1.2.

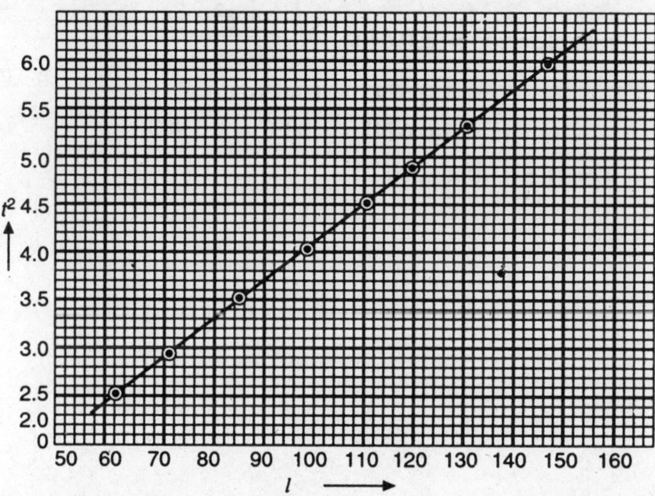

Fig. 1.2

(*ii*) **Determination of standard deviation.** To calculate the standard deviation we subtract the mean value of '*g*' found in the above case from each of the individual calculated value *i.e.* X_1, X_2, X_3 etc.

Mean value of g, $\overline{X} = 979.91$ cm s^{-2}

No.	$X - \overline{X}$	δ	δ^2
1.	$\delta_1 = 980.90 - 979.91$	$+0.99$	0.98
2.	$\delta_2 = 972.21 - 979.91$	-7.70	59.29
3.	$\delta_3 = 980.90 - 979.91$	$+0.99$	0.98
4.	$\delta_4 = 977.34 - 979.91$	-2.57	6.60
5.	$\delta_5 = 984.06 - 979.91$	$+4.15$	17.22
6.	$\delta_6 = 978.92 - 979.91$	-0.99	0.98
7.	$\delta_7 = 983.27 - 979.91$	$+3.36$	11.29
8.	$\delta_8 = 981.69 - 979.91$	$+1.78$	3.17

$$\Sigma\,\delta^2 = \delta_1^2 + \delta_2^2 + \delta_3^2 + \delta_4^2 + \delta_5^2 + \delta_6^2 + \delta_7^2 + \delta_8^2 = 100.51$$

Number of observations $\quad n = 8$

Standard deviation $\sigma = \sqrt{\dfrac{\Sigma \delta^2}{n-1}} = \sqrt{\dfrac{100.51}{7}} = 3.7893$

Standard error. Standard error $\sigma_m = \sqrt{\dfrac{\Sigma \delta^2}{n(n-1)}}$

$\therefore \quad \sigma_m = \sqrt{\dfrac{100.51}{8 \times 7}} = 1.3397$

Probable error. Probable error $= \pm 0.6745 \sqrt{\dfrac{\Sigma \delta^2}{n(n-1)}}$

$= \pm 0.6745 \times \text{standard error}$

$= \pm 0.6745 \times 1.3397 = \pm 0.9036$

$\therefore$ Value of $g = 979.91 \pm 0.9036 \text{ cm } s^{-2}$.

PART I

MECHANICS AND PROPERTIES OF MATTER

2

Measurement of Length, Mass and Time

2.1. Measurement of length. A measurement of length is necessary in almost all experiments in Physics. Where great accuracy is not required a metre rod can be used conveniently to measure the distances correct up to a millimetre. To measure a length correct up to a fraction of a millimetre the divisions must be further subdivided. This, however, is not practicable beyond a certain limit because it is difficult to see the small divisions clearly. Moreover, closer the division on the scale, the greater is the inaccuracy in graduation. Hence for measurements accurate up to 1/10th or 1/100th of a millimetre the various devices are : (*i*) *vernier callipers* (*ii*) *a screw gauge* and (*iii*) *a spherometer*.

2.2. Vernier callipers. A vernier callipers consists of a rectangular steel bar graduated in inches on one edge and in centimetres on the other edge as shown in Fig. 2.1. This is known as the **main scale.** Over this scale slides a small scale called the **vernier scale.**

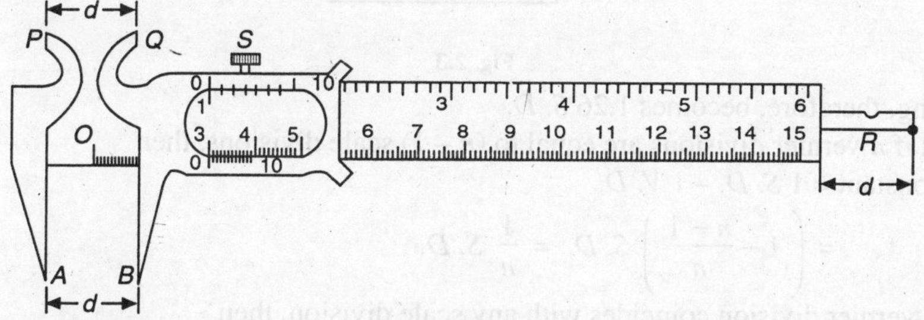

Fig. 2.1

The instrument has got two jaws, *A* and *B*. The jaw *A* is fixed at the end of the rectangular bar on the zero side, while the other jaw *B* is movable and can slide along the main scale. Each jaw is at right angles to the main scale and the movable jaw can be fixed at any position by the screw *S*. The two vernier scales are attached to the movable jaw as shown. Usually when the two jaws are touching each other, the *zero* of the vernier scale coincides with *zero* of the main scale.

In some forms of the instrument the jaws project in the upper part as shown at *P* and *Q*. These projecting jaws are used to measure the internal diameters of the tubes. The movable jaw also carries a thin rectangular rod *R* which is used to measure the depth of vessels. As shown the distance *AB*, *PQ* and the length of projecting rod *R* are the same and can be measured on the scale.

2.3. Vernier constant. To understand how the fraction of the smallest scale division is read accurately with the help of a vernier, consider a vernier scale having 10 equal divisions.

Let these 10 vernier divisions coincide with 9 main scale divisions as shown in Fig. 2.2. Since 10 divisions of the vernier scale coincide with 9 main scale divisions.

$$\therefore \qquad 1\ V.D. = \frac{9}{10} S.D.$$

or
$$1\ S.D. - 1\ V.D. = \frac{1}{10} S.D. = 0.1\ S.D.$$

The difference between one scale division and one vernier division is called the **vernier constant** *or the least count of the vernier.*

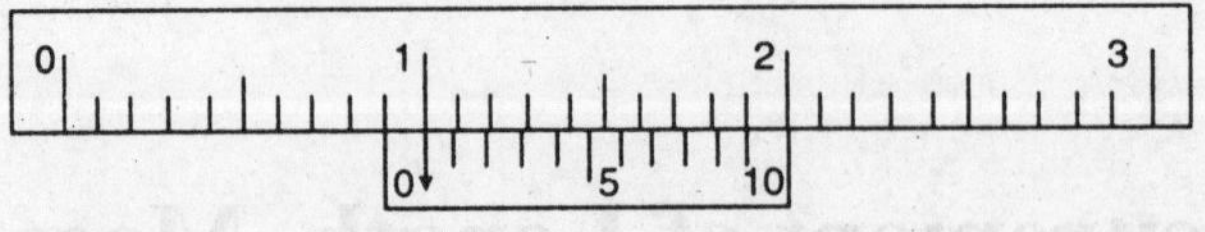

Fig. 2.2

Now suppose that the zero of the vernier lies between 1.2 and 1.3 as shown in Fig. 2.3. It is clear from the figure that the 6th vernier division coincides with a scale division. If we denote the fraction after 1.2 by x, then

$$1.2 + x + 6\ V.D. = 1.2 + 6\ S.D.$$

or
$$x = 6\ S.d. - 6\ V.D. = 6\ (1\ S.D. - 1\ V.D.)$$
$$= 6 \times \text{vernier constant}$$
$$\therefore \qquad x = 0.6\ S.D.$$

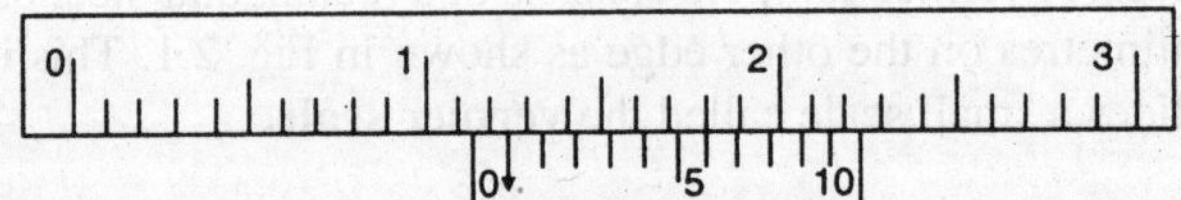

Fig. 2.3

The reading, therefore, becomes 1.26 *S. D.*

In general if n vernier divisions are equal to $(n - 1)$ scale divisions, then

Vernier constant = 1 *S. D.* – 1 *V. D.*

$$= \left(1 - \frac{n-1}{n}\right) S.D. = \frac{1}{n}\ S.D.$$

If the mth vernier division coincides with any scale division, then

The fraction to be added

$$= m \times \frac{1}{n}$$

Hence the rule to read the fraction of a scale division is to multiply the number of the vernier division which coincides with any scale division with the vernier constant.

To find the vernier constant. (*i*) Find the magnitude of the smallest division on the main scale.

(*ii*) Count the total number of divisions on the vernier scale.

(*iii*) Slide the movable jaw so that the zero mark (the first division) of the vernier scale coincides with any one of the main scale divisions.

(*iv*) Find the number of scale divisions, which coincide with the total number of vernier divisions. Calculate the vernier constant as explained above.

Many instruments which are used for measurement are provided with vernier arrangement. The vernier constant of some of the instruments has been calculated in the following examples.

Linear Vernier

(*i*) **Fortin's barometer.** It is used to find the value of atmospheric pressure. In one type of this instrument suppose 10 *V. D.* coincide with 19 *S. D.* and one *S. D.* is 1 millimetre. To find vernier constant each *V. D.* is supposed to be further divided into two, thereby making a total of 20 such *vernier divisions*. The vernier constant of this new vernier scale will be as follows :

$$20 \text{ supposed } V.D. = 19\, S.D.$$

$$1 \text{ supposed } V.D. = \frac{19}{20}\, S.D.$$

$$\text{Vernier constant} = 1\, S.D. - 1\, V.D. = \left(1 - \frac{19}{20}\right) S.D.$$

$$= \frac{1}{20} \text{ mm} = 0.05 \text{ mm.}$$

In fact each vernier division is twice this supposed division.

$\therefore$ Vernier constant = 2 × 0.05 = 0.1 mm = 0.01 cm

In second type of this instrument 20 V.D coincide with 19 S.D and one S.D. is 1 m.m. The vernier constant of such kind of Fortin s barometer will be as follows

$$20 \text{ V.D.} = 19 \text{ S.D}$$

$$1 \text{ V.D} = \frac{19}{20} \text{ S.D}$$

$$\text{Vernier constant} = 1 \text{ S.D} - 1 \text{ V.D} = \left[1 - \frac{19}{20}\right] \text{S.D}$$

$$= \frac{1}{20} \text{ mm} = 0.05 \text{ mm} = 0.005 \text{ cm.}$$

(*ii*) **Travelling microscope.** It is used for finding the length of very small objects. The microscope is first focused at one end and then at the other end of the small object. The distance through which the microscope moves is measured by a vernier rigidly fixed to the microscope and sliding over a centimetre scale.

In this case suppose 25 *V. D* coincide with 24 *S. D.* and each scale division is 0.5 mm.

$$\therefore \quad 25\, V.D. = 24\, S.D.$$

$$\text{or} \quad 1\, V.D. = \frac{24}{25}\, S.D.$$

$$\therefore \quad \text{Vermier constant} = 1\, S.D. - 1\, V.D. = \left(1 - \frac{24}{25}\right) S.D.$$

$$= \frac{1}{25}\, S.D. = \frac{1}{25} \times 0.5 \text{ mm}$$

$$= \frac{1}{50} \text{ mm} = 0.002 \text{ cm}$$

Circular Vernier

Verniers used in spectrometers, sextants, polarimeter, dip circle etc. are called circular verniers. The circle is divided into 360 degrees and each degree is further divided into 2, 5 or 6 parts.

(*i*) **Spectrometer.** A spectrometer is used to measure the refractive index by finding the angle of the prism and the angle of minimum deviation. In this case suppose 30 *V. D.* coincide with 29 *S.D.* and magnitude of each scale division is 0.5 degree.

$$\therefore \quad 30\, V.D. = 29\, S.D.$$

or $$1\ V.D. = \frac{29}{30}\ S.D.$$

$\therefore$ Vernier constant
$$= 1\ S.D. - 1\ V.D. = \left(1 - \frac{29}{30}\right) S.D.$$
$$= \frac{1}{30}\ S.D. = \frac{1}{30} \times 0.5 \text{ degree}$$
$$= \frac{1}{60} \text{ degree} = 1 \text{ minute.}$$

(ii) **Sextant.** A sextant is used to find the height of an object by measuring the angle subtended by it. In this case usually 60 *V. D.* coincide with 119 *S. D.* and each scale division is equal to 10 minutes. To find the vernier constant each vernier division is supposed to be further divided into two thereby making a total of 120 such vernier divisions. The vernier constant of this new vernier scale will be as follows :—

$$120\ V.D. = 119\ S.D.$$

or $$1\ V.D. = \frac{119}{120}\ S.D.$$

$\therefore$ Vernier constant
$$= 1\ S.D. - 1\ V.D. = \left(1 - \frac{119}{120}\right) S.D.$$
$$= \frac{1}{120}\ S.D. = \frac{1}{120} \times 10 = \frac{1}{12} \text{ mint.}$$
$$= 5 \text{ seconds.}$$

In fact each vernier division is twice this value.

$\therefore$ Vernier constant = 2 × 5 = 10 seconds.

(iii) **Polarimeter.** A polarimeter is used for studying polarisation of light and to find the specific rotation of liquids or solutions. In this case each division on the circular scale of the polarimeter is 1 degree and 10 V.D coincide with 9 S.D. Therefore

$$10 \text{ V.D} = 9 \text{ S.D}$$

or $$1 \text{ V.D} = \frac{9}{10} \text{ S.D}$$

$\therefore$ Vernier constant
$$= 1 \text{ S.D} - 1 \text{ V.D} = \left(1 - \frac{9}{10}\right) \text{S.D}$$
$$= \frac{1}{10} \text{ S.D} = \frac{1}{10} \times 60' = 6'.$$

(iv) **Dip circle.** This instrument is used to find the value of angle of dip at a place. It has a horizontal circular scale graduated in degrees and each division in equal to 1 degree. 10 V.D coincide with 9 S.D.

Therefore, in this case also vernier constant is the same as in the case of a polarimeter *i.e.* 6′

2.4. Screw gauge. It is based upon the principle of a screw and consists of a U-shaped frame *F* having a fixed end at *A*. Through the other end *B* passes a fine and an accurately cut screw of a uniform pitch. A cap fits on to the screw and carries on its inner edge *H*, 50 or 100 equal division marks. This is called the *head scale or circular scale* and is used to measure the fraction of a revolution. The number of complete revolutions can be read on the pitch scale *S* which is graduated on the nut parallel to the axis of the screw.

The reading of the circular scale is taken against a line known as the *reference line.*

A screw gauge is used to measure accurately small thicknesses or diameters of the wires.

In some screw gauges the screw head is provided with a ratchet arrangement *R*. When the studs *A* and *C* are in contact either with each other or with some other object placed in between, the ratchet slips over the screw cap without moving the screw forward. *This avoids undue pressure between the studs or on the object.*

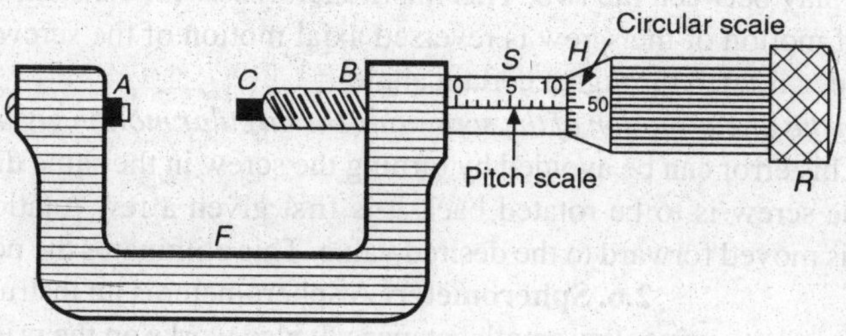

Fig. 2.4

To find the least count. To find the least count of a screw gauge, we first determine the pitch.

Pitch *of a screw is defined as the distance between two consecutive threads taken parallel to its axis and is measured by the distance through which the screw moves forward or backward when one complete rotation is given to the circular cap.*

To find the pitch. (*i*) Rotate the circular scale *H* and coincide the zero mark with the reference line. Note the reading on the pitch scale.

(*ii*) Give four complete rotations to the circular scale and again note the reading on the pitch scale.

(*iii*) Calculate the pitch by the formula

$$\left(\frac{\text{Distance travelled on the pitch scale}}{\text{Number of rotations}}\right)$$

Note. *The pitch of a screw gauge is usually 0.5 mm or one millimetre.*

The **least count** *of the screw gauge is defined as the distance through which the screw moves backward or forward when the cap is rotated through one division on the circular scale.*

To find the least count note the number of divisions on the circular scale *H*, then

$$\text{Least count} = \frac{\text{Pitch of the screw}}{\text{Number of divisions on the circular scale}}$$

Zero correction. If we bring the jaws *A* and *C* in contact without applying any undue pressure, the zero of the circular scale may not coincide with the reference line. In such a case there is a **zero error.** In some instruments the zero mark goes beyond the reference line, while in others it is left behind.

To find zero correction count the number of divisions on the circular scale by which the zero mark has advanced beyond or is left behind the reference line. Multiply this number with the least count. This is the zero correction. *If the zero of the circular scale has advanced beyond the reference line the zero correction is* **positive** *but if it is left behind the reference line it is* **negative**.

(Remember : **zero beyond-add**).

Diameter of the wire. To find the diameter of a wire insert the wire between the jaws *A* and *C* and turn the cap till the wire is **just** held between them *without undue pressure. The ratchet stop R* begins to turn freely when a contact has been established. Note the reading on the circular scale against the reference line. Multiply it by the least count and add it to the number of complete divisions visible on the main scale.

If the pitch is *half millimetre* and there are 50 divisions on the circular scale, care should be taken to note whether the last uncovered divisions is less than half or more than half. If it is more than half, add 50 to the circular scale reading before multiplying it by the least count. Turn the wire through a right angle and note the reading again at the same point.

2.5. Backlash error. In the screw gauge, a spherometer or any other instrument in which the principle of a screw is used to measure small distances, the *accurately* cut screw moves in a nut. There is always a misfit or play between the two. This misfit increases with constant use. Due to this play when the direction of motion of the screw is reversed axial motion of the screw does not take place until the screw has been rotated through a certain angle.

This lag between the linear motion of the screw and the circular motion given to its head is called the backlash error. This error can be avoided by turning the screw in the same direction while taking an observation. If the screw is to be rotated back it is first given a few rotations in the backward direction and then it is moved forward to the desired value. This eliminates the possibility of an error.

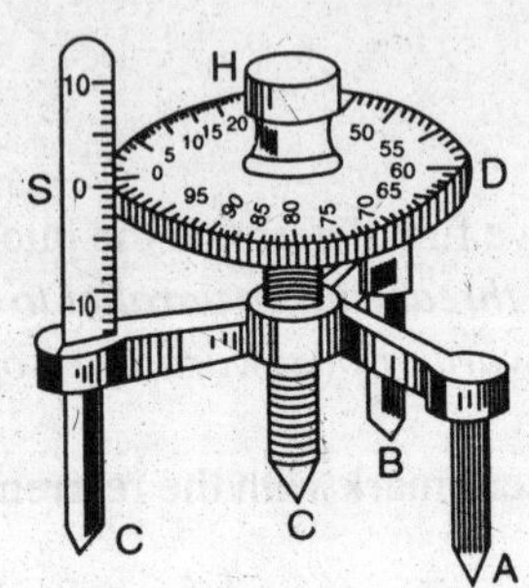

Fig. 2.5

2.6. Spherometer. A spherometer is an instrument used for measuring very small distances. It also works on the principle of a micrometer screw. It is generally used for measuring the thickness of a thin plate and for determining the radius of curvature of spherical surfaces such as lenses or mirrors.

Construction. It consists of a metal framework supported on three fixed legs of equal length. The end of the three legs lie at the corners of an equilateral triangle. An accurately cut screw works through a threaded hole at the centre of the framework. The screw terminates at the top into a milled head and carries a large graduated disc as shown in Fig. 2.5.

The lower end of this screw forms the central leg of the instrument. A small vertical scale marked in millimetres or half millimetres is fixed at one end of the frame with its graduations close to those on the circular disc. The edge of the circular disc is divided into a large number of equal divisions, generally 50 or 100.

Least count. The least count of a spherometer is the smallest distance that can be measured with it. *It is the distance moved by the screw when it is turned through one division on the circular scale.*

To find the least count of the spherometer :

(*i*) Bring the zero of the circular scale against a division mark on the mm scale

(*ii*) Give four complete rotations to the screw and find the distance moved by it on the millimetre scale.

(*iii*) Divide this distance by 4 to get the '*pitch*' of the screw.

(*iv*) Divide the pitch by the total number of divisions on the circular scale to get the '*least count*' of the spherometer.

2.7. Radius of curvature of a spherical surface. A spherometer can be used to determine the thickness of a glass slip and the radius of curvature of a spherical surface. The theory of the method is given below.

Let *AEXFY* be the sphere of which the given spherical surface forms a part. Consider a vertical section passing through one of the legs say *A*, the tip of the screw *E* and the centre of the sphere *O*. It will be a circle as shown in Fig. 2.6. The third leg *C* will not be visible in this position of the spherometer. The tip of the central screw touches the spherical surface at *E* and if the surface were plane it would touch it at *D*. Thus $DE = h$ represents the height of the central screw above the plane of the outer legs, *ABC*. The section of the sphere through *ABC* will be a circle of radius $AD = r$.

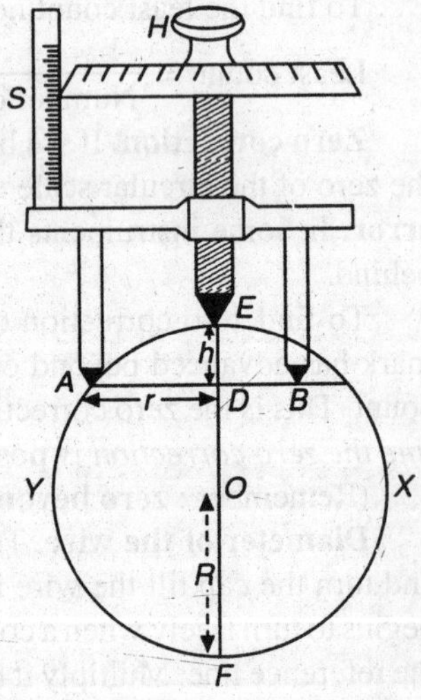

Fig. 2.6

From the geometry of the circle, we have

$$AD^2 = ED.DF$$

or $$r^2 = h(2R - h)$$

or $$2Rh = r^2 + h^2$$

$$\therefore \quad R = \frac{r^2}{2h} + \frac{h}{2} \qquad ...(i)$$

If l is the length of each side of the equilateral triangle ABC formed by joining the tips of the three outer legs, then as shown in Fig. 2.7, we have

$$\frac{l}{2} = r \cos 30 = \frac{r\sqrt{3}}{2}$$

Fig. 2.7

Substituting $r = \frac{l}{\sqrt{3}}$ in (*i*),

We have

$$R = \frac{l^2}{6h} + \frac{h}{2}.$$

The value of R can be calculated from the measurement of l *and* h.

Experiment 2.1. To find the radius of curvature of (*i*) a convex mirror and (*ii*) a concave mirror by a spherometer.

Apparatus. A spherometer, a plane glass slab, a convex and a concave mirror of appropriate diameter (about 8 to 10 cm), a vernier callipers and wooden blocks for supporting the mirrors.

Theory. The radius of curvature R of a convex or a concave surface is given by the relation

$$R = \frac{l^2}{6h} + \frac{h}{2}$$

where l is the mean length of the three sides of the equilateral triangle formed by joining the tips of three outer legs and h represents the height of the central screw above or below the plane of the outer legs.

Procedure (*i*) **Convex mirror** 1. Find the least count of the spherometer as explained in **Art 2.6.**

2. Raise the screw by tuning its head, so that it may be above the plane containing the tips of the three legs. Place the spherometer on the page of your note-book on which the experiment is to be recorded. Press the spherometer to get the impression of the tips of three legs. Mark the position of each of the three points by drawing a small circle round each point. Measure the length of each side of the triangle formed by joining these points *very accurately* and find the mean value. Let the mean length be l cm.

Note. Great care should be taken in measuring l as in the calculations for the radius we have to use l^2 and any error made in the measurement of l will be doubled. It is advisable to measure l by the use of projecting jaws P and Q of the vernier callipers correct to a fraction of a millimetre or with the help of a travelling microscope.

3. Set the convex mirror firmly on a *horizontal surface.* Place the spherometer on the polished surface of convex mirror after raising its central leg through four or five complete rotations of the circular scale. Turn the screw so that the central leg while moving *downwards just* touches the surface of the mirror Note the circular scale reading against the edge of the vertical scale. Repeat this step three times.

4. Place the spherometer on the surface of the plane glass slab. Turn the screw in the same direction as in the previous step till tip of the central leg *just* touches the plane surface. Count the number of complete rotations and the additional number of circular divisions moved. Repeat the step *three times by moving the central leg every time in the upward direction through a sufficient large distance as compared to the value of 'h'.*

(*ii*) **Concave Mirror.** 1. First fix the mirror firmly in position on the blocks of wood as shown in Fig. 2.8.

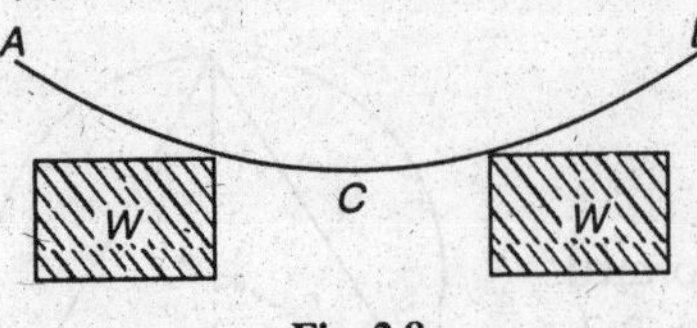

Fig. 2.8

2. Find the circular scale reading by placing the spherometer *first* on the plane glass slab and then on the polished concave surface. Count the number of complete rotations and the additional number of circular scale divisions moved. Repeat it three times and take the mean.

Observations. Distance moved by the screw in four complete rotations d = mm

Pitch of the screw $d/4$ = mm

Total number of divisions on circular scale n =

Least count of the spherometer = $\frac{\text{Pitch}}{n}$ = mm

Distance between the legs l = (*i*) cm

(*ii*) cm

(*iii*) cm

Mean distance between the legs = l = cm

No.	Type of mirror	Number of complete rotations m	Number of additional circular scale divisions moved x	h = m × pitch + x × Least count	Mean value of h in cm
1	Convex				
2					
3					
1	Concave				
2					
3					

Radius of curvature for convex mirror $R = \frac{l^2}{6h} + \frac{h}{2} =$ cm.

Radius of curvature for concave mirror $R = \frac{l^2}{6h} + \frac{h}{2} =$ cm.

Precautions. 1. The circular disc should always be rotated in the same direction to avoid error due to *back-lash*.

2. The reading should be noted when the spherometer *just* begins to rotate about the central screw.

3. The distance between the two legs must be measured accurately, correct up to a fraction of a millimetre.

4. The mirrors should be in perfectly stable positions when readings are being taken.

5. There should be no play between the screw and the nut in which it rotates.

2.8. Mass. *The mass of a body is the quantity of matter contained in it.* It is constant at all places and is found by comparing the weight of the body with that of standard weights by an ordinary balance.

The weight of a body is the pull of the earth and varies from place to place. It is given by mg where m is the mass and g the acceleration due to gravity at the place.

2.9. Measurement of time. We are familiar with a watch, a time piece and a clock which are ordinary means of measuring time. In the laboratory we have to measure accurately the interval of time between two events. This is done with the help of a stop-watch or a stop-clock.

Stop-watch. It is a specially constructed watch having a long second's hand which moves over a circular dial with sixty equal division marks representing seconds. Each division is further divided into 5 or 10 equal parts, thus enabling us to read correct up to 1/5th or 1/10th of a second. A small minute hand moves over a small circular dial graduated into 30 divisions each representing a minute. For every complete rotation of the second's hand the minute hand advances by one division and one complete rotation of the minute hand is completed in 30 minutes.

To begin with both the hands are at zero. At the beginning of an event as the knob at the top is pressed, the second's hand starts. At the end of the event the knob is again pressed when both the second's and the minute's hands stop. The interval of time can thus be read. If the knob is again pressed both the hands fly back to the zero position and the watch is ready for use.

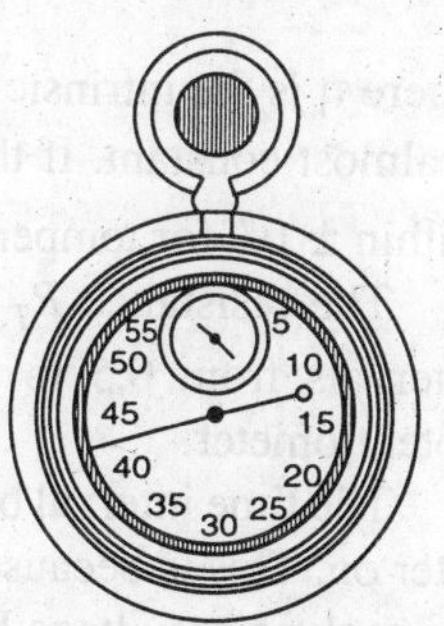

Fig. 2.9

Stop-clock. A stop-clock also works on the same principle as a stop-watch does, but is bigger in size and reads correct up to a second. The fraction of a second has to be estimated by judgement. The starting and stopping is done by a rod fixed to it. When the rod is pushed to the right the clock starts and when it is pushed to the left it stops. In this type of stop-clock, the hands cannot be brought to zero by a fly-back action as in the case of a stop-watch. Instead there is a third hand known as index hand which can be moved from outside and is made to coincide with the second's hand before starting the stop-clock.

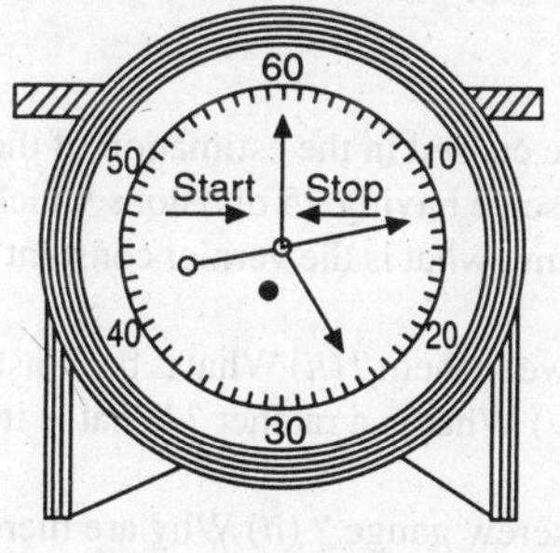

Fig. 2.10

In another type of stop-clock instead of the rod a knob is provided as in a stop-watch. When the knob is pressed the clock starts and it stops when the knob is pressed again. If the knob is now pressed both the hands fly back to the zero position and the clock is ready for use.

2.10. Metronome. A metronome is an electronic device to measure small time intervals by producing periodic audible signals in a speaker or head phone. This instrument is highly reliable and is particularly useful in the study of slowly varying phenomenon like the charging and discharging of a capacitor or growth and decay of current in an inductance-resistance (*LR*) circuit. The main advantage of such a device over that of a stop-watch is that the experimenter can take a reading on a meter or any other instrument and at the same time also measure 'time' by listening to the ticks (signals) produced by the metronome thus relieving him from the tedious task of looking at the stop-watch and the measuring instrument simultaneously.

It consists of a unijunction transistor (UJT) 2N 2646 as a *relaxation oscillator*. The basic circuit is shown in Fig. 2.11. At the beginning of the operating cycle the emitter E is *reverse biased* and, therefore, non-conducting. As the capacitor C_T (which is an electrolytic capacitor of capacity 25 μf) is charged through the resistor R_T the emitter voltage V_E rises exponentially towards the supply voltage $+V$. When the emitter voltage reaches the peak point voltage V_p the transistor becomes forward biased so that the dynamic resistance between the emitter E and the base B_1 drops to a low value of the order of 10-15 ohms. At this stage the capacitor C_T discharges through the emitter till the emitter voltage reaches a low value of about 2 volts known as *Valley voltage* (V_V). The emitter ceases to conduct and the cycle is repeated. As the speaker *LS* is in series with the emitter a pulse passes through the speaker corresponding to every cycle and a tick is produced.

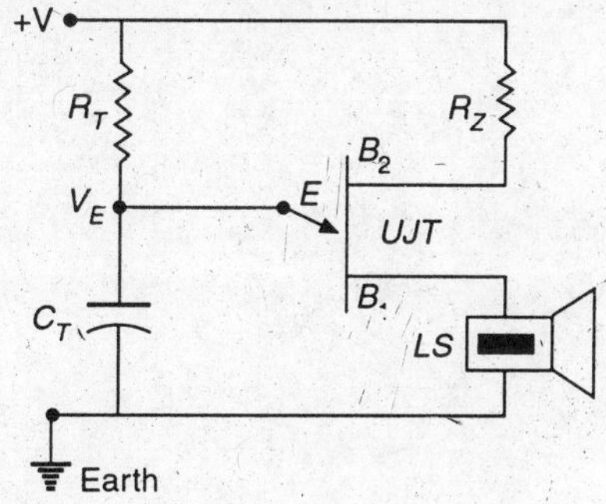

Fig. 2.11

The period of the pulse is given by

$$T = R_T\ C_T\ \log_e \frac{1}{1-\eta}$$

where η is the intrinsic stand off ratio which varies from device to device but for a particular device is almost constant. If the resistance R_2 is about 100 ohm the frequency of the oscillations is stable within ± 1% for temperature upto 70°C and 10% supply voltage variation.

The resistance R_T can be varied in 11 discrete steps by a band switch to produce signals at intervals from 0.5 to 10 sec. The period can also be adjusted at any intermediate value by a potentiometer.

The time interval between switching on and first tick is greater than that between successive ticks later on. This is because in the very beginning the capacitor has to be charged from zero potential to the peak point voltage V_p from which it discharges to give the first tick. After this the capacitor retains about 2 volts and therefore takes a smaller time to reach the peak point voltage V_P. The successive ticks after the first few ticks are ignored to have a variation of less than 1%.

Oral Questions

1. Vernier callipers. (*i*) What is a vernier constant ? (*ii*) What error will be caused in the estimation of the smallest fraction without a vernier scale? (*iii*) You are provided with a vernier scale having 25 divisions which coincide with 49 divisions of the main scale. If each main scale division is 0.5 mm., what is the vernier constant? (*iv*) What is zero error?

2. Screw gauge. (*i*) Define pitch and the least count. What is the relation between them ? (*ii*) What is backlash error ? How is it avoided ? (*iii*) When is the zero error positive or negative ? (*iv*) What is a ratchet ? What is its function ?

3. Spherometer. (*i*) What is the difference between a spherometer and a screw gauge ? (*ii*) Why are there three legs in a spherometer ? (*iii*) What quantity should be measured more accurately to get a correct value of the radius of curvature *R* ?

4. Mass. (*i*) What is the difference between mass and weight ? (*ii*) Is the weight of a body constant at various places on the surface of the earth ? (*iii*) What do you measure with a balance ?

5. Stop-watch and metronome. (*i*) What is the least time that can be measured by a stop-clock and a stop-watch ? (*ii*) Which is more accurate ? Stop-watch or stop-clock. (*iii*) What is a metronome ? (iv) What is its advantage over a stop-watch? (*v*) How can you adjust the duration of pulses produced by it ? (*vi*) What is the order of accuracy of this device ?

PART I

PROPERTIES OF MATTER

3

Moment of Inertia

3.1. Moment of inertia *of a body about an axis is defined as the sum of the product of the mass and the square of the distance of different particles of the body from the axis of rotation.*

If a body of mass M is supposed to be made up of a large number of small masses $m_1, m_2, \ldots$ at distances $r_1, r_2, \ldots$ from the axis of rotation AB, then

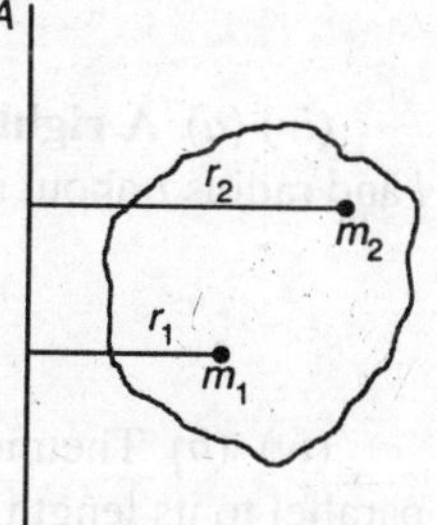

Fig. 3.1.

Moment of inertia $$I = m_1 r_1^2 + m_2 r_2^2 + \ldots\ldots\ldots$$
$$= \Sigma mr^2.$$

The *S.I.* unit of moment of inertia is $kg\text{-}m^2$.

3.2. Radius of gyration *of a body is the square root of the mean square distance of the particles of the body from the axis of rotation.* If the body is divided into n particles each of mass m and they lie at distances $r_1, r_2, \ldots$ from the axis of rotation AB, then

Radius of gyration $$K = \left(\frac{r_1^2 + r_2^2 + \ldots\ldots}{n}\right)^{\frac{1}{2}}$$

It may also be defined as the distance of the point at which the whole mass of the body may be concentrated so as to have the same moment of inertia.

$$\therefore \qquad I = MK^2.$$

The moment of inertia plays the same role in rotational motion as mass does in linear motion. The *S.I.* unit of radius of gyration is a metre.

3.3. Energy of rotation of a body. If a body rotates about an axis AB with an angular velocity ω, all its particles have the same angular velocity but different linear velocities as they lie at different distances from the axis of rotation. Let the linear velocities of the particles of mass $m_1, m_2, \ldots$ from the axis of rotation be $v_1, v_2, \ldots$ respectively, then

Total kinetic energy of the body = Sum of the kinetic energy possessed by the various particles

$$= \frac{1}{2}m_1 v_1^2 + \frac{1}{2}m_2 v_2^2 + \ldots\ldots\ldots$$

$$= \frac{1}{2}m_1 r_1^2\omega^2 + \frac{1}{2}m_2 r_2^2\omega^2 + \ldots\ldots\ldots$$

$$= \frac{1}{2}\left(\sum mr^2\right)\omega^2 = \frac{1}{2} I \omega^2$$

where I is the moment of inertia of the body about the axis AB.

3.4. Values of moment of inertia. Moment of inertia of some regular solids is given below :

(*i*) **Circular ring.** The moment of inertia of a circular ring of mass M and radius r about an axis passing through its centre and perpendicular to its plane is given by

$$I = Mr^2$$

(*ii*) **Circular disc.** The moment of inertia of a circular disc about an axis passing through its $C.G.$ and perpendicular to its plane is given by

$$I = \frac{1}{2} Mr^2$$

where M is the mass of the disc and r its radius.

(*iii*) **A rectangular bar.** The moment of inertia of a rectangular bar about an axis passing through its $C.G.$ and perpendicular to the edges of length l and breadth b is given by

$$I = M\frac{l^2 + b^2}{12}$$

(*iv*) (*a*) **A right circular cylinder.** The moment of inertia of a right circular cylinder of length l and radius r about an axis through its $C.G.$ and perpendicular to the axis of the cylinder is given by

$$I = M\left(\frac{l^2}{12} + \frac{r^2}{4}\right)$$

(*iv*) (*b*) The moment of inertia of the same cylinder about an axis passing through its $C.G$ and parallel to its length is given by

$$I = \frac{1}{2} Mr^2$$

(*v*) **A sphere.** The moment of inertia of a sphere of radius r about a diameter is given by

$$I = \frac{2}{5} Mr^2.$$

(*vi*) **A spherical shell.** The moment of inertia of a thin spherical shell of mass M and radius r about its diameter is given by

$$I = \frac{2}{3} Mr^2$$

3.5. Moment of inertia of a flywheel. A flywheel is simply a heavy wheel with a long axle supported in bearings such that it can rest in any position. In other words, the $C.G.$ lies on the axis of rotation.

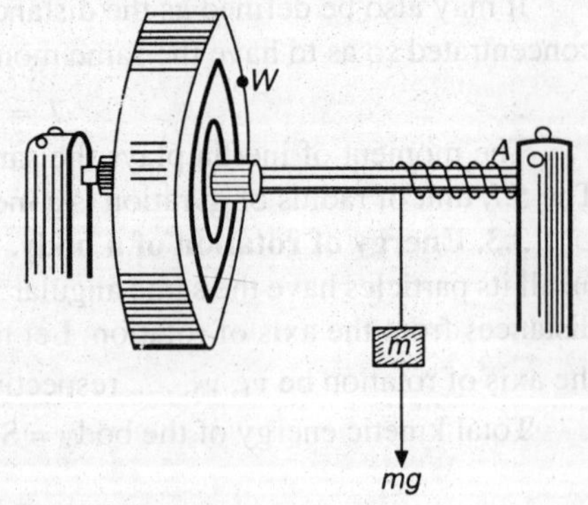

Fig. 3.2

To find the moment of inertia of a flywheel a mass m is attached to the axle of the wheel by a cord which is wrapped several times round the axle as shown in Fig. 3.2. One end of the string is in the form of a loop so that it can easily be attached to or detached from a pin A projecting from the axle. The length of the string is so adjusted that it gets detached from the axle as soon as the bottom of the mass m is just to touch the floor.

When the mass is allowed to fall, its potential energy is partly converted into the kinetic energy due to the velocity gained by it and partly into the energy of rotation of the flywheel.

Let ω be the angular velocity imparted to the wheel at the moment the mass m is detached.

After the string has been detached from the wheel, the wheel continues to revolve for some time. Its angular velocity decreases on account of friction and finally the wheel comes to rest. If n_1 is the number of revolutions that the wheel makes in time t before coming to rest, then

$$\text{Average angular velocity} = \frac{2\pi n_1}{t}$$

If the frictional force is constant the rotation of the wheel is uniformly retarded. It begins with an angular velocity ω and its final angular velocity is zero, so that the initial velocity ω is double the average velocity.

or
$$\omega = 2 \times \frac{2\pi n_1}{t} = \frac{4\pi n_1}{t}$$

According to the principle of conservation of energy, when the string is detached

P.E. of mass m = K.E. of mass m + K.E. of wheel + Work done against friction.

If h is the height through which the mass has fallen, then

P.E. of mass $m = mgh$

If r is the radius of the axle, then

Linear velocity of mass $m = r\omega$ and K.E. of mass $m = \frac{1}{2} m (r\omega)^2$

If I is the moment of inertia of the wheel, then K.E. of the wheel $= \frac{1}{2} I\omega^2$

Let F be the energy per revolution used in overcoming the frictional force. If n is the number of revolutions the wheel makes during the descent of the mass m, then

Total energy used to overcome friction = nF

Hence
$$mgh = \frac{1}{2} mr^2\omega^2 + \frac{1}{2} I\omega^2 + nF \qquad ...(i)$$

The kinetic energy possessed by the wheel is used up in overcoming friction. As the wheel comes to rest after making n_1 revolutions

$$\therefore \quad n_1 F = \frac{1}{2} I\omega^2$$

or
$$F = \frac{1}{n_1}\frac{1}{2} I\omega^2$$

Substituting the value of F in (i), we have

$$mgh = \frac{1}{2} mr^2\omega^2 + \frac{1}{2} I\omega^2 \left(1 + \frac{n}{n_1}\right)$$

or
$$I = \frac{2mgh - mr^2\omega^2}{\omega^2(1 + n/n_1)}$$

Experiment 3.1. To find the moment of inertia of a fly-wheel.

Apparatus. A fly wheel, a few different masses and a mass provided with a hook, a strong and thin string, stop watch, a metre rod, a vernier callipers and a piece of chalk.

Formula. Moment of inertia

$$I = \frac{2mgh - mr^2\omega^2}{\omega^2 (1 + n/n_1)}$$

Procedure. 1. Examine the wheel and see that there is the least possible friction. Oil the bearings, if necessary.

2. Measure the diameter of the axle with a vernier callipers at different points and find the mean. Measure also the circumference of the wheel W with a thread.

3. Take a strong and thin string whose length is less than the height of the axle from the floor. Make a loop at its one end and *slip* it on the pin A on the axle. Tie a suitable mass to the other end of the string. Suspend the mass by means of the string so that the loop is just on the point of slipping from the pin A. Make a chalk mark on the wheel behind the pointer in this position. Also note the position of the lower surface of the mass m on a scale fixed behind on the wall as at C.

4. Now rotate the wheel and wrap the string uniformly round the axle so that the mass m is slightly below the rim of the wheel and the chalk mark is again opposite to the pointer P. Again note the position of the lower surface of the mass on the scale as at B. If now the mass is allowed to fall, it will descend through a height $BC = h$ before being detached from the pin A. Count the number of turns wound round the axle and let it be n. The wheel will thus make n revolutions before the thread is detached.

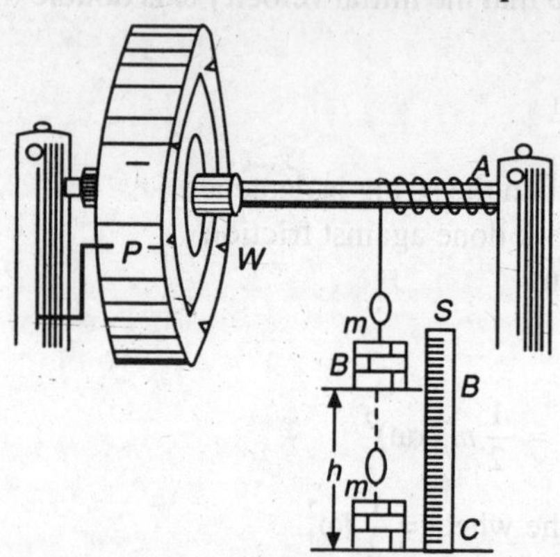

Fig. 3.3

5. Hold a stop-watch in your hand and allow the mass to descend. As soon as the sound of the weight striking the ground is heard, start the stop-watch. Count the number of revolutions n_1 made by the wheel before coming to rest with reference to the chalk mark and note the time t taken for the purpose.

It is possible that in addition to the complete rotations made by the wheel, there may be a fraction of the rotation. To estimate the fraction measure the distance along the circumference by which the chalk mark has advanced beyond the pointer P, by means of a thread. Divide this distance by the circumference of the wheel W. Repeat three times for the same height and load.

6. Repeat the experiment with three different masses, suitably adjusting the height through which the mass falls so that the number of rotations made by the wheel can be easily counted.

Observations. Vernier constant =

Diameter of the axle = (*i*) (*ii*) (*iii*)

Mean diameter of the axle = m

∴ Radius of the axle r = m

Circumference of the flywheel W, S = m

Mass m = kg

Observation 1 Height h = m

Number of turns wound on axle n =

Sl. No.	*No. of complete revolutions made by the wheel x*	*Distance of chalk mark from pointer d*	*Fraction of revolution* $\frac{d}{S} = y$	*No. of revolutions* $n_1 = x + y$	*Time*
1.					
2.					
3.					

Mean n_1 =

Mean time $t =$

$\therefore$ Angular velocity $\omega = \dfrac{4\pi n_1}{t}$

Moment of inertia of the flywheel $I = \dfrac{2mgh - mr^2\omega^2}{\omega^2\left(1+\dfrac{n}{n_1}\right)}$ kg – m^2

Record observations (2) and (3) as above

Mean moment of inertia $I =$ kg – m^2

Precautions. 1. There should be least possible friction in the flywheel. See that the flywheel starts of its own accord and no push is imparted to it. The mass tied to the end of the cord should be of such a value that it is able to overcome friction at the bearings and thus automatically starts falling.

2. The length of the string should be less than the height of the axle of the flywheel from the floor.

3. The loop slipped over the pin should be loose enough to be detached easily.

4. The string should be thin and should be wound evenly.

5. The stop-watch should be started just when the string is detached.

Source of error. (*i*) The angular velocity ω has been calculated on the assumption that the friction remains constant when the angular velocity decreases from ω to zero. In actual practice this is not the case because the friction increases as the velocity decreases.

(*ii*) The instant at which the string is detached cannot be correctly found out.

3.6. Rotation of a rigid body. The equations of motion of a rigid body rotating about a fixed axis, passing through it, with a uniform angular acceleration α are similar to the equations of motion of translation of a particle in a straight line moving with a uniform linear acceleration *a* as stated below.

Rotational motion	**Translational motion**
(*i*) $\omega = \omega_0 + \alpha t$	(*i*) $v = u + at$
(*ii*) $\theta = \omega_0 t + \frac{1}{2}\alpha t^2$	(*ii*) $s = ut + \frac{1}{2}at^2$
(*iii*) $\omega^2 - \omega_0^2 = 2\alpha\theta$	(*iii*) $v^2 - u^2 = 2as$

where the symbols have their usual meaning.

In linear motion force $\vec{F} = m\,\vec{a}$ and correspondingly in rotational motion $\vec{\tau} = I\,\vec{\alpha}$ where *I* is the moment of inertia of the rotating body.

In rotational motion torque is also given by

$$\vec{\tau} = \vec{r} \times \vec{F}$$

Here $\vec{r}$ is the vector distance of the point where the force $\vec{F}$ is applied from the axis of rotation.

Experiment 3.2 (*a*) **To find the angular acceleration α of a flywheel.**

(*b*) **To find the torque τ and hence to find the moment of inertia of the flywheel.**

Apparatus. Same as in **Exp 3.1.**

Procedure 1. Repeat steps 1, 2, 3 and 4 as in **Expt. 3.1.**

5. With the help of a stop-watch note the time taken by the mass *m* to descend through a height *h*. The value of *h* = length of the string wrapped round the axle and is given by $h = 2\pi rn$ where *r* is the radius of the axle.

6. (*i*) Repeat step 5 keeping *m* constant and varying the number of turns *n*. Take 6 readings.

(*ii*) Again repeat step 5 keeping the number of turns n (maximum) and constant and varying the mass m. Take 6 observations with different values of m. Repeat each observation thrice and calculate the average time taken in each observation.

Observations. Vernier constant =

Diameter of the axle = (*i*) (*ii*) (*iii*)

Mean diameter of the axle = m

Radius of the axle r = m

Acceleration due to gravity $g = 9.8$ ms^{-2}

(*a*) Mass attached to the string m = kg

Sl. No.	*Number of turns n*	*Time t*	t^2
1			
2			
3			
4			
5			
6			

(*b*) Number of turns on the axle n =

S. No.	*Mass attached m(kg)*	*time t* (*i*)	(*ii*)	(*iii*)	*mean t*	$\alpha = \frac{4\pi n}{t^2}$	$\tau = mgr$	$I = \tau/\alpha$
2								
3								
4								
5								
6								

(*a*) **To find the angular acceleration.** As the flywheel starts from rest, $\omega_0 = 0$ and equation of motion given by $\theta = \omega_0 t + \frac{1}{2}\alpha t^2$ becomes

$$\theta = \frac{1}{2}\alpha t^2 \qquad \text{Also } \theta = 2\pi n$$

or

$$n = \frac{\alpha}{4\pi} t^2$$

Hence a graph between n and t^2 gives a straight line passing through the origin. The slope of the line $= \frac{\alpha}{4\pi} = \frac{n}{t^2}$

or angular acceleration $\alpha = 4\pi \times$ **slope of the line** $= \frac{4\pi n}{t^2}$

Therefore, taking n along the Y-axis and t^2 along the X-axis plot a graph as shown in Fig. 3.4. Take a point A on the graph and draw AN perpendicular to the X-axis, then

Slope of the graph $\tan AON = \frac{AN}{ON}$

$$\therefore \qquad \frac{\alpha}{4\pi} = \frac{AN}{ON}$$

or $$\alpha = 4\pi \frac{AN}{ON} = \quad \text{rad } s^{-2}$$

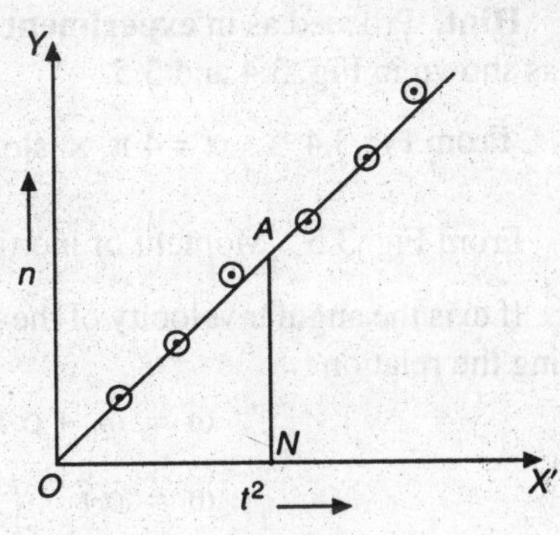

Fig. 3.4

NOTE. The dimensions of torque τ are the same as those of work. Its S.I. unit is a joule or kg-m^2s^{-2}.

(*b*) **To find the moment of inertia of a flywheel.** The moment of inertia of the flywheel can be calculated from the values of angular acceleration α and torque τ.

Suppose the mass *m*, when released starts moving downward with an acceleration *a*. Let *T* be the tension in the string. The tangential force acting on the axle of the wheel = *mg*

$$\therefore \qquad mg - T = ma$$

or $$T = mg - ma = m(g - a)$$

The torque τ on the axle responsible for rotating the fly wheel is given by.

Torque acting on the axle = Tension × radius of axle

or $$\tau = m(g - a)\, r$$

As *m* is very small as compared to the mass of the fly wheel and *a* is very small as compared to *g*, *ma* may be neglected as compared to *mg*

$$\therefore \qquad \tau = mg\, r$$

As $\tau = I\alpha$ where *I* is the moment of inertia of the fly wheel about the axis of rotation, the graph between torque τ and angular acceleration α is a straight line. The slope of the graph gives the moment of inertia of the fly wheel about the axis of rotation.

Hence moment of inertia I = Slope of the $(\tau - \alpha)$ graph $= \dfrac{\tau}{\alpha}$

Therefore, taking $\alpha = \dfrac{4\pi n}{t^2}$ as calculated in (*a*) along *X*-axis and torque $T = mgr$ along *Y*-axis plot a graph as shown in Fig. 3.5.

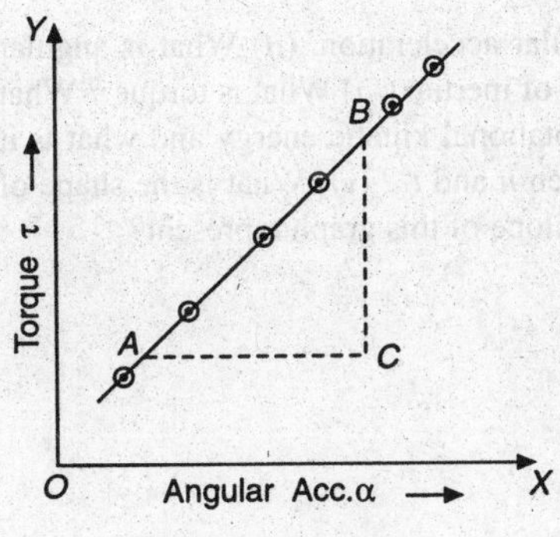

Fig. 3.5

Observations

Slope of the graph $= \dfrac{\tau}{\alpha} = \dfrac{BC}{AC} =$

$\therefore$ Moment of inertia $I = \dfrac{\tau}{\alpha} =$ kg – m^2

NOTE A part of the applied torque is always used up in overcoming friction between the flywheel and the bearings. Therefore, a graph between τ and α does not pass through the origin. It makes an intercept on the τ-axis. Let this intercept be τ_0, then this value gives the frictional torque and the net torque producing acceleratioin is $\tau - \tau_0$

Precautions same as in **Expt 3.1.**

Exerçise. *From the value of the angular acceleration* α *and time t from one set of observations, calculate the angular velocity* ω *and from this calculated value find the kinetic energy of the flywheel corresponding to this angular velocity.*

Hint. Proceed as in **experiment 3.2** and plot a graph between n and t^2 as well as between τ and α as shown in Fig. 3.4 and 3.5.

From Fig 3.4 $\alpha = 4\pi \times$ slope of the line $=$ rad s^{-2}

From Fig. 3.5 Moment of inertia $I = \frac{\tau}{\alpha} = \frac{BC}{AC} =$ kgm^2

If ω is the angular velocity of the fly wheel when the attached mass first touches the ground, then using the relation

$$\omega = \omega_0 + \alpha t \quad \text{we have}$$

$$\omega = \alpha t \quad [\because \omega_0 = 0]$$

$\therefore$ Kinetic energy acquired by the fly wheel $= \frac{1}{2} I \omega^2 =$ Joule

and kinetic energy acquired by the attached mass $= \frac{1}{2} mv^2$

$$= \frac{1}{2} mr^2 \omega^2 = \text{Joule.}$$

Oral Questions

1. Moment of Inertia. (*i*) Define moment of inertia and radius of gyration. (*ii*) What are the units of moment of inertia? (*iii*) State the theorem of parallel axis and perpendicular axis. (*iv*) What is the moment of inertia of a rectangular bar? (*v*) Why is it necessary to mention the axis ? (*vi*) What is the physical significance of moment of inertia? (*viii*) What is the moment of inertia of (*a*) A circular ring about an axis passing through its centre and perpendicular to its plane (*b*) A circular dise about an axis passing through its C.G and perpendicular to its plane (*c*) A right circular cylinder about an axis through its C.G and perpendicular to the axis of the cylinder (*d*) about an axis passing through its C.G and parallel to its length (*e*) A sphere about a diameter and (*f*) A thin spherical shell about a diameter.

2. Flywheel. (*i*) What is a flywheel and what is its practical advantage ? (*ii*) Why should the length of the string be less than the height of the axle from the ground ? (*iii*) About what axis do you find moment of inertia of the flywheel ? (*iv*) What are the methods of reducing friction at the ball bearings ? (*v*) What are ball bearings and roller bearings ? How do they reduce friction ?

3. Rotation of a rigid body. (*i*) Define angular velocity and angular acceleration. (*ii*) What is angular momentum ? State the relation between angular momentum and moment of inertia. (*iii*) What is torque ? What is the relation between torque and angular acceleration ? (*iv*) What is rotational kinetic energy and what is it equal to ? (*v*) How do you find angular acceleration from a graph between n and t^2? (*vi*) What is the shape of the graph between torque and angular acceleration ? (*vii*) What does the slope of this graph represent?

4

Acceleration due to Gravity

4.1. Acceleration due to gravity. Earth attracts bodies lying on or near its surface towards its centre. Bodies left free, move towards the centre of the earth with a constant acceleration g known as *acceleration due to gravity*. The value of this acceleration is 9.8 m/sec^2. The value of g varies from place to place on the surface of the earth, being greater on the poles than on the equator, greater at sea-level than on the mountain tops.

4.2. Simple pendulum. The acceleration due to gravity can be easily determined with a simple pendulum. *It consists of a heavy particle suspended by a weightless, inextensible and perfectly flexible string fixed from a point about which the pendulum oscillates without friction.* In practice, it is not possible to have such an *ideal* pendulum because neither we can get a single material particle nor a weightless and inextensible string. The nearest approach to an ideal simple pendulum consists of a *small metallic sphere, suspended from a fixed support by a very fine flexible cotton thread.*

If the amplitude is small, the time period t of a simple pendulum of length l is given by

$$t = 2\pi \sqrt{\frac{l}{g}}$$

A simple pendulum whose time period is two seconds is called a *second's pendulum.* Its length at a place where g is 9.8 m/sec^2 is 0.994 m.

4.3. Compound pendulum. The conditions of an ideal simple pendulum cannot be realised in actual practice. Most of the defects are removed by using a compound pendulum.

A compound pendulum consists of a rigid body which can oscillate freely about a horizontal axis passing through it.

Time period of a compound pendulum. Consider a rigid body of any shape and mass m capable of oscillating freely about a horizontal axis passing through it perpendicular to its plane.

Let O be the **centre of suspension** of the body and G its centre of gravity vertically below O at a distance l in the position of rest. When the body is displaced through a small angle θ to the dotted position, the centre of gravity is shifted to the position G' and its weight mg acts vertically downward at G'.

If the pendulum is now released a restoring couple acts on it and brings it back to the initial position. But due to inertia, it overshoots the mark and hence starts oscillating about the mean position.

The moment of the restoring couple or torque

Fig. 4.1

$$\tau = -mg \times G'A = -mg\, l \sin\theta = -mg\, l\, \theta \qquad ...(i)$$

since the angle θ through which the pendulum is displaced is small so that $\sin\theta = \theta$.

This restoring couple gives rise to an angular acceleration α in the pendulum. If I is the moment of inertia of the rigid body (pendulum) about an axis passing through its centre of suspension, then restoring couple (torque) is given by

$$\tau = I\alpha = I\,\frac{d^2\theta}{dt^2} \qquad \text{...}(ii)$$

Comparing equations (i) and (ii), we have

$$I\,\frac{d^2\theta}{dt^2} = -\,mg\,l\,\theta$$

or

$$\frac{d^2\theta}{dt^2} = -\,\frac{mg\,l}{I}\,\theta \qquad \text{...}(iii)$$

$\therefore$

$$\frac{d^2\theta}{dt^2} \propto \theta \qquad \left[\because\ -\frac{mgl}{I} = \text{constant}\right]$$

As the angular acceleration is proportional to angular displacement, the motion of the compound pendulum is simple harmonic and its time period T is given by

$$T = 2\pi\sqrt{\frac{\text{Angular displacement}}{\text{Angular acceleration}}} = 2\pi\sqrt{\frac{\theta}{\frac{mg\,l\,\theta}{I}}} = 2\pi\sqrt{\frac{I}{mg\,l}} \qquad \text{...}(iv)$$

If I_G is the moment of inertia of the body (or compound pendulum) about an axis parallel to the given axis but passing through C.G. (G), then according to the theorem of parallel axis, we have

$$I = I_G + ml^2$$

Now, $\quad I_G = mK^2$ where K is the radius of gyration

$\therefore$

$$I = mK^2 + ml^2 = m\,(K^2 + l^2)$$

Substituting the value of I in relation (iv), we get

$$T = 2\pi\sqrt{\frac{mK^2 + ml^2}{mgl}} = 2\pi\sqrt{\frac{K^2 + l^2}{lg}} \qquad \text{...}(v)$$

4.4. Length of an equivalent simple pendulum. Relation (v) can be put in the form

$$T = 2\pi\sqrt{\frac{\frac{K^2}{l} + l}{g}}$$

This relation shows that the time period of a compound pendulum is the same as that of a simple pendulum of length $L = \frac{K^2}{l} + l$. The length L is called **length of an equivalent simple pendulum** or **reduced length** of the pendulum.

Hence

$$L = \frac{K^2}{l} + l \quad \text{and} \quad T = 2\pi\sqrt{\frac{L}{g}}$$

Since K^2 is always a positive quantity, the length of an equivalent simple pendulum is always greater than l.

4.5. Centre of suspension. As stated above, *the point O through which the horizontal axis about which the pendulum vibrates, passes is called the centre of suspension.* If l_1 is the distance of O from the centre of gravity G, then

$$\text{Time period} \qquad T = 2\pi\sqrt{\frac{K^2 + l_1^2}{l_1 g}}$$

4.6. Centre of Oscillation. A point C on the other side of centre of gravity G and at a distance $l_2 = \frac{K^2}{l_1}$ from it is called the *centre of oscillation* as shown in Fig. 4.2.

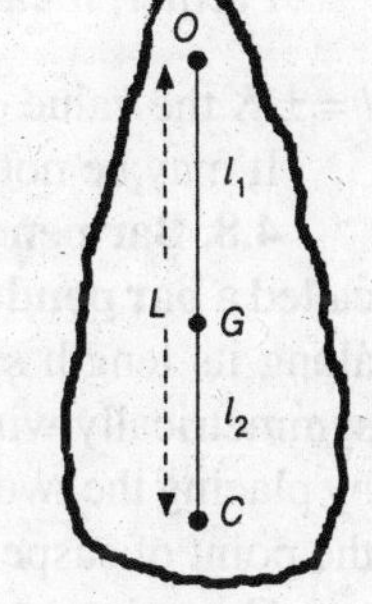

Fig. 4.2

Now
$$OC = OG + GC = l_1 + \frac{K^2}{l_1} = l_1 + l_2 = L$$

$$\therefore \qquad T = 2\pi\sqrt{\frac{l_1 + l_2}{g}} = 2\pi\sqrt{\frac{L}{g}}$$

Thus the time period of the compound pendulum about a horizontal axis through C is the same as about O.

The point C at a distance equal to the length of an equivalent simple pendulum from the point of suspension O on the straight line passing through the centre of gravity G is called the **centre of oscillation.**

4.7. Interchangeability of the centre of oscillation and centre of suspension. *The centre of suspension and the centre of oscillation are interchangeable.*

In fact there are four points two on either side of the centre of gravity and collinear with it about which the time period of the compound pendulum is the same. If we select two points on either side of $C.G$ but not equidistant from it, then one of these corresponds to the centre of suspension and the other to the centre of oscillation. If the distance of one of these from $C.G.$ is l_1 and that of the other is l_2, then

Equivalent length $L = l_1 + l_2$.

The time period of a compound pendulum

$$T = 2\pi\sqrt{\frac{K^2 + l_1^2}{l_1 g}} = 2\pi\sqrt{\frac{L}{g}}$$

$$\therefore \qquad \frac{K^2}{l_1} + l_1 = L = l_1 + l_2$$

$$\text{or} \qquad K^2 = l_1 l_2$$

$\therefore$ Radius of gyration of the compound pendulum $K = \sqrt{l_1 l_2}$

The time period of the compound pendulum is **minimum** when the distance of the point of suspension from $C.G.$ is equal to the radius of gyration.

To prove this put the relation for the time period

$$T = 2\pi\sqrt{\frac{K^2 + l^2}{lg}} \quad \text{as}$$

$$T^2 = \frac{4\pi^2}{g}\left(\frac{K^2 + l^2}{l}\right) = \frac{4\pi^2}{g}\left(\frac{K^2}{l} + l\right)$$

Differentiating this expression with respect to l, we get

$$2T\frac{dT}{dl} = \frac{4\pi^2}{g}\left(-\frac{K^2}{l^2} + 1\right)$$

For the value of time period T to be minimum or maximum

$$\frac{dT}{dl} = 0$$

or
$$1 - \frac{K^2}{l^2} = 0 \quad \text{or} \quad l = \pm K$$

Further, it can be proved that for $l = \pm K$ the value of $\frac{d^2T}{dl^2}$ is positive showing thereby that for $l = \pm K$ the value of **time period is minimum.**

It may be noted that for $l = 0$ or for $l = \infty$ the value of $T = \infty$

4.8. Bar pendulum. To find the value of g a special type of a compound pendulum is used. It is called **a bar pendulum** and is shown in Fig. 4.3. It consists of a uniform metal bar having holes drilled along its length symmetrically on either side of the centre of gravity. Two knife edges are placed symmetrically with respect to the *C.G.* as at *A* and *B*. The time period is determined about each hole by placing the two knife edges symmetrically about *C.G.* The distance of each of the knife edges (*i.e.*, the point of suspension) from the centre of gravity is measured in each case.

Experiment 4.1. To plot a graph between the distance of the knife-edges from the centre of gravity and the time period of a compound pendulum. From the graph find

(*a*) the acceleration due to gravity.

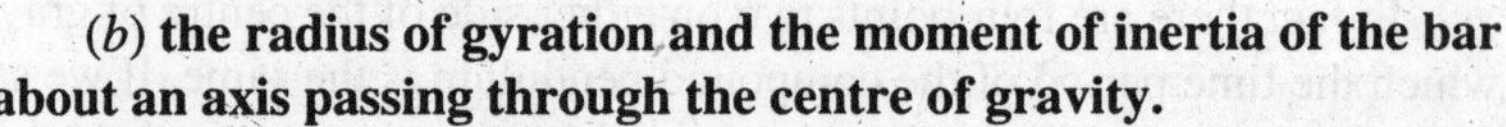

(*b*) the radius of gyration and the moment of inertia of the bar about an axis passing through the centre of gravity.

Apparatus. A compound pendulum, a wedge, a spirit level, a telescope, a stop-watch, a metre rod, a spring balance and a graph paper.

Theory. If a graph is plotted between the distance of the knife-edges from the centre of gravity taken along the *X*-axis and the corresponding time period *t* taken along the *Y*- axis for a bar pendulum, then the shape of the graph is as shown in Fig. 4.4.

If a horizontal line *ABCDE* is drawn, it cuts the graph in points *A*, *B* and *D*, *E* about which the time period is the same. The points *A* and *D* or *B* and *E* lie on opposite sides of the centre of gravity at unequal distances such that the time period about these points is the same. Hence one of these corresponds to the centre of suspension and the other to the centre of oscillation. The distance *AD* or *BE* gives the length of the equivalent simple pendulum *L*. If *t* is the corresponding time period, then

$$t = 2\pi\sqrt{\frac{L}{g}}$$

or
$$g = \frac{4\pi^2 L}{t^2}$$

If l_1 and l_2 are the distances of the point of suspension and the point of oscillation from the centre of gravity, then

Radius of gyration $K = \sqrt{l_1 \, l_2}$

and Moment of inertia $I = MK^2$

where M is the mass of the bar pendulum.

Fig. 4.3

Procedure. 1. Paste a small piece of paper on either end of the compound pendulum as shown in Fig. 4.3. Draw a line parallel to the edge of the pendulum on each paper to serve as a reference mark. Mark on the one side *A* and on the other side *B*.

Place the knife-edges in the first hole on either side parallel to each other and make them tight, so that the sharp edge is pointing towards the centre of gravity.

2. Place a spirit level on the glass plates fixed on the bracket in the wall meant for suspending the pendulum and see that the upper surfaces of the glass plates are in the same level.

3. Suspend the pendulum from the knife-edge on the side A so that the knife-edge is perpendicular to the edge of the slot and the pendulum is hanging parallel to the wall.

4. Adjust the eye-piece of the telescope so that the cross wires are clearly visible through it. Focus the telescope at the reference line of the compound pendulum. Turn the eye-piece so that the cross-wires are in the positions as shown in Fig. 4.3.

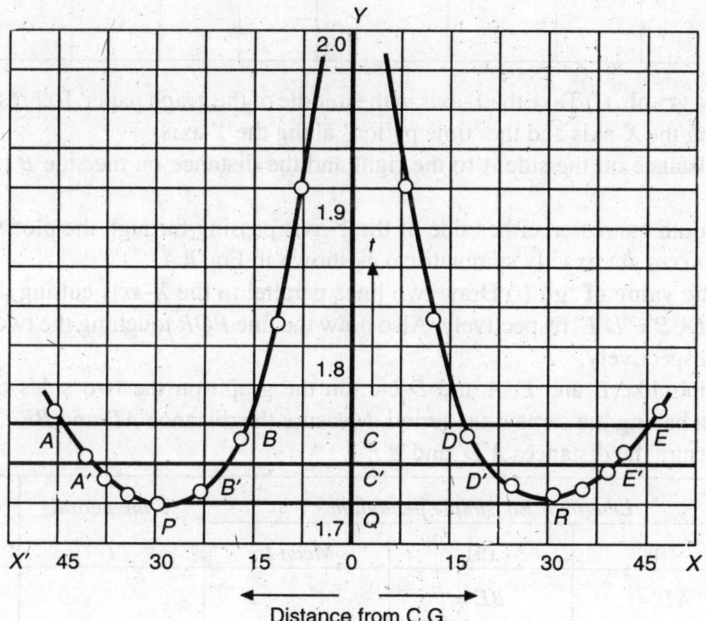

Fig. 4.4

5. Set the pendulum into vibration with a small amplitude of about 5° and allow it to make a few vibrations so that these become regular.

6. Look through the telescope and when the image of the reference mark passes across the point of intersection of the cross-wires, start the stop-watch and **count zero.** Count one when the pendulum is passing through *the same position in the same direction* and so on. Note the time taken for 20 vibrations. Repeat again and take the mean.

7. Measure the distance between the *C.G.* and the inner edge of the knife-edge.

8. Now suspend it on the knife-edge on the side B and repeat the observations.

9. Repeat the observations with the knife-edges in the 2nd, 3rd, 4th etc. holes on either side of the centre of gravity.

See that the knife-edges are always placed symmetrically with respect to C.G.

Observations.

Sl. No.	Side A					Side B				
	Time for 20 vibrations			Time period t	Distance from C.G.	Time for 20 vibrations			Time period t	Distance from C.G
	1	2	Mean			1	2	Mean		
1										
2										
3										
...										
...										
...										

10. **To plot the graph.** (*i*) Take the *Y*-axis in the middle of the graph paper. Represent the distance from the *C.G.* along the *X*-axis and the 'time period' along the *Y*-axis.

(*ii*) Plot the distance on the side *A* to the right and the distance on the side *B* to the left of the origin.

(*iii*) Draw smooth curves on either side of the *Y*-axis passing through the plotted points taking care that the two curves are exactly symmetrical as shown in Fig. 4.4.

11. **To find the value of 'g'.** (*i*) Draw two lines parallel to the *X*-axis cutting the curves at the points *ABCDE* and *A′B′C′D′E′* respectively. Also draw the line *PQR* touching the two portions of the graph at *P* and *R* respectively.

(*ii*) Select points like *B* and *E*, *A* and *D* etc. on the graph on the two sides of the *C.G.*, *not equidistant from it*, having the same time period. Measure the distance *AD* and *BE*.

Similarly measure the distances *A′D′* and *B′E′*.

Sl. No.	Length of Eq. simple pendulum			Time period	$\frac{L}{t^2}$
	(i)	(ii)	Mean L	t	
1	AD =	BE =			
2	A′D′ =	B′E′ =			

Mean $\frac{L}{t^2}$ =

$\therefore \quad g = \frac{4\pi^2 L}{t^2} = \quad ms^{-2}$

Actual value of g = 9.8 ms^{-2}

Percentage error =

12. **To find the radius of gyration.** Measure the distances *AC*, *CD*; *BC*, *CE* and *A′C′*, *C′D′*, *B′C′*, *C′E′*, as well as *PQ* and *QR*.

Sl. No.	l_1	l_2	$K = \sqrt{l_1 l_2}$
1	AC =	CD =	
2	CE =	BC =	
3	A′C′ =	C′D′ =	
4	C′E′ =	B′C′ =	
5	PQ =	QR =	

Mean value of *K* =

13. **Moment of inertia.** Find the mass of the pendulum.

Mass of pendulum M = kg

Radius of gyration K = m

$\therefore$ Moment of inertia MK^2 = kg – m^2

Precautions 1. The knife-edges should be horizontal and the bar pendulum parallel to the wall.

2. Amplitude should be small.

3. The time period should be noted after the pendulum has made a few vibrations and the vibrations have become regular.

4. The two knife-edges should always lie symmetrically with respect to the *C.G.*

5. The distance should be measured from the knife-edges.

6. The graph drawn should be a free-hand curve.

Sources of error. 1. Slight error is introduced due to (*i*) resistance of air, (*ii*) curvature of knife-edges, (*iii*) yielding of support and (*iv*) finite amplitude.

2. The stop-watch may not be very accurate.

Exercise. *Plot a graph between t^2l and l^2 for a compound pendulum and hence find the value of g the acceleration due to gravity and K the radius of gyration.*

Hints. For a compound pendulum.

$$t = 2\pi\sqrt{\frac{K^2 + l^2}{lg}} \quad \text{or} \quad t^2 = \frac{4\pi^2 K^2}{lg} + \frac{4\pi^2 l^2}{lg}$$

or
$$t^2 l = \frac{4\pi^2}{g} \cdot l^2 + \frac{4\pi^2}{g} \cdot K^2$$

Thus if a graph is plotted between t^2l taken along the Y-axis and l^2 taken along the X-axis, it will be a straight line as shown in Fig. 4.5.

The slope of the curve is the co-efficient of l^2 and is, therefore, given by,

$$\tan\theta = \frac{4\pi^2}{g}$$

Hence
$$g = \frac{4\pi^2}{\tan\theta}$$

The intercept on the Y-axis

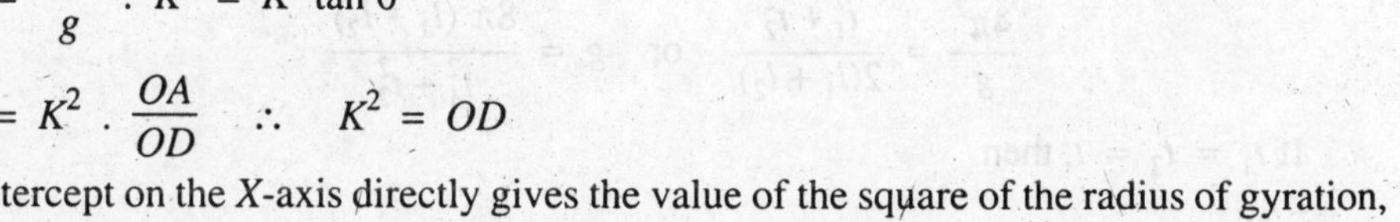

$$OA = \frac{4\pi^2}{g} \cdot K^2 = K^2 \tan\theta$$

or
$$OA = K^2 \cdot \frac{OA}{OD} \quad \therefore \quad K^2 = OD$$

Fig. 4.5

Hence the intercept on the X-axis directly gives the value of the square of the radius of gyration, *i.e.*, K^2.

4.9. Kater's pendulum. It is a compound pendulum constructed on the principle that centre of oscillation and centre of suspension are interchangeable. It consists of a brass or a steel bar capable of vibrating about two adjustable knife edges K_1 and K_2 facing each other. Two metal weights W and w can be made to slide, along the length of the bar and can be clamped in any position. A wooden weight W' exactly similar to the metal weight W can also slide along the bar. The smaller weight w is placed between the two knife-edges and the bigger weight W is fixed at one end while the wooden weight W' is kept symmetrically at the other end of the bar as shown in Fig. 4.6. In this position the centre of gravity lies in-between and near one of the knife-edges.

The position of the two knife edges and the weight w are so adjusted that the time-periods of the pendulum about the two knife-edges are equal. In such a case one knife-edge is at the centre of oscillation with respect to the other and the distance between the two knife-edges is equal to the length of an equivalent simpie pendulum L. If t is the time period, then the acceleration due to gravity g can be found out from the relation

$$t = 2\pi \sqrt{\frac{L}{g}}$$

It is, however, very difficult and tedious to make the time of oscillation about the two knife-edges *exactly* equal. Bassel showed that it was not absolutely necessary to make the time-periods about the two knife-edges exactly equal. The calculations of the value of g can be made if time periods were very *nearly* equal, the difference being not more than .01 second.

If t_1 and t_2 are time-periods about the two knife-edges K_1 and K_2 respectively and l_1 the distance of the knife-edge K_1 and l_2 that of the knife-edge K_2 from the centre of gravity, then

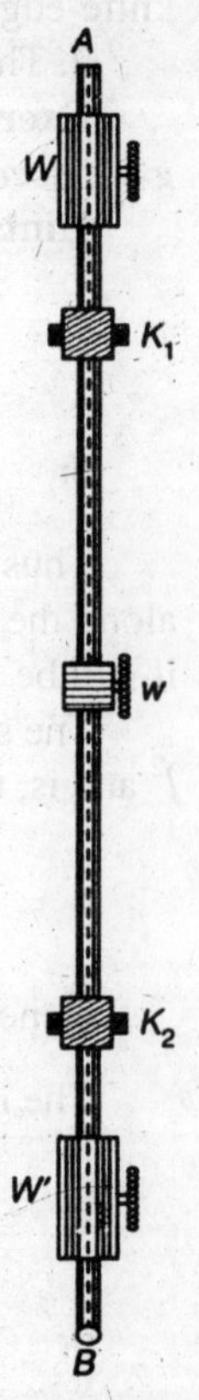

Fig. 4.6

$$t_1 = 2\pi \sqrt{\frac{k^2 + l_1^2}{l_1 g}} \quad \text{and} \quad t_2 = 2\pi \sqrt{\frac{k^2 + l_2^2}{l_2 g}}$$

or
$$t_1^2 l_1 = \frac{4\pi^2}{g}\left(K^2 + l_1^2\right) \qquad \text{...(i)}$$

and
$$t_2^2 l_2 = \frac{4\pi^2}{g}\left(K^2 + l_2^2\right) \qquad \text{...(ii)}$$

Subtracting (*ii*) from (*i*), we get

$$\frac{4\pi^2}{g} = \frac{t_1^2 l_1 - t_2^2 l_2}{(l_1 + l_2)(l_1 - l_2)} = \frac{t_1^2 + t_2^2}{2(l_1 + l_2)} + \frac{t_1^2 - t_2^2}{2(l_1 - l_2)}$$

$$\therefore \quad g = \frac{8\pi^2}{\dfrac{t_1^2 + t_2^2}{l_1 + l_2} + \dfrac{t_1^2 - t_2^2}{l_1 - l_2}}$$

If t_1 and t_2 are very nearly equal and l_1 and l_2 differ by a fairly large amount the second term will be very small and hence if its value is determined with only a fair degree of accuracy, the error in the result will be practically negligible.

If the second term is negligible, then

$$\frac{4\pi^2}{g} = \frac{t_1^2 + t_2^2}{2(l_1 + l_2)} \quad \text{or} \quad g = \frac{8\pi^2(l_1 + l_2)}{t_1^2 + t_2^2}$$

If $t_1 = t_2 = t$, then

$$\frac{4\pi^2}{g} = \frac{t^2}{(l_1 + l_2)}$$

or
$$g = \frac{4\pi^2(l_1 + l_2)}{t^2}$$

Experiment 4.2. To determine the value of acceleration due to gravity at a place with Kater's pendulum.

Apparatus. Kater's pendulum, a telescope, a stop watch and a metre rod.

Formula.
$$g = \frac{8\pi^2}{\dfrac{t_1^2 + t_2^2}{l_1 + l_2} + \dfrac{t_1^2 - t_2^2}{l_1 - l_2}}$$

Procedure. 1. Shift the weight *W* to one end *A* of Kater's pendulum and fix it. Fix the knife-edge K_1 just below it as shown in Fig. 4.6.

The weight W is kept at one end of the rod to shift the centre of gravity of the pendulum to one side. The two points about which the time-period is the same will now lie at unequal distances from the C.G. on either side of it.

2. Keep the knife-edge K_2 at the other end *B* and the small weight *w* nearly in the centre. Fix the wooden weight *W′* at the end *B* in a position symmetrical to the metal weight *W*. Since the pendulum is thus made symmetrical about its middle point, the error due to air drag is minimised.

3. Suspend the pendulum about the knife-edge K_1 and set it in to vibrations with a *small amplitude*. Start the stop-watch when the pendulum is just passing through its equilibrium position and count *zero*.

Count one when the pendulum again passes through its equilibrium position in the same direction and so on. Note the time taken for 20 vibrations.

4. Now suspend the pendulum about the knife-edge K_2 and find the time for 20 vibrations again. Adjust the position of the knife-edge K_2 so that time-period of the pendulum about K_2 is very nearly the same as that about the knife edge K_1.

5. Again suspend the pendulum about the knife-edge K_1 and note the time for 20 vibrations.

The time-period about K_1 will be different from the time period found in step 3. It is because the position of *C.G.* has shifted due to the change in position of the knife-edge K_2.

6. Again suspend the pendulum about the knife-edge K_2 and adjust its position *very slightly till the time-period about K_2 is as nearly equal to the time-period about K_1 as possible.*

7. Adjust the eye-piece of the telescope so that the cross-wires are clearly visible. Focus the telescope at the Kater's pendulum. Find the time for 100 vibrations about K_1 first and then about K_2 three times. The two time-periods may differ by 1% but not more.

8. Balance the pendulum on a sharp wedge and mark the position of its centre of gravity. Measure the distance of the knife-edge K_1 as well as that of K_2 from the centre of gravity.

Observations.

Knife edge	*Time for 100 vibrations*				*Time period*
	(i)	*(ii)*	*(iii)*	*Mean*	
K_1					$t_1 =$
K_2					$t_2 =$

Distance of the knife-edge K_1 from *C.G.*, $l_1 =$ m

Distance of the knife-edge K_2 from *C.G.*, $l_2 =$ m

$\therefore \quad l_1 + l_2 =$ m and $l_1 - l_2 =$ m

Hence
$$g = \frac{8\pi^2}{\dfrac{t_1^2 + t_2^2}{l_1 + l_2} + \dfrac{t_1^2 - t_2^2}{l_1 - l_2}} = \text{ ms}^{-2}$$

Actual value of g = ms^{-2}

∴ % error =

Precautions. 1. The heavy weight should be placed at one end so that the *C.G.* lies near one of the knife-edges and wooden weight *W'* symmetrically at the other end to avoid error due to air drag.

2. The two knife-edges should be parallel to each other.

3. The amplitude of vibration should be small so that the motion of the pendulum satisfies the condition $\sin \theta = \theta$

4. To avoid any irregularity of motion the time-period should be noted after the pendulum has made a few vibrations and the vibrations have become regular.

5. For final observation the time of at least 100 vibrations must be taken with an accurate stop-watch.

Sources of error. Same as in **Experiment 4.1.**

Experiment 4.3. To determine the moment of inertia of a rectangular bar about an axis through its C.G. by the method of coincidence.

Apparatus. A compound pendulum (rectangular bar) or a Kater's pendulum, a simple pendulum, a telescope, a wedge, a metre rod and a vernier callipers.

Theory. The radius of gyration and the moment of inertia of a compound pendulum can be determined by the **method of coincidence.** This method of timing a pendulum consists in suspending a simple pendulum in front of a compound pendulum. If both the pendulums have the same time-period and they start oscillating simultaneously, they will vibrate as one. If however, their time-periods are different, one will vibrate faster than the other and thus they will go out of step.

Suppose once the two pendulums together pass through the same point in the same direction. *After some time when one of the pendulums has made one vibration more than the other, they will again pass through the same point in the same direction and a* **coincidence** *is said to take place.*

Let the time-period of the compound pendulum be t and that of simple pendulum t_1. Suppose the compound pendulum makes $(n - 1)$ vibrations during the time the simple pendulum makes n vibrations *i.e.*, the simple pendulum is vibrating faster than the compound pendulum, then

$$(n-1)\,t = nt_1 \qquad ...(i)$$

But
$$t = 2\pi\sqrt{\frac{K^2+l^2}{lg}}$$

and
$$t_1 = 2\pi\sqrt{\frac{L}{g}}$$

where l, K, g, and L have their usual meanings.

Substituting the values of t and t_1 in equation (i), we have

$$(n-1)\,2\pi\sqrt{\frac{K^2+l^2}{lg}} = n\,2\pi\sqrt{\frac{L}{g}}$$

or
$$\frac{K^2+l^2}{l} = \left(\frac{n}{n-1}\right)^2 L \qquad ...(ii)$$

∴
$$K^2 = \left(\frac{n}{(n-1)}\right)^2 lL - l^2$$

If M is the mass of the rectangular bar, then

Moment of inertia = MK^2

Procedure. 1. (*i*) Paste a long strip of paper on the compound pendulum and draw a pencil line on it passing through the centre and parallel to the length of the pendulum.

(*ii*) Suspend the compound pendulum from a knife-edge at one of its ends from a wall bracket. Also suspend a simple pendulum symmetrically in front of the compound pendulum, from a fine slit in the same wall bracket so that the thread exactly covers the line drawn on the compound pendulum.

2. Adjust the length of the thread to about 10 cm less than the length of the compound pendulum, so that the time- period of the simple pendulum is less than that of the compound pendulum. The simple pendulum will make a greater number of vibrations in the same time than the compound pendulum does and hence it møves faster.

3. Set both the pendulums into vibration simultaneously. Start counting the number of vibrations made by the simple pendulum when it moves together with the compound pendulum through the same point in the same direction. Count the first vibration *zero*. Count the number of vibrations up to the instant when they *again pass simultaneously through the same point in the same direction.*

The number of vibrations between these two coincidences should not be less than 60.

4. If the number of vibrations after which coincidence takes place is less than 60, increase the length of the simple pendulum by one cm and proceed as in the last step. See whether the number of vibrations after which the coincidence takes place increases or decreases. If it increases then go on increasing the length till coincidence takes place after 60 or more vibrations.

If it decreases then decrease the length of the simple pendulum instead of increasing it.

To focus the telescope. (*i*) Adjust the position of the eye- piece so that the cross-wires are clearly visible.

(*ii*) Focus the telescope on the reference line near the lower end of the compound pendulum.

(*iii*) Rotate the telescope to make one of the cross-wires vertical.

(*iv*) Shift the position of the telescope so that the vertical cross-wire, the simple pendulum and the reference line are seen overlapping through the telescope.

(*v*) Set the two pendulums into vibrations with a small amplitude. Find the number of vibrations made by the simple pendulum between two coincidences.

Repeat three times.

Do not disturb the adjustment while counting finally.

Note. *When the coincidence takes place after about 100 vibrations it is not possible to estimate correctly within 2 or 3 vibrations when the coincidence actually takes place. The error caused by making a mistake of ± 2 vibrations in counting 100 vibrations is, however, hardly 1 in 5000.*

Suppose the coincidence takes place after 100 vibrations and it is not possible to judge ± 2 vibrations, then

$$\{(100 \pm 2) - 1\}\, t = (100 \pm 2)\, t_1$$

or

$$\frac{t}{t_1} = \frac{1}{1 - \dfrac{1}{100 \pm 2}} = 1 + \frac{1}{100 \pm 2}$$

$$= 1 + \frac{1}{100}\left(\frac{1}{1 \pm \dfrac{2}{100}}\right) = 1 + \frac{1}{100} \pm \frac{2}{100^2}$$

The error is only 2 in 100^2 or 1 in 5000.

6. Find the length of the simple pendulum from the point of suspension to the top of the bob. Change its length slightly and again find the number of vibrations after which coincidence takes place.

The length should be changed in such a way that the coincidence takes place between 60 and 100 vibrations.

7. Find the diameter of the bob of the simple pendulum with a vernier callipers.

8. Determine the *C.G.* of the compound pendulum by supporting it on a wedge and find the length l between the *C.G.* and the lower edge of the knife-edge. Find the mass M of the compound pendulum.

Observations. Vernier constant = cm

Diameter of the bob = cm

Radius of the bob r = cm = m

Sl. No.	*Number of vibrations of simple pendulum*				*Length of the simple pendulum x*	*Correct length L = x + r (in metres)*	*Length of Compound pendulum l*	$K^2 = \left(\frac{n}{n-1}\right)^2 lL - l^2$
	(i)	(ii)	(iii)	Mean n				
1								
2								
.								
.								
.								
6								

Mass of the compound pendulum M = kg

Moment of inertia of the compound pendulum MK^2 = kg – m²

Note. *In the above experiment it has been supposed that the simple pendulum vibrates faster than the compound pendulum. On the other hand if the simple pendulum vibrates slower than the compound pendulum the formula will be*

$$K^2 = \left(\frac{n}{n+1}\right)^2 lL - l^2$$

Precautions. 1. Same as in **Experiment 4.1.**

2. The coincidence should take place between 60 and 100 vibrations.

Exercise. *Find the length of an equivalent simple pendulum for a given compound pendulum by the coincidence method.*

Hints : We know that

$$\frac{K^2 + l^2}{l} = \left(\frac{n}{n-1}\right)^2 L$$

∴ The length of an equivalent simple pendulum $= \left(\frac{n}{n-1}\right)^2 L$

4.10. Determination of radius of gyration. The time period of vibration of a rigid body is given by

$$T = 2\pi \sqrt{\frac{I}{mgl}}$$

where I is the moment of inertia about the axis of rotation.

If I_0 is the moment of inertia of the body about an axis through *C.G.* and l the perpendicular distance of the axis of rotation from *C.G.*, then

$$I = I_0 + ml^2 = mK^2 + ml^2$$

where K is the radius of gyration.

$$\therefore \qquad T = 2\pi \sqrt{\frac{K^2 + l^2}{lg}}$$

or

$$\frac{T^2 lg}{4\pi^2} = K^2 + l^2$$

$$\therefore \qquad K^2 = \frac{T^2 lg}{4\pi^2} - l^2$$

Experiment 4.4. (*a*). To find the radius of gyration of objects of different geometrical shapes but of the same mass by noting the time period of oscillation.

(*b*). To show that the time period is independent of the mass so long as the shape and size remain the same.

Apparatus. Two thin rectangular plates of same mass and different sizes, an annular disc of the same mass, two circular discs one having the same mass as the other three objects and the second circular disc of the same size but double the mass, a knife edge, stop-watch and telescope.

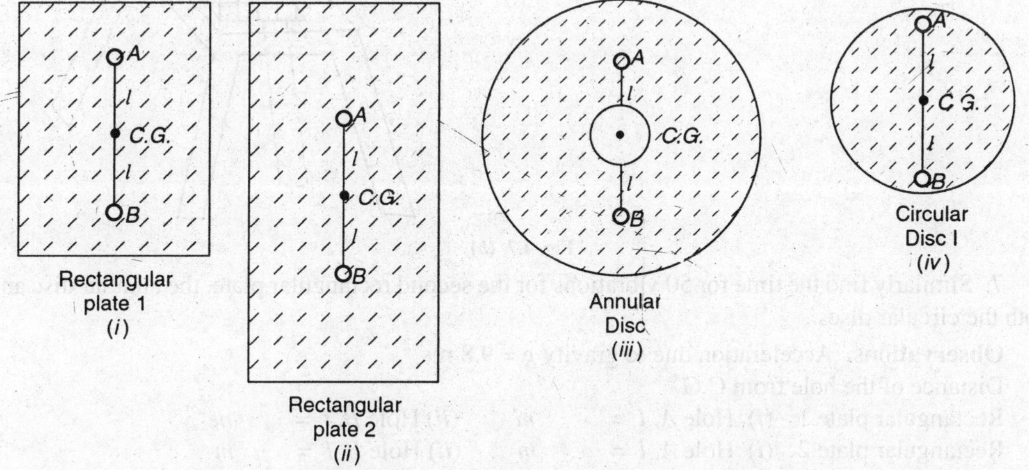

Fig. 4.7 (*a*)

Procedure. 1. Note the position of *C.G* of the two rectangular plates, the annular disc and the two circular discs. See that the centre of the hole in each case is situated at the same distance *l* from the *C.G*.

2. Check the mass of each of the two rectangular plates, the annular disc and one circular disc either by a sensitive physical balance or by a spring balance and see that each of them has the same mass.

3. Fix a knife edge in the hole at the point *A* at a distance *l* from the *C.G* in one rectangular plate and draw a reference line or mark at the lower edge.

4. Suspend the plate on two smooth glass plate supports with the help of the knife edge as shown in Fig. 4.7(*b*). Focus the vertical reference mark through the telescope placed at a distance of about two metres.

5. Set the plate into uniform vibrations of small amplitude and with the help of a stop watch find the time taken for 50 vibrations. Repeat thrice.

6. Invert the plate and fix it at *B*. Again find the time for 50 vibrations. Repeat thrice.

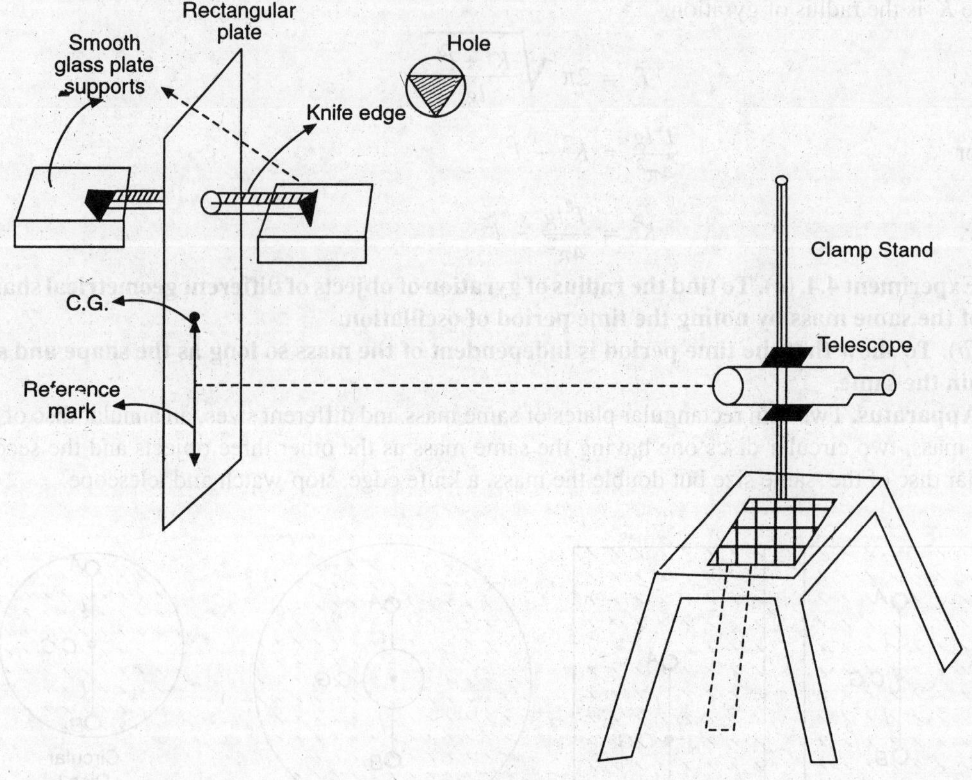

Fig. 4.7 (*b*)

7. Similarly find the time for 50 vibrations for the second rectangular plate, the annular disc and both the circular discs.

Observations. Acceleration due to gravity $g = 9.8 \text{ ms}^{-2}$.

Distance of the hole from C.G

Rectangular plate 1 (*i*) Hole *A*, $l =$ *m* (*ii*) Hole *B*, $l =$ *m*

Rectangular plate 2 (*i*) Hole *A*, $l =$ *m* (*ii*) Hole *B*, $l =$ *m*

Annular disc (*i*) Hole *A*, $l =$ *m* (*ii*) Hole *B*, $l =$ *m*

Circular disc 1 (*i*) Hole *A*, $l =$ *m* (*ii*) Hole *B*, $l =$ *m*

Circular disc 2 (*i*) Hole *A*, $l =$ *m* (*ii*) Hole *B*, $l =$ *m*

Mass of Rectangular plate 1 = kg

Mass of Rectangular plate 2 = kg

Mass of Annular disc = kg

Mass of Circular disc 1 = kg.

Object	*Time for 50 vibrations*							*Time period T*	T^2	$K^2 = \frac{T^2 lg}{4\pi^2} - l^2$
	Hole A			*Hole B*			*Mean*			
	(*i*)	(*ii*)	(*iii*)	(*i*)	(*ii*)	(*iii*)				
Rectangular plate 1										
Rectangular plate 2										
Annular disc										
Circular disc 1										
Circular disc 2										

Calculations and verification. (*i*) *Rectangular plates*

Measure the length *a* and breadth *b* of each of the two rectangular plates, then if *M* is the mass

Moment of inertia $I = \frac{M(a^2 + b^2)}{12}$

or $K^2 = \frac{a^2 + b^2}{12}$

For rectangular plate 1 $a =$ m ; $b =$ m

$\therefore$ $K^2 = \frac{a^2 + b^2}{12} =$ m^2

For rectangular plate 2 $a =$ m ; $b =$ m

$\therefore$ $K^2 = \frac{a^2 + b^2}{12} =$ m^2

(*ii*) *Annular disc.* Measure the internal radius *r* and external radius *R* of the annular disc, then

Moment of inertia $I = M\frac{(R^2 + r^2)}{2}$

or $K^2 = \frac{R^2 + r^2}{2} =$ m^2

(*iii*) *Circular discs.* Measure the radius *R* of each of the circular discs then

Moment of inertia $I = M\frac{R^2}{2}$

or $K^2 = \frac{R^2}{2}$

For circular disc 1 $R =$ m

$\therefore$ $K^2 = \frac{R^2}{2} =$ m^2

For circular disc 2 $R =$ m

$\therefore$ $K^2 = \frac{R^2}{2} =$ m^2

Compare these calculated values of K^2 in each case with the values determined experimentally.

Conclusion. 1. As the time period is different for each body, the moment of inertia does not only depend upon mass as other wise it would have been equal.

2. As *K* is different in all the cases and M.I depends upon *K* being equal to MK^2, we conclude that the M.I of an object depends upon the distribution of mass of the body about the axis of rotation *i.e* on its shape and size.

(*b*) Measure the mass of each of the two circular discs.

The time period of the circular discs is the same although the mass of disc 2 is double that of the disc 1. This is because their shapes and sizes are the same *i.e.* they have the same value of *K*. As the value of *l* has also been kept constant

$$\therefore \quad T = 2\pi\sqrt{\frac{K^2 + l^2}{lg}} = \text{a constant}$$

This shows that time period is independent of the mass so long as the shape and size is the same.

Note. The two circular discs have the same value of *K* so long as these are thin and have the same radius. The mass of one will be double of the other if its thickness is twice the other.

Precautions 1. The knife edges should be sharp and the point of suspension should not slip.

2. The amplitude should be small.

3. The time period should be calculated by observing accurately the time of 50 vibrations.

4. The telescope should be focussed from a distance of about 2 metres.

Sources of error 1. There is always some air resistance to the motion, which is different foı different shapes of the oscillating objects.

2. The masses of the oscillating objects may not be exactly equal.

3. The distance of the point of oscillation from the *C.G* of each object may not be exactly the same.

Oral Questions

1. Gravity and gravitation. (*i*) What is *g*? (*ii*) What is the difference between '*G*' and '*g*' ? (*iii*) How does *g* vary from place to place on the surface of earth ? (*iv*) What is the effect of altitude, depth and rotation on the value of *g* ? (*v*) Is the value of '*g*' the same on the moon as on the earth ?

2. Simple pendulum. (*i*) What is a simple pendulum ? Why can it not be realised in actual practice ? (*ii*) What is a second's pendulum ? (*iii*) What is the value of '*g*' at your place and what are its units ?

3. Compound pendulum. (*i*) What is a compound pendulum ? (*ii*) What do you mean by 'centre of suspension' and 'centre of oscillation' ? (*iii*) What is an 'equivalent length' of a compound pendulum ? (*iv*) What is the distance of the point having the minimum time-period from the centre of gravity? (*v*) What is the time-period of a compound pendulum at its *C.G.* ? (*vi*) About how many points the time-period of a compound pendulum is the same ? (*viii*) Why do you place the knife-edges symmetrically ?

4. Kater's pendulum. (*i*) What is Kater's pendulum ? (*ii*) Why do you place the heavy mass at one end ? (*iii*) Is it necessary to have the two time-periods exactly equal ? How will you find the value of *g* if the two time-periods are not exactly equal ? (*iv*) What is the use of wooden weight ? Where should it be placed ?

5. Coincidence method. (*i*) How will you find out which pendulum moves faster ? (*ii*) Why is it necessary to have the coincidence between 60 and 100 vibrations ?

6. Radius of gyration. (*i*) Define radius of gyration. What are its units ? (*ii*) What is the radius of gyration of (*a*) a rectangular lamina (*b*) an annular ring and (*c*) a circular disc. (*iii*) Does radius of gyration depend upon mass ? (*iv*) What is the effect of shape and size on the radius of gyration ?

5

Elastic Constants

5.1. Elasticity. An external deforming force acting on a body may produce a change in length, volume or shape. On withdrawing the force the body more or less regains its original length, volume or shape.

This property of a body to regain its original length, volume or shape when the deforming forces have been removed is called elasticity.

When an external force acts on a body relative displacement of its various parts takes place. The displaced particles tend to come to their original positions to restore the original length, volume or shape of the body and thus exert a restoring force.

The restoring force per unit area called into play inside the body is called stress.

The reaction set up in the body is equal and opposite to the applied force, so long as there is no permanent change produced in the body. The restoring force is, therefore, equal to the applied force. Hence if a force F is applied normally to the area of cross section a of a wire, then

$$\text{Stress} = \frac{F}{a}$$

S.I. units of stress are Newton per square metre (Nm^{-2})

The external force acting on the body causes a relative displacement of its various parts. A change in length, volume or shape takes place. The body is then said to be under a strain.

Strain *is defined as the ratio of change in length, volume or shape to its original length, volume or shape.* Strain is thus a pure ratio and has no units.

5.2. Young's modulus. It is defined as the ratio of stress to longitudinal strain within elastic limits. The longitudinal strain is measured by observing the change in length per unit length. If L is the length of a wire and an increase in length l is produced by a force F, then

$$\text{Strain} = \frac{l}{L}$$

$$\therefore \quad \text{Young's modulus } Y = \frac{\text{stress}}{\text{strain}} = \frac{F}{a} \div \frac{l}{L} = \frac{FL}{al}$$

In *S.I.* units, young's modulus is expressed in Newton per square metre (N/m^2)

5.3 Cantilever. *It is a beam fixed horizontally at one end and loaded at the other.*

Suppose AB represents the *neutral axis* of the cantilever of length l fixed at A and loaded at the end B with a load W. Let the end B be deflected to the position B' under the action of the load W.

Consider a section of the beam at P at a distance x from A, then moment of the couple due to the load W

$$= W \times PB' = W(l - x)$$

As the point B' lies very close to B, PB' is almost perpendicular to BW.

As the beam is in equilibrium, the bending couple is equal to the restoring couple $\frac{YI}{R}$ where R is the radius of the neutral axis at P, Y the Young's modulus and I the geometrical moment of inertia. Hence

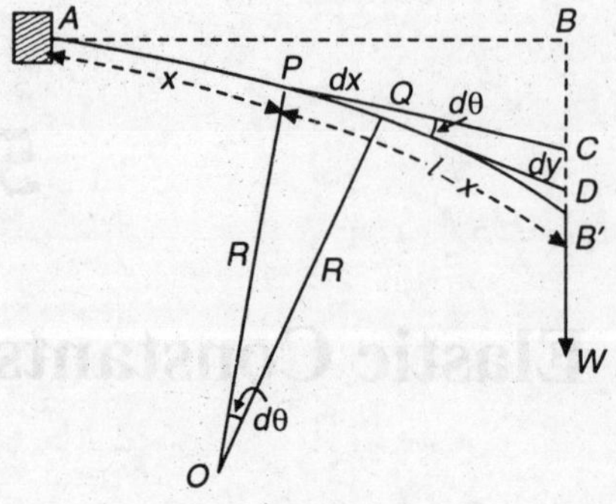

Fig. 5.1

$$W(l-x) = \frac{YI}{R}$$

or

$$\frac{1}{R} = \frac{W(l-x)}{YI}$$

As we move towards the fixed end A the moment of the load increases. Hence the radius of curvature will be different at different points and decreases as we move towards A. Consider another point Q lying very close to P at a distance dx. The radius of curvature at Q is practically the same as dx is very small :

$$\therefore \quad PQ = R \, . \, d\theta = dx$$

or

$$d\theta = \frac{dx}{R} = \frac{W(l-x)dx}{YI}$$

Draw tangents at P and Q meeting the vertical line at C and D respectively, then depression of Q below P, is given by

$$CD = dy = (l-x) \, d\theta$$

where $d\theta$ is the angle between the two tangents. Thus

$$dy = \frac{(l-x) \, W(l-x)dx}{YI} = \frac{W}{YI}(l-x)^2 \, dx$$

$\therefore$ Total depression BB'

$$y = \int_0^l \frac{W}{YI}(l-x)^2 \, dx = \int_0^l \frac{W}{YI}(l^2 + x^2 - 2lx) \, dx$$

$$= \frac{W}{YI}\left[l^3 + \frac{l^3}{3} - l^3\right] = \frac{Wl^3}{3YI}$$

5.4. Geometrical moment of inertia. *The geometrical moment of inertia of a beam is defined as its moment of inertia if it has a unit mass per unit area.*

(*i*) For a beam of rectangular cross section

$$I = \frac{bd^3}{12}$$

where b is the breadth and d the thickness of the beam.

(*ii*) For a beam of circular cross-section

$$I = \frac{\pi r^4}{4}$$

where r is the radius of the beam.

5.5. Double cantilever. If a bar is supported at two knife edges A and B, l metre apart in a horizontal plane so that equal lengths of the bar project beyond the knife edges and a weight W is suspended at the middle point C, then it acts as a double cantilever.

The middle part of the bar is practically horizontal. It is, therefore, equivalent to two inverted cantilevers fixed at the middle point C and loaded at A and B with a load $W/2$ acting upward. The depression y at C is given by

$$y = \frac{W/2\,(l/2)^3}{3YI} = \frac{Wl^3}{48\,YI}$$

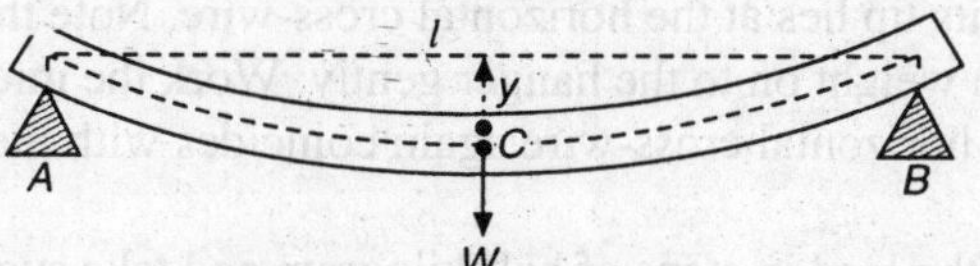

Fig. 5.2

For a rectangular bar of breadth b and thickness d,

$$I = \frac{bd^3}{12}$$

$\therefore$ Depression $y = \dfrac{Wl^3}{4Ybd^3}$

Experiment 5.1. To find the Young's modulus of the material of a rectangular bar by bending.

Apparatus. A uniform rectangular bar of brass or iron about one metre long, 3 cm wide and 0.5 cm thick, two knife-edges fixed on rigid supports, a stirrup with a hook and having a pointer, half kilogram weights, a spirit level, a metre rod, a travelling microscope, a vernier callipers and a screw gauge.

Theory. Consider a bar supported on two knife-edges l m apart in a horizontal plane, so that equal lengths of the bar project beyond the knife-edges. If a weight mg is suspended at the middle point, a depression y is produced.

For a rectangular bar of breadth b and thickness d the depression

$$y = \frac{Wl^3}{4Ybd^3} = \frac{mgl^3}{4Ybd^3}$$

where Y is the Young's modulus of the material of the bar.

$$\therefore \qquad Y = \frac{mgl^3}{4ybd^3}$$

Procedure. 1. Find the centre of gravity of the given bar and draw a sharp transverse line at the position of the *C.G.* Mark points on either side of *C.G.* at equal distance of 0.3 to 0.4 m each. Draw sharp lines at these points as well.

2. Place the bar on two rigid knife-edges as shown in Fig. 5.3 so that it rests on the two marked lines equidistant from *C.G.* Make the bar horizontal and test with a spirit level.

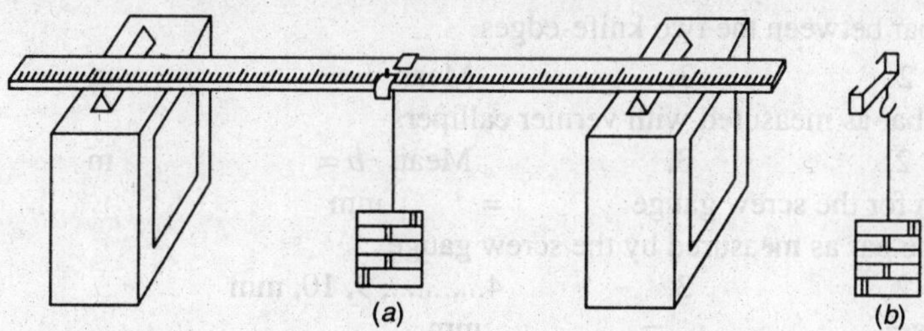

Fig. 5.3

3. Slip the stirrup on the bar and adjust its position so that the tip of the needle lies vertically above the line marked at *C.G.* Suspend the hanger from the hook.

4. Find the vernier constant of the vertical scale of the microscope. Adjust the position of the cross-wires so that the cross-wires are clearly visible and one of the cross-wires is horizontal. See that the microscope tube is also horizontal. Adjust if necessary. Focus the microscope on the pointer fixed to the stirrup so that its tip lies at the horizontal cross-wire. Note the reading on the scale.

5. Slip a half kilogram weight on to the hanger gently. Work the micrometer screw at the top of the microscope so that the horizontal cross-wire again coincides with the tip of the needle. Note the reading.

6. Gradually increase the load in steps of half kilogram and take such eight observations.

7. Now decrease the load in steps of half kilogram and take the readings as before. Thus for a given load there will be two readings, one while the load is increasing and the other while the load is decreasing.

8. Remove the bar carefully without disturbing the position of the knife-edges. Place a metre rod across the knife-edges with its graduated side vertical and measure the distance between the two knife-edges accurately.

9. Measure the breadth of the bar with a vernier callipers and thickness with a screw gauge at least at 10 different points.

10. Plot a graph between the load and the mean of corresponding microscope readings on loading and unloading. Find from the graph the mean depression for two kilograms.

Observations. Vernier constant of the microscope = mm

Sl. No.	*Load in kg*	*Microscope reading*		*Mean*	*Depression for 2 kg (y)*
		Load increasing	*Load decreasing*		
1	0				
2	0.5				
3	1				
4	1.5				
5	2				
6	2.5				
7	3				
8	3.5				
9	4				

Mean depression for 2 kg by calculations from graph

$y =$ mm = m

Length of the bar between the two knife-edges

1. 2. 3. Mean $l =$ m

Breadth of the bar as measured with vernier callipers

1. 2. 3. Mean $b =$ m

Zero correction for the screw gauge = mm

Thickness of the bar as measured by the screw gauge

1. 2. 3. 4.........., 9, 10, mm

Mean thickness = = mm

Corrected thickness $d =$ mm = m

$$\therefore \qquad Y = \frac{mgl^3}{4bd^3y} = \text{N}/m^2$$

Value of Y from tables =

% Error =

Precautions. 1. The knife-edges should be rigid and fixed on a rigid support.

2. The knife-edges should be at equal distances from the centre of the bar. The stirrup should be parallel to the knife-edges and placed exactly at the centre to get a symmetrical loading of the bar.

3. The weights should be placed or removed from the hanger gently and should be increased or decreased gradually in equal steps.

4. The maximum load used should be such that it keeps the bar within elastic limits.

5. The thickness should be measured very accurately at a number of points along the whole length of the bar. The thickness d is a small quantity and since its cube is to be used in the calculations, a small error in its measurement will cause three times large error in the result.

Alternative method for finding depression. The depression in the middle of the bar due to increased load can be determined by supporting a spherometer on a stand just above the middle point of the bar.

In order to find out the exact contact position of the central leg of the spherometer an electrical circuit containing a cell, a tangent galvanometer and a key is set up between the spherometer and the bar as shown in Fig. 5.4. When the key is inserted and the central leg is lowered to bring it just in contact with the rod, a sudden deflection is indicated in the galvanometer as soon as the leg comes in contact with the bar. The total number of divisions, through which the disc is rotated, multiplied by the least count gives the depression corresponding to the increased load.

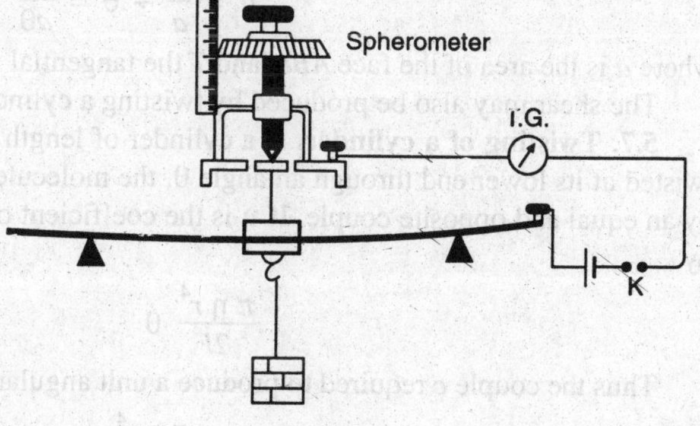

Fig. 5.4

Exercise. *Find the Young's modulus of the material of a bar of circular cross-section by bending.*

Hints. For a circular bar of radius r and length l, the depression y when a mass m is suspended at its mid-point is given by

$$y = \frac{mgl^3}{12\pi Yr^4}$$

or

$$Y = \frac{mgl^3}{12\pi r^4 y}$$

5.6. Rigidity. A solid has its own shape and size, a liquid has its own size, but takes the shape of the vessel in which it is put. A gas has no shape or size of its own. It takes the shape of the vessel in which it is put and fills it completely.

In other words, gases yield to the slightest force tending to change their shape or size, whereas liquids resist all forces tending to change their size (or volume) but yield to even the slightest force tending to change their shape. Solids resist all forces tending to change their shape or size (volume).

The property of a solid to resist the forces tending to change its shape is called **rigidity.** *Liquids and gases have no rigidity.*

Modulus of rigidity. *It is defined as the ratio of the tangential stress to the shearing strain*

To find the value of the shearing strain, consider a solid cube *ABCDefgh*. The lower face *CDgh* is fixed and a tangential force F is applied over the face *ABef* so that it is displaced to the position *A′B′e′f′*. Each horizontal layer of the cube is thus displaced, the displacement being proportional to its distance from the fixed plane. Thus shearing strain

$$= \frac{\text{Displacement of a plane}}{\text{Distance from the fixed plane}}$$

$$\frac{AA'}{AD} = \frac{l}{L} = \tan\theta$$

When θ is small $\tan\theta = \theta$

∴ Shearing strain = θ

Hence the angle through which a line originally perpendicular to the fixed plane is turned is a measure of the shearing strain.

Fig. 5.5

$$\text{Now modulus of rigidity} = \frac{\text{Tangential stress}}{\text{Shearing~strain}}$$

$$\therefore \qquad \eta = \frac{F}{a} \div \theta = \frac{F}{a\theta} = \frac{T}{\theta}$$

where a is the area of the face *ABef* and T the tangential force per unit area.

The shear may also be produced by twisting a cylinder keeping one of its faces fixed.

5.7. Twisting of a cylinder. If a cylinder of length l and radius r is fixed at its upper end and twisted at its lower end through an angle θ, the molecules of the cylinder resist this twisting couple by an equal and opposite couple. If η is the coefficient of rigidity, then the twisting couple is given by

$$\frac{\pi \eta r^4}{2l}\theta$$

Thus the couple c required to produce a unit angular twist (1 radian) is given by

$$c = \frac{\pi \eta r^4}{2l}$$

To prove the above relation consider a cylinder of length l and radius r rigidly clamped at its upper end as shown in Fig. 5.6(*a*). Let it be twisted by applying a couple at its lower end. The solid cylinder can be considered to be made up of a very large number of thin co-axial cylinders. Consider one such cylinder of radius x and thickness dx. Let the twist at the lower end be θ (radian). The angle of shear is φ.

Then $\qquad BB' = x\theta = l\phi$

or $\qquad \phi = \dfrac{x\theta}{l}$

$$\text{Now co-efficient of rigidity} \quad \eta = \frac{\text{Tangential stress}}{\text{Shearing strain } \phi}$$

$$\therefore \quad \text{Tangential stress} = \frac{F}{a} = \eta\phi = \eta\frac{x\theta}{l}$$

$$\text{or Applied force } F = \eta\frac{x\theta}{l} \times a = \eta\frac{x\theta}{l} \times 2\pi x\, dx$$

$$= \frac{2\pi\eta\theta}{l} x^2\, dx$$

since the area of the base of the thin co-axial cylinder of radius x or circumference $2\pi x$ and thickness $dx = 2\pi x\, dx$ as shown in Fig. 5.6(*b*).

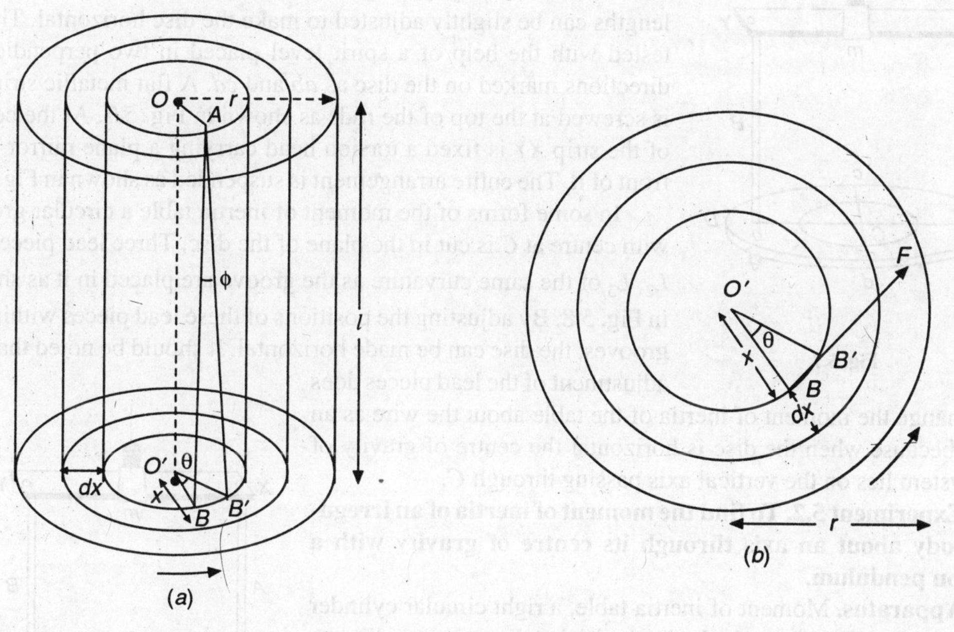

Fig. 5.6

Torque acting on this cylinder

$$d\tau = F.x = \frac{2\pi\eta\theta}{l}\, x^2\, x\, dx = \frac{2\pi\eta\theta}{l}\, x^3\, dx$$

$\therefore$ Total torque acting on the whole solid cylinder

$$\tau = \frac{2\pi\eta\theta}{l}\int_0^r x^3\, dx = \frac{2\pi\eta\theta}{l}\,\frac{r^4}{4} = \frac{\pi\eta r^4}{2l}\,\theta$$

$$= c\,\theta$$

where $c = \dfrac{\pi\eta r^4}{2l}$ *gives the couple per unit angular twist or torsional rigidity.*

5.8. Torsion pendulum. It is a heavy circular disc suspended from one end of a fine wire attached to its centre. The other end of the wire is fixed in a torsion head. When the disc is turned in the horizontal plane so as to twist the wire and then released, it executes torsional vibrations about the wire as an axis. The time period t of the torsional vibration is given by

$$t = 2\pi\sqrt{\frac{I}{c}}$$

where I is the moment of inertia of the disc about the wire as an axis and c the couple per unit angular twist of the wire.

To find the moment of inertia of a regular or an irregular body by a torsion pendulum a **moment of inertia table** is used.

5.9. Moment of inertia table. It consists of a flat circular metallic disc D to which are fixed vertically two equal lengths of metal rods A and B at points equidistant from the centre along a diameter of the disc.

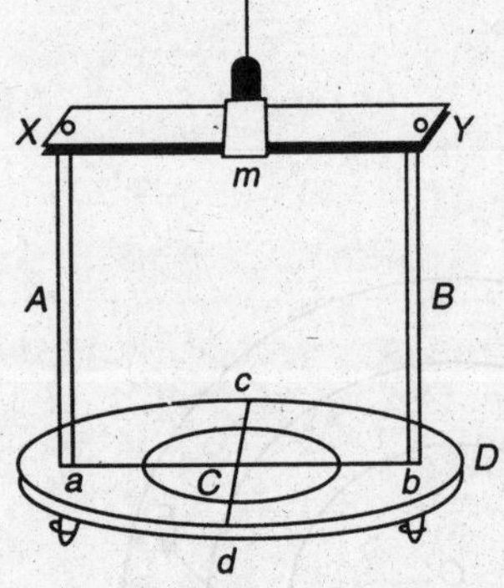

Fig. 5.7

These rods carry nuts at the bottom of the disc D by which their lengths can be slightly adjusted to make the disc horizontal. This is tested with the help of a spirit level placed in two perpendicular directions marked on the disc as ab and cd. A flat metallic strip XY is screwed at the top of the rods as shown in Fig. 5.7. At the centre of the strip XY is fixed a torsion head carrying a plane mirror m in front of it. The entire arrangement is suspended as shown in Fig. 5.7.

In some forms of the moment of inertia table a circular groove with centre at C is cut in the plane of the disc. Three lead pieces L_1, L_2, L_3 of the same curvature as the groove are placed in it as shown in Fig. 5.8. By adjusting the positions of these lead pieces within the grooves, the disc can be made horizontal. It should be noted that the adjustment moment of the lead pieces does not change the moment of inertia of the table about the wire as an axis, because when the disc is horizontal the centre of gravity of the system lies on the vertical axis passing through C.

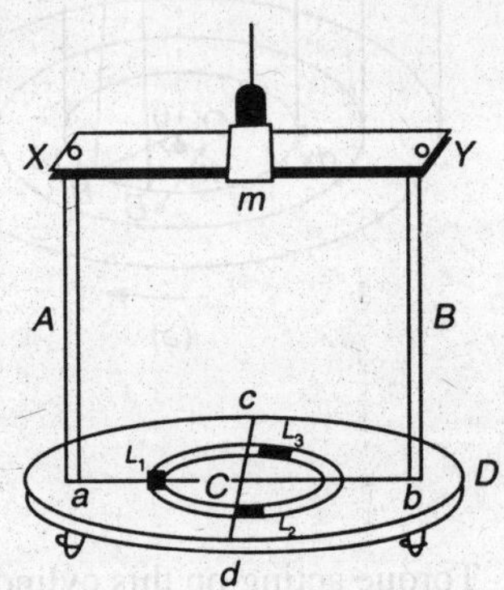

Fig. 5.8

Experiment 5.2. To find the moment of inertia of an irregular body about an axis through its centre of gravity with a torsion pendulum.

Apparatus. Moment of inertia table, a right circular cylinder (a regular body), an irregular body, a spirit level, a vernier callipers, telescope with inverted scale with zero mark in the middle, a stop watch, a balance and a weight box.

Theory. Let t be the time-period of the torsional vibration of the moment of inertia table alone, I its moment of inertia and c the couple per unit angular twist of the wire, then

$$t = 2\pi \sqrt{\frac{I}{c}} \qquad ...(i)$$

Suppose a regular body, of known moment of inertia I_1, about an axis through its $C.G.$ is placed on the table such that the axis of the wire passes through the centre of gravity. If t_1 is the time-period of the torsional vibration of the system, then

$$t_1 = 2\pi \sqrt{\frac{I + I_1}{c}} \qquad ...(ii)$$

If now the regular body is replaced by an irregular body whose moment of inertia I_2 about an axis through its $C.G.$ is required and the time period t_2 of the system is again found out, then

$$t_2 = 2\pi \sqrt{\frac{I + I_2}{c}} \qquad ...(iii)$$

Squaring and dividing (*ii*) by (*i*), we have

$$\frac{t_1^2}{t^2} = \frac{I + I_1}{I}$$

or

$$\frac{t_1^2 - t^2}{t^2} = \frac{I_1}{I} \qquad ...(iv)$$

Similarly squaring and dividing (*iii*) by (*i*), we have

$$\frac{t_2^2 - t^2}{t^2} = \frac{I_2}{I} \qquad ...(v)$$

Dividing (*v*) by (*iv*), we get

$$\frac{I_2}{I_1} = \frac{t_2^2 - t^2}{t_1^2 - t^2}$$

or

$$I_2 = \frac{t_2^2 - t^2}{t_1^2 - t^2} \times I_1$$

Generally a body of known moment of inertia is a right circular cylinder. If M is the mass of the cylinder and r its radius, then the moment of inertia about its own axis is

$$= \frac{1}{2}Mr^2$$

Procedure 1. Suspend the moment of inertia table by means of a copper wire of a suitable length and thickness from a rigid support. Place a spirit level along the diameter *ab* and see that the table is horizontal. If it is not, then adjust the nuts provided at the ends of the rods *A* and *B* below the disc or adjust the position of the lead weights in the groove. Test again by placing the spirit level along the diameter *cd*.

2. Place a telescope with an inverted centimetre scale clamped to its stand at a distance not less than one metre towards the mirror fixed to the torsion head.

Place the eye just above the telescope and adjust the position of the scale so that the images of the scale divisions are seen in the mirror. Now place the telescope in the position of the eye.

Adjust the eye-piece of the telescope so that the cross-wires are clearly visible. Focus the telescope on the centimetre scale divisions as seen in the mirror but not on the mirror. *It may be remembered that the image lies as far behind the mirror as the scale is in front of it.*

Observe the reading of a division mark in the centre of the scale image and adjust the vertical cross-wire with it. Let this be the reference mark.

3. Give a slight twist to the table so that it begins to execute torsional vibrations in the horizontal plane alone. When the image of the reference line on the scale just passes the vertical cross-wire, start the stop-watch and count *zero*. Count one when the image of the same mark passes in the same direction and so on. Find the time for 20 vibrations. Repeat three times.

4. Place the solid cylinder in the centre of the table. Adjust its position so that the table is again horizontal. Test with a spirit level. Find the time for 20 vibrations of the system three times.

5. Remove the cylinder and place the irregular body in the centre of the table. Adjust the position so that the table is horizontal. *In this case the axis of the wire passes through the C.G. of the body.* Find the time for 20 vibrations of the system three times again.

6. Weigh the solid cylinder and find its diameter with a vernier callipers.

Observations

Sl. No.		*Time for 20 vibrations*				*Time period*
		1	*2*	*3*	*Mean*	
1.	Table alone					$t =$
2.	With cylinder					$t_1 =$
3.	With irregular body					$t_2 =$

Mass of the solid cylinder $M =$ kg

Diameter of the solid cylinder = 1. 2. 3.

Mean diameter $d =$ cm

Radius of the cylinder $r =$ cm = m

$\therefore$ Moment of inertia $I_1 = \frac{1}{2}Mr^2 =$ kg–m^2

Calculations.

M.I. of the irregular body $I_2 = I_1 \times \frac{t_2^2 - t^2}{t_1^2 - t^2} =$ kg–m^2

Precautions. 1. The *M.I.* table should always be horizontal.

2. The *M.I.* table should execute torsional vibrations only.

3. The amplitude should be small so that the wire is not twisted beyond the elastic limits.

Experiment 5.3. To compare the moment of inertia of a solid sphere and a hollow sphere (or a solid disc and an annular disc) of same mass using a torsion pendulum and hence to show that the moment of inertia depends upon the distribution of mass.

Apparatus Same as in **Expt. 5.2.**

Procedure. Compare the moment of inertia of the solid sphere I_2 and the hollow sphere of the same mass I_1 as explained in **Exp. 5.2.**

Conclusion. As $\frac{I_2}{I_1} \neq 1$, we conclude that the *M.I* of a body depends not only on its mass but also on the distribution of mass about the axis of rotation.

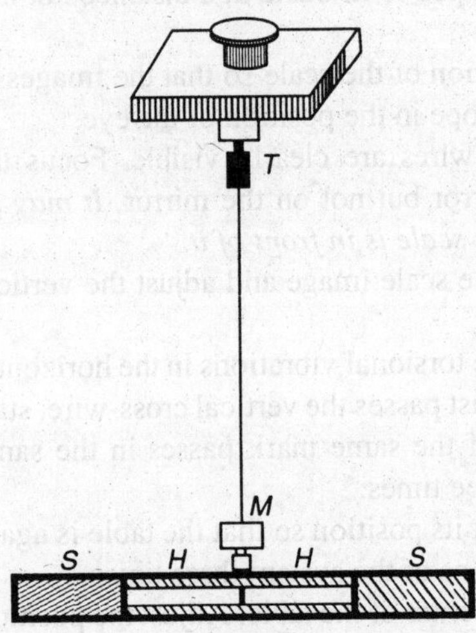

Fig. 5.9

5.10. Maxwell's needle. The modulus of rigidity can be determined with the help of a Maxwell's needle. It consists of a hollow cylindrical tube open at both ends provided with a torsion head in the middle to which is fixed a plane mirror. One end of a wire of uniform radius, modulus of rigidity of the material of which is to be determined is fixed in the torsion head. The other end of the wire is fixed to a second torsion head fixed to a rigid support. The tube is fitted with four cylinders of equal lengths of which two are hollow and two are solid. The length of each cylinder is equal to one quarter of the length of the cylindrical tube so that when they are put end to end they just fill the hollow tube completely. The hollow cylindrical tube along with the cylinders is suspended in a vertical plane as shown in Fig. 5.9. The time period can be determined with the help of a telescope and scale arrangement.

Experiment 5.4. To determine the modulus of rigidity of copper by Maxwell's needle.

Apparatus. A Maxwell's needle, a copper wire of a suitable length and thickness, a fixed support with a torsion head, a telescope with a scale attached to its stand, a stop-watch, a screw gauge, a spring balance, a metre rod and an electric lamp with a holder.

Theory. To find the co-efficient of rigidity, the Maxwell's needle is suspended first with solid cylinders in the outermost position as shown in Fig. 5.10 (*i*). Let t_1 be the time period. If I_1 is the moment of inertia of the loaded tube about the suspended wire as an axis, then

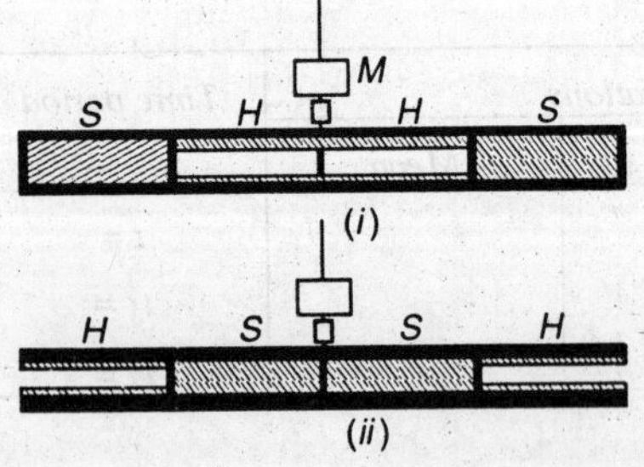

Fig. 5.10

$$t_1 = 2\pi\sqrt{\frac{I_1}{c}} \qquad ...(i)$$

The position of the solid and the hollow cylinders is then changed as shown in Fig. 5.10 (*ii*). Let the time period now be t_2. If I_2 is the moment of inertia of the loaded tube in this position about the wire as an axis, then

$$t_2 = 2\pi\sqrt{\frac{I_2}{c}} \qquad ...(ii)$$

From (*i*) and (*ii*), we get

$$t_1^2 - t_2^2 = 4\pi^2 \frac{(I_1 - I_2)}{c} \qquad ...(iii)$$

or
$$c = 4\pi^2 \frac{(I_1 - I_2)}{(t_1^2 - t_2^2)}$$

But
$$c = \frac{\pi \eta r^4}{2l}$$

where η is the modulus of rigidity of the material of the wire, r the radius and l the length of the wire.

$$\therefore \qquad \frac{\pi \eta r^4}{2l} = 4\pi^2 \frac{(I_1 - I_2)}{(t_1^2 - t_2^2)}$$

or
$$\eta = \frac{8\pi l}{r4} \frac{(I_1 - I_2)}{(t_1^2 - t_2^2)} \qquad ...(iv)$$

Let m_1 be the mass of each hollow cylinder and m_2 that of each solid cylinder. Suppose I_0 is the moment of inertia of the hollow tube, I_H the moment of inertia of each hollow cylinder and I_S the moment of inertia of each solid cylinder about an axis passing through their middle points and perpendicular to their length. If L is *half* the length of the hollow tube of Maxwell's needle, then applying principle of parallel axis, we have

$$I_1 = I_o + 2\left[I_H + m_1\left(\frac{L}{4}\right)^2 + I_s + m_2\left(\frac{3L}{4}\right)^2\right]$$

as the centres of hollow cylinders are at a distance $\frac{L}{4}$ and those of solid cylinders at a distance $\frac{3L}{4}$ from the axis of oscillation. Similarly

$$I_2 = I_0 + 2\left[I_H + m_1\left(\frac{3L}{4}\right)^2 + I_S + m_2\left(\frac{L}{4}\right)^2\right]$$

as now the centres of hollow cylinders are at a distance $\frac{3L}{4}$ and those of solid cylinders at a distance $\frac{L}{4}$ from the axis of oscillation

$$\therefore \qquad I_1 - I_2 = (m_2 - m_1)L^2$$

Substituting in equation (*iv*), we get

$$\eta = \frac{8\pi l}{r^4} \frac{(m_2 - m_1)L^2}{(t_1^2 - t_2^2)} \qquad ...(v)$$

Procedure. 1. (*i*) Put the hollow cylinders of the Maxwell's needle inwards and the solid cylinders outwards symmetrically in the tube as shown in Fig. 5.10 (*i*).

(*ii*) Take a copper wire of suitable length and thickness. Remove the kinks by stretching the wire gently.

(*iii*) Suspend the Maxwell's needle from one end of this wire and fix the other end in a rigid support. See that the needle is horizontal and oscillates freely in a horizontal plane about the wire as an axis.

(*iv*) See that the needle lies parallel to the wall. If it is not, adjust the torsion head at the top.

(*v*) Arrest the motion of the needle just near its mean position by a stand placed in contact with it or tie one end of a thread to the needle and its other end to a fixed point.

2. Place a telescope having an inverted centimetre scale clamped to its stand on a stool, at a suitable distance not less than one metre from the Maxwell's needle, with the scale towards the mirror fixed to the needle.

Place your eye just above the telescope and adjust the position of the scale till its image is visible in the mirror. If the image is not clearly visible illuminate the scale by an electric lamp placed in front of the scale.

3. Place the telescope in the position of the eye. Adjust the eye-piece so that the cross-wires are clearly visible. Focus the telescope on the image seen in the mirror and not on the mirror. *The image lies as far behind the mirror as the scale is in front of it*. Observe the reading of the scale in the centre of the image. Adjust the vertical cross-wire so that it coincides with a cm division mark.

4. Gently remove the stand or apply a burning match stick to the thread, as the case may be. The needle will begin to vibrate in a horizontal plane. *See that the amplitude is small and the needle does not oscillate up and down.* If it does so stop it and release it again gently or touch the wire just above the needle with the fingers. *Protect the apparatus from air currents.*

5. When the image of the reference line on the scale just passes the vertical cross-wire, start the stop watch and count **zero**. Count **'one'** when the image of the same mark passes in the same direction and so on.

Find the time for 15 vibrations. Repeat three times.

6. Now place the solid cylinders inwards and the hollow cylinders outwards as shown in Fig. 5.10 (*ii*). Set the telescope again and repeat the observations.

7. Find the weight of each solid cylinder with a spring balance and find the mean. Similarly find the mean weight of a hollow cylinder. Measure the length of the Maxwell's needle and that of the wire from the point of suspension to the point where it is attached to the needle.

8. Measure the diameter of the wire accurately at different points with a screw gauge. At each point measure the diameter in two mutually perpendicular directions.

Note that the radius 'r' occurs as r^4 in the formula and hence a small error in the measurement of diameter will be magnified four times.

Observations

Position of soild cylinders	*Time for 15 vibrations*				*Time period*
	1	*2*	*3*	*Mean*	
Outside					$t_1 =$
Inside					$t_2 =$

Weight of the hollow cylinders = (*i*) (*ii*)

Mean m_1 = kg

Weight of the solid cylinders = (*i*) (*ii*)

Mean m_2 = kg

Length of the needle $2L$ = cm = m

Half the length of the needle L = cm = m

Length of the wire l = cm = m

Diameter of the wire

Least count of the screw gauge = mm =

Zero correction = ± mm

Position	*1*	*2*	*3*	*4*	*5*	*6*	mean
a							
b							

Mean corrected diameter = mm

Mean corrected radius r = mm = m

Calculations. $\eta = \dfrac{8\pi\, l\,(m_2 - m_1)\, L^2}{(t_1^2 - t_2^2)\, r^4}$ = N/m^2

Actual value from tables = N/m^2

Percentage error =

Precautions. 1. There should be no kinks in the wire.

2. The Maxwell needle should vibrate equally on both sides of the reference mark.

3. The thickness of the wire should be such that it remains taut under the load of the Maxwell needle.

4. The Maxwell needle should remain horizontal and should not vibrate up and down.

5. The amplitude of vibrations should be small so that the wire is not twisted beyond the elastic limit.

6. The telescope should be focused on the image of scale which lies as far behind the mirror as the object is in front of it and not on the mirror.

Exercise 1. *Find the moment of inertia of a cylinder by finding its time period. Co-efficient of rigidity is given.*

Hints : $t = 2\pi\sqrt{\dfrac{I}{c}}$, where $c = \dfrac{\pi\eta r^4}{2l}$.

Exercise 2. *Find the torsional couple per unit angular twist of a wire.*

Hints. Suspend the Maxwell's needle from the wire and find the time periods t_1 and t_2 with solid cylinders outside and then inside respectively.

Now $$t_1^2 - t_2^2 = \frac{4\pi^2 L^2 (m_2 - m_1)}{c}$$

or $$c = \frac{4\pi^2 L^2 (m_2 - m_1)}{t_1^2 - t_2^2}$$

Note. L is half the length of Maxwell's needle.

5.11. Poisson's ratio (σ) *It is defined as the ratio of lateral contraction per unit length* (β) *and longitudinal elongation per unit length* (α) *within elastic limits.*

$$\therefore \quad \sigma = \frac{\text{Lateral contraction per unit length } (\beta)}{\text{Longitudinal elongation per unit length } (\alpha)}$$

$$= \frac{\text{Lateral contraction or Lateral strain}}{\text{Longitudinal elongation or longitudinal strain}}$$

or $$\sigma = \frac{\beta}{\alpha}$$

Experiment 5.5. To find the relationship between the longitudinal strain and lateral strain in the case of rubber tube and hence to find Poisson's ratio for rubber.

Apparatus. Bicycle rubber tube with metal sleeves and rubber stoppers, a metal stand to hold the rubber tube, a graduated tube, a hanger with a few 250 gm slotted weights, a metre scale, measuring cylinder etc.

Theory. If a rubber tube suspended vertically is loaded at its lower end with a small weight, it is slightly stretched. This results in the increase in length and hence the internal volume. Suppose V is the original volume of the tube, A its area of cross-section, L its length and D its internal diameter. Let dV be the slight increase in volume, dL the corresponding increase in length and dA the decrease in the area of cross-section, then

Initial volume $\quad V = A \times L$

Volume after loading

$$(V + dV) = (A - dA)(L + dL)$$
$$= AL + A \, . \, dL - L \, . \, dA$$

neglecting the product of two small quantities $dL \times dA$. Hence

$$dV = A \, . \, dL - L \, . \, dA$$

or
$$\frac{dV}{dL} = A - \frac{L}{dL} \, . \, dA \qquad \text{...}(i)$$

Now area $\quad A = \dfrac{\pi D^2}{4}$

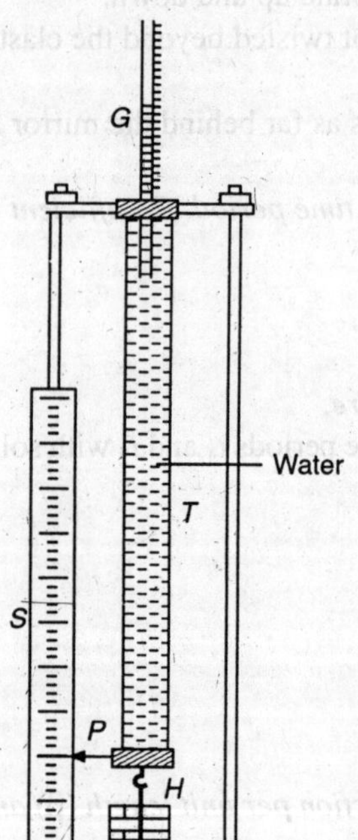

Fig. 5.11

Differentiating, we get

$$dA = \frac{2\pi D}{4} \, . \, dD = 2A \, . \, \frac{dD}{D}$$

Substituting in (i), we get

$$\frac{dV}{dL} = A - \frac{L}{dL} \, . \, 2A \, \frac{dD}{D}$$
$$= A\left(1 - 2\,\frac{dD/D}{dL/L}\right)$$
$$= A\,(1 - 2\,\sigma)$$

$\therefore$ Poisson's ratio $\sigma = \dfrac{\text{Lateral contraction or strain}}{\text{Longitudinal elongation}} = \dfrac{dD/D}{dL/L}$

Hence
$$\sigma = \frac{1}{2}\left(1 - \frac{1}{A}\,\frac{dV}{dL}\right)$$

Thus by measuring the small change in volume dV corresponding to a small change in length dL and the area of cross-section A, Poisson ratio σ can be calculated.

Procedure. 1. Close the upper end of the bicycle rubber tube T with a rubber stopper having a hole in it and pass the graduated tube G through the hole. Surround the rubber stopper by a metal ring which can be fixed rigidly to the upper end of a metal stand as shown in Fig. 5.11. Close the lower end of the rubber tube with another rubber stopper which can be surrounded by another metal sleeve provided with a hook below for loading. The sleeve also carries a light pointer P which can move freely over a vertical metre scale S fixed to the stand.

2. To find the area of cross-section of the tube suspend the hanger H from the hook and fill the tube up to the lower end of the upper rubber stopper with a known volume of water. Level the stand so that the tube stands vertical and the pointer P moves freely on the metre scale. Measure the length of the rubber tube between the two rubber stoppers. Calculate the area of cross-section from the knowledge of volume and length.

3. Introduce the graduated tube and pour water if necessary so that the meniscus of water is visible in the graduated tube. Adjust the position of the scale S so that the pointer lies some where on the middle on some division on the scale. Note the position of the pointer as also the reading of water meniscus in the gruaduated tube while the hanger H is in position. Take this as the zero position.

4. Place gently a load of 250 gm on the hanger H. Wait for 5 minutes to allow the apparatus to settle down. Note down the reading of water meniscus in the tube G and also the position of the pointer P on the scale. The difference between the readings of water meniscus in the tube G and readings of the pointer P on the scale gives the value of dV and dL respectively.

5. Repeat the process by increasing the load in steps of 250 gm and note the observations, after waiting for 5 minutes after each loading.

6. Repeat the process by unloading the hanger in steps of 250 gm till the load on the hanger is reduced to zero.

7. Find the mean of the two observations and calculate $\frac{dV}{dL}$ in each case. Find the mean value of $\frac{dV}{dL}$ and calculate σ.

Observations. Length of the rubber tube $L =$ cm
Internal volume of the tube $V =$ cc

Area of the cross-section $A = \frac{V}{L}$ sq. cm =

Sl. No.	*Load in gm*	*Water level in tube G*			*Change in Vol. dV (c.c.)*	*Position of pointer on scale*			*Change in length dL (cm)*	$\frac{dV}{dL}$
		Loading	*unloading*	*Mean*		*Loading*	*Unloading*	*Mean*		
1	0									
2	250									
3	500									
4	750									
5	1000									
6	1250									
7	1500									
8	1750									
9	2000									

Mean $\frac{dV}{dL} =$

∴ Poisson's ratio $\sigma = \frac{1}{2}\left[1 - \frac{1}{A}\frac{dV}{dL}\right] =$

Graph. Draw a graph between the extension (change in length) dL and increase in volume dV taking dV along Y-axis and dL along X-axis. The graph is a straight line. The slope of the graph

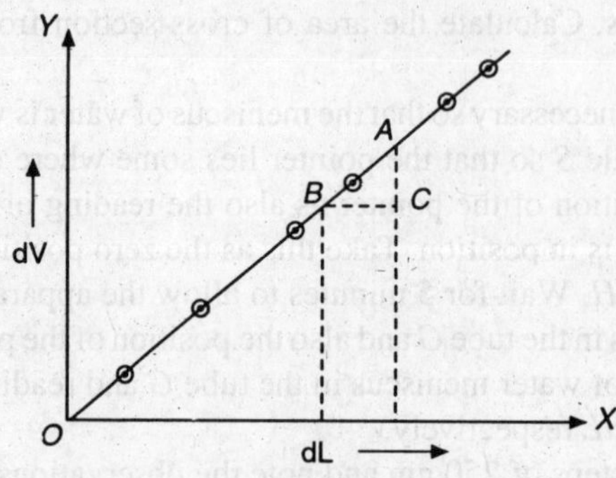

Fig. 5.12

$$\frac{dV}{dL} = \frac{AC}{BC}$$

$$\therefore \quad \text{Poisson's ratio } \sigma = \frac{1}{2}\left[1 - \frac{1}{A}\frac{AC}{BC}\right]$$

(From the graph.)

Precautions. 1. Rubber tube, the graduated tube and the scale should all be vertical.

2. The rubber tube should not be pulled beyond the elastic limits for the validity of Hooke's law.

3. There should be no air bubbles in the tube.

4. The loading and unloading should be very gentle and observations should be recorded after waiting for 5 minutes after each loading or unloading.

Note. As dL is very small, it can be measured more accurately with the help of a travelling microscope.

5.12. Searle's apparatus for determining the elastic constants. The elastic constants like Young's modulus, modulus of rigidity and Poisson's ratio can be determined with the help of Searle's apparatus.

The apparatus consists of two exactly equal rectangular or circular metal bars LM and PQ fastened by means of a short thin wire of the material whose elastic constants are to be determined. The bars are suspended from a rigid support by two small equal lengths of thread so that the wire is straight and the bars are held parallel to each other as shown in Fig. 5.13.

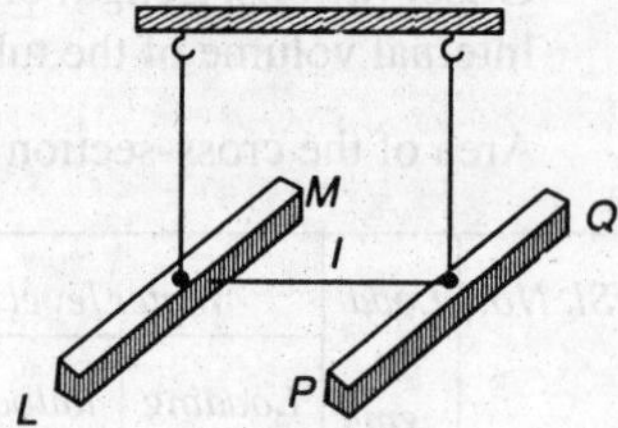

Fig. 5.13

Experiments 5.6. To find Young's modulus, modulus of rigidity and Poisson's ratio for the material of a wire by Searle's method.

Apparatus. Searle's apparatus, thread, a stop watch, a telescope, a metre rod, vernier callipers and screw gauge.

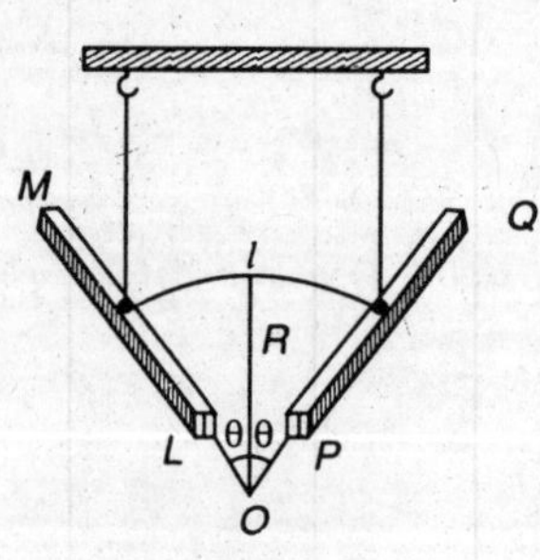

Fig. 5.14

Theory. When the ends P and L of the two bars of the Searle's apparatus for finding the elastic constants are slightly pulled together symmetrically the wire is bent into a circular arc as shown in Fig. 5.14. The length of the arc is given by

$$l = R \times 2\,\theta$$

where R is the radius of the circular arc and $2\,\theta$ the angle subtended by it at its centre 0.

The bending moment of the wires exerts a couple on each bar given by $\frac{Y I_g}{R}$ where $I_g = \frac{\pi r^4}{4}$ is the geometrical moment of inertia of the wire and Y the Young's modulus of elasticity of the material

$$\therefore \quad \text{Bending couple} = \frac{Y}{R}\frac{\pi r^4}{4} = Y\frac{\pi r^4}{4}\cdot\frac{2\,\theta}{l} = Y\frac{\pi r^4}{2l}\,\theta$$

When the bars are released these begin to vibrate in a horizontal plane due to the couple exerted by the wire on the bars. The wire first bends into a circular arc having its centre at O and then into an arc having its centre on the opposite side.

If I is the moment of inertia of the bar about a vertical axis passing through its $C.G$, then the restoring couple is given by

$$I\alpha = I\frac{d^2\theta}{dt^2}$$

where $\alpha = \frac{d^2\theta}{dt^2}$ is the angular acceleration of the bar.

The equation of motion for the angular vibration of the bar is now given by

$$I\frac{d^2\theta}{dt^2} = -Y\frac{\pi r^4}{2l}\theta$$

or
$$\alpha = \frac{d^2\theta}{dt^2} = -\frac{Y\pi r^4}{2lI}\theta\cdot$$

As the angular acceleration α is proportional to displacement θ, the motion is simple harmonic. The time period of oscillation of each bar t_1 is given by

$$t_1 = 2\pi\sqrt{\frac{2lI}{Y\pi r^4}}$$

Hence **Young's modulus** $Y = \frac{8\pi I l}{r^4 t_1^2}$ ...(*i*)

If now the suspension threads are removed and one of the bars is clamped so that the other hangs in the horizontal plane as shown in Fig. 5.15 then the bar if given a little twist will execute torsional vibrations of time period t_2 given by

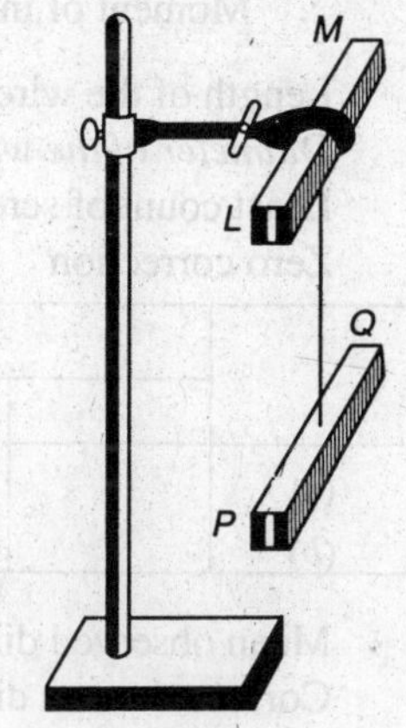

Fig. 5.15

$$t_2 = 2\pi\sqrt{\frac{I}{c}}$$

where c is the couple exerted per unit angular twist.

If η is the modulus of rigidity for the wire, then

$$c = \frac{\pi\eta r^4}{2l}$$

$$\therefore \quad t_2^2 = \frac{4\pi^2 I}{c} = \frac{4\pi^2 I . 2l}{\pi\eta r^4}$$

or **Modulus of rigidity** $\eta = \frac{8\pi I l}{r^4 t_2^2}$ ...(*ii*)

Dividing (*i*) by (*ii*), we get

$$\frac{Y}{\eta} = \frac{t_2^2}{t_1^2}$$

Now **Poisson's ratio** $\sigma = \frac{Y}{2\eta} - 1$

$$= \frac{t_2^2}{2t_1^2} - 1$$

Procedure. 1. Fasten the wire under test to the two identical bars LM and PQ at their middle points as shown in Fig. 5.13 Suspend the two bars from a rigid support, by two threads of suitable lengths so that when the system is at rest, the two bars and the wire lie in the same horizontal plane.

2. Mark a vertical reference line at the centre of the end of each bar. Focus the telescope on this line on one of the bars so that the vertical cross-wire coincides with this mark.

3. Draw the two bars slightly together by a loop of thin cotton thread slipped at the ends L and P and carefully bring the system to rest. Burn the thread of the loop to release the bars which will begin to oscillate with a small amplitude. Look through the telescope and find the time for 25 vibrations. Repeat four times and calculate the mean time period t_1.

4. Remove the bars from the suspended threads and clamp one of the bars as shown in Fig. 5.15 so that the wire is vertical and the second bar is horizontal. Focus the telescope on the reference line on the lower bar. Twist the wire by rotating the lower bar so that it begins to execute torsional vibrations. Find the time for 25 vibrations. Repeat four times and calculate the time of one vibration t_2.

5. Measure the length of the bar with a metre rod and its breadth with a vernier callipers. Find its mass by weighing it. Measure the length of the wire under test with a metre rod and its diameter with a screw gauge accurately at a number of places and at each place in two mutually perpendicular directions.

Observations and calculations.

(*a*) **Determination of Y.**

Time for 25 vibrations = 1. 2. 3. 4.

Mean = seconds

∴ Time period t_1 = sec.

Length of the bar L = m

Vernier constant = mm

Breadth of the bar = 1. 2. 3.

Mean breadth b = cm = m

Mass of the bar m = kg.

∴ Moment of inertia of the bar $I = m \times \dfrac{L^2 + b^2}{12}$ kg – m²

Length of the wire l = m

Diameter of the wire

Least count of screw gauge = mm

Zero correction = ± mm

	Observed diameter							
	1	2	3	4	5	6	7	8
(*a*) (*b*)								

Mean observed diameter = mm

Corrected mean diameter d = mm = m

∴ Corrected radius r = m

Young's modulus $Y = \dfrac{8\pi I l}{r^4 t_1^2}$ = N/m²

(*b*) **Modulus of rigidity** η

Time for 25 vibrations = 1. 2. 3. 4.

Mean = sec

∴ Time period t_2 = sec

Co-efficient of rigidity $\eta = \dfrac{8\pi I l}{r^4 t_2^2}$ = N/m² (*c*) **Poisson's ratio** $\sigma = \frac{t_2^2}{2t_1^2} - 1$

$=$

It is clear from the expression for the Poisson's ratio that the value of this important constant can be determined from the measurement of the two time periods alone, which can be determined with sufficient accuracy. It is not necessary to find the value of Y and η to find the value of σ.

Precautions. 1. The amplitude of vibration need not be very small as in the derivation for the expression of the time period no such assumption is made. Care should be taken that the wire is not strained beyond the elastic limit and the thread remains vertical.

2. The bars should vibrate in a horizontal plane and there should be no up and down motion.

3. The radius of the wire should be measured very accurately. In the calculations the radius occurs as the **fourth** power and a small error in its value will be magnified four times.

4. The time period occurs in the second power in calculations and hence it should also be measured accurately. To achieve this the time of a large number of vibrations should be taken.

Note. If the two bars are not exactly identical the moment of inertia of both the bars should be calculated separately and the mean value of I should be used in the calculations of Young's modulus of elasticity. In the calculations of rigidity, the moment of inertia of the lower bar which is set into vibration must only be considered.

Exercise. *Find the bulk modulus for the material of the given wire using Searle's rigidity apparatus.*

Hint. Find the value of Y and σ as explained in **Expt. 5.6** and calculate bulk modulus K by the relation

$$K = \frac{Y}{3(1-2\sigma)}$$

Oral Questions

1. Elasticity. (*i*) Define stress, strain and Young's modulus of elasticity. (*ii*) What do you mean by elastic limits and breaking stress? (*iii*) What are the dimensions of stress, strain and Young's modulus? (*iv*) Which is more elastic; rubber or steel?

2. Young's modulus by bending. (*i*) What is neutral axis? (*ii*) How should the bar be placed? (*iii*) What will happen if the two knife-edges are not equidistant from the centre of gravity of the bar? (*iv*) How will the depression change for the same load, if

(*a*) the length of the rod is made half,

(*b*) the breadth of the rod is made half; and

(*c*) the thickness of the rod is made half

(*d*) What do you mean by geometrical moment of inertia?

3. Rigidity. (*i*) Define modulus of rigidity. (*ii*) What is shear? (*iii*) What are the units of rigidity? (*iv*) What is the effect of temperature on rigidity?

4. Torsion pendulum. (*i*) What part does moment of inertia play in the rotatory motion? (*ii*) How is moment of inertia of a body related to its period of torsional vibration? (*iii*) What will be the change in periodic time if a thin wire of half the radius is used? (*iv*) Does the moment of inertia depend upon the axis of rotation?

5. Maxwell's needle. (*i*) In what way the Maxwell's needle method is better than an ordinary cylinder? (*ii*) Do you get the same value of rigidity from this method and other methods? If not why? (*iii*) Why should the amplitude be small? (*iv*) Which observations will you take with maximum accuracy and why?

6. Poisson's ratio. (*i*) Define Poisson's ratio? What are its units? (*iii*) What is its value for rubber? (*iv*) Can Poisson's ratio have any value? (*v*) Knowing the value of Y and bulk modulus, can you find the value of σ?

7. Elastic constants. (*i*) What is the advantage of Searle's method of determining elastic constants over other methods of determining Y and η? (*ii*) State the value of Poisson's ratio and Bulk modulus of elasticity in terms of Y and η. (*iii*) While finding the time period in the first part of the experiment, should the amplitude be small? If so, why ? Is this restriction also there in the second part of the experiment? If not why?

6

Surface Tension and Viscosity

6.1. Surface tension. The force of cohesion between molecules of a liquid gives rise to a phenomenon called *surface tension.* According to this aspect of molecular attraction the free surface of a liquid acts at all times as though it has a *thin membrane stretched over it and that this membrane is under tension and trying to contract.* It is for this reason that a thin needle, if placed on the surface of water carefully, makes a small depression in the surface and rests there without sinking even though its density is about eight times that of water. Again if a clean glass tube of a small bore is dipped in water, the water rises in the tube, but if the tube is dipped in mercury, the mercury is depressed.

The surface tension is measured as the force per unit length acting on either side of a line in the liquid surface in equilibrium, the direction of force being tangential to the surface and perpendicular to the line.

In *S.I.* units surface tension is measured in newton/metre. (N/m)

6.2. Angle of contact. The free surface of a liquid like turpentine, in a narrow tube, is concave and that of mercury is convex as shown in Fig. 6.1.

If we draw a tangent plane to the liquid surface and a tangent plane to the surface of the solid, *then the angle between the* two tangent planes within the liquid is called the **angle of contact** between the *liquid and the solid.* Thus α is the angle of contact between mercury and glass and β is the angle of contact between turpentine and glass, as shown in Fig. 6.1.

Note that α is obtuse whereas β is an acute angle.

In general, all liquids which wet glass have an acute angle of contact.

The force of surface tension acts all along the line of contact of the liquid with the solid *i.e. along the direction of the tangent plane to the liquid surface.*

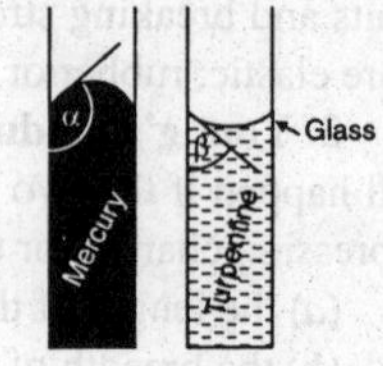

Fig. 6.1

Experiment 6.1. To determine the surface tension of water by Jaeger's method.

Apparatus. Jaeger's apparatus, a glass tube of about 5 mm diameter, microscope, a beaker and an adjustable stand.

Theory. Jaeger's apparatus consists of a Woulf's bottle *W* fitted in one mouth with a dropping funnel *F* containing water. Through the other mouth passes a glass tube *ABCD* to which is fitted a manometer *M* containing a liquid of low density. The end *D* of the tube is drawn into a fine capillary and dips in the liquid, the surface tension of which is to be found out to a depth of 4 to 5 cm contained in a beaker. As the capillary tube dips in the liquid, the liquid rises in it and the shape of the meniscus is hemispherical. If the stop-cock of the dropping funnel is slightly opened, the water slowly falls into the bottle and forces an equal volume of air into the tube *ABCD*, thereby compressing the liquid in the tube *CD*, which moves down and forms a bubble at the end *D*. The radius of the bubble gradually decreases with increasing pressure inside it till it reaches the minimum value when the shape of the bubble is hemispherical of radius r equal to the radius of the opening at D. The difference in the level of the liquid in the two limbs of the manometer is now maximum say h_1. The bubble is unstable

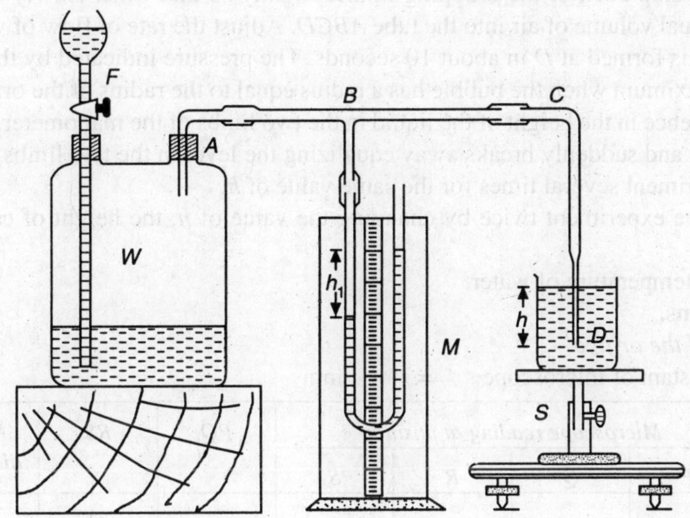

Fig. 6.2

because a slight increase in the radius decreases the internal pressure to produce equilibrium and as the external pressure is constant the bubble breaks away. Just before the bubble breaks away, the pressure inside

$$= P + h_1 \rho_1 g$$

where P is the atmospheric pressure and ρ_1 the density of the liquid in the manometer.

At the same time pressure outside the bubble

$$= P + h\rho g$$

where h is the depth of the open end D below the free surface of the liquid in the vessel and ρ the density of the experimental liquid.

Since the excess pressure inside the bubble is $\frac{2T}{r}$,

$$\therefore \qquad \frac{2T}{r} = (P + h_1 \rho_1 g) - (P + h\rho g)$$

or

$$T = \frac{rg}{2}(h_1 \rho_1 - h\rho)$$

Procedure. 1. Take a thin glass tube, clean it thoroughly and at its one end draw a thin capillary of about 0.2 mm to 0.5 mm diameter. Cut the lower end of the capillary smooth and perpendicular to the axis of the tube. Examine the tip under a microscope and see that the opening is perfectly circular and there is no roughness. Measure the diameter.

2. Clamp the capillary tube in a horizontal position and focus the travelling microscope on its orifice and measure the diameter in two mutually perpendicular directions PQ and RS as shown in Fig. 6.3. Slightly rotate the capillary tube and repeat the observation. Take such three observations.

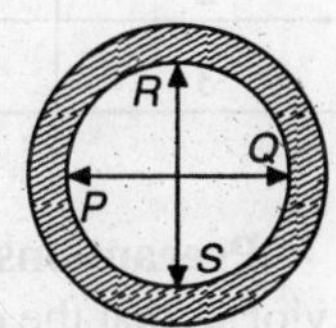

Fig. 6.3

3. Now make a scratch on the tube at a distance of 3 to 4 cm from the orifice. Measure the distance of this scratch from the orifice with the microscope set transversely. Clamp the tube in a vertical position in the Jaeger's apparatus as shown in Fig. 6.2. Place a beaker containing water on an

adjustable stand and let the capillary tube dip in it. Adjust the height of the beaker so that the scratch marked on the capillary coincides with the water surface.

4. Open the stop-cock of the dropping funnel slightly so that water slowly falls into the bottle and forces an equal volume of air into the tube *ABCD*. Adjust the rate of flow of water ino the bottle so that a bubble is formed at *D* in about 10 seconds. The pressure indicated by the manometer rises and becomes maximum when the bubble has a radius equal to the radius of the orifice at *D*. Note the maximum difference in the height of the liquid in the two limbs of the manometer. The bubble at this stage is unstable and suddenly breaks away equalizing the level in the two limbs of the manometer. Repeat the experiment several times for the same value of *h*.

5. Repeat the experiment twice by changing the value of *h*, the height of capillary dipping in water.

6. Note the temperature of water.

Observations.

Diameter of the orifice.

Vernier constant of microscope = mm

Sl. No.	*Microscope reading at position*				*PQ*	*RS*	*Mean diameter*	*Mean radius*
	P	*Q*	*R*	*S*				
1								
2								
3								

Mean radius from all observations *r* = cm

= m

Temperature of water *t* = °C

Density of water at *t*°C ρ = kg/m³

Density of liquid in the manometer ρ_1 = kg/m³

Sl. No.	*Depth of orifice h*	*Manometer readings*			*Mean* h_1	$T = \frac{rg}{2}(h_1 \rho_1 - h\rho)$
		upper level	*lower level*	*Difference*		
1		(*i*) (*ii*) (*iii*) (*iv*)	(*i*) (*ii*) (*iii*) (*iv*)	(*i*) (*ii*) (*iii*) (*iv*)		
2						
3						

Mean surface tension *T* = N/m

Precautions and sources of error. 1. The manometer must contain a liquid of low density like xylol so that the difference in the two columns may be large.

2. The orifice of the capillary should be circular and about 0.2 mm to 0.5 mm in diameter.

3. The diameter of the orifice must be measured at different places and at each place in tw mutually perpendicular directions to reduce the error due to ellipticity of the cross-section.

4. The glass tube should be thoroughly cleaned before drawing the capillary. Any traces of grease will affect the result.

5. The apparatus should be completely air-tight.

6. The bubbles should be formed slowly and one by one at a rate of about one bubble in 10 seconds. This can be achieved by keeping the air space in the Woulf's bottle small.

6.3. Stream line and turbulent flow. Consider the motion of water flowing from left to right through a pipe whose diameter varies as shown in Fig. 6.4. If the motion of water is smooth and steady each small drop of water reaching a point such as *A* follows exactly the same path as is done by its preceding drop which passed through that point. Such a smooth flow of liquid through a pipe is called **stream lined flow** and the lines indicating the path of the particles in the stream of any fluid are called stream lines.

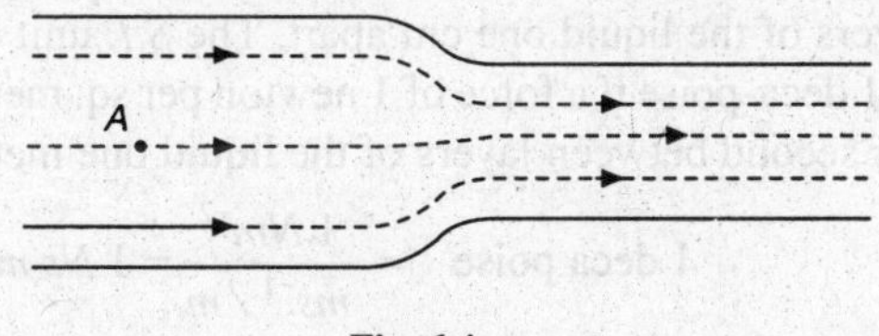

Fig. 6.4

Hence the *orderly, stream line* or *steady flow* is one in which every particle of liquid follows the same path as that of its preceding particle and the velocity of each particle crossing a particular point in a particular layer is exactly the same both in magnitude as well as in direction as the velocity of the preceding particle when it was at that point. But the velocity of the particles at different points in the same layer or at the same point in different layers is not the same. Such a motion does not have eddies, whirlpools or zig-zag path. Three such stream lines have been shown in Fig. 6.4. The stream lines never cross each other and they are more crowded in the part of the tube where the flow of the fluid is faster and farther apart where the motion is slow.

If, however, the velocity of the fluid becomes too large or the bends or constrictions in the pipe cause the fluid to change its direction too abruptly the regular motion gets broken up into eddies and whirls. Such a churning motion of the fluid is called **turbulent** and the flow is called **turbulent flow.**

6.4. Viscosity. *The property of a liquid by virtue of which the liquid offers resistance to its flow is called viscosity.*

When a liquid flows over a flat surface the layer of the liquid in contact with the surface remains stationary because of adhesion. The next upper layer moves slowly over the first, the third over the second and so on. The speed of each layer increases with its distance from the flat surface.

Fig. 6.5

Consider a layer *AB* of a fluid moving with a velocity v and a parallel layer *CD* at a distance dx from it moving with a velocity $v + dv$, then $\frac{dv}{dx}$ is known as the **velocity gradient.**

The difference between the velocities of the two layers is due to a force acting parallel to the upper layer in the direction of its motion. Let the force acting on an area *A* of *CD* be *F*. On the lower layer *AB* this force obviously will act in opposite direction. It is found that the force per unit area

$$\frac{F}{A} \propto -\frac{dv}{dx}$$

or

$$\frac{F}{A} = -\eta\frac{dv}{dx}$$

or

$$\eta = \frac{-F}{A\frac{dv}{dx}} \quad ...(i)$$

where η is a constant depending upon the nature of the liquid and is called the **co-efficient of viscosity.** *It is defined as the tangential force per unit area required to maintain a unit velocity difference between two parallel layers of the liquid a unit distance apart.*

The relation (*i*) is true only if each particle of the liquid moves along the same path as its preceding particle *i.e.* the motion is **streamline motion**. The motion is streamline so long as the velocity of flow of the liquid is below a certain limiting value called the *critical velocity*. When the velocity increases beyond the critical value the motion becomes disorderly or *turbulent*.

The co-efficient of viscosity of a liquid decreases rapidly with rise of temperature and increases with rise of pressure.

Units. The *C.G.S.* unit of viscosity is a **poise**. A liquid has a viscosity of 1 poise if a force of 1 dyne per sq. cm is able to maintain a velocity difference of 1 cm per second between two parallel layers of the liquid one cm apart. The *S.I.* unit of viscosity is a **deca-poise.** A liquid has a viscosity of 1 deca-poise if a force of 1 newton per sq. metre is able to maintain a velocity difference of 1 metre per second between layers of the liquid one metre apart.

$$1 \text{ deca poise} = \frac{1\,Nm^{-2}}{ms^{-1}/m} = 1\,Ns\,m^{-2}$$

$$= 1 \times (10^5\,dyne)\,s\,(100\,cm)^{-2} = 10\,dyne\,s\,cm^{-2} = 10 \text{ poise}$$

or $\quad 1 \text{ poise} = 0.1 \text{ deca–poise.}$

Dimensions of co-efficient of viscosity. From the relation

$$\eta = -\frac{F}{A\,\frac{dv}{dx}}$$

substituting the dimensions of various quantities, we have

$$[\,\eta\,] = \frac{[\,M^1\,L^1\,T^{-2}\,]}{[L^2]\,[L^1\,T^{-1}/L^{-1}]} = M^1\,L^{-1}\,T^{-1}$$

6.5. Viscosity apparatus. To find the viscosity of a liquid by its flow through a capillary tube, the apparatus used is as shown in Fig. 6.6. It is known as viscosity apparatus or Poiseuille's apparatus and consists of a constant level tank *C*, a capillary tube *AB* of uniform bore, a manometer *M* and a graduated cylinder as arranged.

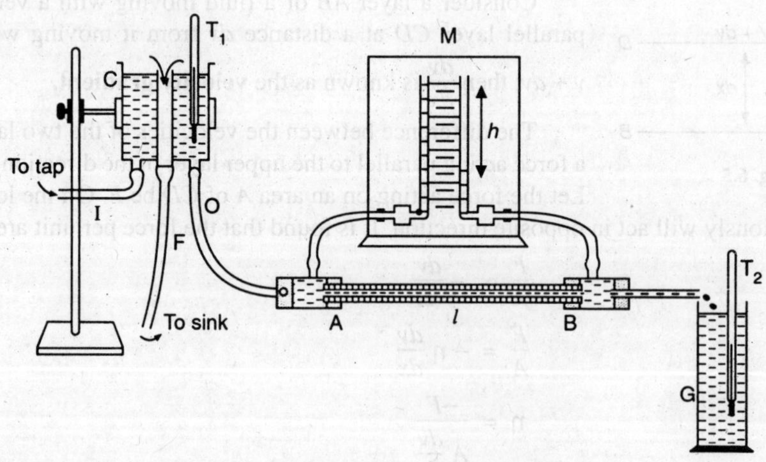

Fig. 6.6

The constant level tank makes the water to enter into the capillary tube at a constant pressure. Water enters the tank through the inlet tube *I* and flows into the capillary tube through the outlet tube *O*. The excess of water flows out through the overflow tube *F* thus maintaining the level of water in the tank constant. The difference of pressure across the ends of the capillary tube and hence the rate of flow of water through it can be changed by raising or lowering the constant level tank.

The manometer M measures the height of the free surface of water on entering and leaving the capillary tube AB. Hence the difference in heights of the water level in the two tubes of the manometer gives the difference in pressure in terms of the height of water column. In this way a steady flow of water can be maintained through the capillary tube.

Experiment 6.2. To find the co-efficient of viscosity of water by noting its flow through a capillary tube of uniform bore.

Apparatus. Viscosity apparatus, a fine capillary tube of uniform bore, a graduated cylinder, a stop watch, a travelling microscope and two thermometers.

Theory. When a liquid which wets the surface is allowed to flow through a uniform capillary tube of length l and radius r, the volume V flowing out per second against a constant difference of pressure across its ends is given by

$$V = \frac{\pi P r^4}{8 l \eta}$$

or

$$\eta = \frac{\pi P r^4}{8 l V}$$

This is known as Poiseuille's relation. If h is the difference in height of the free surface of the liquid at the two ends and ρ the density of the liquid, then

$$P = h\rho g$$

or

$$\eta = \frac{\pi h \rho g r^4}{8 l V}$$

The above formula holds true only under ideal conditions *i.e.* when the pressure difference is so small that the liquid leaves the tube in a slow trickle.

Procedure. 1. Take a capillary tube and clean it by passing dilute nitric acid and then a strong solution of caustic soda. Wash it freely under tap water and allow the water to flow through it for about 5 minutes. Dry it by passing hot air and measure its length.

2. Clamp the tube in a horizontal position and find the internal diameter of the tube at both the ends with the help of a travelling microscope accurately.

Note. Since the radius of the capillary tube occurs as r^4 in the final formula a slight error in its measurement will be magnified four times in the result. Hence in order to find accurately the mean radius of the tube following method may be used.

Suck a column of mercury AB about 10 cm long near one end of the capillary tube. Measure its length accurately with a travelling microscope. Move the thread towards the other end as at $A'B'$ and find the length again. Repeat at different positions. Find the weight of mercury by weighing it in a watch glass and hence calculate the radius r, as given below.

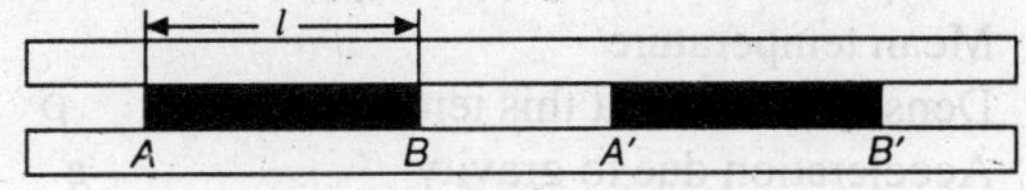

Fig. 6.7

Radius

$$r = \sqrt{\frac{m}{\pi d l}}$$

where m is the mass of mercury d its density and l the mean length of the mercury column.

3. Place the tube in position in the viscosity apparatus. Allow the water to flow and adjust the height of the constant level tank till the water flows out of the capillary tube *drop by drop*. Measure the height h, *i.e.* the difference in the level of water in the two tubes of the manometer M.

4. Place the graduated cylinder below the dropping water and at the same time start the stop watch. Collect about 50 cc of water and note the time taken. Repeat twice.

5. Note the temperature of the incoming and the outgoing water by thermometers T_1 and T_2 placed as shown.

6. Now alter the pressure difference by changing the height of constant level tank and repeat the experiment. Take such three sets of observations.

Note. Do not use a pinch cock to regulate the flow of the liquid. The flow is regulated by the length, the radius of the capillary and the difference of pressure at its two ends.

Observations. Length of the capillary tube l = m

Diameter of the capillary tube

(*i*) **With microscope**

Vernier constant of the microscope = cm

End	*Microscope reading at*				*PQ*	*RS*	*Mean*
	P	*Q*	*R*	*S*			
A							
B							

Mean diameter = cm = m

Mean radius = cm = m

or (*ii*) **With mercury thread.**

Sl. No.	*Microscope reading at*		*Length of mercury thread l*
	A	*B*	
1			
2			

Mean length l = m

Weight of watch glass = gm

Weight of watch glass + mercury = gm

∴ Weight of mercury m = gm = kg

Density of mercury d = kg/m^3

∴ Radius of the tube $r = \sqrt{\frac{m}{\pi d l}}$ = m

Temperature of water (*i*) °C; (*ii*) °C, (*iii*) °C

Mean temperature = °C

Density of water at this temp. ρ = kg/m^3

Acceleration due to gravity g = m/s^2

Sl. No.	*h in metre*	*Volume of water collected*	*Time taken in sec.*	*Volume collected per sec. V in cubic metre*	*Mean V*	$\frac{h}{V}$
1		(*i*)				
		(*ii*)				
2		(*i*)				
		(*ii*)				
3		(*i*)				
		(*ii*)				

Mean $\frac{h}{V}$ = m^{-2}

Calculations. $\eta = \frac{\pi\rho g r^4}{8l} \times \frac{h}{V}$

= Newton sec/m^2 or deca poise

Value from tables =

% error =

Precautions. 1. The bore of the capillary tube should be narrow and uniform.

2. The capillary tube should be placed in a horizontal position.
3. The flow of water through the capillary tube should be streamlined.
4. The water should flow out of the capillary tube drop by drop.
5. The temperature should remain constant during the experiment.
6. The diameter must be measured very accurately observing all precautions for the use of a microscope.

Sources of error. (*i*) The temperature might have changed during the experiment.

(*ii*) The flow may be slightly turbulent.

(*iii*) When the liquid enters the tube its motion is accelerated at the inlet end and as a result of this velocity distribution becomes uniform only when the liquid has travelled a short length of the tube. This necessitates a correction for the length of the tube from l to $(l + 1.64\, r)$.

(*iv*) When the liquid leaves the tube it has some velocity. Hence the entire pressure difference $P = h\rho g$ is not used up in overcoming the viscous resistance but a part of it is spent in imparting kinetic energy to the liquid. The effective pressure is less than the actual pressure by an amount

$$\frac{V^2\rho}{\pi^2 r^4}$$

Hence the effective pressure

$$P_1 = h\rho g - \frac{V^2\rho}{\pi^2 r^4}$$

Substituting the value of P_1 for P in Poiseuille's relation the corrected formula for co-efficient of viscosity becomes

$$\eta = \frac{\pi\, h\, \rho\, g r^4}{8\, V\, (l + 1.64r)} - \frac{V\rho}{8\pi(l + 1.64r)}$$

Exercise. *Find the resistance offered to the flow of water by a capillary tube by analogy with Ohm's law.*

Hint. According to Poiseuille's relation, the rate of flow of a liquid through a capillary tube is given by

$$V = \frac{\pi\, P r^4}{8\, \eta\, l}$$

This relation may be put in the form

$$V = \frac{P}{R}$$

where $R = \frac{8\, \eta\, l}{\pi\, r^4}$ and is known as *fluid resistance i.e* the resistance offered to the free flow of the given liquid by the capillary tube. To find R, determine experimentally the value of l, r and η as in **Expt. 6.2** and calculate the value of

$$R = \frac{8\, \eta\, l}{\pi\, r^4}.$$

Experiment 6.3. To study the flow of water through a capillary tube (without manometer) as a function of pressure head using two tubes of (*a*) same length but different radii and (*b*) different lengths but same radii.

Apparatus. Constant pressure head arrangement, three capillary tubes *A, B* and *C*; *A* and *B* of the same length but different radii *B* and *C* of the same radius but different lengths, a measuring cylinder, stop watch, spirit level, travelling microscope, metre rod, a thermometer and pieces of rubber tubing.

Theory. The volume of liquid flowing per second through a uniform capillary tube of length l and radius r under a constant pressure head P is given by

$$V = \frac{\pi P r^4}{8\eta l}$$

where η is the co-efficient of viscosity.

(*i*) For a given capillary tube its length l and radius r are constant and η the co-efficient of viscosity for the liquid is also constant.

$$\therefore \qquad V \propto P$$

Now pressure $P = \rho g h$ where ρ is the density of liquid. As ρ and g are constant, hence

$$V \propto h.$$

If a graph between V along X-axis and h along Y-axis is plotted, it will be a straight line upto certain values of 'h' and beyond this point, it deviates from straight line path. The straight line path shows the *stream-line* motion and beyond that *turbulant motion.*

(*ii*) For the same value of P and l for a liquid, the volume

$$V \propto r^4$$

Hence for capillary tubes of radii r_1 and r_2 having the same length, the volumes V_1 and V_2 of water collected in the same time are given by

$$\frac{V_1}{V_2} = \left(\frac{r_1}{r_2}\right)^4$$

(*iii*) For the same value of P and r the volume

$$V \propto \frac{1}{l}$$

Thus for capillary tubes of lengths l_1 and l_2 having the same radii the volumes V_1 and V_2 collected in the same time are given by

$$\frac{V_1}{V_2} = \frac{l_2}{l_1}$$

Procedure. 1. Measure the lengths of the capillary tubes *A, B* and *C*. Measure the internal diameter of each capillary tube at either end with the help of a travelling microscope by clamping the tube in horizontal position, taking two readings in mutually perpendicular directions as *PQ* and *RS*.

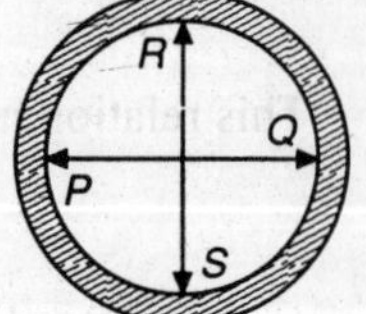

Fig. 6.8

2. Place the capillary tube *A* in position. Allow the water to flow and adjust the height of the constant level tank (h = 5 cm) till the water starts dripping out of the capillary tube. If there is a tendency of water to flow along the outer surface of the capillary tube, apply a little of grease on the tube or attach a piece of thread near the open end with a small length hanging below.

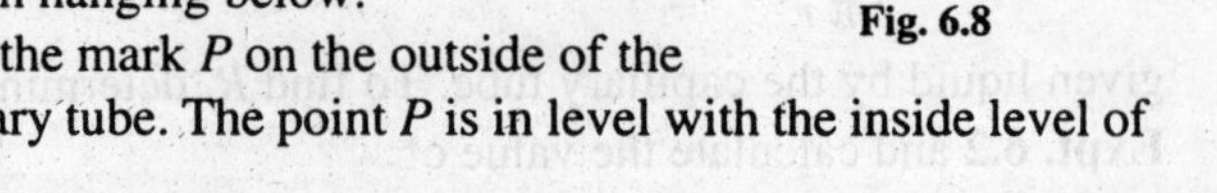

The head of pressure h is measured from the mark P on the outside of the constant level tank upto the level of the capillary tube. The point P is in level with the inside level of water in the tank.

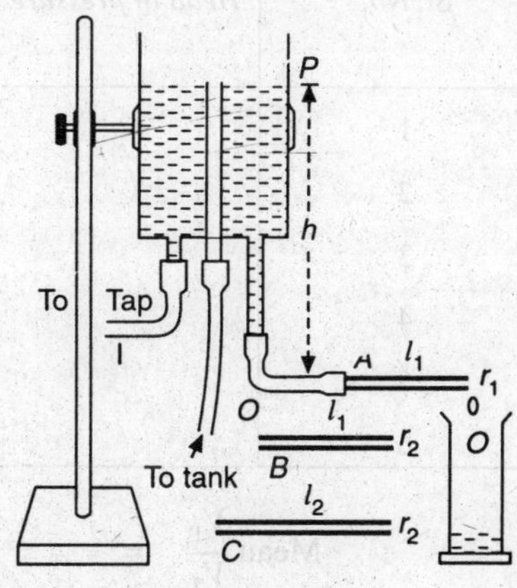

Fig. 6.9

3. Measure the head of pressure h and place the graduated cylinder below the end of the capillary from which water is dropping and at the same time start the stop watch. Collect about 50 cc of water and note the time taken. Repeat twice.

4. Repeat the observation for different values of h varying in steps of 5 cm. Note the temperature of water.

5. Similarly repeat the observations with capillary tubes B and C.

Observations.

Length of tube A, l_1 = m

Length of tube B, l_2 = m

Length of tube C, l_3 = m

Diameter of the capillary tubes.

Tube	*End*	*Microscope reading at*				*PQ*	*RS*	*Mean diameter*
		P	*Q*	*R*	*S*			
A	1 2							
B	1 2							
C	1 2							

Mean radius of A, r_1 = cm = m

Mean radius of B, r_2 = cm = m

Mean radius of C, r_3 = cm = m

Tube A

Sl. No.	Head h in metre	*Volume of water collected*	*Time taken t*	*Volume per second*	*Mean rate of flow V*	$\frac{h}{V}$
1		*(i)* *(ii)*				
2						

Similarly tabulate data for tubes B and C for the same value of h.

NOTE : The mean rate of flow V is in cubic metre per second and h is in metre so that h/V is in m^{-2}.

Stream line and turbulent flow.

Plot a graph between the mean rate of flow 'V' taken along X-axis and corresponding values of 'h' taken along Y-axis for the three capillary tubes A, B and C on the same graph paper. It is seen that upto certain values of 'h' the graph is a straight-line and beyond this point the graph deviates from straight line. The straight line portion of each graph represents **stream-line** flow and beyond it, the **turbulent** flow.

(*a*) For the capillary tubes A and B having the same length l_1 but different radii r_1 and r_2, tabulate the following data within the value of 'h' for which the flow is *stream-line*.

Sl. No.	Head of pressure h	Rate of flow for		$\frac{V_1}{V_2}$
		Tube A V_1	Tube B V_2	
1				
2				
3				
4				
5				
6				

Mean $\frac{V_1}{V_2}$ =

Calculated value of $\left(\frac{r_1}{r_2}\right)^4$ =

As the value of $\frac{V_1}{V_2} = \left(\frac{r_1}{r_2}\right)^4$ within experimental error for stream-line flow

$$V \propto r^4$$

for the capillary tubes having the same length and same head of pressure.

From the graph. 1. Plot a graph taking pressure head (h) along X-axis and volume of water collected per second (Rate of flow) V_1 for the tube A and V_2 for the tube B (of the same length l_1 but of radii r_1 and r_2 respectively) along the Y-axis. We will get two straight lines not parallel to each other. The slope of the graph for tube of larger radius will be greater than that of the graph for tube of smaller radius as shown in Fig. 6.10.

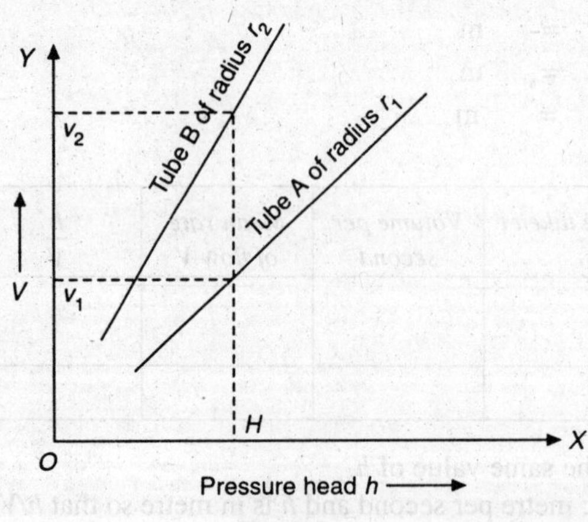

Fig. 6.10

2. For the same pressure head H note the valumes V_1 and V_2 of water collected per second (rate of flow) for the tube A and B respectively.

Record. From the graph pressure head H = m

Volume of water collected per second for the tube A, V_1 = cubic metre

Volume of water collected per second for the tube B, V_2 = cubic metre

Mean radius of the tube $A = r_1$ = m

Mean radius of the tube $B = r_2$ = m

For the tube A, $\frac{V_1}{r_1^4}$ =

For the tube B, $\frac{V_2}{r_2^4}$ =

Difference between $\frac{V_1}{r_1^4}$ and $\frac{V_2}{r_2^4} = \frac{V_1}{r_1^4} - \frac{V_2}{r_2^4}$ =

Conclusion. (*i*) The fact that the graph between *h* and *V* is a straight line shows that the rate of flow is directly proportional to pressure head *h*

(*ii*) As $$\frac{V_1}{r_1^4} = \frac{V_2}{r_2^4} \text{ or } \frac{V_1}{V_2} = \left(\frac{r_1}{r_2}\right)^4$$

$$\therefore \quad V \propto r^4$$

i.e. the volume of liquid collected per second is directly proportional to the fourth power of the radius of the capillary tube, length remaining the same.

(*b*) For the capillary tubes *B* and *C* of the same radius r_2 and lengths l_1 and l_2 tabulate the following data within the values of *h* for which the flow is stream line

Sl. No.	*Head in metres h*	*Rate of flow of water in cubic metre per sec for capillary* *B* (V_1)	*C* (V_2)	$\frac{V_1}{V_2}$

Calculated value of $\frac{l_2}{l_1}$ =

As the value of $\frac{V_1}{V_2} = \frac{l_2}{l_1}$ within experimental error for stream line flow $V \propto \frac{1}{l}$ for the capillary tubes of the same radius and the same head of pressure.

From the graph. 1. Plot a graph taking pressure head (*h*) along *X*-axis and volume of water collected per second (rate of flow) V_1 for the tube *B* and V_2 for the tube *C* (of the same radius r_2 but of lengths l_1 and l_2 respectively) along the *Y*-axis. We will get two straight lines not parallel to each other as shown in Fig. 6.11.

2. For the same pressure head *H* note the volume V_1 and V_2 of water collected per second (rate of flow) for the tubes *B* and *C* respectively.

Record. From the graph, pressure head

H = m

Volume of water collected per second for the tube *B*

V_1 = cubic metre

Volume of water collected per second for the tube *C*

V_2 = cubic metre

Length of the tube

B, l_1 = cm = m

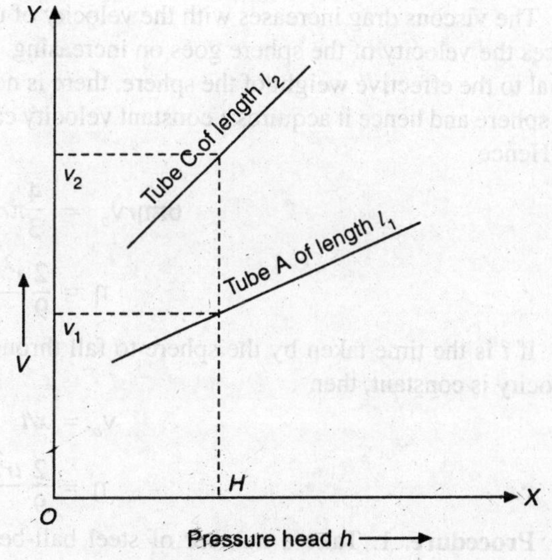

Fig. 6.11

Length of the tube C, l_2 = cm = m

For the tube B, $V_1 l_1$ =

For the tube C, $V_2 l_2$ =

Difference between $V_1 l_1$ and $V_2 l_2 = V_1 l_1 - V_2 l_2$ =

Conclusion. 1. The fact that the graph between h and V is a straight line shows that the rate of flow V is directly proportional to the pressure head h

2. As $V_1 l_1 = V_2 l_2$; $\frac{V_1}{V_2} = \frac{l_2}{l_1}$ or $V \alpha \frac{1}{l}$.

Therefore, for the capillary tubes of the same radius and for the same head of pressure the volume of the liquid collected is inversely proportional to the length of the tube.

Precautions. Same as in **Expt 6.2.**

6.6. Falling sphere viscometer. Falling sphere viscometer consists of a glass tube A about 50 cm long and 3.5 cm wide closed at the bottom with a rubber plug. The top end is closed with a rubber cork having a hole through which passes a narrow glass tube funnel B of about 0.5 cm diameter. This arrangement ensures that the motion of spheres falling through the liquid is along the axis of the wide tube A, containing the liquid. To ensure constant temperature of the liquid, the tube A containing the liquid is surrounded by a wider glass cylinder C which is filled with water. A stirrer S is also provided. The temperature of the bath is measured with a thermometer T.

Experiment 6.4. To determine the co-efficient of viscosity of glycerine or castor oil by Stoke's method.

Apparatus. Falling sphere viscometer, steel ball-bearings, a thermometer, a stop watch, a screw gauge, a metre scale, forceps, rubber bands etc.

Theory. The method is based on the application of Stoke's law to the fall of small spheres through a viscous liquid. If a small sphere of radius r and density ρ is allowed to fall freely from rest through a viscous liquid of density σ and co-efficient of viscosisy η, then at any instant the sphere is acted upon by the following forces :

(*i*) Viscous drag $F = 6\pi\eta r v$ directed vertically upward where v is the velocity of the sphere and

(*ii*) the effective weight of the sphere $= \frac{4}{3}\pi r^3 (\rho - \sigma) g$ directed vertically downward

The viscous drag increases with the velocity of the sphere and so long it is less than the upward forces the velocity of the sphere goes on increasing. When the viscous drag is equal to the effective weight of the sphere, there is no resultant force acting on the sphere and hence it acquires a constant velocity called the *terminal velocity* v_o. Hence

$$6\pi\eta r v_o = \frac{4}{3}\pi r^3 (\rho - \sigma) g$$

$$\therefore \quad \eta = \frac{2}{9}\frac{r^2 (\rho - \sigma) g}{v_o}$$

If t is the time taken by the sphere to fall through a distance x, when the velocity is constant, then

$$v_o = x/t$$

$$\therefore \quad \eta = \frac{2}{9}\frac{t r^2 (\rho - \sigma) g}{x}$$

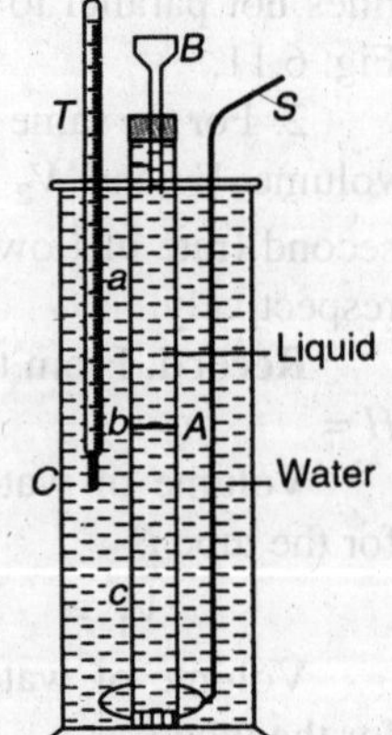

Fig. 6.12

Procedure. 1. Take a number of steel ball-bearings and measure their diameters accurately with a screw gauge. Thoroughly clean the spheres so that no grease and dirt is sticking to their surface. Place these balls in a clean beaker or watch glass and soak their surfaces thoroughly with the experimental liquid.

2. Set the viscometer as shown in Fig. 6.12 and tie three rubber bands as at *a, b* and *c* equidistant from each other. *i.e., ab = bc*. The distance between two consecutive bands should be 10 - 15 cm.

3. Transfer one of the balls with the help of forceps into the funnel *B*. The ball after falling through the tube of the funnel will fall centrally through the liquid along the axis of the tube *A*. When the ball reaches the band *a* start the stop watch and note the time taken by it to cover the distance *ab* and then *bc*. If the two time intervals are equal the ball has acquired the terminal velocity. Find the mean of the time of transit between *ab* and *bc*.

4. Repeat the experiment with other balls and calculate the value of r^2t for each ball separately. Find the mean value of r^2t.

5. Measure the density of the liquid with the help of a hydrometer and note the temperature of the water bath surrounding the tube *A*.

Observations.

Density of the liquid σ = kg/m^3

Density of ball-bearings (Steel) ρ = kg/m^3

Temperature of the bath = °C

Measurement of diameter of ball-bearings

Pitch of the screw gauge = cm

Least count = cm

Zero correction = ± cm

Balls	*Diameter of the ball*			*Radius*		
	One direction	*Perpendicular direction*	*Mean d*	*cm*	*m*	r^2 *(in m^2)*
A						
B						
C						
...						
...						
...						
H						

Mean r^2 = m^2

Measurement of the time of fall

Balls		*Distance between consecutive bands*		*Time of transit*	*Mean time t*	r^2t
		x	Mean *x*			
A	(*i*)					
	(*ii*)					
B						
..						
..						
G						
H	(*i*)					
	(*ii*)					

Mean r^2t =

Calculations. $\eta = \frac{2}{9}\frac{r^2 t(\rho - \sigma)g}{x} =$ deca poise

Precautions and sources of error. 1. The ball-bearings should be small in size so that the time of transit between two consecutive bands is large for accurate measurements.

2. To eliminate the wall effect on the value of terminal velocity the radius of the tube containing the liquid should be large as compared to the radius of the ball.

3. The balls should be thoroughtly soaked in the experimental liquid before making them fall through the liquid and the soaked balls should be picked up with the help of forceps.

4. The balls should fall along the axis of the wider tube.

5. The viscosity changes with temperature rapidly, hence the experiment with various balls should be carried out in quick succession.

6. The diameters of the balls should be measured accurately and in two mutually perpendicular directions.

7. The temperature of the liquid should be measured with a sensitive thermometer.

Note. For accurate work to eliminate the wall effect on the terminal velocity Landenberg has shown that the corrected terminal velocity is given by

$$v_0 = v\left(1 - 2.4\frac{r}{R}\right)$$

where v is the observed velocity and R the radius of the container tube A.

6.7. Effect of temperature on viscosity. The viscosity of liquids decreases with rise of temperature. No satisfactory relation between viscosity of a liquid at different temperatures has been obtained. The following table gives the co-efficient of viscosity of water from 0 - 100°C.

Temp. °C	*Viscosity in deca poise*	*Temp. °C*	*Viscosity in deca poise*
0	0.001793	40	0.000657
5	0.001522	50	0.000550
10	0.001311	60	0.000469
15	0.001142	70	0.000406
20	0.001006	80	0.000356
25	0.000893	90	0.000316
30	0.000800	100	0.000284

It should, however, be noted that viscosity of gases increases with rise of temperature.

6.8. Effect of pressure on viscosity. Pressure has practically no effect on the viscosity of liquids which are fairly mobile *i.e.*, possess a small co-efficient of viscosity. For example, the co-efficient of viscosity of ether at 20°*C* increases by about 60% for an increase of about 500 atmosphere pressure. The viscosity of water *decreases* with pressure for the first few hundred atmospheres. For liquids having a large co-efficient of viscosity, the value increases appreciably with pressure.

Oral Questions

1. Surface tension. (*i*) Define surface tension. (*ii*) What are its dimensions and the units in which it is measured? (*iii*) What is the angle of contact? (*iv*) Why does a liquid rise in a capillary tube? Why is mercury depressed? (*vi*) What will happen if the height of the capillary tube is less than the height to which water can rise into it? Will the water overflow? If not, Why? (*vii*) What is the effect of temperature and impurities on the value of surface tension? (*viii*) What part does surface tension play in our daily life? (*ix*) What is the value of the angle of contact in case of (*a*) mercury and glass (*b*) water and glass? (*x*) Where from does the energy come when the liquid rises in a capillary tube against gravity? (*xi*) Should the top of capillary be open or closed? (*xii*) Why don't you use distilled water instead of tap water?

2. Jaeger's method. (*i*) State the formula for surface tension for Jaeger's method. (*ii*) Why is the pressure inside the air bubble greater than that outside it? (*iii*) Why does the liquid in the manometer rise and fall during the formation and breaking away of the bubble?

3. Viscosity. (*i*) Define viscosity, (*ii*) What is stream line flow? (*iii*) What is meant by turbulent flow? (*iv*) What is critical velocity and on what factors does it depend? (*v*) What are the units of viscosity? (*vii*) How does viscosity of water vary with temperature? (*viii*) Which has greater viscosity; honey or water? (*ix*) What is the effect of pressure on viscosity? (*x*) What do you mean by fluid resistance?

4. Flow method. (*i*) On what factors does the rate of flow of a liquid through a capillary tube depend? (*ii*) Does the relation hold good for a tube of wide bore? If not, why? (*iii*) How will you test whether the capillary tube is of a uniform bore or not? (*iv*) What arrangement do you use to keep the difference of pressure at the two ends constant? (*v*) Which is the better method for finding the diameter of the capillary tube; mercury column method or the microscope method?

5. Flow through tubes of different length and radius. (*i*) What type of graph do you get between h and V for a given capillary of constant length and radius? (*ii*) Is the graph a straight line for all values of h? (*iii*) What is the relation between V_1 and V_2 *i.e.* volumes of water collected per second for the capillaries of same length but different radii under the same head of pressure? (*iv*) What is the relation between V_1 and V_2 *i.e.* volumes of water collected per second for the capillaries of same radius but different lengths under the same head of pressure.

6. Stoke's method. (*i*) What is Stoke's law and what are its limitations? (*ii*) What is terminal velocity? (*iii*) Why does the falling sphere acquire a terminal velocity? (*iv*) Why are the spheres soaked with the liquid before allowing them to fall through the liquid? (*v*) Can you study the variation of viscosity with temperature with this apparatus?

7

One Dimensional Collisions

7.1. Collision. *A collision is said to have taken place whenever two bodies in their independent motion strike against each other or come so close to each other that in a very short interval of time the magnitude and the direction of their velocity changes.*

7.2. Types of collision. Collisions are generally of two types

(*i*) **Elastic collision.** *A collision is said to be perfectly elastic if the total linear momentum and the total kinetic energy of the system (colliding bodies) remains conserved i.e.* the law of conservation of linear momentum and the law of conservation of energy hold good.

Collisions between atomic and sub-atomic particles are perfectly elastic. Collisions between two glass or ivory or polished steel balls are also *nearly* elastic. In an elastic collision the relative velocity of approach before the collision is equal to the relative velocity of separation after the collision.

(*ii*) **Inelastic collision.** *A collision is said to be perfectly inelastic if the total linear momentum of the system remains conserved but not the total kinetic energy of the system i.e.* there is a definite loss of kinetic energy during the impact. In general, all collisions are of this type. Familiar examples are that of a collision of a bat and ball or between two billiard balls. In inelastic collision the two colliding bodies move together after the collision and relative velocity after collision is always less than the relative velocity before collision.

7.3. Law of conservation of linear momentum. Linear momentum is defined as the product of mass and linear velocity of a body.

$$\vec{p} = m\vec{v}$$

It is a *vector* quantity and its direction is the same as that of velocity of the body. It is measured in gm cm s^{-1} in C.G.S. system and its S.I unit is Kg m s^{-1}.

The law of conservation of linear momentum states that *the total linear momentum of the system before collision is equal to the total momentum of the system after collision.* This law holds good for elastic as well as inelastic collisions.

If two bodies of masses m_1 and m_2 moving with velocities u_1 and u_2 in the same direction ($u_1 > u_2$) suffer elastic collision and their velocities after collision are v_1 and v_2 respectively, then according to the law of conservation of linear momentum.

$$m_1u_1 + m_2u_2 = m_1v_1 + m_2v_2 \quad ...(i)$$

7.4. Law of conservation of energy. *This law states that the total kinetic energy of the system before collision is equal to the total kinetic energy of the system after collision.* This law holds good only for elastic collisions. According to this law

$$\frac{1}{2}m_1 u_1^2 + \frac{1}{2}m_2 u_2^2 = \frac{1}{2}m_1 v_1^2 + \frac{1}{2}m_2 v_2^2 \quad ...(ii)$$

7.5. Velocities of the bodies after collision. Relation (*i*) can be stated in the form

$$m_1 (u_1 - v_1) = m_2 (v_2 - u_2) \quad ...(iii)$$

Relation (*ii*) can be stated in the form

$$m_1 (u_1^2 - v_1^2) = m_2 (v_2^2 - u_2^2) \quad ...(iv)$$

Dividing (*iv*) by (*iii*), we get

$$u_1 + v_1 = v_2 + u_2$$

$$\therefore \quad v_2 = u_1 + v_1 - u_2$$

Substituting in Eq. (*i*), we get

$$m_1 u_1 + m_2 u_2 = m_1 v_1 + m_2 u_1 + m_2 v_1 - m_2 u_2$$

or $$2 m_2 u_2 + (m_1 - m_2) u_1 = (m_1 + m_2) v_1$$

or $$v_1 = \frac{2 m_2 u_2 + (m_1 - m_2) u_1}{m_1 + m_2} \quad ...(v)$$

Similarly $$v_2 = \frac{2 m_1 u_1 - (m_1 - m_2) u_2}{m_1 + m_2}$$

7.6. Co-efficient of restitution. No collision is perfectly elastic or perfectly inelastic. It is only the degree of elasticity of the collision that varies. The degree of elasticity of a collision is measured by the *co-efficient of restitution.*

It is defined as the ratio of the relative velocity of seperation of the two bodies after collision to the relative velocity of their approach before collision

$$\therefore \text{ Co-efficient of restitution } e = \frac{\text{Relative velocity of separation of two bodies after collision}}{\text{Relative velocity of approach before collision}}$$

$$= \frac{v_2 - v_1}{u_1 - u_2}$$

For perfectly elastic collision, $e = 1$ and for a perfectly inelastic collision $e = 0$ or $v_2 - v_1 = 0$ or $v_2 = v_1$ *i.e.* in a perfectly inelastic collision, the two colliding bodies *move together* after collision.

Generally, the collisions are neither perfectly elastic nor perfectly inelastic. Therefore, for a practical collision the value of *e* lies between *zero* and *one.*

7.7. Dependence of fraction of kinetic energy transferred on masses of colliding bodies. Let a body of mass m_1 moving with a velocity u_1 suffer a collision with a body of mass $m_2 = xm_1$ supposed to be at rest for simplicity *i.e.* $u_2 = 0$. If v_1 and v_2 respectively are their velocities after collision, then

$$v_2 = \frac{2 m_1 u_1 - (m_1 - m_2) u_2}{m_1 + m_2}$$

$$= \frac{2 m_1 u_1}{m_1 + m_1 x} \quad [\because u_2 = 0]$$

$$= \frac{2u_1}{1 + x}$$

$$\therefore \quad \text{K.E of mass } m_2 = x\, m_1 \text{ after collision} = \frac{1}{2} m_2 v_2^2$$

$$= \frac{1}{2} x\, m_1 \frac{4 u_1^2}{(1 + x)^2} = \frac{4x}{(1 + x)^2} \cdot \frac{1}{2} m_1 u_1^2$$

$$\therefore \quad \frac{\text{K.E of stationary mass } m_2 \text{ after collision}}{\text{K.E of moving mass } m_1 \text{ before collision}} = \frac{\frac{4x}{(1+x)^2}\frac{1}{2}m_1 u_1^2}{\frac{1}{2}m_1 u_1^2} = \frac{4x}{(1+x)^2}$$

or Percentage of K.E transferred by the moving body to the stationary body after collision

$$= \frac{4x}{(1+x)^2} \times 100\%$$

Thus the fraction or percentage of kinetic energy transferred by the moving body to the stationary body after collision depends upon the masses of the two bodies.

7.8. Verification of the law of conservation of linear momentum. Let a small spherical ball of mass m_1, moving with a velocity u_1 in a horizontal plane collide with a small spherical ball of mass m_2 at rest and let v_1 and v_2 respectively be their velocities after collision. Suppose these balls fall through a height h,

Then $$h = ut + \frac{1}{2}gt^2 = \frac{1}{2}gt^2$$

as the velocity of each ball in the vertical direction $u = 0$

$$\therefore \quad t = \sqrt{\frac{2h}{g}}$$

It is because the time taken by all bodies to fall from the same height is equal irrespective of their masses.

The horizontal distance travelled by these balls while falling through the height h is given by

$$s_1 = v_1 t \quad \therefore \quad v_1 = \frac{s_1}{t} = s_1\sqrt{\frac{g}{2h}}$$

and $$s_2 = v_2 t$$

$$\therefore \quad v_2 = \frac{s_2}{t} = s_2\sqrt{\frac{g}{2h}}$$

Knowing m_1, m_2, u_1, v_1 and v_2 the law of conservation of momentum can be verified.

Experiment 7.1. To verify (*a*) The law of conservation of linear momentum and (*b*) The law of conservation of kinetic energy in the case of elastic collision.

Apparatus. The apparatus used to perform the experiment is shown in Fig. 7.1. It consists of a grooved ruler *ABC*. The part *AB* is inclined and the part *BC* is horizontal. *S* is a *set screw* having a slight depression on the top of it. The height of this screw can be adjusted. The grooved ruler is fixed on the top of a table. The stationary ball which is to be struck by the moving ball is placed in the depression on the set screw top. A drawing board with a sheet of white paper fixed on it with drawing pins is placed on the floor. The height of the struck ball, when at rest; from the drawing board can be measured with the help of a plumb line.

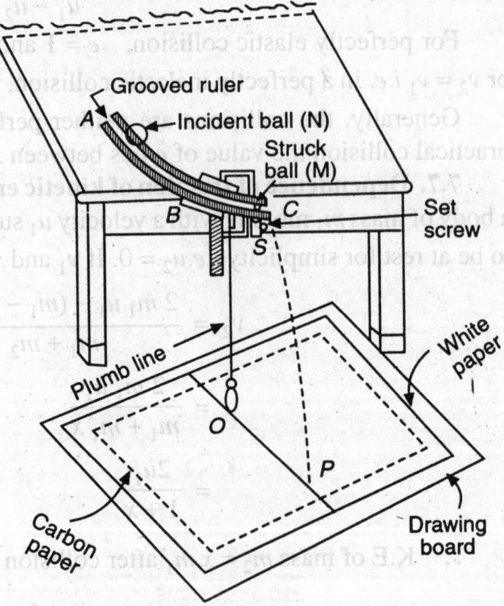

Fig. 7.1

Procedure. 1. Fix the grooved ruler on the top of the table with a clamp. With the help of the *set screw* make the portion *BC* horizontal.

2. Place a drawing board on the floor. Fix a carbon paper with its carbon side up on the drawing board. Place a white paper sheet on the carbon paper and fix it in position with the help of drawing pins.

3. Place the light ball (*M*) on the *set screw*. Mark a point *O* on the white sheet of paper vertically below *M* with the help of a plumb line. The point *O* gives the position of the incident ball at the instant of the collision. Adjust the position of the board in such a way that the point *O* lies on the centre of the shorter side of the sheet.

4. Remove the ball (*M*). Place the heavier ball (*N*) at the top of the grooved ruler and release it. Mark the point *X* where the incident ball (*N*) falls on the sheet. (Fig. 7.2)

5. Now place the *struck ball* (*M*) on the *set screw* and release the *incident ball* (*N*) from the top of the grooved ruler. The two balls will collide and then fall and strike the sheet at certain points. Mark these points *Y* and *Z*.

6. Find the weight of the incident ball *N* and the struck ball *M*. Let their masses be represented as m_1 and m_2 respectively.

7. Join the points *O* and *X*, *O* and *Y* and *O* and *Z*.

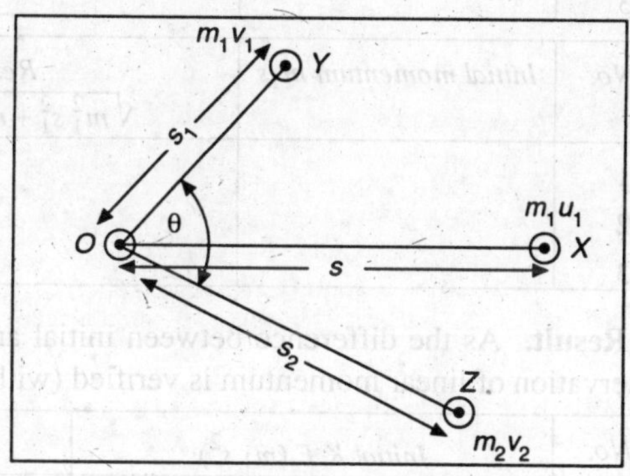

Fig. 7.2

Let $OX = s$; $OY = s_1$ and $OZ = s_2$ and $< YOZ = \theta$

Now $$v_1 = \frac{s_1}{t} = k\,s_1 \text{ where } k \text{ is a constant} = \frac{1}{t}$$

$$v_2 = \frac{s_2}{t} = k\,s_2 \quad \text{and} \quad u_1 = \frac{s}{t} = ks$$

$\therefore$ Initial momentum $= m_1\,u_1 = m_1\,ks$

$$\text{Resultant final momentum} = \sqrt{(m_1 v_1)^2 + (m_2 v_2)^2 + 2m_1 v_1 m_2 v_2 \cos\theta}$$

$$= \sqrt{(k\,m_1 s_1)^2 + (k\,m_2\,s_2)^2 + k^2\,2\,m_1 s_1\,m_2 s_2 \cos\theta}$$

$$= k\sqrt{m_1^2\,s_1^2 + m_2^2\,s_2^2 + 2m_1\,m_2\,s_1\,s_2 \cos\theta}$$

$$\therefore \quad m_1\,ks = k\sqrt{m_1^2\,s_1^2 + m_2^2\,s_2^2 + 2m_1 m_2\,s_1 s_2 \cos\theta}$$

or $$m_1\,s = \sqrt{m_1^2\,s_1^2 + m_2^2\,s_2^2 + 2m_1 m_2\,s_1 s_2 \cos\theta}$$

This is the law of conservation of linear momentum.

Again $$\frac{1}{2}m_1\,u_1^2 = \frac{1}{2}m_1 v_1^2 + \frac{1}{2}m_2\,v_2^2$$

or $$\frac{1}{2}m_1 k^2\,s^2 = \frac{1}{2}m_1\,k^2\,s_1^2 + \frac{1}{2}m_2 k^2\,s_2^2$$

or $$m_1 s^2 = m_1\,s_1^2 + m_2\,s_2^2$$

This is the law of conservation of kinetic energy.

8. Repeat the experiment three times on new sheets of paper.

Observations. Mass of the incident ball (*N*), m_1 = gm.

Mass of the struck ball (*M*), m_2 = gm.

S. No.	*Distance between O and X = s cm*	*Distance between O and Y = s_1 cm*	***Distance between O and Z = s_2 cm***	*Angle θ*	*cos θ*
1					
2					
3					

S.No.	*Initial momentum m_1s*	*Resultant momentum $\sqrt{m_1^2 s_1^2 + m_2^2 s_2^2 + 2m_1 m_2 s_1 s_2 \cos\theta}$*	*Difference*
1			
2			
3			

Result. As the difference between initial and resultant momentum is very small, the law of conservation of linear momentum is verified (within experimental error).

S.No.	*Initial K.E ($m_1 s^2$)*	*Final K.E. ($m_1 s_1^2 + m_2 s_2^2$)*	*Difference*

Result. As the difference between initial kinetic energy and final kinetic energy is small the law of conservation of energy is verified. (within experimental error).

Precautions. 1. The grooved ruler should be very smooth.

2. The portion *BC* of the grooved ruler should be horizontal.

3. The grooved ruler should be clamped tightly to the table.

4. The height of the apparatus should be such that the balls after collision fall on the sheet fixed on the drawing board and make a mark on the carbon paper.

5. The incident ball should be just released. It should not be pushed.

Source of error. 1. The portion *BC* of he grooved ruler may not be exactly horizontal.

2. While releasing the incident ball it may get pushed slightly.

Experiment 7.2. To study one dimensional elastic collision using two hanging spheres.

Apparatus. Two metallic spheres of equal masses, a metre rod, light and uniform thread, Fixed, unyielding support with suspension head to suspend the spheres.

Theory. When a suspended sphere is moved horizontally with a small amplitude say 5 to 7 cms, its path is almost a straight line. When released the velocity attained by it is directly proportional to the distance moved by it, if the air resistance is negligible. If *v* is the velocity attained and *s* the distance moved, then

$$v \propto s$$

or $$v = ks \quad \text{where } k \text{ is a constant.}$$

Procedure. 1. Find the weight of the two metallic spheres *P* and *Q* separately. See that the two are as nearly equal as possible.

2. Suspend the two spheres with threads of *exactly* equal lengths in the suspension clamps in such a way that they are not more than 5 cm above the floor.

3. Place a metre scale horizontally below the spheres.

4. Note the initial positions of the spheres on the metre scale when at rest.

5. Now displace the sphere P towards the left through a distance of 4 to 5 cm to the position P' and then release it. The sphere P will strike against the sphere Q. As their masses are equal, the sphere P will come to rest after collision (which is elastic for all practical purposes) while, the sphere Q would move to wards the right. Note the position Q' upto which the sphere Q moves to the right before retracing its path. If the two spheres are of exactly the same mass, then the sphere Q will travel through exactly the same distance as travelled by the sphere P from the point of release to the position of sphere Q while at rest.

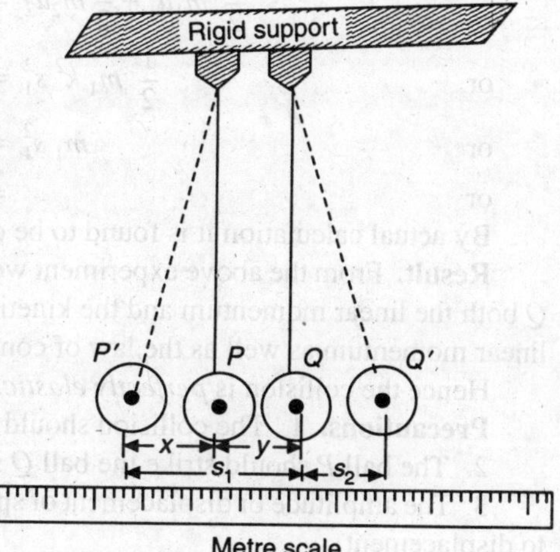

Fig. 7.3

6. Repeat the experiment 6 to 7 times to get the exact distance travelled by the sphere P and Q before and after collision.

Observations Mass of sphere P, m_1 = gm

Mass of sphere Q, m_2 = gm

Distance between spheres P and Q, when at rest, y = cm

S.No.	*Distance through which sphere P is displaced to the left* x = cm	*Distance travelled by sphere P before colliding with sphere Q* $s_1 = x + y$ = cm	*Distance travelled by sphere Q to the right after collision* s_2 = cm.
1			
2			
3			
4			
5			
6			
7			

Note. The distance x is kept the same in all the observations and corresponding value of s_2 is noted in each case.

Law of conservation of momentum. Since velocity is proportional to horizontal distance

$\therefore$ Initial velocity of P before collision $u_1 \propto s_1$ or $u_1 = ks_1$

Final velocity of P after collision $v_1 = 0$

Initial velocity of Q before collision $u_2 = 0$

Final velocity of Q after collision $v_2 \propto s_2$ or $v_2 = ks_2$

According to the law of conservation of momentum

$$m_1u_1 + m_2u_2 = m_1v_1 + m_2v_2$$

$$\therefore \quad m_1 ks_1 + m_2 \times 0 = m_1 \times 0 + m_2 ks_2$$

or $$m_1s_1 = m_2s_2$$

or $$=$$

By actual calculation it is found to bé correct (within experimental error)

Law of conservation of energy. According to the law of conservation of energy

$$\frac{1}{2} m_1 u_1^2 + \frac{1}{2} m_2 u_2^2 = \frac{1}{2} m_1 v_1^2 + \frac{1}{2} m_2 v_2^2$$

or
$$\frac{1}{2} m_1 k^2 s_1^2 = \frac{1}{2} m_2 k^2 s_2^2$$

or
$$m_1 s_1^2 = m_2 s_2^2$$

or
$$=$$

By actual calculation it is found to be correct (within experimental error)

Result. From the above experiment we find that during the collision of the sphere P with sphere Q both the linear momentum and the kinetic energy remain conserved *i.e* the law of conservation of linear momentum as well as the law of conservation of energy are obeyed.

Hence the collision is *perfectly elastic.*

Precautions. 1. The collision should be one dimensional and elastic.

2. The ball P should strike the ball Q in a manner so as to have a head-on collision.

3. The amplitude or displacement of sphere P should be small, so that the velocity is proportional to displacement.

4. The masses of the two spheres should be exactly equal.

Sources of error. 1. The masses of the two spheres may not be exactly equal. In such a case both the spheres shall have some displacement after the collision so that the displacement of the sphere P will also have to be measured. Therefore, the equations of motion will have to be modified accordingly.

2. There is always some air resistance to the motion of the balls.

Experiment 7.3. To study (*i*) The law of conservation of linear momentum (*ii*) The law of conservation of kinetic energy and (*iii*) to calculate the co-efficient of restitution using one dimensional collision apparatus of two hanging spheres.

Apparatus. One dimensional collision apparatus using two hanging spheres, a battery and a tapping key.

One dimensional collision apparatus. It consists of two spheres A and B of masses m_1 and m_2 respectively suspended from the suspension clamps C_1 and C_2 attached to the rigid ceiling of the box with the help of two threads of exactly the same length. The box is fitted with glass windows. The spheres oscillate parallel to a scale graduated in m.m and placed in position slightly above the spheres.

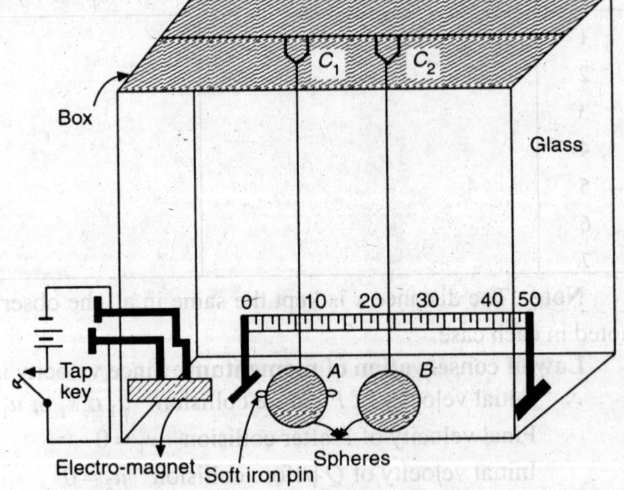

Fig. 7.4

On the left side of the apparatus there is an electromagnet having an iron core and a coil. The electromagnet is activated when an electric current is passed through its coil with the help of the battery by pressing the tapping key. The sphere close to the magnet is made of a magnetic material with a *soft iron pin* embedded into it.

Theory. Let the metre scale be placed at a distance l below the points of suspension O and O'. Let L be the length of the thread between the points of suspension O and O' and the centre of oscillation of the spheres (Fig. 7.5). The sphere A is now pulled towards the left so that it comes in contact with the electro magnet.

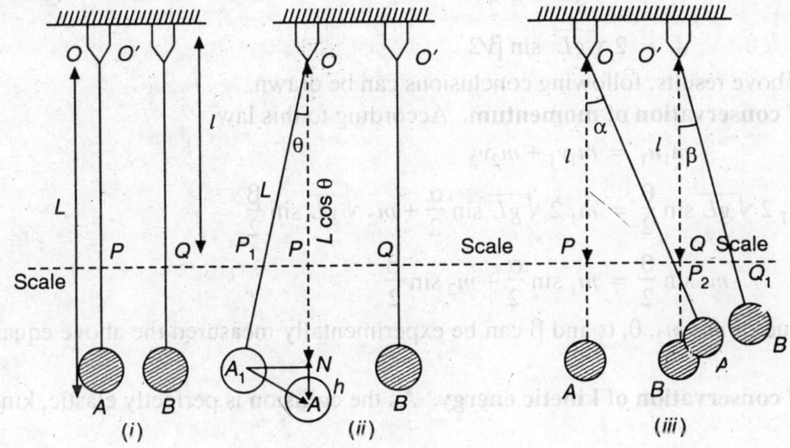

Fig. 7.5

Let its angular displacement be θ and the position of thread on the scale be P_1. Vertical height gained by the sphere $A = AN = h = OA - ON = L - L\cos\theta = L\,(2\sin^2\theta/2)$

$$[\because\ 1 - \cos 2A = 2\sin^2 A/2]$$

or $$h = 2\,l\sin^2\theta/2 \qquad \text{...}(i)$$

Again $$\tan\theta = \frac{P_1P}{OP} = \frac{P_1P}{l}$$

As P_1P and l are known the value of θ can be calculated. Substituting this value of θ in Eq. (i), we can find the value of h

$\therefore$ Potential energy of sphere A in the displaced position

$$E_p = m_1gh \qquad \text{...}(ii)$$

When the sphere A is released (by releasing the tap key in the circuit of the electromagnet) it comes down and collides with sphere B. Let the velocity of the sphere A, when it comes back to its mean position be u_1. Then the kinetic energy gained by it before colliding with sphere B is given by

$$E_k = \frac{1}{2}m_1u_1^2 \qquad \text{...}(iii)$$

As we are assuming that there are no energy losses, the energy is conserved.

$$\therefore \qquad \frac{1}{2}m_1u_1^2 = m_1gh$$

or $$u_1 = \sqrt{2gh}$$

$$= \sqrt{2g \times 2L_1\sin^2\frac{\theta}{2}} = 2\sqrt{gL}\,\sin\frac{\theta}{2} \qquad \text{...}(iv)$$

Let the velocity of sphere A after collision with B be v_1 and that of sphere B be v_2. Also

Let the angular displacement of A after collision $= \alpha$

and the angular displacement of B $= \beta$

The position P_2 of A and Q_1 of B after collision is noted on the scale. Then

$$\tan\alpha = \frac{PP_2}{OP} = \frac{PP_2}{l} \text{ and } \tan\beta = \frac{QQ_1}{O'Q} = \frac{QQ_1}{l}$$

Using equation (*iv*), we have

$$v_1 = 2\sqrt{gL}\,\sin\frac{\alpha}{2} \quad ...(v)$$

and

$$v_2 = 2\sqrt{gL}\,\sin\beta/2 \quad ...(vi)$$

From the above results, following conclusions can be drawn.

1. **Law of conservation of momentum.** According to this law

$$m_1u_1 = m_1v_1 + m_2v_2$$

or

$$m_1\, 2\sqrt{gL}\,\sin\frac{\theta}{2} = m_1\, 2\sqrt{gL}\,\sin\frac{\alpha}{2} + m_2\sqrt{gL}\,\sin\frac{\beta}{2}$$

or

$$m_1\sin\frac{\theta}{2} = m_1\sin\frac{\alpha}{2} + m_2\sin\frac{\beta}{2} \quad ...(vii)$$

As the value of m_1, m_2, θ, α and β can be experimentally measured the above equation can be verified.

2. **Law of conservation of kinetic energy.** As the collision is perfectly elastic, kinetic energy is conserved.

or

$$\frac{1}{2}m_1u_1^2 = \frac{1}{2}m_1v_1^2 + \frac{1}{2}m_2v_2^2$$

Substituting the values of u_1, v_1 and v_2 we get

$$\frac{1}{2}m_1(4\,gL)\sin^2\frac{\theta}{2} = \frac{1}{2}m_1(4gL)\sin^2\frac{\alpha}{2} + \frac{1}{2}m_2(4gL)\sin^2\frac{\beta}{2}$$

or

$$m_1\left(\sin^2\frac{\theta}{2} - \sin^2\frac{\alpha}{2}\right) = m_2\sin^2\frac{\beta}{2} \quad ...(viii)$$

This equation can also be verified by measuring the values of m_1, m_2, θ, α and β experimentally.

3. **Fraction of K.E transferred by moving sphere A to stationary sphere B.** As already proved in **Art 7.7.**

$$\frac{\text{K.E of stationary sphere } B \text{ of mass } m_2 \text{ after collision}}{\text{K.E of moving sphere } A \text{ of mass } m_1 \text{ before collision}} = \frac{4x}{(1+x)^2}$$

Where $x = \dfrac{m_2}{m_1}$

$$\therefore \quad \frac{\text{K.E of } B \text{ after collision}}{\text{K.E of } A \text{ before collision}} = \frac{4m_2}{m_1\left(1+\dfrac{m_2}{m_1}\right)^2}. \quad ...(ix)$$

Also

$$\frac{\text{K.E of } B \text{after collision}}{\text{K.E of } A \text{ before collision}} = \frac{\frac{1}{2}m_2v_2^2}{\frac{1}{2}m_1u_1^2} = \frac{m_2}{m_1}\frac{\left[2\sqrt{gL}\,\sin\frac{\beta}{2}\right]^2}{\left[2\sqrt{gL}\,\sin\frac{\theta}{2}\right]^2}$$

$$= \frac{m_2}{m_1}\,\frac{\sin^2\frac{\beta}{2}}{\sin^2\frac{\theta}{2}} \quad ...(x)$$

By actual calculation the value of (*ix*) and (*x*) comes out to be the same, thus proving the result.

4. **Co-efficient of restitution.** The co-efficient of restitution is given by

$$e = \frac{v_2 - v_1}{u_1 - 0} = \frac{v_2 - v_1}{u_1} = \frac{\sin\frac{\beta}{2} - \sin\frac{\alpha}{2}}{\sin\frac{\theta}{2}}.$$

Procedure 1. Take the two spheres A and B. The sphere A should have a soft iron pin embedded into it so that it gets attracted by the electromagnet. Weigh them and record their masses.

2. Suspend the spheres with light uniform threads of the same length L from the suspension clamps and see that they are perfectly vertical.

3. Note down the length of the threads from the points of suspension O and O' to the points P and Q respectively where the threads come parallel to the scale.

4. Also note down the positions of the threads on the scale corresponding to the points P and Q.

5. Press the tap key so that the circuit of the electromagnet is completed and it gets activated. Pull the sphere A to the left and bring it in contact with the electromagnet. The electromagnet holds the sphere A in the position A_1. Note the position of the suspension thread on the scale at the position P_1 where the thread comes to rest in the position A_1 of the sphere A.

6. Release the sphere A by releasing the tap-key. The sphere A will collide with sphere B and the two will move in the same direction. Note the position P_2 on the scale upto which the thread A is displaced after the collision. Also note the position Q_1 of the thread of B on the scale and find the maximum displacement QQ_1.

Note. Generally a light *rider* is attached to the thread on the right side of each of the spheres A and B, which moves freely on the scale and helps to read accurately the position P_2 and Q_1 on the scale.

7. With the same value of displacement for the sphere A repeat the experiment at least three times and find the mean of various displacements of spheres A and B.

8. Adjust the position of the electromagnet by moving it forward or backward and repeat the experiment. The velocity and hence the kinetic energy of the sphere A will change and therefore the positions of P_1 and as a result that of P_2 and Q_1 will also change. For each position of the electromagnet repeat the experiment at least three times.

9. Also repeat the experiment by changing one or both the spheres A and B.

Observations. Mass of sphere A, m_1 = gm

Mass of sphere B, m_2 = gm

Initial position of the thread of A on the scale P = cm

Initial position of the thread of B on the scale Q = cm

Length of the thread from the point of suspension to the scale $OP = O'Q = l$ = cm.

S.No.	Reading of scale at P_1	Reading of scale at P_2				Reading of scale at Q_1			
		(i)	(ii)	(iii)	mean	(i)	(ii)	(iii)	mean
1									
2									
3									
4									
5									

S.No.	$\tan\theta = \frac{P_1P}{l}$	$\tan\alpha = \frac{P_2P}{l}$	$\tan\beta = \frac{Q_1Q}{l}$	$\theta = \tan^{-1}\frac{P_1P}{l}$	$\alpha = \tan^{-1}\frac{P_2P}{l}$	$\beta = \tan^{-1}\frac{Q_1Q}{l}$	$\sin\theta/2$	$\sin\alpha/2$	$\sin\beta/2$	$\sin^2\theta/2$	$\sin^2\alpha/2$	$\sin^2\beta/2$
1												
2												
3												
4												
5												

Verification

1. **Conservation of linear momentum**

S. No.	$m_1 \sin\frac{\theta}{2}$	$m_1 \sin\frac{\alpha}{2} + m_2 \sin\frac{\beta}{2}$
1		
2		
3		
4		
5		

Conclusion. As $m_1 \sin\frac{\theta}{2}$ is very nearly equal to $m_1 \sin\frac{\alpha}{2} + m_2 \sin\frac{\beta}{2}$ the collision obeys the law of conservation of linear momentum and is thus one dimensional.

2. **Conservation of kinetic energy**

S. No.	$m_1 \sin^2\frac{\theta}{2}$	$m_1 \sin^2\frac{\alpha}{2} + m_2 \sin^2\frac{\beta}{2}$
1		
2		
3		
4		
5		

Conclusion. As $m_1 \sin^2\frac{\theta}{2} = m_1 \sin^2\frac{\alpha}{2} + m_2 \sin^2\frac{\beta}{2}$ the kinetic energy is conserved during the collisions and thus the collisions are one dimensional.

3. **Co-efficient of restitution *e***

S. No.	$e = \dfrac{\sin\frac{\beta}{2} - \sin\frac{\alpha}{2}}{\sin\theta/2}$	*Mean e*
1		
2		
3		
4		
5		

4. **Kinetic energy transferred by m_1 to m_2.**

S. No.	$\frac{m_2}{m_1} \times \frac{4}{\left(1+\frac{m_2}{m_1}\right)^2}$	$\frac{m_2}{m_1} \times \frac{\sin^2 \frac{\beta}{2}}{\sin^2 \frac{\theta}{2}}$
1		
2		
3		
4		
5		

Conclusion. As the results in the two columns are almost equal, the relation is verified.

Precautions. 1. Same as in **Expt. 7.2.**

2. The collision should be head-on.

3. A light rider should be used to measure the maximum displacement.

Oral Questions

1. One dimensional collision. (*i*) What do you understand by the term collision? (*ii*) Define elastic and inelastic collision. (*iii*) Define linear momentum of a body. (*iv*) Is momentum a scalar or a vector quantity? (*v*) What are the units of momentum. (*vi*) State the law of conservation of linear momentum. (*vii*) State the law of conservation of energy. (*viii*) What is the relation between momentum and force? (*ix*) Give the characteristics of an elastic collision. (*x*) Give the characteristics of an inelastic collision. (*xi*) Give the examples of elastic and inelastic collisions. (*xii*) When two bodies of equal masses collide elastically, what is the relation between initial and final velocities of the colliding bodies. (*xiii*) In Q (*xii*) if the second body is at rest before collision what happens to the first after collision?

PART II

HEAT AND STATISTICAL PHYSICS

8

Expansion of Gases

8.1. Expansion of gases. Gases expand much more than solids and liquids do. If a gas is heated but its volume is kept constant, the pressure of the gas will increase. We shall, therefore, study,

(*i*) Variations in volume of a gas when the temperature is changed, keeping *pressure constant.*

(*ii*) Variations in pressure of a gas when temperature is changed keeping *volume constant.*

(*iii*) The variations in the volume of a gas when the pressure is changed keeping the temperature constant. This has been studied under Boyle's law.

8.2. Co-efficients of expansion. *The co-efficient of* **increase of volume** *of a gas is defined as the increase in volume per unit volume at 0°C, per degree centigrade rise of temperature, pressure remaining constant.*

If V_0 is the volume at 0°*C* and V_t the volume at *t*° *C*, then the co-efficient of increase of volume at constant pressure γ_p is given by

$$\gamma_p = \frac{V_t - V_0}{V_o \times t}$$

or

$$V_t = V_0 (1 + \gamma_p t)$$

The value of co-efficient of increase of volume at constant pressure for all gases is $\frac{1}{273}$ or 0.00367. It is given by Charles' law which states that *the volume of a gas changes by* $\frac{1}{273}$ *of its volume at 0°C for every 1°C change of temperature, pressure remaining constant.*

The co-efficient of **increase of pressure** *at constant volume is defined as the increase in pressure per unit pressure at 0°C, per degree centigrade rise of temperature, volume remaining constant.*

If P_0 is the pressure at 0°*C* and P_t the pressure at *t*°*C*, then the co-efficient of increase of pressure at constant volume γ_v is given by

$$\gamma_v = \frac{P_t - P_0}{P_0 \times t}$$

or

$$P_t = P_0 (1 + \gamma_v t)$$

The value of co-efficient of increase of pressure at constant volume for all gases is $\frac{1}{273}$ or 0.00367.

While determining the co-efficient of linear expansion of a solid or co-efficient of apparent expansion of a liquid the lower temperature was taken to be room temperature instead of 0°*C*. *In the case of gases the lower temperature must be 0°C and not the room temperature* as an appreciable error will be caused due to the very large value of the co- efficient of expansion of a gas, *viz.*, $\frac{1}{273}$ or 0.00367, which is approximately 10 times the co-efficient of cubical expansion of water and 100 times the co-efficient of cubical expansion of copper.

8.3. Constant pressure air thermometer. The constant pressure air thermometer is used to determine the co-efficient of increase of volume at constant pressure. It consists of a vertical wooden board fixed to a horizontal iron base fitted with levelling screws as shown in Fig. 8.1. A metre scale is fixed on the vertical board with its zero mark at the lower most position. Along the scale is attached a wide-bore glass tube fitted with corks. Through the upper cork passes an inlet tube *I* and a thermometer *T*. Through the lower cork passes an outlet tube *O* and a graduated glass tube closed at one end and having a zero graduation at the closed end. This tube has graduations from 0-50 cc. The lower end of this tube is connected by means of a thick walled rubber tubing to a wider glass tube *D* open at both ends. The graduated tube is generally fixed while the tube *D* can slide along the vertical metre scale. A certain volume of dry air is enclosed in the graduated tube by filling the rubber tube and some portions of both these tubes with clean and dry mercury.

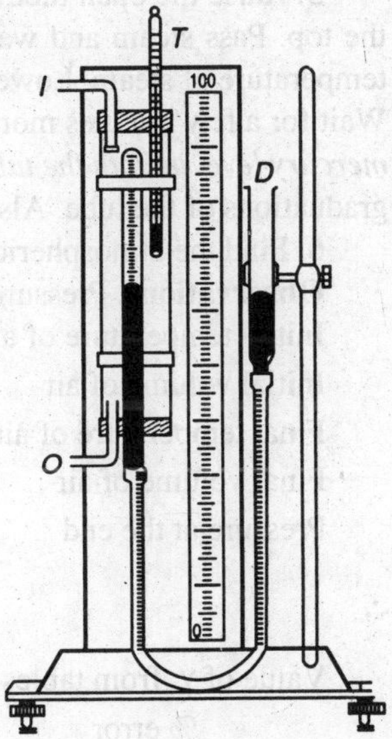

Fig. 8.1

Experiment 8.1. To determine the co-efficient of increase of volume of air at constant pressure.

Apparatus. A constant pressure air thermometer, a plumb line, a thermometer, a steam generator, a bunsen burner, a tripod stand and a set square.

Theory. If V_0, V_1 and V_2 are the volumes of the enclosed gas at temperature 0°C, t_1 °C, and t_2 °C respectively and γ_p is the co-efficient of increase of volume at constant pressure, then

$$V_1 = V_0(1+\gamma_p t_1) \quad ...(i)$$

$$V_2 = V_0(1+\gamma_p t_2) \quad ...(ii)$$

Dividing equation (*ii*) by (*i*), we have

$$\frac{V_2}{V_1} = \frac{1+\gamma_p t_2}{1+\gamma_p t_1}$$

or

$$\gamma_p = \frac{V_2 - V_1}{V_1 t_2 - V_2 t_1}$$

Procedure. Put enough water in the steam generator and start heating it. *Do not connect it with the inlet tube in the very beginning.*

2. *To set the apparatus vertical.* Suspend a plumb line in front of the scale. Adjust the levelling screws at the base so that the plumb line is parallel to the edge of the scale and equidistant from it at all points. *Adjust the central screw to make it parallel to the edge and the two side screws to make it equidistant.*

3. To find the atmospheric pressure : (*i*) Find the vernier constant of the Fortin's barometer.

(*ii*) Adjust the screw at the base of the barometer so that the mercury surface just touches the ivory tip.

(*iii*) Place the eye at the same level as the meniscus of mercury to avoid error due to parallax and note the reading.

4. Adjust the position of the open tube D so that the upper surface of mercury in it and in the closed graduated tube stands at the same level. Note the volume of air enclosed and also the temperature as recorded by the thermometer T.

Note. A convenient method of finding whether mercury in the open and the closed tubes is at the same level is to focus a telescope on the apparatus and adjust its horizontal cross-wire in level with the upper meniscus of mercury in both the tubes.

5. Raise the open tube to its highest point and connect the steam generator to the inlet tube I at the top. Pass steam and wait for about 15 minutes so that the air in the graduated tube attains the temperature of steam. Lower the tube D till mercury stands at the same level in both the tubes again. Wait for a few minutes more. *If the air in the graduated tube has attained the temperature of steam, mercury level in both the tubes will remain stationary*. Note the reading of the volume of air from the graduations of the tube. Also note the temperature.

6. Find the atmospheric pressure again.

Observations. Pressure in the beginning = cm

Initial temperature of air $t_1 =$ °C

Initial volume of air $V_1 =$ cc

Final temperature of air $t_2 =$ °C

Final volume of air $V_2 =$ cc

Pressure at the end = cm

$$\therefore \quad \gamma_p = \frac{V_2 - V_1}{V_1 t_2 - V_2 t_1}$$

Value of γ_p from tables = 0.00367

% error =

NOTE (*i*) The pressure is noted from a Fortin's barometer directly in cm of mercury column and is not to be used in calculations. Therefore it has not been converted into metres.

(*ii*) The volume is noted directly in cc. It is not converted into cubic metres as only ratio of volumes is involved.

Precautions. 1. The enclosed air should be completely dry.

2. The volume should be measured when the temperature remains constant for several minutes and mercury in the two tubes stands at the same level.

3. The pressure in the beginning and at the end of the experiment should be the same.

4. Steam should be passed from above.

5. The mercury should not be sticking to the sides of the glass tube.

Exercise. *Using a constant pressure air thermometer find the room temperature.*

Hints. Find the volume V_0 of the enclosed air at 0°C by surrounding it with ice, then V_1 the volume at room temperature t_1 and finally V_2 the volume at the temperature of steam. Find the temperature of steam from the barometer reading either from the tables of constants or from the formula.

Steam temperature = 100 – 0.00367 (760-p),

where p is the barometer reading in millimetres of mercury.

Assuming the temperature of steam to be t_2°C the room temperature is calculated from the relation

$$t_1 = \frac{V_1 - V_0}{V_2 - V_0} \times t_2$$

8.4. Joly's constant volume air thermometer. A simple form of a constant volume air thermometer due to Joly consists of a thin walled glass bulb C blown at the end of a capillary tube about 1 mm in diameter. The tube is bent twice at right angles and is joined by a thick walled rubber tubing to an open glass tube *D*. The part of the capillary tube attached to the bulb beyond the second bend is fixed vertically on a vertical wooden board which is, in turn, fixed on a horizontal iron base provided with three levelling screws. A millimetre scale is fixed on the vertical board with its zero mark in the lowermost position. The open tube which is also in front of the scale can slide up or down along the vertical board and can be clamped at any position. The rubber tube and the open tube contain mercury and during the experiment the level of mercury is kept at the fixed mark *F* in the capillary tube.

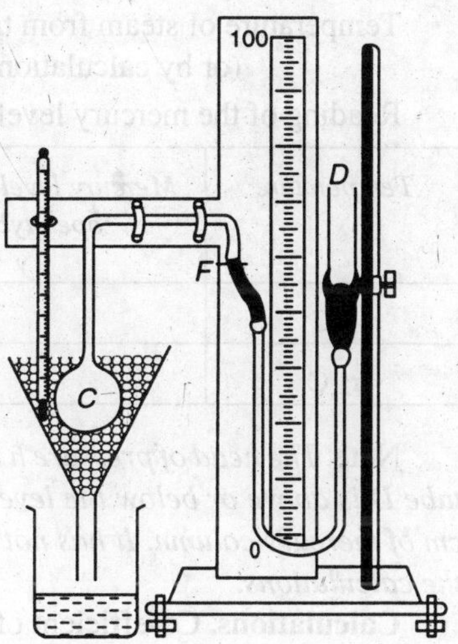

Fig. 8.2

Experiment 8.2. To determine the co-efficient of increase of pressure of air at constant volume.

Apparatus. Joly's constant volume air thermometer, a plumb line, a beaker, a thermometer, ice, a funnel, a pair of set squares and steam bath.

Theory. If P_0 is the pressure of the enclosed gas at 0 °C and P_t the pressure at the temperature of steam *t* °C, then co-efficient of increase of pressure at constant volume is given by

$$\gamma_v = \frac{P_t - P_0}{P_0 \times t}$$

Procedure. 1. Set the apparatus vertical with the help of plumb line and levelling screws.

2. Make a fixed mark *F* below the bend on the capillary tube by pasting a piece of gum paper on it. *The mark is made below the bend so that mercury may not enter the bulb C* due to negligence.

3. Note the atmospheric pressure from Fortin's barometer.

4. Bring the open tube *D* to the lowest position. Place a glass funnel below the bulb *C*. Wash ice in running water and pound it in a *clean duster*. Surround the bulb *completely* with pounded ice and suspend from a clamp a thermometer with its bulb dipping in ice near the glass bulb. Wait for about 10 minutes so that the air in the bulb attains the temperature of ice. Raise the tube *D* till the mercury level in the capillary tube stands at the fixed mark *F*.

5. Wait for a few minutes. If the air in the bulb has attained the temperature of ice and there is no leakage in the apparatus, then the mercury level will remain stationary at *F*.

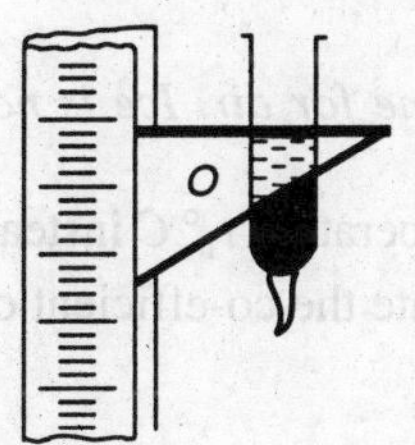

Fig. 8.3

In order to note the reading of the upper meniscus of mercury in the open tube *D* place a set square such that one of its perpendicular sides is horizontal and the other presses against the side of the metre scale, as shown in Fig. 8.3. Move it till the horizontal edge is at the level of mercury. Place the eye at the same height in order to avoid error due to parallax and note the scale reading against the horizontal edge of the set square. Similarly, note the reading of mercury level on the scale against the fixed mark *F*.

6. Remove the funnel containing ice surrounding the bulb *C*. Place the bulb in the steam bath as shown in Fig. 8.4. When the bulb remains surrounded with steam for about 10 minutes the air in the bulb will attain the temperature of steam. See that the thermometer records a constant temperature. Raise the tube *D* so that mercury in the capillary tube again stands at the fixed mark *F*. Wait for a few minutes more. If the air in the bulb has attained the temperature of steam the mercury level will remain stationary at *F*. Note the scale reading against the level of mercury in the tube *D*, with the help of a set square. *Bring the tube D to its lowest position and then remove the burner.*

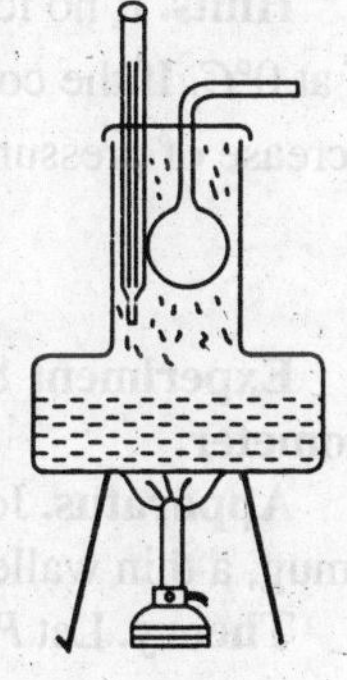
Fig. 8.4

7. Note the atmospheric pressure again and find the temperature of steam at this pressure.

Observations. Atmospheric pressure in the beginning = cm
Atmospheric pressure at the end = cm
Mean atmospheric pressure P = cm
Temperature of steam from tables
(or by calculation) t_2 = °C
Reading of the mercury level at the fixed mark F x = cm

Temperature	*Mercury level reading in open tybe D* y	*Head of pressure y-x = h (+ or –)*	*Pressure P ± h*
0°C			$P_0 =$
t_2°C			$P_2 =$

Note. *The head of pressure h is positive or negative according as the level of mercury in the open tube D is above or below the level of mercury at the fixed mark F. The value of pressure is taken in cm of mercury column. It has not been converted into metres as the ratio of pressures is involved in the calculations.*

Calculations. Co-efficient of increase of pressure at constant volume

$$\gamma_v = \frac{P_2 - P_0}{P_0 \times t_2}$$

Value from tables = 0.00367
% error =

Precautions. 1 and 5 are same as in **Experiment 8.1.**

2. The reading of mercury level should be taken when the temperature remains constant for several minutes and the mercury remains at the fixed mark F.

3. The open tube should be considerably lowered after the reading at the steam temperature has been taken otherwise mercury will rush into the bulb when the burner is removed.

4. The bulb should be surrounded by steam and should not be immersed in boiling water.

Sources of error. 1. Glass bulb expands and the volume of air increases, *the volume, therefore, does not remain constant.*

2. The air in the capillary tube is not at the same temperature as the air in the bulb.

3. The enclosed air may not be completely dry.

Exercise. *Find the co-efficient of increase of pressure at constant volume for air. Ice is not provided.*

Hints. If no ice is available the first reading may be taken at the room temperature t_1° C instead of at 0°C. If the corresponding value of the pressure of air is P_1 we can calculate the co-efficient of increase of pressure by the formula

$$\gamma_v = \frac{P_2 - P_1}{P_1 t_2 - P_2 t_1}$$

Experiment 8.3. To find the melting point of wax using Joly's constant volume air thermometer.

Apparatus. Joly's constant volume air thermometer, a plumb line, a thermometer, ice, a funnel, a mug, a thin walled test tube filled half with wax, set squares.

Theory. Let P_0 be the pressure of the air in the bulb of a Joly's constant volume air thermometer at 0°C, the temperature of melting ice and P_2 the pressure at the temperature of steam t_2 °C. If t_1 °C is the melting point of wax or any unknown temperature and P_1 the pressure corresponding to it, then

$$P_2 = P_0(1 + \gamma_v t_2)$$

or $$P_2 - P_0 = P_0 \gamma_v t_2 \quad ...(i)$$

and $$P_1 = P_0(1 + \gamma_v t_1)$$

or $$P_1 - P_0 = P_0 \gamma_v t_1 \quad ...(ii)$$

Dividing (*ii*) by (*i*), we get

$$\frac{t_1}{t_2} = \frac{P_1 - P_0}{P_2 - P_0}$$

or $$t_1 = \frac{P_1 - P_0}{P_2 - P_0} \times t_2$$

The temperature of steam t_2 can be calculated from the knowledge of the atmospheric pressure.

Procedure. Steps 1 to 5 same as in **Experiment 8.2.**

6. Remove the funnel containing ice surrounding the bulb *C*. Place the bulb in water contained in a mug.

Suspend the thin-walled test tube containing wax in the mug so that portion containing wax dips in water. Heat the bath slowly and note the head of pressure when the wax just begins to melt. Remove the flame when the whole of wax has melted. Stir the bath and allow the water to cool slowly. Note the head of pressure again when the melted wax just begins to solidify. The mean of these two values for the head of pressure gives the correct value of head of pressure at the temperature of melting wax.

7. Remove the test tube containing wax and the water bath. Place the bulb in steam bath and find the head of pressure corresponding to the temperature of steam.

8. Note the atmospheric pressure again and find the temperature of steam at this pressure.

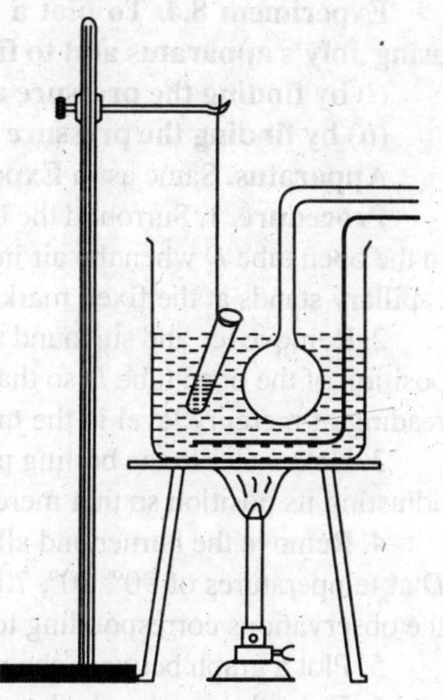

Fig. 8.5

Observations. Atmospheric pressure in the beginning

= cm

Atmospheric pressure at the end = cm

Mean atmospheric pressure P = cm

= mm

Temperature of steam $t_2\,^0C = 100 - 0.00367\,(760 - P)$

where the atmospheric pressure P is in millimetres of mercury.

Reading of mercury level at the fixed mark x = cm

Temperature	*Mercury level reading in open tube D* y	*Head of Pressure* $y - x = \pm h$	*Pressure* $P \pm h$
Melting ice 0°C			P_0 =
Melting wax t_1 °C			P_1 =
Steam t_2 °C			P_2 =

Calculations.

(i) Melting point of wax $t_1\ ^0C = \dfrac{P_1 - P_0}{P_2 - P_0} \times t_2 =$

Precautions and sources of error. Same as in **Experiment 8.2.**

(ii) **Exercise.** *Find the room temperature using Joly's constant volume air thermometer.*

Hints. Find the pressure P_1 corresponding to the room temperature without surrounding the bulb. If P_0 and P_2 are the pressures corresponding to the temperature of melting ice and temperature of steam respectively, then

$$t_1 = \frac{P_1 - P_0}{P_2 - P_0} \times t_2$$

Experiment 8.4. To plot a graph between temperature and pressure at constant volume using Joly's apparatus and to find the co- efficient of increase of pressure at constant volume.

(*i*) by finding the pressure at any two temperatures and

(*ii*) by finding the pressure at 0°C and 100°C.

Apparatus. Same as in **Experiment 8.2** and a brass mug.

Procedure. 1. Surround the bulb *C* in ice and note the reading of the upper meniscus of mercury in the open tube *D* when the air in the bulb attains the temperature of ice and the mercury level in the capillary stands at the fixed mark *F*.

2. Remove ice and surround the bulb with water in the brass mug at room temperature. Raise the position of the open tube *D* so that mercury stands at the fixed mark *F* in the capillary tube. Note the reading of mercury level in the tube *D* at room temperature.

3. Heat water to the boiling point and again note the reading of mercury in the open tube *D* after adjusting its position so that mercury again stands at *F* in the capillary tube.

4. Remove the burner and allow the water to cool. Note the reading of mercury level in the tube *D* at temperatures of 90°, 80°, 70° etc. by keeping the temperature constant at these points. Record the observations corresponding to all the temperatures.

5. Plot a graph between the pressure *P* and the temperature *t*.

6. From the graph calculate the values of P_1 and P_2 corresponding to any two temperatures, t_1 and t_2 respectively and then calculate the co-efficient of increase of pressure from the relation

$$\gamma_v = \frac{P_2 - P_1}{P_1 t_2 - P_2 t_1}$$

7. Also find from the graph the pressure of the gas P_0 corresponding to 0°*C* and P_{100} corresponding to 100°*C*. Calculate the co-efficient of increase of pressure at constant volume from the relation

$$\gamma_v = \frac{P_{100} - P_0}{P_0 \times 100}$$

Observation and calculations. Same as in **Expt. 8.2.**

Exercise. *Find the value of absolute zero by taking observations with Joly's constant volume air thermometer.*

Hints. Plot a graph between the pressure *P* and the temperature *t* representing pressure along *X*-axis and temperature along *Y*-axis taking zero value at the origin in both the cases. The position of *X*-axis is selected in such a manner that it is possible to go up to 100 °*C* on the positive side and up to 300 °*C* on the negative side. The graph is a straight line as shown in Fig. 8.6. Produce the curve to meet the *Y*-axis. The point at which it meets the *Y*-axis gives the value of the absolute zero. This is because corresponding to this temperature the pressure of the enclosed gas is zero.

Graph between temperature and pressure

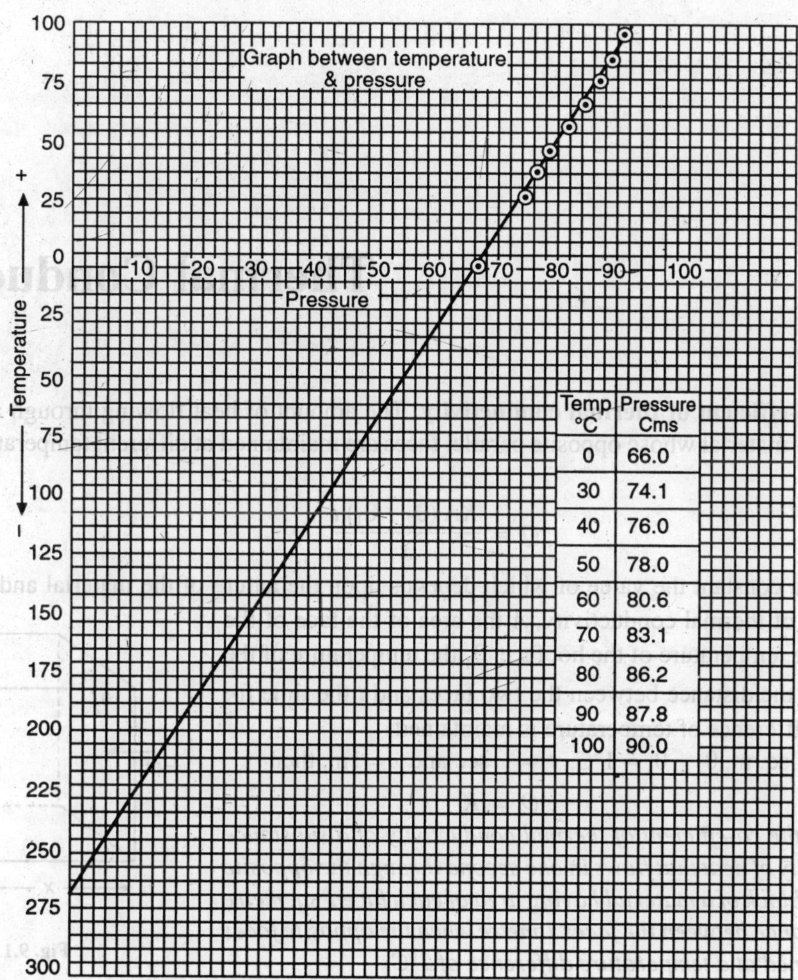

Temp °C	Pressure Cms
0	66.0
30	74.1
40	76.0
50	78.0
60	80.6
70	83.1
80	86.2
90	87.8
100	90.0

Fig. 8.6

Oral Questions

1. Expansion of gases. (*i*) State Boyle's law, Charle's law, and Regnault's law. (*ii*) What is the co-efficient of increase of volume and the co-efficient of increase of pressure for air? (*iii*) What are their values for hydrogen? (*iv*) Is it the same for all gases? (*v*) What is a perfect gas? Is there any such gas in nature? (*vi*) What is an absolute zero and what is its value on the Centigrade and Fahrenheit scales? (*vii*) Is it possible to reach this temperature? If not what is the minimum temperature reached so far? (*viii*) What is a gas thermometer? Why is it preferred to ordinary thermometers? (*ix*) Why is the mercury drawn in when the bulb cools? How will you prevent it?

9

Thermal Conductivity

9.1. Co-efficient of thermal conductivity. The amount of heat flowing through a rectangular block of any material whose opposite parallel faces are maintained at different temperatures is given by

$$Q = \frac{KA(\theta_2 - \theta_1)\,t}{x}, \qquad (i)$$

where K is a constant the value of which depends upon the nature of the material and is known as co-efficient of thermal conductivity, A the area of the face of the block, θ_2 the temperature of the hot face, θ_1 the temperature of the cold face, x the distance between the two faces and t the time for which the difference of temperature is maintained.

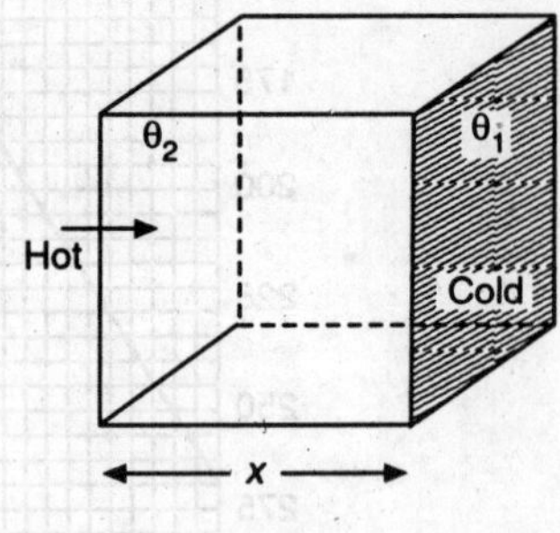

Fig. 9.1

If $A = 1$ sq. m, $\theta_2 - \theta_1 = 1°C$, $t = 1$ sec and $x = 1$m, then

$$Q = K$$

Hence the co-efficient of thermal conductivity of a substance is the quantity of heat that flows in one second through the opposite faces of the block of a material having area of each face 1 sq. metre and the distance between the faces 1 metre when the opposite faces are maintained at a temperature difference of 1°C.

Since a block each side of which is 1 metre is called a metre cube the co-efficient of thermal conductivity may also be defined as follows :—

It is the quantity of heat that flows in one second through a metre cube of a substance the opposite faces of which are maintained at a temperature difference of 1°C.

9.2. Temperature gradient. The quantity $\frac{\theta_2 - \theta_1}{x}$ in equation (*i*) represents the *temperature gradient* i.e the rate of fall of temperature with distance. In cases where the rate of fall of temperature with distance is not uniform, we replace $\frac{\theta_2 - \theta_1}{x}$ by $\frac{d\theta}{dx}$ · Therefore equation (*i*) becomes

$$Q = -KA\frac{d\theta}{dx}t \qquad ...(ii)$$

The *negative* sign indicates that the temperature θ decreases in the direction of increasing x.

9.3. Dimensions of K. From Eq. (*ii*)

$$K = \frac{Q\,dx}{A\,d\theta\,t}$$

$\therefore$ Dimensions of $K = \dfrac{[M^1 L^2 T^{-2}][L^1]}{[L^2][K][T]}$

where $[K]$ represents the dimension of $d\theta$ as degree Kelvin.

or Dimensions of $K = [M^1 L^1 T^{-3} K^{-1}]$

Units of K. The S.I. unit of K is *Watt per metre per degree kelvin* or $Wm^{-1} K^{-1}$.

9.4. Flow of heat along a bar. When one end of a metal bar AB is placed in a bath at a high temperature, heat travels from the hot end to the cold end of the bar along its length. If we consider a small portion XY of the bar, it receives heat from the portion immediately behind it towards the hot end. It absorbs a part of the heat received, thereby raising its temperature. It loses another part by radiation from its exposed surface and convection of gases around it. The rest of heat it transmits to the portion immediately in front of it towards its cold end. As the portion XY is absorbing heat its temperature goes on rising.

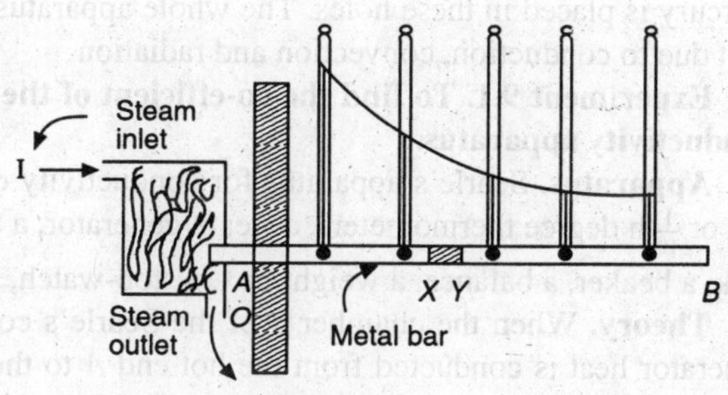

Fig. 9.2

This state during which the temperature of each portion of the bar is rising is known as the **variable state.**

After a sufficiently long time the temperature of each portion rises to a steady value. At this stage a small portion like XY of the rod receives heat from the portion immediately behind it, radiates a part of it and transmits the rest to the portion immediately in front of it. In other words no heat is retained by it to raise its own temperature.

This state when there is no more absorption of heat by the bar to raise the temperature of any part of it is called the **stationary** *or* **the steady state.**

Before this stage is reached the temperature of any portion like XY of the bar depends upon

(*i*) The co-efficient of thermal conductivity of the material and

(*ii*) Specific heat.

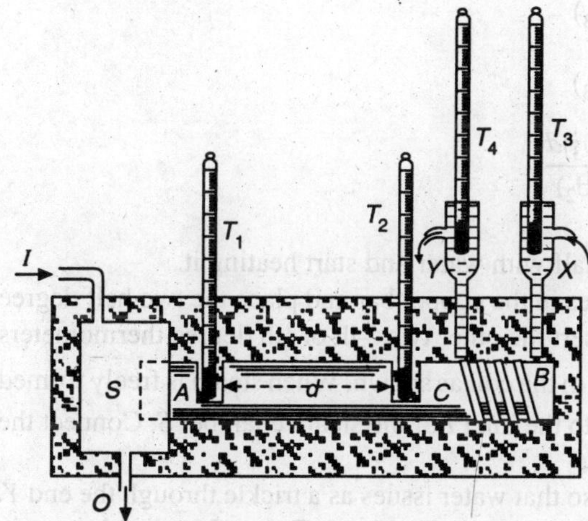

Fig. 9.3

In the steady state since there is no rise of temperature, the temperature of any portion will be independent of the specific heat and depends only on the conductivity of the material. Further if the bar is well insulated so that the loss of heat due to radiation and convection is reduced to a minimum, the amount of heat entering at the hot end is equal to the amount of heat leaving at the cold end.

9.5. Searle's conductivity apparatus. The co-efficient of thermal conductivity of good conductors like copper, brass etc., can be determined with the help of **Searle's conductivity apparatus.**

The essential parts of the apparatus are a rod AB of the material whose co-efficient of thermal conductivity is to be determined. The end A of the rod is placed in a chamber S having an inlet tube I and an outlet tube O. When the pipe leading steam from the steam generator is connected to I steam enters into the chamber and thus the end A is heated to the steam temperature. A coil C of a brass tube is wound round the end B and a steady flow of cold water is maintained through it. The thermometer T_3 placed in position records the temperature of the water that enters at X and the thermometer T_4 records the temperature of the water that leaves the coil at Y. Two other thermometers T_1 and T_2 are placed a known distance apart in two holes drilled in the rod. To ensure good contact of the thermometer bulbs with the rod a little of mercury is placed in these holes. The whole apparatus is covered with felt or cork to reduce loss of heat due to conduction, convection and radiation.

Experiment 9.1. To find the co-efficient of thermal conductivity of copper using Searle's conductivity apparatus.

Apparatus. Searle's apparatus for conductivity of metals, two half degree thermometers, two $\frac{1}{5}$th or $\frac{1}{10}$th degree thermometers, a steam generator, a tripod stand, a bunsen burner, a constant level tank, a beaker, a balance, a weight box, a stop-watch, a pair of dividers and a metre rod.

Theory. When the chamber S of the Searle's conductivity apparatus is connected to a steam generator heat is conducted from the hot end A to the cold end B of the bar whose co-efficient of thermal conductivity is to be determined. The flow of water through the coil C is regulated so that a *steady state of temperature* is reached. In this state as the bar is *thermally* insulated, it does not lose heat by radiation and further as there is no absorption of heat anywhere along the bar (the temperature of every point remaining constant) the quantity of heat flowing through any section of the bar in a given time is the same.

Let θ_1 and θ_2 be the temperatures indicated by the thermometers T_1 and T_2 respectively. If A is the area of cross-section of the bar, d the distance between the two thermometers and K the co- efficient of thermal conductivity, then the amount of heat Q passing through these points in one second is given by

$$Q = \frac{KA(\theta_1 - \theta_2)}{d}$$

This heat is absorbed by water flowing per second through the coil C. To find this let m be the mass of water flowing per second and θ_4 and θ_3 be the temperatures of the outgoing and incoming water respectively, then

$$Q = m(\theta_4 - \theta_3)$$

$$\therefore \quad \frac{KA(\theta_1 - \theta_2)}{d} = m(\theta_4 - \theta_3)$$

or
$$K = \frac{m\,(\theta_4 - \theta_3)d}{A\,(\theta_1 - \theta_2)}$$

Hence K can be calculated.

Procedure. 1. Fill the steam generator about half with water and start heating it.

2. Put a little of mercury in each of the two holes in the copper bar and place the two half degree thermometers T_1 and T_2 in position in the holes. Insert the other two $\frac{1}{10}$th or $\frac{1}{5}$th degree thermometers T_3 and T_4 at the entrance and exit points of the water in the coil as shown. When steam is freely formed in the steam generator, connect it by a rubber tube to the inlet I of the steam chamber S. Connect the end X of the coil C to the constant level tank.

3. Adjust the flow of water through the coi! C so that water issues as a trickle through the end Y.

4. Wait till a steady state is reached. This will take about half an hour. To confirm that the steady state has reached note and record the readings of the four thermometers after intervals of about 5 minutes. When the thermometers indicate constant temperatures for about 10 minutes, the steady state has reached. The last observation of temperatures should be used for calculations.

5. Take a clean and dry beaker and weigh it. Collect about 100 cc of water in it and note the time during which the water is collected with a stop-watch. Repeat the observation three times.

6. Repeat the experiment by changing the flow of water.

7. Measure the diameter of the bar with a vernier callipers at different points and the distance between the centres of the two holes in the bar with a pair of dividers.

Observations.

Distance between the centres of the two holes d = cm = m

Diameter of the bar

Vernier constant = cm

	1	2	3	4	5	6	Mean
Readings along one diameter (*a*)							
Rending in perpendicular direction (*b*)							

Mean radius r = cm = m

Area of cross section of the bar $A = \pi r^2$ = sq. metre

Readings of temperature during variable and steady state

Thermometer	*30 mt.*	*35 mt.*	*40 mt.*	*45 mt.*	*50 mt.*	*55 mt.*	*60 mt.*
T_1							
T_2							
T_3							
T_4							

Sl. No.	*Mass of water collected*	*Time taken*	*Mass collected per second*		*Steady final readings of thermometers*			
			in gm	*in kg*	T_1 θ_1	T_2 θ_2	T_3 θ_3	T_4 θ_4
1.								
2.								
3.								

Calculations.

$$K = \frac{m(\theta_4 - \theta_3)d}{A(\theta_1 - \theta_2)} = \quad \text{Kilo-calories/sec/sq.m/unit temp. gradient}$$

Value from tables =

% error =

Precautions. 1. Water should be allowed to flow at a uniform rate and a mere trickle of water should issue from the outlet *Y* of the copper spiral so that the rise of temperature may be constant and sufficient.

2. The temperatures should be recorded only when a steady state has been reached.

3. The diameter of the bar should be measured at a number of places and at every place in two mutually perpendicular directions.

4. The rod should be insulated thermally to avoid loss of heat due to radiation.

Sources of error. In deriving the relation used in the calculation of K, it has been supposed that flow of heat takes place in a direction parallel to the length of the bar. The conditions of linear flow of heat usually do not prevail in the bar and in the form of the apparatus described above it is not possible to ascertain it.

9.6. Thermal conductivity of bad conductors. The quantity of heat flowing through bad conductors is very small. In measuring their conductivity long bars should not be used as in the case of metals because it will not be possible to measure the small rate of flow accurately even with a reasonably high temperature difference between the ends. Again the flow of heat Q by conduction is directly proportional to the cross-sectional area A and inversely proportional to the thickness d. It is, therefore advantageous to use the substance in the form of a thin plate of large area and small thickness so that the quantity of heat flowing is conveniently measured. One of the various methods employed is Lee's disc method.

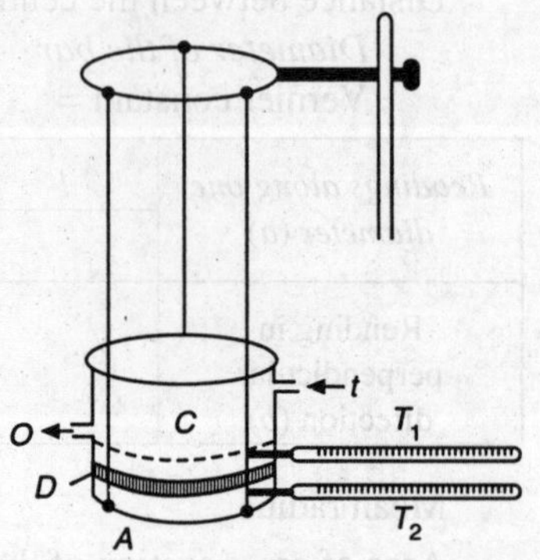

Fig. 9.4

Lee's disc method. The apparatus consists of a metal slab A of copper which is suspended by means of three strings from a heavy stand. On this disc rests a disc D of the material of the same diameter whose co-efficient of thermal conductivity is required. Above this rests a hollow cylindrical vessel C with inlet and outlet for steam. A thermometer T_1 is placed near the bottom of the cylindrical vessel and the thermometer T_2 is placed in the lower slab in which a hole is bored. To ensure good contact of thermometer bulb with the disc a little of mercury is put in the hole. The disc and the cylindrical vessel are nickel plated to give them the same emissivity.

Experiment 9.2. To find the co-efficient of thermal conductivity of a bad conductor by Lee's method.

Apparatus. Lee's disc apparatus, two 1/10°C thermometers, circular disc of the specimen of a bad conductor, (ebonite or card board), a stop watch, a screw gauge, vernier callipers etc.

Theory. On passing steam through the cylindrical vessel a steady state is reached soon. In this condition the rate at which heat is conducted across the specimen disc is equal to the rate at which heat is emitted through the exposed surface of the lower disc. If K is the co-efficient of thermal conductivity of the material of bad conductor, d its thickness and r its radius; θ_1 and θ_2 the constant readings of the thermometers T_1 and T_2 in the steady state, then rate at which heat is conducted across the disc of the material

$$Q = K\frac{\pi r^2(\theta_1 - \theta_2)}{d}$$

If M is the mass of the metal disc, s the specific heat of its material, then rate of cooling at θ_2 is equal to

$$= Ms\frac{d\theta}{dt}$$

where $\frac{d\theta}{dt}$ is the rate of fall of temperature at θ_2.

$$\therefore \qquad \frac{K\pi r^2(\theta_1 - \theta_2)}{d} = Ms\frac{d\theta}{dt}$$

or

$$K = \frac{Msd}{\pi r^2(\theta_1 - \theta_2)}\frac{d\theta}{dt}$$

The rate of cooling is found by heating the metal disc to a temperature about 10°C above the steady temperature θ_2, it is then allowed to cool and temperature is noted after every 30 seconds till the temperature falls to about 10°C below θ_2. A graph is then plotted between the temperature and time. A tangent is drawn at a point P corresponding to θ_2. The slope of the tangent gives the value of $\frac{d\theta}{dt}$ corresponding to temperature θ_2.

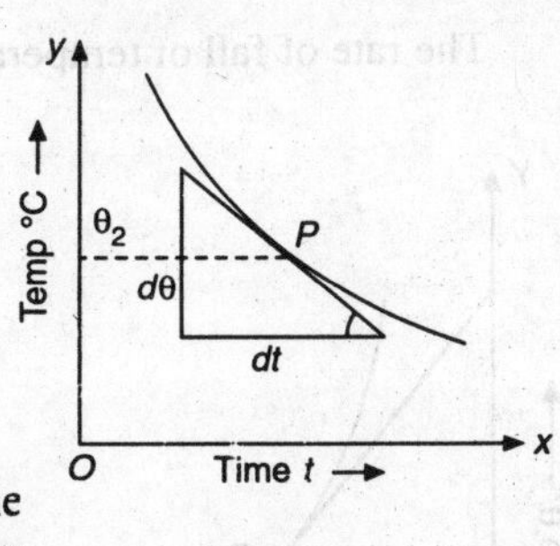

Fig. 9.5

Procedure. 1. Set the apparatus as shown in Fig. 9.4 so that the flat surface of the disc is horizontal. Insert the disc of the material in between this disc and the cylindrical vessel. Place the thermometers T_1 and T_2 in position.

2. Pass steam from the inlet of the cylindrical vessel and wait till the steady state is reached. This will take 30-40 minutes. When the temperatures indicated by the thermometers T_1 and T_2 are steady note down the temperatures. Interchange the thermometers T_1 and T_2 and again note down the temperatures.

3. Remove the cylindrical vessel as also the disc of the material and heat the disc A till its temperature is about 10°C above the steady temperature indicated by T_2. Allow it to cool and note the temperature at intervals of about 30 seconds, till its temperature falls to about 10°C below θ_2. Plot a graph between temperature and time.

4. Measure the diameter of the disc with a vernier callipers along two diameters mutually perpendicular to each other and measure its thickness with a screw gauge at different points. Also find the mass of the disc when cooled.

Observations.

Mass of the metallic disc M = gm = kg

Specific heat of metal s = kilo-cal/kg

Diameter of the disc = 1. 2.

Mean diameter D = cm = m

∴ Radius of the disc r = cm = m

Thickness of the disc d = 1., 2..., 3..., 4...

Mean thickness = cm = m

Steady temperature of the thermometers

T_1 = (*i*) (*ii*)

T_2 = (*i*) (*ii*)

Mean temperature θ_1 = °C

Mean temperature θ_2 = °C

Readings for cooling curve.

No. of Obs.	1	2	3	4	5	6	7	8	9
Time in seconds	30	60	90	120	150	180	210	...	...
Temp. of disc									

To find the rate of cooling $\frac{d\theta}{dt}$ · Draw a graph taking time t along the X-axis and temperature θ along the Y-axis. Draw a tangent to the graph at a point P corresponding to the steady temperature θ_2 as shown in Fig. 9.6 cutting the X-axis at A. Draw PM perpendicular to OA.

The rate of fall of temperature $\frac{d\theta}{dt}$ at the temperature θ_2 is given by

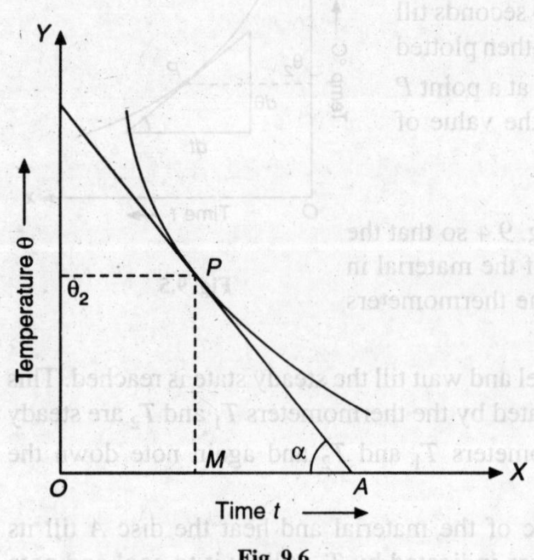

Fig. 9.6

$$\frac{d\theta}{dt} = \tan\alpha = \frac{PM}{MA}$$

$\therefore$ Rate of cooling $\frac{d\theta}{dt}$ from the graph corresponding to steady temperature θ_2 = °C/sec.

Calculations. $K = \frac{Msd}{\pi r^2(\theta_1 - \theta_2)} \times \frac{d\theta}{dt} =$

Precautions. 1. Thickness d of the disc of the material should be measured at a number of places on its surface.

2. The diameter of the disc should be equal to that of the cylindrical vessel and the metallic disc and should be measured in two perpendicular directions.

3. The thermometers should be placed close to the face of the disc of the specimen.

4. There should be a good thermal contact between the disc of material and the lower surface of the cylindrical surface and the upper surface of the circular metallic disc. If necessary glycerine may be applied between the surfaces.

5. The steady state temperatures should be recorded only when the readings of T_1 and T_2 remain constant after an interval of about five minutes.

Oral Questions

1. Conductivity. (*i*) Define co-efficient of thermal conductivity. (*ii*) What are the units in which it is expressed? (*iii*) What are the dimensions of thermal conductivity? (*iv*) What do you mean by temperature gradient (*v*) On what factors does the rise of temperature of any portion of the rod depends (*a*) in the variable state and (*b*) in the steady state. (*vi*) Distinguish between steady and variable states. (*vii*) Why is it necessary to obtain the steady state before taking observations? (*viii*) Which material is the best conductor of heat? (*ix*) Can Searle's apparatus be used for finding the conductivity of a bad conductor? If not why? (*x*) In Lee's method why should the experimental disc be thin? (*xi*) Why the two thermometers are placed close to the disc?

10

Mechanical Equivalent of Heat

10.1. Mechanical equivalent of heat. *It is defined as the amount of work done in order to produce a unit quantity of heat.* It is denoted by J.

On the C.G.S. system $J = 4.18 \times 10^7$ ergs/calorie

In S.I. units $J = 4.18$ Joules/calorie

$= 4180$ Joules/kilocal.

If W units of work produce H units of heat, then

$$W = JH$$

This is known as **Joule's law** or **first law of thermodynamics.**

10.2. Constant flow calorimeter. A number of methods are available for determining mechanical equivalent of heat. An accurate method of determining J is the constant flow method due to Callendar and Barnes. The apparatus used is known as **constant flow calorimeter.** It consists of a constant level tank C which maintains a constant head of pressure at the ends of the tube AB. Water enters the tank through the inlet tube I and flows into the tube AB through the outlet tube O. The excess of water flows out through the overflow tube F thereby maintaining the level of water in the tank constant.

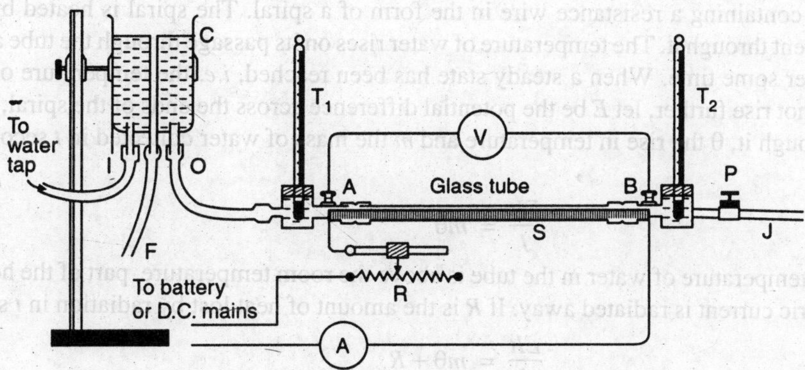

Fig. 10.1

The flowing water is heated by passing current through the heating coil S placed in the tube AB between the two terminals. The thermometers T_1 and T_2 record the temperature of the water entering and leaving the tube AB. A screw type pinch-cock P regulates the flow of water through the tube.

The apparatus can be worked on battery, $D.C.$ mains or $A.C.$ mains. While working on a battery or $D.C.$ mains a suitable rheostat R is used to adjust the value of current passing through the heating coil *S*, as shown in Fig. 10.1. The *D.C.* voltmeter placed across the spiral *S* measures the potential difference across its ends and the *D.C.* ammeter *A* measures the current flowing through it.

While working on *A.C.* mains a suitable variable step down transformer is used to bring down the voltage to the required value as shown in Fig. 10.2. The *D.C.* voltmeter and the ammeter are replaced by the *A.C.* instruments of the required ranges. This method is very convenient and has the advantage that there is very little wastage of energy.

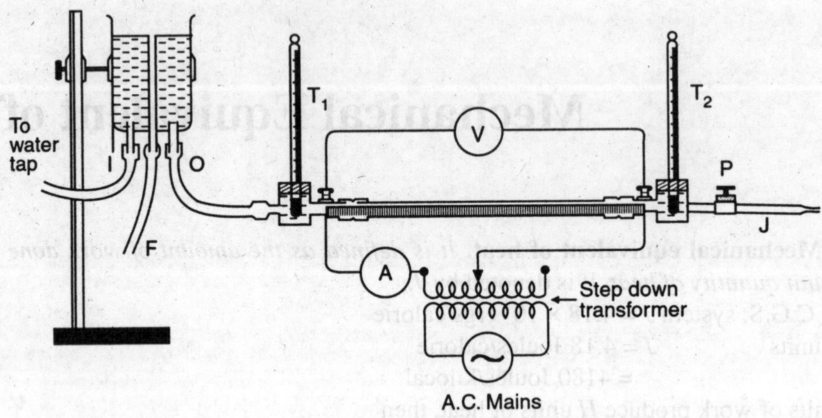

Fig. 10.2

Experiment 10.1. To determine the mechanical equivalent of heat by Callender and Barne's constant flow method.

Apparatus. Constant flow calorimeter, two $\frac{1}{5}$th degree thermometers, an *A.C.* ammeter of range 1.5 amp., a voltmeter of range 50 volts (any other suitale range depending upon the design of the apparatus may be used), a step-down transformer, a beaker, a balance, a weight box, a graduated cylinder, a screw type pinch cock, a stop-watch and well insulated connecting wires.

Theory. A steady stream of water is made to flow through a glass tube of the constant flow calorimeter containing a resistance wire in the form of a spiral. The spiral is heated by passing an electric current through it. The temperature of water rises on its passage through the tube and becomes constant after some time. When a steady state has been reached, *i.e.* the temperature of outflowing water does not rise further, let *E* be the potential difference across the ends of the spiral, *I* the current flowing through it, θ the rise in temperature and *m* the mass of water collected in *t* seconds, then we have

$$\frac{EIt}{J} = m\theta$$

As the temperature of water in the tube is above the room temperature, part of the heat produced by the electric current is radiated away. If *R* is the amount of heat lost by radiation in *t* seconds, then

$$\frac{EIt}{J} = m\theta + R \quad ...(i)$$

The quantity *R* cannot be determined directly and hence must be eliminated. To eliminate *R* another reading is taken by varying the value of *E, I* and the flow of water in such a manner that the rise of temperature θ is again the **same.** For the second observation, we have

$$\frac{E_1 I_1 t}{J} = m_1\theta + R \quad ...(ii)$$

where m_1 is the mass of water flowing in t seconds, E_1 the potential difference across the ends of the spiral and I_1 the current flowing through it.

Subtracting (*ii*) from (*i*), we have

$$\frac{(EI - E_1I_1)t}{J} = (m - m_1)\theta$$

or

$$J = \frac{(EI - E_1I_1)t}{(m - m_1)\theta}$$

Procedure. 1. Draw a diagram showing the scheme of connections as in Fig. 10.2. Connect the various instruments accordingly and check the connections.

2. Connect the overflow pipe F through a rubber tube to the sink the outlet tube O to the tube containing spiral coil and the inlet tube I to a water tap. Open the pinch-cock and place a bucket to collect the outflowing water. Open the tap so that water flows *freely and steadily* through the tube containing the heating coil. Adjust the rate of flow by means of the pinch-cock so that it is about 60 *cc* per minute.

Show the connections and the arrangement of the apparatus to the instructor before passing current.

3. Note the temperature indicated by the thermometers T_1 and T_2 and find out the correction to be applied, if any. Also clean and dry a beaker and weigh it.

4. Pass the current through the spiral resistance coil so that the ammeter and the voltmeter show a deflection. Adjust the value of the current till the temperature difference indicated by the two thermometers lies between 8 to 10°*C* when a *steady state* has been reached.

Note. *For a rate of flow of 1 c c per second and a rise of temperature of 10°C, 10 calories of heat per second are required.*

Now 10 cal/sec = 10 × 4.2 = 42 joules/sec.
= 42 watts

If the resistance of the spiral lies between 40 and 45 ohm a current of about 1.2 ampere will be obtained for a *P. D.* of 35 volts across it and a suitable rise of temperature will be obtained.

If any air bubbles appear in the tube, remove them by gently tapping the tube. When the temperature of thermometer T_2 remains constant for about 5 minutes, note the temperature of thermometers T_1 and T_2 and the readings of the ammeter and the voltmeter.

Collect water in the beaker for two to three minutes. Note the time accurately and weigh the beaker with water again.

5. Increase the current by about 0.1 to 0.2 ampere and adjust the rate of flow of water so that the difference in temperatures of T_1 and T_2 when a steady state has been reached is the *same* as in the first case. Note the readings of the thermometers T_1 and T_2 and also of the ammeter and the voltmeter. Find the weight of water collected and the time again.

Note. Slight variations of flow are brought about by raising or lowering the constant level tank to make the temperature difference the same as in the previous observation.

6. Repeat the observation by increasing the value of the current, and adjusting the rate of flow, keeping the temperature difference to the same value as in the previous observation.

Observations. Initial temperature indicated by T_1 = °*C*

Initial temperature indicated by T_2 = °*C*

Correction to be applied to the difference = $T_1 - T_2 =$ °*C*

Weight of dry beaker = gm

Sl. No.	Current I	P.D. E	Temperature of water on		Difference of Temperature θ		Wt. of water collected	Time t	Rate of flow $m = w/t$
			Entering	Leaving	Obs.	Corr.			
1									
2									
3									

Calculations. $J = \dfrac{EI - E_1I_1}{(m - m_1)\theta}$ Joule/cal

(*i*) From Obs. (1) and (2) $J =$

(*ii*) From Obs. (2) and (3) $J =$

(*iii*) From Obs. (1) and (3) $J =$

Mean values of $J =$ J/cal

Actual value of $J =$ J/cal

Percentage error =

Precautions. 1. Show the connections to the instructor before passing current.

2. When alternating current is used *A.C.* ammeter and voltmeter should be used.

3. While using direct current the positive of the voltmeter and the positive of the ammeter should be connected to the positive of the mains.

4. Water should be passed through the tube before passing the current.

5. The connecting wires should not be touched without switching off the current.

6. The turns of the spiral should not touch each other or the sides of the glass tube.

7. The tube should be tapped gently to remove air bubbles.

Oral Questions

Mechanical equivalent of heat. (*i*) Define Joule's mechanical equivalent of heat. (*ii*) What is its value ? (*iii*) What other methods are available for finding the value of *J* and why do you prefer the constant flow method? (*iv*) How do you apply a correction for the loss of heat due to radiation ? (*v*) How will you increase the difference in temperature without changing the current ? (*vi*) What material is used for making the spiral and why ? (*vii*) From the reading of the voltmeter and the ammeter calculate the resistance of the coil ? (*viii*) Is the resistance same when the coil is heated ? If not, why?

11

Introduction to Theory of Probability

11.1. Probability. Suppose we toss a coin. Either the *head* can come upward or the *tail* i.e. the event can happen in a total number of two equally likely ways. The number of ways in which the *head* can come up is only one. Therefore, the probability that the *head* may come up is $\frac{1}{2}$. Similarly the probability that the *tail* may come up is also $\frac{1}{2}$.

Let us consider another example of a dice. A dice is a homogeneous cube with six faces marked respectively with number of dots from 1 to 6. It is supposed that the dice cannot fall on its edges. When the dice is thrown it falls with one of its faces upward. The face can have any number (of dots) from 1 to 6. Thus there are six equally likely outcomes in a single throw of the dice. In other words, the probability of any one face (say with number 2) to come up is $\frac{1}{6}$.

The probability of dice coming up with an *even* number is $\frac{3}{6}$ as there are only three faces on the dice with even numbers i.e. 2, 4 and 6.

Similarly, suppose we are asked to draw *king of hearts* from a well shuffled pack of cards. We know there are 52 cards in a pack. When we draw a card, there is a chance that we might get the king of hearts in the very first trial but then there is also a possibility that we may not get it in 51 trials. In any case, we are sure to get the king of hearts at least once in 52 trials. So the required probability is $\frac{1}{52}$.

However, if we are to draw a king from a well shuffled pack of cards, the probability is $\frac{4}{52}$ as there are 4 kings in a pack of 52 cards and if we make 52 trials, we get all the four kings.

From the above examples we find that in the case of throwing a coin the event can happen in two mutually exclusive and equally likely ways, out of which only one is favourable to the *head* coming up. In the case of throwing a dice, the event can happen in six mutually exclusive and equally likely ways, out of which only one is favourable to a particular face (say with number 2) coming up and the number of ways favourable to an even numbered face coming up is three. In the case of a pack of cards, there are 52 mutually exclusive and equally likely ways, out of which only one is favourable to a particular card say king of hearts being drawn and the number of ways favourable to a king being drawn is four. Therefore, we define probability as under.

If an event A can occur in a total number of N exhaustive, mutually exclusive and equally likely ways and if a certain number m of these are considered favourable to the event A, then the ratio $\frac{m}{N}$ is called its probability of success.

∴ Probability of an event

$$P = \frac{\text{Number of ways which are favourable to the event}}{\text{Total number of equally likely possible ways.}}$$

11.2. Zero probability. In the case of throwing up of a six faced dice with faces marked from 1 to 6, suppose we want to know the possibility of the dice coming up with a face marked with number 8. As the dice has only six faces marked with numbers 1 to 6, there is no face marked with number 8.

∴ Probability of dice coming up with number 8 $= \frac{0}{6} = 0$

Hence the probability of an impossible event is zero.

11.3. Probability one. Suppose in the above case we want to know the probability that the dice may come up with a face marked with a number less than 7. As all the faces of the dice are marked with a number less than 7 the probability of dice coming up with number $< 7 = \frac{6}{6} = 1.$

Hence the probability of a sure or certain event is one.

11.4. Total probability. If an event can occur or happen in m ways and fail to happen in n ways, then probability of happening of the event $= \frac{m}{m+n}$ and probability of not happening of the event $= \frac{n}{m+n}$.

As the event must either happen or not happen (fail to happen) the sum of the two probabilities

$$= \frac{m}{m+n} + \frac{n}{m+n} = \frac{m+n}{m+n} = 1.$$

11.5. Principle of equal a priori probability. When we toss a coin we are clear in our mind that the coin will fall either with its *Head up* or *Tail up*. Similarly if we throw a six faced cubical dice, we are sure that the dice will fall with *one* of its six faces upward. In the same way, if we have an open box divided into two equal sized compartments, as shown, and a small particle is thrown from a large distance in such a way that it must fall in either of the two compartments, then the probability of the particle to fall in the compartment marked 1 is equal to the probability that it may fall into the compartment marked 2.

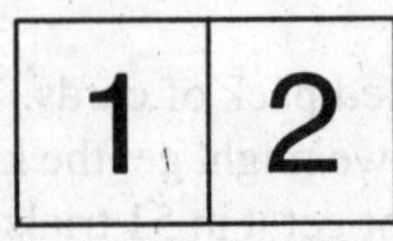

Fig. 11.1

This principle of assuming equal probability for events which are equally likely is known as the principle of equal a priori probability.

A priori really means something which exists in our mind prior to and independently of the observation we are going to make.

The principle of equal a priori probability will not hold good, if in the above example, the two compartments are of unequal size.

11.6. Tossing of two coins. Suppose we have two coins and we toss them simultaneously. There are two possible situations.

(*i*) The coins are unlike or distinguishable

and (*ii*) The coins are similar or indistinguishable.

(*i*) **Coins distinguishable.** Suppose the two coins are unlike. For example, one of the coins is of copper and the other is of silver. Let a_1, b_1 represent the head and tail respectively of the first and a_2, b_2 that of the second coin.

In a number of tossings of two unlike coins, four events are possible, which are shown in a tabular form.

No.	Event	Symbol	Probability
1.	Heads of two face up	$a_1 a_2$	1/4
2.	Head of first tail of second up	$a_1 b_2$	1/4
3.	Head of second tail of first up	$b_1 a_2$	1/4
4.	Tail of two faces up	$b_1 b_2$	1/4

Thus the total number of ways in which the event is possible $n = 4 = 2^2$.
Number of favourable ways for any one of the event $m = 1$.

$\therefore$ Probability of any one of the event to happen $p = \frac{m}{n} = \frac{1}{4}$.

(*ii*) **Coins indistinguishable.** When the coins are similar

$$a_1 = a_2 = a \qquad b_1 = b_2 = b \qquad \text{(say)}$$

In this case the second and the third event become identical and we are left with only three situations

Now, the probability of getting the head of one up and tail of second up is $2 \times \frac{1}{4} = \frac{1}{2}$. In other words the weight of this event is 2. This can be shown in a tabular form as under :

No.	Event	Symbol	Weight	Probability
1.	Heads of two face up	$a . a = a^2$	1	$\frac{1}{4}$
2.	Head of one and tail of other up	ab	2	$\frac{1}{2}$
3.	Tail of two face up	$b . b = b^2$	1	$\frac{1}{4}$

(*iii*) **Tossing of three distinguishable coins.** The heads of the three coins are now represented as a_1, a_2, a_3 and tails as b_1, b_2, b_3.

The possible events represented in a tabular form are as under :

No.	Event	Symbol	Probability
1.	Heads of three face up	$a_1\ a_2\ a_3$	$\frac{1}{2} \times \frac{1}{2} \times \frac{1}{2} = \frac{1}{8}$
2.	Head of first and second and tail of third up	$a_1\ a_2\ b_3$	$\frac{1}{8}$
3.	Head of first and third and tail of second up	$a_1\ b_2\ a_3$	$\frac{1}{8}$
4.	Head of second and third and tail of first up	$b_1\ a_2\ a_3$	$\frac{1}{8}$
5.	Head of first and tail of second and third up	$a_1\ b_2\ b_3$	$\frac{1}{8}$
6.	Head of second and tail of first and third up	$b_1\ a_2\ b_3$	$\frac{1}{8}$
7.	Head of third and tail of first and second up	$b_1\ b_2\ a_3$	$\frac{1}{8}$
8.	Tail of three face up	$b_1\ b_2\ b_3$	$\frac{1}{8}$

Thus the total number of ways in which the event is possible $n = 8 = 2^3$.
Number of favourable ways for any one of the event $m = 1$.

$\therefore$ Probability of any one event to happen $p = \frac{m}{n} = \frac{1}{8}$.

(*iv*) **Coins indistinguishable.** If the three coins are indistinguishable (similar)

$$a_1 = a_2 = a_3 = a \qquad b_1 = b_2 = b_3 = b \qquad \text{(say)}$$

Then the probability of getting heads of two and tails of one upward or tails of two and head of one upward is three time *i.e.*, $3 \times \frac{1}{8} = \frac{3}{8}$ *i.e.*, the weight of these events is 3. This is a shown in tabular form as under :

No.	*Event*	*Symbol*	*Weight*	*Probability*
1.	Heads of three up	a^3	1	$\frac{1}{8}$
2.	Heads of two and tail of one up	$a^2 b$	3	$\frac{3}{8}$
3.	Head of one and tails of two up	$a b^2$	3	$\frac{3}{8}$
4.	Tails of three up	b^3	1	$\frac{1}{8}$

11.7. Macrostate. Suppose we have 4 distinguishable particles and we wish to distribute them into two exactly similar compartments in an open box. Since the particles are distinguishable we denote them as *a, b, c* and *d.* Any particle thrown into the box must fall into one of the compartments. Since both the compartments are exactly alike, the particles have the same *a priori* probability of going into either of them. In other words the probability of a particle falling into any compartment is $\frac{1}{2}$. The possible ways in which 4 particles can be distributed in two compartments is as under.

Compartment,	*No. of particles*				
	*Dist** *1*	*Dist* 2	*Dist* 3	*Dist* 4	*Dist* 5
1	0	1	2	3	4
2	4	3	2	1	0

i.e. there are 5 different distributions (0, 4), (1, 3), (2, 2), (3, 1), (4, 0).

Each compartmentwise distribution of a system of particles is known as a macrostate.

For a system of *n* particles distributed among two compartments the various macrostates are $(0, n)$ $(1, n-1)$, $(2, n-2)$... $(n-1, 1)$ and $(n, 0)$. Thus total number of macrostates for *n* particles is $(n+1)$.

Microstate. In the above example, since the particles are *distinguishable* each compartmentwise distribution can have a number of different arrangements as shown below.

Macrostate	Possible arrangement		No. of microstates
	Compart 1	Compoartt 2	
0.4	0	abcd	1
1.3	a b c d	bcd cda dab abc	4
2.2	ab ac ad bc bd cd	cd bd bc ad ac ab	6
3.1	bcd cda dab abc	a b c d	4
4.0	abcd	0	1

Each distinct arrangement is known as the **microstate** *of the system.*

The distribution (0, 4) can have only one arrangement, distribution (1, 3) can have 4 distinct arrangements, distribution (2, 2) can have 6, distribution (3, 1) four and (4, 0) only one. Thus a given macrostate may consist of a number of microstates. In the above case of 4 particles the total number of microstates is $16 = 2^4$.

The total number of microstates for n particles = 2^n.

The number of microstates corresponding to a given macrostate is known as its frequency.

The postulate of equal a priori probability may also be stated as

'All accessible microstates corresponding to possible macrostates are equally probable.

According to this postulate the probability of occurrence of a given macrostate is proportional to the number of microstates that correspond to the given macrostate.

Hence the probability of occurrence of a macrostate is defined as the number of microstates in it to the total number of possible microstates.

11.8. Probability of distribution of *n* identical particles in two compartments. If there are n identical particles to be distributed in two equal compartments, then the total number of ways in which the event is possible or the total number of microstates is 2^n. If it is proposed to have r particles in one compartment and the remaining $(n - r)$ in the other compartment, then the number of meaningful arrangements or the number of microstates in the macrostate $(r, n - r)$

$$= {}^nC_r = \frac{n!}{r!\,(n-r)!}$$

$\therefore$ Probability of the macrostate $(r, n - r)$ is given by

$$P(r) = \frac{\text{Number of microstates in the given macrostate}}{\text{Total number of microstates of the system}}$$

$$= \frac{n!}{r!\,(n-r)\,!}\,\frac{1}{2^n}\,.$$

In the above discussion by compartment, we mean a particular situation. Thus if there are n coins which are thrown N number of times, then one compartment means *heads up* and the other compartment means *tails up.*

For example, if 2 identical coins are thrown, then $n = 2$ and the probability that one coin may come with head up ($r = 1$) and the other may come with tail up $(n - r) = 1$ will be

$$\frac{2!}{1!\,1!}\frac{1}{2^2} = \frac{1}{2}$$

which is the same as calculated in **Art. 11.6** (*i*)

Similarly if 3 identical coins are thrown, then $n = 3$ and the probability that 2 coins may come with head up $r = 2$ and one coin may come with tail up $(n - r) = 1$ will be

$$\frac{3!}{2!\,1!}\frac{1}{2^3} = \frac{3}{8}$$

which is the same as calculated in **Art 11.6** (*iii*)

Again if 4 identical coins are thrown up $n = 4$ and the probability that 2 coins may come with head up $r = 2$ and 2 coins may come with tail up $(n - r) = 2$ will be

$$\frac{4!}{2!\,2!}\frac{1}{2^4} = \frac{6}{16}$$

which is the same as calculated in **Art 11.7.**

11.9. Theoretical and observed frequency. Suppose we have *n* identical coins. If we throw them in such away that *r* coins fall with their *heads up* and the remaining $(n - r)$ with their *tails up* then the probability of the event

$$P(r) = \frac{n!}{r!\,(n-r)!}\frac{1}{2^n}$$

If the throw is repeated *N* times, *N* being very large, then for this event.

Theoretical expected frequency $F(r) = N \times P(r)$.

We denote the actual observed frequency of *r heads up* state as $f(r)$.

For a very large value of *N* i.e. for a very large number of trials of tossing of two coins, the theoretical expected frequency $F(r)$ will almost be the same as actual observed frequency $f(r)$ and the deviation if any will be very small.

Experiment 11.1. (*i*) To verify certain laws of probability distribution (*ii*) To verify laws of probability by throwing one coin, two coins and 10 (or more) coins.

Apparatus. A few identical coins (say 10 or 20 coins of the same denomination and small thickness), a plastic smooth mug, or container or tumbler and a wooden tray.

Procedure 1. Place the tray on the table. See that it is clean and there are no dust particles in it. Take a single coin and put it in the given container (mug or tumbler). Shake it well and throw the given coin in the tray. Note whether the head of the coin is up (*a*) or the tail is up (*b*). Repeat the process *at least 100 times.*

2. Now take two identical coins and put them in the cylindrical container. Shake the container well and throw the coins in the tray. Note down each time the combination produced. If the heads of both are up record it as (*a a*). If the head of one and tail of the other is up record it as (*a, b*) and if the tails of both are up record it as (*b b*). *Repeat at least 100 times.*

3. Now take 10 identical coins and put them together in the cylindrical container. Shake them well and again throw them in the tray. Record the number of coins which show with *heads up* and also the number of coins which show with *tails up. Repeat the process 100 times.*

Observations 1. **With one coin**

Number of throws, $N = 100$

Number of coins, $n = 1$

Record the observations as shown in the table.

No	1	2	3	4	5	6	7	8	9	10	....	98	99	100
Heads up	*a*	*a*	×	*a*	×	×	..	..	..	..	....	..	*a*	×
Tails up	×	×	*b*	×	*b*	*b*	..	..	..	..	....	..	×	*b*

Number of times head is up = A =

Theoretical probability of getting head up = $\frac{1}{2}$

∴ Number of times head should be up theoretically (for 100 trials) = $\frac{1}{2} \times 100 = 50$.

Percentage variation $= \left(\frac{50 - A}{50}\right) \times 100 =$

2. **With two coins**

Number of trials $N = 100$

Number of coins $n = 2$

Record the observations as shown in the table below:

No.	1	2	3	4	5	6	7	8	9	10	....	98	99	100
Heads of two together	*aa*	×	×	×	*aa*	×	×	×	×	×	....	*aa*	×	×
Tails of two together	×	*bb*	×	×	×	×	*bb*	×	×	*bb*	....	×	×	*bb*
Heads of one and tail of other	×	×	*ab*	*ab*	×	*ab*	×	*ab*	*ab*	×	....	×	*ab*	×

Experimental count of Head-Head combination = A =

Experimental count of Tail-Tail combination = B =

Experimental count of Head-Tail combination = C =

Theoretical probability of Head-Head or Tail-Tail combination $= \frac{1}{4}$

Theoretical count of Head-Head or Tail-Tail combination $= 100 \times \frac{1}{4} = 25$.

Percentage deviation of Head-Head combination $= \left(\frac{25 - A}{25}\right) \times 100 =$

Percentage deviation of Tail-Tail combination $= \left(\frac{25 - B}{25}\right) \times 100 =$

Theoretical probability of Head-Tail combination $= \frac{1}{2}$

Theoretical count of Head-Tail combination $= \frac{1}{2} \times 100 = 50$

Percentage deviation of Head-Tail combination $= \left(\frac{50 - C}{50}\right) \times 100 =$

3. **With any number of coins**

Number of trials, $N = 100$

Number of coins, $n = 10$

Record the observations as shown in the table below:

No.	1	2	3	4	5	6	7	8	9	10		98	99	100
No of Heads up (r)	4	7	8	6	5	4	6	5	3	5		4	2	3
Number of Tails up $(n-r)$	6	3	2	4	5	6	4	5	7	5		6	8	7

Combination *(r)* *(n − r)* *(Head)* *(tail)*	*Observed frequency f(r)*	*Probability* $P(r) = \frac{n!}{r!\,(n-r)\,!} \times \frac{1}{2^n}$	*Theoretical expected frequency F(r)* $= P(r) \times N$	*Deviation*
0, 10				
1, 9				
2, 8				
3, 7				
4, 6				
5, 5				
6, 4				
7, 3				
8, 2				
9, 1				
10, 0				

4. Graph (*i*) Draw a graph between theoretical expected frequency $F(r)$ and (r). A typical graph for number of coins $n = 10$ and number of trials $N = 100$ is shown in Fig. 11.2.

(*ii*) Draw a graph between observed frequency $f(r)$ and r.

(*iii*) Also draw a graph between observed frequency $f(r)$ and r for 200 trials.

Result. It is observed that as the number of trials is increased the deviation decreases.

Conclusions. 1. With one coin the number of times we get *heads up* is *nearly* the same as theoretical expected value given by the product of the total number of throws and the probability of *heads up* i.e $N \times P$ (heads). This verifies the principle of *a priori probability*.

2 With two coin system

(*a*) Number of times the heads of both are up is *nearly* equal to the product of total number of throws and probability of both heads being up *i.e.* $N \times P$ (head-head).

(*b*) Number of times the tails of both are up is *nearly* equal to the product of total number of throws and probability of both tails being up *i.e.* $N \times P$ (tail-tail).

(*c*) Number of times the tail of one and head of the other is up together is *nearly* equal to the product of the total number of the throws and probability of the event (head-tail) i.e $N \times P$ (head-tail).

(*d*) Number of times of (head-head) + Number of times of (head-tail) + Number of times of (tail-tail) is always equal to the total number of trials or throws.

All these conclusions lead to the validity of the law of composite events in terms of probability of two independent events.

3. With large number of coins the actual observed frequencies for different combinations [*r*, (*n*–*r*)] match closely with theoretically expected frequencies. This matching improves as the number of trials is increased.

Precautions 1. The tray should be clean and smooth.

2. The mug or tumbler should be perfectly smooth from inside.

3. The coin should be thin and identical with perfectly smooth edges.

4. The coins should not be able to stand on their edges.
5. The tumbler and the coins should be shaken well before throwing.
6. The total number of throws or trails should be as large as possible.

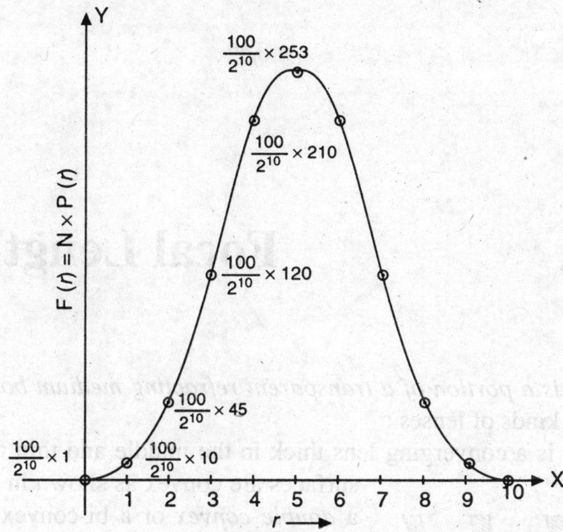

Fig. 11.2

Oral Questions

Probability (*i*) What is the meaning of the term probability? (*ii*) What is 'a priori probability'? (*iii*) What is the law of equal a priori probability? (*iv*) What is the probability of getting a king from a pack of cards? (*v*) What is addition theorem of probability? (*vi*) What is the multiplication theorem of probabilities? (*vii*) What is meant by frequency of occurrence? (*viii*) A coin is tossed 100 times. What is the frequency of occurrence of heads up? (*ix*) Two coins are thrown together for 100 times. What is the frequency of occurrence of heads of both together? (*x*) 10 coins are thrown 1024 times calculate the theoretical expected frequency for the combination (0 head, 10 tails), (1 head, 9 tails), (2 heads, 8 tails), (3 heads, 7 tails), (4 heads, 6 tails), (5 heads, 5 tails).

PART III
OPTICS

12

Focal Length of Lenses

12.1. Lens. *A lens is a portion of a transparent refracting medium bounded by two spherical surfaces*. There are two kinds of lenses :

(*i*) **Convex lens.** It is a converging lens thick in the middle and thin at the edges. If both the surfaces are convex as shown in Fig. 12.1 (*a*) it is called a *double convex* or a bi-convex lens. If one surface is convex and the other plane as shown in Fig. 12.1 (*b*) it is called a *plano-convex* lens. If one surface is concave and the other convex as shown in Fig. 12.1 (*c*) it is called a *concavo-convex lens*.

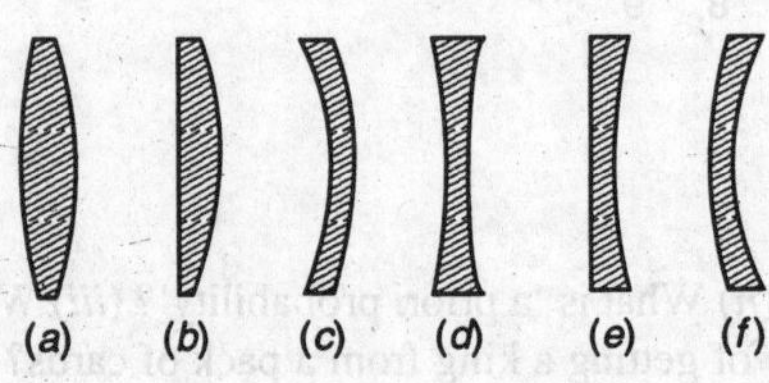

Fig. 12.1

(*ii*) **Concave lens.** It is a diverging lens thin in the middle and thick at the edges. If both the surfaces are concave as shown in Fig. 12.1 (*d*) it is called a *double concave* or a biconcave lens. If one surface is concave and the other plane as shown in Fig. 12.1 (*e*) it is called a *plano-concave* lens. If one surface is convex and the other concave as shown in Fig. 12.1 (*f*) it is known as a *convexo-concave* lens.

Optical centre. A ray of light always suffers some deviation in passing through a lens but there is a point inside or outside the lens such that the rays passing through it suffer no deviation. Such a point is called the *optical centre*.

Principal axis. It is a straight line joining the centres of curvature of the two surfaces of a lens.

Principal focus. If a parallel beam of light falls on the surface of a lens in a direction parallel to the principal axis, the rays after passing through the lens converge to or appear to diverge from a point on the axis. This point is called the **principal focus.** In the case of a convex lens, the rays after refraction actually meet at F as shown in Fig. 12.2 (*a*). The principal focus is, therefore, a real point. In the case of a concave lens the rays after refraction become divergent and when produced backward appear to meet at F. The principal focus is thus a virtual point as shown in Fig. 12.2 (*b*)

Focal length. It is the distance between the optical centre and the principal focus of a lens.

12.2. Cartesian sign convention. According to this sign convention

(*i*) Incident rays are drawn from left towards the right of the lens.

(*ii*) All distances are measured from the optical centre of the lens.

(*iii*) Distances measured in the direction of incident light are taken as positive and those measured in a direction opposite to that of the incident light are taken as negative.

(*iv*) Distances measured perpendicular to the X-axis in the upward ($+Y$) direction are taken as positive and those measured in the downward ($-Y$) direction are taken as negative.

According to this sign convention the focal length of a convex lens is positive and that of a concave lens negative.

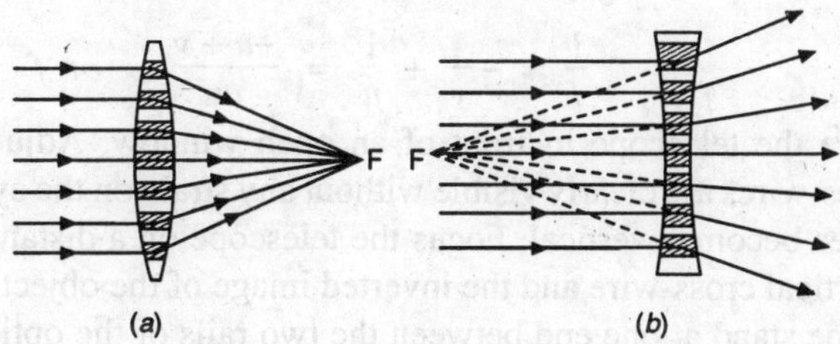

Fig. 12.2

Lens formula. If u is the distance of the object and v that of the image from the optical centre, then

$$\frac{1}{f} = \frac{1}{v} - \frac{1}{u}$$

where f is the focal length of the lens.

12.3. Determination of focal length of a concave lens. A concave lens always produces a virtual image of an object. Hence its focal length cannot be determined directly by the parallax method but indirect methods have to be used. We shall discuss the method of finding the focal length of a concave lens by using a telescope.

Experiment 12.1. To determine the focal length of a concave lens by a telescope using the relation $\frac{1}{v} - \frac{1}{u} = \frac{1}{f}$

Apparatus. A concave lens, a convex lens of focal length longer than the focal length of the concave lens, a telescope, an optical bench with three uprights, a needle, a knitting needle, a cork, a clamp-stand, a half metre rod etc.

Theory. If an object O is placed at the principal focus of a convex lens lying just in front of a telescope *set for parallel rays*, the image of O is formed at the cross-wires in the focal plane of the objective. It is because the rays after passing through the convex lens are rendered into a parallel beam. The distance between O and the optical centre C of the convex lens gives the focal length of the convex lens.

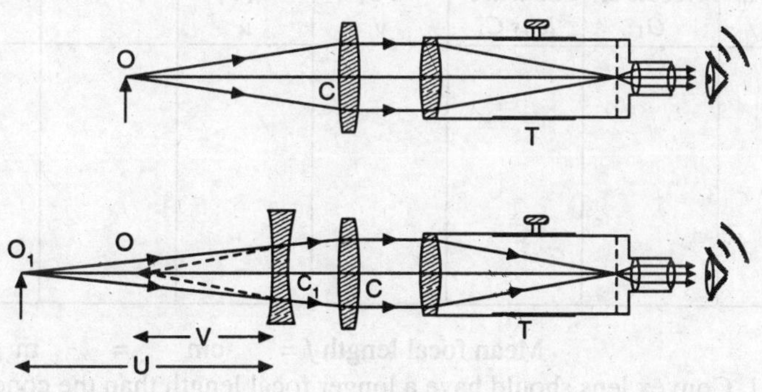

Fig. 12.3

Now if a concave lens is placed between the object O and the convex lens without disturbing the position of the convex lens and the position of the object needle O is adjusted as at O_1 so that there is no parallax between the image of O_1 and the vertical cross wire, then in this position the virtual image of O_1 is formed at O by the concave lens. Hence

$$O_1C_1 = -u \text{ and } OC_1 = -v$$

$$\therefore \quad \frac{1}{f} = -\frac{1}{v} + \frac{1}{u} = \frac{-u+v}{uv} \quad or\ f = \frac{uv}{v-u}$$

Procedure. 1. Place the telescope in front of an open window. Adjust the eye-piece of the telescope so that the cross-wires are clearly visible without any strain on the eye. Rotate the eye-piece till one of the cross-wires becomes vertical. Focus the telescope on a distant object, till there is no parallax between the vertical cross-wire and the inverted image of the object.

2. Place the telescope stand at one end between the two rails of the optical bench with the axis of the telescope parallel to the rails. In this position the image of a distant object should be visible through the telescope without any obstruction.

3. Clean the convex lens and find its rough focal length. Hold it *in an upright just in contact with the objective of the telescope*. Place the needle in an upright on the optical bench at a distance *slightly less than the rough focal length of the convex lens* from the lens and look through the telescope. A magnified inverted image of the needle will be visible. Adjust the position and height of the needle so that there is no parallax between the tip of the image and the vertical cross-wire. Note the position of the index mark of the needle upright.

4. Mount the concave lens on another upright and interpose between the convex lens and the needle O close to the convex lens with its centre at the same height as that of the centre of the convex lens. Look through the telescope and adjust the position of the needle O as at O_1 so that again there is no parallax between the tip of the needle and the vertical cross-wire. Note the positions of the index marks of the uprights containing the concave lens and new position of the needle upright as at O_1.

5. Repeat the observation 3-4 times after disturbing the arrangement.

6. Find the index correction between the needle upright and the concave lens.

Observations. Rough focal length of convex lens = cm

Length of the knitting needle x = cm

Observed distance between the concave lens upright and the needle upright y = cm

Index correction $(x - y)$ = ± cm

No	Position of			Observed		Corrected		$f = \frac{uv}{v-u}$
	Needle at O	Needle at O_1	Concave lens C_1	OC_1 v	O_1C_1 u	v	u	
1								
2								
3								
4								
5								

Mean focal length f = cm = m

Precautions. 1. Convex lens should have a longer focal length than the concave lens.

2. The tip of the needle, the centre of convex lens, concave lens and the objective of the telescope should be at the same height.

3. To check that the focal length of the convex lens is greater than that of the concave lens, see that the combination when the two lenses are placed in contact is concave. In other words, the power of the concave lens is greater than that of the convex lens.

12.4. Nodal slide assembly. It consists of a heavy optical bench about a metre and a half in length and provided with four uprights. One of the uprights is placed at the extreme end of the bench and on it is mounted a lamp housing provided with a circular opening for light to pass through it from a **frosted lamp.** On the second upright adjacent to it is mounted a metal plate with cross-slits. Away from the cross slits is placed the third upright which carries a nodal slide.

Nodal slide. It consists of a carriage on which two lens holders can be fixed in the desired positions. The distance between centres of the lenses can be directly read on the scale provided on the carriage. The carriage can be made to slide in the horizontal plane along the scale with the help of a rack and pinion arrangement. It is also capable of rotation about a vertical axis *XY* and if desired its angle of rotation can be noted from a graduated scale on the circular base. The axis of rotation of the nodal slide which remains fixed can be made to pass through any point on the axis of the lens system which is parallel to the axis along which the carriage can slide. To find the position of the point on the axis of the lens system, through which the axis of rotation is passing, an index mark *P* is made on the slide, past which the carriage scale moves. The point *P* also marks the axis of rotation of the nodal slide.

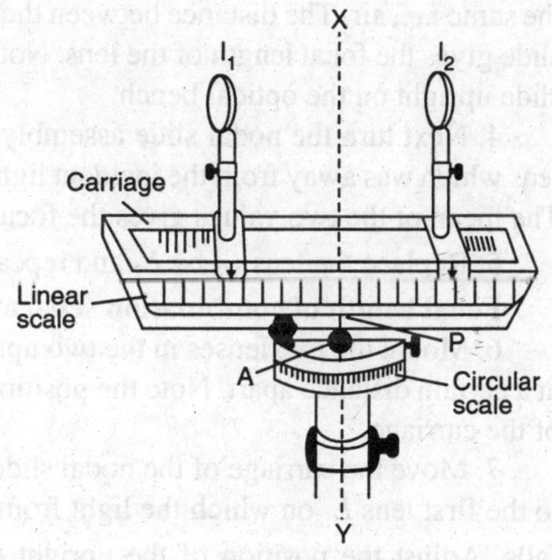

Fig. 12.4

The fourth upright on the optical bench carries a plane mirror which is capable of rotation about a horizontal axis perpendicular to the bed of the bench.

Experiment. 12.2. To determine the focal length of a combination of two convergent lenses separated by a distance *d* using nodal slide assembly and to verify the relation

$$\frac{1}{F} = \frac{1}{f_1} + \frac{1}{f_2} - \frac{d}{f_1 f_2}$$

where the letters have their usual significance.

Apparatus. Nodal slide assembly, two convex lenses of nearly the same focal lengths, cross-slits, a plane mirror etc.

Theory. If f_1 and f_2 are the focal lengths of the component lenses separated by a distance d and F is the focal length of the combination, then

$$\frac{1}{F} = \frac{1}{f_1} + \frac{1}{f_2} - \frac{d}{f_1 f_2}$$

The observed value of the focal length F should be nearly equal to the calculated value from the knowledge of f_1, f_2 and d.

Procedure. *Focal length of individual lenses.* 1. Set the lamp housing assembly on the upright placed at one extreme end of the optical bench and the cross-slits in a metal plate on the second upright adjacent to the first. Fix the nodal slide assembly on the third upright next to the cross-slit upright and the plane mirror on the fourth upright placed next to the nodal slide assembly. Adjust the heights of the four uprights to be nearly the same.

2. Mount one lens in one of the lens holders on the carriage and shift its position so that the index mark coincides with the index mark of *P*.

3. Move the nodal slide upright carrying the lens L_1 along the optical bench till a well defined image of the cross-slits is formed besides the cross-slits on the metal plate. Rotate the nodal slide through a small angle about the vertical axis *XY*. The image will ordinarily shift its position sideways. Move the carriage of the nodal slide assembly by a mm or so from the index mark *P* and focus the image again. Test by rotating the nodal slide assembly through a small angle. The image may shift now to the other side. Adjust the position of the upright carrying lens from the index mark *P* and focus the image again. Test again by rotating the nodal assembly. Repeat if necessary till there is no shift of the image. When this position is achieved, the metal plate with cross-slits coincides with the focal plane of the lens and the axis of rotation of the nodal slide passes through the corresponding nodal point and hence through the corresponding principal point as the medium on both sides of the lens is the same *i.e.*, air. The distance between the metal plate cross-slits and the axis of rotation of the nodal slide gives the focal length of the lens. Note the position of the cross-slits plate upright and the nodal slide upright on the optical bench.

4. Next turn the nodal slide assembly through 180° on the circular scale so that the face of the lens which was away from the incident light now faces it. Repeat the process and find its focal length. The mean of the two values gives the focal length f_1 of the first lens.

5. Replace the lens L_1 by L_2 and repeat the above process to find the mean focal length f_2.

Focal length of combination separated by a distance.

6. Mount the two lenses in the two uprights on the nodal slide assembly and fix the two uprights at a certain distance apart. Note the position of the index mark of each lens holder on the linear scale of the carriage.

7. Move the carriage of the nodal slide so that the index mark *P* of the axis of rotation lies close to the first lens L_1 on which the light from the cross-slits falls. Adjust the position of the upright carrying nodal slide assembly till a well focused image of the cross-slits is obtained on the metal plate. Rotate the nodal slide assembly about *XY* through a small angle. The image is found to be displaced to one side. Note the direction of displacement relative to the near end of the nodal slide. Next shift the nodal slide carriage so that the point *P* is now close to the second lens. Focus the image by moving the nodal slide and observe the shift of the image by rotating the nodal slide about its axis *XY*. The shift of the image would be opposite to the one observed with the index mark *P* close to the first lens. It is thus evident that for the axis of rotation *XY* somewhere between these two positions there will be no shift of the image on rotation of the nodal slide about the axis *XY*. By trial find the position of the index mark *P* where the shift of the image of the cross slits does not take place on slight rotation of the nodal slide about its axis *XY*. In this position the metal plate with cross-slits is situated on the first focal plane and the axis of rotation passes through the first focal point N_1 as in Fig. 12.5.

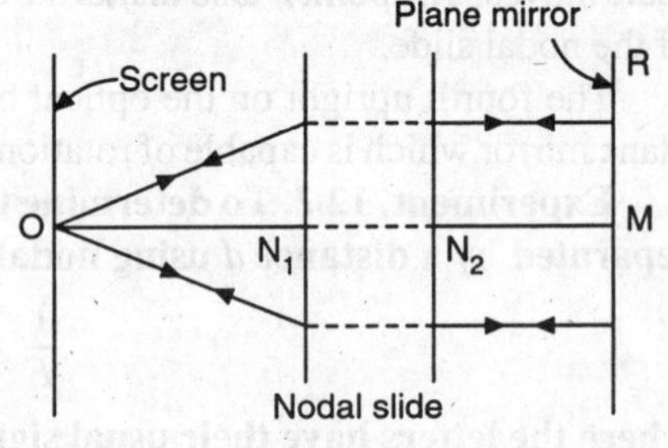

Fig. 12.5

Note the positions of the uprights carrying cross-slits and the nodal slide on the optical bench. The difference of the two readings gives the focal length of the combination.

8. Rotate the nodal slide through 180° without changing the distance between the two lenses L_1 and L_2 and repeat the above adjustments. The position of the cross-slits gives the position of the second principal focus and the position of the axis of rotation the second nodal point. Note down the positions of the uprights and find the distance between them. The result will be found to be nearly the same. Find the mean of the two results. Let this be *F*.

9. Repeat the observation by changing the distance *d* between the two lenses

Observations.

Focal length of individual lenses

Lens	No.	Light incident on	Position of		Focal length $= (b - a)$
			Cross slit a	Axis of nodal slide b	
L_1	1	(i) Ist face (ii) 2nd face			
	2				Mean $f_1 =$
L_2	1	(i) Ist face (ii) 2nd face			
	2				Mean $f_2 =$

Focal length of combination

Sl. No.	Distance between lenses d	Light incident on	Position of		Focal length $= (b - a) = F$
			Cross slit a	Axis of nodal slide b	
1		Lens L_1 Lens L_2			
2					
3					

Mean $F =$ cm

Calculations.

Focal length of first lens $f_1 =$

Focal length of 2nd lens $f_2 =$

Distance between the lenses $d =$

$$\therefore \quad \frac{1}{F} = \frac{1}{f_1} + \frac{1}{f_2} - \frac{d}{f_1 f_2}$$

Result. The calculated and the experimental values of the focal length of the combination being approximately equal, this verifies the result.

$$\frac{1}{F} = \frac{1}{f_1} + \frac{1}{f_2} - \frac{d}{f_1 f_2}$$

Precautions. 1. The cross-slits should be properly illuminated by the light coming from a frosted lamp.

2. The rotation of the nodal slide about the axis XY should not exceed 4–5°, for testing.

3. The distance between the nodal assembly and the cross-slits is adjusted every time for a well focused image of the cross-slits when testing.

4. The plane mirror should be truly plane and perpendicular to the axis of the nodal slide.

5. Sometimes a bright image is formed on the metal plate screen by reflection from the surface of the lens. This image should not be confused with the desired image formed by the rays which suffer refraction through the lens and retrace their path after reflection from the plane mirror. The genuine image can be detected by placing the hand in front of the plane mirror when it will disappear.

Oral Questions

1. Lenses. (*i*) What is the difference between a convex lens and a concave lens? (*ii*) How will you find the rough focal length of (*a*) a convex lens; (*b*) a concave lens? (*iii*) What is the nature of the image formed in a convex and a concave lens? (*iv*) What is the power of your spectacle lens? Calculate its focal length. (*v*) Why are the images formed by a lens usually curved and coloured? (*vi*) Why is it not possible to determine the focal length of a concave lens directly? (*vii*) In the concave mirror method where should the image be formed by the concave lens. (*viii*) Will the telescope method succeed if the focal length of the concave lens is greater than that of the convex lens?

2. Cardinal points and nodal slide. (*i*) What are cardinal points in an optical system? (*ii*) Define focal points, principal points and nodal points. (*iii*) How will you determine the nodal points of a lens system? (*iv*) What is a nodal slide? (*v*) How will you determine the focal points? (*vi*) How will you distinguish between the correct image formed after refraction and false image formed by reflection alone? (*viii*) How will you get a well defined sharp image of the cross slits?

13

Magnifying Power and Resolving Power of a Telescope

13.1 Telescope. A Telescope is an optical instrument used for seeing distant objects very clearly. There are two types :

(*i*) Astronomical telescope and

(*ii*) Terrestrial telescope.

The **astronomical** telescope of the refracting type in its simple form essentially consists of two convex lens systems one of which serves the purpose of an *objective* and the other that of an *eye piece*.

The **objective** is a convex lens of large focal length and large aperture. It is usually a combination of two lenses in contact to reduce spherical and chromatic aberrations to a minimum. It is mounted at one end of a brass tube.

The **eye-piece** is also a convex lens but has a short focal length and a small aperture. In order to increase the field of view and also to minimise aberrations it consists of two plano-convex lenses separated by a suitable distance. The lens towards the objective is called the *field lens* and the other through which the eye views the object is known as the *eye lens*. The combination is mounted in a small tube which can slide inside a bigger tube carrying the objective.

A cross-wire W is mounted in front of the eye-piece and is used as a reference mark. The distance between the objective and the eye-piece can be adjusted with the help of a rack and pinion arrangement R. In a **terrestrial telescope** an additional lens is used in between the objective and the eye-piece. This lens forms a re-inverted image of the inverted image of the object formed by the objective. The final image as seen by the eye-piece is, therefore, erect.

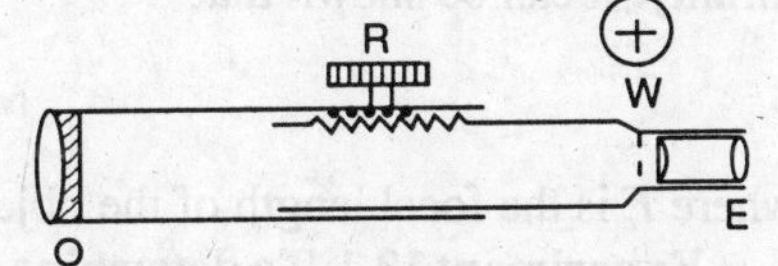

Fig. 13.1

13.2. Eye-pieces. The eye-pieces commonly used are of two kinds :

(*i*) Ramsden's eye-piece and (*ii*) Huyghen's eye-piece.

(*i*) **Ramsden's eye-piece.** It consists of two plano-convex lenses each of focal length f separated by a distance $\frac{2}{3}f$ with convex surfaces facing each other as shown in Fig. 13.2. The focal length of the equivalent lens is $\frac{3}{4}f$ and its position is $\frac{1}{2}f$ behind the field lens.

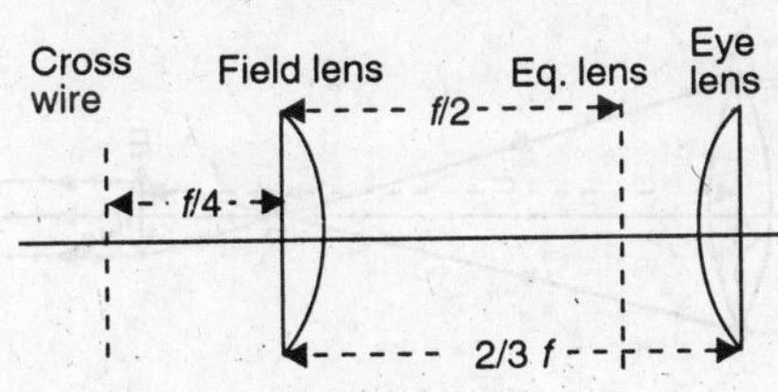

Fig. 13.2

The cross-wires are, therefore, placed at a distance

$$\frac{3}{4}f - \frac{1}{2}f = \frac{1}{4}f$$

in front of the field lens.

The cross-wires can be replaced by a scale, if necessary, for purpose of measurement. This is known as a *positive* eye- piece.

In this arrangement the field lens is not very efficient but the field of view is fairly wide. The arrangement is not entirely achromatic. The chromatic effects are minimised by using achromatic combinations for both the field and the eye-lenses.

The spherical aberration is minimised as the image is seen through two lenses and hence the deviation is small as it takes place at **four** different surfaces. It is further minimised by making both the lenses plano-convex, their convex surfaces facing each other.

This eye-piece is very commonly used in all the optical instruments where measurements are to be made.

(*ii*) **Huyghens' eye-piece.** It consists of two plano-convex lenses, the field lens having a focal length ($3f$) three times that of the eye lens (f). The distance between them is $2f$. The convex surfaces of both the lenses face towards the objective. The focal length of the equivalent lens is $\frac{3}{2}f$ and its position is $3f$ behind the field lens.

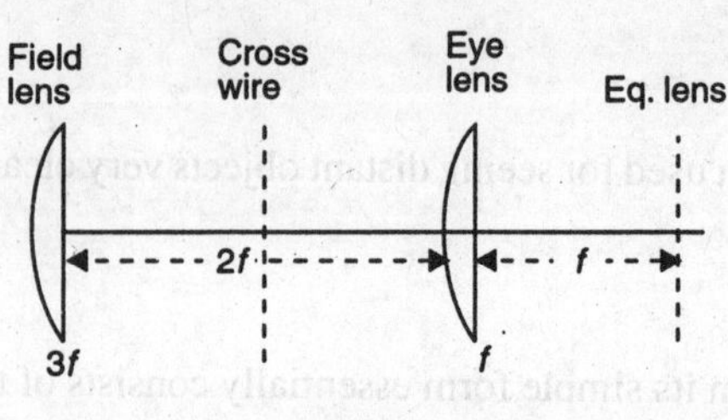

Fig. 13.3

The cross-wires are placed midway between the two lenses. This is known as a *negative* eye-piece. This eye-piece has been so designed that there is minimum spherical aberration and it is free from chromatic aberration.

In Huyghens' eye-piece the scale or the cross-wires have to be placed midway between the field lens and the eye lens. This does not give trustworthy measurements as the image undergoes refraction in both the lenses of the eye-piece whereas light from the cross-wires suffers refraction only at the surface of eye lens. This gives rise to unequal magnification of the image and the cross-wires.

13.3 Magnifying power of a telescope. *It is defined as the ratio of the angle subtended by the image as seen through the telescope to the angle subtended by the object at the eye as seen directly when both are at infinity.*

In normal adjustment when the telescope is set for parallel rays and the image is also formed at infinity, it can be shown that

$$\text{Magnifying power} = \frac{F}{f}$$

where F is the focal length of the objective and f that of the eyepiece.

Experiment 13.1. To determine the magnifying power of a telescope by the slit method using a microscope.

Apparatus. A telescope, a cardboard piece, a sharp blade, gum paper, one needle in an upright, a travelling microscope and brass slits to fit the tube on the objective side.

Theory. If a telescope having an objective O and an eyepiece E is set for infinity (parallel rays), then a beam of light parallel to the principal axis after passing through the objective of focal length F meets at its principal focus P. If the eye-piece is so adjusted that its distance from P is equal to its focal length f, the light again emerges as a parallel beam. Thus in the normal adjustment the distance between the objective and eye-piece $C_1 C_2 = F + f$.

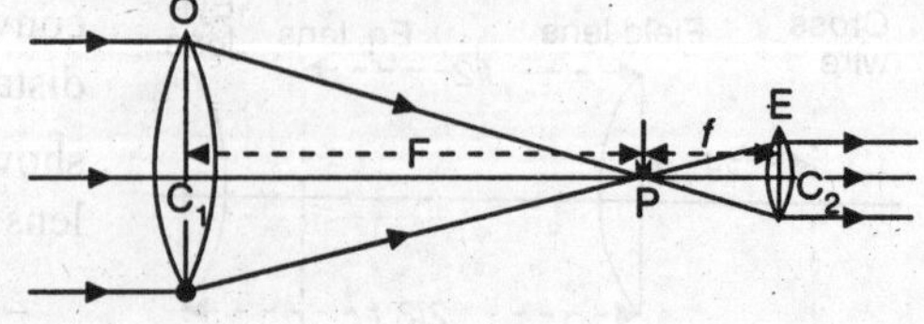

Fig. 13.4

If now the objective is replaced by a slit AB, its real image A_1B_1 is formed by the eye-piece as shown in Fig. 13.5

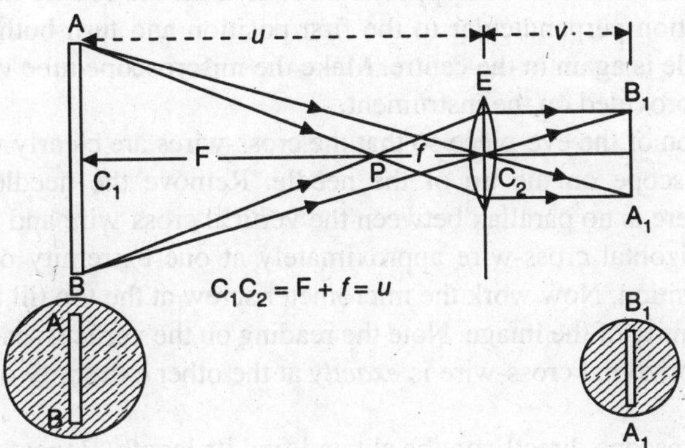

Fig. 13.5

$$\therefore \quad \frac{AB}{A_1B_1} = -\frac{u}{v}$$

But $$\frac{1}{v} = \frac{1}{f} + \frac{1}{u}$$

or $$\frac{u}{v} = \frac{u}{f} + 1 = \frac{u+f}{f}$$

$$\therefore \quad \frac{AB}{A_1B_1} = -\left(\frac{u+f}{f}\right)$$

Now $$-u = F+f$$

$$\therefore \quad \frac{AB}{A_1B_1} = -\left[\frac{-F-f+f}{f}\right] = \frac{F}{f} = m$$

Thus we see that when the telescope is set for parallel rays the magnifying power m is given by

$$m = \frac{\text{Length of the slit as seen directly}}{\text{Length of its image as seen in eye–piece}} = \frac{AB}{A_1B_1}$$

Procedure. 1. *To focus the telescope for parallel rays.* Turn the telescope to an open window. Adjust the eye-piece so that the cross-wires are clearly visible. Rotate the eye-piece to set one of the cross-wires in a vertical position. Focus the telescope on a distant object, till there is no parallax between the cross-wires and the image of a distant object.

2. *To prepare the slit.* Cut a circular disc, from a piece of card-board, of diameter exactly equal to that of the objective. Draw a narrow rectangle along a diameter of the disc with a sharp pencil and cut it away neatly with a sharp blade. Paste small pieces of paper near the extremities of the slit to get a sharp straight edge.

If a brass slit is provided, which can fit the telescope tube, paste small pieces of paper at the extremities of the slit.

3. Remove the objective and replace it by the slit placed in a vertical position.

4. Place the eye at a distance not less than 30 cm from the eye-piece and locate *the image of the slit*. Place a needle in an upright near the eye-piece in line with the image of the slit and adjust its position till there is no parallax between the needle and the image of the slit.

5. Place the travelling microscope on a level table near a window so that sufficient light is available. Place the spirit level on the stage of the microscope parallel to the line joining the two levelling screws and turn the screws in opposite directions till the bubble lies in the centre. Turn the spirit level in a direction perpendicular to the first position and turn both the screws in the same direction till the bubble is again in the centre. Make the miscroscope tube vertical by coinciding the two reference marks provided on the instrument.

Adjust the position of the eye-piece so that the cross-wires are clearly visible.

Focus the microscope on the tip of the needle. Remove the needle and finally adjust the microscope so that there is no parallax between the vertical cross-wire and the image of the slit.

6. Bring the horizontal cross-wire approximately at one extremity of the image. Clamp the microscope in this position. Now work the micrometer screw at the top till the horizontal cross wire is *exactly* at one extremity of the image. Note the reading on the vertical scale. Work the micrometer screw again till the horizontal cross-wire is *exactly* at the other extremity of the image and note the reading. Repeat twice.

7. Focus the microscope directly on the slit and find its length. Repeat twice.

8. Cover a small portion of the slit at each extremity with a gum paper and take such three observations.

Observations.

No.	*Reading at the image*			*Mean*	*Reading at the slit*			*Mean*	$m = \frac{AB}{A_1B_1}$
	A_1	B_1	*Difference*	A_1B_1	A	B	*Difference*	AB	
1. (*i*)									
(*ii*)									
2. (*i*)									
(*ii*)									
3. (*i*)									
(*ii*)									

Mean $m =$

Precautions. 1. The ends of the slit should be sharp and straight.

2. The slit should be sufficiently and uniformly illuminated.

3. While locating the image of the slit the eye should be kept at a distance more than 30 cm from the eye-piece of the telescope.

4. The objective should be replaced in its position after completing the experiment.

13.4. Resolving power. *It is defined as the reciprocal of the smallest angle subtended at the objective of the telescope by the two point objects which can just be distinguished as separate.*

The image of a point source is not a point but is a diffraction pattern consisting of a central bright disc surrounded by alternate dark and bright rings. If there are two nearby point sources the diffraction discs of the two patterns may overlap and the two images may not be distinguished. According to Rayleigh *the two nearby images are said to be resolved if the position of central maximum of one coincides with the first secondary minima of the other and vice versa. Using Rayleigh criterion,* we have

$$\text{Resolving power of a telescope} = \frac{1}{d\theta} = \frac{a}{\lambda}$$

where $d\theta$ is the smallest angle that can be resolved by the telescope objective, a its aperature (diameter) and λ the wave length of light.

Laboratory talescopes use a convex lens as an objective but large telescopes used in astronomical observatories also use concave mirrors with large apertures to have a greater resolving power. The resolving power of a mirror telescope is also given by $\frac{a}{\lambda}$.

Experiment 13.2. To verify the expression for the resolving power of a telescope.

Apparatus. A telescope, a monochromatic source of light, a variable rectangular slit, a sheet of plane glass coated with a tin foil and containing two fine parallel lines 2 to 3 mm apart cut with a fine razor blade, a travelling microscope and a measuring tape.

Theory. If two point objects P and Q (not shown) lying close to each other having an angular separation $d\theta$ are viewed through the objective of a telescope, then their images are formed at P_1 and Q_1 in the focal plane of the objective of the telescope. Since light from object P is incident normally

$$\therefore \quad AP_1 = BP_1$$

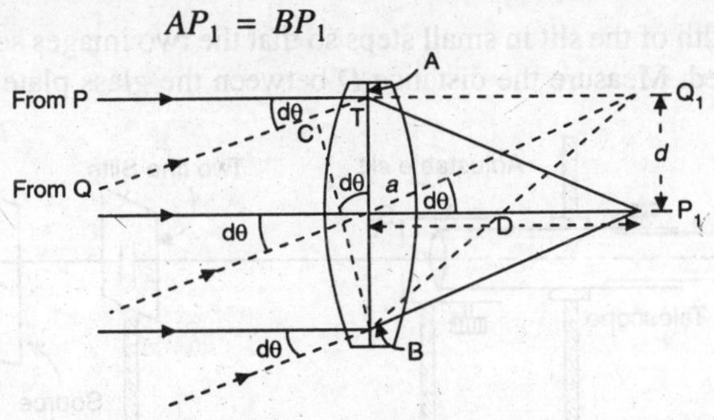

Fig. 13.6

Hence the first maximum of P is formed at P_1. In order that the image Q_1 of Q may appear as separate, the first maximum of the image Q_1 should fall on the first secondary minimum of the image P_1. This condition is achieved if the path difference between the extreme rays from Q and those from P is λ *i.e.*,

$$AC = \lambda$$

If a is the diameter of the objective, then

$$AC = AB \,.\, d\theta$$

or

$$d\theta = \frac{AC}{AB} = \frac{\lambda}{a} \qquad \text{... } (i)$$

The angle $d\theta$ subtended by the two point objects is given by

$$d\theta = \frac{d}{D}$$

where d is the separation between P_1 and Q_1 the images of the two point objects and D their distance from the objective

Hence

$$d\theta = \frac{d}{D}$$

Now resolving power $= \frac{1}{d\theta} = \frac{D}{d} = \frac{a}{\lambda}$

This relation can be satisfied if λ is known as all other quantities can be measured.

Note The relation $d\,\theta = \frac{\lambda}{a}$ holds good for a rectangular aperture. For a circular aperture like that of a telescope this relation can be written as

$$d\,\theta = 1.22\,\frac{\lambda}{a}$$

and Resolving power $= \frac{1}{d\theta} = \frac{a}{1.22\,\lambda}$

Procedure. 1. Hold the glass plate coated with tinfoil and having two fine parallel lines drawn on it with a sharp blade 2 to 3 mm apart, in a stand so that the plane is vertical and so also the two parallel lines. Place a powerful source of sodium light behind it and focus the light on the lines if necessary.

2. Mount the telescope on a stand. Place it at a convenient distance and adjust, such that the axis lies horizontally and meets the plate. Focus the telescope on the two lines which act as two sources. Mount the adjustable slit on a stand and place it in front of the telescope objective as close to it as possible. Adjust the position of the telescope by its horizontal movement at right angle to its axis, so that the two slits lie symmetrically with respect to the intersection of the cross-wires in the field of view.

3. Decrease the width of the slit in small steps so that the two images seen through the telescope **just** cease to be resolved. Measure the distance D between the glass plate and the objective of the telescope. Also measure the minimum width of the slit a with the help of a microscope.

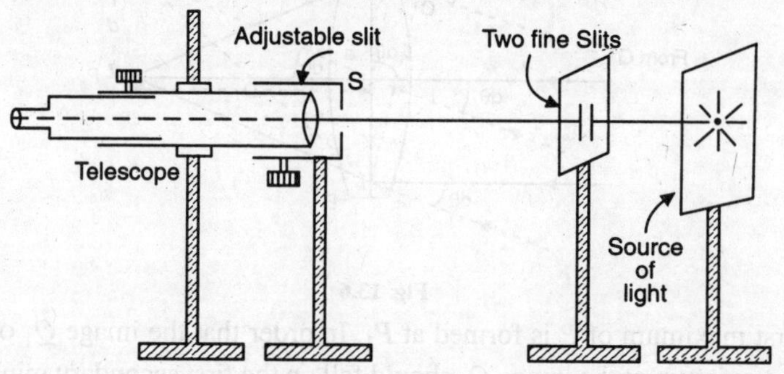

Fig. 13.7

4. Repeat at least six times for different distances of glass plate with tin foil, from the objective of the telescope. Finally measure the distance d between the two lines on the glass plate with the help of the travelling microscope.

Observations. Vernier constant of microscope =

Wavelength of light used $\lambda =$ m.

No.	*Microscope reading on*			*Distance of glass plate from telescope objective D (metre)*	$\frac{D}{d}$	*Miscroscope reading on*			$\frac{a}{\lambda}$
	First line	*Second line*	*d (metre)*			*one edge of slit*	*other edge of slit*	*a (metre)*	
1									
2									
3									
4									
5									
6									

As $\frac{D}{d} = \frac{a}{\lambda}$ the relation is verified.

Precautions. 1. The two lines drawn on the glass plate with tin foil should be very narrow and close to each other.

2. The plane of glass plate should be vertical as also the two lines marked on it.

3. The axis of the telescope should be horizontal and normal to the plane of the glass plate.

4. While measuring the width of the slit and also the distance between the two lines with travelling microscope, avoid back lash error.

5. The observation should be noted when the images of the two slits in the field of view of telescope just cease to be separated.

6. The minimum width of the aperture for resolution of two lines should be found by *opening* as well as *closing* the slit.

7. The slit should not be closed too tightly, otherwise its edges will get damaged.

Oral Questions

1. **Magnifying power.** (*i*) Define magnifying power. (*ii*) On what factors does the magnifying power of a telescope depend? (*ii*) When a telescope is set for infinity what is the distance between the objective and the eye-piece? (*iv*) What is the difference between the magnifying power and the resolving power? (*v*) On what factors does the resolving power depend? (*vi*) What is (*a*) an ultra- microscope and (*b*) an electron microscope? In what respects are these better than the ordinary microscope? (*vii*) What is the aperture of the biggest telescope of the world? (*viii*) What are the different types of eye-pieces used in a telescope? (*ix*) What are their relative advantages and disadvantages? (*x*) What type of an eye-piece is used in your (*a*) telescope and (*b*) microscope and why? (*xi*) What are the defects in the images formed by lenses? Are these defects present in the objective of your telescope? If not, how have these been removed?

2. **Resolving power of a telescope.** (*i*) What do you mean by the resolving power of a telescope? (*ii*) How do you measure the resolving power of a telescope? (*iii*) What is Rayleigh's criterion for resolution? (*iv*) Why the objective of a telescope is of large diameter? (*v*) Does the resolving power of a telescope depend on the distance of the telescope from the object? (*vi*) Explain why sometimes the theoretical resolving power is greater than the value obtained experimentally? (*vii*) What is the resolving power of a normal eye?

14

Sextant

14.1. Sextant. This is an instrument which enables us to determine the angle subtended by two distant point objects at the place of observation. If the distance of one of the points from the sextant is known the distance between the two points can be calculated.

It consists of a graduated circular arm *AB* supported by two fixed radial arms *AC* and *BC* which meet at a point *C* which is the centre of the circle of which the arms form a part.

An *index arm CD* which is free to rotate round an axis passing through *C* carries a plane mirror M_1 *called the index glass* at the end *C* fixed in a plane perpendicular to the plane of the scale. At the other end of this arm at *D* there is a vernier scale *V* which can slide along the main scale. Ordinarily this vernier scale is fixed in position, but its motion can be made free by pressing the lever at *L*. A tangent screw *S* is provided for fine adjustments of the vernier.

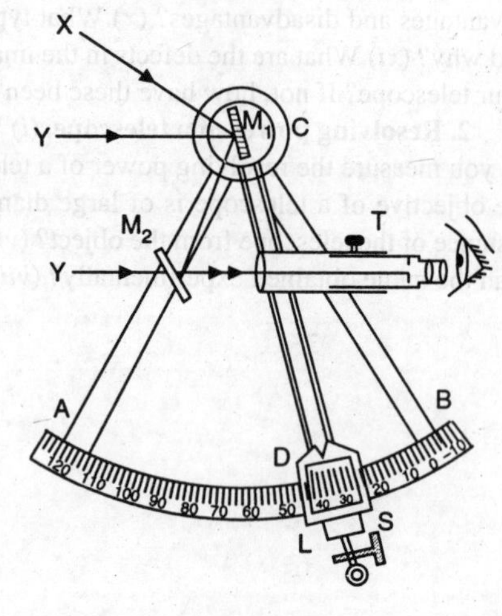

Fig. 14.1

A glass plate M_2 called the *horizon glass* is fixed on the arm *AC* with *its plane perpendicular to the plane of the scale in such a manner that when the index arm is at zero this is parallel to the mirror* M_1. Lower half of this glass plate is silvered whereas the upper half is transparent. A telescope *T* is fixed to the arm *BC* with its axis parallel to the plane of the scale and passing through the centre of the horizon glass. A screw is provided at the side of the telescope (not shown) to move it in a plane perpendicular to the plane of the scale.

Working. To find the angular distance between two distant points like *X* and *Y*, the reading of the vernier is noted when the image of *Y* as seen direct through the clear portion of M_2 overlaps the image formed after reflection from M_1 and then from the silvered part of M_2. The arm *CD* is rotated till the image of *X* as seen after reflection from M_1 and the silvered portion of M_2 coincides with the image of *Y* as seen direct. The reading of the vernier scale is again noted. The angular separation between *X* and *Y* is equal to twice the angle through which the mirror M_1 has been rotated. To facilitate a direct reading from the scale, the arm is graduated in half degrees although markings are in degrees. Such markings are called **sextant degrees.** Thus the arm *AB* which has 60° is marked 120°.

Micrometer type sextant. *In some sextants a micrometer screw is provided in place of the vernier*. In such a case in general the smallest division on the scale is *one degree*. The number of divisions on the circular scale of the micrometer or drum is 60. Thus the least count of the micrometer

$$= \frac{1^\circ}{60} = 1'$$

The micrometer or drum is further provided with a vernier having 5 divisions which coincide with 4 divisions of the micrometer scale. Thus

$$\text{Vernier constant} = \left(1 - \frac{4}{5}\right) = \frac{1}{5} \text{ minute}$$

$$= 12''$$

In some instruments the micrometer is provided with a vernier having 6 divisions which coincide with 5 divisions of the micrometer scale. In such a case

$$\text{Vernier constant} = \left(1 - \frac{5}{6}\right) \text{ of a scale division}$$

$$= \frac{1}{6} \text{ minute} = 10''$$

Experiment 14.1. To find the vertical distance between two points using a sextant.

Apparatus. A sextant, chalk pieces of two different colours, a rigid clamp stand, a measuring tape, a plumb line and a spirit level.

Theory. If h is the vertical distance between two points X and Y and θ is the angle subtended by XY at the index glass M_1 of the sextant at a distance d as shown in Fig. 14.2, then

$$\frac{h}{d} = \tan\theta$$

or

$$h = d\tan\theta$$

This is true only when M_1Y is perpendicular to XY.

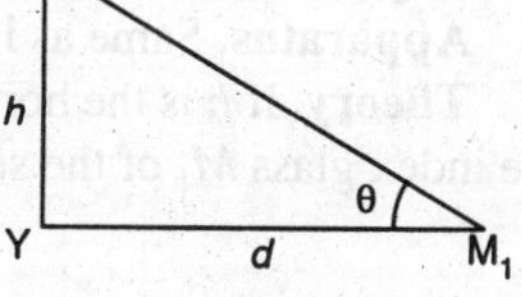

Fig. 14.2

Procedure. 1. Find the vernier constant or the least count of the sextant.

2. Mark a horizontal arrow nearly 10 cm long on the wall with a coloured chalk at a height of about l metre from the floor. Mark another horizontal arrow vertically above it near the ceiling.

3. Place a rigid clamp stand at a distance of nearly 5 metre from the wall. Clamp the sextant in the stand with the scale downwards and the objective of the telescope pointing horizontally towards the lower mark. Adjust the height of the sextant so that the axis of the telescope is horizontal and is at the same level as the lower mark Y from the ground. Test with a spirit level. With the help of a plumb line adjust the plane of the scale to be vertical. Move the telescope sideways by the screw at its side so that the line of demarcation of the silvered and unsilvered portions of the horizon glass M_2 lies in the centre of telescope objective.

4. Focus the telescope on the lower mark Y and adjust the position of the sextant so that the image of the horizontal arrow is in the centre of the field of view. *This is the direct image as seen in the telescope.*

5. Move the arm CD so that the image of the same horizontal arrow is also visible in the telescope after reflection from the two mirrors. This will happen when the vernier reading is nearly zero and the instrument is in proper adjustment. If the reflected image is not in the field of view or does not coincide with the direct image, the horizon glass is not perpendicular to the plane of the scale. *To adjust this, work the small screw provided on the carriage of the horizon glass till the two images lie one above the other.*

If the two images seen are not of the same brightness, it can be adjusted by moving the telescope sideways with the help of the screw provided at the side of the telescope.

6. When the two images of the mark Y overlap, note the reading on the scale.

7. Keeping the eye in the same position move the arm *CD*. One-half of the field of view will appear to slide past the other half and ultimately the image of the upper arrow will appear to approach the direct stationary image. Note the scale reading when the two images again overlap.

8. *Measure the distance from the lower arrow to the index mirror by a measuring tape.*

9. Repeat the experiment with three different distances of the sextant from the lower arrow.

Observations. Vernier constant of the sextant =

Sl. No.	*Scale reading*		*Difference*	*tan* θ	*Distance*	Height
	Initial	*Final*	θ		*d cm*	h = *d* tan θ
1						
2						
3						

Mean height = cm = m

Actual height by direct measurement = cm = m

% error =

Precautions. 1. Telescope should be at the same height from the floor as the bottom of the object.

2. The scale should be vertical and the axis of the telescope horizontal.

3. The two images should overlap and should have the same intensity.

Experiment 14.2. To find the horizontal distance between two points using a sextant.

Apparatus. Same as in **Experiment 14.1.**

Theory. If *h* is the horizontal distance between two points *X* and *Y* and θ the angle subtended at the index glass M_1 of the sextant at a distance *d* lying in the horizontal position, then

$$\frac{h}{d} = \tan\theta$$

or

$$h = d\tan\theta$$

Procedure. The method is exactly the same as explained in **Experiment 14.1.** In this case two vertical arrows are marked in a horizontal line at the same height from the ground as that of the telescope or the sextant. The sextant is fixed in a horizontal position on a stand in such a manner that the stand is not shaky. The plane of the scale is adjusted to be horizontal with the help of a spirit level. The sextant is placed in such a position that the axis of the telescope is perpendicular to the horizontal line on the wall and points towards the left mark.

Observations. Same as in **Experiment 14.1.**

Precautions. 1. Same as in **Experiment 14.1.**

2. The scale should be horizontal and the axis of the telescope should be perpendicular to the line joining the two points.

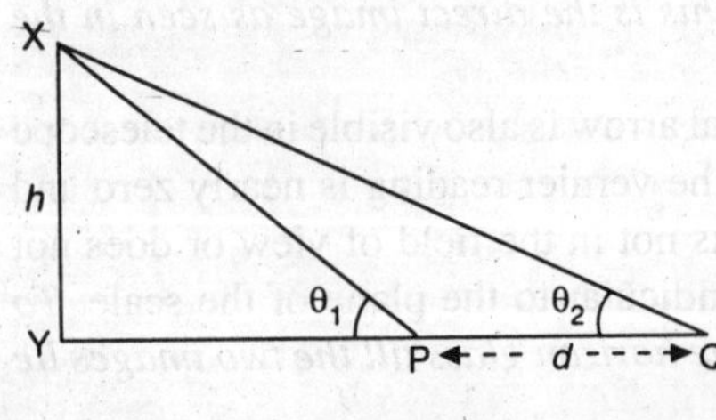

Fig. 14.3

Experiment 14.3. To find the height of an inaccessible object using a sextant.

Apparatus. Same as in **Experiment 14.1.**

Theory. When the distance from the sextant to the foot of the distant object cannot be measured directly the object is said to be *inaccessible*. To find the height *XY* of such an object the angles θ_1 and θ_2 subtended by it at two different points *P* and *Q* in the same straight line as the foot of the distant object *Y*, are determined.

Now $$\cot\theta_1 = \frac{YP}{h} \quad \text{and} \quad \cot\theta_2 = \frac{YQ}{h}$$

$$\therefore \quad \cot\theta_2 - \cot\theta_1 = \frac{YQ - YP}{h} = \frac{d}{h}$$

or

$$h = \frac{d}{\cot\theta_2 - \cot\theta_1}$$

Procedure. 1. Mark an arrow-head in the same horizontal line as the telescope, at a large distance. This serves as the lower reference point for the inaccessible object. The height of the top of this inaccessible object is determined with respect to this point.

2. Set the sextant so that the direct and the reflected images of the arrow overlap. Note the reading. When the reflected image of the top of the object coincides with the direct image of the arrow, note the reading again. Repeat by placing the sextant in three different positions *P, Q* and *R* in the same straight line by moving the sextant from 2 to 3 metres at a time. Find the value of *h* from any pair of observations and calculate the mean value.

Observations.

Vernier constant =

Sextant at	*Scale reading*		*Difference* θ	*cot* θ
	Initial	*Final*		
P			θ_1	
Q			θ_2	
R			θ_3	

(*i*) Distance *PQ*, d_1 = cm

$$\therefore \quad h = \frac{d_1}{\cot\theta_2 - \cot\theta_1} = \quad \text{cm} \quad = \text{m}$$

(*ii*) Distance *QR*, d_2 = cm

$$\therefore \quad h = \frac{d_2}{\cot\theta_3 - \cot\theta_2} = \quad \text{cm} \quad = \text{m}$$

(*iii*) Distance *PR*, d_3 = cm

$$\therefore \quad h = \frac{d_3}{\cot\theta_3 - \cot\theta_1} = \quad \text{cm} \quad = \text{m}$$

Mean height h = cm = m.

Precautions. Same as in **Exp. 14.1.**

Exercise. *To find the area of a laboratory window.*

Hints. The area of the window = length × breadth.

Mark two points *A* and *B* on the extremities of length and measure the distance *AB* (vertical). In a similar manner measure the breadth (horizontal distance) and hence find the area.

Experiment 14.4. To determine the angular diameter of the sun with the help of a sextant.

Apparatus. Same as in **Expt. 14.1.**

Procedure. 1. Clamp the sextant in a rigid clamp stand in the vertical position. Look at the sun through the unsilvered portion of glass plate M_2 (Horizon glass) with the telescope using *dark red coloured filters* to reduce the intensity of light. This is denoted by *S* in Fig. 14.4.

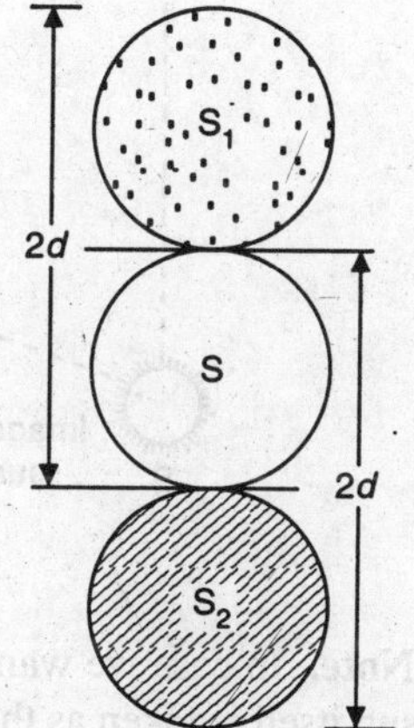

Fig. 14.4

2. Turn the moveable index arm *CD* of the sextant such that the image S_1 obtained by double reflection from the two mirrors (M_1) and (M_2) just touches the rim of *S*. Note the reading.

3. Now again rotate the arm such that the image S_2 is on the other side of *S* again in the position just touching the rim of *S*. Note the reading.

4. The difference of these two readings gives $2d$ (twice the angular diameter of the sun). Half this gives the angular diameter of the sun. Repeat the experiment three times.

Observations. Vernier constant of the sextant =

S. No.	*Scale reading*			
	S_1 touching the rim of S (a)	*S_2 touching the rim of S (b)*	*2d = (a–b)*	*Angular diameter d*
1				
2				
3				

Precautions 1. Same as in **Expt. 14.1.**

2. The filters (dark red glasses) must be used to reduce the intensity of light.

Experiment 14.5. To determine the altitude or the angular elevation of the Sun using a sextant.

Apparatus. Same as in **Expt. 14.1** and mercury in a flat dish or trough to act as artificial horizon.

Theory. To find the altitude or the angular elevation of the sun we require an artificial horizon. For this purpose we use a flat dish or trough full of mercury. We use mercury as it is highly reflective and quite heavy so that it is not liable to be easily disturbed by wind etc. The angle subtended at the sextant telescope by the sun and its image in mercury is measured. One-half this value gives the altitude of the sun.

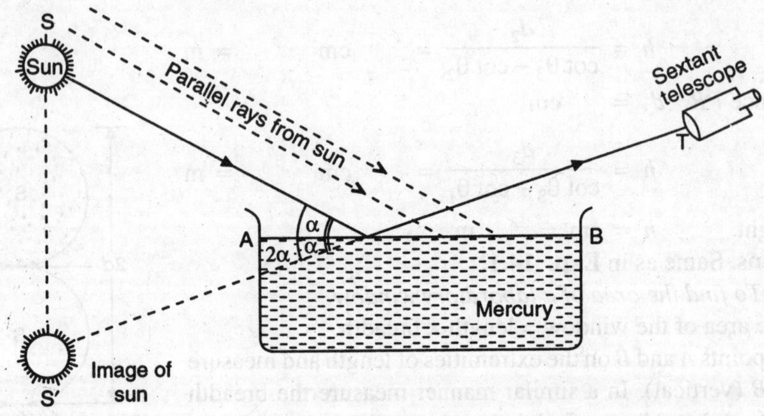

Fig. 14.5

Note. In case we want to find the altitude of the sun at sea, the horizon is taken as one object and sun itself is taken as the second object.

Procedure. 1. Put mercury in the flat dish (or trough) and place it on a stool in the sun in the morning or evening when the sun is quite low. See the image of the sun in mercury. It will be seen in the direction *TS′* on the mercury surface *AB*. *AB* serves as the *artificial horizon.*

2. Take the sextant and fix it in the rigid clamp-stand in such a way that the telescope is directed towards S'. Use sun glasses (red and blue) to avoid glare. Adjust the telescope to see the direct image of S' through the unsilvered portion of the *horizon glass* M_2.

3. Now gradually move the moveable *index arm* till the image of S' formed by reflection from both the mirrors M_1 and M_2 (index glass and silvered portion of horizon glass) concides with the direct one. Note the reading. Let is be θ_1.

4. Again rotate the *index arm* to bring the image of the sun S formed by reflection from both the mirrors M_1 and M_2 to coincide with the direct image S'. Note the reading. Let it be θ_2.

5. The angle of elevation of the sun $\alpha = \frac{\theta_2 - \theta_1}{2}$. Repeat three times.

Observations. Vernier constant of the sextant =

S. No.	*Scale reading*		*Angle of elevation*
	Initial θ_1	*Final* θ_2	$\alpha = \frac{\theta_2 - \theta_1}{2}$

Mean α =

Precautions. Same as in **Expt. 14.1.**

Oral Questions

1. **Sextant** (*i*) What is the principle of a sextant? (*ii*) What are sextant degrees and why are they so marked? (*iii*) What precautions will you take for the proper setting of the telescope? (*iv*) Name the various parts of the sextant. (*v*) What are the various uses to which a sextant is put? (*vi*) What type of telescope is used in a sextant? (*vii*) Name the type of eye-piece used in it. (*viii*) What is artificial horizon? Why do we use mercury when finding the elevation of the sun? (*ix*) How will you find the height of an inaccessible object? State the formula used. (*x*) What is the use of coloured glasses in the sextant? (*xi*) What is the name given to fixed mirror M_2? (*xii*) What is the name given to the mirror M_1 fixed to the index arm?

15

Photometry

15.1. Photometry. This branch deals with the measurement of intensity of illumination of a surface, illuminating power of a source of light and comparison of illuminating powers of two sources. To understand these terms, we shall first define *luminous flux.*

Luminous flux. The luminous flux ϕ of a given source of light is defined as the time rate of flow of light energy from the source with reference to the visual sensation produced by it.

The unit of luminous flux is a '*lumen*'. It is defined as the luminous flux emitted by a uniform point source having a luminous intensity of *one candela* in a solid angle of *one steradian.*

Luminous intensity (or illuminating power). The luminous intensity L of a source of light is defined as the time rate of flow of light energy through a unit solid angle.

If $d\phi$ is the luminous flux across a surface which subtends a solid angle $d\omega$ at the given point source of light, then

$$\text{Luminous intensity } L = \frac{d\phi}{d\omega}$$

If ϕ is the total flux from the point source of light, then since the total solid angle subtended by a closed surface at a point inside it is 4π, we have

$$L = \frac{\phi}{4\pi}$$

or

$$\phi = 4\pi L$$

Luminous intensity may also be defined as the amount of visual light energy falling per unit time on the surface of a unit area placed at a unit distance from the source when the surface is held normally to the direction of light.

The unit of luminous intensity is a '*Candela*'. One Candela is $\frac{1}{60}$th part of the luminous intensity of 1 cm^2 surface of a perfect black body maintained at the melting point of pure platinum 1773° C or 2046 K.

Intensity of illumination (or Illuminascence). The intensity of illumination at a point on a given surface is defined as the luminous flux falling on a unit area of the surface surrounding the given point.

If $d\phi$ is the luminous flux falling on an area ds surrounding a point on a surface, then

$$\text{Intensity of illumination at the point } I = \frac{d\phi}{ds}$$

The S.I. unit of, intesity of illumination is a '*lux*' *or* '*metre candle*'.

The intensity of illumination at a point on a surface is one lux if the luminous flux falling on it is one lumen per square metre.

The C.G.S. unit of intensity of illumination is a '*phot*'. It is the intensity of illumination of a surface which receives a luminous flux of one lumen per square cm.

$\therefore$ 1 lux = 10^{-4} phot or 1 phot = 10^4 lux

15.2. Inverse square law. Let O be a point source of light, and ds a small element of area at a distance r from it. If θ is the angle that normal to the surface ds makes with the direction of r, then the solid angle $d\omega$ subtended by the area ds at O is given by

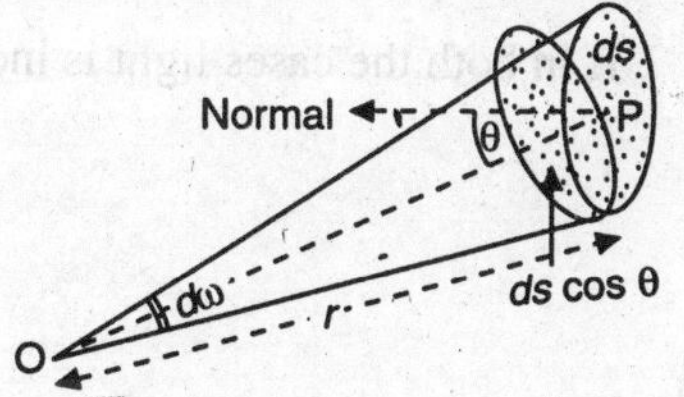

Fig. 15.1

$$d\omega = \frac{ds \cos\theta}{r^2}.$$

If L is the luminous intensity of the source O in the direction of r, then

Luminous flux received by the area $ds = d\phi = L\, d\omega$

$$= L\frac{ds \cos\theta}{r^2}$$

If I is the intensity of illumination at a point on the surface ds, then

$$I = \frac{d\phi}{ds} = \frac{L \cos\theta}{r^2}$$

According to this relation the intensity of illumination of a surface is

(*i*) directly proportional to the luminous intensity of the light source.

(*ii*) Inversely proportional to the square of the distance between the source and the surface, and

(*iii*) directly proportional to the cosine of the angle between the normal to the surface and direction of r. This is known as *Lambert's cosine law*.

For normal incidence $\theta = 0$

$$\therefore \quad I = \frac{L}{r^2}$$

This is **inverse square law.** It states that *the intensity of illumination is inversely proportional to the square of the distance between the light source and the given surface.*

15.3. Principle of a photometer. A photometer is used to compare the illuminating powers, or luminous intensities of two sources of light.

Let L_1 and L_2 be the luminous intensities of two sources of light respectively. Suppose light from the source L_1 falls on a surface at a distance r_1 from it and produces an intensity of illumination I_1 on the surface, then

$$I_1 = \frac{L_1 \cos\theta_1}{r_1^2}$$

where θ_1 is the angle between the direction of incident light and normal to the surface.

Suppose B_1 is the brightness of the surface as judged by the eye, then

$$B_1 = \alpha I_1$$

where α is a constant of proportionality known as reflection factor.

Let I_2 be the intensity of illumination of the same surface when placed at a distance r_2 from the source L_2 and B_2 the corresponding brightness, then

$$I_2 = \frac{L_2 \cos\theta_2}{r_2^2}$$

and $$B_2 = \alpha I_2$$

If, as judged by the eye or any other method $B_1 = B_2$, then

$$\alpha I_1 = \alpha I_2$$

or
$$I_1 = I_2$$

or
$$\frac{L_1 \cos \theta_1}{r_1^2} = \frac{L_2 \cos \theta_2}{r_2^2}$$

If in both the cases light is incident normally on the surface, then $\cos \theta_1 = \cos \theta_2 = 1$ and

$$\frac{L_1}{r_1^2} = \frac{L_2}{r_2^2}$$

or
$$\frac{L_1}{L_2} = \frac{r_1^2}{r_2^2} = \left(\frac{r_1}{r_2}\right)^2$$

This is the principle underlying photometry.

15.4. Lummer—Brodhun photometer. This photometer makes use of a combination of two right angled glass prisms P_1 and P_2. The base (hypotenuse) of the prism P_2 is made convex by grinding and then the two prisms are cemented together with canada balsam such that the two prisms are in contact only at the central part with an air gap all round. Thus the light rays falling on this *cube* will be transmitted through the central part and the rays falling on the outer part will be totally internally reflected as shown. This combination of prisms is known as **Lummer Brodhun cube.**

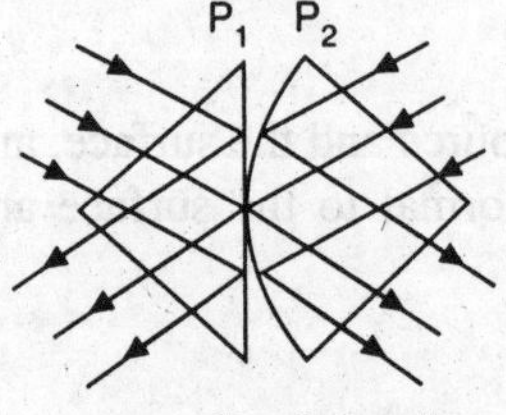

Lummer Brodhun cube

Fig. 15.2

The photometer consists of a box B blackened from inside on the top of which is fitted a screen S in a vertical position. The screen is an opaque sheet with both sides coated with a diffusing material like plaster of paris or magnesium carbonate.

The Lummer Brodhun cube is fitted in the centre of the box and two totally reflecting right angled prisms P_3 and P_4 are placed symmertrically on the walls on either side so that these deviate the incident beam of light through a right angle to make it incident normally on the two prisms forming the Lummer Brodhun cube.

Light after passing through the LB cube is received in a low power telescope T in which we make a simultaneous visual comparison of the illumination on the two sides of the screen each of which is illuminated by one of the two light sources.

Working. The two sources of light S_1 and S_2 whose luminous intensities are to be compared are placed on either side of the screen S, so that light falls normally on its respective faces. As the screen is coated with a diffusing material light falling on it is scattered in all possible directions. Suppose the parallel rays marked 1, 2, 3 scattered by the left face of the screen are totally reflected by the prism P_3 and those marked 4, 5, 6 scattered by the right-face of the screen are totally reflected by the prism P_4 respectively as shown. Thereafter these rays fall on the Lummer Brodhun cube which transmits the rays marked 2 and 5 falling on its central part and totally internally reflects the rays 1 and 3 or 4 and 6 falling on its outer parts. The light rays reaching the tele-

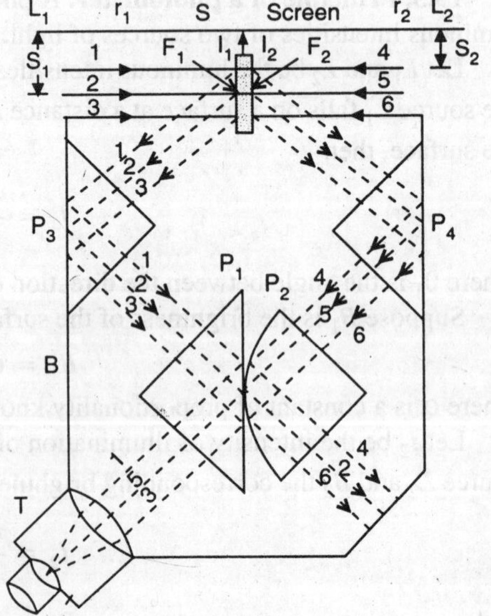

Fig. 15.3

scope consist of central ray marked 5 from the source S_2 and outer rays marked 1 and 3 from the source S_1. The central ray marked 2 from the source S_1 and outer rays marked 4 and 6 fall on the inner blackened wall of the box and are fully absorbed.

In general, the intensity of illumination produced by the sources on the two sides of the screen are not equal and there will be difference in the brightness of the central and outer parts of the field of view as seen through the telescope. If the intensity of illumination I_1 is less than I_2, the central part will be brighter (Fig. 15.4. *a*) and if I_1 is greater than I_2 the outer part will be brighter (Fig 15.4. *b*).

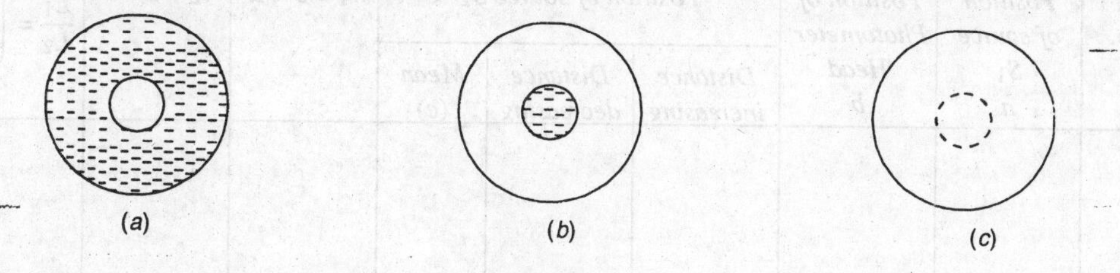

Fig. 15.4

The distances of the sources S_1 and S_2 are so adjusted that the whole field of view is uniformly illuminated as shown in Fig 15.4 (*c*).

When this adjustment has been made

$$I_1 = I_2$$

$$\therefore \quad \frac{L_1}{L_2} = \left(\frac{r_1}{r_2}\right)^2$$

where r_1 and r_2 are the distances of the sources S_1 and S_2 from the screen respectively.

Experiment 15.1. To compare the illuminating powers (or luminous intensities) of two given sources of light using a Lummer- Brodhun photometer.

Apparatus. A heavy optical bench 175 cm long, a Lummer Brodhun photometer, two sources of light etc.

Procedure 1. Mount the Lummer Brodhun photometer in an upright and place it in the middle of the optical bench. Rotate the box so that the screen S (coated with a diffusing material) is perpendicular to the length of the optical bench bed. Also mount the two sources of light S_1 and S_2 (suitably enclosed in their casings) on two separate uprights and place them on the opposite sides of the photometer. Adjust their heights equal to that of the screen S and see that the line joining the sources S_1 and S_2 is also parallel to the bed of the optical bench. Focus the telescope on the *contact surface* of the *Lummer Brodhun cube* P_1 P_2.

2. Adjust the source S_1 at a suitable distance r_1 from the photometer and the source S_2 very close to it. Switch on the two sources so that the lamps begin to emit light. Allow the light to fall normally on the opposite faces of the screen S. The light scattered by the two faces of the screen after suffering total internal reflection at the prisms P_3 and P_4 and transmitted or totally reflected by the *L.B cube* is received by the telescope. On seeing through the telescope it will be observed that the central portion which receives light from the source S_2, is brighter than the outer portion which receives light from the source S_1. Move the source S_2 away from the photometer and keep on looking through the telescope till the *whole* field of view appears to be uniformly illuminated. Note the positions of the sources S_1 and S_2 as also of the photometer on the scale of the optical bench.

3. Remove the source S_2 to a far off position. It will be observed that the outer portion is brighter than the central portion. Slowly bring S_2 closer to the photometer till again the whole field of view appears to be uniformly illuminated as seen through the telescope. Again note the position of S_2.

4. Rotate the head of the Lummer Brodhun photometer through 180° so that the face F_1 which was in front of source S_1 now lies on the side of source S_2 and the face F_2 which was in front of source S_2 now lies on the side of source S_1. Repeat observations of step 2 and 3 keeping the distance of the source S_1 from the screen S the same say r_1.

5. Repeat the experiment for different values of r_1.

Observations.

Sl. No.	*Position of source* S_1 *a*	*Position of Photometer Head* *b*	*Position of source* S_2			$r_1 = b - a$	$r_2 = c - b$	$\frac{L_1}{L_2} = \left(\frac{r_1}{r_2}\right)^2$
			Distance increasing	*Distance decreasing*	*Mean (c)*			
1								
2								
3								
4								

Mean $\frac{L_1}{L_2}$ =

Precautions. 1. The heights of the two sources of light should be suitably adjusted so that light from them falls normally on the diffusing screen.

2. The diffusing screen should be adjusted normal to the length of the bed of the optical bench.

3. The line joining the two sources should be parallel to the length of the bed.

4. To eliminate error due to difference in the reflecting powers of the two surfaces of the diffusing screen, readings should also be taken after rotating the photometer head through 180° so as to interchange the surfaces exposed to the two sources of light.

Oral Questions

I. Photometry (*i*) Define photometry. (*ii*) Define Luminous flux, Luminous intensity and intensity of illumination. (*iii*) Give the S.I. units of luminous flux, luminous intensity and intensity of illumination. (*iv*) What is inverse square law? (*v*) State the factors on which the intensity of illumination at a surface depend. (*vi*) State Lambert's cosine law of photometry. (*vii*) What is the principle of a photometer? (*viii*) In a Lummer Brodhun photometer why do you repeat the observation after rotating the diffusing screen through 180°? (*ix*) State the limitations of L.B. photometer. (*x*) What other methods are available for comparing the illuminating powers of two sources?

16

Spectrometer

16.1. Spectrometer. It is a compact apparatus for obtaining a pure spectrum. It is used for the study of spectra and for finding the refractive index of a material in the form of a prism. In its simplest form it has the following main parts :—

(*i*) The collimator, (*ii*) the telescope and (*iii*) the prism table.

(*i*) **The collimator C.** Its purpose is to produce a parallel beam of light. It consists of a tube mounted horizontally on the arm of the spectrometer. The tube has a converging achromatic lens at one end and a sliding tube having an adjustable vertical slit at the other end. The focal length of the lens is almost equal to the length of the collimator tube. The distance between the slit and the lens can be altered by a rack and pinion arrangement or by sliding the inner tube to obtain parallel rays. The tube rests on two screws C_1 and C_2 by which it can be slightly tilted up or down if necessary.

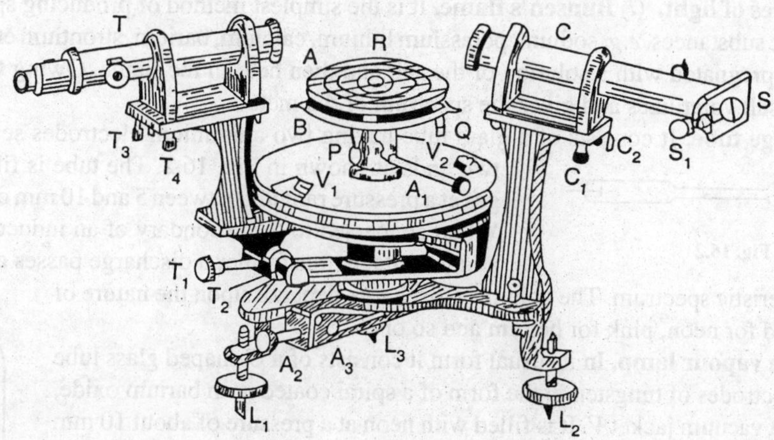

Fig. 16.1

The slit. It consists of two sharp edges. One of the edges is fixed while the other can be moved parallel to it by working the screw S_1 provided at its side.

(*ii*) **The telescope T.** It is an astronomical telescope having an achromatic objective and a Ramsden's eye-piece. It is mounted on another arm fixed rigidly to the circular scale graduated in half degrees. The telescope along with the scale can be turned round a vertical axis passing through the centre of the spectrometer. It can be fixed in any position by the screw T_1 and then can be given a slow rotation or fine adjustment by a tangent screw T_2. The position of the telescope can be read by the two verniers V_1 and V_2, 180° apart fixed to the prism table. The telescope rests on the screws

T_3 and T_4 by which it can be slightly tilted up or down for adjustment of its axis, if necessary. A rack and pinion arrangement is provided on the side of the telescope for focussing.

(iii) **The prism table.** It consists of an upper plate *A* and a lower plate *B* separated by three springs through which pass the levelling screws *P, Q* and *R*. A set of parallel equidistant lines are engraved on the upper plate as shown in Fig. 16.1. These lines are parallel to the line joining any two of the screws, say *P* and *Q*. *The prism is always placed with one of its reflecting faces perpendicular to these lines.* A series of circles concentric with the axis of rotation of the prism are also ruled on this plate. These help in placing the prism correctly. The height of the prism table can be adjusted by a clamping screw A_1 which fixes the prism table to the verniers V_1 and V_2. The table can rotate about a vertical axis and may be fixed in any desired position by means of the screw A_2. When fixed by the screw A_2 it can be turned very slowly by a tangent screw A_3 placed at the base of the spectrometer. The position of the table can be accurately read by the verniers moving on the circular scale.

16.2. Levelling the spectrometer. The telescope and the collimator of most of the spectrometers are already set by the manufacturers so that their axes are parallel to the plane of the graduated circular scale. **The collimator and the telescope, therefore, ordinarily require no setting.**

If it becomes necessary to level the telescope and the collimator it may be done as follows :—

(i) Place the spirit level on the graduated circular scale parallel to the line joining any two levelling screws provided at the base say L_1 and L_2. Adjust by turning the screws in opposite direction till the air bubble is in the centre. Now place the spirit level perpendicular to this position and adjust the third screw L_3 if necessary. (in some of the spectrometers levelling screws L_1, L_2, L_3, are not provided).

(ii) Place the spirit level on the tube of the telescope and adjust the screws T_3 and T_4 to make it horizontal.

(iii) Similarly adjust the levelling of the collimator with the help of screws C_1, and C_2.

16.3. Sources of light. *(i)* **Bunsen's flame.** It is the simplest method of producing spectrum of various inorganic substances. *e.g.* sodium, potassium lithium, calcium, barium, strontium etc. A brush of asbestos is impregnated with a solution of the salt and then held in the flame. Owing to the high temperature the salt volatilizes and gives the spectrum of the metal.

(ii) **Discharge tube.** It consists of a glass tube having two aluminium electrodes sealed at the two ends as shown in Fig. 16.2. The tube is filled with a gas at a pressure ranging between 5 and 10 mm of mercury. When connected to the secondary of an induction coil or a high voltage transformer a discharge passes and the gas gives its characteristic spectrum. The colour of discharge depends upon the nature of the gas, being red for neon, pink for helium and so on.

Fig. 16.2

(iii) **Sodium vapour lamp.** In its usual form it consists of a *U* shaped glass tube *L* having two electrodes of tungsten in the form of a spiral coated with barium oxide, enclosed inside a vacuum jacket *V*. It is filled with neon at a pressure of about 10 mm of mercury and some specks of metallic sodium are deposited on the sides of the tube. To start the discharge a potential of about 400 volts is required which is obtained from a step up transformer. The lamps are available in 45 watts, 80 watts and 140 watts sizes. The 140 watt lamp takes 0.9 amp. at 165 volts. The discharge first passes through the neon gas which gives its characteristic red colour. The heat generated causes the metallic sodium to evaporate and gradually the tube is filled with the persistent golden yellow light of sodium. The sodium lamp is connected to the mains through a special transformer called the **auto-leak transformer.**

Fig. 16.3

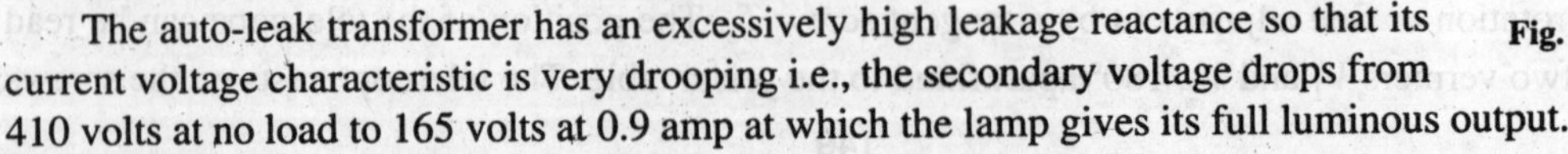
The auto-leak transformer has an excessively high leakage reactance so that its current voltage characteristic is very drooping i.e., the secondary voltage drops from 410 volts at no load to 165 volts at 0.9 amp at which the lamp gives its full luminous output.

(*iv*) **Mercury vapour lamp.** The basic principle of a mercury vapour lamp is the same as that of sodium vapour lamp. A high pressure lamp of the modern form essentially consists of an outer bulb *A* and an inner bulb *B* the space in between being highly evacuated. The inner bulb contains a small quantity of argon and mercury, two main electrodes E_1 and E_2 and a starting electrode *S* placed very close to E_1. The auxiliary electrode *S* is connected to the other electrode E_2 through a high resistance *R* of about 500 ohm. Each of the main electrodes E_1 and E_2 consists of a tungsten wire spiral containing a rod of barium compound which when heated gives out a copious supply of electrons.

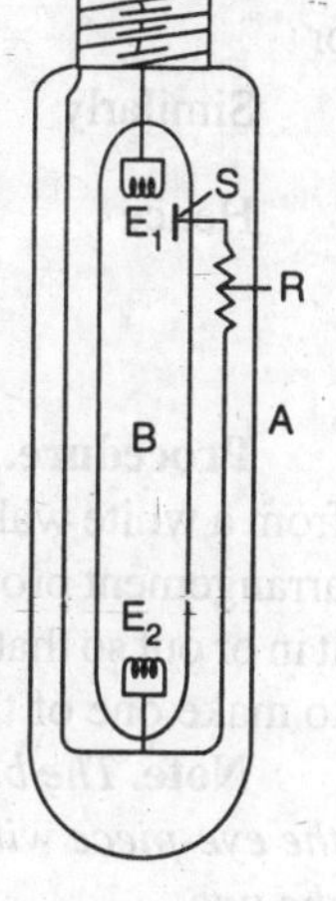

Fig. 16.4

When the lamp is switched on, full mains voltage is applied across the electrodes E_1 and *S* so that a discharge passes between the two, thereby starting the ionisation of the gas. As the lamp warms up, the mercury gradually vaporises and the pressure increases. The mercury discharge between E_1 and E_2 fills the whole tube. The argon discharge becomes ineffective because the resistance of the starting circuit is very high as compared to that of the parallel circuit.

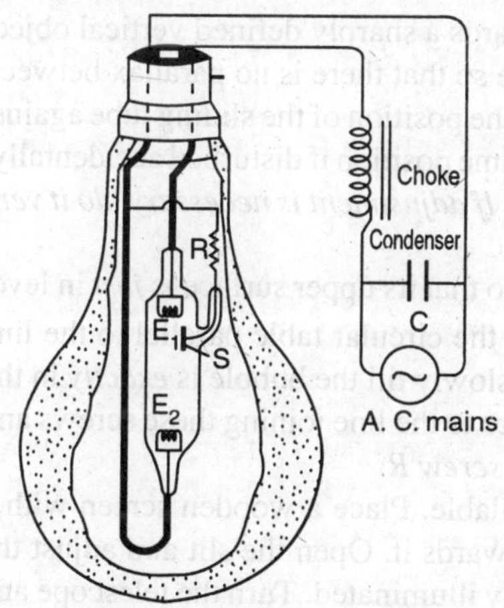

Fig. 16.5

Such lamps are available in 250-500 watts capacity. Special mercury vapour lamps are also being made in 80 and 125 watts capacity. These operate at an extra high pressure of 5-10 atmospheres. The discharge tube is of quartz instead of glass so that it can withstand the high pressure. The tube is about 5 cm long and is housed inside a pearl glass bulb similar to an ordinary 100 watt lamp as shown in Fig. 16.5. Such lamps are generally used in the laboratory.

The lamp is worked on *A.C.* mains with the help of a choke which serves as a stabilizer. A condenser is also placed across the mains to improve the efficiency of the lamp. If the lamp is switched off it cannot be started again at once because mercury must condense before the starting electrodes can function. This requires about 10 minutes after which the lamp can be started as usual.

Experiment 16.1. To find the angle of the prism by rotating the telescope method.

Apparatus. A spectrometer, a spirit level, a source of monochromatic light (sodium lamp), a glass prism, a wooden screen with a circular aperture, an eye-piece and an electric lamp.

Theory. If a narrow parallel beam of light from the collimator is allowed to fall on the edge *A* of the prism as shown in Fig. 16.6 it is partly reflected from the face *AB* and partly from the face *AC*. If a telescope is placed with its axis parallel to *KG* or *AE* an image of the slit will be seen in it. If the telescope is rotated so that its axis lies parallel to *HF* or *AD*, the image of the slit will again be seen in it. The angle of the prism is equal to half the angle through which the telescope is rotated.

This can be proved as follows :

As *MK* is the incident ray and *KG* the reflected ray

$$\angle MKA = \angle GKC = \angle AKO$$

But $$\angle MKA = \angle KAO$$

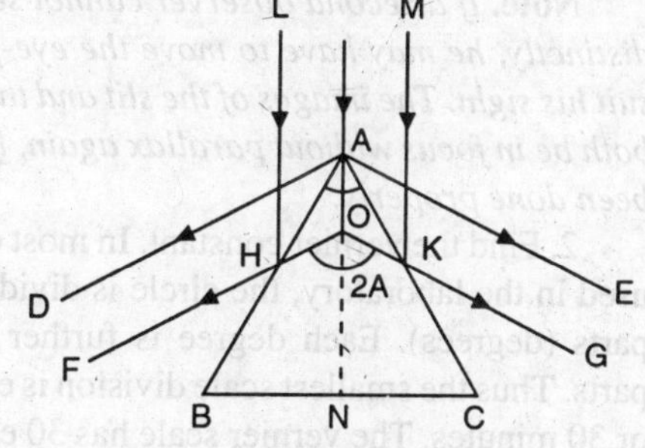

Fig. 16.6

$\therefore$ $\angle KAO = \angle AKO = \dfrac{A}{2}$

or $\angle KON = \angle KAO + \angle AKO = A$

Similarly $\angle HON = A$

Hence $\angle A = \dfrac{1}{2}\ (\angle KON + \angle HON)$

$= \dfrac{1}{2}\ \angle HOK = \dfrac{1}{2}\ \angle FOG = \dfrac{1}{2}\ \angle DAE.$

Procedure. 1. To set the spectrometer. (*i*) Place the spectrometer at a distance of a few metres from a white wall and adjust the distance of the eye-piece from the objective with the rack and pinion arrangement provided to make the field of view to appear white. Now focus the eye-piece by moving it in or out so that the cross-wires are *clearly visible without any strain on the eye*. Rotate the eye-piece to make one of the cross-wires vertical.

Note. *The best method to focus the eye-piece without any strain is to see the cross-wires through the eye-piece with one eye and the wall directly by the other eye so that there is no parallax between the two.*

(*ii*) Now direct the telescope through an open window towards a sharply defined vertical object at a very large distance say an electric pole. Focus the telescope so that there is no parallax between the vertical cross-wire and the image of the distant object. Mark the position of the sliding tube against the edge of the outer tube so that the telescope may be set in the same position if disturbed accidentally.

(*iii*) Check the levelling of the telescope and the collimator. *If adjustment is necessary, do it very carefully.* **Ordinarily this levelling is not required.**

Release the clamping screw A_1 and raise the circular table so that its upper surface is *just* in level with the lower edge of the collimator. Place the spirit level on the circular table parallel to the line joining the two levelling screws P and Q. Adjust their position slowly till the bubble is *exactly* in the centre of the spirit level. Now place the spirit level *at right angles* to the line joining these screws and bring the bubble again in the centre *by adjusting only the third screw R*.

(*iv*) Prepare a sodium flame or use a sodium lamp, if available. Place a wooden screen with a circular hole in front of the source and direct the collimator towards it. Open the slit and adjust the position of the collimator so that the slit is uniformly and brightly illuminated. Turn the telescope and bring it in line with the collimator so that the image of the slit is visible and is in the centre of the field of view. Make the slit as narrow as possible. Fix the telescope in this position by the screw T_1.Adjust the distance of the slit from the lens of the collimator with the rack and pinion arrangement, till there is no parallax between the vertical cross-wire and the *sharp* image of the slit. If the image of the slit is not symmetrical with respect to the horizontal cross wire the levelling of the telescope or the collimator may be wrong.

Note. *If a second observer cannot see the image of the cross-wires and the slit in the telescope distinctly, he may have to move the eye-piece in or out to suit his sight. The images of the slit and the cross-wires will both be in focus without parallax again, if the focusing has been done properly.*

2. Find the vernier constant. In most of the instruments used in the laboratory, the circle is divided into 360 equal parts (degrees). Each degree is further divided into two parts. Thus the smallest scale division is equal to 0.5 degree or 30 minutes. The vernier scale has 30 equal divisions which coincide with 29 divisions of the main scale as shown in Fig. 16.7. The vernier constant in such a case is one minute.

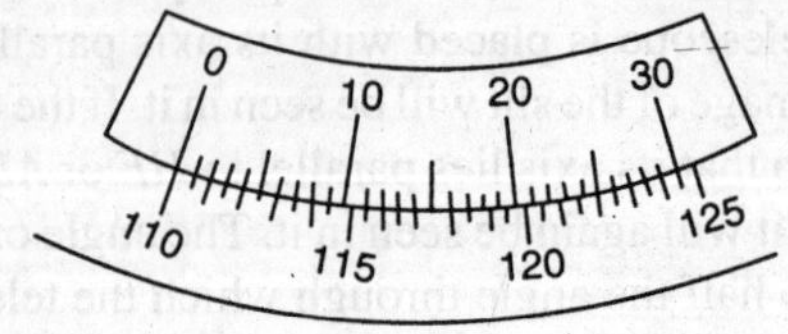

Fig. 16.7

3. **Optical levelling and adjustment of the prism.** Clean the prism and place it on the prism table with its refracting edge *A* at the centre of the table. Turn it so that one of its polished faces is perpendicular to the lines drawn on the table and parallel to the line joining the two screws *P* and *Q*. Observe the image of these lines in the base and turn the prism slowly so that the image of the lines appears to be in continuation with the lines drawn on the table. The face *AB* is now exactly at right angles to the line joining the two screws *P* and *Q* as shown in Fig. 16.8. Rotate the table so that the refracting edge points towards the collimator and light falls on both the polished faces of the prism. Observe the reflected image on one side first with naked eye and then bring the telescope in position so that the image is seen in its field of view. See that this image is symmetrical with respect to the horizontal cross wire. If not, adjust one of the screws *P* or *Q*.

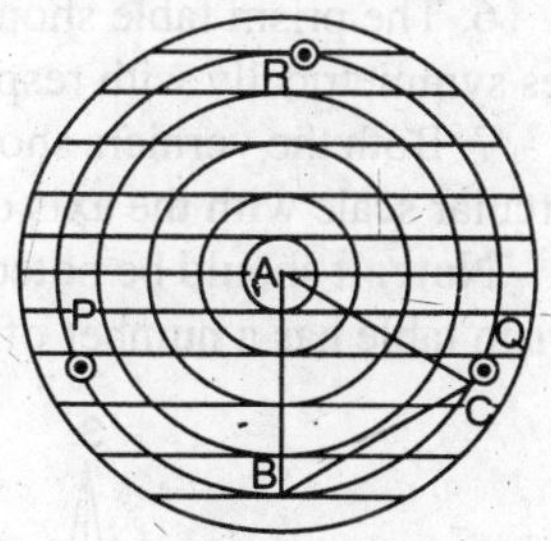

Fig. 16.8

Now look from the other face and place the telescope in position and see that the image is again symmetrical with respect to the horizontal cross-wire. If not, adjust with the screw *R*. Turn back the telescope into the first position and examine if the image is in the right position. Adjust if necessary.

These operations are repeated till the image as seen from either face is at the right height. The edge of the prism is then parallel to the axis of the spectrometer.

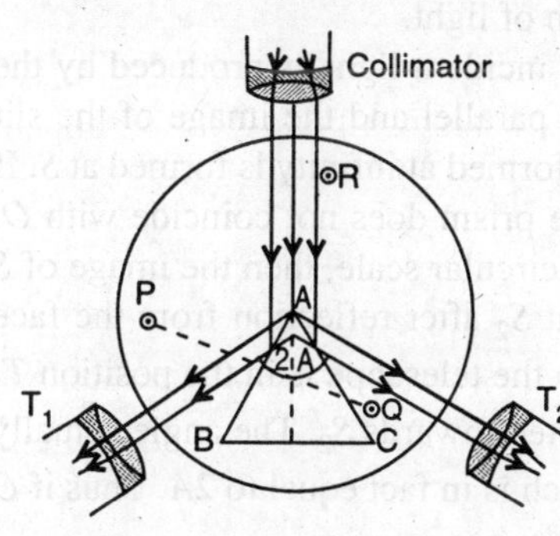

Fig. 16.9

4. Direct the telescope to receive light after reflection from the face *AB* so that the image of the slit is in the centre of the field of view. Clamp the telescope in this position and work its tangent screw to bring the centre of the image exactly on the vertical cross-wire. Note the reading on the scale against both the verniers V_1 and V_2.

5. Now direct the telescope to receive light after reflection from the face *AC*. Make the necessary adjustments and note the reading on both the vernier scales again.

6. Repeat the observation three times.

Observations.

Vernier constant =

Sl. No.	*Vernier V_1* Telescope reading		*Difference* θ	$A = \frac{\theta}{2}$	*Vernier V_2* Telescope reading		*Difference* θ	$A = \frac{\theta}{2}$
	1st face	*2nd face*			*1st face*	*2nd face*		
1								
2								
3								

Mean A =

Precautions. 1. The axis of the telescope, the collimator and the plane of the prism table should be horizontal.

2. The position of the eye piece should be adjusted so that the cross wires are clearly visible without any strain.

3. The telescope should be focused for infinity and the collimator should be adjusted to give a parallel beam of light.

4. The slit should be narrow.

5. The prism should be placed with its refracting edge at the centre of the prism table when finding the angle by rotating the telescope.

6. The prism table should be levelled so that the reflected image from both the reflecting faces lies symmetrically with respect to the horizontal cross-wire.

7. Both the verniers should be read to avoid error due to the non-coincidence of the centre of the circular scale with the axis of rotation of the telescope or the table.

Note. It should be noted that placing of the prism with the refracting edge A at the centre of the prism table has a number of advantages.

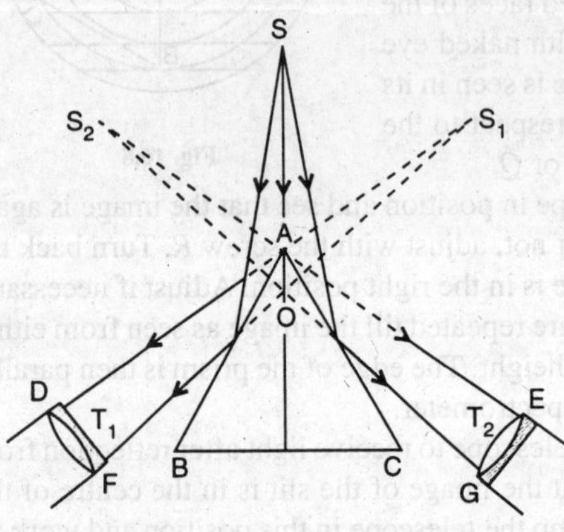

Fig. 16.10

(*i*) It makes the maximum amount of light available from the collimator to enter into the telescope, thereby increasing the brightness of the image.

(*ii*) It increases the resolving power of the instrument as the full length of the base of the prism is utilized.

(*iii*) It corrects for any want of parallelism in the incident beam of light.

Suppose the incident light as produced by the collimator is not parallel and the image of the slit instead of being formed at infinity is formed at S. If the edge A of the prism does not coincide with O, the centre of the circular scale, then the image of S will be formed at S_2 after reflection from the face AC and at S_1 after reflection from the face AB of the prism. When the telescope is in the position T_1 it is directed towards S_1 and when it is in the position T_2 it is directed towards S_2. The angle actually measured is FOG which is evidently less than the angle DAE which is in fact equal to $2A$. Thus if O coincides with A, this error will be eliminated.

Experiment 16.2. To find the angle of the prism by rotating the table method.

Apparatus. Same as in **Experiment 16.1.**

Theory. If the telescope is placed with its axis parallel to FG and light after reflection from the face AB is received by it the image of the slit will be seen. If the telescope is kept fixed and the prism table is rotated through an angle θ so that *the face AC occupies exactly the same position* as was occupied by AB, the image of the slit will again be seen. Evidently the point C has been rotated through an angle θ and the angle A of the prism is given by

$$A = 180 - \theta$$

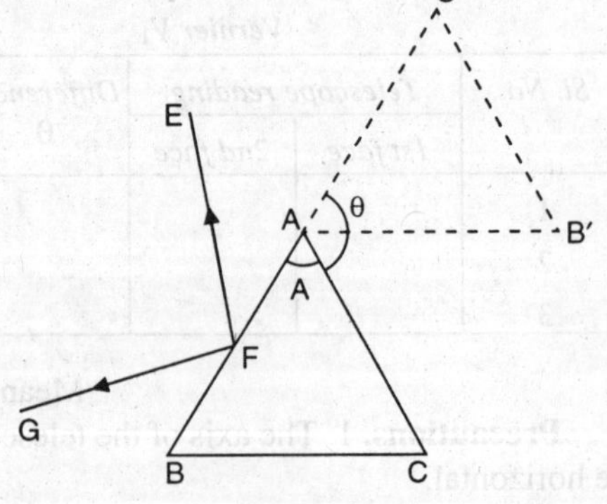

Fig. 16.11

Procedure. 1. Set the spectrometer as already explained.

2. Place the prism on the prism table with its *centre coinciding with the centre of the table and the reflecting face AB perpendicular to the line joining the screws P and Q.*

3. Turn the telescope so that it is approximately at right angle to the collimator axis. Clamp it in this position. Release the screw A_2 and rotate the table so that the image of the slit after reflection from the face AB falls on the cross-wire as shown in Fig. 16.12. Clamp the table in this position. Work the tangent screw A_3 of the table so that the image coincides exactly with the vertical cross-wire. Read the position of the table on both the vernier scales V_1 and V_2.

4. Rotate the table again so that light after reflection from the second polished face AC of the prism is received by the telescope. Make adjustments with the tangent screw so that the image of the slit again falls on the cross-wire as shown in Fig. 16.13. Read the position of the table on both the vernier scales again. If the scale has passed the 360° mark during rotation add 360 to the reading.

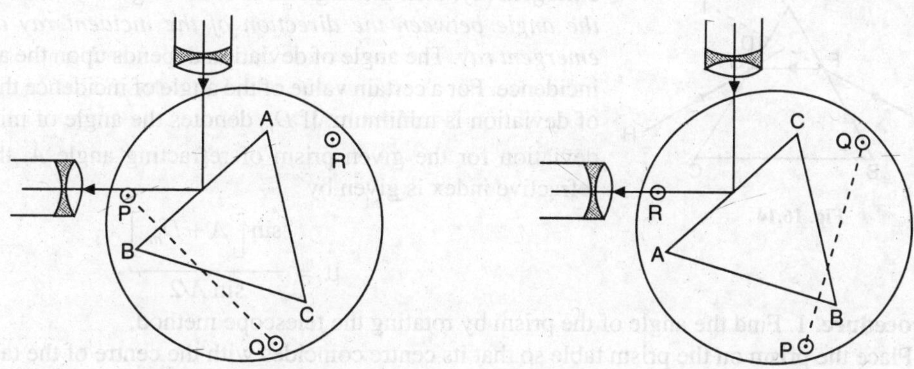

Fig. 16.12 **Fig. 16.13**

5. Repeat the observation three times.

Note. The table should be rotated in the anti-clockwise direction so that the face AC turns through an angle $\theta = (180 - A)$ in order to occupy the position previously occupied by the face AB.

Observations. Vernier constant =

	Vernier V_1				Vernier V_2			
Sl. No.	*Reading of the table*		*Difference* θ	$A = 180 - \theta$	*Reading of the table*		*Difference* θ	$A = 180 - \theta$
	Face AB	*Face AC*			*Face AB*	*Face AC*		
1								
2								
3								

Mean value of A =

Precautions. 1. Same as in **Experiment 16.1.**

2. The prism should be placed with its centre coinciding with the centre of the table.

16.4. Refractive index. When a ray of light travels from one medium into another it suffers refraction. If i is the angle of incidence in the first medium and r the corresponding angle of refraction in the second medium, then the refractive index μ of the second medium with respect to the first medium is given by.

$$\mu = \frac{\sin i}{\sin r}$$

According to the wave theory of light

$$\mu = \frac{\text{Velocity of light in vacuum}}{\text{Velocity of light in the medium}}$$

The refractive index of a medium also varies with the wavelength of light. It is greater for violet rays of shorter wavelength and smaller for red rays of longer wavelength.

The refractive index is usually denoted for the yellow light or mean wavelength of white light.

Experiment 16.3. To find the refractive index of the material of given prism using a spectrometer.

Apparatus. Same as in **Experiment 16.1.**

Theory. When a ray of light passes through a prism it suffers refraction as shown in Fig. 16.14.

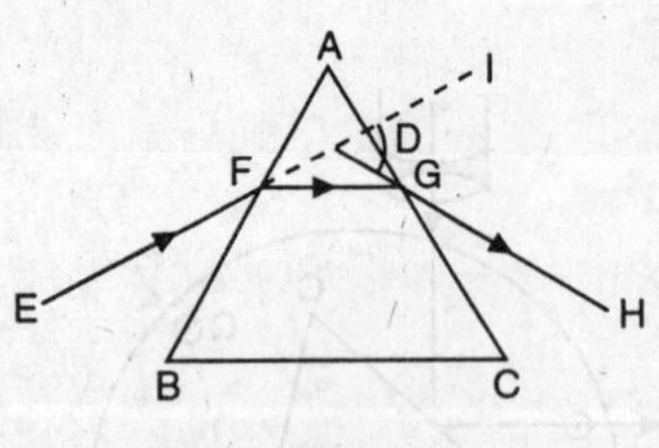

Fig. 16.14

If EF is the incident ray, FG the refracted ray and GH the emergent ray, then the angle IDH is the angle of deviation. *It is the angle between the direction of the incident ray and the emergent ray.* The angle of deviation depends upon the angle of incidence. For a certain value of the angle of incidence the angle of deviation is minimum. If D_m denotes the angle of minimum deviation for the given prism of refracting angle A, then the refractive index is given by

$$\mu = \frac{\sin\left[\frac{A + D_m}{2}\right]}{\sin A/2}.$$

Procedure. 1. Find the angle of the prism by rotating the telescope method.

2. Place the prism on the prism table so that its centre coincides with the centre of the table and one of the reflecting faces say AB is perpendicular to the line joining the two screws P and Q as shown in Fig. 16.15. Rotate the table so that light from the collimator is incident on the reflecting face AB and after refraction passes out of the face AC as shown.

3. Look through the face AC and locate the position of the refracted image with the naked eye. Bring the telescope in this position. Adjust its position till the image of the slit falls on the cross-wire. *Now turn the prism table in such a direction that the telescope has to be moved towards the line of the collimator axis in order to keep the image in the centre of the field of view.* As the prism table is rotated in this way the angle of deviation decreases. When the position of minimum deviation is reached the image becomes stationary and further rotation of the table makes it move in the backward direction. The direction of the telescope is finally adjusted till the image is on the cross wire in the stationary position from which it turns back in whichever direction the prism is rotated. Clamp the telescope and the table. The minimum deviation position is correct if the image remains stationary when the tangent screw of the table is given one rotation either way. Note the reading on both the vernier scales V_1 and V_2.

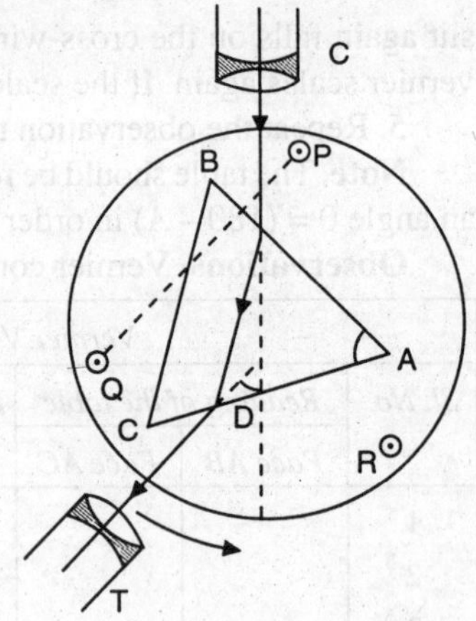

Fig. 16.15

4. Keep the table fixed and remove the prism gently. Turn the telescope and bring it in line with the collimator and clamp it. Work the tangent screw of the telescope so that the image of the slit falls exactly on the vertical cross-wire. Take the reading on both the vernier scales.

5. Repeat the experiment and take three such observations.

Observations. Vernier constant =

Wave length of light used =

Angle of the prism.

Sl. No.	*Vernier V_1*				*Vernier V_2*			
	Telescope reading		*Difference*	$A = \frac{\theta}{2}$	*Telescope reading*		*Difference*	$A = \frac{\theta}{2}$
	1st face	*2nd face*	θ		*1st face*	*2nd face*	θ	
1								
2								
3								

Mean A =

Angle of minimum deviation

Sl. No.	*Vernier V_1*			*Vernier V_2*		
	Telescope reading		*Difference D_m*	*Telescope reading*		*Difference D_m*
	Min. Dev.	*Direct*		*Min. Dev.*	*Direct*	
1						
2						
3						

Mean D_m =

$$\text{Refractive index for sodium light } \mu = \frac{\sin\left(\dfrac{A + D_m}{2}\right)}{\sin A/2}$$

Precautions. 1. Same as in **Experiment 16.1.**

2. The position of minimum deviation should be noted when the image of the slit remains stationary even when the tangent screw of the table is given one rotation either way.

Experiment. 16.4. To find the refractive index of a transparent liquid using a hollow prism and spectrometer.

Apparatus. Same as in **Expt. 16.1** but a hollow glass prism in place of a solid glass prism.

Theory. Same as in **Expt. 16.3.**

Procedure 1. Clean the hollow glass prism and fill it with the given transparent liquid. Wipe it dry from outside and see that there are no air bubbles.

2. Find the refracting angle of the prism and the angle of minimum deviation.

NOTE. *While determining the refracting angle of the hollow prism two or more images of the slit are generally seen. For purposes of measurement of angle of prism always use the brightest image.*

Observations, Calculations and **Precautions.** Same as in **Expt. 16.3.**

Experiment 16.5. To plot a graph between the angle of incidence and the corresponding angle of deviation and to find the refractive index of the material of the prism.

Apparatus. Same as in **Experiment 16.1.**

Procedure. Set the spectrometer and find the angle of the prism by the rotating telescope method.

2. Place the prism with its centre coinciding with the centre of the prism table and set it approximately in the position of minimum deviation, so that light falls on the face AB and emerges out from the face AC as shown in Fig. 16.16. Clamp the table.

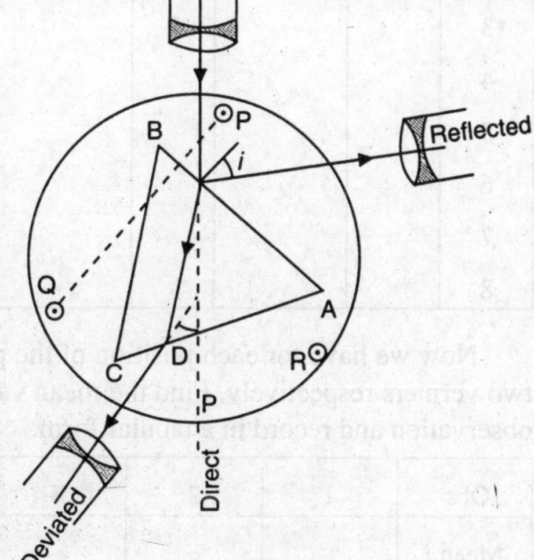

Fig. 16.16

3. Turn the telescope to receive the emergent light and adjust its position, so that the image of the slit is formed on the cross-wire. Clamp the telescope and note its reading on both the verniers V_1 and V_2.

4. Now turn the telescope to receive the reflected light from the face AB as shown in Fig. 16.16. Adjust the position of telescope till the image of the slit falls on the vertical cross-wire. Clamp it and note the reading on both the verniers.

5. Bring the telescope back to receive the deviated ray. Turn the prism table **without disturbing the circular scale** in the clockwise direction so that the deviated ray is displaced by about one degree.

Adjust the telescope so that the image is formed on the vertical cross-wire again. Note the reading on both the vernier scales.

6. Turn the telescope again to receive the reflected light from the face *AB*. Make the necessary adjustments and note the reading on both the vernier scales.

7. Turn the table in the clockwise direction again and take three or four observations as explained.

8. Rotate the prism table back to its starting position so that the prism is again in the minimum deviation position approximately. Turn the table in the anti-clockwise direction and take four observations.

9. Remove the prism and turn the telescope so that the direct light is received and the image of slit falls on the vertical cross-wire. Note the reading of both the verniers.

Note. *If θ is the angle between the direct and the reflected rays, then the angle of incidence is given by*

$$i = \frac{180 - \theta}{2}$$

The angle of deviation *D* is given by the difference between the reading of deviated ray and direct ray.

10. Plot a graph between *i* and *D* and from the graph find the angle of minimum deviation.

Observations. Vernier constant =

Record observation for the angle of prism, as in **Expt. 16.1.**

	Vernier V_1 *Direct reading =*					*Vernier V_2* *Direct reading =*				
Sl. No.	*Telescope reading*		θ	$i = \frac{180-\theta}{2}$	*D*	*Telescope reading*		θ	$i = \frac{180-\theta}{2}$	*D*
	Dev.	*Refl.*				*Dev.*	*Refl.*			
1										
2										
3										
4										
5										
6										
7										
8										

Now we have for each position of the prism two values of *i* and two values of *D* observed with two verniers respectively. Find the mean value of *i* and the corresponding mean value of *D* for each observation and record in a tabular form.

Obs	1	2	3	4	5	6	7	8
Mean *i*								
Mean *D*								

Graph between *i* and *D* for a prism

Plot a graph taking angle of incidence *i* along the *X*- axis and angle of deviation *D* along the *Y*-axis. A typical graph is shown in Fig. 16.17.

Graph between i and D for a prism

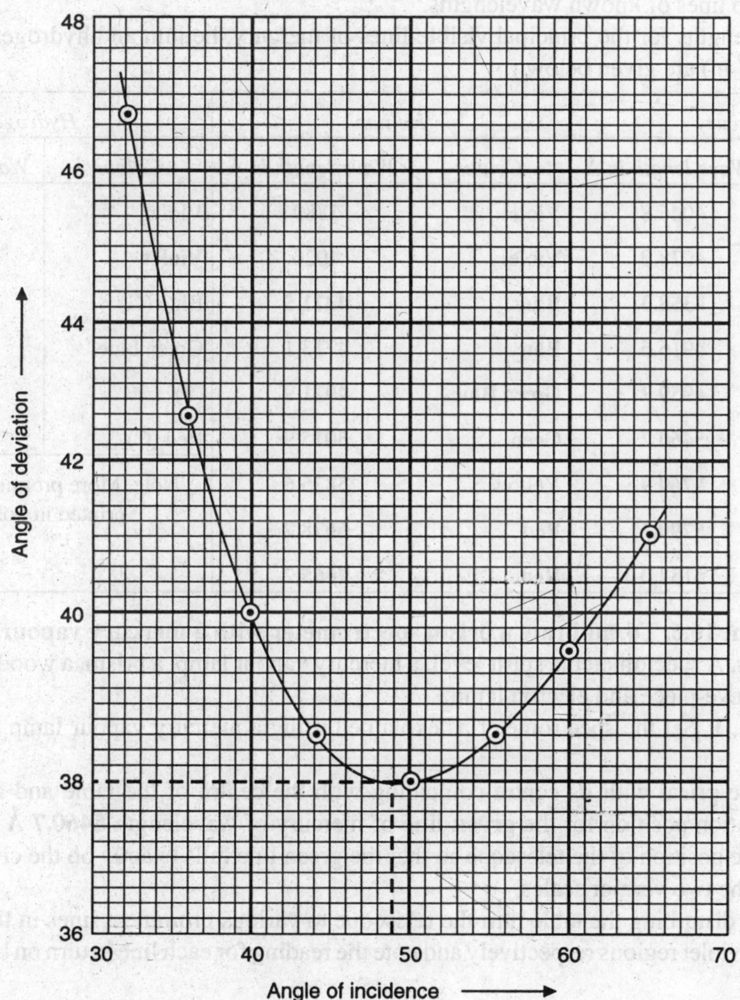

Fig. 16.17

Angle of minimum deviation from the graph D_m =

$$\therefore \qquad \mu = \frac{\sin\left(\frac{A + D_m}{2}\right)}{\frac{A}{2}}$$

Precautions. 1. Same as in **Experiment 16.1.**

2. The circular scale table should be kept fixed during the experiment. *Only the prism table should be rotated to change the observation.*

16.5. Calibration of a prism spectrometer. For a given prism the deviation D of a ray is inversely proportional to the square of the wavelength λ. Hence a graph between D and $\frac{1}{\lambda^2}$ is a straight line. From such a graph the unknown wavelength can be determined. For purposes of calibration the sources commonly used are :

(*i*) Mercury vapour lamp, (*ii*) Helium tube, (*iii*) Hydrogen tube etc. These sources give a large number of sharp lines of known wavelengths.

The wavelengths for the principal visible lines of mercury, helium and hydrogen in Angstrom units (Å = 10^{-10} m) are given below :

Mercury		*Helium*		*Hydrogen*	
Colour	*Wave-length in Å*	*Colour*	*Wave-length in Å*	*Colour*	*Wave-length in Å*
Violet	4046.8	Violet	3889	Violet	3970
Violet	4078.0	Violet	4026	**Violet**	**4102**
Blue	**4358.3**	**Blue**	**4371.5**	Blue (G)	4340
Blue green	4916.6	Blue	4713.1	Green blue	
Green	4959.7	Green Blue	4921.9	(F)	4861
Green	**5460.7**	Geen	5015.9	Red (C)	6563
Yellow	**5769.9**	Yellow	5875.6	Note. More prominent lines are printed in bold type.	
Yellow	**5790.7**	Red	6678.2		
Orange	6152.0	**Red**	**7065.2**		

Experiment 16.6. To calibrate a prism spectrometer with a mercury vapour lamp.

Apparatus. A spectometer, a spirit level, a mercury vapour lamp, a prism, a wooden screen with an aperture, an eye-piece and electric lamp.

Procedure. 1. Set the spectrometer as explained using a mercury vapour lamp in place of the sodium lamp.

2. Place the prism with its centre coinciding with the centre of the table and adjust it in the minimum deviation position for the **green line** of mercury of wavelength **5460.7 Å** and clamp the table. Adjust the position of the telescope so that the green line falls exactly on the cross-wire. Note the reading of the two vernier scales.

3. Without disturbing the table turn the telescope to various prominent lines in the red, yellow, green, blue and violet regions respectively and note the reading for each line in turn on both the vernier scales.

4. Remove the prism without disturbing the position of the table and note the direct reading on both the vernier scales.

Observations. Vernier constant =

Direct reading with vernier V_1 =

Direct reading with vernier V_2 =

Sl. No.	*Colour*	*Wave-length* λ	$\frac{1}{\lambda^2}$	*Telescope reading*		*Deviation with*		
				Vernier V_1	*Vernier* V_2	V_1	V_2	Mean D
1								
2								
3								
4								

5. Plot a graph between $\frac{1}{\lambda^2}$ and the corresponding mean value of the deviation D.

Precautions. 1. Same as in **Experiment 16.1.**

2. The table should be kept fixed after setting the prism in the minimum deviation position for the mean green wavelength in the case of mercury light.

Exercise 1. *Calibrate the given spectrometer with a helium tube.*

Hints. Set the prism in the minimum deviation position for the green wavelength and proceed as explained above using a helium discharge tube.

16.6. Dispersive power. When a ray of white light passes through a prism it splits up into its constitutent colours.

This phenomenon of splitting up of white light into its constitutent colours is called dispersion.

The refractive index for the material of a prism is given by

$$\mu = \frac{\sin\left(\frac{A + D_m}{2}\right)}{\sin\frac{A}{2}}$$

For a small angled prism the angle of deviation is also small and almost constant. Therefore, for a small angled prism

$$\mu = \frac{\frac{A+D}{2}}{\frac{A}{2}} \qquad [\because \ \sin\theta = \theta \text{ when } \theta \text{ is small}]$$

or $$D = A(\mu - 1)$$

As refractive index for light of red colour is smaller than the refractive index for light of violet colour, the *angle of deviation* for red colour is the least and the deviation for violet colour is maximum. For other colours of the spectrum the deviation lies in between.

The total angle through which the colours of a spectrum are spread is a measure of dispersion. *Hence angular dispersion is defined as the difference between the deviations produced in the extreme colours of the spectrum.*

If D_v and D_r are the deviations for violet and red colours respectively, then

Angular dispersion $\alpha = D_v - D_r = (\mu_v - 1)A - (\mu_r - 1)A$

$= (\mu_v - \mu_r)A$

The mean deviation is the deviation produced for the mean colour of the spectrum. The mean colour is generally taken as the yellow colour and the mean refractive index as the refractive index for yellow light μ_y or the average of the refractive indices of the two extreme colours μ_v and μ_r.

$\therefore$ Mean deviation $D = (\mu_y - 1)A = (\mu - 1)A$

Where $\mu_y = \mu = \frac{\mu_v + \mu_r}{2}$

The **despersive power** *of the material of a prism is defined as the ratio of angular dispersion to mean deviation.*

$\therefore$ Dispersive power $\omega = \frac{(\mu_v - \mu_r)A}{(\mu - 1)A} = \frac{\mu_v - \mu_r}{\mu - 1}$

It is clear from the above expression that dispersive power depends upon the nature of the material of the prism and is independent of the angle of the prism.

Experiment 16.7. To find the dispersive power of a given prism using a hydrogen discharge tube.

Apparatus. A spectrometer, a spirit level, a glass prism, an eye-piece, a hydrogen discharge tube, a sodium lamp etc.

Theory. The dispersive power ω for the material of a prism is given by the relation

$$\text{Dispersive power} \quad \omega = \frac{\mu_v - \mu_r}{\mu - 1}$$

where μ_v is the refractive index for violet light, μ_r for red light and μ is the mean refractive index For a source like the hydrogen discharge tube, dispersive power is defined by the relation.

$$\omega = \frac{\mu_F - \mu_C}{\mu_D - 1}$$

where μ_F, μ_C and μ_D are the refractive indices for *F* (4861 Å), and *C* (6563 Å) lines of hydrogen spectrum and *D* (5893 Å) line of sodium respectively.

Procedure 1. Set the spectrometer. using sodium lamp and find the angle of the prism.

2. Find the angle of minimum deviation for the mean wavelength of *D* line of sodium light.

3. Using hydrogen discharge tube find the angle of minimum deviation for the *F* and *C* lines.

4. Calculate the refractive index corresponding to each wavelength using the relation

$$\mu = \frac{\sin\left(\frac{A + D_m}{2}\right)}{\sin\left(\frac{A}{2}\right)}$$

Observations. Record observations separately for wavelength of *D, F* and *C* lines as in **Expt. 16.3.**

Calculations. Knowing the refractive index for *F, C* and *D* lines, find the dispersive power for the material of the given prism

$$\omega = \frac{\mu_F - \mu_C}{\mu_D - 1}$$

Precautions. Same as in **Expt. 16.3.**

Exercise 1. *Find the dispersive power of the given transparent liquid using a spectrometer.*

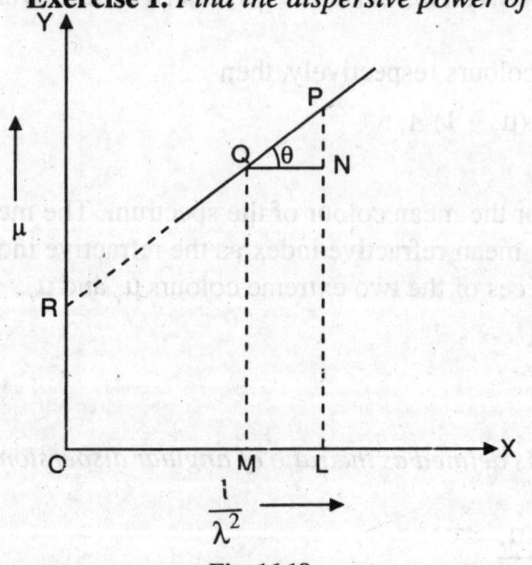

Fig. 16.18

Hints. Using a hollow prism find the refractive index of the given liquid for the *F, C* and *D* lines and hence calculate the dispersive power of the liquid.

Exercise 2. *Using mercury lamp find the dispersive power of the material of the prism.*

Hint. Find the refractive index for violet (λ = 4358.3 Å), yellow (λ = 5790.7 Å) and green (λ = 5460.7 Å) lines, then

$$\text{Dispersive power} \quad \omega = \frac{\mu_v - \mu_y}{\mu_g - 1}$$

16.7. Cauchy's contants. The refractive index of a material depends upon wavelength. Cauchy gave an empirical relation between μ and λ which is given as below :

$$\mu = A + \frac{B}{\lambda^2}$$

where *A* and *B* are known as *Cauchy's constants*. This relation is of the type $y = mx = c$. Hence a graph between μ *and* $\frac{1}{\lambda^2}$ is a straight line as shown in Fig. 16.18. The value of Cauchy's constant *A* is given by the *intercept* on the *Y*-axis and the value of *B* is given by the slope of the line.

Experiment 16.8. To find the value of Cauchy's constants A and B for the material of the given prism using a mercury vapour lamp.

Apparatus. Same as in **Expt. 16.3** with extra-dense glass prism and mercury vapour lamp.

Theory. According to Cauchy's relation between refractive index μ and wave length λ, we have

$$\mu = A + \frac{B}{\lambda^2}$$

Where *A* and *B* are Cauchy's constants. If μ_1 and μ_2 are the refractive indices corresponding to wavelength λ_1 and λ_2 respectively, then

$$\mu_1 = A + \frac{B}{\lambda_1^2} \quad ...(i)$$

$$\mu_2 = A + \frac{B}{\lambda_2^2} \quad ...(ii)$$

From (*i*) and (*ii*), we have

$$\mu_1 - \mu_2 = B\left[\frac{l}{\lambda_1^2} - \frac{l}{\lambda_2^2}\right] = B\,\frac{\lambda_2^2 - \lambda_1^2}{\lambda_1^2\lambda_2^2}$$

or $$B = \frac{(\mu_1 - \mu_2)\,\lambda_1^2\,\lambda_2^2}{\lambda_2^2 - \lambda_1^2}$$

Substituting the value of *B* in relation (*i*), we have

$$\mu_1 = A + \frac{(\mu_1 - \mu_2)\,\lambda_2^2}{\lambda_2^2 - \lambda_1^2}$$

or $$A = \mu_1 - \frac{(\mu_1 - \mu_2)\,\lambda_2^2}{(\lambda_2^2 - \lambda_1^2)}$$

Similarly $$A = \mu_2 - \frac{(\mu_1 - \mu_2)\,\lambda_1^2}{(\lambda_2^2 - \lambda_1^2)}$$

Procedure 1. Using sodium vapour lamp set the spectrometer and find the angle of the prism.

2. Replace sodium vapour lamp by mercury lamp and find the angle of minimum deviation for violet line of wave length 4358.3 Å and green line of wavelength 5460.7 Å.

3. Find the refractive index μ_1 and μ_2 for the violet and green wavelengths respectively.

Observations. Record observations for the angle of prism and for the refractive index of two lines separaly as in **Expt. 16.3.**

Calculations.

$$A = \mu_1 - \frac{(\mu_1 - \mu_2)\,\lambda_2^2}{\lambda_2^2 - \lambda_1^2} = \mu_2 - \frac{(\mu_1 - \mu_2)\,\lambda_1^2}{\lambda_2^2 - \lambda_1^2}$$

$$B = \frac{(\mu_1 - \mu_2)\,\lambda_1^2\lambda_2^2}{\lambda_2^2 - \lambda_1^2}$$

Precautions : Same as in **Expts. 16.1 and 16.3.**

Experiment 16.9. Plot a graph between μ and $\frac{1}{\lambda^2}$ using a mercury vapour lamp and hence find the values of Cauchy's constants *A* and *B*.

Apparatus. Same as in **Expt. 16.6.**

Procedure. 1. Using sodium vapour lamp set the spectrometer and find the angle of the prism.

2. Replace sodium vapour lamp by mercury vapour lamp and find the angle of minimum deviation for the **violet** line of wave length 4046.8Å.

3. Similarly find the angle of minimum deviation for the **blue** line (λ = 4358.3Å), **blue green** line (λ = 4916.6Å), **green** line (λ = 5460.7Å), **yellow** lines (λ = 5769.9Å) and the **orange** line (λ = 6152Å).

Observations Record the observations for the angle of prism as in **Expt. 16.1.**

Mean angle of prism A =

No.	Colour of spectral line	Wavelength of line $\lambda = \times 10^{-10}$ m	$\frac{1}{\lambda^2}$ m^{-2}	Vernier V_1			Vernier V_2			Mean D_m	$\mu = \frac{\sin \frac{A+D_m}{2}}{\sin \frac{A}{2}}$
				Telescope reading							
				Min. deviation position	Direct position	Min. dev. D_m	Min. deviation position	Direct position	Min. dev. D_m		
1	Violet	4046.8									
2	Blue	4358.3									
3	Blue green	4916.6									
4.	Green	5460.7									
5.	Yellow	5769.9									
6	Orange	6152.0									

Graph. Plot graph taking $\frac{1}{\lambda^2}$ (in m^{-2}) along X-axis and μ along Y-axis. The graph is a straight line as shown in Fig. 16.18. Produce the line PQ so as to meet the Y-axis at a point R. Measure the intercept OR. This gives the value of Cauchy's constant A.

Draw PL and QM perpendicular to the X-axis and QN perpendicular to PL. Measure PN and QN. The slope of the line $\tan\theta = \frac{PN}{QN}$ gives the value of second Cauchy's constant B.

Record OR =

$\therefore$ Value of Cauchy's constant A for the material of the given prism =

PN = QN = $\frac{PN}{QN} = \tan\theta$ = slope of the line =

$\therefore$ Value of Cauchy's constant B for the material of the given prism =

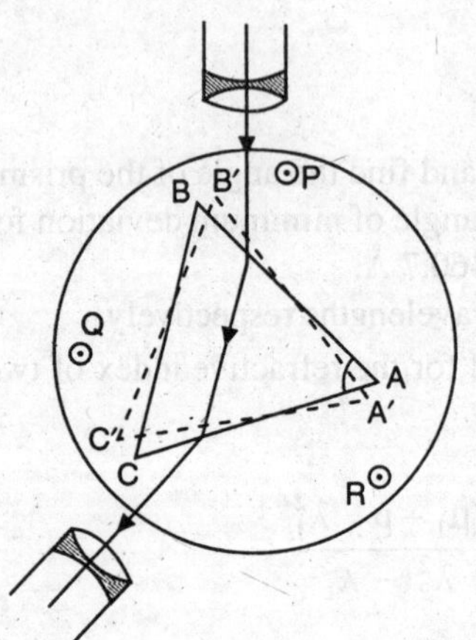

Fig. 16.19

16.8. Schuster's method of setting a spectrometer for parallel rays. This method is suitable when a distant object for focussing is not available.

Level the spectrometer table with a spirit level. Illuminate the slit of the collimator and place a prism on the table. Set the prism in the minimum deviation position. Now rotate the prism table slightly so that the refracting edge of the prism moves towards the telescope as shown by dotted lines $A'B'C'$ in Fig. 16.19.

The deviation will slightly increase. Now turn the telescope so that the image falls on the cross-wire. Fix the telescope in this position and focus it so that a sharp image of the slit is seen in the field of view.

Now rotate the prism table in the opposite direction so that the refracting edge of the prism moves towards the collimator. The image of the slit will move, come to the minimum deviation position, then will move backwards and again coincide with the cross-wire. Stop rotating the prism table. The position of the prism is now shown by thick lines *ABC*. Now bring the slit sharply in focus by adjusting the collimator.

Repeat this process of alternately focusing the telescope and the collimator 3 or 4 times, till the image of the slit remains sharply in focus for both the positions of the prism. When this is so, the collimator produces a parallel beam of light and the telescope is set to receive it.

16.9 Resolving power of a prism spectrometer. *It is defined as the ratio of the wavelength λ to the smallest difference in wavelength d λ of the next line, that can just be distinguished as separate by the prism spectrometer.*

$$\therefore \qquad \text{Resolving power} = \frac{\lambda}{d\lambda}.$$

For a prism of base of thickness *t*, the resolving power is given by

$$\frac{\lambda}{d\lambda} = t\,\frac{d\mu}{d\lambda}$$

where $d\mu$ is the change in refractive index for a small change in wavelength $d\lambda$.

16.10. Experimental determination of resolving power. To find the resolving power we first find $\frac{d\mu}{d\lambda}$ using Cauchy's formula

$$\mu = A + \frac{B}{\lambda^2}$$

$$\therefore \qquad \frac{d\mu}{d\lambda} = -\frac{2B}{\lambda^3}$$

The value of $B = \dfrac{(\mu_1 - \mu_2)\,\lambda_1^2\,\lambda_2^2}{\lambda_2^2 - \lambda_1^2}$

where μ_1 and μ_2 are the refractive indices of the material of the prism corresponding to wavelengths λ_1 and λ_2 respectively and λ is the mean wavelength.

As the whole thickness *t* of the base of the prism is not generally used, the effective thickness of the prism used is calculated as follows.

The prism is placed in the position of minimum deviation for wavelength λ_1 and the angle of minimum deviation D_m is determined. An adjustable slit is placed in front of the telescope objective and its width is gradually reduced so that the two lines corresponding to the wavelengths λ_1 and λ_2 are *just* resolved.

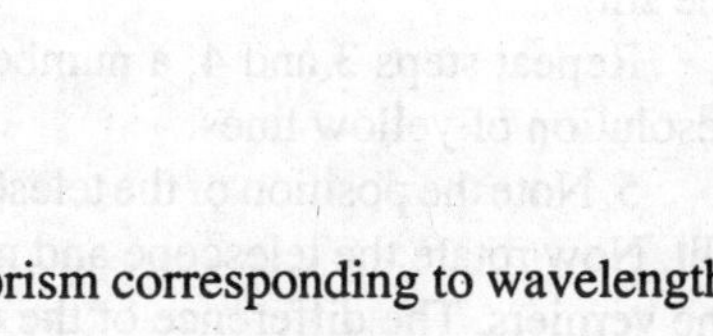

Fig. 16.20

If *a* is the width of the slit, *i* the angle of incidence, then

$$AD = AF = \frac{a}{\cos i}$$

and $$t = DF = DH + HF = AD\,\sin\frac{A}{2} + AF\,\sin\frac{A}{2}$$

or $$t = \frac{2\,a\,\sin\frac{A}{2}}{\cos i}$$

Now $i + e = 2i = A + D_m$

$\therefore$ $i = \dfrac{A + D_m}{2}$

Hence $t = \dfrac{2\,a \sin A/2}{\cos\left(\dfrac{A+D_m}{2}\right)}$

Experiment 16.10. To determine the resolving power of a prism

Apparatus. Same as in **Expt. 16.9** and an adjustable micrometer slit.

Procedure 1. Find the value of Cauchy's constant B using violet line of wavelength 4358.3 Å and green line of wavelength 5460.7 Å as explained in **Expt. 16.9.**

2. Mount the adjustable micrometer slit on the objective of the telescope keeping it wide open so that full aperture of the objective of the telescope may be used.

3. Set the spectrometer in the minimum deviation position for yellow doublet (of wavelength 5769.9 Å and 5790.7 Å) of mercury.

3. Slowly decrease the width of the slit with the help of the micrometer screw till the two yellow lines **just** appear as separate. On slightly further reducing the width of the slit the two lines will just merge into each other. Note the reading of the micrometer. Continue to turn the screw till the light is totally cut off. Again note the reading of the micrometer. The difference of the two readings gives the width of the slit.

4. Now turn the micrometer screw in the backward direction and note the reading as the light **just** appears. Continue turning the screw till the two lines appear to be **just** resolved and note the reading of the micrometer. The difference of these two readings also gives the value of the width of the slit.

Repeat steps 3 and 4, a number of times to find the mean value of the width of the slit for just resolution of yellow lines.

5. Note the position of the telescope on both the verniers and remove the prism and the micrometer slit. Now rotate the telescope and note the direct reading of the light from the collimator slit on both the verniers. The difference of the two readings gives the angle of minimum deviation.

Observations. Angle of prism A =

For Cauchy's constant B. Record observations as in **Expt.** 16.3 for the wavelengths 4358.3 Å and 5460.7 Å separately.

For width of the slit

Sl. No.	*Micrometer Reading*						*Mean width of operture for just resolution (a) cm*
	Aperture Decreasing			*Aperture Increasing*			
	Lines just resolved x	*Intensity just zero y*	$a = x - y$	*Light just appear x′*	*Lines just resolved y′*	$a = x' - y'$	
1							
2							
3							
4							
5							

Mean a = cm

Angle of minimum deviation for yellow doublet

D_m =

Calculations.

Effective thickness of the prism $t = \dfrac{2\,a \sin A/2}{\cos\left(\dfrac{A + D_m}{2}\right)} = \text{cm}$

Mean wavelength of yellow doublet $\lambda = \text{Å} = \text{cm}$

Resolving power $= \dfrac{\lambda}{d\lambda} = \dfrac{2B}{\lambda^3}\, t \quad =$

Precautions 1. Observe all precautions for use of spectrometer and determination of refractive index.

2. The slit should be properly adjusted for just resolution of two yellow lines of mercury spectrum while decreasing the width and again while increasing the width.

3. The angle of minimum deviation should be determined for the mean wavelength of the yellow doublet of mercury.

16.11. Ordinary and extra-ordinary refractive indices.

The refractive index of a substance is defined as the ratio of the velocity of light in vacuum to the velocity of light in the medium. In the case of a doubly refracting crystal the ray is split-up into *O* and *E* components which travel with different velocities. The velocities of *O* and *E* components are the same along the direction of optic axis but are different in a direction perpendicular to the direction of the optic axis. Hence there are two principal refractive indices one corresponding to the *O*-ray denoted by μ_o and the other corresponding to the *E*-ray denoted by μ_e in a direction perpendicular to the direction of optic axis.

The velocity of *O*-ray is constant in all directions and μ_o is, therefore, constant. Thus

$$\mu_o = \frac{\text{Velocity in vacuum}}{\text{Velocity of } O\text{–ray}}.$$

The velocity of *E*-ray is different in different directions and hence the refractive index is not constant.

In a *negative* crystal like calcite the velocity of *E*- ray is *maximum* in the direction perpendicular to the direction of optic axis and hence the refractive index has a *minimum* value. This refractive index is known as the **principal refractive index** and is denoted by μ_e.

$$\mu_e = \frac{\text{Velocity in vacuum}}{\text{Maximum normal velocity of } E\text{–ray}}.$$

The value of μ_o for sodium light for calcite is 1-658 and μ_e is 1.486.

In *positive crystals like quartz the velocity of* *E*-ray is *minimum* in a direction perpendicular to the direction of optic axis and hence the refractive index has a maximum value in this direction. This is known as the **principal refractive** index for *E*-ray and is denoted by μ_e.

$$\mu_e = \frac{\text{Velocity in vacuum}}{\text{Minimum normal velocity of } E\text{–ray}}.$$

The value of μ_o for sodium light for quartz is 1.544 and μ_e is 1.553.

Experiment 16.11. To determine the principal refractive indices for the *O*-ray and the *E*-ray when refraction takes place through calcite or quartz.

Apparatus. A spectrometer, a spirit level, a source of sodium light (sodium lamp), a calcite or quartz prism cut with its optic axis perpendicular to the edge and lying in the principal section or parallel to the edge and perpendicular to the principal section, an eye-piece, an electric lamp etc.

Theory. When a ray of light passes through a prism of uniaxial crystal cut with its optic axis perpendicular to the edge and lying in the principal section or parallel to the edge and perpendicular to the principal section *ABC* as shown in Fig. 16.21 (*a*) and (*b*) respectively, it is broken into *O* and *E* rays. If the prism is adjusted in the minimum deviation first for *E*-ray and then for *O*-ray, then

$$\mu_e = \frac{\sin\left(\frac{A+\delta_e}{2}\right)}{\sin A/2} \quad \text{and } \mu_o = \frac{\sin\left(\frac{A+\delta_o}{2}\right)}{\sin A/2}$$

where A is the angle of the prism and δ_e and δ_0 the corresponding angles of minimum deviation of the *E*-ray and the *O*-ray respectively.

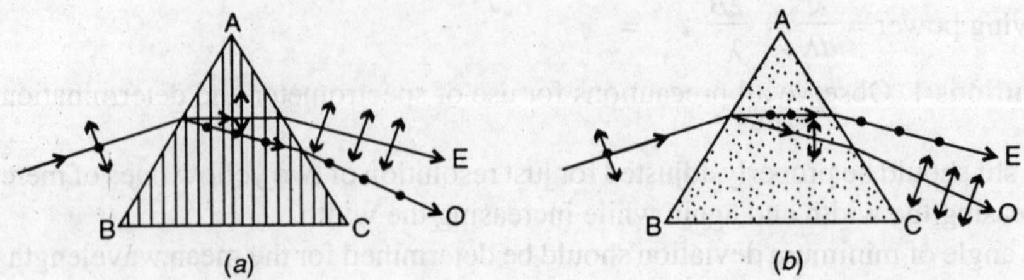

Fig. 16.21

Procedure. 1. Find the angle of the prism by rotating telescope method.

2. Place the prism on the prism table of the spectrometer as explained in **Expt. 16.3.**

3. Illuminate the slit with sodium light and allow the light to fall on the face *AB* so that it emerges out from the face *AC* after refraction. The light inside the prism travels at right angle to the optic axis and hence is broken up into the *E*-ray and the *O*-ray. Hence two refracted images of the slit are seen in the field of view of the telescope. Obtain the *E*-ray image on the cross-wire and make adjustments so that the prism is set in the minimum deviation position for the *E*-ray. Note the position of *E*-ray. Remove the prism and note down the position of direct ray and find the value of δ_e.

4. Repeat the experiment and find the value of δ_0.

5. Repeat the experiment and take at least three such observations each for *E*-ray and *O*-ray.

Observations. *Angle of the prism.*

Make a record of observations as in **Expt. 16.1.**

Angle of minimum deviation.

Sl. No.		*E-ray*		*Difference* δ_e	*O-ray*		*Difference* δ_o
		Telescope reading			*Telescope reading*		
		Min. Dev. position	*Direct*		*Min. Dev. position*	*Direct*	
1	V_1						
	V_2						
2	V_1						
	V_2						
3	V_1						
	V_2						

Mean δ_e = Mean δ_o =

Calculations. Angle of the prism A =

$$\mu_e = \frac{\sin\left(\dfrac{A+\delta_e}{2}\right)}{\sin A/2} =$$

$$\mu_o = \frac{\sin\left(\dfrac{A+\delta_o}{2}\right)}{\sin A/2} =$$

Precautions. Same as in **Expt. 16.1 and 16.3.**

Oral Questions

1. Spectrometer. (*i*) What is a pure spectrum? What are the conditions for obtaining it? (*ii*) What is a spectrometer and how the conditions for obtaining a pure spectrum are realized in it? (*iii*) What is the function of the collimator? (*iv*) Explain how will you level the spectrometer table? (*v*) Why do you set the telescope for parallel rays? (*vi*) How do you place the prism when determining its angle by (*a*) rotating table, and (*b*) rotating telescope and why? (*viii*) What is minimum deviation? How does a ray of light pass through the prism in this position? (*viii*) What is the relation between the angle of incidence and the angle of emergence when the prism is in the minimum deviation position?

2. Determination of refractive index (*i*) Define refractive index. (*ii*) How does refractive index change with wavelength of light? (*iii*) What is dispersive power of a prism? (*iv*) You are provided with three prisms of the following materials: (*a*) Crown glass (*b*) Dense flint and (*c*) Extra dense flint. Which will have the highest refractive index and which the highest dispersive power? (*v*) Does the angle of deviation depend upon the size and the refracting angle of the prism? If so, how?

3. Sources of light. (*i*) Which source of light are you using? Is it monochromatic? (*ii*) What is the wavelength of light given by the source? (*iii*) What is the construction of a sodium lamp? Are you working it on A.C. or D.C.? (*iv*) What are prominent wavelengths of mercury light? (*v*) Why is the light red in the beginning when you switch on the sodium lamp?

4. Dispersion. (*i*) What is meant by angular dispersion and dispersive power of a medium? (*ii*) What is the angle of deviation and on what factor does it depend? (*iii*) When is the angle of deviation minimum? What is the practical importance of the minimum deviation position of the prism?

5. Cauchy's constants (*i*) What is Cauchy's relation? (*ii*) What is the shape of graph between μ and $\frac{1'}{\lambda^2}$?

(*iii*) How will you find the values of Cauchy's constants from the graph?

6. Resolving power of prism (*i*) What is the resolving power of a prism ? (*ii*) Give the resolving power of a prism in terms of the thickness of the base. (*iii*) What is the effective thickness of the base of a prism? (*iv*) How does resolving power depend upon the thickness of the base ? (*v*) Distinguish between dispersive power and resolving power.

7. Ordinary and extra-ordinary refractive indices. (*i*) What is a doubly refracting crystal? (*ii*) What is optic axis? (*iii*) What do you mean by *O* and *E* component of light? (*iv*) What is the relation between velocities of *O* and *E* components along the direction of optic axis? (*v*) What is the relation between the velocities of *O* and *E* components in a direction perpendicular to the optic axis? (*vi*) What do you mean by a negative crystal and a positive crystal? (*vii*) In a doubly refracting crystal, how does the refractive index for ordinary ray u_o change with direction of light? (*viii*) In a doubly refracting crystal, how does the refractive index for extra-ordinary ray u_e change with direction of light? (*ix*) In what direction the velocity of *E*-ray is maximum in a negative crystal? (*x*) What do you mean by principal refractive index of a negative crystal for *E*-ray? (*xi*) Is its value maximum or minimum in a direction perpendicular to the direction of optic axis. (*xii*) What do you mean by principal refractive index of a positive crystal for *E*-ray? (*xiii*) Is it value maximum or minimum in a direction perpendicular to the direction of optic axis.

17

Interference

17.1. Interference. Light is a transverse wave-motion. When two waves of light of the same wavelength and amplitude travel along the same straight line, they re-inforce each other at points where the crest of one falls over the crest of the other or the trough of one falls over the trough of the other and destroy each other's effect if the crest of one falls over the trough of the other and *vice versa. This modification in the distribution of light intensity due to superposition of two waves is called* **interference.**

The phase relation between the waves at the time of emission rapidly changes with time not only in different sources but even in different parts of the same source. As a result of this the brightness and darkness at a point changes rapidly giving rise to general illumination. This is why it is not possible to produce interference with two independent sources. On the other hand two **coherent sources** *i.e.*, two sources which have been derived originally from the same source are required to produce stationary interference. Any change in phase at the time of emission will effect the two sources equally and there will exist a constant phase difference between the light emitted by them.

For *re-inforcement or for maximum intensity* the two waves should differ in path by an even multiple of half a wavelength. For *destructive interference or for minimum intensity* the path difference should be an odd multiple of half a wavelength.

17.2. Phase change at reflection. When a ray of light is reflected at the surface of a medium which is optically denser than the medium through which the ray is travelling, a change of phase equal to π or a path difference of $\frac{\lambda}{2}$ is introduced.

Hence if interference takes place between direct light coming from one of the coherent sources and light coming from the other coherent source after suffering one reflection at a denser medium, then the conditions for maximum and minimum intensity will be reversed. In such a case for maximum intensity the two waves should differ in path by an *odd* multiple of half a wavelength and for minimum intensity the path difference should be zero or an *even* multiple of half a wavelength.

17.3. Thin films. When light is incident on a thin film, a part of it is reflected at the upper surface and the rest is transmitted into the medium. It is again partly reflected at the lower surface and emerges out of the upper surface. The two beams of light, one reflected at the upper surface and the other reflected at the lower surface give rise to the phenomenon of interference. The path difference is given by

$$2\,\mu\,t\cos r$$

where μ is the refractive index of the medium of the film with respect to the surrounding, t its thickness and r the angle of refraction in the film.

In addition the ray of light reflected at the upper surface undergoes an additional path difference of $\frac{\lambda}{2}$ due to reflection if $\mu > 1$. On the other hand if $\mu < 1$, the ray of light reflected at the lower surface undergoes an additional path difference of $\frac{\lambda}{2}$. Hence

$$\text{Net path difference} = 2\,\mu\, t \cos r \pm \frac{\lambda}{2}$$

In the case of a thin air film enclosed between two glass surfaces and for normal incidence $i = r = 0$ and $\cos r = 1$

$$\therefore \quad \text{Path difference} = 2\,\mu\, t + \frac{\lambda}{2}$$

For maximum intensity

$$2\,\mu t + \frac{\lambda}{2} = n\,\lambda$$

or

$$2\,\mu t = (2n-1)\frac{\lambda}{2}$$

For minimum intensity

$$2\,\mu t = n\lambda$$

17.4. Newton's rings. Circular interference fringes produced by enclosing a very thin air film of varying thickness between the surface of a convex lens of a large radius of curvature and a plane glass plate are known as Newton's rings.

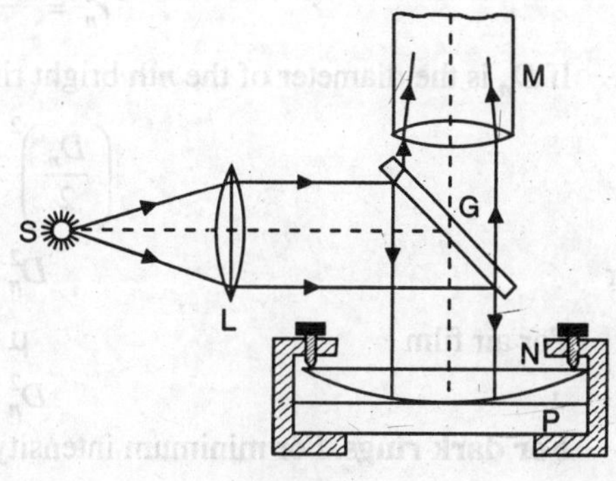

Fig. 17.1

In order to produce these fringes light from an extended monochromatic source S is rendered parallel by a convex lens L. It falls on the glass plate G inclined at an angle of 45° to the vertical and is reflected normally on to the lens N. An air film of gradually increasing thickness is thus enclosed between the lower surface of this lens and the glass plate P. The light reflected from the upper and the lower surfaces of the air film produces interference fringes. At the centre the lens is in contact with the glass plate and the thickness of the air film is zero. *The centre will be dark as a phase change of π is introduced due to reflection at the lower surface of the air film.* As we proceed outwards from the centre the thickness of the air film gradually increases being the same all along the circle with centre at the point of contact. Hence the fringes produced are concentric circles, and are localised in the air film. The fringes are viewed by means of a low power microscope M as shown in Fig. 17.1.

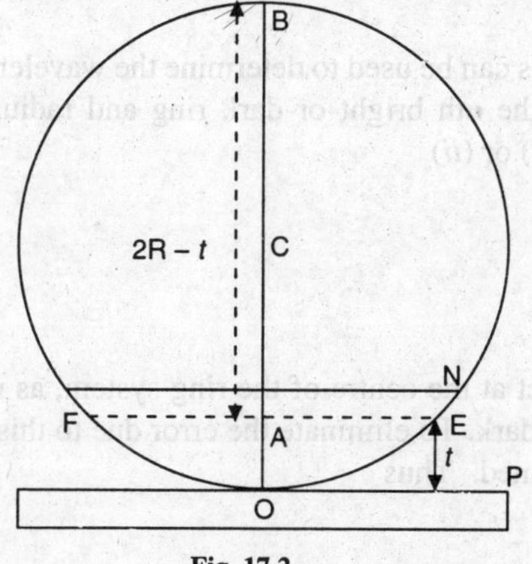

Fig. 17.2

17.5. Diameter of dark and bright rings. The plano convex lens N placed on the glass plate P is shown separately in Fig. 17.2. O is the point of contact of the lens N with plate P. Let R be the radius of the curved surface of the lens and t the thickness of the air film at a certain point E. From the property of a circle

$$AE \times AF = OA \times AB$$

But $\quad AE = AF = r$ the radius of the circular fringe passing through E.

and $$OA = PE = t$$

$\therefore$ $$r^2 = t\,(2R - t) = 2Rt - t^2$$

In actual practice R is quite large and t is very small. Therefore t^2 may be neglected as compared to $2Rt$.

$\therefore$ $$r^2 = 2\,Rt$$

or $$t = \frac{r^2}{2R}$$

For bright rings. For maximum intensity or for a bright ring the condition is

$$2\,\mu t = (2n - 1)\frac{\lambda}{2}$$

or $$2\,\mu\,\frac{r^2}{2R} = (2\,n - 1)\frac{\lambda}{2}$$

$\therefore$ $$r^2 = \frac{(2n - 1)\,\lambda\,R}{2\mu}$$

Writing r_n for r we get the radius of the nth bright ring

$$r_n^2 = \frac{(2\,n - 1)\,\lambda\,R}{2\,\mu}.$$

If D_n is the diameter of the nth bright ring, we have

$$\left(\frac{D_n}{2}\right)^2 = \frac{(2n - 1)\,\lambda\,R}{2\,\mu}$$

or $$D_n^2 = \frac{2\,(2n - 1)\,\lambda\,R}{\mu}.$$

For air film $$\mu = 1$$

$\therefore$ $$D_n^2 = 2\,(2\,n - 1)\,\lambda\,R \qquad ...(i)$$

For dark rings. For minimum intensity or for a dark ring, the condition is $2\,\mu t = n\,\lambda$

$\therefore$ $$2\mu\,\frac{r^2}{2R} = n\,\lambda \text{ or } r_n^2 = \frac{n\,\lambda\,R}{\mu} \text{ or } D_n^2 = \frac{4\,n\,\lambda\,R}{\mu}$$

For air film $\mu = 1$ $\therefore$ $D_n^2 = 4\,n\,\lambda\,R$ ...(ii)

17.6. Wavelength of light. Newton's rings apparatus can be used to determine the wavelength of monochromatic light by measuring the diameter of the nth bright or dark ring and radius of curvature of the surface of the lens by applying relation (*i*) or (*ii*)

From relation (*i*) $$\lambda = \frac{D_n^2}{2\,(2\,n - 1)\,R}$$

and from relation (*ii*) $$\lambda = \frac{D_n^2}{4\,nR}$$

If the lens and the glass plate are not quite in contact at the centre of the ring system, as may occur if the surfaces are not clean, the centre may not be dark. To eliminate the error due to this the diameter of any two dark rings say nth and mth is determined. Thus

$$D_n^2 = 4nR\lambda$$

and $$D_m^2 = 4mR\lambda$$

$\therefore$ $$D_n^2 - D_m^2 = 4(n - m)R\lambda$$

or $$\lambda = \frac{D_n^2 - D_m^2}{4(n-m)R}$$

This formula involves the difference of the squares of the diameters of two rings and is independent of the thickness of the air film at the so called point of contact.

The diameter of the *n*th bright ring is given by the relation

$$D_n^2 = 2\,(2n-1)\,R\,\lambda.$$

Hence the diameter of the ring depends upon the wavelength of light used. With white light the rings are coloured, violet being on the innermost side and red on the outermost edge as the wavelength of violet light is the least and that of red light the highest in the visible region. Each colour gives rise to its own system of rings and these coloured rings soon superimpose and overlap thereby resulting in almost uniform illumination after a few rings. It is for this reason that when Newton's rings are obtained with white light we get only a few rings. The central spot is dark.

When a mirror is used in place of the glass plate below the lens no interference rings are seen because the reflection is complete. The reflected and the transmitted system of rings get superimposed, the two systems being complementary give rise to uniform illumination.

Experiment 17.1. To find the wavelength of sodium light by measuring the diameters of Newton's rings.

Apparatus. A travelling microscope, a sodium lamp, Newton's rings apparatus consisting of an optically plane glass plate and a convex lens of about 100 centimetre focal length placed in a box having an optically plane glass plate inclined at an angle of 45°, a spherometer, a convex lens of short focal length, etc.

Theory. Circular interference fringes produced by enclosing a thin air film of varying thickness between the surface of a convex lens of large radius of curvature and a plane glass plate are known as Newton's rings. The wavelength of monochromatic light which produces these rings is given by

$$\lambda = \frac{D_n^2 - D_m^2}{4R\,(n-m)}$$

where R is the radius of curvature of the surface of the lens in contact with the glass plate and D_n and D_m the diameters of the *n*th and the *m*th dark or bright rings.

Procedure. 1. Level the microscope table and set the microscope tube in a vertical position. Find the vernier constant of the horizontal scale.

2. Clean the surface of the glass plate *P*, the lens *N* and the glass plate *G*. Place them in position as shown in Fig. 17.1. Place the arrangement in front of a sodium lamp so that the height of the centre of the glass plate *G* is the same as that of the centre of the sodium lamp. Place a screen in between having a hole of about one inch square in it at the same height. Place the convex lens in between and adjust its position so that a parallel beam of light is made to fall on the glass plate *G* inclined at an angle of 45°.

3. Adjust the position of the microscope so that it lies vertically above the centre of the lens *N*. Focus the microscope, so that alternate dark and bright rings are clearly visible.

4. Adjust the position of the microscope till the point of inter-section of the cross-wires coincides with the centre of the ring system and one of the cross-wires is perpendicular to the horizontal scale.

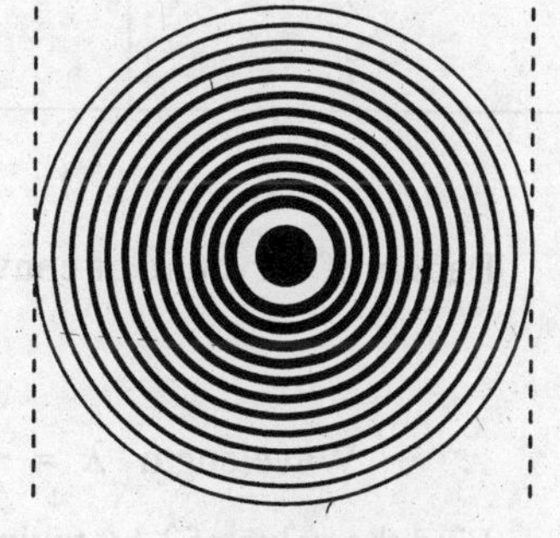

Fig. 17.3

5. Slide the microscope to the *left* till the cross-wire lies tangentially at the centre of the 20th dark ring. The position of the cross-wire when the microscope is focused on the 10th dark ring is shown in Fig. 17.3. Note the reading on the vernier scale of the

microscope. Slide the microscope backward with the help of the slow motion screw and note the readings when the cross-wire lies tangentially at the centre of the 16th, 12th, 8th and 4th dark rings respectively.

Keep on sliding the microscope to the *right* and note the reading when the cross-wire again lies tangentially at the centre of the 4th, 8th, 12th, 16th, and 20th dark rings respectively.

6. After reaching the 20th ring slide the microscope backwards and again note the readings corresponding to the same rings on the *right* and then on the *left* of the centre of the ring system.

7. Remove the lens N and find the radius of curvature of the surface of the lens in contact with the glass plate P accurately using a spherometer.

Observations. Venier constant = cm

Ring No.	*Microscope reading*		*Diameter*	*Microscope reading*		*Diameter*	*Mean Diameter*	
	Left	*Right*		*Right*	*Left*		*cm*	*m*
20								
16								
12								
8								
4								

Pitch of the spherometer screw = mm
Number of divisions on circular scale =
Least count = mm
Distance between the two legs = 1....... 2. ...... 3. ......
Mean l = cm

No.	*Spherometer reading on*		*h*
	Convex surface	*Plane surface*	
1			
2			
3			

Mean h = cm

Radius of curvature of convex surface $R = \dfrac{l^2}{6h} + \dfrac{h}{2}$

= cm = m

$$\therefore \quad \text{Wavelength } \lambda = \frac{D_n^2 - D_m^2}{4(n-m)R}$$

Find the value of λ by taking the various combinations of n and m as for example, (20, 12), (16, 8), (12, 4).

λ = 1. 2. 3.

Mean wavelength of sodium light λ = m

Note. *The radius of curvature of the surface of the lens, in contact with the glass plate should be measured accurately. In addition to a careful measurement of h, the distance between the legs l should also be measured very carefully as in the formula for R, l^2 is involved. It is desirable to measure l with the help of a travelling microscope, instead of a pair of dividers.*

Precautions. 1. The lens and the glass plate should be cleaned.

2. A lens of large radius of curvature, preferably plano-convex, should be used.

3. The point of intersection of the cross-wires should coincide with the centre of the ring system.

4. The micrometer screw should always be moved in the same direction to avoid error due to back-lash.

5. The radius of curvature of the surface of the lens in contact with the glass plate should be measured accurately.

6. The amount of light from the source should be adjusted for maximum visibility. Too much light increases the general illumination and decreases the contrast between bright and dark rings.

Exercise 1. *Plot a graph between* n and D_n^2 *in the case of Newton's rings and hence find the radius of curvature of the surface of the lens. Given that mean wavelength of sodium light is 5893* $\times 10^{-10}$ *m.*

Hints. Measure the diameter of the alternate dark rings starting from 20th i.e. 20th, 18th, 16th, 14th, 12th, 10th, 8th, 6th, 4th and second. Plot D_n^2 along the *Y*-axis, and the number of rings along *X*-axis as shown in Fig. 17.4.

The graph is a straight line. Take two points *P* and *Q* on it and draw perpendiculars *PM* and *QN* on the *X*-axis, then $D_n^2 = QN$ and $D_m^2 = PM$ where *n* and *m* are the numbers of the corresponding rings.

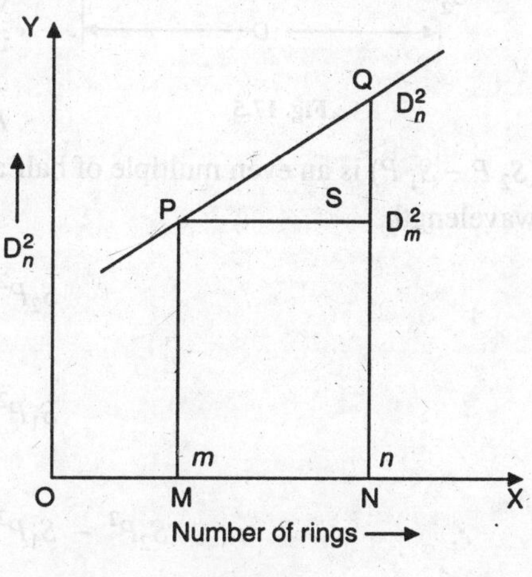

Fig. 17.4

$$\text{Now} \qquad QS = QN - PM$$

$$= D_n^2 - D_m^2$$

$$\text{Hence} \qquad = D_n^2 - D_m^2$$

$$= QS = 4\,(n-m)\,R\,\lambda$$

$$\therefore \qquad R = \frac{QS}{4\,(n-m)\,\lambda}.$$

Exercise 2. *Find the refractive index of the given liquid, using Newton's rings apparatus.*

Hints. The diameter of the *n*th dark ring in Newton's apparatus having air film is given by

$$(D_n^2)_{\text{Air}} = 4\,n\,R\,\lambda \qquad \text{...(i)}$$

and with a liquid film of refractive index μ it is given by

$$(D_n^2)_{\text{Liquid}} = \frac{4\,n\,R\,\lambda}{\mu} \qquad \text{...(ii)}$$

Therefore, to find the refractive index of a liquid, measure the diameters of various dark rings first when there is an air film between the lens and the glass plate. Then, introduce a film of the liquid and again measure the diameter of the rings. If we select the *n*th and the *m*th dark ring, then

$$(D_n^2 - D_m^2)_{\text{Air}} = 4\,(n-m)\,R\,\lambda$$

$$\text{and} \qquad (D_n^2 - D_m^2)_{\text{Liquid}} = \frac{4\,(n-m)\,R\,\lambda}{\mu}$$

$$\therefore \quad \text{Refractive index} \qquad \mu = \frac{(D_n^2 - D_m^2)_{\text{Air}}}{(D_n^2 - D_m^2)_{\text{Liquid}}}$$

This relation also holds true for bright rings.

From relations (*i*) and (*ii*), we get

$$\frac{(D_n^2)_{\text{Liquid}}}{(D_n^2)_{\text{Air}}} = \frac{1}{\mu}$$

or $$\frac{(D_n)_{\text{Liquid}}}{(D_n)_{\text{Air}}} = \frac{\text{Diameter of } n\text{th ring with liquid}}{\text{Diameter of } n\text{th ring with air}} = \frac{1}{\sqrt{\mu}}$$

For $$\mu > 1, \quad (D_n)_{\text{Liquid}} < (D_n)_{\text{Air}}$$

Thus when the liquid is introduced between the lens and the glass plate the diameter of the rings decreases i.e. the rings contract.

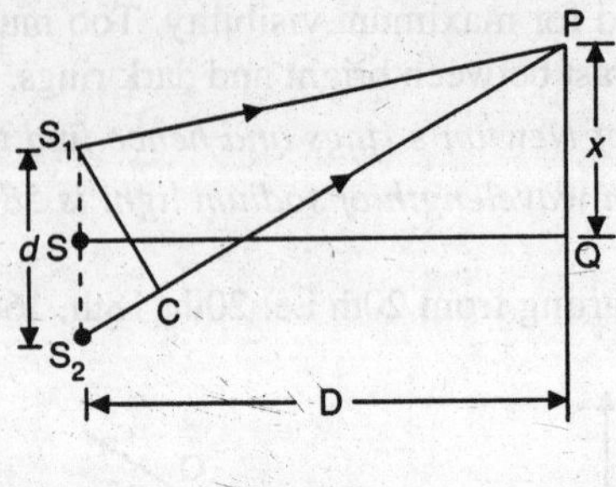

Fig. 17.5

17.7. Interference bands with biprism. When a source S is placed parallel to the refracting edge of a biprism, it gives rise to two virtual images as at S_1 and S_2 a certain distance d apart. The two sources S_1 and S_2 give out light waves parallel to each other in the same phase having the same amplitude. As the point O is equidistant from S_1 and S_2 the displacement will be in the same phase and thus will produce maximum intensity. The intensity at any point P distant x from O depends upon the path difference between the rays reaching P from S_1 and S_2. It will be maximum if the path difference $(S_2 P - S_1 P)$ is an even multiple of half a wavelength and minimum if it is an odd multiple of half a wavelength.

$$S_2P^2 = D^2 + \left(x + \frac{d}{2}\right)^2 \quad \text{and}$$

$$S_1P^2 = D^2 + \left(x - \frac{d}{2}\right)^2$$

$$\therefore \quad S_2P^2 - S_1P^2 = \left(x + \frac{d}{2}\right)^2 - \left(x - \frac{d}{2}\right)^2 = 2dx$$

Path difference $$S_2P - S_1P = \frac{2dx}{S_2P + S_1P} = \frac{2dx}{2D} = \frac{d}{D}x$$

For maximum intensity $\frac{d}{D}x = n\lambda$

or $$x = \frac{D}{d}n\lambda$$

Fringe width $$\beta = x_n - x_{n-1}$$

$$= \frac{D}{d}n\lambda - \frac{D}{d}(n-1)\lambda = \frac{D}{d}\lambda$$

or $$\lambda = \beta\frac{d}{D}$$

By measuring β, d and D, the wavelength of monochromatic light can be found out.

Experiment 17.2. To find the wavelength of sodium light using Fresnel's biprism.

Apparatus. A good quality optical bench about 150 cm long, four upright stands with transverse motion, an adjustable slit, a biprism, a lens holder, a micrometer Ramsden eye-piece provided with a divided head reading upto .001 cm, a short focus lens, variable aperture and a sodium lamp.

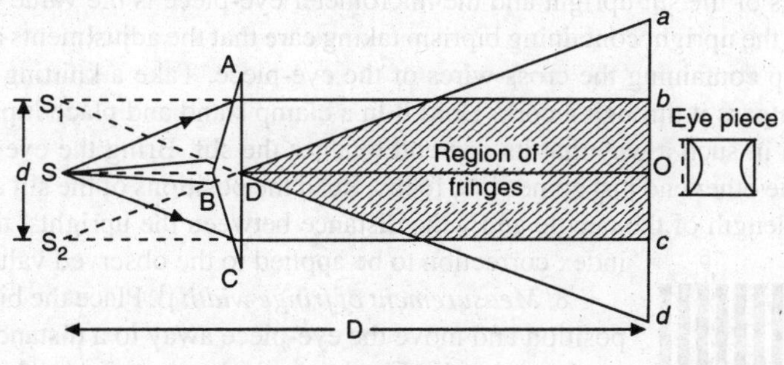

Fig. 17.6

Theory. Fresnel's biprism consists of a glass prism two faces of which make an angle of nearly 180° with one another, and are equally inclined to the third face making the other two angles each equal to 30″ or so.

When a source of light S is placed such that the biprism ABC is illuminated, the edge B of the biprism divides the incident light into two portions. One of the rays after suffering refraction through the portion ABD appears to come from S_1 and the other after suffering refraction through the portion CBD appears to come from S_2. In this manner two virtual coherent sources S_1 and S_2 are obtained. Interference bands can be observed on the screen which coincide with the plane of cross wires of the micrometer eye-piece. The fringe width β is given by the relation

$$\beta = \frac{D}{d}\lambda$$

where d is the distance between the two sources S_1 and S_2 and D is the distance between the slit S and the cross-wires at O of the eye-piece.

Procedure. 1. *Adjustment*. Level the bed of the optical bench with the help of levelling screws and a spirit level.

2. Mount the slit on the upright and focus the light from the sodium lamp on it with a lens.

3. Focus the micrometer eye-piece on the cross-wires and mount it on the upright meant to carry it. Turn it so that one of the cross-wires becomes vertical. Move it up to the upright carrying the slit and adjust its height equal to the height of the slit. Make the slit vertical with the help of the tangent screw provided for the purpose. To test the correctness mount a lens of short focal length on the spare upright and get the image of the slit on the vertical cross-wire.

4. Mount the biprism on the upright next to the slit and adjust its height equal to the height of the slit. Adjust the plane of the biprism, so that its plane is perpendicular to the plane of the bed.

5. Place the eye on the side of the biprism away from the slit, two virtual images will be seen. Move the eye across the bench, one of the images will appear to cross the edge of the biprism from one side to the other. If the refracting edge of the biprism is parallel to the slit, then the image as a whole will appear to cross the edge, otherwise the adjustment is faulty. Adjust the biprism by rotating it in its own plane by the tangent-screw provided for the purpose to affect the sudden transition of the full image. When this is done, the slit, the cross-wire and the edge of the biprism are all parallel.

6. Look through the eye-piece, a set of fringes will be seen. If these are not visible, then move the biprism upright at right angle to the bed of the optical bench till the fringes appear in the field of view of the eye-piece. Set the vertical cross-wire on the central bright fringe and move the eye-piece

away from the bi-prism and note if there is any lateral displacement of the fringes. In case there is any lateral shift, move the biprism at right angle to the bed of the bench till the shift ceases.

7. *Measurement of the distance D between the slit and focal plane of the eye-piece*. The difference between the readings of the slit upright and the micrometer eye-piece is the value of D. To find the actual value remove the upright containing biprism taking care that the adjustments are not disturbed. Remove also the cap containing the cross-wires of the eye-piece. Take a knitting needle and after measuring its length pass it through a cork. Hold it in a clamp stand and place it parallel to the bed of the optical bench in such a way that its one end touches the slit. Bring the eye-piece closer and move it slowly till the other end just comes into focus. Read the positions of the slit and the eye-piece uprights. If x is the length of the needle and y the distance between the uprights, then $(x - y)$ is the index correction to be applied to the observed value of D.

0 5 10 15

Fig. 17.7

8. *Measurement of fringe width* β. Place the bi-prism upright in position and move the eye-piece away to a distance of about 80-90 cm from the slit. Fix the vertical cross wire on the centre of one of the bright fringes and note down the reading of the micrometer screw of the eye-piece. Move the screw gradually till 8-10 fringes have moved across the field of view. Note again the reading of the screw of the micrometer eye- piece, then

$$\text{Fringe width } \beta = \frac{\text{Total distance moved}}{\text{Number of fringes moved across the cross-wire}}$$

Note the position of the eye-piece upright on the optical bench. Without disturbing the position of the slit and the biprism, change the position of the eye-piece upright and repeat the observation twice to measure β.

9. *Measurement of distance d between the two virtual sources*. Introduce a convex lens of short focal length mounted on an upright between the eye-piece and the biprism uprights. Keeping the positions of the slit, biprism and the eye-piece uprights fixed, move the lens upright away from the eye-piece upright to the position L_1 till the images of the two virtual sources are seen distinctly. Fix the vertical cross-wire on one of the images and note the reading of the micrometer screw. Move the screw till the vertical cross-wire falls on the second image. Again note the reading of the micrometer screw. The difference between the two readings gives the distance C_1 between the two images formed.

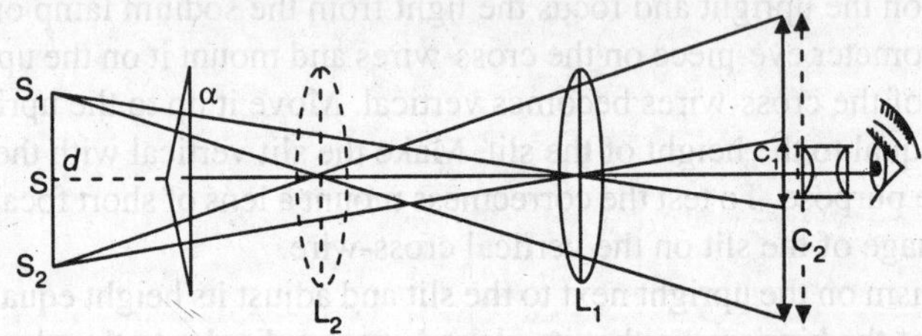

Fig. 17.8

Next move the lens towards the biprism upright to the position L_2 till again the two images are seen in the field of view of the eye-piece. Measure the distance C_2 between the two images again, then the distance

$$d = \sqrt{C_1 C_2}$$

Repeat this observation thrice.

Observations. Pitch of the screw = mm

Number of divisions on the circular scale =

Least of count = mm = cm

Length of the knitting needle x = cm

Observed reading between the uprights y = cm
Index correction $(x - y)$ = cm
Least count of the vernier provided on the upright = cm

Measurement of fringe width

No.	*Position uprights*			*Micrometer reading on*			$\beta = \frac{b-a}{n}$
	slit	*eye-piece*	*D*	*1st fringe a*	*nth fringe b*	*difference (b – a)*	
1							
2							
3							

Measurement of distance *d*

No.	*Lens position L_1*		C_1	*Lens position L_2*		C_2	$d = \sqrt{C_1 C_2}$
	Micrometer reading on			*Micrometer reading on*			
	1st image	*2nd image*		*1st image*	*2nd image*		
1							
2							
3							

Mean value of d = cm

Calculations.

Corrected value of D (*i*) ... (*ii*) ..... (*iii*) .....

(*i*) Wavelength of sodium light $\lambda = \beta \frac{D}{d}$ = cm = m

(*ii*)

(*iii*)

Mean value of wavelength λ = m

Precautions. 1. The slit should be as narrow as possible.

2. The refracting edge should be parallel to the slit.

3. The line joining the slit and the edge of the biprism should be parallel to the bed of the optical bench.

4. Index correction should be found out and applied to the observed distance D between the slit and the micrometer eye-piece uprights.

5. While using the micrometer screw backlash error must be avoided.

6. The vertical cross-wire should be placed in the middle of the bright fringe to find the fringe width.

7. A lens of short focal length should be used to measure the distances C_1 and C_2.

8. To avoid spherical and other aberrations a variable circular aperture must be used in front of the convex lens. The axis of the lens should be parallel to the bed of the optical bench.

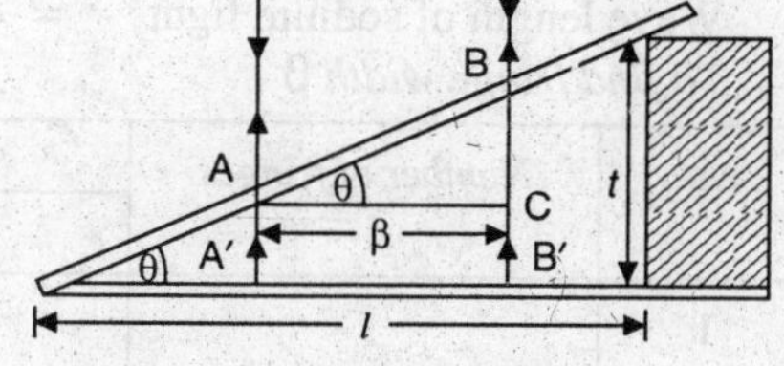

Fig. 17.9

9. As many fringes as possible should be taken to find the mean fringe width β.

Experiment 17.3. To determine the thickness of a thin plate or a thin paper by measurement of width of interference fringes in an air wedge.

Apparatus. A travelling microscope, a sodium lamp, an optically plane glass plate, a microscope slide, an arrangement having an optically plane glass plate inclined at an angle of 45°, a piece of thin paper *e.g.* cigarette paper whose thickness is to be determined.

Theory. A wedge-shaped air film can be obtained by inserting a thin mica plate or a thin paper between two plane parallel glass plates. A highly magnified view is shown in Fig. 17.9. The fringe at A is produced by the light beam reflected at A, the point at the bottom of the microscope slip and at the point A' immediately below A across the air film. The next fringe is produced when the air film has increased in thickness by $\lambda/2$. If θ is the angle of the air wedge, then

$$\tan\theta = \frac{BC}{AC} = \frac{BC}{A'B'} = \frac{\lambda}{2} \times \frac{1}{\beta}$$

where β is the fringe width. If t is the thickness of paper, then

$$\tan\theta = \frac{t}{l}$$

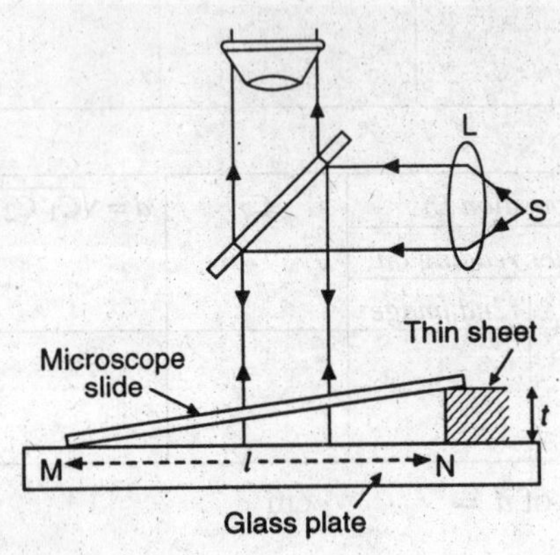

Fig. 17.10

$$\therefore \quad \frac{t}{l} = \frac{\lambda}{2\beta}$$

or

$$t = \frac{l\lambda}{2\beta}$$

Procedure 1. Level the microscope table, and adjust the microscope tube in a vertical position and find the vernier constant of the horizontal scale.

2. Arrange the apparatus as shown with the edge of the thin plate inserted under one end of the microscope slide placed on the glass plate.

3. Adjust the source and the microscope tube until parallel interference fringes are formed in the air wedge and are seen in the field of view of the microscope and the vertical cross-wire coincides with the centre of one of the straight fringes. Note the reading of the microscope vernier on the scale in this position.

4. Slide the microscope to one side with the help of the slow motion screw till the cross-wire moves on to the 20th fringe. Note the reading of the microscope vernier on the scale. Find the traverse of the microscope to cover a given number of interference fringes and find the fringe width β.

5. Repeat the observation thrice selecting different number of fringes.

6. Now measure the distance l of the edge of thin plate from the line of contact of the microscope slide with the glass plate at M and N with the help of the travelling microscope. Repeat three times.

Observations. Vernier constant = cm

Wave length of sodium light = 5893×10^{-10} m

To find fringe width β

No.	*Number of fringes*	*Microscope reading*		*Traverse*	*Fringe width* β
		(i)	*(ii)*		
1					
2					
3					

Mean β = cm

= m

Measurement of l

No.	Microscope reading on		l
	Thin plate N	Line of contace M	
1			
2			
3			

Mean l = cm
= m

Calculations.

Thickness $t = \frac{\lambda l}{2\beta} =$ m

Precautions. 1. The two glass plates must be cleaned and should be optically worked.
2. The quantity of light should be adjusted for maximum visibility.

Oral Questions

1. Interference. (*i*) What is interference of light ? (*ii*) Can two independent sources produce interference of light ? If not why ? (*iii*) What is the condition ordinarily for (*a*) darkness and (*b*) brightness ? (*iv*) What is the phase change on reflection ? (*v*) Does this change take place always or under some particular conditions ? (*vi*) What are the various methods of getting interference fringers ?

2. Newton's rings. (*i*) How does interference take place in the case of Newton's rings ? Is it possible to have interference with a lens of short focal length ? (*iii*) What will happen if the lens is cylindrical ? (*iv*) Where are the fringes located ? (*v*) Why are the fringes circular ? (*vi*) Why is the centre dark ? Can it be bright ? If so, when ? (*vii*) Why do the fringes get closer and finer as we move away from the centre ? (*viii*) On what factors does the diameter of a ring depend ? (*ix*) What will happen if the glass plate is replaced by a mirror ? (*x*) What is the difference between rings seen by reflected light and those seen by transmitted light ? (*xi*) Would you prefer a lens of long focal length or that of a short focal length ? (*xii*) Where are the fringes formed ? (*xiii*) Why do you get only a few fringes with white light ? (*xiv*) Can you find the refractive index of a liquid by this method ? (*xv*) How does the diameter of the fringes change on the introduction of a liquid ?

3. Biprism. (*i*) What is a biprism and how is it made ? (*ii*) What are coherent sources and how are they produced ? (*iii*) Explain how the fringes are produced with biprism ? (*iv*) What conditions are necessary for the success of the experiment ? (*v*) What adjustments are necessary for the success of the experiment ? (*vi*) How is the bench correction determined for the measurement of *D* ? (*vii*) How do you measure the distance *d* between the two images of the slit ? (*viii*) What is the nature of the two images formed by the biprism ? (*ix*) Why do we use lens of short focus ? (*x*) What is the use of a circular stop of a variable aperture placed before the lens ? (*xi*) Describe the mechanical parts of the optical bench that you are using.

4. Thickness of a thin plate. (*i*) What type of air film do you get by inserting a thin plate between two plane parallel glass plates? (*ii*) What is the shape of interference fringes? (*iii*) On what factors does the fringe width depend?

18

Diffraction

18.1. Diffraction. Ordinarily light appears to travel in straight line. A very careful observation, however, shows that light does suffer some deviation from its straight path when it passes close to the edge of opaque obstacles or when it passes through narrow slits or holes. The phenomenon is more pronounced when the dimensions of the obstacle or the aperture are comparable to the wavelength of light. It is found that not only light bends into the region of geometrical shadow, but also the illumination outside the geometrical shadow is non-uniform and shows alternate variations in intensity. *The bending of light round corners and spreading of light waves in the geometrical shadow of an object is called* **diffraction.** The phenomenon of diffraction is due to the interference of secondary wave-lets which originate from a single wave front.

There are two types of diffraction phenomenon, known as (*i*) *Fresnel diffraction* and (*ii*) *Fraunhofer diffraction.*

Fresnel diffraction. In this type of diffraction the source of light and screen are at finite distances from the obstacle or aperture at which diffraction takes place. No lenses are used to make the rays parallel or convergent. The incident wave front is either spherical or cylindrical. Such a diffraction takes place when light suffers diffraction at a straight edge, a narrow slit, a thin wire etc.

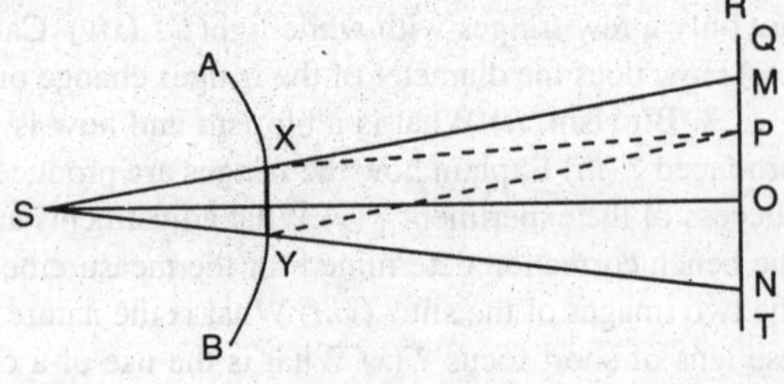

Fig. 18.1

Fraunhofer diffraction. In this type of diffraction the source of light and screen are effectively at infinity with respect to the aperture or obstacle causing diffraction. This is achieved by using two convex lenses, one to make the light from the source parallel before it falls on the aperture and the other to focus the light on the screen after diffraction. The incident wave front is plane. The most important example of this type is met with in a diffraction grating.

18.2. Fresnel diffraction at a wire. Let *S* represent a narrow rectangular slit placed parallel to the wire of thickness *XY* and perpendicular to the plane of the paper. This will give rise to a geometrical shadow *MN* on the screen *RT*.

The effect at a point *Q* outside the geometrical shadow is the same as that due to a straight edge at *X* and hence diffraction bands of unequal width and intensity will be seen above *M* and similarly below *N*. These bands are independent of the thickness of the wire, as on either side the effect of the other half of the wave is negligible because most important half-period elements are cut off due to the finite width of the wire.

Within the geometrical shadow interference fringes are observed. The effect due to the portion *AX* of the cylindrical wave front at any point *P* within the geometrical shadow is due entirely to a few half-period elements at X so that it may be regarded as a small luminous source at X. Similarly the effect due to the portion BY of the wave-front is equal to a source at Y.

The effect at any point P within the geometrical shadow MN will depend upon the path difference ($PY–PX$). The point P will be bright if path difference $(PY–PX) = n\lambda$ and dark if it is equal $(2n + 1) \frac{\lambda}{2}$. These fringes are of equal width and the fringe width

$$\beta = \frac{D}{d} \lambda$$

where D is the distance of the screen from the obstacle and d the thickness of the obstacle, *i.e.*, the diameter of the wire.

The centre O will be bright because at O the waves from X and Y will always meet in phase.

The variation of intensity of the interference fringes produced within the geometrical shadow of a thin wire and diffraction bands outside the geometrical shadow is shown in Fig. 18.2 (*a*).

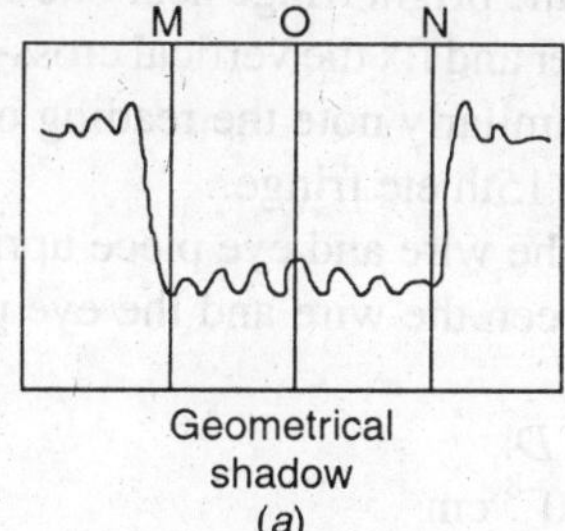

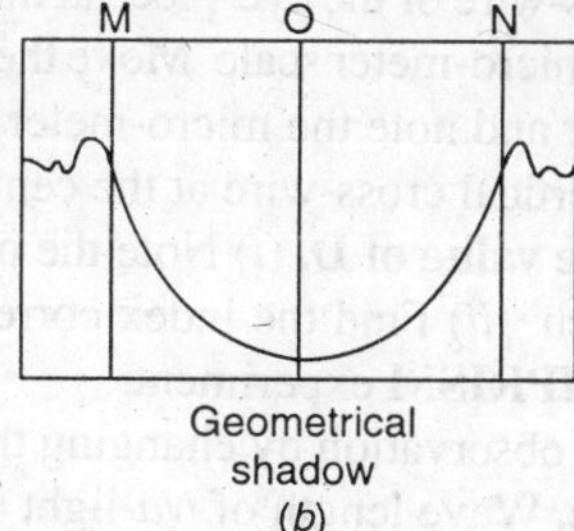

Fig. 18.2

As the diameter of the wire is increased the fringe width decreases and when the thickness of the wire is sufficiently large the interference fringes will disappear and only the diffraction bands outside the limits of the geometrical shadow will be visible as shown in Fig. 18.2 (*b*).

To find the diameter of the wire. Focus the micrometer eye-piece on the interference fringes obtained within the geometrical shadow of the wire. The fringes are of equal width. Find the mean fringe width by measuring the shift of the cross-wire for a known number of fringes. The diameter of the wire can be found out from the expression

$$\beta = \frac{D}{d} \lambda \quad \text{or} \quad d = \frac{D\lambda}{\beta}$$

where d is the diameter of the wire, D the distance between the slit and the cross-wires and λ the wavelength of monochromatic light used.

Experiment 18.1. To determine the diameter of a thin wire by studying the diffraction (and interference) pattern produced by it.

Apparatus. A good quality optical bench about 150 cm long, four upright stands with transverse motion, an adjustable slit, a thin wire mounted in a frame and fitted with a tangent screw for rotation in a vertical plane, a micrometer Ramsden eye-piece, provided with a divided head reading upto 0.001 cm, a convex lens of short focal length, variable aperture, a sodium vapour lamp and a spirit level.

Procedure. 1. Level the bed of the optical bench with the help of levelling screws and spirit level.

2. Mount the adjustable slit on the upright and focus the light from the sodium lamp on it with a convex lens.

3. Focus the micrometer eye-piece on the cross-wires and mount it on the upright meant to carry it. Turn it so that one of the cross-wires becomes vertical. Move it up to the upright carrying the slit and adjust its height equal to the height of the slit. Make the slit vertical with the help of the tangent screw provided for the purpose.

4. Mount the thin wire held in its frame in an upright and place it between the slit and the eye-piece. Adjust its height to that of the slit. Make it vertical by seeing through the eye-piece and rotating it with the help of the tangent screw, till the wire coincides with the vertical cross-wire.

5. Move the straight thin wire close to the slit and bring the micrometer eye-piece close to the wire. Make the slit narrow till diffraction fringes appear in the field of view. If the fringes are not sharp, slightly rotate the wire in its own plane with the help of the tangent screw.

6. Move the eye-piece away from the wire, the fringes may shift laterally to the left or to the right. This means that the line joining the slit and the wire is not parallel to the bed of the optical bench. To remove this defect, move the upright carrying the wire in a direction perpendicular to the optical bench with the help of the screw at its base, till the fringes come back to their original position.

7. **To find the fringe width** β (*i*) Find the least count of the micrometer of the eye piece. (*ii*) Keeping the wire upright very close to the slit, move the eye-piece upright to a convenient distance, so that interference fringes are visible in the geometrical shadow of the wire. (*iii*) Adjust the position of the vertical cross-wire of the eye-piece at the centre of the bright fringe near one extreme end. Note the reading on the micro-meter scale. Move the micrometer and fix the vertical cross-wire at the centre of 5th bright finger and note the micro-meter reading. Similarly note the reading of the micrometer by adjusting the vertical cross-wire at the centre of 10th, 15th etc fringe.

8. **To find the value of D.** (*i*) Note the position of the wire and eye piece uprights on the scale of the optical bench. (*ii*) Find the index correction between the wire and the eye piece uprights, as explained in the **BIPRISM** experiment.

9. Repeat the observation by changing the value of *D*.

Observations. Wave length of *Na*-light = 5893×10^{-8} cm

Least count of micro-meter = mm = cm

Length of knitting needle x = cm

Observed distance between the wire and eye-piece uprights y = cm

Index correction $(x - y)$ = cm

Sl. No.	Position of Uprights				Micrometer Reading				$\beta = \frac{b-a}{n}$
	Wire	Eye-piece	D Observed	D Corrected	first fringe (a)	nth fringe (b)	n	b − a	
1.					(i) (ii) (iii)			Mean β	
2.					(i) (ii) (iii)			Mean β	

Calculations Diameter of the wire $d = \frac{D\lambda}{\beta}$ (*i*) .... (*ii*) ....

Mean Diameter d = cm = m

Precautions 1. The slit should be as narrow as possible.

2. The slit, the wire and the cross-wire of the eye-piece should all be vertical and in the same line, parallel to the bed of the optical bench.

3. The wire should be placed very close to the slit.

4. While using the micrometer, error due to backlash must be avoided.

5. While measuring the fringe width the vertical cross wire should be fixed in the centre of the bright fringe.

6. Index correction must be applied to find the correct value of *D*.

18.3. Diffraction grating. A plane diffraction grating consists of an optically plane glass plate on which are ruled a number of equidistant, parallel straight lines. The lines divide the glass plate into opacities and transparencies, the thickness of which is of the order of the wavelength of visible light. The region where a line is drawn becomes opaque whereas the space between the two lines is transparent. The number of lines in a plane transmission grating is of the order of 6000 lines per cm.

In laboratories an actual grating is not used as it is very costly. The original grating is usually ruled on a plate of *speculum* metal by a sharp diamond point fitted to a dividing engine. For ordinary use *replicas* of the grating are obtained by depositing a very thin film of gelatine on it. The film is removed when dry and is pasted on an optically plane glass plate.

ABC ..... *H* in Fig. 18.3 represents the section of a plane transmission grating supposed perpendicular to the plane of the paper. Let the width of the clear space or *transparency* be equal to *AB* = *a* and that of the ruling or *opacity* be *BC* = *b*. The distance $(a + b)$ is called the *grating element*. Prints like *A* and *C* separated by a distance $(a + b)$ are called *corresponding points*.

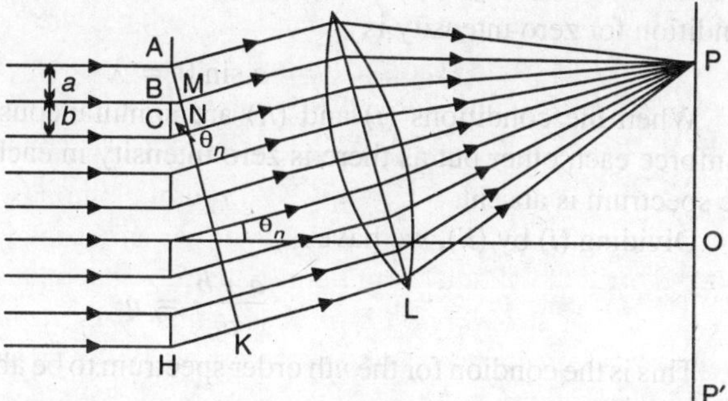

Fig. 18.3

When a parallel beam of monochromatic light is incident normally on the grating it suffers diffraction. As shown in Fig. 18.3 *ABC* .... *H* represents the incident plane wave front and *AMN* ...... *K* represents the transmitted wave front after diffraction at the grating through an angle θ_n. There is no path difference between the rays before reaching the plane *ABC* .... *H*. The only phase change that occurs between the various rays is due to the difference in paths traversed by the rays from the plane *ABC* .... *H* to the plane *AMN* .... *K*. It is clear that the path difference between the rays from two corresponding points *A* and *C* is $CN = (a + b) \sin \theta_n$. The same is true for any two corresponding points. The transmitted beam of parallel light after falling on the lens *L* is brought to focus at the point *P*. The point *P* will be bright or dark according as the rays *reinforce* or *interfere* with one another. They would reinforce and give bright ness or a *maximum* if

$$(a + b) \sin \theta_n = n \lambda$$

They would interfere and produce darkness or a *minima* if

$$(a + b) \sin \theta_n = (2n + 1) \frac{\lambda}{2}$$

When $n = 0$, $\theta_n = 0$. This gives the *central bright maximum*

When $n = 1$, $(a + b) \sin \theta_1 = \lambda$ and we get the *first bright maximum* or first order spectrum

∴ For *first order spectrum*, $n = 1$

and $$(a + b) \sin \theta_1 = \lambda$$

Similarly, For the *second order* spectrum $n = 2$

and $$(a + b) \sin \theta_2 = 2 \lambda$$

Thus $$\sin \theta_2 = 2 \sin \theta_1 .$$

This relation helps us to verify that the order of spectrum after the first is second as under certain conditions second order spectrum is missing and we may get the third order spectrum after the first.

If light is not monochromatic, then each wavelength gives rise to its own primary maxima in each order. Hence each order will consist of a spectrum.

The number of lines *N per inch* are marked on the grating. The value of the grating element $(a + b)$ is given by

$$(a+b) = \frac{2.54}{N} \text{ cm}$$

When number of lines per cm is given $a + b = \frac{1}{N}$

18.4 Absent spectra. In a diffraction grating the condition for the primary maxima is

$$(a+b) \sin \theta = n \lambda \qquad \text{...}(i)$$

If the path difference between rays from the extreme ends of one single slit is λ, then the intensity of light is zero. It is because in such a case the slit can be divided into two halves and the path difference between the corresponding points in the two halves will be $\frac{\lambda}{2}$ which gives rise to zero intensity. Hence condition for zero intensity is

$$a \sin \theta = \lambda \qquad \text{...}(ii)$$

When the conditions (*i*) and (*ii*) are simultaneously obeyed the beams from all the slits do reinforce each other but as there is zero intensity in each slit, the resultant intensity is also zero and the spectrum is absent.

Dividing (*i*) by (*ii*), we have

$$\frac{a+b}{a} = n.$$

This is the condion for the *n*th order spectrum to be absent. In order that the **second** order spectrum may be absent.

$$\frac{a+b}{a} = 2$$

or

$$a = b.$$

In other words the width of a transparency must be equal to that of an opacity.

Experiment 18.2. To determine the wavelength of sodium light using a plane diffraction grating.

Apparatus. A spectrometer, a spirit level, a sodium lamp, an eye-piece, a diffraction grating with clamping arrangement etc.

Theory. When a parallel beam of monochromatic light is incident normally on a grating, the transmitted light gives rise to primary maxima in certain directions given by the relation

$$(a+b) \sin \theta_n = n \lambda$$

where a is the width of a transparency, b that of an opacity, θ_n *the angle of diffraction for the* nth order maximum and λ the wavelength of light.

Procedure. 1. **Setting.** Adjust the position of the eye-piece of the telescope so that the cross-wires are clearly visible. Focus the telescope on a distant object and set it for parallel rays. Level the spectrometer by the levelling screws and then the prism table with the help of a spirit level.

2. Fix the grating stand on the circular table with two screws in the holes drilled on one of the lines parallel to the line-joining two of the screws meant for the purpose, say P and Q. The face of the stand to which the clamps are attached should be at the centre of the table. *Take out the grating carefully from the box, holding it from the edge and without touching its surface*, fix it very carefully to the frame with its **ruled surface** towards the telescope.

3. **Optical levelling of the grating table.** Rotate the table so that the plane of the grating is approximately inclined at an angle of 45° to the axis of the collimator. Rotate the telescope to receive the reflected light from the grating surface. If the image of the slit is not symmetrical with respect to the horizontal cross-wire, adjust with the help of the third screw R. In this position the plane of the grating will be vertical.

Rotate the table carrying the grating so that the plane of the grating is approximately perpendicular to the axis of the collimator. Look for the first order spectrum on one side of the direct image of the slit. Turn the telescope so that vertical cross-wire coincides with the first order diffracted image. If this image is not symmetrical with respect to the horizontal cross-wire, adjust it with the help of one of the screws P or Q. In this position the grating lines are parallel to the axis of the spectrometer. Now turn the telescope to the other side so that the vertical cross-wire again coincides with the first order diffracted image. If the adjustments are carefully done then the diffracted images of the slit will be symmertical with respect to the horizontal cross-wire in all positions.

4. **Setting the grating normal to the incident light.** Place the telescope in line with the collimator so that the vertical cross-wire falls exactly in the centre of the image of the slit. Note the scale reading. Add 90 to the reading and place the telescope at this reading to set it perpendicular to the axis of the collimator. Clamp it in this position.

Rotate the grating table till the plane face of the grating is facing both the collimator and the telescope. Look through the telescope and turn the table very slowly till the centre of the slit falls exactly on the vertical cross-wire as shown in Fig. 18.4.

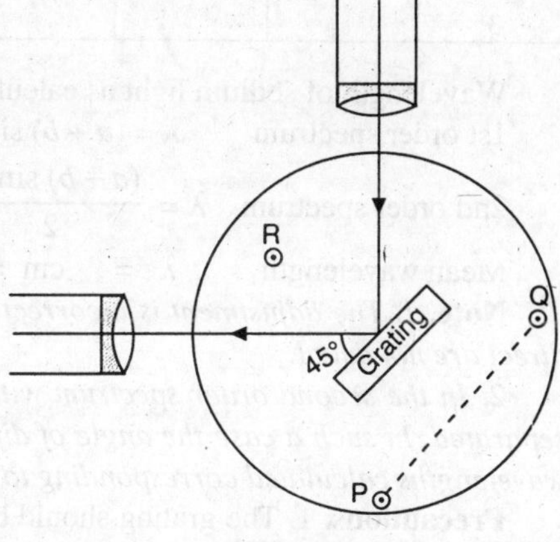

Fig. 18.4

In this position the plane of the grating is inclined at an angle of 45° to the incident light. Note the reading. Turn the table through 45° from this position so that the plane of the grating is normal to the incident light with its plane face towards the collimator. *The grating is now set normal to the incident light with its ruled surface away from the collimator.* Clamp the table in this position.

5. Place the eye in front of the collimator and move it gradually towards the telescope till the first order diffracted image is visible. Bring the telescope in this position and observe the image through it. Clamp the telescope in this position. If the resolving power of the grating is sufficiently high two distinct narrow lines corresponding to the wavelength 5890 Å and 5896 Å will lie side by side in the field of view. Ordinarily the two lines will appear as one in the first order spectrum. Turn the tangent screw of the telescope till the vertical cross-wire coincides with the centre of the image of the slit. Note the reading of the scale on both the verniers. Similarly observe the first order spectrum on the other side of the direct image and note the reading on both the verniers.

6. Similarly note the reading of the verniers by setting the telescope on the second order diffracted image on either side of the direct light.

7. Repeat the above observations three times.

8. Note the number of lines per inch as marked on the grating and replace it carefully in the box with ruled surface *upwards*.

Observations. Vernier constant =

Number of lines per inch on the grating N =

∴ Grating element $(a + b) = \dfrac{2.54}{N}$ = cm

Direct reading of telescope =

Sl. No.	*Order of spectrum*	*Vernier*	*Telescope reading*			*Angle of diffraction*		
			Left	*Direct*	*Right*	*Left*	*Right*	*Mean*
1	1st order	V_1						
2		V_2						
3								θ_1
1	2nd order	V_1						
2		V_2						
3								θ_2

Wavelength of sodium light as calculated from

1st order spectrum $\lambda = (a + b) \sin \theta_1 =$ cm

2nd order spectrum $\lambda = \dfrac{(a+b)\sin\theta_2}{2} =$ cm

Mean wavelength $\lambda =$ cm $=$ m

Note. *1. The adjustment is incorrect if the angles of diffraction on the left and the right of the direct are not equal.*

2. In the second order spectrum with a good grating the two D-lines generally appear to be separated. In such a case the angle of diffraction should be found separately for the two lines and wavelengths calculated corresponding to each.

Precautions. 1. The grating should be held from the edges and the ruled surface should not be touched.

2. The telescope should be focused on the brightest image of the slit while seeing the reflected image.

3. The ruled surface should face away from the collimator.

4. The light should fall on the whole of the grating surface.

Exercise 1. *Find the wavelength of yellow line of helium.*

Exercise 2. *Determine the wavelength of various lines in the spectrum of a mercury vapour lamp.*

Experiment 18.3. To find the number of lines per centimetre of the given grating. Given mean wavelength of sodium light to be 5893 × 10^{-10} m.

Apparatus. Same as in **Expt. 18.2.**

Theory. When a parallel beam of monochromatic light of wavelength λ is incident normally on a grating, the light after diffraction gives rise to principal maxima in certain directions given by

$$(a + b) \sin \theta_n = n \lambda$$

where $(a + b)$ is the grating element and θ_n he angle of diffraction for the nth order maximum.

$$\therefore \qquad a + b = \frac{n\lambda}{\sin \theta_n}$$

If we measure θ_1 and θ_2 the angles of diffraction in the *first* and the *second* order spetra respectively, then

$$(a + b) = \frac{\lambda}{\sin\theta_1} = \frac{2\lambda}{\sin\theta_2}$$

The number of lines per cm $= \dfrac{1}{a+b}$

Procedure. Setting and measurement of θ_1 and θ_2 same as in **Expt. 18.2,** Steps 1–7.

Observations. Wavelength of sodium light $= 5893 \times 10^{-10}$ m

Sl. No.			Telescope reading			Angle of diffraction			Grating element $(a + b)$
			Left	Direct	Right	Left	Right	Mean	
1	First order	V_1							
2		V_2							
1	Second order	V_1							
2		V_2							

Mean $(a + b)$ = m

Number of lines per metre = $\dfrac{1}{a+b}$

∴ Number of lines per cm =

Precautions. Same as in **Expt. 18.2.**

18.5. Hydrogen spectrum. Long before Bohr formulated his theory of hydrogen spectrum, Balmer, Lyman, Paschen and others made experimental studies of hydrogen spectrum in the visible, ultra violet, infra red and far infra red regions respectively. They gave an empirical formula representing the wave number or reciprocal of wave length λ of different spectral lines as

$$\bar{\nu} = \frac{1}{\lambda} = R\left(\frac{1}{n_1^2} - \frac{1}{n_2^2}\right)$$

where R is a constant known as **Rydberg's constant** the value of which is 1.097×10^{-3} Å^{-1} (or 1.097×10^{7} m^{-1})

Lyman series of spectral lines lying in the ultra violet region are given when

$$n_1 = 1 \text{ and } n_2 = 2, 3, 4 \text{ etc.}$$

or $$\bar{\nu} = \frac{1}{\lambda} = R\left[\frac{1}{1^2} - \frac{1}{n_2^2}\right]$$

Balmer series of spectral lines in the *visible region* are given when

$n_1 = 2$ and $n_2 = 3, 4, 5$ etc

or $$\bar{\nu} = \frac{1}{\lambda} = R\left[\frac{1}{2^2} - \frac{1}{n_2^2}\right]$$

Paschen series of spectral lines in the *infra–red region* are given when

$$n_1 = 3 \text{ and } n_2 = 4, 5, 6, \textit{etc.}$$

or $$\bar{\nu} = \frac{1}{\lambda} = R\left[\frac{1}{3^2} - \frac{1}{n_2^2}\right]$$

Brackett series of spectral lines in the *far infra–red* region are given when

$$n_1 = 4 \text{ and } n_2 = 5, 6, 7, \text{ etc.}$$

or $$\bar{\nu} = \frac{1}{\lambda} = R\left[\frac{1}{4^2} - \frac{1}{n_2^2}\right]$$

Pfund series of spectral lines also in the *far infra–red* region are given when

$$n_1 = 5 \text{ and } n_2 = 6, 7, 8 \text{ etc.}$$

or $$\bar{\nu} = \frac{1}{\lambda} = R\left[\frac{1}{5^2} - \frac{1}{n_2^2}\right]$$

The formation of various series in the hydrogen spectrum is shown by energy level diagram in Fig 18.5 (*a*) and (*b*) The energy corresponding to $n = 1$ is shown along the lowermost line in Fig. 18.5 (*b*).

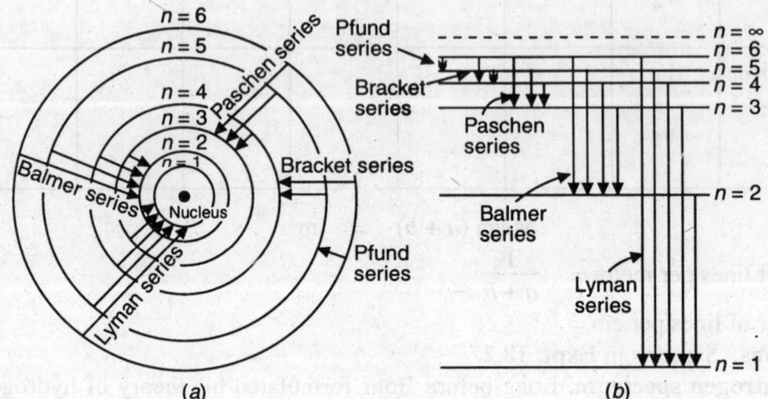

Fig. 18.5

According to Bohr's theory the value of Rydberg's constant in S.I. *units* is given by the relation

$$R = \frac{me^4}{8c\,\varepsilon_0^2\,h^3}$$

where m is the mass and e the charge on the electron, c the velocity of light. ε_0 permittivity of free space and h Planck's constant.

Substituting the various values, we get

$$R = 1.097 \times 10^7 \text{ m}^{-1}$$

which shows remarkable agreement with the experimental value.

18.6. Rydberg's constant. If we study the spectrum produced by a hydrogen gas discharge tube in the visible region with the help of a grating spectrometer, we observe four prominent lines as under which form Balmer series of hydrogen spectrum

(*i*) H_α line in the *red* region $\lambda = 6563$ Å corresponding to $n_1 = 2$ and $n_2 = 3$

(*ii*) H_β line in the *greenish blue* region $\lambda = 4861.3$ Å corresponding to $n_1 = 2$ and $n_2 = 4$

(*iii*) H_γ line in the *blue* region $\lambda = 4340.5$ Å corresponding to $n_1 = 2$ and $n_2 = 5$ and

(*iv*) H_δ line in the *violet region* λ= 4101.7 Å corresponding to $n_1 = 2$ and $n_2 = 6$. The wave length of each line is determined with the help of the grating spectrometer using the relation $(a + b) \sin \theta = n\lambda$ where θ is the angle of diffraction, $(a + b)$ the grating element and n the order of the spectrum. Substituting the value of λ for H_α, H_β, H_γ, and H_δ spectral lines of hydrogen in the relation

$$\frac{l}{\lambda} = R\left[\frac{l}{n_1^2} - \frac{l}{n_2^2}\right]$$

where $n_1 = 2$ and $n_2 = 3, 4, 5, 6$ respectively, we can calculate the value of R.

Experiment 18.4. To determine the value of Rydberg's constant with the help of a grating spectrometer using hydrogen gas discharge tube.

Apparatus. A hydrogen gas discharge tube with a suitable transformer, a spectrometer, a plane transmission grating, spirit level, eye-piece, electric lamp etc.

Procedure. 1. Set the spectrometer and the hydrogen gas discharge tube. Find the angle of diffraction in the first and the second order spectrum corresponding to the spectral lines H_α, H_β, H_γ and H_δ as explained in **Expt. 18.2.**

Observations. Number of lines per cm of the grating N =

Grating element $(a+b) = \dfrac{1}{N} =$ cm

H_α line.

Sl. No.	*Order of spectum*	*Vernier*	*Telescope reading*			*Angle of diffraction*		
			Left	*Direct*	*right*	*Left*	*right*	*mean*
	1st	V_1						
		V_2						θ_1
	2nd	V_1						
		V_2						θ_2

Wavelength of H_α line from

1st order spectrum $\lambda = (a+b)\sin\theta_1$ = cm = m

2nd order spectrum $\lambda = \dfrac{(a+b)\ \sin\theta_2}{2}$ = cm = m

Mean wavelength of H_α line λ = cm = m

For H_α line $n_1 = 2$ and $n_2 = 3$

$$\therefore \qquad \frac{1}{\lambda} = R\left[\frac{1}{2^2} - \frac{1}{3^2}\right]$$

or $$R = \frac{36}{5} \times \frac{1}{\lambda} = \quad m^{-1}$$

H_β line. Record the observations as for H_α line and calculate λ.

For H_β line $n_1 = 2$ and $n_2 = 4$

$$\therefore \qquad \frac{1}{\lambda} = R\left[\frac{1}{2^2} - \frac{1}{4^2}\right]$$

or $$R = \frac{16}{3} \times \frac{1}{\lambda} \qquad = \quad m^{-1}$$

Similarly for H_γ line

$$\frac{1}{\lambda} = R\left[\frac{1}{2^2} - \frac{1}{5^2}\right] \qquad = \quad m^{-1}$$

or $$R = \frac{100}{21} \times \frac{1}{\lambda} \qquad = \quad m^{-1}$$

and for H_δ line

$$\frac{1}{\lambda} = R\left[\frac{l}{2^2} - \frac{l}{6^2}\right] \qquad = \quad m^{-1}$$

or $$R = \frac{9}{2} \times \frac{1}{\lambda}$$

Mean value of Rydberg's constant R = cm^{-1}
= m^{-1}
= $Å^{-1}$

Standard value of $R = 1.097 \times 10^7\ m^{-1}$
Difference =
% error = %

Precautions. Same as in **Expt. 18.2.**

Experiment 18.5. To determine the resolving power of a plane trasmission diffraction grating.

Apparatus. A spectrometer, a sodium lamp, an eye-piece, a diffraction grating with clamping arrangement, a variable rectangular slit provided with micrometer arrangement to measure its width.

Theory. Resolving power of a diffraction grating is defined as its ability to show the two neighbouring lines in a spectrum as separate. It is measured as $\frac{\lambda}{d\lambda}$ where $d\lambda$ is the smallest difference in wavelength of the two lines. According to Rayleigh's criterion the *two lines of wavelengths* λ *and* $\lambda + d\lambda$ *are said to be resolved in a certain order grating spectrum if the principal maximum of one falls on the first secondary minimum of the other.* For the nth order principal maximum for a wavelength λ for normal incidence and a grating element $(a + b)$, we have

$$(a+b)\sin\theta = n\lambda \qquad ...(i)$$

If the angle of diffraction θ changes by a small amount $d\theta$ corresponding to a change in wavelength $d\lambda$, then differentiating, we have

$$(a+b)\cos\theta \,.\, d\theta = n\, d\lambda$$

or

$$d\theta = \frac{n\, d\lambda}{(a+b)\cos\theta} \qquad ...(ii)$$

If light after diffraction enters the objective of a telescope of diameter *d*, then the smallest angl $d\theta'$ that it can resolve for a wavelength λ is given by

$$d\theta' = \frac{\lambda}{d}$$

Thus the principal maxima in a certain order spectrum corresponding to wavelengths λ and $(\lambda + d\lambda)$ having angular separation of $d\theta$ will be just resolved by the telescope, if

$$d\theta = d\theta'$$

or

$$\frac{n\, d\lambda}{(a+b)\cos\theta} = \frac{\lambda}{d}$$

$\therefore$ Resolving power $$\frac{\lambda}{d\lambda} = \frac{n\,.\,d}{(a+b)\cos\theta} \qquad ...(ii)$$

If the total width of the grating AB has N lines, then

$$AB = N(a+b)$$

Now diameter of the objective of the telescope

$$AC = d = AB\cos\theta$$
$$= N(a+b)\cos\theta$$

Hence from (ii), we have

Resolving power $$\frac{\lambda}{d\lambda} = \frac{n\,N(a+b)\cos\theta}{(a+b)\cos\theta}$$
$$= nN.$$

Fig. 18.6

In practice it is difficult to find two spectral lines in a grating spectrum which just fulfill this requirement. Moreover it is essential that the entire grating surface should be illuminated uniformly. Hence to find the resolving power of the grating, the smallest portion of the grating is found out with an adjustable slit which just resolves the two neighbouring lines of wavelength λ and $(\lambda + d\lambda)$ and then the maximum resolving power of the grating is calculated.

Procedure. 1. Set the spectrometer, mount the grating on the turn table and after doing the optical levelling set the grating normal to the incident light.

2. Illuminate the slit with sodium light and turn the telescope from the normal position to one side to obtain the *first order* spectrum in which the two *D*-lines of sodium of wavelengths 5890 Å and 5896 Å are seen. Clamp the telescope so that the vertical cross-wire coincides with the middle of the two lines. Find the angle of diffraction θ_1 in the first order.

3. Mount an adjustable slit on to the objective of the telescope and adjust its width. The two lines will be found to approach each other. Continue reducing the width of the slit to reduce the part of grating producing the spectrum, until the two lines just cease to be resolved.

4. To measure the width of the slit note the reading on the micrometer scale. Carefully remove the slit and turn the screw slowly till the slit just closes and no light is visible. Note the reading of the micrometer screw and find the width of the slit and hence the portion of the grating which just resolves the two wavelengths of the *D*-lines of sodium in the first order. Make several trials and repeat the experiment for the first order on the other side of the direct image.

Note : If the micrometer screw is not provided then measure the width of the slit with the help of the travelling microcope.

5. Measure the full width of the grating and the diameter of the objective of the collimator as also that of the telescope lenses.

6. Repeat the experiment using the second order spectrum.

Observations.

Vertier constant of the spectrometer =

Number of lines on the grating (per inch) N =

Grating constant $(a+b) = \dfrac{2.54}{N}$

Wavelength of D_1 line of sodium $= 5896 \times 10^{-10}$ m

Wavelength of D_2 line of sodium $= 5890 \times 10^{-10}$ m

Mean wavelength $\lambda = 5893 \times 10^{-10}$ m

Difference in wavelength $d\lambda = 6 \times 10^{-10}$ m

Least count of the micrometer screw = cm

Full width of the grating = cm

Diameter of the collimator lens = cm

Diameter of the objective of telescope = cm

(b) Reading of the angle of diffraction

	Vernier	*Telescope reading*			*Angle of diffraction*		Mean θ
		Left	*Direct*	*Right*	*Left*	*Right*	
First order	V_1						
	V_2						$\theta_1 =$
Second order	V_1						
	V_2						$\theta_2 =$

(b) Reading of width of slit when lines are just resolved

			Reading of micrometer			
			Lines just resolved (p)	*Slit closed (q)*	*Width of slit p − q*	*Mean width (d)*
First order	Left	*(i)*				
		(ii)				
		(iii)				
	Right	*(i)*				
		(ii)				
		(iii)				
Second order	Left	*(i)*				
		(ii)				
		(iii)				
	Right	*(i)*				
		(ii)				
		(iii)				

Calculations.

First order.

Width of grating used AB $= \dfrac{d}{\cos\theta_1} =$

Number of lines in width AB of grating $= \dfrac{AB}{(a+b)} = N'$

$\therefore$ Resolving power $r = nN' =$

If the diameter of the collimator lens is equal to or greater than the width of grating, then

Maximum resolving power $= \dfrac{\text{Full width of grating}}{AB} \times r$

Theoretical resolving power $= N\,n =$

Actual resolving power $\dfrac{\lambda}{d\lambda} = \dfrac{5893}{6} = 982$

Second order. Make similar calculations taking the corresponding values of θ and *d*.

Precautions. 1. The width of the slit should be adjusted so that the two lines are just resolved.

2. The screw should be turned only in one direction while adjusting the width of the slit, to avoid backlash error.

3. The final reading of the micrometer should be taken when the two jaws of the slit are just in contact.

4. The adjustment for the width of the slit should be done a number of times and the mean thus obtained should be used in calculations.

18.7. Dispersive power of a grating. *The dispersive power of a grating is defined as the change in the angle of diffraction corresponding to a unit change in the wavelength.*

The expression for the principal maxima in the *n*th order for a diffraction grating is

$$(a+b)\sin\theta = n\lambda \qquad \text{...(i)}$$

The expression shows that θ varies with wavelength. If the wavelength changes from λ to (λ + *d*λ), then the angle of diffraction will change from θ to (θ + *d*θ). The ratio $\dfrac{d\theta}{d\lambda}$ is called the **dispersive power of the grating.**

Differentiating relation (*i*), we get

$$(a+b)\cos\theta\, d\theta = n\, d\lambda$$

or Dispersive power $\dfrac{d\theta}{d\lambda} = \dfrac{n}{(a+b)\cos\theta}$

Experiment 18.6. To prove that for a plane transmission diffraction grating the dispersive power $\dfrac{d\theta}{d\lambda} = \dfrac{n}{(a+b)\cos\theta}$

Apparatus Same as in **Expt. 18.2.**

Theory. Dispersive power of a grating

$$\frac{d\theta}{d\lambda} = \frac{n}{(a+b)\cos\theta}$$

If θ_1 is the angle of diffraction for wavelength λ_1, θ_2 for a close wavelength λ_2 and θ is the mean angle of diffraction for the mean wavelength in a given order, then

$$\frac{d\theta}{d\lambda} = \frac{\theta_1 - \theta_2}{\lambda_1 - \lambda_2} = \frac{n}{(a+b)\cos\theta}$$

Procedure. 1. Find the angles of diffraction in the first and second order for the wavelengths λ_1 and λ_2 corresponding to D_1 and D_2 lines of sodium light.

Observations. Tabulate as **in Expt 18.2** separately for wavelengths λ_1 and λ_2 to find the mean values of the angle of diffraction in the first and second order.

Grating element $(a+b) = \dfrac{2.54}{N} =$ cm = m

Wavelength of D_1 line $\lambda_1 = 5896 \times 10^{-10}$ m

Wavelength of D_2 line $\lambda_2 = 5890 \times 10^{10}$ m

$d\lambda = \lambda_1 - \lambda_2 = 6 \times 10^{-10}$ m

Order of spectrum (n)	*Angle of diffraction for* λ_1 θ_1	λ_2 θ_2	*Mean angle of diff.* $\theta = \dfrac{\theta_1+\theta_2}{2}$	$d\theta = \theta_1 - \theta_2$	Dispersive power $\dfrac{d\theta}{d\lambda}$
1st					
2nd					

For the 1st order $\dfrac{n}{(a+b)\cos\theta} =$

$\dfrac{d\theta}{d\lambda} =$

For the 2nd order $\dfrac{n}{(a+b)\cos\theta} =$

$\dfrac{d\theta}{d\lambda} =$

We find that in any order $\dfrac{d\theta}{d\lambda} = \dfrac{n}{(a+b)\cos\theta}$

Precautions Same as in **Expt. 18.2.**

18.8. Corona plate. It is a metal plate having a small hole about 0.25 to 0.5 mm in diameter. Along two perpendicular lines passing through the centre of the hole, pin holes are made at distance of 1 cm apart. The function of the pin-holes is to have equal distance from the centre of the hole. The plate is mounted on an upright of an optical bench and light from the sodium lamp is focussed on the central hole. It is used to find the diameter of lycopodium powder by the corona ring method or knowing the diameter, the wavelength of light can be found out. For this purpose a glass plate is taken and is dusted with lycopodium powder. This plate is adjusted at a distance from the corona plate and on looking through the glass plate circular diffraction fringes of yellow colour can be seen.

Theory. When light from a small circular hole is made to fall on small lycopodium particles, it is diffracted and a diffraction pattern *i.e.*, a bright spot surrounded by dark and bright rings can be seen.

If λ is the wavelength of light and d the diameter of the lycopodium powder, then the angle of diffraction θ is given by the relation

$$\sin\theta = \frac{1.22\,\lambda}{d}$$

The angle of diffraction can be found out by adjusting the position of the glass plate from the corona plate so that the first dark ring coincides with the first set of pin holes. If D is the distance of the glass plate from the corona plate and x the distance of first set of pin-holes from the centre of the circular hole, then as θ is small

$$\sin\theta = \tan\theta = \frac{x}{D}$$

or $$\frac{1.22\,\lambda}{d} = \frac{x}{D}$$

$\therefore$ $$d = \frac{1.22\,\lambda\,D}{x}$$

If the first dark ring coincides with the 2nd or 3rd etc set of pin holes and x is the distance of the corresponding set of pin holes from the centre, then

$$d = \frac{1.22\,\lambda\,D}{x}$$

Experiment 18.7. To determine the diameter of lycopodium powder particle using corona rings.

Apparatus. A corona plate, a glass plate, lycopodium powder, a high-power microscope an optical bench with two uprights, source of sodium light etc.

Procedure. 1. Sprinkle a thin uniform layer of lycopodium powder on a glass plate.

2. Mount the corona plate on an upright of the optical bench placed at the extreme end and mount the glass plate with thin layer of lycopodium powder on the second upright and find the index correction between the two. Illuminate the central hole of the corona plate with sodium light and look through the glass plate on the central hole. Diffraction circular fringes of yellow colour will be visible.

3. Adjust the distance of the glass plate upright, keeping the eye close to the glass plate, so that the first dark ring coincides with the first set of pin-holes distant x from the centre of the circular hole. Note the positions of the two uprights.

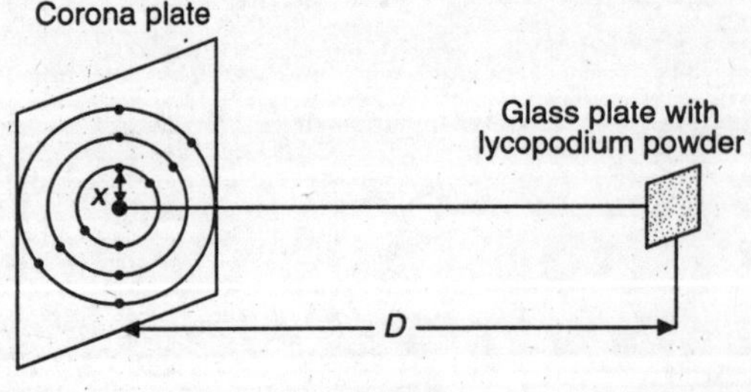

Fig. 18.7

4. Alter the position of the glass plate so that the first dark ring coincides with the second set of pin-holes. Repeat the experiment for third set of pin-holes.

5. Remove the corona plate and measure the distance of different sets of pin-holes from the centre of the circular hole with a travelling microscope. Find the mean value of $\frac{D}{x}$ and calculate the value of the diameter of lycopodium powder particle taking $\lambda = 5893$ Å.

Observations.

Wavelength of light λ = 5893×10^{-8} cm

Index correction for D = ± cm

No.	Position of		Value of $D = (a - b)$		x	$\frac{D}{x}$	Mean $\frac{D}{x}$
	Corona plate (a)	glass plate (b)	obs.	correct			
1							
2							
3							
:							

Calculations.

Diameter of lycopodium powder particle

$$d = 1.22\,\lambda \frac{D}{x} = \quad \text{cm}$$

Precautions 1. There should be a uniform thin layer of lycopodium powder on the glass plate.

2. The light from sodium lamp should illuminate all the holes in the corona plate.

3. The middle portion of the dark ring should coincide with the centre of the pin-holes.

4. Index correction should be found out and applied to find the correct value of D.

Oral Questions

1. Diffraction (*i*) What is diffraction? (*ii*) What is difference between interference and diffraction? (*iii*) What are the two kinds of diffraction? (*iv*) Give examples of Fresnel and Faunhaufer diffraction. (*v*) How do you find the diameter of a wire using Fresnel diffraction?

2. Diffraction grating. (*i*) What is diffraction grating and how is it constructed? (*ii*) Do you use the original grating in the laboratory? (*iii*) How is a replica prepared? (*iv*) How do you set the grating for normal incidence? (*v*) How many orders of spectra are possible with a grating of 15000 lines per inch? (*vi*) What is an absent spectra? (*vii*) What is the condition that the second order spectra may be absent? (*viii*) Is the width of the transparency of your grating equal to the width of the opacity? If not, how do you know it? (*ix*) Why is prism spectrum more intense than the grating spectrum?

3. Resolving power and dispersive power of a grating. (*i*) What is resolving power of a grating and on what factors does it depend? (*ii*) What is Rayleigh criterion of resolution? (*iii*) The two sodium lines appear separate in the second order spectrum but not in the first order, why? (*iv*) With light from sodium lamp, in addition to the prominent yellow lines some other lines are also visible, what are these due to? (*v*) What do you mean by dispersive power of a grating? (*vi*) What are the factors on which (*a*) resolving power (*b*) dispersive power depend? (*vii*) What is the difference between resolving power and dispersive power?

4. Rydberg's constant (*i*) What are the various spectral series of hydrogen spectrum? (*ii*) Give Bohr's formula for each of the series of hydrogen spectrum (*iii*) What is Rydberg's constant? State its value and units. (*iv*) What is the value of Rydberg's constant according to Bohr's theory in terms of fundamental constants like *h, e, m* etc.? (*iv*) Which of the series lies in the ultra viotet, visible, infrared and far-infrared regions? (*vi*) Distinguish between frequency and wave number. What is the relation between them?

5. Corona rings and diameter of lycopodium powder. (*i*) What is the principle of corona rings? (*ii*) What type of fringes do you get and why? (*iii*) Why should the hole in the middle of the metal plate be small? (*iv*) What is the function of the pin-holes along two mutually perpendicular lines drawn on the metal plate? (*v*) Can you use corona plate to find the diameter of the lycopodium powder? (*vi*) Can white light be used in this experiment?

19

Polarisation

19.1. Polarisation. Light is a kind of transverse wave motion. In ordinary light the vibrations take place in all directions perpendicular to the direction of propagation. Such light has the same properties in all directions and is known as *unpolarised light*. When light is passed through crystals, like tourmaline, quartz etc., it acquires the property of one-sidedness *i.e.*, the vibrations take place only in one plane. In such a case light is said to be *plane-polarised*. An arrangement for producing polarised light is called a *polariser*. A nicol prism is generally used for the purpose.

19.2. Nicol prism. It is an optical device made from calcite and is used in many instruments for producing and analysing plane polarised light.

When light is passed through a doubly refracting crystal it is split up into *an ordinary and an extraordinary ray*. Both these rays are plane polarised. In the nicol prism one of these rays is cut off by total internal reflection.

Construction. A calcite crystal about three times as long as it is wide is taken. The principal section *ABCD* is shown in Fig. 19.1. Its end faces are cut down so as to reduce the angles at *B* and *D* from 71° to 68° with principal section. The crystal is then cut apart along *A′C′* perpendicular both to the principal plane and the end faces such that *A′C′* makes an angle of 90° with the ends *A'B* and *C′ D*. The two cut surfaces are ground, polished optically flat and cemented together with canada balsam which is a clear transparent cement whose refractive index lies midway between the refractive indices of calcite for the ordinary and the extraordinary rays. The sides of the prism are blackened to absorb the totally reflected ray.

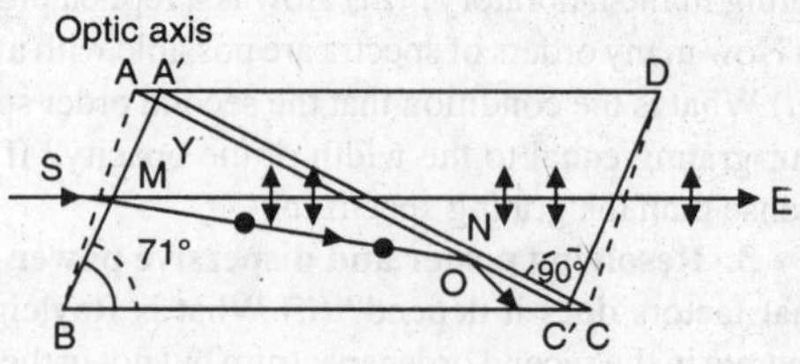

Fig. 19.1

A ray *SM* incident on the face of the prism is split up into the ordinary and the extraordinary rays. The ordinary ray is totally reflected at *N* whereas the extraordinary ray passes through giving rise to plane polarised light having vibrations in the principal plane.

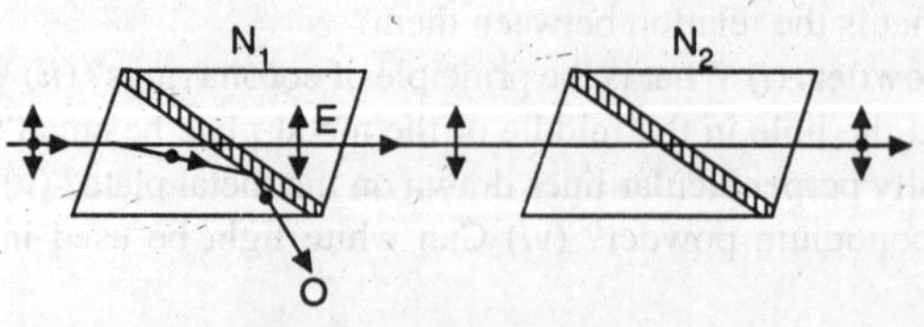

Fig. 19.2

The nicol prism can be used both as a polariser as well as an analyser. When the principal section of the two nicols are parallel as shown in Fig. 19.2, the extraordinary ray is transmitted through the analyser. If one of the nicols is rotated the intensity of the transmitted beam decreases and finally no light is transmitted. The extraordinary ray for the first nicol forms an ordinary ray for the second and is thus totally reflected. In this position the two nicols are said to be *crossed* as shown in Fig. 19.3. The first nicol polarises the light and is called the **polariser**, whereas the second nicol analyses the light and is called the **analyser**.

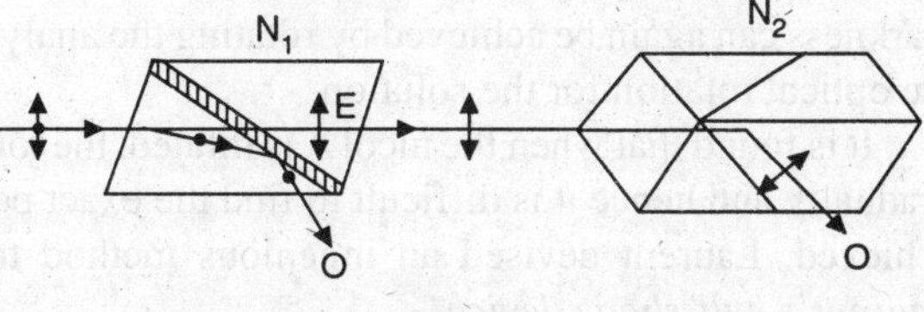

Fig. 19.3

Again when the analyser is further rotated the two nicols are in parallel position and E-ray is again transmitted as shown in Fig. 19.4.

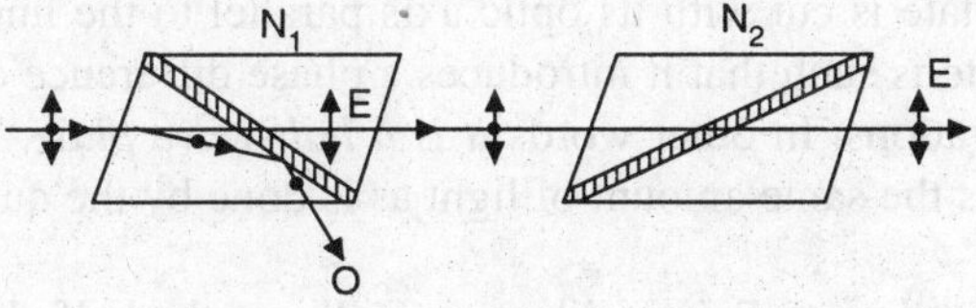

Fig. 19.4

19.3. Specific rotation. Certain substances like quartz, turpentine and sugar solution rotate the plane of polarisation of a plane polarised light passing through them. Such substances are called *optically active*. There are two types of substances.

(*i*) *Dextro-rotatory or right-handed.* These produce the rotation of the plane of polarisation towards the right *i.e.* in the clockwise direction on looking towards the source.

(*ii*) *Laevo-rotatory or left-handed.* These produce the rotation of the plane of polarisation towards the left *i.e.*, in the anti-clockwise direction on looking towards the source.

The angle through which the plane of polarisation is rotated depends upon

(*i*) thickness of the substance,

(*ii*) concentration of the solution or the density of the material,

(*iii*) wavelength of light, and

(*iv*) temperature.

Specific rotation for the given wavelength of light at a given temperature is defined as the rotation produced by one decimetre (10 centimetre) length of the solution of unit concentration.

If θ is the rotation produced by l decimetere of a solution and c is the concentration in gram per c.c., then specific rotation S at a given temperature and corresponding to a wavelength λ is given by

$$[S]_\lambda^t = \frac{\theta}{lc} = \frac{\text{Rotation in degrees}}{\text{Length in decimetres} \times \text{concentration in gm/c.c.}}$$

19.4. Laurent's half-shade polarimeter. It is an instrument used for finding the optical rotation of certain solutions. When used for finding the optical rotation of sugar it is called a saccharimeter. If the specific rotation of sugar is known, the concentration of the sugar solution can be determined.

Construction. The optical parts of a polarimeter are shown in Fig. 19.5 in which S is a source of monochromatic light usually a sodium lamp.

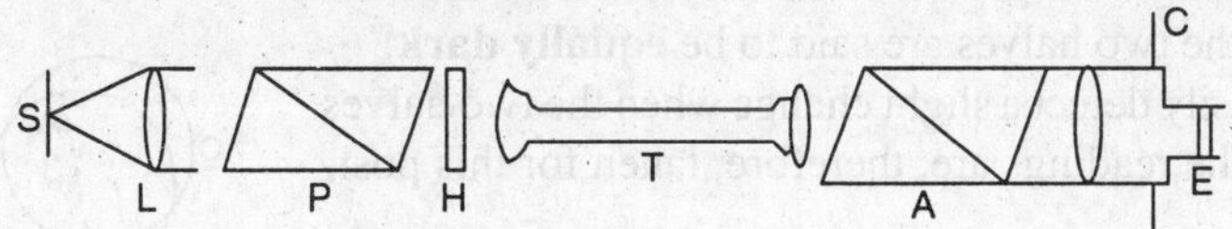

Fig. 19.5

The light from the source after passing through a narrow slit is rendered into a parallel beam by the lens L. The light is made plane polarised by the nicol prism P and after passing through the half-shade device H (described later), a glass tube T containing the solution is made to fall on the analysing nicol A. The light is viewed through a Galilean telescope E. The analysing nicol A can be rotated about the axis of the tube and its rotation can be measured on the graduated circular scale C divided in degrees, with the help of a vernier.

The position of the analyser is adjusted so that the field of view is completely dark. The tube T is filled with the required solution and placed in position. The field now becomes illuminated. Darkness can again be achieved by rotating the analysing nicol A through a certain angle which gives the optical rotation for the solution.

It is found that when the nicol A is rotated, the total darkness of the field of view is attained rather gradually and hence it is difficult to find the exact position correctly for which complete darkness is achieved. Laurent devised an ingenious method to achieve this. The arrangement is known as *Laurent's half-shade device*.

Laurent's half-shade device. It consists of a semi-circular plate ADB of glass cemented to a semi-circular plate ACB of quartz. The quartz plate is cut with its optic axis parallel to the line of separation AOB. The thickness of the quartz plate is such that it introduces a phase difference of π between the ordinary and the extraordinary vibrations. In other words *it is a half-wave plate*. The thickness of the glass plate is such that it absorbs the same amount of light as is done by the quartz plate.

Suppose that light after passing through the polariser P, is incident normally on the half-shade plate and has vibrations along OP. On passing through the glass half the vibrations will remain along OP but on passing through the quartz half, these will be split up into E and O components. The vibrations of the O component are along OD and those of E-component along OA.

Fig. 19.6

On passing through the quartz plate a phase difference of π is introduced between the two vibrations. The O-vibration will advance in phase by π and will take place along OC instead of OD. The resultant vibration on emergence from the quartz plate will be along OQ, so that

$$\angle POA = \angle QOA.$$

If the analysing nicol is fixed with its principal plane parallel to OP, the plane polarised light through glass will pass and hence it will appear brighter than the quartz half from which light is partially obstructed.

If the principal plane of the nicol is parallel to OQ the quartz half will appear brighter than the glass half due to the same reason.

When the principal plane of the analysing nicol is parallel to AOB, the two halves will appear **equally bright.** It is because the vibrations emerging out of the two halves are equally inclined to its principal plane and hence the two components have equal intensities.

When the principal plane of the analyser is at right angles to AOB, again the components of OP and OQ are of equal intensity. The two halves are again equally illuminated but as the intensity of the component passing through is small as compared to those in the previous case, the two halves are said to be **equally dark**.

The eye can easily detect a slight change when the two halves are equally dark. The readings are, therefore, taken for this position.

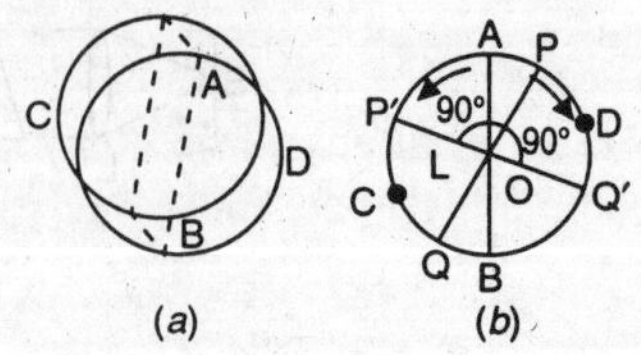

Fig. 19.7

19.5. Bi-quartz A biquartz is a simple and sensitive device which can be used in place of a half shade arrangement in a polarimeter. It consists of two semi-circular quartz plates ACB and ADB each 3.75 mm thick, one of left handed L and the other of right handed R quartz cut with their optic axes perpendicular to their refracting faces. These are cemented together along AB to form a circular plate as shown in Fig. 19.7 (*a*).

When plane polarised white light is incident normally on the plate each half of this plate will rotate each colour equally in opposite directions. For yellow light this rotation is about 90°. If POQ

is the direction of incident vibration, then for yellow light it will be along OQ' in the right handed half and along OP' in the left handed half. Hence the vibration of yellow light will be along the direction $P'OQ'$ perpendicular to the direction POQ. Thus if the principal plane of the analysing nicol is parallel to POQ yellow light will be quenched in both halves of the field, while other colours will be present in the same proportion thereby giving the same tint of greyish violet colour to both the halves. This is known as the tint of passage. If the analyser is rotated to one side from this position one half of the field of view appears blue while the other half appears red. If the analyser is rotated in the opposite direction, the first half which was blue earlier appears red and the second half which was red earlier appears blue. As the transition from red to blue is very rapid, the zero position can be obtained very accurately.

Experiment 19.1. To determine the specific rotation of sugar using Laurent's half-shade polarimeter.

Apparatus. Laurent's half-shade polarimeter, a sodium lamp, sugar, a balance, a weight box, a graduated cylinder, two beakers, filter paper, a funnel, an eye-piece, a pipette and a glass rod.

Theory. If θ is the optical rotation produced by l decimetres of a solution and c the concentration in gram per *c.c.*, then specific rotation S at a given temperature t and corresponding to a wavelength λ is given by

$$[S]_\lambda^t = \frac{\theta}{lc} = \frac{\text{Rotation in degreess}}{\text{Length in decimetres} \times \text{concentration in gm/c.c.}}$$

Procedure. 1. *Preparation of 20% solution.* Taken a clean dry beaker and weigh it. Add about 20 gram of sugar in it and weigh again. Calculate the volume of solution to have a 20% strength as follows :—

$$\text{Volume required} = \frac{m \times 100}{20} \; c.c.$$

where m is the mass of sugar in the beaker.

Add nearly half this volume of water in the beaker and stir well till the whole of sugar is dissolved. Add more water, if necessary taking care that the volume of solution is *less* than the calculated value of the volume.

Transfer the solution into a graduated cylinder. Rinse the beaker with a small quantity of water and add to the cylinder. Make the required volume by adding more water little by little with a pipette. Filter the solution in a clean beaker and cover it.

2. **Setting.** Find the vernier constant of the circular scale. Place the polarimeter so that the aperture is in front of the sodium lamp (or a sodium flame). Look through the telescope and adjust the position of the eye-piece so that the two halves of the half shade device are clearly in focus.

3. Remove the brass caps of the polarimeter tube. Clean the tube as well as the glass windows. Now replace one of the caps in position *taking care that no strain is exerted on the glass window.* Hold the tube in a vertical position and fill it with water. Slip the second glass window gently on the tube taking care *that no air bubbles are left underneath it*. Screw the cap gently *taking care that no strain is exerted on the glass window.*

4. Place the tube in position in the polarimeter and cover it. Rotate the analysing nicol by rotating the circular scale till the *two halves of the half shade device are in* **equally dark** *position. In this position there will be an abrupt change in the intensity of the two halves when slight rotation on either side is given*. Note the reading on the scale.

Turn through 180°, set the analyser again for equally dark position and note the reading on the scale.

Note. *If the instrument is properly set, there is no strain on the glass windows, and water is free from optically active impurities, the scale reading will be 0* and *180* in the two positions respectively.

5. Now fill the tube with 20% sugar solution prepared as explained above. Place it in position and note the scale readings when the two halves are in equally dark position. Repeat by turning through 180°.

6. Take 30 *c.c.* of the 20% sugar solution and make the volume to 60 *c.c.* so as to reduce the strength to *one-half* (10%). Fill the tube with this 10% solution and take the readings as explained above.

7. Similarly repeat with 5% and 2.5% solutions.

Observations. Weight of the beaker = gm

Weight of beaker + sugar = gm

Weight of sugar $m =$ gm

Volume of solution required $= \frac{m}{20} \times 100$ *c.c.* =

Vernier constant =

Sl. No.	*Strength of solution in gram per 100 c.c.*	*per c.c.(c)*	*Scale readings through solution 1st pos.*	*2nd pos.*	*Rotation 1st pos.*	*2nd pos.*	*Mean* θ	$\frac{\theta}{c}$
1	Zero							
2	20							
3	10							
4	5							
5	2.5							

Length of the tube l = decimetre

Mean θ/c =

∴ Specific rotation $\frac{\theta}{lc}$ =

Temperature of water t = °C

Value of specific rotation from tables at t°C =

Percentage error =

Laevo-rotatory or dextro rotatory =

Precautions. 1. The tube and the glass windows should be clean.

2. There should be no air bubble in the glass tube when the tube is filled with the solution.

3. The caps should be screwed in such a manner that there is no leakage. These should not be made very tight so as to strain the glass windows. Explain why?

4. The reading should be taken in the equally dark position and not in the equally bright position.

5. Before filling the tube with a solution the tube should be rinsed with the same solution.

Experiment 19.2. Plot a graph between the concentration and rotation for various strengths of sugar solution and hence find (*a*) the specific rotation and (*b*) the concentration of the given sugar solution.

Apparatus Same as in **Expt. 19.1.**

Theory. If we plot a graph between concentration of sugar solution c taken along X-axis and corresponding rotation of plane of polarisation θ along the Y-axis, the graph is a straight line as shown in Fig. 19.8.

The slope of the graph gives the value of θ/c from which we have specific rotation $= \frac{1}{l}\frac{\theta}{c}$ where is the length of the polarimeter tube in decimetre. To find the concentration of the given solution the tube is filled with this solution and the corresponding value of rotation θ is observed. The value of concentration against this value of θ is noted from the graph.

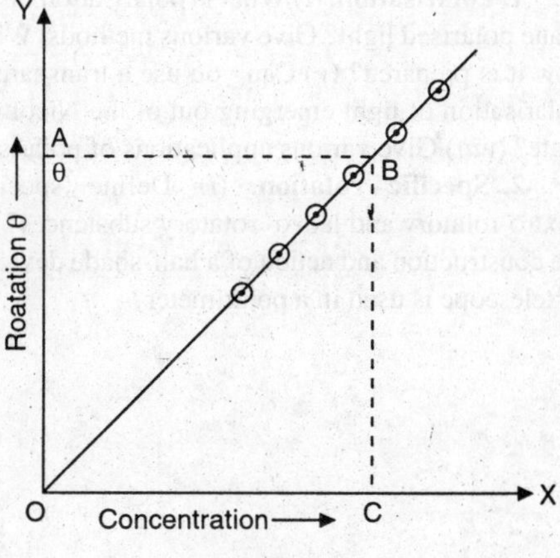

Fig. 19.8

Procedure 1. Find the rotation for 20%, 10% and 5% solution of sugar as explained in **Expt. 19.1.**

2. Now take 40 c.c of the original 20% sugar solution and add water to make the volume 50 c.c. This will make a 16% solution. Now find the angle of rotation for 16%, 8% and 4% as explained in **Expt. 19.1.**

3. Taking concentration in gm/c.c. along the *X*-axis and rotation θ along *Y*-axis, plot a graph. The graph is a straight line.

4. To find the concentration of the given sugar solution take a point *A* on the *Y*-axis corresponding to the value of rotation θ produced by this solution. Draw *AB* parallel to *X*-axis cutting the graph at the point *B*. Draw *BC* perpendicular to *X*-axis, then the value of concentration corresponding to the point *C* on the *X*-axis is the concentration of the given sugar solution.

5. The graph passes through the origin *O*. The slope of the graph $\frac{BC}{OC}$ gives the mean value of $\frac{\theta}{c}$.

Observations Vernier constant =

S.No.	*Strength of solution in gm*		*Scale reading through solution*		*Rotation*		*Mean rotation θ*
	per 100 cc	*per cc(c)*	*1st position*	*2nd position*	*1st position*	*2nd position*	
1	Zero						
2	20						
3	10						
4	5						
5	16						
6	8						
7	4						
8	Unknown	*x*					

Value of θ corresponding to unknown concentration $x = \theta =$

∴ [value of concentration x from the graph corresponding to the point C] = gm/cc. =

Slope of the graph $= \frac{BC}{OC} = \frac{\theta}{c}$

Length of the polarimeter tube l = decimetre

∴ Specific rotation $= \frac{1}{l}\frac{\theta}{c} = \frac{1}{l}\frac{BC}{OC} =$

Precautions : Same as in **Expt. 19.1.**

Exercise 2. *Find the specific rotation of turpentine.*

Oral Questions

1. Polarisation. (*i*) What is polarisation of light? (*ii*) What is plane polarised light? (*iii*) How do you obtain plane polarised light? Give various methods. Which method is the best and why? (*iv*) What is a Nicol prism and how it is prepared? (*v*) Can you use a transparent cement other than canada balsam? (*vi*) What is the plane of polarisation of light emerging out of the Nicol prism? (*vii*) What is a quarter wave plate? What is a half wave plate? (*viii*) Give various applications of polarisation. (*ix*) Explain the working of a biquartz.

2. Specific rotation. (*i*) Define specific rotation. On what factors does it depend? (*ii*) What are dextro-rotatory and laevo- rotatory substances? (*iii*) What are the various parts of the polarimeter? (*iv*) Explain the construction and action of a half-shade device. In which position is it most sensitive and why? (*v*) What type of telescope is used in a polarimeter?

PART IV

SOUND

20

Vibrations of stretched Strings

20.1. Vibrations of stretched strings. In sound a *string means a cord of any material whose length is very large as compared to its diameter and which is perfectly flexible and uniform in cross-section.* A thin metal wire fulfils these conditions quite satisfactorily.

When a string stretched between two fixed points is plucked in a direction at right angles to its length, *stationary transverse vibrations* are set up in it. The string may vibrate as a whole in one segment as shown in Fig. 20.1 (*i*). The frequency of sound produced by the string in this case is the lowest and is called the *fundamental*. It may vibrate in two or more segments as shown in Fig. 20.1 (*ii*) and (*iii*). In these cases the frequency of sound produced will be double and three times the fundamental frequency respectively. These frequencies are known as *overtones* or *harmonics*.

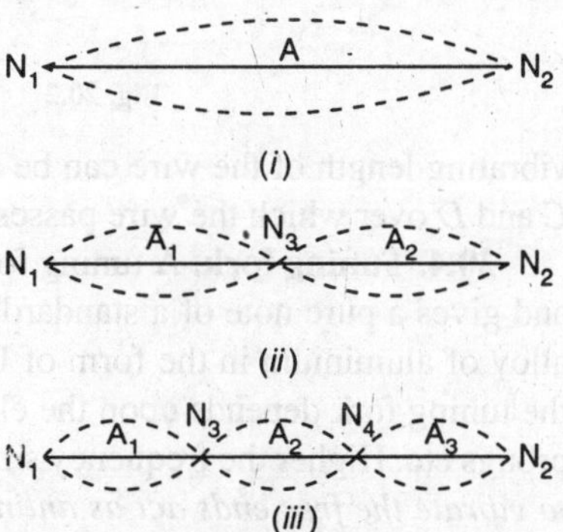

Fig. 20.1

Node. *The fixed points like N_1, N_2 ...etc. where the amplitude is zero but the strain is maximum are called* **nodes.**

Antinode. *The points like A, A_1, A_2...etc., where the amplitude is maximum but the strain is the least are called* **antinodes.**

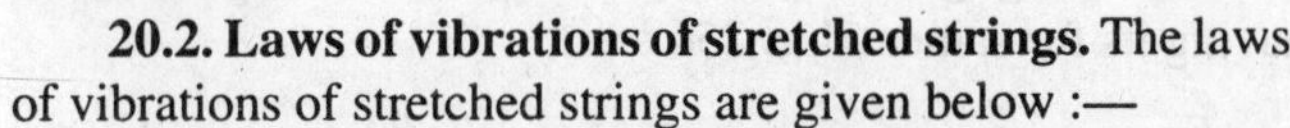

20.2. Laws of vibrations of stretched strings. The laws of vibrations of stretched strings are given below :—

The frequency of the fundamental note *n* of a stretched string is

(*i*) inversely proportional to the length *l* *i.e.* $n \propto \frac{1}{l}$;

(*ii*) directly proportional to the square root of the stretching force *T i.e.*,

$$n \propto \sqrt{T}$$

(*iii*) inversely proportional to the square root of the density ρ of the material of the wire *i.e.*,

$$n \propto \sqrt{1/\rho}$$

(*iv*) inversely proportional to the diameter *D* of the wire *i.e.*

$$n \propto 1/D$$

Combining all these laws, wet get

$$n = \frac{1}{lD}\sqrt{\frac{T}{\pi\rho}}$$

$$= \frac{1}{2l}\sqrt{\frac{T}{\frac{\pi D^2 \rho}{4}}} = \frac{1}{2l}\sqrt{\frac{T}{m}}$$

where $m = \frac{\pi D^2 \rho}{4}$ is the mass per unit length of the wire.

In other words the third and the fourth laws can be combined into one which states that the frequency of the fundamental note of a stretched string

(*iii*) is inversely proportional to the square root of the mass per unit length of the wire *i.e.*,

$$n \propto \frac{1}{\sqrt{m}}$$

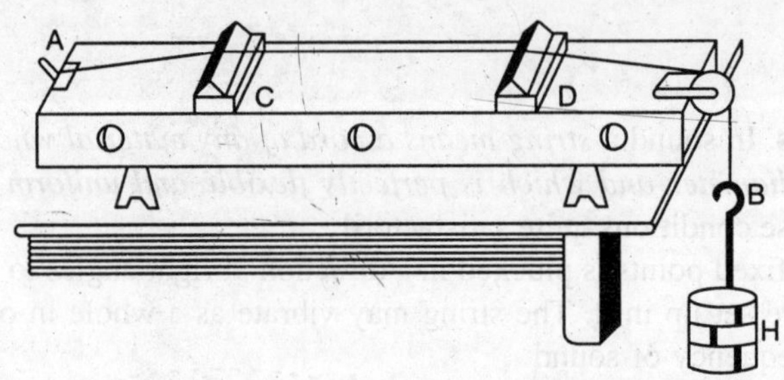

Fig. 20.2

20.3. Sonometer or monochord. The laws of vibrations of stretched string can be verified by a sonometer as shown in Fig. 20.2.

It consists of a hollow sounding box, having two movable bridges. One end of the string is fixed to a peg at one end of the sounding box while the second end passes over a pulley fixed at the other end of the sounding box, and carries a hanger on which slotted weights can be slipped. The vibrating length of the wire can be adjusted by means of two sharp knife edges *C* and *D* over which the wire passes.

20.4. Tuning fork. A tuning fork is a valuable instrument in the laboratory and gives a pure note of a standard frequency. It is usually made of steel or an alloy of aluminium in the form of U provided with a handle. The frequency of the tuning fork depends upon the elasticity of the material and the length of the prongs etc. Higher the frequency smaller is the length of the prongs. *When made to vibrate the free ends act as antinodes and the middle point as the node.* The changes in frequency with temperature are almost negligible. The frequencies of the various notes on the diatonic scale having its fundamental as 256 are :

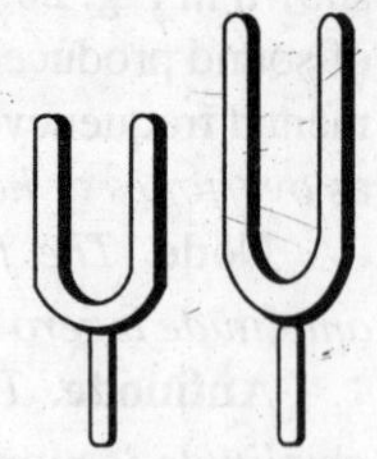

Fig. 20.3

C	*D*	*E*	*F*	*G*	*A*	*B*	*C*
256	288	320	341.3	384	426.6	480	512

The tuning forks in the laboratory generally have these values marked on them.

Experiment 20.1. To determine the frequency of a tuning fork using a sonometer.

Apparatus. A sonometer, kilogram weights, two wedges, wooden blocks, a pair of scissors, sensitive balance, a weight box, a metre rod and a rubber hammer.

Theory. The frequency of the fundametal note of a stretched string is given by

$$n = \frac{1}{2l}\sqrt{\frac{T}{m}}$$

In S.I. units l, the length of string is in metres, T the stretching force in newtons and m the mass of the string in kg/metre.

Procedure. 1. *To set the sonometer.* Place two sharp wedges under the wire of the sonometer. Load the hanger with a load of about 4 kilogram including the hanger. *See that the wire is properly* *stretched between the wedges and touches both of them.* Wooden blocks may be placed under the wedges, if necessary. *See that there are no kinks in the wire.*

2. (*i*) Bring two wedges *close to each other*. Cut a piece of paper about 2 cm long and about 2 mm wide. Fold it into a *V* shape. Place this paper '*rider*' on the wire between the wedges.

(*ii*) Take the tuning fork of the frequency provided. Strike gently at the end of one of its prongs with a rubber hammer to set it into vibrations. Press the stem against the sonometer board keeping the tuning fork *vertical*. At the same time move one of the wedges outwards *very slowly* till the rider, *kept always on the wire, midway between the two wedges*, begins to flutter. Strike the tuning fork again and again during this adjustment and press against the board to keep it vibrating. If the adjustment is now made very carefully the '*rider*' flies off at a certain fixed distance between the wedges. The wire now vibrates in resonance with the frequency of the tuning fork.

(*iii*) Measure the length of the wire between the wedges by a metre rod. Increase the distance between the wedges by a few centimetres. Again adjust the length of the wire for resonance by slowly decreasing the distance between the wedges.

3. Take such four observations by changing the load.

4. Adjust the two wedges to be exactly 50 cm apart. Mark with ink a point on the wire at the position of each wedge and cut the wire at these points. Find its weight accurately by a sensitive balance.

Observations. Length of the wire = cm = m

Mass of the wire = gm = kg

Mass per metre of the wire = gm = kg

Sl. No.	*Load in kg (W)*	*Tension in Newton* $T = W \times 9.8$	*Length giving a note of frequency n*			$n = \frac{1}{2l}\sqrt{\frac{T}{m}}$
			increasing	*decreasing*	*Mean l (metre)*	
1						
2						
3						
4						

Mean frequency n =

Frequency marked on the tuning fork =

Percentage error =

Precautions. 1. The wire should be of a uniform area of cross- section, free from kinks and should be taut.

2. The observations should start with *minimum* distance between the two wedges as otherwise it is possible to get double the length of wire.

3. The resonance position should be obtained by first slowly increasing the distance between the wedges and then by slowly decreasing it.

4. The rubber hammer should be struck gently on the prong. Never strike the tuning fork on the table.

5. The tuning fork should be pressed gently on the sonometer board.

6. The weight of the hanger should always be included in the load.

7. The pulley should be free from friction

Sources of error. 1. The wire may not be perfectly *flexible*.

2. The tension may not be the same at the two sides of a bridge.

3. Due to friction of the pulley the tension in the string may be less than the actual load applied.

4. Slight error may be caused due to the yielding of the supports.

5. The wire may not be of a uniform area of cross-section.

Experiment 20.2. To verify the laws of transverse vibrations of stretched strings using a sonometer.

Apparatus. A sonometer, 3 different wires, a hanger, half kilogram weights, two wedges, wooden blocks, 3 tuning forks of different frequencies, a pair of scissors, sensitive balance, a weight box, a metre rod and a rubber hammer.

Procedure. *To verify the law of length i.e.,*

$$n \propto \frac{1}{l} \text{ or } nl = \text{constant.}$$

(*i*) Proceed as in **Experiment 20.1.** Keeping the same tension find the length of the wire which vibrates in resonance with each of the three tuning forks.

Observations. Load including the hanger = kg

Sl. No.	*Frequency n*	*Distance between wedges*			$n \times l$
		Length increasing	*Length decreasing*	*Mean l (metres)*	
1					
2					
3					

Verification. Since $n \times l$ = constant, the law of length is verified.

2. *To verify the law of tension i.e.,*

$$n \propto \sqrt{T} \text{ or } \frac{\sqrt{T}}{n} = \text{constant}$$

(*i*) Take a load of 4 kg including the hanger. Obtain the length of the wire which vibrates in resonance with one of the tuning forks as in **Experiment 20.1.**

(*ii*) Increase the load by 1 kg and repeat the experiment with the *same* tuning fork.

(*iii*) Similarly repeat with three different loads.

It is not possible to find directly the frequency with which the same length of wire will vibrate under different loads. If l_1 and l_2 are the lengths of the wire vibrating with the same frequency n_1 under the loads T_1 and T_2 respectively, then the frequency of sound n_2 produced by the same length l_1 under the load T_2 can be calculated as follows :—

Under a load T_2 the length l_2 of the wire produces a frequency n_1.

Applying the law of length.

Under a load T_2 the length l_1 will produce a frequency

$$n_2 = \frac{n_1 \times l_2}{l_1}$$

Observations. Frequency of the tuning fork n_1 =

Sl. No.	*Tension in Newton (W × 9.8)*	*Length of the wire giving the frequency n_1*			*Calculated frequency for a length l_1*	$\frac{\sqrt{T}}{n}$
		Increasing	*Decreasing*	*Mean l (in metres)*		
1	$T_1 =$			$l_1 =$	$n_1 =$	$\frac{\sqrt{T_1}}{n_1} =$
2	$T_2 =$			$l_2 =$	$n_2 = \frac{l_2}{l_1} \cdot n_1$	$\frac{\sqrt{T_2}}{n_2} =$
3	$T_3 =$			$l_3 =$	$n_3 = \frac{l_3}{l_1} \cdot n_1$	$\frac{\sqrt{T_3}}{n_3} =$

Verification. Since $\frac{\sqrt{T}}{n}$ is a constant quantity the law of tension is verified.

3. *To verify the law of mass i.e., $n \propto \sqrt{\frac{1}{m}}$ or $n\sqrt{m}$ is constant.*

(*i*) Stretch the wire with a load of 4 kg including the hanger and find the length of the wire which vibrates in resonance with a tuning fork as explained in **Experiment 20.1.**

(*ii*) Repeat with the same load and the same tuning fork for three different wires.

Note. *The wires may be of different materials and different diameters.*

(*iii*) Cut a wire 0.5 m long in each case and find the weight.

Observations. Frequency of the tuning fork n_1 =

Sl. No.	*Length of the wire giving the frequency n_1*			*Calculated frequency for a length l_1*	*Mass per metre of the wire m (in kg)*			*$n\sqrt{m}$*
	Increasing	*Decreasing*	*Mean l (in metres)*		*Total mass*	*Length*	*m*	
1			$l_1 =$	$n_1 =$			$m_1 =$	$n_1\sqrt{m_1} =$
2			$l_2 =$	$n_2 = \frac{l_2}{l_1} . n_1$			$m_2 =$	$n_2\sqrt{m_2} =$
3			$l_3 =$	$n_3 = \frac{l_3}{l_1} . n_1$			$m_3 =$	$n_3\sqrt{m_3} =$

Verification. Since *$n\sqrt{m}$ is constant the law of mass is verified.*

Precautions and sources of error. Same as in **Experiment 20.1.**

Exercise 1. *Plot a graph between log l and log m and hence verify the law of mass in the case of a stretched string.*

Hints. Find the length of the sonometer wire of mass m_1 per unit length which when stretched by a certain load T vibrates in resonance with a tuning fork of frequency n_1. Similarly for a wire of mass m_2 per unit length find the length l_2 which vibrates in resonance with the same tuning fork under the same load. According to the law of length the frequency of the second wire n_2 for a length l_2 under the same load is given by

$$n_2 = \frac{n_1 l_2}{l_1}$$

If the law of mass is true, then

$$n_1\, m_1^{1/2} = n_2\, m_2^{1/2}$$

or
$$n_1\, m_1^{1/2} = n_1 \frac{l_2}{l_1} m_2^{1/2}$$

or
$$l_1 m_1^{1/2} = l_2\, m_2^{1/2}$$

or
$$l\, m^{1/2} = \text{constant}$$

$\therefore$
$$\log l + \frac{1}{2} \log m = \text{constant.}$$

In other words, a graph between log l and log m is a straight line whose slope is $-\frac{1}{2}$. To plot the graph take at least six wires of different diameters of the same material.

Exercise 2. *Determine the density of the material of the given wire.*

Hints. Using the given tuning fork find the value of m the mass per unit length of the wire.

Now $$m = \pi D^2 \frac{\rho}{4}$$

$\therefore$ $$\rho = \frac{4m}{\pi D^2}$$

Exercise 3. *Determine the diameter of the given wire using a sonometer and a tuning fork of known frequency.*

Hints. Find the value of m, then

$$m = \frac{\pi D^2}{4} \rho$$

or $$D = \sqrt{\frac{4m}{\pi \rho}}$$

Exercise 4. *Determine the weight of the given body i.e., brick using a sonometer.*

Hints. Suspend the given body in place of the hanger. With the help of a tuning fork of known frequency determine the length of the wire which vibrates in resonance with it. Find m the mass per unit length of the wire as already explained. Calculate the value of T in Newton from the relation

$$n = \frac{1}{2l} \sqrt{\frac{T}{m}}$$

$$\text{The weight of the body} = \frac{T \text{ in newton}}{9.8} \text{ kg}$$

Exercise 5. *Compare the frequencies of two given tuning forks using a sonometer.*

Hints. For the same wire under the same tension, we have

$$n_1 l_1 = n_2 l_2$$

$\therefore$ $$\frac{n_1}{n_2} = \frac{l_2}{l_1}$$

20.5. Electrically maintained tuning fork. As shown the stem of the tuning fork is clamped rigidly to a non-metallic stiff base. To one prong of the tuning fork a short metal strip A having a platinum tipped point contact is fixed. The contact with the strip is made with a platinum tipped screw S. An electromagnet M is placed between the two prongs of the fork. One end of the coil of the electromagnet is connected to the screw S and the other to the stem through a key K, a battery B and a rheostat. As the current is passed through the coil of the electromagnet, the prongs are attracted inwards, the contact is broken and the current stops. The electromagnet loses its magnetism and the prongs return back to their mean positions. The circuit is again completed due to which the electro-magnet exerts attracting force on the prongs, but due to inertia, the prongs continue to move outwards till the outward velocity is reduced to zero by the elastic force. Thus the fork does work against the magnetising force. The prongs then return and as they pass the mean position the contact is again broken, but due to inertia they contiune to move inwards and the whole process is repeated.

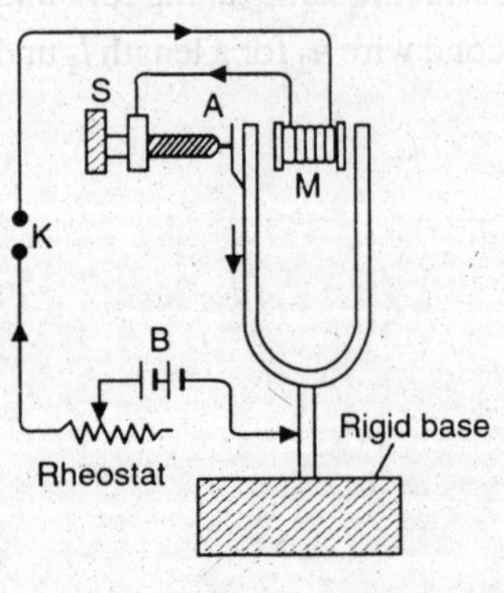

Fig. 20.4

20.6. Melde's experiment. The apparatus consists of a light string AB of uniform diameter and about one metre long, one end of which is connected to the prong of an electrically maintained tuning fork and the other end after passing over a friction less pulley is connected to a light scale pan P. The tension in the string can be varied by placing different weights in the pan. Stationary waves are produced due to the superimposition of the direct waves sent by the tuning fork and the reflected waves from the pulley. There are two modes of vibrations.

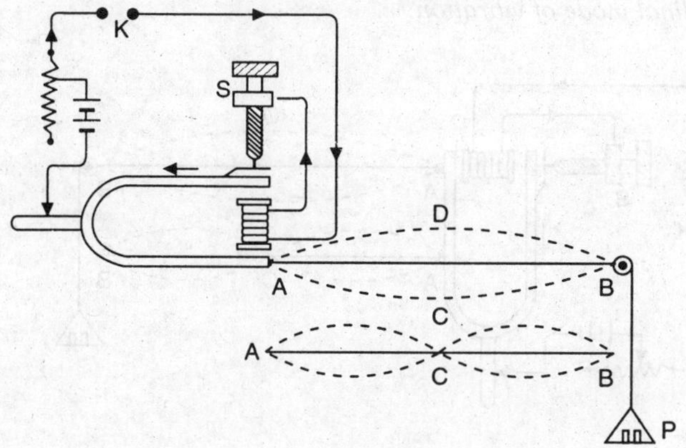

Fig. 20.5

(*a*) **Transverse mode of vibration.** In this arrangement the vibrations of the prong of the tuning fork are in a direction perpendicular to the length of the string as shown in Fig. 20.5. The time during which the tuning fork completes one vibration, the string also completes one vibration. If the thread of length l vibrates in one loop, the frequency of the fundamental note under a tension T is given by

$$n = \frac{1}{2l}\sqrt{\frac{T}{m}} \qquad ...(i)$$

where m is the mass per unit length of the thread. If the same length vibrates in p-loops under a tension T_p, then

$$n = \frac{p}{2l}\sqrt{\frac{T_p}{m}} \qquad ...(ii)$$

From (*i*) and (*ii*), we have

$$\frac{T_p}{T} = \frac{1}{p^2} \text{ or } T_p = \frac{T}{p^2}$$

Thus if we wish to have p-loops in the same length of the string, the tension should be reduced to $1/p^2$ of its previous value.

(*b*) **Longitudinal mode of vibration.** In this arrangement the tuning fork is set in such a manner that the vibrations of the prongs are parallel to the length of the string as shown in Fig. 20.6. In this case the time during which the tuning fork completes one vibration the string completes *half* a vibration. Hence the frequency of the thread is half that of the tuning fork. If the same length l of the thread vibrates in one loop under a tension T, then

Frequency of fork $n = 2 \times$ frequency of the string

$$= 2 \times \frac{1}{2l}\sqrt{\frac{T}{m}}$$

$$= \frac{1}{l}\sqrt{\frac{T}{m}}$$

where m is the mass per unit length of the string. For p- loops the frequency

$$n = \frac{p}{l}\sqrt{\frac{T}{m}}$$

Thus if the same length of the thread vibrates under the same tension in the two modes of vibration, the number of loops in the transverse arrangement is double as compared to the number of loops in the longitudinal mode of vibration.

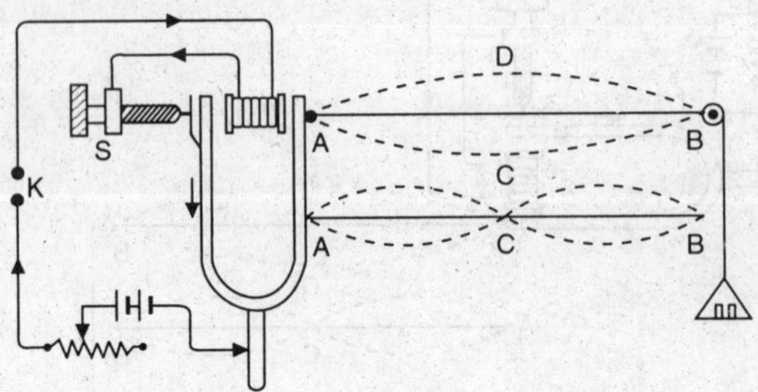

Fig. 20.6

Experiment 20.3. To determine the frequency of an electrically maintained tuning fork by Melde's experiment.

Apparatus. Electrically maintained tuning fork, a stand with clamp, pulley, a light weight pan, a weight box, a balance, a battery of two cells, a key, a rheostat, connecting wires etc.

Theory. A string can be set in to vibrations by means of an electrically maintained tuning fork, thereby producing stationary waves due to reflection of waves at the pulley. The end of the string where it is fixed to the prong of the tuning fork and the position where it touches the pulley are *nodes*.

(*i*) For the **transverse arrangement** the frequency n is given by

$$n = \frac{1}{2l}\sqrt{\frac{T}{m}}$$

where l is the length of the thread in fundamental mode of vibration, T the tension applied to the thread and m the mass per unit length of the thread.

If p-loops are formed in the length l of the thread, then

$$n = \frac{p}{2l}\sqrt{\frac{T}{m}}$$

(*ii*) For the **longitudinal arrangement.** When p-loops are formed the frequency n is given by

$$n = \frac{p}{l}\sqrt{\frac{T}{m}}$$

Procedure. Transverse arrangement.

1. Find the weight of the pan P and arrange the apparatus as shown in Fig. 20.5.

2. Place a load of 4 to 5 gm in the pan attached to the end of the string passing over the pulley. Excite the tuning fork by putting the plug in the key K.

3. Adjust the position of the pulley so that the string is set into resonant vibrations and well defined loops are obtained. If necessary, adjust the tension by adding weights in the pan slowly and gradually. For finer adjustment add milligram weights so that the nodes are reduced to points.

4. Measure the length of say 4 loops formed in the middle part of the string. If L is the distance in which 4 loops are formed then distance between two consecutive nodes $l = \frac{L}{4}$.

5. Note down the weight placed in the pan and calculate the tension T. Tension T = (wts. in the pan + wt. of pan) g

6. Repeat the experiment twice by changing the weight in the pan in steps of one gram and altering the position of the pulley each time to get well defined loops.

7. Measure one metre length of the thread and find its mass to find the value of m, the mass per unit length.

8. **Longitudinal arrangement.** Set the apparatus as shown in Fig. 20.6. and proceed with the adjustments and measurements as explained above.

Observations.

Mass of the pan w = gm = kg

Mass of one metre of thread = gm

Mass per metre m = gm = kg

Mode of vibration	*No.*	*No. of loops (p)*	*Length of corresponding thread L (in metre)*	*Length of each loop* $\frac{L}{p} = l$ *metre*	*Mass in the pan W in kg*	*Tension T = (w + W)g*	*Frequency n*
Transverse (*A*)	1						
	2						
	3						
Longitudinal (*B*)	1						
	2						
	3						

Mean frequency n = vib./sec

Calculations. A. *Transverse arrangement.*

$$n = \frac{1}{2l}\sqrt{\frac{T}{m}}$$

B. *Longitudinal arrangement*

$$n = \frac{1}{l}\sqrt{\frac{T}{m}}$$

Precautions. 1. The thread should be uniform and inextensible.

2. The thread should be horizontal and in alignment with the tuning fork. In transverse arrangement the thread should be stretched in line with the length of the prong so that the vibrations of the tip of the prong are at right angle to it. In the longitudinal arrangement the thread should be in line with the line of vibration of the prong.

3. Well defined loops should be obtained by adjusting the tension with milligram weights.

4. Friction in the pulley should be least possible as otherwise it causes the tension to be less than the actual applied tension.

5. The loops in the central part of the thread should be counted for measurements. The nodes at the tip of the prong and at the pulley should be neglected as these have some motion.

Sources of error. 1. The friction at the pulley and sparking at the platinum points cannot be totally eliminated.

2. There is a change in frequency due to the clamping screw at the tip of the prong of the tuning fork.

Experiment 20.4. To verify the laws of vibrating strings by Melde's experiment that is to show that λ^2/T = constant.

Apparatus. Same as in **Experiment 20.3.**

Theory. When a string is set into resonant vibrations by the electrically maintained tuning fork, stationary waves are produced. If l is the distance between the two consecutive nodes of a loop, then *In transverse arrangement.*

$$n = \frac{p}{2l}\sqrt{\frac{T}{m}} = \frac{1}{\lambda}\sqrt{\frac{T}{m}} \quad \left(\because \frac{2l}{p} = \lambda\right)$$

$$\therefore \quad \frac{\lambda^2}{T} = \left(\frac{1}{n^2 m}\right) = k_1 \text{ (constant)}$$

In longitudinal arrangement.

$$n = \frac{p}{l}\sqrt{\frac{T}{m}} = \frac{2}{\lambda}\sqrt{\frac{T}{m}}$$

or

$$\frac{\lambda^2}{T} = \left[\frac{4}{n^2 m}\right] = k_2 \text{ (constant)}$$

Procedure. Same as in **Experiment. 20.3.**

Observations. Mass of the pan $P = w$ gm = kg.

Mode of vibration	*No.*	*No. of loops p*	*Length of corresponding thread l (in metre)*	*Wave length $\lambda = \frac{2l}{p}$ metre*	*Mass in the pan W in kg*	*Tension T = (W + w)g*	$\frac{\lambda^2}{T}$
Transverse (*A*)	1						
	2						
	3						
Longitudinal (*B*)	1						
	2						
	3						

Verification. Within the limits of experimental error λ^2/T is constant. Hence the law is verified.

Precautions and sources of error. Same as in **Expt 20.3.**

Experiment 20.5. To compare mass per unit length of two strings by Melde's experiment.

Apparatus. Same as in **Expt. 20.3.** Two strings whose mass per unit length is to be compared.

Theory. When a string is set into resonant vibrations by the electrically maintained tuning fork, stationary waves are produced. If T_1 is the tension applied to the first string of mass per unit length m_1, then

(*i*) *In transverse arrangement*

$$n = \frac{1}{2l_1}\sqrt{\frac{T_1}{m_1}} \qquad ...(i)$$

where l_1 is the length of **one loop** of the string i.e. $l_1 = \frac{L_1}{p_1}$, L_1 being the length of the first string and p_1 the number of loops.

Similarly for the second string of mass per unit length m_2, if T_2 is the tension applied and l_2 the length of **one loop** of the string *i.e.* $l_2 = \frac{L_2}{p_2}$, L_2 being the length of the second string and p_2 the number of loops with the **same** tuning fork of frequency n, then we have

$$n = \frac{1}{2\,l_2}\sqrt{\frac{T_2}{m_2}} \qquad ...(ii)$$

From equations (*i*) and (*ii*), we get

$$\frac{1}{2l_1}\sqrt{\frac{T_1}{m_1}} = \frac{1}{2l_2}\sqrt{\frac{T_2}{m_2}}$$

or

$$\frac{1}{l_1^2}\frac{T_1}{m_1} = \frac{1}{l_2^2}\frac{T_2}{m_2}$$

$$\therefore \qquad \frac{m_2}{m_1} = \frac{T_2}{l_2^2}\frac{l_1^2}{T_1} = \frac{l_1^2}{l_2^2}\times\frac{T_2}{T_1}$$

Thus knowing the values of l_1, l_2, T_1 and T_2 we can find the value of $\frac{m_2}{m_1}$.

The same relation holds true in *longitudinal arrangement.*

Procedure (*i*) **Transverse arrangement.** Set the apparatus as shown in Fig. 20.5 and proceed as in **Expt. 20.3.** Steps 1-6.

7. Repeat steps 1-6 by using the second string.

(*ii*) **Longitudinal arrangement.** 8. Set the apparatus as shown in Fig. 20.6 and repeat steps 1-6.

Observations Mass of the pan = w =

Mode of vibration	*String*	*No.*	*No. of loops p*	*Length of corresponding string L*	*Length of each loop* $\frac{L}{p} = l$	*Mass in the pan* W	*Tension* $T = (W + w)g$	$\frac{T}{l^2}$
(A) Transverse	First No. 1	(*i*)						
		(*ii*)						
		(*iii*)						
	Second No.2	(*i*)						
		(*ii*)						
		(*iii*)						
(B) Longitudinal	First No.1	(*i*)						
		(*ii*)						
		(*iii*)						
	Second No. 2	(*i*)						
		(*ii*)						
		(*iii*)						

Calculations (*A*) **Transverse arrangement**

$$\frac{m_2}{m_1} = \frac{T_2}{l_2^2}\times\frac{l_1^2}{T_1} = (i)$$

(*ii*)

(*iii*)

$$\text{mean } \frac{m_2}{m_1} =$$

(*B*) **Longitudinal arrangement**

$$\frac{m_2}{m_1} = \frac{T_2}{l_2^2}\times\frac{l_1^2}{T_1} = (i)$$

(*ii*)

(*iii*)

$$\text{mean } \frac{m_2}{m_1} =$$

Mean $\frac{m_2}{m_1}$ from both the modes =

Verification. Actual mass of 1 metre length of first string

$$M_1 = \qquad g$$

$\therefore$ Mass per unit length of the first string $\frac{M_1}{100} = m_1 =$ g/cm

Actual mass of 1 metre length of second string

$$M_2 = \qquad g$$

$\therefore$ Mass per unit length of the second string $\frac{M_2}{100} = m_2 =$ g/cm

Actual ratio $\frac{m_2}{m_1} =$

Percentage error =

Precautions and sources of error. Same as in **Expt. 20.3.**

Experiment 20.6. To determine the frequency of A.C. mains using a sonometer and an electro-magnet.

Apparatus. Sonometer with steel wire, an electro-magnet, a step down transformer, balance, weight box, metre scale and a clamp stand.

Theory. If a sonometer wire of length l stretched between two knife-edges under a tension T and mass per unit length m is once plucked and then released it executes transverse vibrations and emits a note of fundamental frequency n Hz given by

$$n = \frac{1}{2l}\sqrt{\frac{T}{m}}$$

To find the frequency of A.C. mains using an electro-magnet and a sonometer, the A.C. is passed through the primary of a step-down transformer (220–230 to 4–6 volts). The two ends of the secondary coil of the step-down transformer are connected to the two ends of the windings of the electro-magnet which consists of a coil of insulated copper wire wound over a soft iron core provided with an insulated handle.

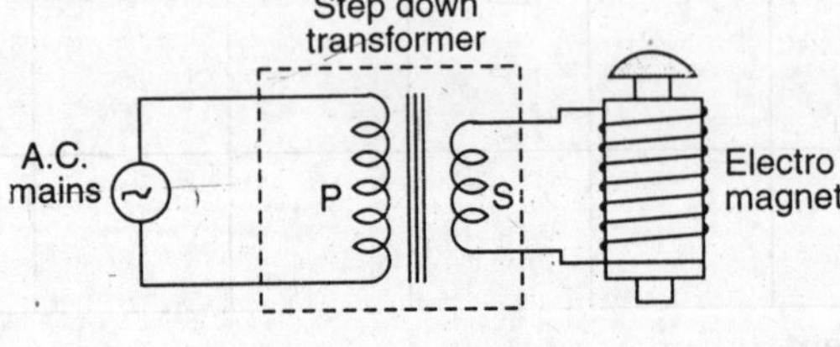

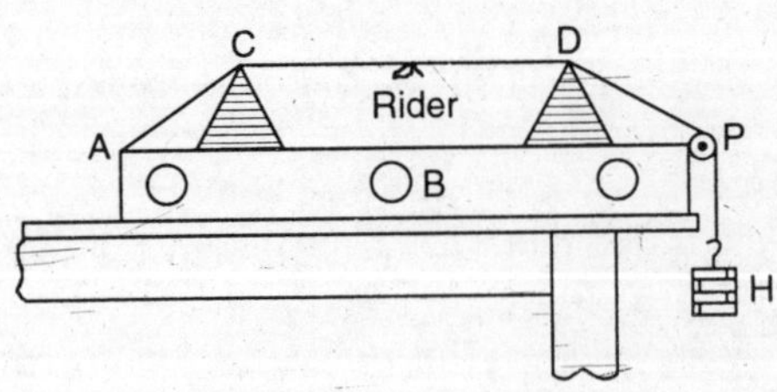

Fig. 20.7

As the A.C. from the secondary of the step-down transformer passes through the electro-magnet it gets magnetised *twice* in each cycle, first with one of its faces as *a north pole* and then with the same face as the south pole. The electro-magnet is kept close to and vertically above the sonometer steel wire.

The position of the knife edges C and D is so adjusted that the sonometer wire vibrates in resonance with the A.C. supply (In India A.C. supply is 50 Hz). The wire is attracted and pulled twice in each cycle of the A.C. mains supply; once when the end of the electro-magnet just above the wire is a north pole and again after half a cycle when this end is a south pole. In other words, the natural frequency (n) of the sonometer wire is double the frequency (f) of the A.C. mains.

$$\therefore \quad n = 2f = \frac{1}{2l}\sqrt{\frac{T}{m}}$$

or

$$f = \frac{n}{2} = \frac{1}{4l}\sqrt{\frac{T}{m}}$$

Procedure. 1. Set the sonometer as explained.

2. Arrange the electro-magnet in a clamp stand and hold it 2–3 mm vertically above the centre of the steel wire of the sonometer. Connect the electro-magnet to the secondary of the stepdown transformer. Switch on the A.C. mains and test the magnetisation of the electro-magnet with the help of an iron needle.

3. Cut a V-shaped light paper rider about one cm long and 2 mm wide. Bring the two knife edges close to each other and place the paper rider on the wire in between the knife-edges. See that the pole of the electro-magnet is *just* above the centre of the sonometer wire.

Now gradually increase the distance between the two knife edges till the rider begins to *flutter*. At this stage, make the adjustment very carefully till the rider flies off at a certain fixed distance between the two knife-edges. The wire is now in resonance with the frequency of the A.C. mains supply. Measure the length of the wire between the two knife edges with a metre rod.

4. Increase the distance between the two knife-edges by a few centimetre. Repeat the above process by decreasing slowly the distance between the two knife-edges till the rider again flies off. Measure the length of the wire between the two knife-edges again. The mean of the two lengths is the true resonant length.

5. Weigh the weights suspended including the hanger with a trip-scale balance.

6. Increase the weight by half a kilogram and repeat the observation to find the length of the wire vibrating in resonance with A.C. mains supply.

Take such four sets of observations by changing the load by $\frac{1}{2}$ kg each time.

7. Switch off the A.C. mains and remove the electro-magnet. To find the mass per unit length of wire adjust the distance between the two knife-edges to be exactly 50 cm apart. Mark with ink a point on the wire at the position of each knife edge and cut the wire at these points. Find its weight accurately by a sensitive balance.

Observations. Length of wire = cm = m

Mass of the wire = gm = kg

Mass per unit length of wire m = gm/cm = kg/m

Sl. No.	*Load in Kgm* M	*Tension* $T = Mg$	*Length of the wire in metre at resonance*			$f = \frac{n}{2} = \frac{1}{4l}\sqrt{\frac{T}{m}}$
			Load increasing l_1	*Load decreasing* l_2	*Mean l* (in metre)	
1						
2						
3						
4						
5						

Mean frequency of A.C. mains = *Hz*

Standard value of frequency of A.C. mains = 50 *Hz*

$\therefore$ Percentage error =

Precautions. 1. 1, 2, 3, 6 and 7 same as in **Expt. 20.1.**

2. The magnetisation of the electro-magnet should be checked with an iron needle before starting the experiment.

3. The electro-magnet should be clamped 2–3 mm vertically above the centre of the sonometer wire.

4. The sonometer wire should be made of steel so that it is attracted by the electro-magnet.

Oral Questions

1. Sonometer. (*i*) What is a sonometer? Name any instrument working on the same principle used in everyday life. (*ii*) What is resonance? (*iii*) Does the sonometer board vibrate in resonance with the tuning fork? If not what phenomenon is it? (*iv*) Why does the sound become loud when the stem of the tuning fork is pressed on the board? (*v*) What are nodes and antinodes? How do you locate their positions? (*vi*) How will the frequency change if the length of the wire is doubled and the tension is made four times? (*vii*) How is the frequency of the tuning fork affected when the prongs are (*a*) loaded (*b*) filed, (*viii*) What are overtones? (*ix*) If the frequency of the string differs slightly from the frequency of the tuning fork. What will happen? (*x*) Why does the rider fall down when the frequency of vibration of the wire is equal to that of the tuning fork? (*xi*) Can you test the position of resonance without using a rider? (*xiii*) For finding the fundamental frequency why should we start with minimum distance between the wedges?

2. Melde's experiment. (*i*) What is an electrically maintained tuning fork? Explain its working (*ii*) What are the two arrangements of the strings with respect to the tuning fork in which the experiment can be performed? (*iii*) How is the tension related to each other for obtaining the same number of loops in the two modes of vibration? (*iv*) How will the number of loops alter as the string is changed over from transverse to longitudinal arrangement, keeping all factors constant? (*v*) Why should the loops in the central part of the thread be counted for measurements? (*vi*) How will the friction in the pulley effect the result?

3. Frequency of A.C. mains. (*i*) What do you mean by the term A.C. (*ii*) What do you mean by the frequency of A.C. mains? (*iii*) Why do you use a step down transformer to supply current to the electromagnet? (*iv*) What is the construction of an electromagnet? (*v*) Why do you use a steel wire when finding the frequency of A.C. mains with the electromagnet? (*vi*) Can you use a brass wire? (*vii*) What is the relation between the frequency of A.C. mains and frequency of sonometer wire when A.C. is passed through the electromagnet.

21

Kundt's Tube

21.1. Vibration of rods. Consider a rod of metal rigidly clamped at its middle point as shown in Fig. 21.1.

If the rod is rubbed along its length at one free end or is stroked, longitudinal vibrations travel along its length. The rod is alternately elongated and compressed and stationary waves are set up in it in the same manner as in the case of an open end pipe. The two free ends act as *anti- nodes* and the middle point which is clamped acts as a *node.* Thus the whole length l of the rod is equal to half the wavelength

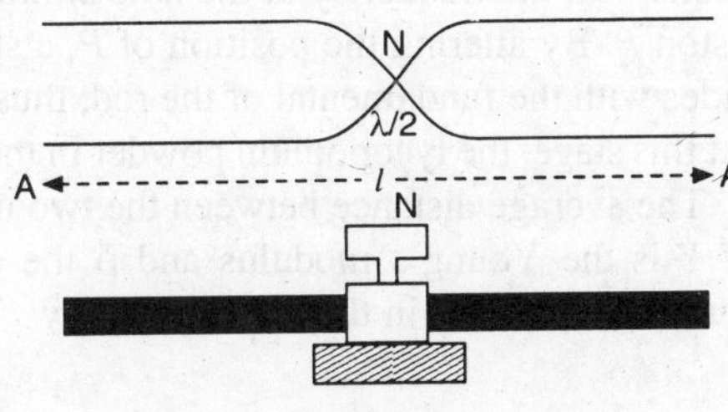

Fig. 21.1

$$\therefore \qquad l = \frac{\lambda}{2}$$

$$\text{or} \qquad \lambda = 2l$$

If V is the velocity of the wave in the material of the rod and n is the frequency of vibration, then

$$n = \frac{V}{2l}$$

When, however, the rod is clamped at one end and is set into longitudinal vibrations at the free end it behaves like a closed end pipe. The stationary wave set up has a node at the clamped end and an antinode at the free end. The whole length of the rod in this case is equal to one quater of wave-length

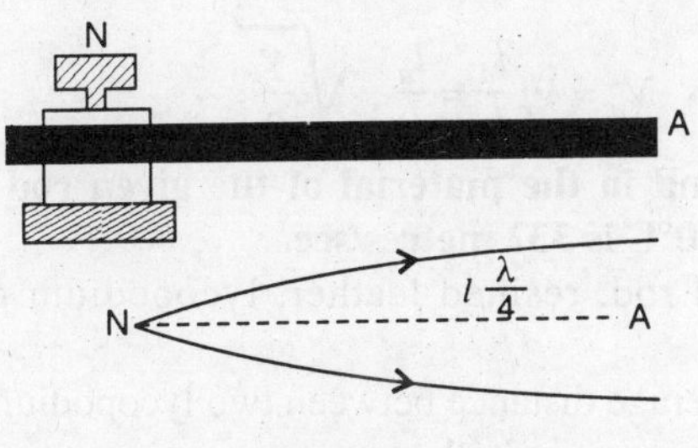

Fig. 21.2

$$\therefore \qquad l = \frac{\lambda}{4}$$

$$\text{or} \qquad \lambda = 4l$$

$$\text{and} \qquad n = \frac{V}{4l}$$

The value of V remains the same as long as the material of the rod is the same.

21.2. Kundt's tube. The apparatus consists of a long glass tube G provided at the ends with metal fittings. Through one end passes a rod AB of uniform radius, about one or two metres long. This rod is clamped in the middle at C and carries a circular disc D rigidly fixed at one end. The diameter of the disc is slightly less than the diameter of the tube. The rod is inserted a few centimetres within the glass tube, without touching it, as shown in Fig. 21.3. The other end of the tube is closed by a metallic piston P, the position of which can be adjusted by moving the rod, carrying the piston, in or out. The

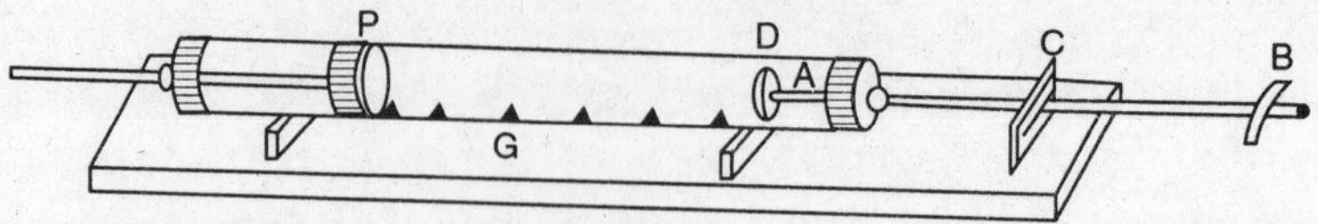

Fig. 21.3

whole apparatus is rigidly clamped on a table, so that there are no jerks on the tube during the experiment.

The tube is thoroughly dried and fixed in position, after spreading a thin layer of evenly laid lycopodium powder in it. The rod is rubbed at the point B, lengthwise with a piece of resined leather and is thus set into longitudinal vibrations, the ends of the rod acting as antinodes and the middle point C as the node. The length of the rod is equal to half the wavelength of sound in the material of the rod. The disc D begins to vibrate backward and forward. The air inside the tube is set into forced vibrations with the frequency of the note emitted and stationary waves are produced by reflection at the piston P. By altering the position of P, a stage is reached when any overtone of the air column coincides with the fundamental of the rod, thus producing resonance within the tube.

At this stage, the lycopodium powder in the tube is violently agitated and collects in heaps at the nodes. The average distance between the two nearest heaps is equal to half the wavelength in air.

If Y is the Young's modulus and ρ the density of material of the rod, then the velocity of longitudinal vibrations in the rod is given by

$$V_r = \sqrt{\frac{Y}{\rho}}$$

Let l_a be the mean distance between two consecutive nodes (lycopodium heaps) in air and l_r the length of the rod, then

Wavelength in air $\qquad \lambda_a = 2\, l_a$

and wavelength in the rod $\qquad \lambda_r = 2l_r$

If n is the frequency of vibrations produced by the rod, then velocity of sound in the material of the rod

$$V_r = 2\, n\, l_r$$

and velocity in air $\qquad V_a = 2n\, l_a$

$$\therefore \qquad \frac{V_a}{V_r} = \frac{l_a}{l_r}, \quad \text{or} \quad V_a = V_r \frac{l_a}{l_r} = \frac{l_a}{l_r}\sqrt{\frac{Y}{\rho}}$$

Experiment 21.1. To find the velocity of sound in the material of the given rod with a Kundt's tube given that velocity of sound in air at 0°C is 332 metres/sec.

Apparatus. Kundt's tube fitted with a metal rod, resined leather, lycopodium powder and a metre rod.

Theory. If l_r is the length of the rod and l_a the average distance between two lycopodium heaps in air in a Kundt's tube when stationary vibrations are set up in it, then

$$\frac{V_r}{V_a} = \frac{l_r}{l_a}$$

where V_r is the velocity of sound in the material of the rod and V_a the velocity of sound in air at room temperature. Hence

$$V_r = V_a \frac{l_r}{l_a}$$

Procedure. 1. Dry the Kundt's tube by passing hot air and clamp it in position on the table.

2. Remove the rod along with the disc and the metal cap from the tube. Spread a thin layer of lycopodium powder at the bottom of the tube along its length. To do it conveniently spread a thin layer of lycopodium powder on a metre rod, insert it into the tube and turn it upside down. Remove the metre rod.

3. Insert the metal rod with the disc into the tube and replace the cap. Clamp the rod exactly at its middle point. Excite longitudinal vibrations in the rod by stroking it along its length near the free end with a piece of resined leather. Adjust the position of the piston P on the other side by moving it in or out very slowly so that the air column between the piston P and the disc D is set into resonant vibrations. When this has been achieved the lycopodium powder will be violently agitated and will settle down in the form of small heaps at and near the nodes where the air is in state of least vibration.

4. Select very carefully the position of two extreme nodes, one near the piston P and the other near the disc D and measure the distance between them. Also find the number of spaces and calculate the average distance between two consecutive nodes.

5. Repeat the experiment at least thrice.

6. Measure the length of the rod and note the room temperature.

Observations. Temperautre t = °C

Length of the rod l_r = cm = m

No.	*Distance between extreme nodes* L	*Number of speaces* m	*Average distance between two consecutive nodes* $l_a = \frac{L}{m}$
1.			
2.			
3.			

Mean l_a = cm = m

Calculations

Velocity of sound in air at 0°C, V_0= 332 metres/sec.

∴ Velocity of sound at t°C, $V_a = V_0 \sqrt{\frac{273+t}{273}}$

Velocity of sound in the material of the rod

$V_r = V_a \frac{l_r}{l_a}$ metre / sec . =

Verification

Young's modulus of the rod Y = Newton / sq. metre

Density of the material of the rod ρ = kg/m^3

∴ Velocity of sound in the material of the rod

$$V_r = \sqrt{\frac{Y}{\rho}} \text{ ms}^{-1}$$

Precautions. 1. The tube and the lycopodium powder should be prefectly dry as the presence of moisture affects the velocity of sound.

2. The rod should be rigidly clamped exactly at the middle point and its axis should coincide with the axis of the tube.

3. A thin and evenly spread layer of lycopodium powder should be used.

4. The rod should be stroked with the resined leather gently and longitudinally near the free end otherwise overtones will be produced.

Exercise 1. *Find the velocity of sound in air with Kundt's tube given the Young's modulus of the material of the rod and its density.*

Hints. The velocity of sound in the material of the rod

$$V_r = \sqrt{\frac{Y}{\rho}}$$

If l_r is the length of the rod and l_a the average distance between two lycopodium heaps in air, then

$$\frac{V_a}{V_r} = \frac{l_a}{l_r} \quad \text{or} \quad V_a = V_r \frac{l_a}{l_r}$$

Oral Questions

Kundt's tube. (*i*) What type of vibrations are produced in the rod ? (*ii*) Why do you clamp the rod in the middle ? (*iii*) What is the function of the piston ? (*iv*) What type of vibrations are formed in the air colume of the tube ? (*v*) What is the effect of temperature, moisture and pressure on the velocity of sound in a gas ? (*vi*) What is the change in velocity of air for 1°C rise in temperature ? (*vii*) Can you use the Kundt's tube for finding the velocity of sound in some other gas ? If so, how ? (*viii*) Can you use the Kundt's tube to find the velocity of sound in a liquid ? If so, how?

22

Ultrasonics

22.1. Ultrasonic waves. Human ear is sensitive to sounds of frequency lying between 20 and 20,000 cycles per second. The ear is unable to hear sounds of frequency less than 20 *Hz* and more than 20,000 *Hz*. The sounds of frequency less than 20 are called *infrasonics* whereas those of frequency more than 20,000 are called *ultrasonics.* In other words ultrasonics are longitudinal mechanical waves of frequency beyond the highest audible frequency (*i.e.* 20,000 *Hz*). As their frequency is very high, the wave length is very low (about 1.6 cm).

The important methods of producing ultrasonic waves are (*i*) magnetostriction oscillator and (*ii*) Piezo electric oscillator. The Piezo electric oscillator makes use of piezo-electric effect exhibited by crystals of quartz, zinc blende, Rochelle salt etc. We shall, however, first describe a quartz crystal in detail.

22.2. Quartz crystal. The quartz crystal found in nature is in the form of *hexagonal prisms* with *hexagonal pyramids* at each end (Fig. 22.1 *a*) The straight line joining the points at the ends or any line parallel to it is called the *optic axis* (or *Z-axis*) of the crystal. Any section cut by a plane at right angles to the optic axis is a hexagon and the lines passing through the corners of the hexagon E_1 E_1, E_2 E_2, and E_3 E_3 are called *electric axis* (or *x-axis*). The three axis perpendicular to the *x*-axis such as M_1-M_1, M_2 M_2 and M_3 M_3 are known as *mechanical axis* or (*y-axis*) of the crystal (Fig. 22.1 *b*).

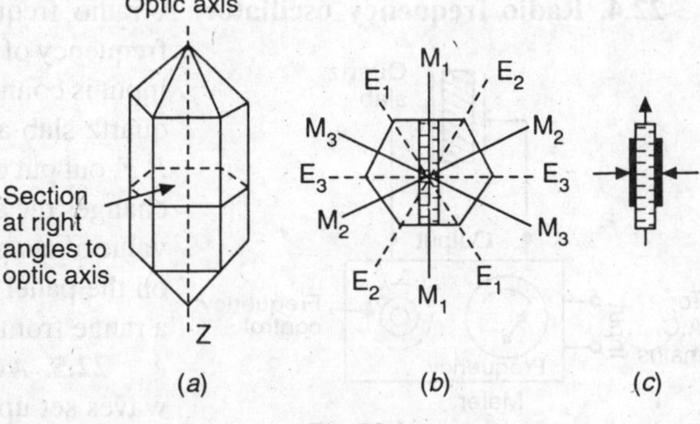

Fig. 22.1

The crystal is cut in the form of a slab with its

Length — parallel to mechanical axis
Breadth — parallel to the optic axis and
Thickness — parallel to the electric axis.

Such a crystal slab sand-witched between two metal plates is shown in Fig. 22.1. (*c*) with the faces perpendicular to the electric axis.

22.3. Piezo-electric effect. When a slab of quartz or tourmaline as described above is subjected to a mechanical pressure on one pair of parallel faces, electric charges are developed on the other pair of parallei faces perpendicular to the first pair. The quantity of charge developed is proportional to the applied pressure. The sign of charge changes when the pressure is changed into tension.

This phenomenon is known as *piezo-electric effect.*

On the other hand when an electric potential difference is applied across the faces of a crystal, the crystal under goes a strain. If the direction of applied potential difference is reversed the direction of strain is also reversed *i.e.* if in the first case it is a compression strain now it will be an extension strain and vice versa. Hence if an *alternating voltage* is applied to the crystal slab rapid alteration of compression and extension takes place in two perpendicular directions resulting in the forced vibrations of the crystal.

This phenomenon is known as *inverse piezo-electric effect.*

When the frequency of these forced vibrations becomes equal to the natural frequency of the slab *resonance* takes place and the amplitude of vibration becomes sufficiently large.

When an alternating voltage is applied to the slab along its *thickness* there are alternating changes in its length and thickness *i.e.* when the length increases, the thickness decreases and *vice-versa* with the same frequency as that of the applied *A.C.* voltage. When the frequency of the applied *A.C.* voltage becomes equal to the natural frequency of the slab, resonant vibrations take place in the slab.

The natural frequency of vibration of the slab is given by

$$n = \frac{1}{2d}\sqrt{\frac{E}{\rho}}$$

where d = distance between the opposite faces of the crystal slab

E = Bulk modulus of quartz (material of the slab) = 8×10^{10} N/m^2

ρ = density of quartz = 2.65×10^3 kg/m^3

$$\therefore \quad n = \frac{1}{2d}\sqrt{\frac{8 \times 10^{10}}{2.65 \times 10^3}} = \frac{2.8 \times 10^3}{d}\ Hz$$

If d = 1 mm = 10^{-3} m, then

$$n = \frac{2.8 \times 10^3}{10^{-3}} = 2.8 \times 10^6\ Hz = 2.8\ MHz$$

22.4. Radio frequency oscillator. A radio frequency oscillator produces *A.C.* voltages of frequency of the order of a few megacycles per second. Its input is connected to the *A.C.* mains and the out put to the quartz slab as shown in Fig. 22.2. The frequency of the *R.F.* out put can be varied by rotating a knob which in turn changes the capacitance of the oscillating *LC* circuit. The value of out put frequency is read on a calibrated dial fixed on the panel known as *Frequency meter.* It has normally a range from 0–10 *MHz.*

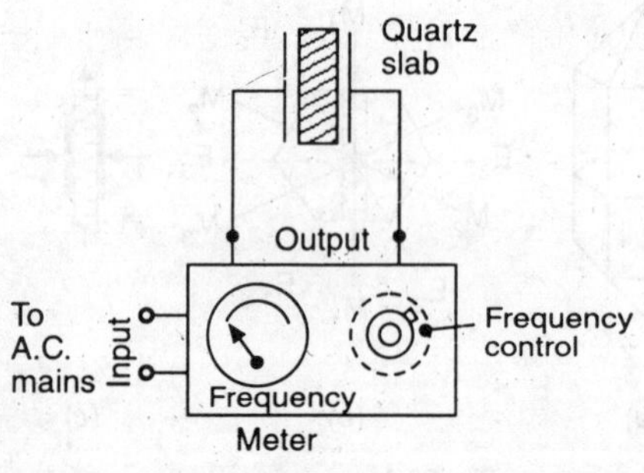

Fig. 22.2

22.5. Acoustic grating. The velocity of ultrasonic waves set up in a liquid can be determined with the help of the *acoustic grating* formed by them. When plane ultrasonic waves are produced in a liquid the pressure varies periodically with the distance from the ultrasonic source being higher at a compression and lower at a rarefaction. The density of the medium and the refractive index also show a periodic variation with distance from the source along the direction of propagation of the wave. Under such a condition, if light from a slit is passed through the liquid in a direction at right angles to the wave propagation the liquid behaves as a diffraction grating with grating element (width of opacity + width of transparency) equal to the wavelength of ultrasonic wave λ_u. Such a grating is known as *acoustic grating.* The light which suffers diffraction at this grating shows a number of diffraction images of the slit when seen through a telescope formed at the position of the cross-wire of the telescope eyepiece.

If θ is the angle of diffraction for the nth order maxima, then

$$\lambda_u \sin\theta = n\,\lambda$$

where λ is the wavelength of light used.

$\therefore$ Wavelength of ultrasonic waves $\lambda_u = \dfrac{n\,\lambda}{\sin\theta}$

If D is the distance between the telescope objective and eye piece cross-wire and d_n the distance of the nth order diffraction maxima from the central image (zero order diffraction maximum), then

$$\sin\theta = \frac{d_n}{D} \qquad [d_n << D]$$

$$\therefore \quad \lambda_u = \frac{n\lambda D}{d_n}$$

If ν is the frequency of ultrasonic waves, then velocity of ultrasonic wave $V = \dfrac{\nu\, n\,\lambda\, D}{d_n}$. To find the velocity of ultrasonic waves, we use *an ultrasonic spectrometer*.

22.6. Ultrasonic spectrometer. The ultrasonic spectrometer is almost similar to an optical spectrometer. It has the following three parts.

(i) **A collimator.** It is a brass tube having a fine adjustable slit at one end and an achromatic convex lens at the other end. It is fitted with a rack and pinion arrangement.

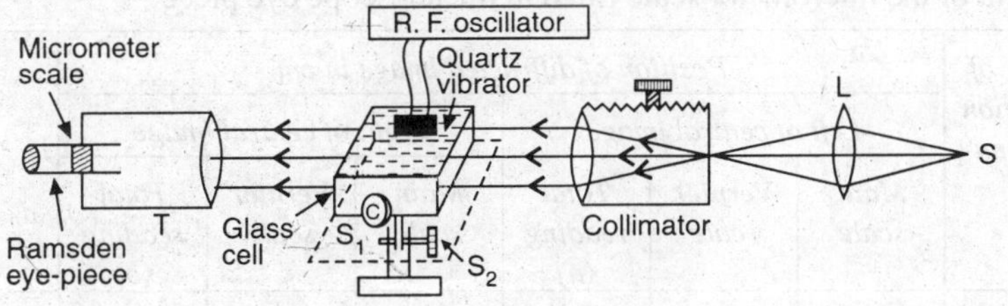

Fig. 22.3

(ii) **A telescope.** It has an objective of large focal length and a Ramsden eye piece provided with a horizontal micrometer scale.

(iii) **A table.** The cell containing the experimental liquid is placed on the table. The table is fitted with a screw S_1 to give it a horizontal motion and a screw S_2 to give it a vertical motion.

Experiment 22.1. To determine the velocity of ultrasonic waves in a given liquid (say kerosene oil).

Apparatus. A radio frequency oscillator fitted with a frequency meter, a quartz crystal slab fitted with two leads, ultrasonic spectrometer, a glass cell with experimental liquid, a sodium lamp, a convex lens, a spirit level etc.

Procedure. 1. Set the telescope and collimator of the ultrasonic spectrometer in a way similar to that of *optical spectrometer*.

2. Place the glass cell containing the experimental liquid on the *spectrometer table* with its opposite parallel sides facing the telescope and the collimator.

3. Mount the quartz crystal slab in its holder and dip it in the liquid near a wall of the glass cell so that the ultrasonic waves produced by the crystal travel in the liquid in a direction perpendicular to that of the incident light. Connect the leads of the crystal slab with the out put terminals of the *R.F.* oscillator.

Fig. 22.4

4. Switch on the sodium lamp and focus its light on the slit of the collimator. See through the telescope eye piece so that a sharp well defined image of the slit is seen in the field of view in the centre of the micrometer scale fitted in the eye piece.

5. Now switch on the *R.F* oscillator. Ultrasonic waves are set up in the liquid. Adjust the frequency of the oscillator so that it becomes equal to the natural frequency of the crystal slab. At this stage resonance takes place and diffraction images of the slit will be seen in the telescope. An enlarged view of the diffraction pattern is shown in Fig. 22.4. Note the frequency of the *R.F* oscillator. Maintain it constant at this value through out the experiment.

NOTE. If the diffraction images are not seen by adjusting the frequency of the oscillator, then readjust the position and orientation of the crystal in the liquid and the position and level of the glass cell.

6. Measure the distance of various order diffraction images on both sides from the central zero order image on the micrometer scale provided in the telescope eye-piece.

7. Measure the distance *D* between the objective lens and the cross-wire (micrometer scale) of the telescope. This is also equal to the focal length of the objective lens and is generally given by the makers.

Observations. Wavelength of sodium light $\lambda = 5893 \times 10^{-10}$ m

Room temperature = °C

Distance D = cm = m

Frequency of the oscillator used ν = *Hz*

Least count of the micrometer scale fitted to the telescope eye piece = mm = m

Sl. No.	*Order of diffraction maxima n*	*Position of diffracted image in cm*						$d_n = \frac{a-b}{2}$	$\frac{d_n}{n}$ *cm*
		Left of central image			*Right of central image*				
		Main scale	*Vernier scale*	*Total reading (a)*	*Main scale*	*Vernier scale*	*Total reading (b)*		
1									
2									
.									
.									
10									

Mean value of $\frac{d_n}{n}$ = cm = m

Wavelength of ultrasonic waves

$$\lambda_u = \frac{n}{d_n}\lambda D = \frac{\lambda D}{\left(\frac{d_n}{n}\right)} = \quad \text{m}$$

Velocity of ultrasonic waves

$$V = \nu\,\lambda_u \quad \text{m/s}$$

Precautions 1. The glass cell should be thoroughly cleaned and filled with the liquid filtered a number of times.

2. The crystal slab should be immersed completely in the liquid and should touch neither the bottom nor the side walls.

3. The crystal slab should be set in such a way that the ultrasonic waves produced by it travel at right angles to the incident light.

4. The diffraction pattern should be sharp and narrow.

5. The frequency of the *R.F.* oscillator should be properly adjusted for resonance and should not be disturbed during the experiment.

Oral Questions

1. Ultrasonics (*i*) **What are ultrasonics?** (*ii*) **What is the audible frequency range of human ear?** (*iii*) **What name is given to frequencies below the audible range ?** (*iv*) **Are ultrasonic waves longitudinal or transverse?** (*v*) **What is piezo-electric effect?** (*vi*) **What is inverse piezo electric effect?** (*vii*) **State the formula for frequency of a quartz crystal.** (*viii*) **Calculate the frequency of a quartz crystal of thickness 1mm when $E = 8 \times 10^{10}$ Nm^{-2} and $\rho = 2.65 \times 10^3$ kg m^{-3}** (*ix*) **What do you understand by acoustic grating?** (*x*) **What is the grating element of such a grating.** (*xi*) **Define optic axis, electric axis and mechanical axis of a quartz crystal.** (*xii*) **How is a quartz crystal cut for producing high frequency oscillations and mounted?**

PART V

MAGNETISM

23

Magnetic Measurements

23.1. Horizontal component of earth's magnetic field. Earth is a huge magnet. *The horizontal component of earth's magnetic field at a place is the component of the total magnetic field of the earth along a horizontal plane.*

It is denoted by B_H. The S.I. unit of magnetic field B is a **Tesla** or **Weber per square metre** (Wb/m^2). The C.G.S. unit is a **Gauss.**

$$1 \text{ Wb/m}^2 = 10^4 \text{ Gauss.}$$

23.2 Magnetic moment. When a magnetic dipole (or a bar magnet) is placed in a uniform magnetic field B with its dipole axis making an angle θ with the direction of the magnetic field, a couple $MB \sin \theta$ acts on the dipole tending to align it with the direction of the magnetic field.

If $B = 1$, $\theta = 90°$ so that $\sin \theta = 1$, then

The moment of the couple = M

Hence the magnetic moment of a dipole (or a bar magnet) is defined as the moment of the couple which tends to align its dipole axis parallel to a magnetic field of unit strength when placed in a direction perpendicular to the field.

The S.I. unit of magnetic moment is Weber-metre (Wb-m) or Ampere metre^2 (A-m^2).

23.3. The value of horizontal component of earth's magnetic field B_H and the magnetic moment of the bar magnet M can be determined with the help of

(*a*) A deflection magnetometer and (*b*) A vibration magnetometer.

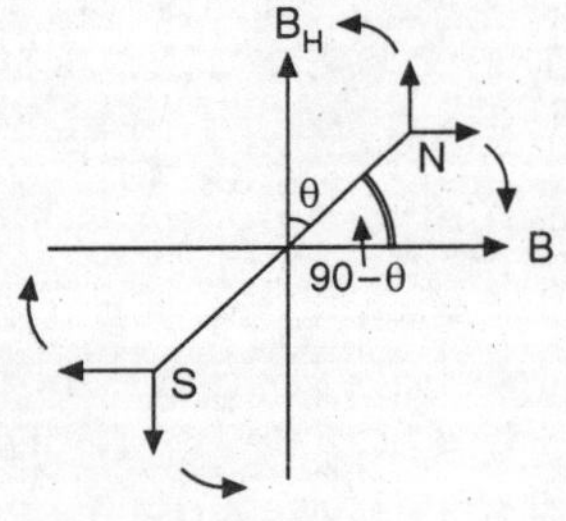

Fig. 23.1

(*a*) **Deflection magnetometer.** When a magnetic dipole (say a magnetic needle *NS*) is placed on a vertical axis in the horizontal plane under the influence of the two uniform magnetic fields B and B_H acting perpendicular to each other, it will be in equilibrium in a direction making an angle θ with the field B_H. At this stage the moment of the couple due to the field B_H acting on the dipole is equal and opposite to the moment of the couple acting on the magnet due to the magnetic field B.

$$\therefore \qquad MB_H \sin \theta = MB \sin (90 - \theta) = MB \cos \theta$$

$$\text{or} \qquad \frac{B}{B_H} = \frac{\sin \theta}{\cos \theta} = \tan \theta .$$

or $$B = B_H \tan \theta.$$

This is known as **tangent law.** The working of a deflection magnetometer is based on this principle.

Construction. It consists of a small magnetic needle pivoted on a vertical axis at the centre of a circular scale enclosed in a *brass* or *bakelite* box covered with a glass lid. The magnetic needle carries a long and light aluminium pointer rigidly attached perpendicular to its length.

The scale is divided into *four* quadrants by the 0–0 and the 90–90 lines. A plane mirror is fixed at the bottom of the box which enables to take the readings of the position of the pointer without error due to parallax. This is achieved by placing the eye in such a position that the pointer covers its image in the plane mirror.

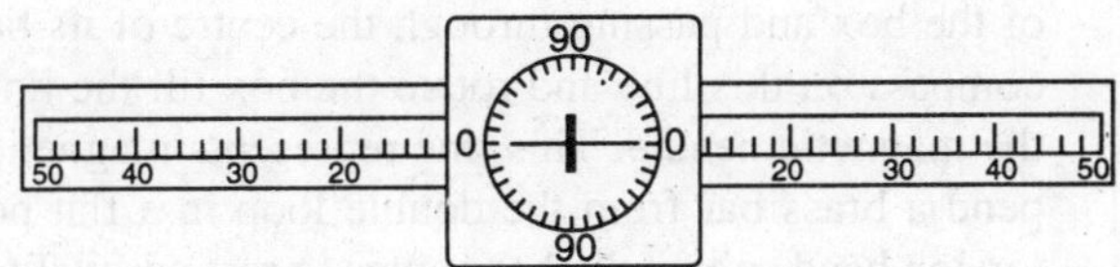

Fig. 23.2

The compass box is placed at the middle of a wooden base about one metre long on which two *millimetre* scales are fixed on either side with their zero marks coinciding with the centre of the circular scale.

Setting the magnetometer in the end on position or tangent 'A' position of Gauss. To set the magnetometer fix a fine thread on the magnetometer arms parallel to the scales fixed on the arms. Rotate the magnetometer box so that the thread lies on the 0–0 line when looked upon vertically.

Now rotate the wooden arms along with the magnetometer box so that the pointer lies on the 0–0 line. As the pointer is fixed at right angles to the magnetic needle and is now parallel to the arms, the arms are set in the east-west direction.

If a magnet is now placed on the arms with its axis parallel to the *millimetre* scale it will create a magnetic field along the axial line perpendicular to the earth's field B_H. The magnetic needle of the magnetometer will therefore, be under the action of two uniform fields at right angles to each other. The tangent law will, therefore, be applicable.

Errors in a magnetometer and their elimination. A magnetometer suffers from various errors which can be eliminated by taking the following **eight** observations in the deflection method.

(*i*) *Error due to the magnetic needle not being pivoted at the centre of the circular scale*. This error is eliminated by reading both ends of the pointer.

(*ii*) *Error due to the poles being situated at different distances from the centre of the magnet.* This error is eliminated by reversing the magnet in its position so that the north pole takes the place of the south pole and *vice versa* and again reading both ends of the pointer. This also eliminates the error if the pointer is not exactly lying on the 0—0 line.

(*iii*) *Error due to the centre of the circular scale not being exactly at the zero of the mm scale marked on the arms*. The error is eliminated by placing the magnet at the corresponding position on the other side of the compass box and repeating the above *four* observations.

The mean of these *eight* observations gives the correct deflection.

(*b*) **Vibration magnetometer.** A freely suspended magnetic dipole always points in the north and south direction. If it is displaced through an angle θ a couple having a moment $MB_H \sin \theta$ acts on it, which tends to bring it back to its initial position of rest. As the magnet reaches the mean position it acquires kinetic energy and over shoots the mark and thus begins to vibrate just as a simple pendulum does. The motion of the magnet is simple harmonic and hence time-period is given by the formula

$$t = 2\pi \sqrt{\frac{I}{MB_H}}$$

where *I* is the moment of inertia of the magnet.

Construction. A vibration magnetometer consists of a wooden box having glass sides. A long brass tube is fixed vertically at the top of the box and has got a torsion head from which hangs an *unspun* silk fibre terminating into a double loop or a light aluminium or brass clamp. There is a slit at the top of the box. The magnetic dipole is placed in a flat position in the double loop of an unspun silk thread or in the clamp parallel to the bottom.

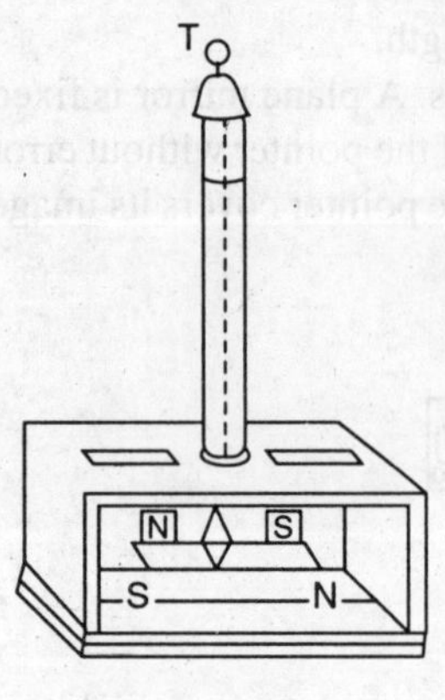

Fig. 23.3

Setting. To set the magnetometer draw a line *ns* parallel to the length of the box and passing through the centre of its base. Place a magnetic compass on this line and rotate the box till the line becomes parallel to the magnetic needle. This line represents magnetic meridian. Now suspend a brass bar from the double loop in a flat position and adjust the torsion head so that the bar comes to rest parallel to the line *ns* as shown in Fig. 23.3. This removes the torsion in the silk fibre and now if the brass bar is replaced by a magnet placed with its north pole towards earth's north it will lie in the magnetic meridian without any other force acting on it.

23.4. Determination of the horizontal component of earth's magnetic field. The value of B_H can be determined in the laboratory using a deflection and a vibration magnetometer. If θ is the deflection produced by a magnet placed at a distance *d* when the magnetometer is set in the end-on position, then

$$\frac{\mu_0}{4\pi}\frac{2Md}{(d^2-l^2)^2} = B_H \tan\theta$$

or

$$\frac{B_H}{M} = \frac{\mu_0}{4\pi}\frac{2d}{(d^2-l^2)^2 \tan\theta} \quad ...(i)$$

If *t* is the time-period of the same magnet determined by a vibration magnetometer, then

$$t = 2\pi\sqrt{\frac{I}{MB_H}} \quad ...(ii)$$

or

$$MB_H = \frac{4\pi^2 I}{t^2} \quad ...(iii)$$

Multiplying (*i*) and (*ii*), we get

$$B_H^2 = \frac{\mu_0}{4\pi}\frac{8\pi^2 Id}{t^2(d^2-l^2)^2 \tan\theta}$$

or

$$B_H = \frac{2\pi}{t\,(d^2-l^2)}\sqrt{\frac{\mu_0}{4\pi}\frac{2Id}{\tan\theta}}$$

where *I* is the moment of inertia of the bar magnet given by

$$I = \text{mass} \times \frac{\text{length}^2 + \text{breadth}^2}{12}$$

The value of $\frac{\mu_0}{4\pi} = 10^{-7}$ Wb/A-m

Experiment 23.1. To find the value of B_H the horizontal component of earth's magnetic field in the laboratory using a deflection and a vibration magnetometer.

Apparatus. A deflection magnetometer, a vibration magnetometer, a telescope, a small magnetic dipole, a rectangular brass bar, a stop watch, a half metre rod, a magnetic compass and a piece of chalk.

Procedure. 1. Find the magnetic length of the magnetic dipole.

2. Set the magnetometer in the end on position. Mark its boundary with a piece of chalk. Mark the centre of the magnetic dipole by a pencil line drawn all round. Find the mean deflection θ by placing the magnetic dipole at a suitable distance on the arms. Repeat for three different distances.

3. Set the vibration magnetometer and find the time for 20 vibrations for the same magnetic dipole. Repeat three times.

4. Find the length, breadth and the mass of the magnetic dipole.

Observations. Magnetic length of the magnet $2l$ = cm

l = cm = m

No.	*Distance d* (in metres)	*Deflection*								*Mean deflection* θ
		Magnetic dipole on east arm				*Magnetic dipole on west arm*				
		N towards needle		*S towards needle*		*N towards needle*		*S towards needle*		
		1	*2*	*3*	*4*	*5*	*6*	*7*	*8*	
1										
2										
3										

Time for 20 vibrations = 1. 2. 3.

Mean time for 20 vibrations = seconds

∴ Time period t = seconds

Mass of the magnetic dipole = gm = kg

Length of the magnetic dipole a = cm = m

Breadth of the magnetic dipole b = cm = m

∴ Moment of inertia $I = \frac{a^2 + b^2}{12} \times$ mass = kg– m²

Now $$B_H = \frac{2\pi}{t\,(d^2 - l^2)} \sqrt{\frac{\mu_0}{4\pi} \frac{2Id}{\tan\theta}}$$

∴ B_H = (i) (ii) (iii)

Mean B_H = Tesla (Wb/m²)

Precautions. 1. No magnet or magnetic substance should lie on the working table. Even the given magnetic dipole should be kept at a safe distance while setting the magnetometer.

2. See that the needle of the magnetometer rotates freely and the arms are not shaky.

3. The boundary of the magnetometer should be marked after setting it in the end-on position and should not be disturbed during the experiment.

4. The deflection should lie between 30° and 60°.

5. The box should be tapped every time gently before noting the reading of the pointer.

6. The middle point of the magnetic dipole should be marked and distance should be measured from this marked point.

7. While noting the observations keep the eye in such a position that the pointer lies over its image in the mirror below to avoid error due to parallax.

8. Unspun silk thread should be used for suspending the magnetic dipole in a vibration magnetometer.

9. Amplitude of vibration must be small otherwise the formula for the time-period $t = 2\pi\sqrt{\frac{I}{MB_H}}$ will not hold good.

10. The line drawn on the base of the vibration magnetometer should be set in the magnetic meridian and torsion in the silk thread should be removed by using a rectangular brass bar so that the magnetic dipole remains in the magnetic meridian without any torsion acting on it.

11. While replacing the brass bar by a magnetic dipole the north pole of the magnetic dipole should be kept towards earth's north.

12. Magnetic dipole should be horizontal and as near the base of the magnetometer as possible.

13. To avoid the effect of air currents on the vibration the door of the magnetometer box should be closed.

Experiment 23.2. To find the value of M in the laboratory using deflection and vibration magnetometers.

Apparatus. Same as in **Expt. 23.1.**

Theory. If θ is the deflection produced by a magnetic dipole placed at a distance d when the magnetometer is set in the end-on position, then

$$\frac{\mu_0}{4\pi}\frac{2Md}{(d^2 - l^2)^2} = B_H \tan\theta$$

$$\therefore \quad \frac{M}{B_H} = \frac{4\pi}{\mu_0}\frac{(d^2 - l^2)^2 \tan\theta}{2d} \qquad ...(i)$$

If t is the time period of the same magnet as determined with vibration magnetometer, then

$$t = 2\pi\sqrt{\frac{I}{MB_H}}$$

$$\text{or} \quad MB_H = \frac{4\pi^2 I}{t^2} \qquad (ii)$$

Multiplying (i) and (ii), we get

$$M^2 = \frac{4\pi^2 I}{t^2}\left(\frac{4\pi}{\mu_0}\right)\frac{(d^2 - l^2)^2 \tan\theta}{2d}$$

$$\text{or} \quad M = \frac{2\pi(d^2 - l^2)}{t}\sqrt{\left(\frac{4\pi}{\mu_0}\right)\frac{I\tan\theta}{2d}}$$

Procedure and record. Same as in **Expt 23.1.**

Calculations. $M = \frac{2\pi(d^2 - l^2)}{t}\sqrt{\left(\frac{4\pi}{\mu_0}\right)\frac{I\tan\theta}{2d}}$

(*i*) (*ii*) (*iii*)

Mean M = Wb–m or (Amp-m^2)

Oral Questions

1. Magnetic field. (*i*) What is meant by the magnetic moment of a magnet? (*ii*) What is magnetic meridian and how do you mark its position? (*iii*) What is the difference between the magnetic length and the length of the magnet? (*iv*). How will you find the position of the poles? (*v*) Define a unit pole. (*vi*) What is the unit of magnetic intensity?

2. Deflection magnetometer. (*i*) What is the tangent law of magnetic fields? What are the essential conditions for its application? (*ii*) Why do you take eight observations in a deflection magnetometer? (*iii*) What is the material of which the magnetometer box is made? (*iv*) How do you set the magnetometer in the end on position? (*v*) How do you place the magnetic dipole in this position? (*vi*) What is the use of a mirror in a deflection magnetometer? (*vii*) Define B_H. What is its value at your place?

3. Vibration magnetometer. What is the principle of a vibration magnetometer? (*ii*) Why is the magnetic dipole enclosed in a box? (*iii*) What is the suspension made of? Can it be of cotton (*iv*) How is the twist removed? (*v*) Can you use an iron bar instead of a brass bar? (*vi*) How do you set a magnetic dipole into vibrations? (*vii*) How do you compare the magnetic moments of two magnetic dipoles which are equal in size? (*viii*) What precautions will you observe in the experiment with a vibration magnetometer?

PART VI

ELECTRICITY

24

Electrical Instruments

24.1. Cells. A cell is used as a source of electric current in all electricity experiments. There are two types of cells which are commonly used in the laboratories.

(*a*) **Primary cells.** In a primary cell the electrical energy is obtained from the chemical reaction taking place within the cell. There are many forms of primary cells. Of these Daniell cell and Leclanche cell are widely used due to their low cost of working. Dry cell is a modified form of a Leclanche cell.

(*b*) **Secondary cells.** The primary cells do not give a continuous heavy current at a constant potential difference for a long time and need a continuous replacement of the zinc plate and the solution. For this reason *secondary cells* are most commonly used in laboratories.

(*i*) *Lead storage cell* (acid accumulator). The *e.m.f* of this cell, when freshly charged is 2.1 volts, but falls to 2.0 volts when used. It should be recharged when the *e.m.f.* falls below 1.8 volts. The internal resistance of such a cell is very low. The two poles of this cell should never be connected directly, without introducing a resistance.

(*ii*) *Edison cell.* (Nickel-iron alkali accumulator). It consists of a steel vessel containing a solution of 20% KOH and 1% $LiOH$ in distilled water. *The positive plate is a steel grid stuffed with nickel hydroxide and flaked nickel* and the *negative* is also a *steel grid containing iron hydroxide* having traces of mercuric oxide which lowers the internal resistance. The *e.m.f.* of such a cell is 1.45 volts and remains constant for a fairly long time. (Fig. 24.1)

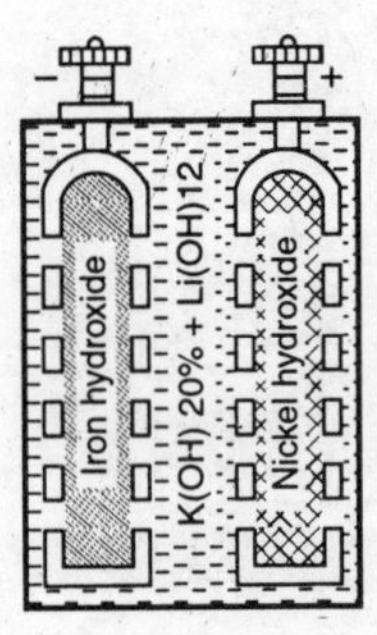

Fig. 24.1

Charging of secondary cells. Electric current from an external D.C. source brings about some chemical changes inside the cell when passed through it, which set up a potential difference between the two plates. The process is known as *charging the cell.*

To charge from A.C. mains, a transformer and a rectifier are used to step down and then to convert into D.C. voltage of a desired value.

(*iii*) *Standard cell.* Two standard cells, in common use, are (*i*) *Weston cadmium cell* and (*ii*) *Clarke cell.*

Weston cadmium cell. It consists of two tubes connected as shown in Fig. 24.2. The positive pole consists of mercury in contact with a paste of pure mercurous sulphate which acts as a depolariser. The negative pole consists of cadmium mercury amalgam. The electrolyte is a saturated solution of cadmium sulphate in contact with cadmium sulphate crystals to mantain the concentration constant.

The *e.m.f.* of the cell at 20°C is 1.0184 volts. At any temperature *t*°C the *e.m.f.* is given by the relation $E = 1.0184–4.06×10^{-5}\ (t–20)–9.5 \times 10^{-7}\ (t–20)^2 + 10^{-8}\ (t–20)^3$ volts.

The *e.m.f.* of the cell remains constant if the temperature does not change. This cell should never be used to draw very heavy currents.

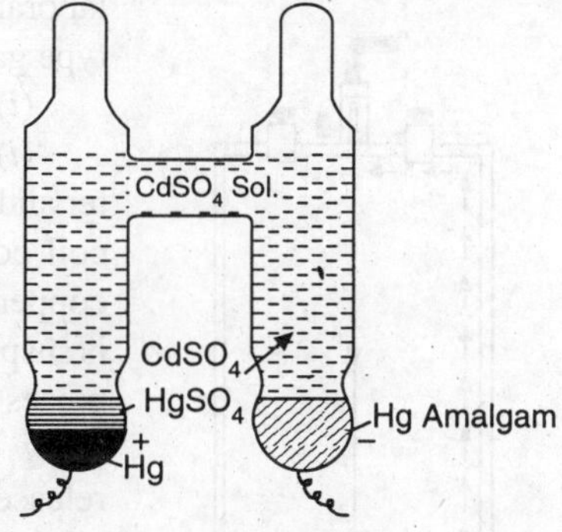

Fig. 24.2

24.2. Resistance. To measure the resistance of a conductor its resistance is compared with a known standard resistance made in the form of a coil.

Resistance coil. The resistance of a standard coil should not vary with temperature. Such coils are made from insulated wires like *constantan, eureka, manganin, german silver* etc. which have a very low temperature co-efficient of resistance. To prepare a *non-inductive resistance* the calculated length of the wire is double folded and then wound carefully on a reel. The free ends of the wire are soldered to the two terminals as shown in Fig. 24.3. The terminals are fixed into an ebonite plate and the whole arrangement is enclosed inside a box.

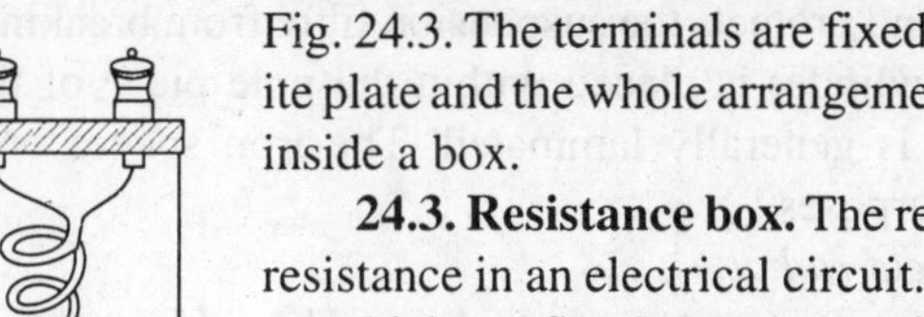
Fig. 24.3

24.3. Resistance box. The resistance box is a device for introducing any known resistance in an electrical circuit. It consists of a wooden box having an ebonite top, on which are fixed stout brass pieces separated by a small distance. A conical hole is drilled in between the two stout brass piece into which fits a plug key. Standard non-inductive resistance coils are placed in the box. The two ends of a coil are soldered to the two adjoining brass piece and the value of this resistance is marked at the top. The various coils having different resistances are arranged in the following order.

1	2	2	5	10	20	20	50
							Infinity
5000	2000	2000	1000	500	200	200	100

When a resistance is to be introduced the plug key from in between the two brass pieces to which the ends of this resistance coil are connected, is removed. When the plug key is in position the current passes through the brass pieces which have practically no resistance. If it is desired to introduce a resistance of 88 ohms, the plug keys of the resistance, 50, 20, 10, 5, 2 and 1 ohms are removed.

24.4. Rheostat. It is an adjustable resistance which is placed in an electrical circuit to regulate the flow of electric current. It consists of a porcelain tube on which is wound a certain length of a resistance wire connected at either end to a terminal. The winding is done in such a manner that the various turns do not touch each other. Over the porcelain tube is fixed a bar carrying a terminal T_1 at one end and a sliding contact S as shown in Fig. 24.4.

The length and diameter of the wire used depend respectively upon the maximum resistance to be introduced and the maximum current to be passed. To use the rheostat as a variable resistance one contact is made with the terminal T_1 fixed to the bar while the other contact is made with one of the two terminals say T_2 to which the ends of the wire are connected. The resistance in the circuit can be varied by moving the sliding contact along the bar.

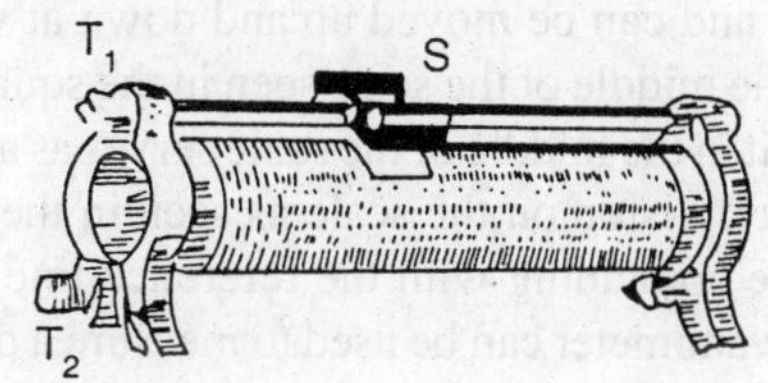

Fig. 24.4

The maximum resistance that can be introduced with the rheostat and the highest instantaneous current that can be passed through it is generally marked on the instrument.

24.5. Galvanometer. A galvanometer is used for the detection and measurement of electric current in a given circuit. There are two types of galvanometers : (*i*) *moving magnet* and (*ii*) *moving coil*.

A tangent galvanometer is the familiar form of a moving magnet type galvanometer used in the laboratory. It has been described in a later chapter. The moving coil type galvanometers are

(*i*) D'Arsonval galvanometer and (*ii*) Weston galvanometer.

(*i*) **D' Arsonval galvanometer.** It is a reflecting type galvamometer and consists of a rectangular metallic box having a glass front. The coil consisting of a large number of turns of a very fine insulated copper wire, is wound over a metallic ring or a rectangular frame and is suspended by a fine **phosphor bronze** strip within the concave pole pieces *N* and *S* of a permanent magnet. The coil can be clamped by a screw fixed at the back when the instrument is not in use and can be released when required. This arrangement checks the unnecessary vibrations of the coil and protects the suspension fibre from breaking. An iron sphere or a cylinder is placed within the pole piece of the magnet. The magnet is generally laminated. The iron sphere or a cylinder serves two purposes :

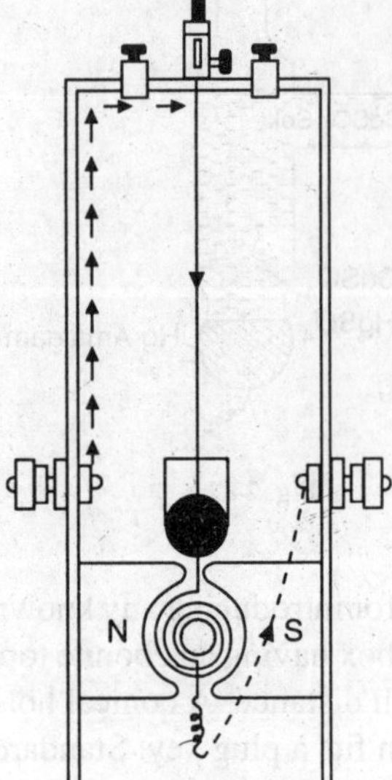

Fig. 24.5

(*i*) *It makes the field radial.*

(*ii*) *It makes the galvanometer dead beat. The eddy currents induced in the soft iron tend to oppose the movement of the coil. The induced currents are also set up in the copper ring on which the coil is would and help to make the galvanometer dead beat.* In a ballistic galvanometer the coil is wound over an ivory ring and the soft iron sphere is not used.

There is a screw at the top called the torsion head which helps to raise or lower the coil and also to rotate it about a vertical axis.

The coil is held in position at the lower end by a light phosphor bronze spring connected to a screw. The screw is connected by a short piece of an insulated copper wire to a binding screw fixed at the side and insulated from the body of the galvanometer. A second binding screw is fixed on the other side and is in contact with the body of the galvanometer. The current enters through one end and leaves through the other binding terminal.

At the base there are three levelling screws. The coil is made symmetrical about the metallic sphere with the help of these screws.

The deflection is read by a mirror fixed on the suspension fibre. A small square piece of the glass plate at the front is silvered in such a manner that it covers a part of the mirror.

A scale printed on a paper and pasted on a horizontal metal strip as shown in Fig. 24.6. is fixed in front of the mirror fixed on the suspension wire, on a vertical rod and can be moved up and down at will. The image of a mark at the middle of the scale seen in the squared silver portion through a slit in the middle of the scale serves as a *reference* line. The image of a division on the scale as seen in the mirror on the suspension wire coinciding with the reference line gives the deflection. The galvanometer can be used to measure a deflection with a lamp and scale arrangement. When the lamp and scale arrangement is used to measure deflection the metallic strip carrying the paper scale and the front fixed squared mirror are removed.

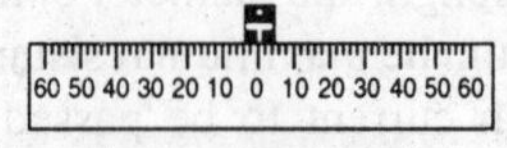

Fig. 24.6

The current *I* passing through the coil of the galvanometer is given by

$$I = \frac{c}{nAB} \quad \frac{\theta}{\cos\theta}$$

where *c* is the couple per unit angular twist of the suspension wire, *n* the number of turns, *A* the area of the coil and *B* the magnetic field due to the permanent magnet and θ the deflection produced.

In the above case the current is not proportional to the deflection θ. The value of the constant of proportionality increases as the deflection increases. To obtain the deflection proportional to the current radial magnetic fields are employed.

Radial magnetic field. A cylindrical piece of soft iron is fixed within the coil between the concave pole pieces of the magnet. The magnetic lines of force crowd in to the soft iron piece and have their directions along the radius of the cylinder, thus resulting in an intense *radial field* in the in-between air space. The vertical sides of the rectangular coil move in this space, the plane of the coil coinciding with the radial lines of force in any position. The formula is then reduced to

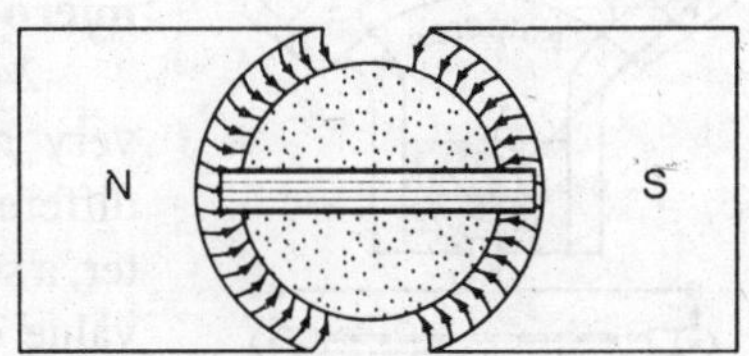

Fig. 24.7

$$I = \frac{c}{nAB}\,\theta$$

and hence the deflection θ is proportional to the current. This arrangement is used in the pointer type galvanometer. Thus in such a moving coil galvanometer current sensitivity or current required to produce a unit deflection is given by

$$\eta = \frac{I}{\theta} = \frac{c}{nAB}$$

Weston type galvanometer. A common from of a moving coil galvanometer used in the laboratory is called a *Weston galvanometer.* It consists of a coil of insulated thin copper wire mounted on pivots. The coil is kept in position between the concave pole pieces of a strong permanent magnet by two phosphor bronze springs. The two springs are soldered to the two ends of the coil and serve as leads for the current. A light pointer is fixed rigidly to the coil and its movement over a scale measures the deflection as shown in Fig. 24.8. This galvanometer has a number of advantages.

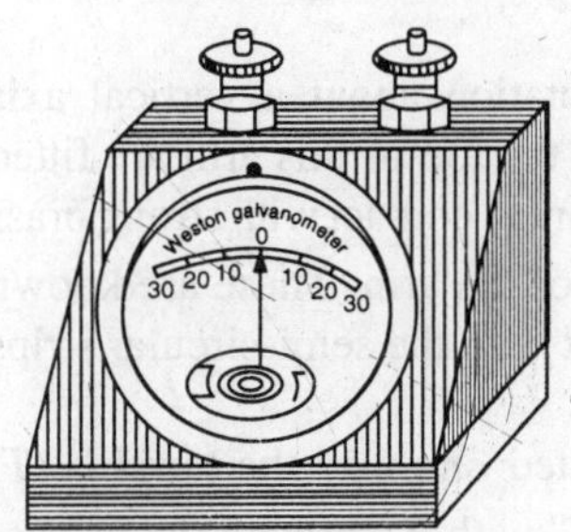

Fig. 24.8

(*i*) As the coil is placed in a strong magnetic field, the magnetic field due to the earth has no effect on it. Thus the galvanometer can be placed in any position unlike a *tangent galvanometer* and remains unaffected due to external magnetic fields.

(*ii*) The strength of current is proportional to the deflection hence the scale used is linear.

(*iii*) It is a *dead beat instrument, i.e.,* it comes to its position of rest instantaneously. This is achieved by placing a cylindrical soft iron core in between the pole pieces of the magnet leaving a **narrow** air gap for the **free** motion of the coil. The motion of the coil due to the flow of current in it produces an induced current in the core, which according to Lenz's law, opposes the motion of the coil and makes it *dead beat.* The soft iron core also makes the magnetic field *radial and strong* and, therefore, the deflection of the coil is proportional to the current flowing through it.

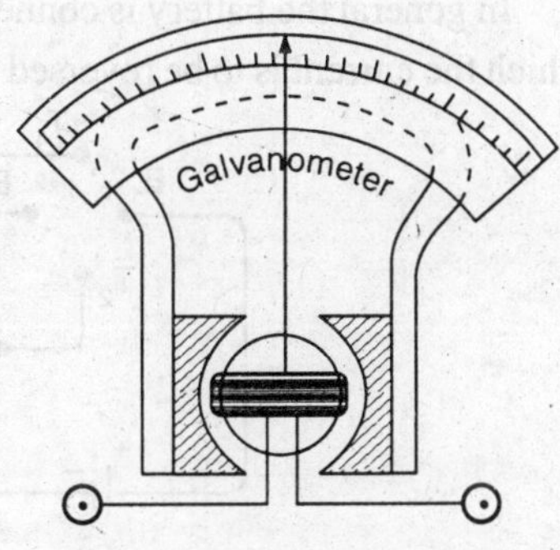

Fig. 24.9

24.6. Ammeter. A moving coil galvanometer gives a full scale deflection for a very low current. In order to measure large currents a suitable low resistances S has to be placed in parallel with the galvanometer coil. This low resistance is known as a *shunt* resistance the value of which depends upon the range of current to be measured. The scale is then calibrated to read the current directly in amperes, and the instrument is called an **ammeter** as shown in Fig. 24.10. If the scale is calibrated to read the current in milliamperes the instrument is called a **milliammeter,** and if the scale is calibrated in microamperes it is known as a **micro-ammeter.**

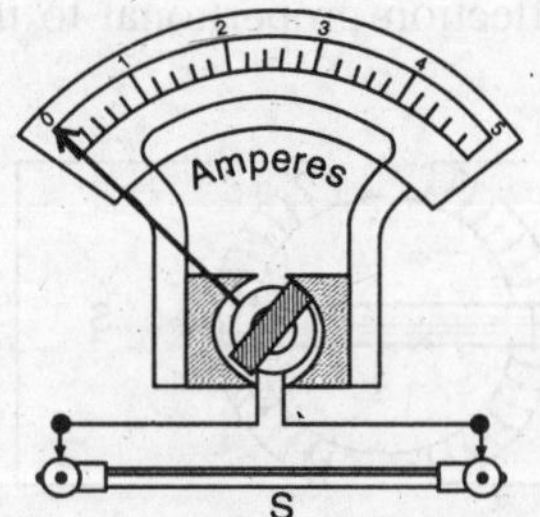

Fig. 24.10

24.7. Voltmeter. A voltmeter is a moving coil galvanometer of a very high resistance. Its scale is calibrated to read the potential difference directly in volts. To convert a galvanometer into a voltmeter, a suitable high resistance R is connected in series with its coil. The value of the series resistance depends upon the range for which the instrument is meant.

To measure the difference of potential between two points on a conductor through which the current is flowing, the voltmeter is connected to the two points in *parallel* with the conductor. As the voltmeter has a very high resistance only a very small fraction of the total current passes through it and the current in the main circuit remains practically unaltered.

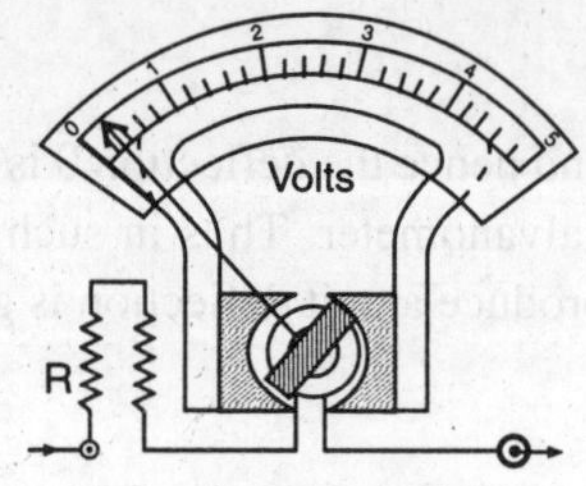

Fig. 24.11

24.8. Reversing key. It is used for reversing the direction of current through an electrical circuit without disconnecting and connecting the wires. A simple form of the key is shown in Fig. 24.12. It consists of two semicircular brass strips fixed on an ebonite plate. Each strip is fitted with a terminal marked B_1 and B_2. It is also provided with an ebonite arm capable of rotation about a vertical axis passing through its centre. On the top of this arm are fitted two terminals T_1 and T_2 which are in contact with strong brass strips fitted on the lower side of the arm. These are known as *brushes* and come in contact with the semi-circular strips when the arm is rotated.

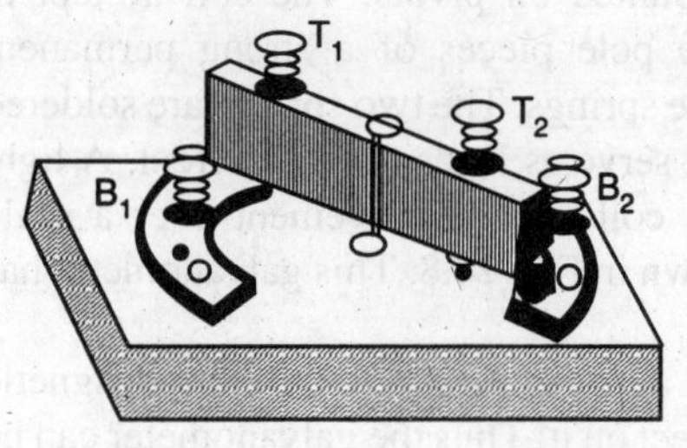

Fig. 24.12

If the ebonite arm is rotated *one way* the terminal T_1 comes in contact with B_1 and T_2 with B_2 through the respective brushes. When rotated in the opposite direction T_1 comes in contact with B_2 and T_2 with B_1.

In general the battery is connected to the lower terminal B_1 and B_2 and the galvanometer through which the current is to be reversed to the top terminals T_1 and T_2. The action of reversing key is shown

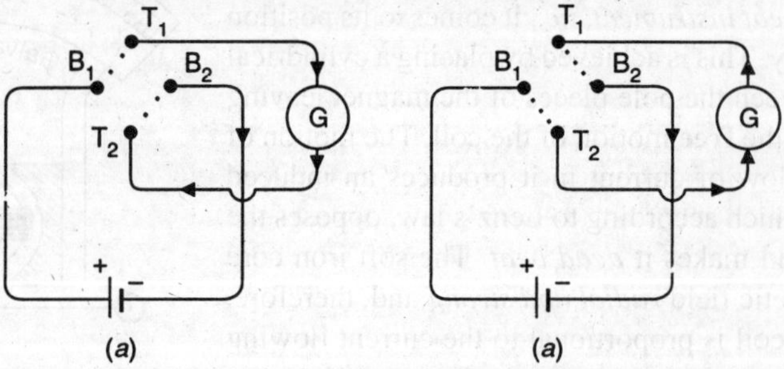

Fig. 24.13

in Fig. 24.13. When T_1 is in contact with B_1 and T_2 with B_2 current through the galvanometer flow as shown in Fig. 24.13. (*a*) but when T_1 is in contact with B_2 and T_2 with B_1 the current in the galvanometer flows in the opposite direction as shown in Fig. 24.13 (*b*).

Oral Questions

1. Cells. (*i*) What are the various types of cells used in the laboratory? (*ii*) What is the difference between the primary and the secondary cells? (*iii*) What are the various parts of (*a*) a Leclanche cell and (*b*) a Daniell cell? (*vi*) What is local action and polarisation? How are they removed in the above cells! (*v*) How is current produced in (*a*) a primary cell (*b*) a secondary cell? (*vii*) Why is a Leclanche cell not used when constant current is required? (*vii*) Name the various parts of a lead accumulator and an Edison cell. (*viii*) What is the e.m.f. of the cell that you are using? Is it a primary or a secondary cell? (*ix*) Which cell is used in (*a*) *your torch* (*b*) in motor car? (*x*) What type of current do you use for charging the secondary cell? (*xi*) Which of the two secondary cells you will prefer and why? (*xii*) What is the minimum e.m.,f. beyond which a lead accumulator should not be used and why? (*xiii*) What do you mean by the ampere hour capacity of a cell? (*xiv*) What is a standard cell? (*xv*) What precautions should be observed in using a standard cell? (*xvi*) What is the formula for the e.m.f. of the Weston cadmium cell?

2. Resistances. (*i*) Define specific resistance. (*ii*) What are the factors on which the resistance of a wire depend ? (*iii*) What is the effect of temperature on the resistance of a conductor? (*iv*) Which wire do you prefer in the construction of a standard resistance coil and why? (*v*) Which wire is used in the coil of a heater? (*vi*) Is the resistance of a wire the same on D.C. and on A.C.? (*vii*) What is a non-inductive resistance? How is it made? (*viii*) How is the infinite resistance provided in a resistance box? (*ix*) What is a rheostat? (*x*) What do the following marking on a rheostat signify : 20 ohms, 4.5 amps. inst., 3 amps. cont?

3. Galvanometers. (*i*) What is the principle of a moving coil galvanometer ? (*ii*) What material is used for the suspension wire of this galvanometer and why? (*iii*) Why is the magnet laminated and provided with concave pole-pieces? (*iv*) What is a radial field and how is it obtained? (*v*) How is the galvanometer made dead beat? (*vi*) What change will you make to convert it into a ballistic galvanometer? (*vii*) What are the advantages of the moving coil type galvanometer? (*viii*) How do you convert a moving coil galvanometer into (a) an ammeter and (*b*) a voltmeter? (*ix*) What is the function of a reversing key?

25

Magnetic Effects

25.1. Magnetic field along the axis of a circular coil. The intensity of magnetic field at a point P lying on the axis of a circular coil AB of radius a having n turns at a distance x from the centre O of the coil in S.I. units, is given by

$$B = \frac{\mu_0}{4\pi} \cdot \frac{2\pi n I a^2}{(a^2 + x^2)^{3/2}}$$

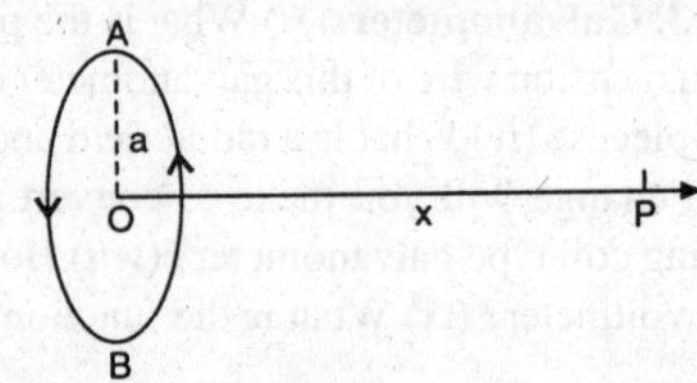

Fig. 25.1

The units of B are Tesla or Wb/m^2

The direction of the magnetic intensity at P is along OP produced if the current flows through the coil in the anti-clock-wise direction as seen from P. If the direction of the current is clockwise the field at P is along PO.

The value of the magnetic intensity is maximum at the centre O of the coil and is given by

$$B = \frac{\mu_o}{4\pi} \frac{2\pi n I}{a} = \frac{\mu_o n I}{2a}$$

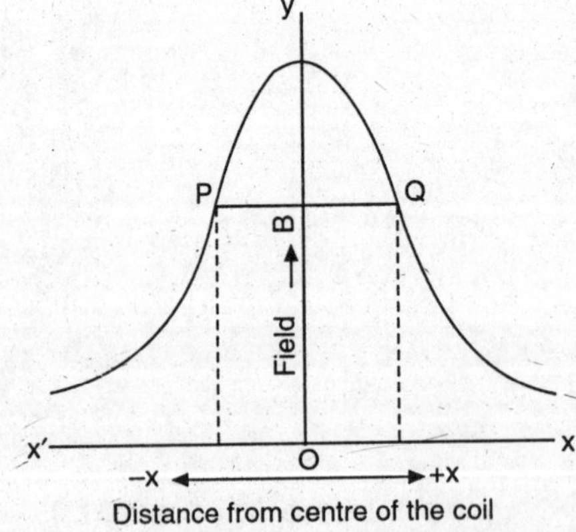

Fig. 25.2

If we move away from O towards the right or left, the intensity of the magnetic field decreases. A graph showing the relation between the intensity of the magnetic field B and the distance x is given in Fig. 25.2. The curve is first concave towards O but the curvature becomes less and less, quickly changes sign at P and Q and afterwards becomes convex towards O. It can be shown that the points of inflexion P or Q where the curvature changes its sign lie at distances $\frac{a}{2}$ from the centre. Hence the distance between P and Q is equal to the radius of the coil.

25.2. Stewart and Gee's tangent galvanometer. To study the variation of magnetic field along the axis of a circular coil Stewart and Gee's type of tangent galvanometer shown in Fig. 25.3 is used. It consists of a circular wooden or brass frame on which are wound a number of turns of insulated copper wire.

The coil is fixed on a horizontal base with its plane vertical and the free ends of the wire are connected to two terminals T_1 and T_2 provided at the base. A deflection magnetometer is fitted on the two pillars V_1 and V_2 provided on the horizontal board in such a manner that it can slide along the horizontal direction with the centre of the needle always lying on the axis of the coil. The distance of the needle from the centre of the coil can be read on the centimetre scale provided on the magnetometer arms with reference to fixed marks A and B on the vertical pillars.

Experiment 25.1. To study the variation of magnetic field with distance along the axis of a circular coil carrying current by plotting a graph.

Apparatus. Stewart and Gee's type tangent galvanometer, a battery, a rheostat, an ammeter, a one-way key, a reversing key, connecting wires, and sand paper etc.

Theory. The intensity of magnetic field at a point on the axis of a circular coil of radius a having n turns, at a distance x from the centre of the coil, in S.I. units is given be

$$B = \frac{\mu_0}{4\pi} \times \frac{2\pi n I a^2}{(a^2 + x^2)^{3/2}}$$

where I is the current in amperes flowing through the coil.

If the field B is perpendicular to the horizontal component of earth's field B_H and θ is the deflection produced in a deflection magnetometer, then

$$B = B_H \tan\theta$$

or $$B \propto \tan\theta$$

Hence a graph between $\tan\theta$ and x will be similar to the graph between B and x.

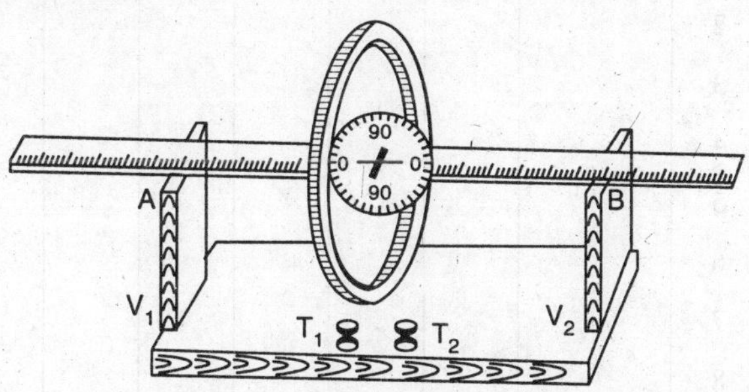

Fig. 25.3

Procedure. 1. Place the instrument on the table so that the arms of the magnetometer lie roughly east and west and the magnetic needle lies at the centre of the vertical coil. Place the eye a little above the coil and rotate the instrument in the horizontal plane till the coil, the needle and its image in the mirror provided at the base of the compass box all lie in the same vertical plane. The coil is thus set roughly in the magnetic meridian. Rotate the compass box so that the pointer lies on the 0–0 line.

2. Connect the galvanometer to a battery, a rheostat, a one way key and an ammeter through a reversing key as shown in Fig. 25.4.

3. Adjust the value of the current so that the magnetometer gives a deflection of the order of 60°–70°. Reverse the current and again note the deflection. If the mean deflection in the two cases agrees closely, the coil lies exactly in the magnetic meridian. If the mean deflection in the two cases does not agree closely, slightly turn the instrument till the deflection with the direct and the reversed currents agree closely.

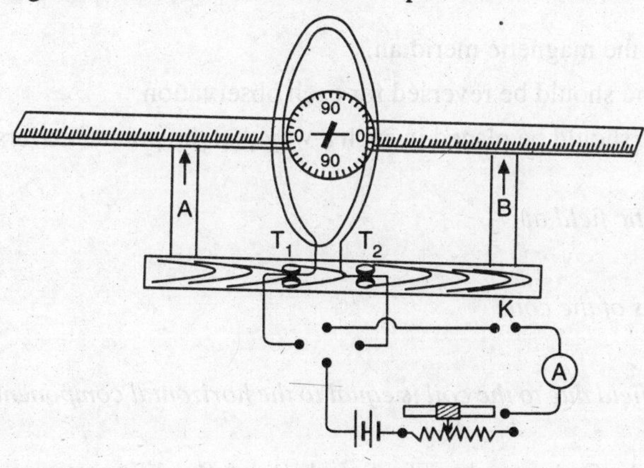

Fig. 25.4

4. Now slide the magnetometer along the axis and find the position where the maximum deflection is obtained. In this position the centre of the needle coincides with the centre of the vertical coil.

5. Note the position of the arm against the reference mark A and also the value of the current as indicated by the ammeter. Read both ends of the pointer in the magnetometer, reverse the current and

again read both ends. Shift the magnetometer by 2 cm and note the reading of the magnetometer keeping the current constant at the same value both for direct and reversed current. In this position the centre of the magnetic needle is at a distance of 2 cm from the centre of the vertical coil. Take a number of observations by shifting the magnetometer by 2 cm at a time.

6. Similarly repeat the observation by shifting the magnetometer in the opposite direction and keeping the current constant at the same value.

Observations. Current $I =$ amp

No.	*Distance from the centre x*	*Left side*				*Mean θ*	*tan θ*	*Right side*				*Mean θ*	*tan θ*
		Direct		*Reversed*				Direct		*Reversed*			
		1	2	3	4			*1*	2	3	4		
1													
2													
3.													
4													
5													
6													
7													
8													
9													

7. **To plot the graph.** To show how the field varies along the axis of the coil, plot a graph between tan θ and x. It will be similar to the graph between B and x, as shown in Fig. 25.2.

Precautions. 1. There should be no magnet, magnetic substances and current carrying conductors near the apparatus.

2. The plane of the coil should be set in the magnetic meridian.

3. The current should remain constant and should be reversed for each observation.

4. To avoid error due to parallax, the eye should be placed in such a way that the pointer covers its image in the mirror below.

Exercise 1. *Find the value of the magnetic field at*

(*i*) *the centre of the coil,*

(*ii*) *a distance a/2 where a is the radius of the coil.*

(*iii*) *a distance a.*

Exercise 2. *Find the distance where the field due to the coil is equal to the horizontal component due to earth's field.*

Hints. From the graph between tan θ and x find the value of x for which tan θ = 1.

Exercise 3. *Plot a graph between tan θ and x in the case of a circular coil and hence find the radius of the coil.*

Hints. Locate the position of the points of inflection P and Q as shown in Fig. 25.2. The distance PQ gives the radius of the coil.

Oral Questions

Magnetic effects. (*i*) What is the magnitude and direction of the magnetic field at the centre of a coil? (*ii*) Define electromagnetic unit of current. (*iii*) What is the use of studying the variation of magnetic field along the axis of a coil? (*iv*) What is a point of inflection and where is it obtained in this case? (*v*) Name the instrument in which use is made of the positions of the point of inflection. (*vi*) What is the value of magnetic field at a point on the axis where the deflection is 45° at a place where $B_H = 0.32$ oersted?

26

Measurement of Resistance

26.1. Wheatstone bridge. It is one of the accurate methods for the measurement of resistance. As shown in Fig. 26.1. four resistances P, Q, R, and X are connected to form a quadrilateral $ABCD$. A galvanometer G is connected across B and D through a tap key K_2 and a cell is connected across A and C through another tap key K_1. The values of the resistances P, Q and R are so adjusted that the galvanometer shows no deflection when the tap key K_1 is pressed first and then the tap key K_2. In such a case the points B and D are at the same potential and it can be shown that

$$\frac{X}{R} = \frac{Q}{P}$$

or

$$X = R\frac{Q}{P}$$

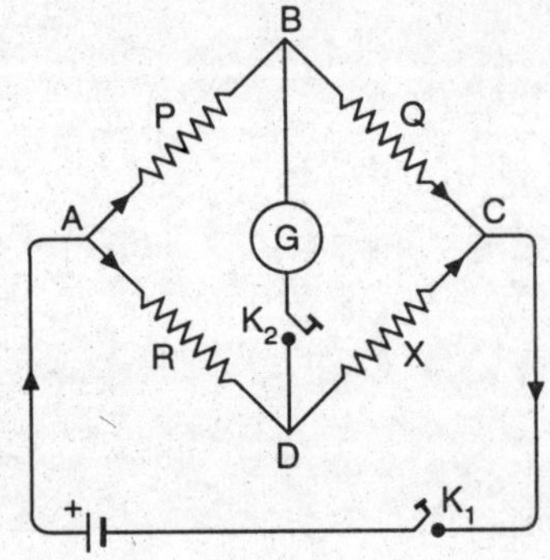

Fig. 26.1

If P, Q and R are known the value of X, the unknown resistance can be calculated. Usually P and Q are fixed resistances known as the ratio arms and R is a resistance box from which any resistance from 1 to 11110 ohm can be introduced in the circuit.

26.2. Post office box. A post office box is a compact form of the Wheatstone bridge. In a plug type instrument each of the arms AB and BC contains three resistances of 10, 100 and 1000 ohm respectively. These are called the *ratio arms*. The arm AD is a complete resistance box containing resistances from 1 to 5000 ohm. The tap keys K_1 and K_2 are also provided as shown. The key K_1 is connected to the point A and the key K_2 to the point B. These connections are shown by lines drawn on the ebonite plate. The unknown resistance X is connected between C and D, the battery between C and the key K_1 and the glavanometer between D and the key K_2 as shown in Fig. 26.2. The circuit is now exactly the same as that of the Wheatstone bridge shown in Fig. 26.1.

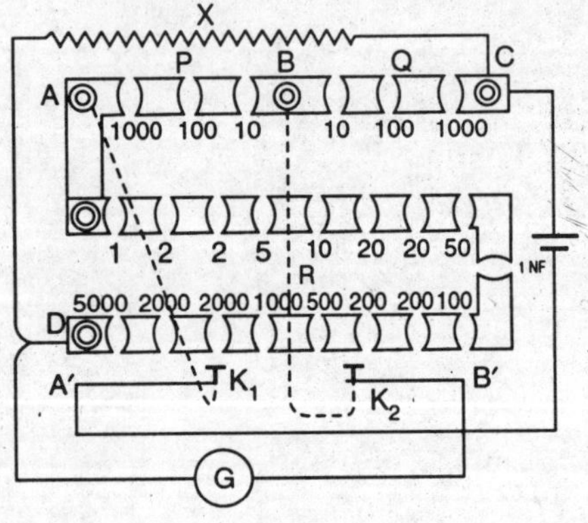

Fig. 26.2

Dial type. Dial type post office boxes are now being widely used in test laboratories. The ratio arms P and Q again consist of three coils each containing resistance of 10,100 and 1000 ohm. By rotating the dial knob any resistance can be brought into the circuit. The resistance R consists of three sets of dials placed in series. The *units* dial consists of 10 coils each of one ohm, the *tens* dial contains 10 coils each of 10 ohm and the *hundreds dial* has 10 coils each of 100 ohm. Sometimes a fourth dial of 10 coils of 1000 ohm each is also provided. Any resistance in steps of one ohm, can be introduced in the arm R. For example, if the knob in the hundreds coil is at the mark 4, tens at 6 and units at 1, then the total resistance of the arm R will be $(400 + 60 + 1) = 461$ ohm. The unknown resistance X is connected between C and D, the galvanometer between D and key K_2 and the battery between C and key K_1 as shown in Fig. 26.3.

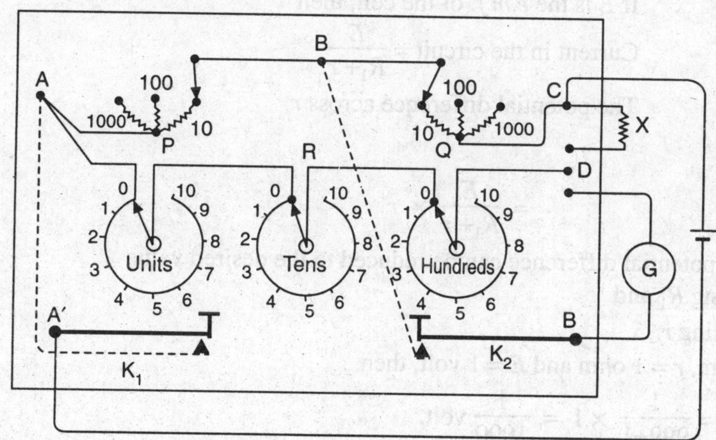

Fig. 26.3

A post office box is a very convenient arrangement for finding the resistance of a coil and to find the specific resistance of the material of a wire. These experiments have been performed by the readers in their lower classes.

26.3. Resistance of a galvanometer. The resistance of galvanometer can be found with a post office box. The method is known as **Kelvin's method.** In this method the same galvanometer is used as an indicator for obtaining the balance point.

The galvanometer of which the resistance is to be determined is placed in the arm CD and a key K_2 is introduced between B and D where usually the galvanometer is connected in a Wheatstone bridge arrangement. The galvanometer acts as its own indicator for obtaining the balance point.

On pressing the key K_1 the galvanometer gives a deflection. To keep the deflection within scale a resistance box is placed in series with the cell, in case of a weston type galvanometer.

If on pressing the key K_2 the *deflection remains unchanged the points* B and D are at the same potential. Applying the usual Wheatstone bridge relation, we have

Fig. 26.4

$$\frac{G}{R} = \frac{Q}{P}$$

or
$$G = R\frac{Q}{P}$$

26.4 D'Arsonval American type galvanometer. It is more sensitive than a Weston type galvanometer. The deflection goes out of scale even when the maximum resistance is taken out from the resistance box.

To obtain the deflection on the scale when the key K_1 is pressed either a high resistance box of the meg ohm type is placed in series with the cell or a potential dividing arrangement is introduced in the cell circuit to reduce the potential difference applied across *A* and *C* in Kelvin's method.

26.5. Potential dividing arrangement. The potential dividing arrangement is shown in Fig. 26.5.

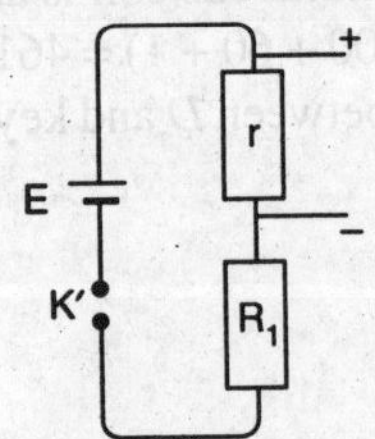

Fig. 26.5

R_1 and r are two resistance boxes connected in series with a cell through a key *K*.

If *E* is the *e.m.f.* of the cell, then

$$\text{Current in the circuit} = \frac{E}{R_1 + r}$$

The potential difference across *r*

$$= \frac{E}{R_1 + r} \times r$$

Evidently the potential difference can be reduced to the desired value

(*i*) by increasing R_1 and

(*ii*) by decreasing *r*.

e.g. if $R_1 = 999$ ohm, $r = 1$ ohm and $E = 1$ volt, then

$$P.D. \text{ across } r = \frac{1}{999+1} \times 1 = \frac{1}{1000} \text{ volt.}$$

A rheostat having a resistance of about 100 ohm can also be used as a potential dividing arrangement.

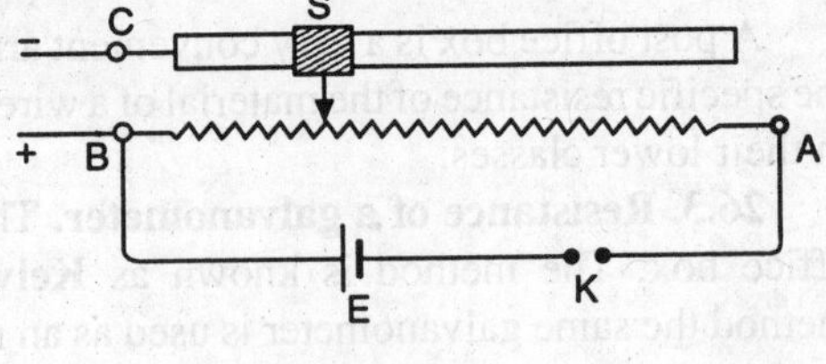

Fig. 26.6

The cell is connected through the key *K* to the fixed terminals *A* and *B* of the rheostat. The small potential difference is obtained by taking connections from *B* and the variable terminal *C*.

By moving the sliding contact *S* towards *C* the potential difference to be applied to the post office box can be decreased and by moving it in the opposite direction the potential difference can be increased. (Fig. 26.6)

Experiment 26.1. To determine the resistance of D'Arsonval American type galvanometer by using Thomson's or Kelvin's method.

Apparatus. A post office box, an American type galvanometer, a cell, two resistance boxes or a rheostat of 100 ohms resistance, a key, connecting wires and a piece of sand paper.

Theory. In Kelvin's method the galvanometer whose resistance is to be measured is placed in the arm *CD* of the Wheatstone bridge arrangement as shown in Fig. 26.4. where normally the unknown resistance is connected. When the bridge is balanced i.e. the galvanometer shows *no change* in deflection

$$\frac{G}{R} = \frac{Q}{P}$$

or

$$G = R\frac{Q}{P}$$

Procedure 1. Draw a diagram showing the scheme of connections as in Fig. 26.7. Mark the points *A, B, C, D*, K_1, and K_2 on the post office box.

2. Clean the ends of the connecting wires with sand paper. Connect the galvanometer between *C* and *D* and the cell through the potential dividing arrangement between *C* and K_1. Connect *D* to *B* through the key K_2.

3. **To set the galvanometer.** (*i*) Release the clamp at the back of the galvanometer and adjust the levelling screws at the base till the coil is symmetrically placed between the pole pieces of the magnet and turns freely.

(*ii*) Adjust the scale so that its extremities are equally distant from the mirror.

(*iii*) Place the eye in front of the slit of the scale to observe the image of the scale in the mirror attached to the coil. Adjust the height of the scale so that the image of the scale is clearly visible in the mirror fixed to the suspension fibre.

(*iv*) Turn the torsion head *very slowly* till zero mark on the image coincides with the image of reference mark observed in the fixed mirror.

4. Press the key K_1 and adjust the values of resistances r and R_1 to obtain the deflection within the scale. Take out 10 ohm plugs from P and Q keeping R zero. Now press the key K_2. *If the deflection decreases the connections are correct.*

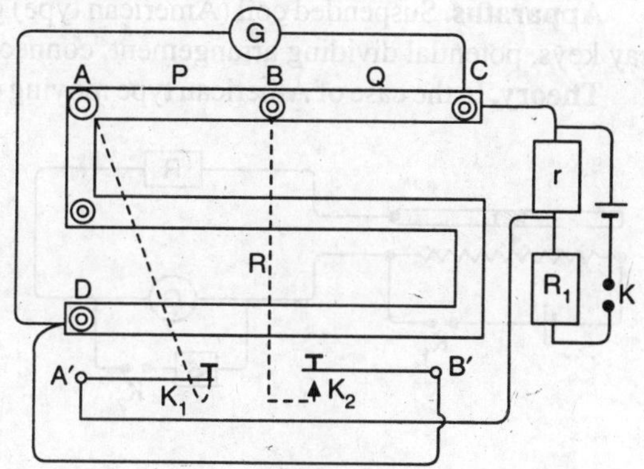

Fig. 26.7

5. Take out the 10 ohm plug from R, press K_1 and note the deflection. Also press K_2, and note the deflection again. If the deflection still decreases, go on repeating with values of R increasing in steps of 10 ohm till on pressing the key K_1 first and then the key K_2 the deflection increases. The resistance of the galvanometer lies between the last two values of R.

6. Start with the lower value of R and increase it in steps of one ohm till on pressing K_1 first and then K_2 the deflection *increases* instead of decreasing.

The resistance of the galvanometer lies between the two final values of R. This gives us the approximate value of the galvanometer resistance.

7. Make P equal to 100 ohm, keeping Q equal to 10 ohm. Starting with 10 times the lower value of R obtained in step 6 adjust it till on pressing K_1 first and then K_2 either the deflection increases instead of decreasing or remains unchanged. Note the values or value of R.

Note. *In this case the galvanometer always shows some deflection when the key K_1 is pressed. When a balance point is obtained there is no change of deflection on pressing the key K_2. It is not possible to adjust for zero deflection.*

8. Similarly repeat with P equal to 1,000 ohm keeping Q equal to 10 ohm and find the value of R for no change of deflection. The value of G is to be calculated from this **last** reading.

Observations.

Sl. No.	*P*	*Q*	*R lies between*		$G = \frac{RQ}{P}$	
			(*i*)	(*ii*)	(*i*)	(*ii*)
1						
2						
3						

Value of G from observation 3 = ohm

Precautions. 1. The galvanometer should be connected between C and D and the points B and D should be connected through the key K_2.

2. Only one cell should be used and the resistances r and R_1 should be so adjusted that the deflection is large and lies within scale.

3. When the key K_2 is pressed after the key K_1 there should be no change in deflection.

Experiment 26.2. To determine the resistance of a suspended coil type galvanometer by equal deflection.

Apparatus. Suspended coil (American type) galvanometer, a cell, two resistance boxes, two one way keys, potential dividing arrangement, connecting wires and a sand paper.

Theory. In the case of American type moving coil galvanometer the deflection is not proportional to the current. The value of constant of proportionality increases as the deflection increases. Hence half deflection method is not suitable. To find its resistance we can use equal deflection method. To obtain the deflection within scale a potential dividing arrangement is used as shown in Fig. 26.8.

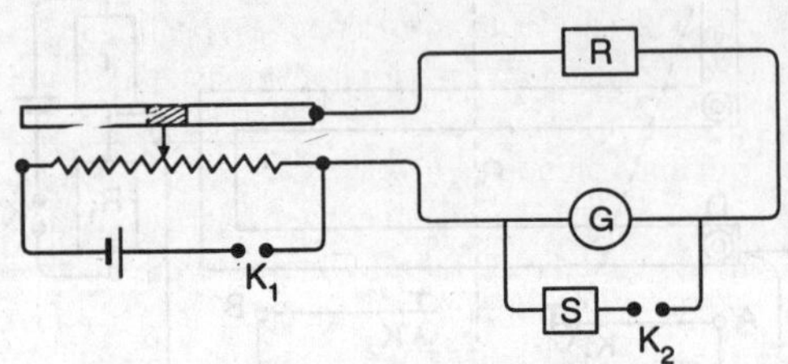

Fig. 26.8

Keeping the key K_2 open, close the key K_1 and adjust the potential difference as also the resistance R from the resistance box so that the deflection is within scale and large. The current through the galvanometer is given by

$$I_g = \frac{E}{R+G} = K\theta \qquad ...(i)$$

where E is the potential difference applied, θ the deflection and G the resistance of the galvanometer.

If now the value of R is reduced sufficiently to the value R_1, the deflection will go out of the scale. Close the key K_2 and adjust the value of shunt resistance S so that the same deflection is obtained, then

Current through galvanometer

$$I_g = \frac{E}{R_1 + \dfrac{GS}{G+S}} \times \frac{S}{G+S} = K\theta$$

or
$$\frac{ES}{R_1(G+S)+GS} = K\theta \qquad ...(ii)$$

Comparing (*i*) and (*ii*), we get

$$R_1(G+S)+GS = S(R+G)$$

or
$$R_1G + R_1S = SR$$

$$\therefore \qquad G = \frac{S(R-R_1)}{R_1}$$

Procedure. 1. Draw a diagram showing the scheme of connections as in Fig. 26.8 and make the connections accordingly.

2. Introduce a resistance R of the order more than 5000 ohm. Adjust the potential difference till the deflection is within scale and maximum.

3. Close the key K_2 and decrease the resistance R to R_1 by about 1000 ohm. Adjust the value of shunt so that the deflection is the same as in the first case. Note the value of shunt.

4. Repeat the experiment three times for different values of R_1.

Observations.

Sl. No.	Resistance R	R_1	Deflection θ	Shunt resistance for same deflection (S)	$G = \frac{S(R - R_1)}{R_1}$
1					
2					
3					
4					

Mean valve of $G =$ ohm

Precautions. 1. The potential difference in one set of observations *i.e.* for R and R_1 should remain the same.

2. The value of R should be sufficiently high.

26.6. Sensitiveness or figure of merit of a galvanometer. In the case of a **reflecting type** galvanometer the **sensitiveness** is defined as the *deflection in millimetres produced on a scale placed at a distance of one metre from the galvanometer mirror by a current of one micro-ampere.*

The **figure of merit** is defined *as the current required in micro-amperes to produce a deflection of one millimetre on a scale placed at a distance of one metre from the galvanometer mirror*. It is denoted by η. It is also called the **current sensitivity.**

If a current i μA produces a deflection of d mm, then

Current sensitivity $\eta = \frac{i}{d}$ μ A/mm

Experiment 26.3. To determine the current sensitivity of a reflecting mirror type moving coil galvanometer.

Apparatus. American type galvanometer, a lamp and scale arrangement, a cell of constant *e.m.f.*, two high resistance boxes and one standard low resistance box (0–10 ohm), a one way key, a reversing key, connecting wires, a metre rod etc.

Theory. To find the current sensitivity a cell of constant *e.m.f.* and negligible internal resistance is connected in series with two resistance boxes Q and P through a key. The resistance box P is a standard low resistance box. This forms a potential dividing circuit. The galvanometer G, connected in series with the third resistance box R, is connected across P through a reversing key. Now

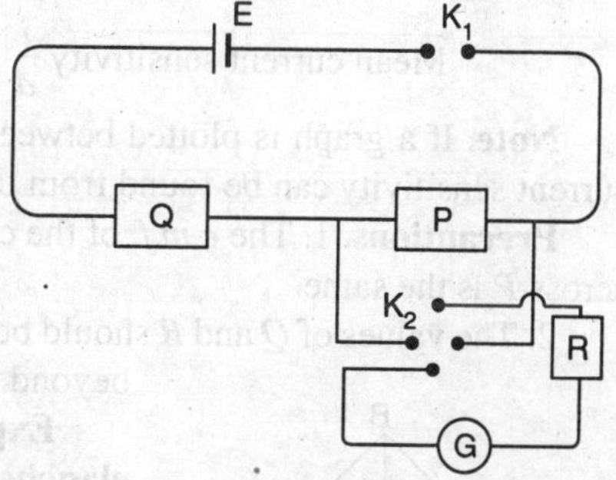

Fig. 26.9

Potential difference across $P = \frac{EP}{P+Q}$

If G is the resistance of the galvanometer, and R the resistance in series with it, then

Current through the galvanometer

$$i = \frac{EP}{(P+Q) \times (R+G+P)} \text{ ampere}$$

$$= \frac{EP \times 10^6}{(P+Q)(R+G+P)} \text{ micro ampere.}$$

If d is the deflection produced on the scale, then

Current sensitivity $\eta = \frac{i}{d}$ μA/mm

Procedure. 1. Determine the resistance of the galvanometer by any of the methods already explained for American type galvanometer.

2. Set up the reflecting mirror galvanometer and adjust the lamp and scale arrangement so that the spot of light is obtained on a scale placed at a distance of one metre from the mirror of the galvanometer. See that the spot of light moves freely and its rest position is on the zero division mark.

3. Draw the scheme of connections as shown in Fig. 26.9. and make the connection accordingly. See that the connections are neat, clean and tight.

4. Take out a resistance of the order of 2000 ohm from the resistance box R. Adjust the value of P to 2 ohm and Q to the desired value so that when the galvanometer circuit is completed nearly a full scale steady deflection is obtained. Note the deflection. Reverse the current and again note the deflection. Find the mean value.

5. Keeping the value of P and Q the same increase the value of R in steps of 1000 ohm and find the mean deflection. Repeat at least for five different values of R.

6. Measure the *e.m.f.* of the cell with an accurate voltmeter.

Observations.

E.M.F. of the cell E = volt

Galvanometer resistance G = ohm

Resistance P = ohm

Resistance Q = ohm

Sl. No.	Resistance R	Deflection			Current in μ A	Current senstivity (i/d)
		Right	Left	Mean (d)	$i = \dfrac{EP \times 10^6}{(P+Q)(R+G+P)}$	
1						
2						
3						
4						
5						

Mean current sensitivity $\dfrac{i}{d}$ = μ A/mm

Note. If a graph is plotted between deflection and current it will be a straight line. The meam current sensitivity can be found from the graph as well.

Precautions. 1. The *e.m.f.* of the cell should be checked again and again so that fall of potential across P is the same.

2. The values of Q and R should be suitably adjusted for a large deflection. If the deflection goes beyond the scale, value of P should be reduced.

Experiment 26.4. To find the internal resistance of a Leclanche cell by Mance's method using post office box.

Apparatus. A post office box, Leclanche cell, a galvanometer (Weston type preferred), a high resistance box, connecting wires, and sand paper.

Theory. The cell of which the resistance is to be determined is introduced in the arm CD of the Wheatstone bridge arrangement in which the unknown resistance is usually connected.

The galvanometer is connected between B and D and a key is introduced between A and C. On pressing the key K_2 the galvanometer gives a deflection. If on pressing the key K_1 the deflection remains unchanged, the points B and D are at the same potential. Appling the usual Wheatstone bridge relation we find

Fig. 26.10

$$X = R\frac{Q}{P}$$

where X is the internal resistance of the cell.

Procedure. 1. Draw a diagram showing the scheme of connections as in Fig. 26.11. Mark the points A, B, C, D, K_1 and K_2 on the post office box with a piece of chalk.

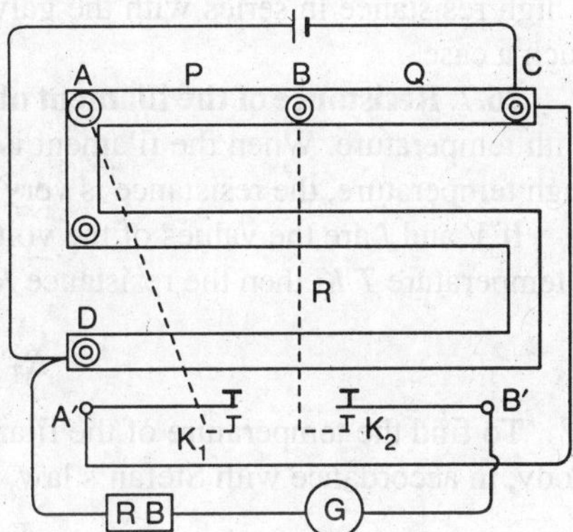

Fig. 26.11

2. Clean the ends of the connecting wires with sand paper. Connect the Leclanche cell between C and D. Connect B and D through a galvanometer, a resistance box and the key K_2. Join A and C through the key K_1. Take out 500 ohm plug from the resistance box and make all other plugs tight by giving them a slight twist.

3. Check the connections again and show to the instructor before passing current.

4. **To test the connections.** Take out the 10 ohm plug from P and Q, keeping R zero. Press the galvanometer key K_2. Adjust the resistance in the resistance box so that the deflection is within scale and is maximum. On pressing the key K_1 if the deflection decreases, the connections are correct.

5. Keeping P and Q each 10 ohm adjust the value of R in steps of 1 ohm, beginning from zero, till on pressing the key K_2 first and then the key K_1 the deflection increases instead of decreasing. The resistance of the cell lies between the last two values of R.

6. Make $P = 100$ ohm keeping $Q = 10$ ohm. The value of R should now lie between ten times the values obtained in step 5. Starting from ten times the lower value of that observation increase R in steps of 1 ohm till on pressing the key K_2 first and then the key K_1 the deflection either increases instead of decreasing or does not change.

Adjust the resistance from the resistance box for maximum deflection and test the balance point again. The resistance of the cell is equal to $\frac{1}{10}$ R.

7. Similarly repeat increasing P equal to 1000 ohm and keeping Q equal to 10 ohm

Observations.

Sl. No.	*P*	*Q*	*R lies between*		$X = \frac{RQ}{P}$	
			(i)	*(ii)*	*(i)*	*(ii)*
1						
2						
3						

Resistance of the cell from obs. 3 = ohm

Precautions 1. Always use a Leclanche cell to find the internal resistance.

2. Take care to press the galvanometer key K_2 first and then the key K_1

3. To bring the deflection of galvanometer within scale either a high resistance should be placed in series with the galvanometer or it may be shunted by a fractional resistance box or a wire of low resistance.

4. Never put a resistance in series with the cell.

Lodge's modification. Lodge suggested that the experiment may also be performed by putting a capacitor of capacitance 0.3 microfarad in series with the galvanometer.

Just after pressing the key K_2 a steady current begins to flow in the circuit but the galvanometer shows no deflection. If the key K_1 is now pressed the galvanometer will show a momentary kick except when $\frac{X}{R} = \frac{Q}{P}$. Hence the balance point can be obtained. This arrangement is better than using a high resistance in series with the galvanometer. It is preferable to use a ballistic galvanometer in such a case.

26.7. Resistance of the filament of a bulb. The resistance of the filament of a bulb varies widely with temperature. When the filament is cold, the resistance is low and when it is glowing and has a high temperature, the resistance is very high.

If V and I are the values of the voltage applied and current through the filament of the lamp, at a temperature T K, then the resistance R_T of the filament is given by

$$R_T = \frac{V}{I}$$

To find the temperature of the filament, we assume that the filament radiates energy as a black body, in accordance with Stefan's law, *i.e.*

$$E = \sigma (T^4 - T_0^4)$$

where T_0 is the room temperature and σ the Stefan's constant. If $T >> T_0$, then, T_0 can be neglected, as compared to T and

$$E = \sigma T^4$$

If the filament of the bulb has a length l and diameter d (both given by the manufacturers), then the total quantity of energy emitted by the filament per second

$$= \pi d l \sigma T^4$$

where $\sigma = 5.67 \times 10^{-8}$ W m^{-2} K^{-4}

The energy being supplied to the filament per second = $V I$ Joules

$$\therefore \quad VI = \pi d l \sigma T^4$$

or

$$T = \left[\frac{VI}{\pi d l \sigma}\right]^{1/4} = K (V I)^{1/4}$$

where $K = \left[\frac{1}{\pi d l \sigma}\right]^{1/4}$

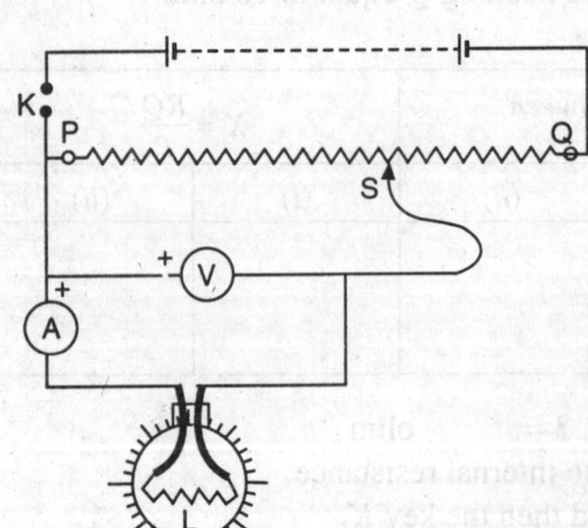

Fig. 26.12

Experiment 26.5. To study the variation of resistance of filament of a bulb with its temperature

Apparatus. 12 volt D.C. battery, a high resistance rheostat, a voltmeter of range 0—15 V, an ammeter of range 0—3 A, a 36 watt 12 volt lamp whose length and diameter of the filament are given by the makers, connecting wires etc.

Procedure. 1. Draw a diagram showing the scheme of connection and connect up accordingly as shown in Fig. 26.12.

2. Adjust the position of the sliding contact S near the end P so that the voltmeter and ammeter give a small initial reading. Note the reading of the voltmeter which measures the potential difference V applied to the lamp and that of the ammeter which measures the current I through it.

3. Go on moving the sliding contact towards Q and take about 8 readings.

Observations. Length of the lamp filament wire $l =$ cm = m

Diameter of the lamp filament wire $d =$ cm = m

(As given by the makers)

Stefan constant $\sigma =$ W m^{-2} K^{-4}

$$\therefore \quad K = \left[\frac{1}{\pi d l \sigma} \right]^{1/4} =$$

Sl. No.	*Voltmeter reading V*	*Ammeter reading I*	*Resistance* $R_T = \frac{V}{I}$	*Temperature* $T = K\,(VI)^{1/4}$

4. Plot a graph taking the absolute temperature T along the X-axis and the resistance R_T along the Y-axis. Interpret the graph.

5. Also plot a graph between log I taken along the Y axis and log V taken along the X-axis. The graph is a straight line as shown in Fig. 26.13.

Discussion. The relation between the current I through the heated filament of a lamp and the applied voltage V is in general given by

$$I = kV^n$$

where k and n are constants for a particular lamp.

$$\therefore \quad \log I = n \log V + \log k$$

Hence a graph between log I (taken along Y-axis) and log V (taken along X-axis) is a straight line.

From the graph $n = \dfrac{BC}{AC}$ = slope of the graph

$$\log k = OA$$

the intercept on the log I axis, from which k can be found out.

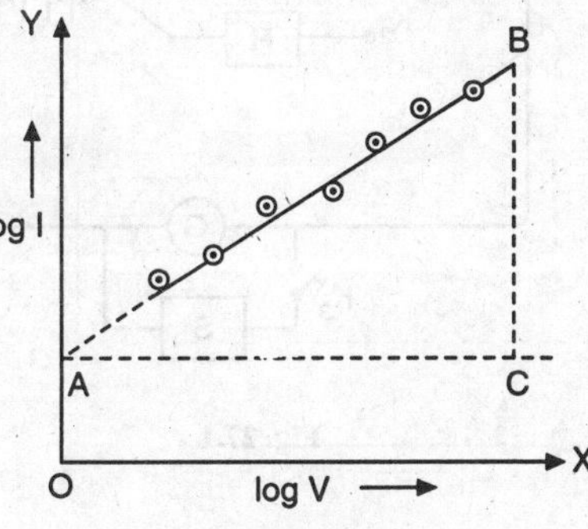

Fig. 26.13

Precautions. *A* 12 volt lamp should be used.

2. The rheostat used should have a very high reistance as compared to the resistance of the bulb when glowing.

Oral Questions

1. Wheatstone bridge. (*i*) What are the advantages of the Wheatstone bridge method of measuring resistance over other methods? (*ii*) What condition must be satisfied to get a balance point? (*iii*) Why is the battery key pressed first and the galvanometer key afterwards? (*iv*) What is the maximum resistance that you can measure with a post office box correct up to an ohm? (*vi*) Is it possible to interchange the position of the battery and the galvanometer? (*vi*) When is the Wheatstone bridge more sensitive? (*vii*) Which of the two methods (*a*) Kelvin's and (*b*) equal deflection, for measuring the resistance of a galvanometer will you prefer and why? (*viii*) In Kelvin's method do you get the null point at zero deflection; if not, why?

2. Internal resistance. (*i*) What is internal resistance of a cell? (*ii*) Which cell do you use for finding the internal resistance of a cell? (*iii*) Why don't you use a secondary cell for the purpose? (*iv*) What is Lodge's modification in Mance's method and what is the action of the capacitor used? (*v*) What is the difference between *e.m.f.* and the potential difference? (*vi*) Why does the potential difference across the terminals of a cell fall when current is drawn from it?

3. Resistance of glowing bulb. (*i*) How does the resistance of a glowing bulb change with temperature of the filament? (*ii*) Is the reistance of hot filament greater than that of the cold filament? (*iii*) How do you calculate the temperature of the fitament ? (*iv*) What is the relation between the voltage and current of a heated filament (v) What is the shape of graph between log V and log I?

27

Measurement of High Resistance

27.1. Measurement of high resistance. The Wheatstone bridge method is suitable only when the order of resistance in the various arms of the bridge is the same. It becomes insensitive when high resistance of the order of 100,000 ohm or more is to be measured. It is because the value of *P, Q, R* and *X* will be 10 ; 1,000 ; 10,000 and 100,000 which are widely different. For the measurement of such high resistances the *substitution method* is simple and quite accurate.

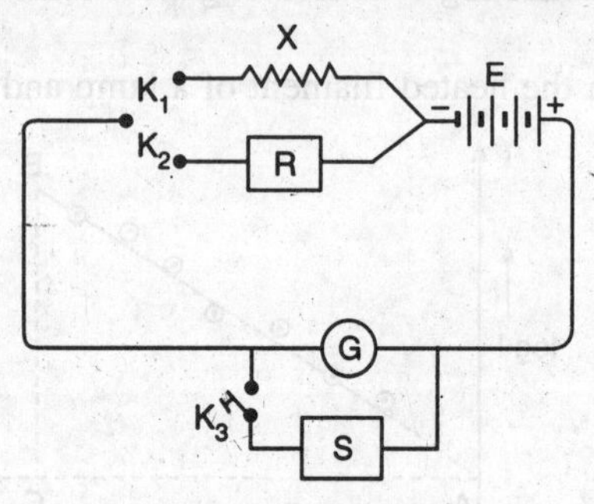

Fig. 27.1

Substitution method. The connections for the substitution method are shown in Fig. 27.1. A battery of four or five cells of *e.m.f.* E is connected to a high resistance box R and an unknown resistance X and then to a high resistance galvanometer G through a two-way key. The galvanometer is shunted by a resistance box S through a key K_3.

When the key K_1 is inserted, the unknown resistance X is introduced in the circuit. A current I_g flowing through the galvanometer gives a deflection θ on the scale. Thus

$$I_g = \frac{E}{X+G} \quad \text{...}(i)$$

When the key K_1 is removed and K_2 is inserted a resistance R of known value is introduced in the circuit. Since the magnitude of R is less than X the deflection of the galvanometer is large and may even go out of scale. The key K_3 is now closed and the value of the shunt resistance S is adjusted so that the same deflection θ is obtained. Now I_g the current through the galvanometer is the same and is given by

$$I_g = \frac{E}{R + \dfrac{GS}{G+S}} \times \frac{S}{G+S} = \frac{ES}{R(G+S)+GS} \quad \text{...}(ii)$$

From (*i*) and (*ii*), we have

$$\frac{E}{X+G} = \frac{ES}{R(G+S)+GS}$$

or $$XS = R(G+S)$$

∴ $$X = R\,\frac{(G+S)}{S}$$

Note. *The high resistance box R is generally a megohm box having steps of 0.1 megohm (1 megohm =* 10^6 *ohm). If X is less than 1 megohm the deflection with resistance R in circuit is less than* *the deflection with X in the circuit. In such a case the shunt is applied when the unknown resistance is in circuit.*

Thus in the first case when R is in circuit and no shunt is applied,

$$I_g = \frac{E}{R+G} \quad ...(iii)$$

In the second case when X is in circuit and shunt S is applied so that the deflection is the same, then

Current through the galvanometer

$$I_g = \frac{E}{X + \frac{GS}{G+S}} \times \frac{S}{G+S} = \frac{ES}{X(G+S)+GS} \quad ...(iv)$$

Comparing (*iii*) and (*iv*), we have

$$\frac{E}{R+G} = \frac{ES}{X(G+S)+GS}$$

or

$$RS = X(G+S)$$

$$\therefore \quad X = R\,\frac{S}{G+S}$$

Experiment 27.1. To determine the given high resistance by the substitution method.

Apparatus. A high resistance of the order of 0.5 megohm, an American type high resistance galvanometer, a megohm box having steps of 0.1 megohm, a two-way key, a one-way key, a resistance box, a battery of 4 to 5 cells, connecting wires and a piece of sand paper.

Formula. When $X > R$

Uunknown high resistance $X = R\,\frac{G+S}{S}$

Note. If the galvanometer does not posses a sufficiently high resistance, a resistance box R_1 is placed in series with the galvanometer and the shunt S is applied across both as shown' in Fig 27.2. Note that one terminal of the shunt resistance box is connected to the galvanometer through a key K and the other to one terminal of the resistance R_1.

If G_1 is the resistance of the galvanometer, then the resistance R_1 from the box is such that $(G_1 + R_1)$ is equal to 1,000 or 2,000 ohm.

The formula in this case will also be modified by putting $(G_1 + R_1)$ for G in the relation

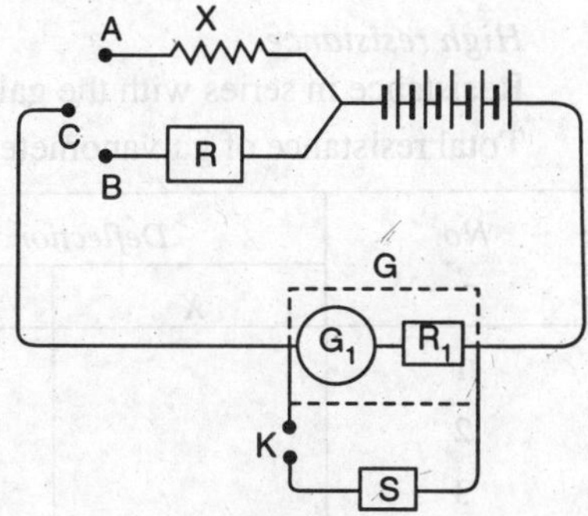

Fig. 27.2

$$X = R\,\frac{(G+S)}{S}$$

$$\therefore \quad X = R\,\frac{(G_1+R_1+S)}{S}$$

Procedure. 1. Draw a diagram showing the scheme of connections as in Fig. 27.2. Clean the ends of the connecting wires with sand paper. Connect one pole of the battery to one terminal of the galvanometer through a resistance box R_1 and the other pole of the battery through an unknown resistance X to one side terminal A and through the megohm resistance box to the other side terminal B of the two-way key. Connect the common terminal C of the two-way key to the other terminal of the galvanometer G_1. Shunt the galvanometer G_1 and the resistance box R_1, together by a resistance box S through one-way key K. Make all the plugs of the resistance boxes R_1 and S tight.

2. Set the galvanometer as already explained. Show the connections to the instructor before passing current.

3. **To find the resistance of the galvanometer.** As in a moving coil American type galvanometer the deflection is not proportional to the current, find its resistance by **equal deflection method.**

4. **To find the value of high resistance *X*.** (*i*) Calculate the resistance of the galvanometer G_1. Introduce a resistance from the resistance box R_1, so that the value of $(G_1 + R_1) = 2000$. This will now be the **effective** resistance of the galvanometer for purpose of calculations. Put in the plug between *A* and *C*. Note the deflection in the galvanometer. Remove the plug from *A* and *C*.

(*ii*) Introduce a resistance of 0.1 megohm from the resistance box *R* and insert the plug key between *B* and *C*. *It will be seen that the deflection is more than that in the first case and may go out of the scale.* Put in the key *K* and adjust the value of the shunt resistance *S* so that the deflection is *exactly* the same. Note the value of and *R* and *S*.

Note *Smaller the resistance of the shunt lesser is the deflection in the galvanometer.*

(*ii*) Increase *R* in steps of 0.1 megohm and repeat the observations so long as the deflection with *R* is more than that with *X*.

Observations.

Resistance of the galvanometer.

Sl. No.	*Resistance R*	*Deflection θ*	*Resistance R_1*	*Shunt for same deflection S*	$G_1 = \frac{S(R - R_1)}{R_1}$
1					
2					
3					

Mean resistance of the galvanometer G_1 = ohm

High resistance.

Resistance in series with the galvanometer R_1 = ohm

Total resistance of galvanometer $(R_1 + G_1) = G$ = ohm

No.	*Deflection with*		*R*	*S*	$X = R\frac{G+S}{S}$
	X	*R and S*	*megohm*		
1					
2					
3					
4					
5					

Mean value of resistance *X* = meg ohm

Precautions. 1. The high resistance and the galvanometer should be placed on a glass or on an ebonite sheet to avoid leakage of current.

2. The connecting wires also should not touch the table.

3. The galvanometer must have a high resistance. To achieve this a resistance R_1 is placed in series with the galvanometer.

Exercise. *Find the resistance of a carbon line drawn on a ground glass strip.*

Hints. A carbon line resistance of any value is made by drawing a pencil line of a suitable thickness with a soft pencil on the rough-surface of a ground glass strip. The strip is held between two terminals making good contact with the carbon line with the help of tinfoils.

27.2. Colour code for radio resistances. For finding the value of high resistance generally carbon resistances used in the radio work are provided in the laboratory. The value of these resistances is indicated by means of a colour code in which the numbers from 0 to 9 are represented by the following colours.

Colour	*Black*	*Brown*	*Red*	*Orange*	*Yellow*	*Green*	*Blue*	*Violet*	*Grey*	*White*
Number	0	1	2	3	4	5	6	7	8	9

Two systems of marking of colours are employed.

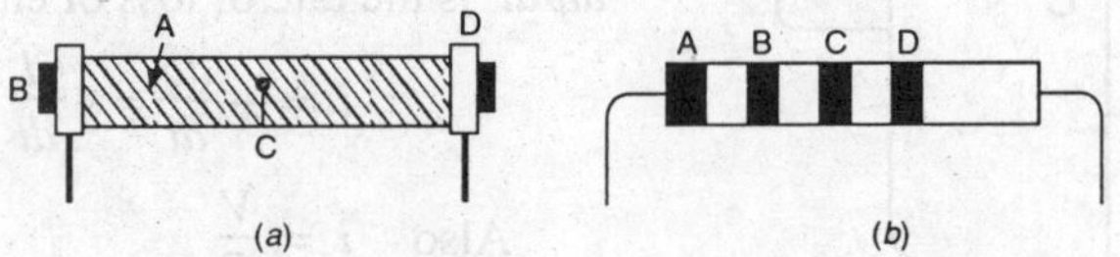

Fig. 27.3

(*i*) The body *A*, the end *B* and the dot *C* are given colours which represent the resistance as will be clear from the following example in which the **body** is green, **end** is red and **dot** is orange. To find the resistance, the colour of the *body* gives the first significant figure, the *end* the second significant figure and the *dot* the number of zeros following it. Thus the above resistance comes out to be

Body	End	Dot	
green	red	orange	
5	2	000	= 52000 ohm.

(*ii*) In the second system the colour of the first ring *A* on the extreme left represents the first significant figure, second ring *B* the second significant figure and the third ring *C*, the number of zeros following it. For example, if the first ring is brown, second black and the third green, the given resistance will be

Brown	Black	Green	
1	0	00000	= 1000000 ohm
			= 1 megohm.

The percentage accuracy or tolerance is represented by the colour of the ring *D* in both the cases. If it is **silver** in colour the accuracy is ± 10% and if it is **gold** in colour, the tolerance is ± 5%.

Experiment 27.2. To determine the high resistance by leakage method.

Apparatus. A ballistic galvanometer, a tap key, a lamp and scale arrangement, a morse key, a capacitor of capacity 0.2 to 0.5 microfarads, a two-way key, a one-way key, a cell, a high resistance of the order of about 100 megohm (carbon line on glass) connecting wires etc.

Note For ballistic galvanometer and lamp and scale arrangement See **Chapter 31.**

Theory. If a capacitor *C* after being charged to a potential difference V_0 is instantaneously discharged through a ballistic galvanometer giving rise to a first throw θ_0 then

Charge on the capacitor

$$q_0 = CV_0 = K\theta_0 \left(1 + \frac{\lambda}{2}\right)$$

where *K* is the constant of the ballistic galvanometer and λ the logarithmic decrement.

The capacitor is again charged to the same potential and the charge is allowed to leak through a high resistance *R* for a known time *t* so that the potential falls from V_0 to V_t.

If the remaining charge q on the capacitor is allowed to pass through the galvanometer so that a first throw θ_1 is produced, then

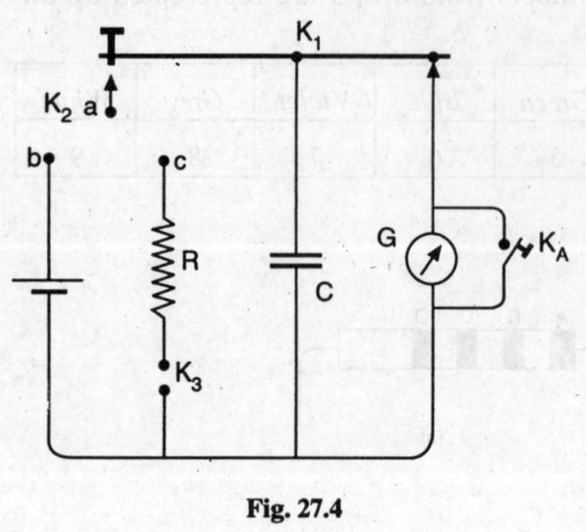

Fig. 27.4

$$q = CV_1 = K\theta_1\left(1 + \frac{\lambda}{2}\right)$$

Hence $$\frac{q_0}{q} = \frac{V_0}{V_t} = \frac{\theta_0}{\theta_1}$$

If V is the potential difference across the capacitor at any instant during the time the charge is leaking and dq/dt is the rate of loss of charge, then

$$i = -\frac{dq}{dt} = -\frac{d}{dt}(CV) = -C\frac{dV}{dt}$$

Also $$i = \frac{V}{R}$$

$$\therefore \quad \frac{V}{R} = -C\frac{dV}{dt}$$

or $$\frac{dt}{RC} = -\frac{dV}{V}$$

Integrating for the time t, we have

$$\frac{1}{RC}\int_0^t dt = -\int_{V_0}^{V_t}\frac{dV}{V}$$

$$\therefore \quad R = \frac{t}{C\log_e\frac{V_0}{V_t}} = \frac{t}{2.3026\,C\log_{10}\frac{\theta_0}{\theta_1}}$$

Procedure. 1. Draw a diagram showing the scheme of connections as in Fig. 27.4.

2. Level the ballistic galvanometer and set the lamp and scale arrangement. Adjust the position of the spot of light on the scale at zero. Put a plug between a and b of the key K_2 and press the morse key K_1. The capacitor is charged to to a constant potential difference equal to the *e.m.f.* of the cell.

3. Release the morse key, the cell capacitor circuit is broken and at the same time the gavanometer capacitor circuit is completed so that the capacitor discharges through the galvanometer. Note the first throw θ_0 corresponding to the charge q_0.

4. Again press the knob of the morse key so that the capacitor is charged to the same potential. Remove the plug from ab and immediately start the stop-watch. When the charge on the capacitor has leaked, if at all, for a known time say 5 seconds release the morse key and observe the first throw θ'. Repeat this self-leakage experiment for different values of leakage time noting the value of the throw θ' for each value of time t.

5. Next close the key K_3 and put the plug into the position ab in the key K_2. Press the morse key and when the capacitor is charged to the same potential turn the key K_2 from the position ab to ac keeping the knob of the Morse key pressed and start the stop watch immediately. When the charge on the capacitor has leaked through the resistance R for a known time t, say 5 seconds, release the morse key and note the throw θ_1. Repeat this experiment for different times observing the throw θ_1 every time for each value of time t.

6. Plot a graph taking the various values of the time t along the X-axis and the corresponding values of $\log_{10}\frac{\theta_0}{\theta}$ along the Y-axis. Find the slope of the curve and calculate the self-leakage resistance r from the relation.

$$r = \frac{t}{2.3026\ C\ \log_{10}\frac{\theta_0}{\theta'}}$$

7. Similarly plot a graph between t and $\log_{10}\frac{\theta_0}{\theta_1}$ and find the combined parallel resistance R' of the combination r and R, from which the resistance R can be calculated.

Observations. Capacitance of the capacitor C = μF = F

Initial position of the spot on the scale =

Position of the spot for full charge q_0, =

∴ First throw due to charge q_0, θ_0=

Sl. No.	*Leakage time t sec.*	*First throw after the charge has leaked*						$\log_{10}\frac{\theta_0}{\theta'}$	$\log_{10}\frac{\theta_0}{\theta_1}$
		Self-leakage			*Leakage through R*				
		Initial	*Final*	θ'	*Initial*	*Final*	θ_1		
1	5								
2	10								
3	15								
4	20								
5	25								
6	30								

Calculations. (*i*) From the graph between t and $\log_{10}\frac{\theta_0}{\theta'}$, we have

Slope of the curve = $\frac{1}{t}\log_{10}\frac{\theta_0}{\theta'}$

∴ Self-leakage resistance $r = \frac{t}{2.3026\ C\ \log_{10}\frac{\theta_o}{\theta'}}$

(*ii*) From the graph between t and $\log_{10}\frac{\theta_0}{\theta_1}$, we have

Slope of the curve = $\frac{1}{t}\log_{10}\frac{\theta_0}{\theta_1}$

Combined resistance $R' = \frac{t}{2.3026\ C\ \log_{10}\frac{\theta_0}{\theta_1}}$

The unknown resistance R can be found from the relation

$$\frac{1}{R} = \frac{1}{R'} - \frac{1}{r}$$

Note. 1. This method is fairly accurate and is used to find the value of high resistance more than 20 megohm.

2. The graph between t and $\log \frac{\theta_0}{\theta'}$ as well as between t and $\log_{10} \frac{\theta_0}{\theta_1}$ is a straight line. To find the slope we shall take the case of graph between t and $\log \frac{\theta_0}{\theta_1}$ as shown in Fig. 27.5.

Take two points A and B on the graph. From B draw BN perpendicular to X-axis and from A, draw AC perpendicular to BN. Then slope of the graph

$$= \frac{BC}{AC} = \frac{\log_{10} \frac{\theta_0}{\theta_1}}{t}$$

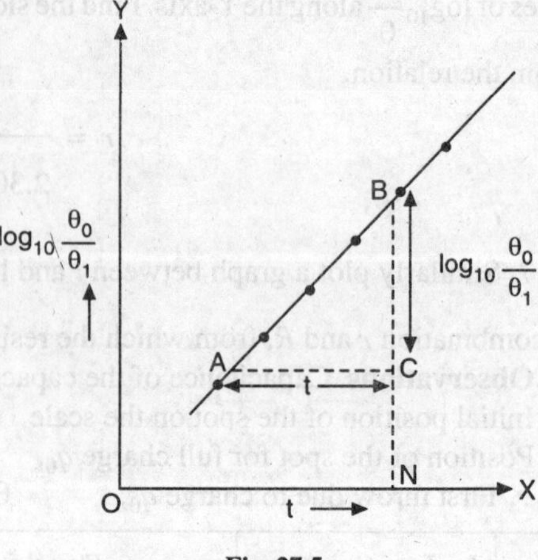

Fig. 27.5

and $$R' = \frac{t}{2.3026\, C \log_{10} \frac{\theta_0}{\theta_1}}$$

$$= \frac{1}{2.3026\, C} \frac{AC}{BC} = \text{meg ohm}$$

If C is taken in microfarad, R′ is calculated directly in meg-ohm.

Precautions. 1. The galvanometer should be carefully levelled so that the spot of light moves freely on the scale.

2. The keys used should be free from leakage.

3. The cell used should have a constant *e.m.f.*

4. All connecting wires should be kept clear of each other and should be prevented from touching the bench, instrument cases etc.

5. The capacitor used should be of such a capacity that there is a satisfactory rate of leakage.

Note. The high resistance to be used can be prepared by drawing a line with lead pencil on glass or by sealing two electrodes in a glass tube about 1 to 2 cm in diameter and filling the tube with a mixture of xylol and amyl alcohol.

Oral Questions

1. High resistance. (*i*) What do you mean by a high resistance? (*ii*) Why can you not use a post office box for measuring a high resistance? (*iii*) Can you use a Weston type galvanometer in the experiment for measuring high resistance? (*iv*) Why are the various wires not allowed to touch the table? (*v*) What is a megohm? (*vi*) Why do you prefer a high resistance galvanometer in this experiment? If the resistance is low what modification will you make? (*vii*) You are provided with a carbon resistance of which the body colour is red, end green and dot yellow. What is its value? (*viii*) What is the value of a carbon resistance in which the colour of the ring on extreme left is blue, next ring black and the third orange?

2. High resistance by leakage (*i*) Why is the method known as leakage method? (*ii*) Can this method be used to measure a low resistance? (*iii*) What is the unit of capacitance? (*iv*) What is the relation between farad and microfarad? (*v*) What is the shape of the graph between t and $\log_{10} \frac{\theta_0}{\theta_1}$?

28

Measurement of Low Resistance

28.1. Measurement of low resistance. A post office box cannot be used for the measurement of a very low resistance because the resistance of the connecting wires becomes comparable and causes an appreciable error. The arrangement also becomes very insensitive because the four arms of the bridge contain widely different resistances.

Carey Foster bridge. It is a modification of the metre bridge designed to measure the difference between two resistances of nearly equal values.

A wire *ab* one metre long of eureka or manganin, having a uniform area of cross-section, is fixed on a metre scale fixed on a wooden board. The ends *a* and *b* are soldered to two thick copper strips. Three more copper strips are fixed parallel to the wire as shown in Fig. 28.1., thus providing four gaps marked 1, 2, 3 and 4.

Two resistance boxes *P* and *Q* or two nearly equal resistances of about 5 or 10 ohm are connected in the inner gaps and the resistances *X* and *Y* whose difference is to be determined, are connected in the outer gaps. The battery is connected across the resistance *P* and *Q i.e.* between points *A* and *C* and the galvanometer between the points *B* and the jockey *J*. The jockey can be made to slide along the wire *ab*.

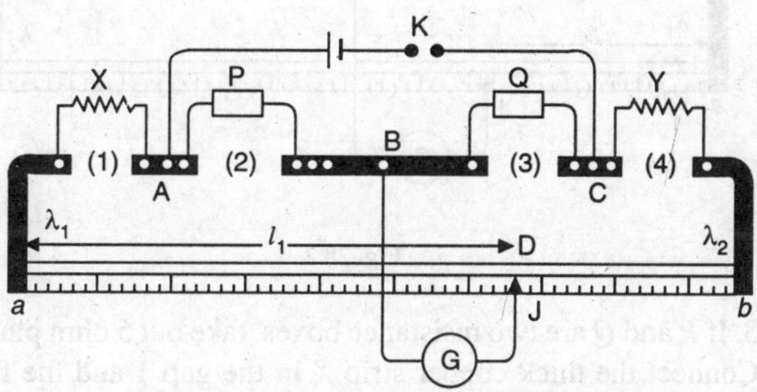

Fig. 28.1

Experiment 28.1. To calibrate the wire of Carey-Foster's bridge and hence to find the resistance of the two turns coil of a tangent galvanometer, or any other small resistance.

OR

To find the low resistance by Carey Foster bridge after calibrating the bridge wire.

Apparatus. Carey-Foster bridge, two equal resistances, thick copper strips, a fractional resistance box (or standard fractional resistance coils), a cell, connecting wires, a sensitive galvanometer, a jockey and a key.

Theory. The arrangement of Carey-Foster bridge is similar to Wheatstone bridge. As shown in Fig. 28.1., *P* and *Q* are the two ratio arms, *X* along with the resistance of wire *aD* and *Y* along with the resistance of the wire *bD* form the other two arms.

If ρ is the resistance per cm of the wire, λ_1 and λ_2 the end resistances at *a* and *b* respectively, and a balance point is obtained at *D*, where $aD = l_1$, *then*

$$\frac{P}{Q} = \frac{X + l_1\rho + \lambda_1}{Y + (100 - l_1)\,\rho + \lambda_2} \qquad ...(i)$$

If now, X and Y are interchanged and the balance point is obtained at D' as shown in Fig. 28.2 such that $a\,D'$ is equal to l_2, then

$$\frac{P}{Q} = \frac{Y + l_2\,\rho + \lambda_1}{X + (100 - l_2)\,\rho + \lambda_2} \qquad ...(ii)$$

Comparing (*i*) and (*ii*) and adding one to both sides, we have

$$\frac{X + Y + 100\rho + \lambda_1 + \lambda_2}{Y + (100 - l_1)\,\rho + \lambda_2} = \frac{X + Y + 100\rho + \lambda_1 + \lambda_2}{X + (100 - l_2)\,\rho + \lambda_2}$$

or $$Y + (100 - l_1)\,\rho + \lambda_2 = X + (100 - l_2)\,\rho + \lambda_2$$

or $$Y - X = (l_1 - l_2)\,\rho$$

or $$X = Y - (l_1 - l_2)\,\rho$$

Thus, the difference between the two resistances is equal to the resistance of the bridge wire between the two balance points. Further the above relation is free from the end correction of the total length of the bridgewire.

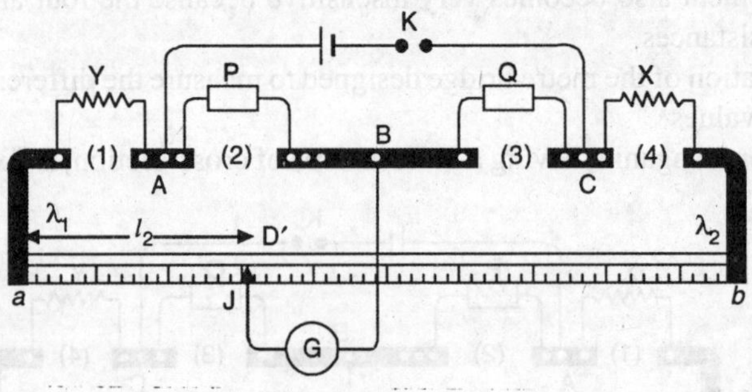

Fig. 28.2

Procedure. 1. *Connections.* (*i*) Draw a diagram showing the scheme of connection as in Fig. 28.1 Mark the gaps 1, 2, 3 and 4 on the bridge.

(*ii*) Clean the ends of the connecting wires and the thick copper strips with a sand paper.

(*iii*) Connect the two equal resistances P and Q in the inner gaps 2 and 3. If P and Q are two resistance boxes, take out 5 ohm plug from either and make all other plugs tight. Connect the thick copper strip X in the gap 1 and the fractional resistance box by means of thick copper rods in the gap 4. Connect one terminal of the galvanometer to the central terminal B and the other to a jockey J. Connect the cell through a key K between the points A and C. Show the connections to the instructor before passing the current.

2. *To test the connections.* Make the plugs of the fractional resistance box tight by giving to each plug a slight twist. Put in the key K and touch the jockey on the wire at the end a. Note the direction of deflection. Now touch the jockey at the end b. If the direction of deflection is reversed the connections are correct.

3. **To find ρ, the resistance per cm of the wire.** (*i*) Without taking out any plug from the fractional resistance box, adjust the position of the jockey to obtain a balance point and note the reading. Take out the key K. Interchange X and Y, put in the key K and obtain the balance point again. Note the reading.

The difference in the two positions gives us the resistance of the lead wires connected to the fractional resistance box (Y) and the resistance of the plugs, brass blocks etc. This shift is known as the correction and must be subtracted from the shift obtained in the subsequent readings to get the correct shift. Let this be denoted by δl.

Take out the key K and again interchange X and Y.

(*ii*) Take out 0.1 ohm plug from the fractional resistance box Y and after putting the key K obtain the balance point. Note the reading and interchange X and Y. Again obtain the balance point and note it.

(*iii*) Repeat the observation with the values of Y equal to 0.2, 0.3, 0.4 and 0.5 ohm etc.

4. **To find the low resistance.** Disconnect the copper strip and connect either the given low resistance in the form of a small piece of wire W soldered to two thick copper strips S_1 and S_2 as shown in Fig. 28.3 or the two turns coil of the tangent galvanometer in its place by pieces of short thick copper wires. Keeping the resistance in the fractional resistance box zero, find the balance point. Interchange the tangent galvanometer and the fractional resistance box and find the balance point again. Similarly repeat the observation by taking out 0.1, 0.2, 0.3, 0.4 and 0.5 ohm resistance from the fractional resistance box.

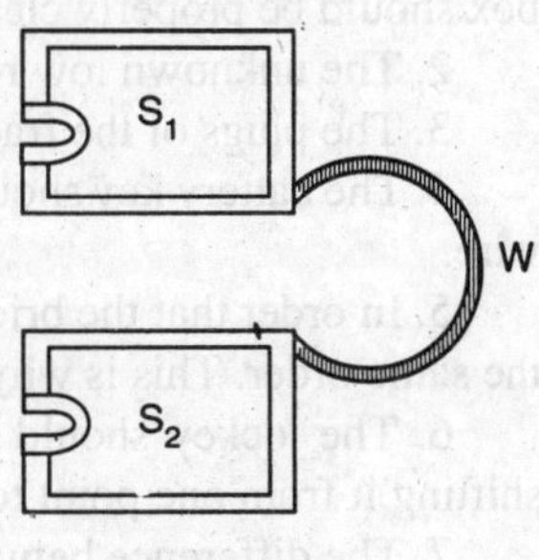

Fig. 28.3

Observations. (*a*) *Determination of* ρ, the resistance per cm of the bridge wire.

Balance point with Y zero and X in left gap $= x$ cm

Balance point with Y zero and X in right gap $= y$ cm

Correction to be applied $(x-y)$ $= \delta l$ cm

Sl. No.	*X ohm*	*Y ohm*	*Position of balance point with copper strip in*		*Shift* $(l_1 - l_2)$ *cm*	*Corrected Shift* $d = (l_1 - l_2) - \delta l$	*Resistance per cm* $\rho = \frac{Y}{d}$
			Left gap (l_1)	*Right gap* (l_2)			
1	0	0					
2	0	0.1					
3	0	0.2					
4	0	0.3					
5	0	0.4					
6	0	0.5					

Mean ρ = ohm/cm

Note. To find the corrected shift subtract from $(l_1 - l_2)$ the shift $\delta\, l$ obtained when both X and Y are zero.

Determination of the unknown low resistance

Sl. No.	*Y ohm*	*Position of balance point with unknown resistance in*		*Shift* $(l_1 - l_2)$ *cm*	*Correct Shift d* $= (l_1 - l_2) - \delta\, l$	*Unknown resistance* $X = Y - \rho\, d$
		Left gap l_1	*Right gap* l_2			
1	0					
2	0.1					
3	0.2					
4	0.3					
5	0.4					
6	0.5					

Mean $X =$ ohm

Note. If the unknown resistance X is greater than the known resistance Y taken from the fractional resistance box, then l_1 is less than l_2. Hence the shift $(l_1 - l_2)$ is a negative quantity.

If X is less than Y, l_1 is greater than l_2 and hence the shift $(l_1 - l_2)$ is positive.

Precautions. 1. The ends of connecting wires, thick copper strip and the leads for the resistance box should be properly cleaned.

2. The unknown low resistance should be connected to the bridge by thick copper leads.

3. The plugs of the fractional resistance box should be tight.

4. The battery key should be taken out while interchanging X and Y to avoid heating of the bridge wire.

5. In order that the bridge may be sensitive, the resistances of the four arms should be nearly of the same order. This is why P and Q should not be very large.

6. The jockey should be touched gently and should not be kept pressed on to the wire while shifting it from one point to the other.

7. The difference between X and Y should not be more than the resistance of the bridge wire.

Experiment 28.2. To find a low resistance using Carey Foster bridge without calibrating the bridge-wire.

Apparatus. Same as in **Experiment 28.1.**

Theory. Let X be the unknown resistance and d_1 and d_2 the shifts obtained with resistances Y_1 and Y_2 from the fractional resistance box respectively, then

$$X - Y_1 = d_1 \rho$$

and

$$X - Y_2 = d_2 \rho$$

$$\frac{X - Y_1}{X - Y_2} = \frac{d_1}{d_2}$$

or

$$d_2 X - d_2 Y_1 = d_1 X - d_1 Y_2$$

or

$$X (d_2 - d_1) = Y_1 d_2 - Y_2 d_1$$

$$\therefore \quad X = \frac{Y_1 d_2 - Y_2 d_1}{d_2 - d_1}$$

It is, therefore, not necessary to find out the value of ρ for the determination of an unknown low resistance, when a fractional resistance box is available.

Procedure. 1. Make the connections as shown in Fig. 28.1. Keeping X and Y both equal to zero, find the balance point. Interchange X and Y and again find the balance point. The shift in the balance point gives the value of the correction δl to be applied to all observations.

2. Replace the thick copper strip X by the unknown low resistance and find the shift in the balance point keeping Y equal to 0.1, 0.2...etc.

Observations.

Correction to be applied $\delta l =$ cm

Sl. No.	*Y ohm*	*Position of balance point with unknown resistance in*		*Shift* $(l_1 - l_2)$ *cm*	*Corrected Shift* $d = (l_1 - l_2) - \delta l$
		Left gap l_1	*Right gap* l_2		
1	0				
2	0.1				
3	0.2				
4	0.3				
5	0.4				
6	0.5				

Value of X from observations 1 and 2 = ohm

2 and 3 = ohm
3 and 4 = ohm
4 and 5 = ohm
5 and 6 = ohm
Mean value of X = ohm

Exercise. *Determine the specific resistance of copper using a Carey-Foster's bridge.*

Hints. Take No. 30 *S.W.G.* copper wire about 1 metre long with its ends soldered to two thick copper strips. Find its resistance by Carey-Foster's bridge and calculate the specific resistance.

28.2. Low resistance by Kelvin's method. It is an accurate method for comparing two low resistances. If one of them is a standard known resistance then the value of the other can be calculated. The general arrangement of the network of resistances forming a Kelvin's bridge is shown in Fig. 28.4. The resistances r_1 and r_2 as well as r_3 and r_4 are equal and are so arranged that total resistance in the arm *EL* = Total resistance in the arm *LH*; and total resistance in the arm *FM* = Total resistance in the arm *MG*.

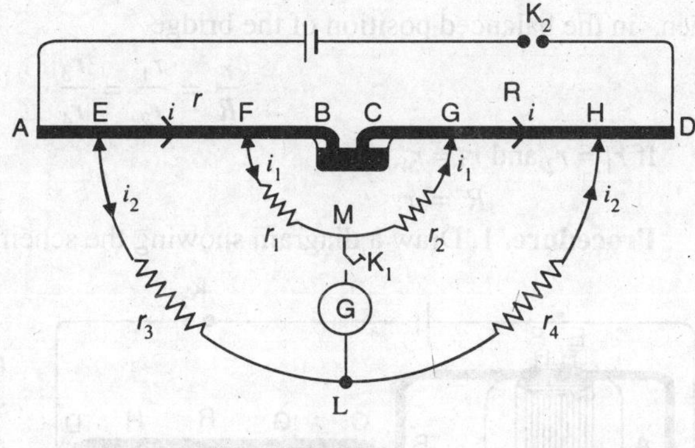

Fig. 28.4

The two pairs of equal resistances r_1, r_2 and r_3, r_4 are each about one or two ohm, and are made out of a wire of the same material having a low temperature co-efficient. These resistances are mounted close to each other to ensure that all of them are at the same temperature.

AB and *CD* are two thick wires whose resistances are to be compared. The ends *B* and *C* of these wires dip in a cup containing mercury which provides a low resistance junction. The contacts *E, F* and *G, H* are so arranged that when the key K_1 in the galvanometer circuit is pressed the galvanometer shows no deflection. In this position potential at *M* is the same as at *L*.

$$\therefore \quad \frac{ir + i_1 r_1}{iR + i_1 r_2} = \frac{i_2 r_3}{i_2 r_4} = \frac{r_3}{r_4}$$

where i, i_1 and i_2 are the currents in the various branches as shown in Fig. 28.4. r is the resistance of portion *EF* and *R* that of the portion *GH*.

$$\text{or} \quad irr_4 + i_1 r_1 r_4 = iRr_3 + i_1 r_2 r_3$$

$$\text{or} \quad i\,(rr_4 - Rr_3) = i_1\,(r_2 r_3 - r_1 r_4)$$

But according to the condition of the bridge

$$\frac{r_1}{r_2} = \frac{r_3}{r_4}$$

$$\text{or} \quad r_1 r_4 = r_2 r_3$$

$$\therefore \quad i\,(rr_4 - Rr_3) = 0$$

$$\text{or} \quad rr_4 = Rr_3$$

$$\text{Hence} \quad \frac{r}{R} = \frac{r_3}{r_4} = \frac{r_1}{r_2}$$

$$\text{As} \quad r_1 = r_2$$

$$\therefore \quad R = r.$$

Experiment 28.3. To determine the specific resistance of copper using a standard low resistance and Kelvin's bridge.

Apparatus. Kelvin's bridge, standard low resistance of 0.01 ohm, copper rod (S.W.G. No. 14) about 40–50 cm long, a low resistance galvanometer, connecting wires, constantan wire, sand paper etc.

Theory. If an unknown resistance R and a known low resistance r are connected so as to form a Kelvin's bridge with resistances r_1 r_2 and r_3 r_4 such that

$$\frac{r_1}{r_2} = \frac{r_3}{r_4}$$

then, in the balanced position of the bridge

$$\frac{r}{R} = \frac{r_1}{r_2} = \frac{r_3}{r_4}.$$

If $r_1 = r_2$ and $r_3 = r_4$, then

$$R = r.$$

Procedure. 1. Draw a diagram showing the scheme of connections as in Fig. 28.5.

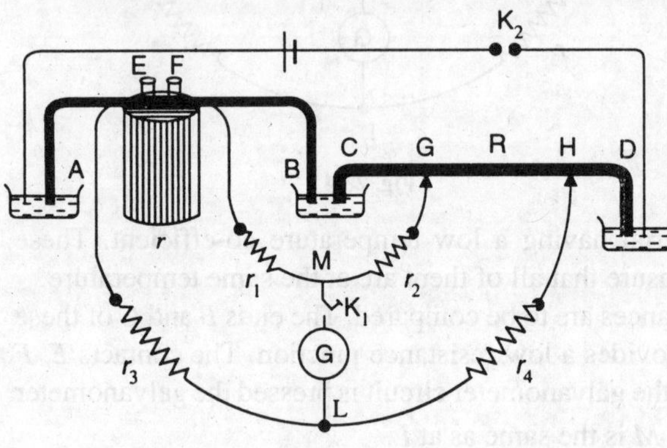

Fig. 28.5

2. Place the copper rod CD as also the standard low resistance with their ends dipping in cups of mercury as shown. Prepare 4 equal resistances r_1, r_2, r_3 and r_4, by taking equal lengths of constantan wire. Connect the free end of r_1 to one terminal of the standard low resistance and the free end of r_2 to a point G on the copper rod near the end C by two pieces of connecting wires of equal lengths. Similarly connect the free end of r_3 to the other terminal of the standard low resistance and the free end of r_4 to a sliding contact H on the copper rod by means of two equal lengths of copper wires. Connect the galvanometer through a tap key K_1 between the junctions of r_1, r_2 and r_3, r_4. Connect the battery through a key K_2 at A and D.

3. Put in the key K_2 and adjust the contact H so that when the key K_1 is also pressed the galvanometer shows no deflection. Measure the length of the rod between the points G and H. Repeat three times.

4. Measure the diameter of the rod with a screw gauge at six different positions and in each position in two mutually perpendicular directions.

Observations.

Standard low resistance = 0.01 ohm

Length of the copper rod GH l = 1. 2. 3.

Mean l = cm = m

Diameter of the rod

Least count of the screw gauge = mm

Zero correction = ± mm

Diameter	*Screw gauge readings*						
	1	*2*	*3*	*4*	*5*	*6*	*Mean*
a							
b							

Corrected diameter $d =$ mm $=$ m

$$\therefore \quad \text{Specific resistance} = \frac{0.01 \times \pi d^2}{4\,l}$$

Precautions. 1. The current should be passed only for a small times so as to avoid excessive heating.

2. The galvanometer should be of low resistance.

3. The contact on the rod at G and H should be sharp to enable accurate measurement of length.

Exercise. *Compare the specific resistances of copper and brass.*

Hints. Replace the standard resistance by the brass rod and obtain the balance point. Measure the length l_1 of the brass rod between E and F and l_2 of copper rod between G and H as in Fig. 28.5.

If a_1 and K_1 are the area of cross-section and specific resistance for brass and a_2 and K_2 the corresponding values for copper, then as $r = R$.

$$\therefore \quad \frac{K_1 l_1}{a_1} = \frac{K_2 l_2}{a_2}$$

or

$$\frac{K_1}{K_2} = \frac{a_1 l_2}{a_2 l_1}$$

28.3. Kelvin's double bridge improved pattern. In the method described in **Art. 28.2**, it is not possible to find the given fixed low resistance or the resistance of a fixed length of the wire of low resistance. This is possible with the improved form of the bridge in which instead of the fixed standard resistance r between E and F a variable standard low resistance in steps of 0.05 ohm along with a length of a manganin wire having a resistance of 0.05 ohm is provided. A self-gripping jockey is provided which can be placed on any point on the wire so that a variation of resistance up to 0.0005 ohm is possible.

Fig. 28.6

To increase the sensitiveness in the measurements, the four fixed resistance r_1, r_2 r_3 and r_4 of equal magnitude are replaced with a double ratio-box, which gives multiplying ratios of 0.01, 0.1, 1 and 10 when the galvanometer is connected between the corresponding marked terminals. This arrangement alters the ratio

$$\frac{r_2}{r_1} = \frac{r_4}{r_3} = n$$

without changing the current in the main circuit. If

$$\frac{r_2}{r_1} = \frac{r_4}{r_3} = 1, \text{ then}$$

$$R = r$$

The value of r is known from the position of knob E and the length of wire FB_1 between the self-gripping jockey F and the terminal B_1.

If the ratio is altered to 0.1 by connecting the galvanometer between the marked terminals, then

$$\frac{r_2}{r_1} = \frac{r_4}{r_3} = 0.1$$

$$\therefore \quad \frac{R}{r} = 0.1$$

or $\quad R = 0.1 \times r$

Again if the ratio is altered to 0.01, then

$$R = 0.01 \times r$$

This gives the most accurate value of the unknown resistance R.

The circuit diagram of the compact form of the Kelvin bridge is as given in Fig. 28.7.

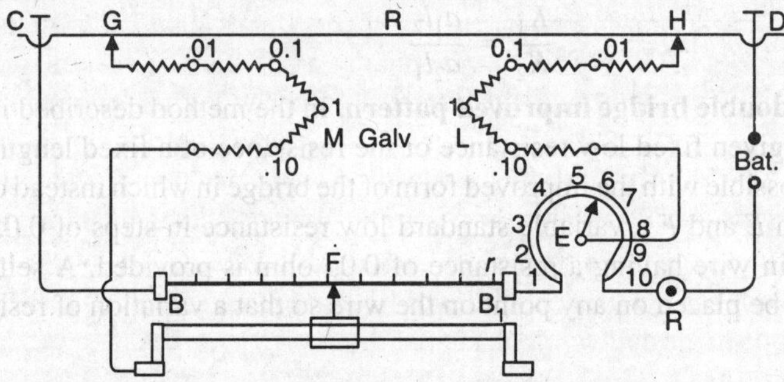

Fig. 28.7

Example. When the galvanometer is connected between the terminals marked 0.01 and 0.01, the adjustable knob E is on the mark 4 and the self gripping jockey F on a division mark 30, so that the galvanometer shows no deflection, then as each step of E is 0.05 ohm.

Unknown resistance $\quad R = (4 \times 0.05 + 30 \times 0.0005) \times 0.01$

$$= (0.20 + 0.015) \times 0.01$$

$$= 0.00215 \text{ ohm.}$$

because resistance of 1 div. mark on the wire $BB_1 = 0.0005$ ohm

Experiment 28.4. To determine the resistance and specific resistance of copper with the help of Kelvin's double bridge.

Apparatus. Kelvin's double bridge, a low resistance moving mirror galvanometer, a battery of a few cells, a copper rod about 40–50 cm long, connecting wires, a beam compass, a screw gauge and metre rod.

Theory. In the balanced position of the Kelvin's bridge

$$\frac{r}{R} = \frac{r_1}{r_2} = \frac{r_3}{r_4}$$

where r_1, r_2 and r_3, r_4 are the two ratio arms, r the known low resistance and R the unknown resistance.

If

$$\frac{r_2}{r_1} = \frac{r_4}{r_3} = 1, \text{ then}$$

$$R = r$$

If $$\frac{r_2}{r_1} = \frac{r_4}{r_3} = 0.1, \text{ then}$$

$$R = 0.1 \times r$$

Again if the ratio is 0.01, then

$$R = 0.01 \times r$$

Procedure. 1. Draw a diagram showing the scheme of connections as in Fig. 28.7.

2. Place the rod (copper or brass) whose resistance is required in the clamp terminals at *C, G, H*, and *D* and make them tight.

3. Connect the battery to the two marked terminals and also the galvanometer to the terminals *L* and *M* through a key *K* (not shown) so that the ratio

$$\frac{r_2}{r_1} = \frac{r_4}{r_3} = 1$$

4. Adjust the knob of the varible resistance *E* on the zero mark and also the self-gripping jockey on the wire at B_1. Press the key *K* and note the direction of the deflection in the galvanometer. Move the self-gripping jockey to the extreme position *B* on the wire. Press the key *K* and again note the direction of deflection. If the direction of deflection is the same, then the unknown resistance *R* is greater than the total resistance of the wire BB_1. Move the knob of the variable resistance to the mark 1 and check the direction of deflection in the galvanometer. If the direction is reversed, then move the self-gripping jockey on the wire to some position such as *F*, so that the galvanometer shows no deflection on pressing the key *K*. The unknown resistance *R* is given by

$$R = (0.05 + B_1\, F \times .0005) \text{ ohm}$$

If the direction of deflection is not reversed when the self-gripping jockey is moved from B_1 to *B*, then adjust the knob *E* to 2, 3, 4, etc., positions till on moving the self gripping jockey from B_1 to *B*, the direction of deflection in the galvanometer is reversed. Find the position such as at *F* on the wire for no deflection, then

$$R = (\text{No. of coils} \times 0.05 + B_1\, F \times 0.0005) \text{ ohm}$$

5. Connect the galvanometer between the terminals marked 0.1 and 0.1, so that the ratio $\frac{r_2}{r_1} = \frac{r_4}{r_3} = 0.1$ and adjust the position of knob *E* and find the point *F* on the wire for no deflection. The value of resistance *r* found in this case will be 10 times the previous value.

6. Connect the galvanometer between the terminals marked 0.01 and 0.01. Adjust the position of knob *E* and also that of the self-gripping jockey for no deflection. Again note the value of the resistance *r*.

7. Measure the length of the rod between the clamp terminals *G* and *H* with the help of a beam compass.

8. Find the least count and zero error in the screw gauge. Measure the diameter of the copper rod at six different positions taking two mutually perpendicular readings at each point.

9. Repeat the experiment with a different copper wire.

Observations.

Resistance of each step of the variable resistance = δ ohm

Resistance of 1 div length of the slide wire = ρ ohm

Length of the copper rod *l* = cm = m

Sl. No.	*Galv. between terminal*	*Position of knob E* x	*Position of jockey at F* $B_1 F = y$	*Resistance* $r = x\delta + y\rho$	*Resistance R*
1	1, 1 0.1, 0.1 0.01, 0.01			–	r $r \times 0.1$ $r \times 0.01$
2					

Resistance of the wire $R = r \times 0.01$ ohm

= ohm

Diameter of the rod.

Least count of the screw gauge = mm

Zero correction = ± mm

Diameter	*1*	*2*	*3*	*4*	*5*	*6*	*Mean*
a							
b							

Mean corrected diameter d = mm

= cm = m

Specific resistance $= \dfrac{R \times \pi d^2}{4l}$ ohm–metre

Precautions. Same as in **Expt. 28.3.**

Oral Questions

1. Low resistance by Carey Foster bridge. (*i*) Why ordinary Wheatstone bridge is not suitable for measuring low resistance ? (*ii*) What do you mean by end correction in a slide wire bridge ? (*iii*) What is Carey-Foster's bridge? (*iv*) What do you measure with it? (*v*) What is the maximum difference in the resistance that you can measure with a Carey-Foster's bridge? (*vi*) Is it essential to have the value of resistance P and Q equal? (*vii*) What should be the order of the resistance P and Q? (*viii*) What will happen if the bridge wire is non-uniform? (*ix*) Can you find the temperature co-efficient of resistance with the help of Carey-Foster's bridge, if so, how? (*x*) How do you make the resistance in a gap zero?

2. Kevin bridge. (*i*) What is the relation for the measurement of low resistance in a Kelvin's bridge? (*ii*) Is it necessary that r_1, r_2 and r_3, r_4 may by equal? (*iii*) What is the material of which the standard resistance is made (*iv*) What is the advantage of mercury cups? (*v*) In the Kelvin's double bridge that you are using what is the resistance of the bridge wire and what is the resistance of each step?

29

Electrolysis

29.1. Electrolysis. The process of decomposition of a liquid or a solution due to the passage of electric current through it is called *electrolysis*. The liquid or the solution through which current is passed is known as an *electrolyte*. The plates placed in the electrolyte are called the *electrodes*. The plate connected to the positive of the battery is called the *anode* while the one connected to the negative is called the *cathode*. The vessel containing the electrolyte and the electrodes is known as a *voltameter*.

The products of decomposition of the electrolyte are called the *ions*. The ions liberated at the cathode carry a *positive* charge and are called *cations*, whereas those liberated at the anode carry a *negative* charge and are called *anions*.

29.2. Faraday's law of electrolysis. (*i*) *The weight of an ion liberated in the process of electrolysis is directly proportional to the current I and the time t for which the current flows.* Thus

$$w \propto I t$$

or

$$w = I \times t \times z$$

where z is a constant. Its value depends upon the nature of the ion and is known as the **electrochemical equivalent.**

It is defined as *the weight of an ion deposited when a current of one ampere flows for one second or when one coulomb of charge passes through the electrolyte*.

(*ii*) *When the same current flows for the same time through several electrolytes, the weights of the various ions deposited are proportional to their chemical equivalents.*

For example, when the same current is passed through acidulated water, copper sulphate solution and silver nitrate solution placed in series the weight of hydrogen, copper and silver liberated are in the ratio of

$$1 : 31.5 : 108$$

where 1, 31.5 and 108 are the equivalent weights of hydrogen, copper and silver respectively.

Note. *Hydrogen and all metals are liberated at the cathode.*

29.3. Copper voltameter. The electro-chemical equivalent of copper can be determined by the electrolysis of copper sulphate solution. The apparatus used is known as a **copper voltameter.** It consists of a glass vessel containing a concentrated solution of copper sulphate and three copper plates. The *two outer copper plates are connected together and form the* **positive pole** (*anode*) *and are connected to the positive of the battery. The central copper plate forms the* **negative pole** *of the voltameter (cathode) and is joined to the negative of the battery* through a key and a rheostat, as shown in Fig. 29.2. When current is passed through the electrolyte, copper is deposited on the cathode, while copper dissolves from the anode into the solution, maintaining the concentration of the solution.

29.4 Tangent galvanometer. It is a moving-magnet type galvanometer. It consists of a circular coil of insulated copper wire of several turns, wound on a vertical non-magnetic wooden or a brass frame. Usually there are three separate coils of 2, 50, and 500 turns. The ends of these coils are connected to the binding screws fixed at the base of the instrument. The frame can revolve about a vertical axis. The base of the galvanometer is provided with three levelling screws to level the instrument. At the centre of the vertical frame is placed a circular horizontal box, containing a scale graduated in degrees from 0 to 90 in four quadrants. At the centre of this scale is placed a small magnetic needle, carrying a light aluminium pointer fixed at right angles to its length on a vertical axis. The needle is free to move in the horizontal plane. The ends of the pointer move over the circular scale. At the base of the circular box is placed a plane mirror, in which the reflected image of the pointer is observed, to avoid error due to parallax.

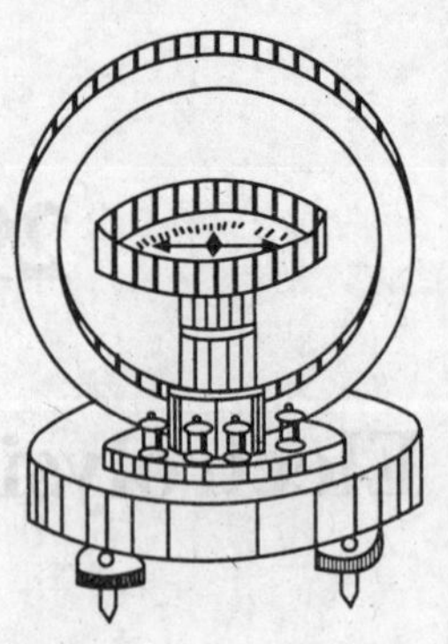

Fig. 29.1

When the coil is in the magnetic meridian, a current I passing through it produces a magnetic field B at the centre, at right angles to the plane of the coil, given by

$$B = \frac{\mu_0}{4\pi} \cdot \frac{2\pi n I}{r} \text{ Web/met}^2$$

where $\frac{\mu_0}{4\pi} = 10^{-7}$ Web/Amp-metre, n the number of turns, and r the radius of the coil in metre. Thus the field B due to the current and the field B_H due to the horizontal component of the Earth's field act at right angles on the magnetic needle and produce a deflection θ in it, given by

$$B = B_H \tan\theta$$

or

$$\frac{\mu_0}{4\pi} \cdot \frac{2\pi n I}{r} = B_H \tan\theta$$

$$\therefore \quad I = \frac{4\pi}{\mu_0} \cdot \frac{r}{2\pi n} B_H \tan\theta$$

$$= \frac{B_H}{G} \tan\theta$$

where $G = \frac{\mu_0}{4\pi} \cdot \frac{2\pi n}{r}$ is a constant for a galvanometer coil and is known as *Galvanometer constant.*

At a given place B_H is also constant and

$$\frac{B_H}{G} = K$$

where K is called the *reduction factor* of the tangent galvanometer.

Hence $$I = K \tan\theta \qquad .(i)$$

The value of $$K = \frac{B_H}{G} = \frac{4\pi}{\mu_0} \cdot \frac{rB_H}{2\pi n}$$

In relation (*i*), if $\theta = 45°$, $\tan\theta = 1$ and $I = K$

Hence reduction factor K of a tangent galvanometer is numerically equal to the current which produces a deflection of 45°.

Experiment 29.1. To find the reduction factor of the two turns coil of a tangent galvanometer using a copper voltameter.

Apparatus. A copper voltameter, four copper plates, a battery of 4 to 5 cells, a rheostat, a one-way key, connecting wires, pieces of sand paper, a balance, a weight box and a stop watch.

Theory. The weight of copper disposited in a copper voltameter is given by

$$w = I z t$$

where I is the current, t the time for which it flows and z the electrochemical equivalent of copper. Thus

$$I = \frac{w}{zt}$$

The current passing through the coil of the tangent galvanometer in amperes is given by

$$I = K \tan \theta$$

where K is the reduction factor and θ the deflection.

$$\therefore \quad K \tan \theta = \frac{w}{zt}$$

or Reduction factor $\quad K = \frac{w}{zt} \times \frac{1}{\tan \theta}$

Procedure. 1. Clean the four copper plates with nitric acid and wash them thoroughly in running water. Dry one of the plates and clean it thoroughly with a fine sand paper. Wash it and let it dry. This is the *experimental plate*.

2. Draw a diagram showing the scheme of connection as in Fig. 29.2.

3. Remove the insulation from the ends of copper wires and clean their tips with sand paper. Remove the plug from the key K and connect the battery, the key, the rheostat the remaining three plates placed in the copper voltameter to the bottom terminals B_1 and B_2 of the reversing key in series as shown in Fig. 29.2. The negative of the battery should be connected to the central copper plate and the outer two plates should be connected together and then connected to the positive of the battery. Place the tangent galvanometer at a distance from the rest of the apparatus and connect the terminal containing the two turns coil to the top terminals T_1 and T_2 of the reversing key by long parallel wires so that they cancel each other's magnetic field.

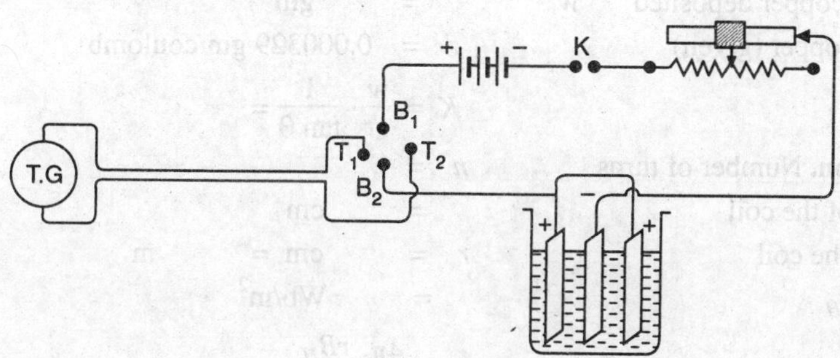

Fig. 29.2

4. **To set the tangent galvanometer.** (*i*) *Remove all magnetic substances from the proximity of the galvanometer.*

(*ii*) Set the plane of the coil vertical by adjusting the levelling screws at the base. When it is so the glass top of the box containing the scale will become horizontal. Test by a spirit level placed in various directions on the glass top.

(*iii*) Turn the coil till it is parallel to the magnetic needle.

(iv) *Tap the box containing the needle gently* to ensure smooth working of the needle. Place the eye such that the pointer *just* covers its own image in the mirror at the bottom of the scale and turn the box till one end of the pointer stands at zero of the scale. Test by tapping the box again.

5. Put the plug in the key *K* and adjust the value of the current so that the galvanometer gives a deflection nearly equal to 45°. Reverse the direction of the current and again note the deflection.

Note. *The tangent galvanometer is most sensitive when the deflection is 45°. But in no case the deflection should be less than 30° or more than 60°.*

6. Weigh the experimental plate accurately using a sensitive balance. Replace the central plate by the weighed copper plate. Put in the key *K* and start the stop watch. Reverse the current after every five minutes and note the deflection by reading both ends of the pointer. Stop the stop watch after about 30 minutes and at the same time take out the key *K*. Note the time taken.

7. Take out the cathode plate, rinse it *immediately in dilute sulphuric acid* and wash it thoroughly in running water. Dry it and again find its weight accurately.

Observations. Initial weight of cathode plate = gm

Sl. No.	*Time*	*Deflection*		*Mean deflection θ*	*tan θ*
		Direct	*Reversed*		
1					
2					
3					
4					
5					
6					

Number of turns used = 2

Time for which current is passed t = secs

Final weight of the cathode plate = gm

Weight of copper deposited w = gm

E.C.E. of copper (given) z = 0.000329 gm/coulomb

$$\therefore \quad K = \frac{w}{zt} \cdot \frac{1}{\tan\theta} =$$

Verification. Number of turns n =

Diameter of the coil = cm

Radius of the coil r = cm = m

Value of B_H = Wb/m^2

$$\therefore \quad K = \frac{4\pi}{\mu_0} \frac{rB_H}{2\pi n} =$$

Difference =

Precautions. 1. The cathode plate must be cleaned, thoroughly dried, and then weighed accurately.

2. The current should be constant throughout the experiment and should be adjusted at the rate of one ampere for every 100 sq. cm. of the area of the copper plate dipping in copper sulphate solution.

3. The plates should not touch each other.

4. The electrolyte should be fresh and free from impurities. If necessary, a few drops of sulphuric acid should be added.

5. The weighed plate must be connected to the negative pole of the battery.

6. The plate must be rinsed in dilute sulphuric acid as soon as it is removed from the electrolyte otherwise the deposit will turn black due to oxidation.

7. Edges and corners of the electrodes must be rounded off otherwise there is a tendency for the deposit to be very thick and rough at the edges.

8. The central plate should be removed very carefully without rubbing it against the slit of the wooden cover.

9. The deflection should be near about 45°.

10. The current should be reversed quickly to avoid loss of time.

11. The tangent galvanometer should be placed at a distance from the rest of the apparatus to avoid magnetic effects.

12. When taking the reading, error due to parallax should be avoided.

Exercise 1. *Find the reduction factor of a tangent galvanometer using 50 turn coil.*

Exercise 2. *Determine the value of B_H at a place using a tangent galvanometer and a copper voltameter.*

Hints. Determine K experimentally as in the above experiment. Measure r and n and apply the formula :

$$B_H = \frac{2\,\pi\,n\,K}{r} \cdot \frac{\mu_0}{4\pi}$$

Exercise 3. *Find the reduction factor of the tangent galvanometer using an ammeter.*

29.5. Determination of electro-chemical equivalent of hydrogen. The electro-chemical equivalent of hydrogen can be determined by the electrolysis of water. The apparatus used is known as a **water voltameter.**

(*i*) In its simple form a water voltameter consists of a glass-vessel in which two platinum electrodes P_1 and P_2 are fused. The electrodes are connected to two terminals at the base as shown in Fig. 29.3. The vessel contains water to which a few drops of sulphuric acid are added to make it conducing, pure water being poor conductor of electricity. Two graduated tubes A and B filled with acidulated water are placed in position over the two electrodes. When an electric current is passed electrolysis takes place and hydrogen collects in the tube A which is inverted over the cathode.

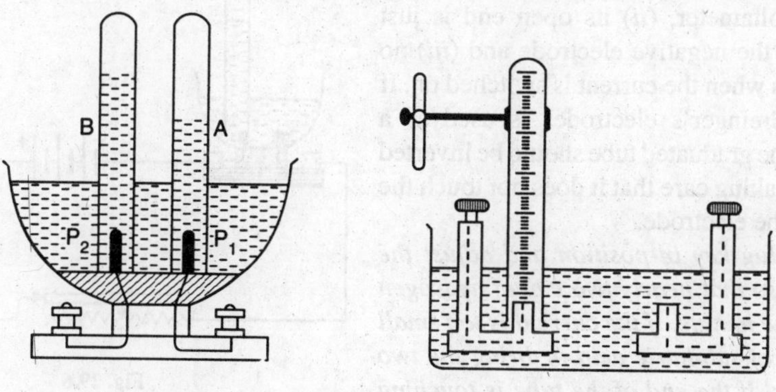

Fig. 29.3 **Fig. 29.4**

(*ii*) In an other form of water voltameter two specially made platinum electrodes known as **Greinger's electrodes** are placed in a glass trough containing acidulated water as shown in Fig. 29.4.

A Greinger's electrode consists of an ebonite base to which are fixed two ebonite rods one of which is about one inch and the other about four inches in height. A thick copper wire passes through the ebonite rods and is connected to a terminal at the end of the longer limb. A platinum strip passes round the shorter limb touching the copper rod. A graduated tube containing acidulated water is inverted over the cathode. When electric current is passed hydrogen collects in this tube.

(*iii*) A third form of the water voltameter known as Hoffman's water voltameter is sometimes used. It consists of graduated U-shaped glass tube calibrated to measure the volume of gases produced in electrolysis. The extremities of the two limbs of the U-tube are provided with stop cocks. A platinum strip about half an inch wide and two inches long with a platinum wire welded to it passes through cork closing the lower end of each arm of the U-tube. To fill the tubes with water a third tube ending in a bulb is provided as shown in Fig. 29.5. When a current is passed, gases are evolved, hydrogen at the cathode and oxygen at the anode. These are collected in respective tubes.

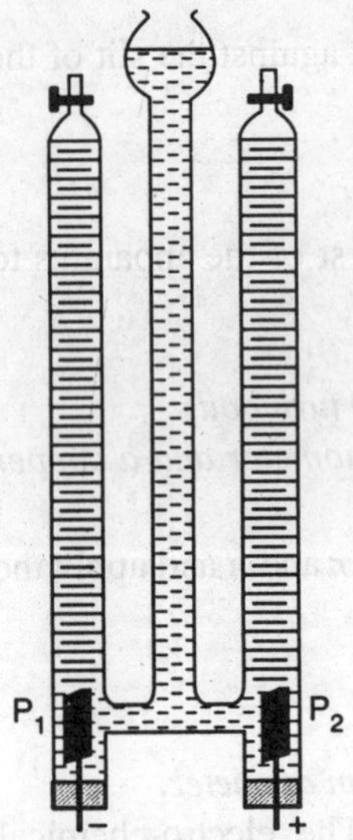

Fig. 29.5

Experiment 29.2. (*a*) To find the electro-chemical equivalent of hydrogen using an ammeter and a water voltameter.

(*b*) Given that one gm equivalent of hydrogen is 1.008 gm and charge on the electron is 1.59×10^{-19} coulomb : find the value of the Faraday and Avogadro number.

Apparatus. A trough containing acidulated water and a pair of Greinger's electrodes or a water voltameter, a graduated tube, a battery of 4 to 5 cells, an ammeter, a rheostat, a key, connecting wires, a thermometer, a half-metre rod, a clamp stand a beaker and a stop watch.

Procedure. 1. Draw a diagram showing the scheme of connections as in Fig. 29.6. Connect the water voltameter, the battery, the key, the ammeter and the rheostat in series so that the positive of the battery is connected to the positive marked terminal of the ammeter. Fill the graduated tube completely with acidulated water placed in the voltameter. Close the open end with thumb and invert it over the electrode connected to the negative of the battery. Clamp the tube carefully keeping it vertical. Adjust the height of the tube so that (*i*) it is not touching the bottom of the voltameter, (*ii*) its open end is just below the top of the negative electrode and (*iii*) no hydrogen escapes when the current is switched on. If a trough with Greinger's electrodes is used as a voltameter then the graduated tube should be inverted on the electrode taking care that it does not touch the base or sides of the electrode.

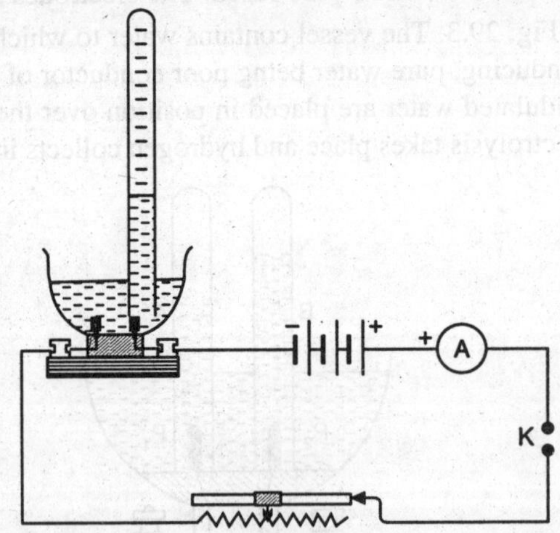

Fig. 29.6

2. *Put the plug key in position and adjust the value of current so that nearly two c.c. of hydrogen are evolved in one minute. If the current is too small add a few drops of sulphuric acid or bring the two electrodes closer. If the end of the tube is touching the base of the electrode the current will be too small as the ions are not free to move from one electrode to the other. When such is the case the current can be increased by slightly raising the tube.*

Test finally by collecting the gas for about 5 minutes. Remove the plug from the key.

3. Fill the tube completely again with acidulated water and replace it in position. Put the key in position and at the same time start the stop watch. Note the value of the current a number of times during the experiment, say after every 3 minutes. When about 40 c.c. of the gas have been collected, take out the plug from the key and at the same time stop the stop watch. Note the time taken.

4. Wait for some time till the water in the voltameter attains the room temperature. Note the volume of the gas collected and also the height of the column of water in the graduated tube above the level of water in the voltameter as shown in Fig. 29.7. Find the atmospheric pressure from the Fortin's barometer. Note the room temperature and find the vapour pressure of water corresponding to the temperature.

Note. *To directly find the volume of the gas at the atmospheric pressure close the tube by the thumb and take it out of the voltameter vessel. Immerse it in a cylinder containing water and remove the thumb. Raise or lower it so that the level of water inside the tube is the same as that in the cylinder. The volume of the gas in the tube in this position is at atmospheric pressure.*

Volume of gas

Height of water

h

Fig. 29.7

Observation. Strength of the current passed.

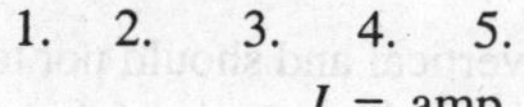

1. 2. 3. 4. 5.

Mean current $I =$ amp.

Time for which the current is passed $t =$ minute = seconds

Observed volume of the gas collected $V_1 =$ c.c.

Height of water colum in the tube $h =$ cm

Pressure of water column $p_1 = \frac{h}{13.6}$ cm =

Room temperature $= °C = T_1 \quad K =$

Maximum vapour pressure of water at room temperature $p_2 =$ cm

Atmospheric pressure $P =$ cm

$\therefore$ Pressure of gas at room temperature

$$P_1 = P - (p_1 + p_2) \text{ cm} =$$

Volume of the gas at *N.T.P.* $V = \frac{P_1 V_1}{T_1} \times \frac{273}{76} =$ c.c.

Density of hydrogen at *N.T.P.* $= .00009$ gm/c.c

$\therefore$ Weight of hydrogen collected $w = V \times .00009$ gm =

Hence *E.C.E.* of hydrogen $z = \frac{w}{It}$ gm/coulomb

(*b*) **Faraday.** *The Faraday is the amount of charge which liberates one gram. equivalent of any substance in electrolysis.* The standard value of Faraday is 96500 coulomb.

As one coulomb of charge liberates an amount equal to electro-chemical equivalent (*E.C.E.*) of any substance and the charge required to liberate one gram equivalent is one Faraday

$$\therefore \qquad \text{Faraday} = \frac{\text{gram equivalent}}{\text{Electro–chemical equivalent}}$$

Now, *E.C.E* of hydrogen from experiment $z =$ gm/coul.

Gram equivalent of hydrogen (given) = 1.008 gm

$$\therefore \qquad \text{Faraday} = \frac{1.008}{z} = \text{coulomb}$$

Difference between experimental and standard values =

Avogadro's number *Avogadro's number is the number of atoms in one gram atom of a substance.*

Now equivalent weight $= \dfrac{\text{Atomic weight}}{\text{Valence}}$

As hydrogen is monovalent

$\therefore$ Atomic weight of hydrogen = Equivalent weight of hydrogen.

Hence *Faraday* is the charge required to liberate one gram atom of hydrogen.

Charge on each atom $e = 1.59 \times 10^{-19}$ Coulomb

$\therefore$ Avogadro's number $N = \dfrac{\textit{Faraday (Experimental value)}}{1.59 \times 10^{-19}} =$

Standard value of Avogadro's number $= 6.0192 \times 10^{23}$ per gm atom

Difference =

Precautions. 1. The tube should be vertical and should not touch the bottom of the electrode.

2. The current should be so adjusted that about 2 c.c. of the gas are collected per minute.

3. When the current is stopped, the tube should be gently tapped so that bubbles sticking to the electrode rise above the water surface.

Exercise. *Compare the E.C.E. of Hydrogen and Copper.*

Hints. Connect the water voltameter and the copper voltameter in series with a battery, a rheostat and a key. Adjust the value of the current in order to collect about 2 c.c. of hydrogen per minute. Pass the current till about 40 or 45 c.c. of the gas have been collected. Find the weight of copper deposited and the weight of hydrogen collected. The ratio of the two weights is equal to the ratio of the respective electro-chemical equivalents. *In this experiment time and current need not be recorded and it is not necessary to keep the current constant as the same current is always passing* through both the voltameters.

29.6. Resistance of an electrolyte. The usual methods of finding resistance are not used to find the resistance of an electrolyte. It is because when a direct current is passed free ions are produced at the electrodes due to electrolysis and hence a back e.m.f. is set up in the circuit. Moreover, changes in concentration also occur near the electrodes which result in the change of resistance.

The most convenient method used to eliminate the effect of *polarisation and migration* of ions is to use an *alternating current*. When an alternating current is passed through an electrolyte the deposits due to flow of current in one direction will be removed when the current flows in the opposite direction. *Hence if the frequency of the alternating current is high and the area of the electrodes is large, the back e.m.f. due to polarisation becomes negligible.*

Although 50 cycle A.C. mains through a step-down transformer are a convenient source of alternating current, the results obtained are not very accurate due to low frequency. An *induction coil* without a condenser and an iron core, known as a '*buzzar*' or more preferably an audio-frequency oscillator of frequency nearly 1000 cycles per second may be used for the purpose. A low resistance telephone earpiece or a vibration galvanometer is used for detection purposes. Ordinary galvanometer is unsuitable to obtain a balance point.

It is not advisable to introduce any inducatance or capacitance in the circuit, therefore, a slide wire bridge is preferable to the post office box.

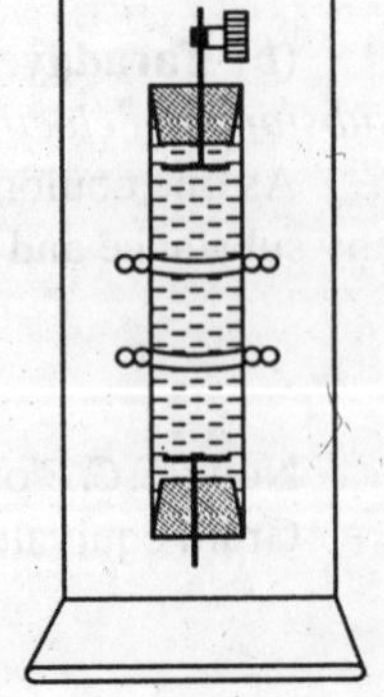
Fig. 29.8

29.7. Electrolytic cell. The electrolytic cell in its usual form consists of a glass tube of uniform area of cross-section about 20 cm long and 2 to 3 cm in diameter, mounted vertically as shown in Fig. 29.8. The tube is closed at its lower end by a rubber cork through which passes a platinum wire carrying a platinum disc which forms one of the electrodes. The disc is coated with platinum black to increase the surface area. The other electrode consists of a similar disc carried by a wire which can be clamped to a binding screw attached to the stand above the tube and thus the distance between the two electrodes can be varied. The distance is measured by a scale fixed behind the tube. While finding the *resistance of $CuSO_4$ solution copper electrodes may be used.*

Experiment 29.3. To find the resistance of an electrolyte.

Apparatus. An electrolytic cell with an elctrolyte, a metre bridge, resistance box, a low impedance headphone, a small induction coil without a capacitor and iron core (or a buzzar) or an audio-oscillator of frequency 1000, a battery, connecting wires and a piece of sand paper.

Theory. The electrolyte contained in an electrolytic cell is connected in one gap of a slide wire bridge and a resistance box R in the other gap. If a balance point is obtained at a distance l from the zero end of the metre bridge, then

$$\frac{X}{R} = \frac{\rho l}{\rho(100 - l)}$$

where X is the resistance of the electrolyte and ρ the resistance per centimetre length of the bridge wire. Hence

$$X = R\frac{l}{100 - l}$$

Procedure. 1. Clean the electrolytic cell and fill it with the given electrolyte taking care that there are no air bubbles between the two electrode.

2. Draw a diagram showing the scheme of connections as in Fig. 29.9.

3. Remove the insulation from the ends of the connecting copper wires and clean their tips with a sand paper. Connect the eletrolytic cell in the left gap and a resistance box R in the right gap of the bridge. Connect one end of a low impedance headphone to the terminal D and other end to a jockey. Connect the ends A and C to the secondary of the induction coil or a source of low voltage high frequency alternating current.

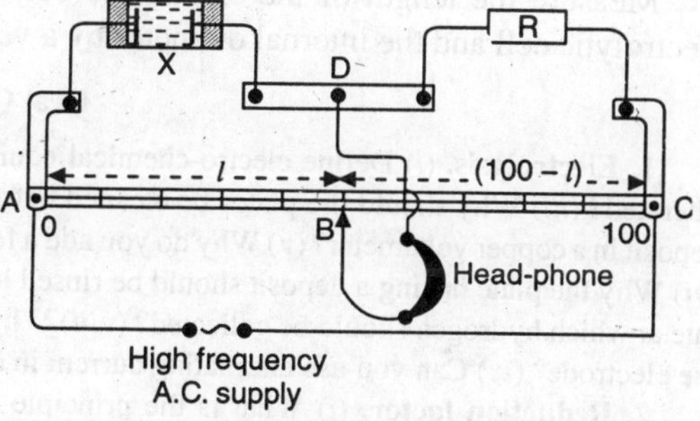

Fig. 29.9

4. **To test the connections.** Start working the induction coil and take out a suitable resistance from the resistance box R. Press the jockey near the end A, sound will be heard in the head phone. Press it now near the end C, a sound will again be heard. Adjust the position of the jockey on the bridge wire where the sound is minimum. Note the length $AB = l$. Calculate the approximate resistance of the electrolyte by the formula

$$X = \frac{R \times l}{100 - l}$$

5. Take out from the resistance box R a resistance very nearly equal to the calculated value. Adjust the position of the jockey to get accurately the position of *minimum sound. If no definite point of minimum sound is obtained the centre of the range over which sound remains minimum is taken as the true balance point.* It will be seen that it is not possible to get a position of *no sound* similar to the position of *no deflection* in a galvanometer. This is due to the self- capacitance of the resistance box which remains unbalanced because there is no capacitance in the corresponding arm. For a perfect balance in alternating current not only resistance but also the inductance and capacitance in the various arms have to be adjusted.

6. Repeat thrice by changing the value of R by a small amount so that the balance point remains between 30 and 70 cm of the bridge wire.

Observations.

Sl. No.	R	l	$100-l$	$X = \frac{R\,l}{100-l}$
1				
2				
3				

Mean resistance of the elctrolyte = ohm

Precautions. 1. There should be no air bubbles in-between the electrodes.

2. The resistance should be non-inductive and of as small self- capacitance as possible.

3. The connecting wires should be straight and not coiled.

4. The frequency of the *A.C.* supply should be high as with a low frequency *A.C.* 50 cycles mains, the balance point will be flat over a wide range.

5. The temperature of the electrolyte should be kept constant as the resistance of electrolyte varies considerably with change of temperature.

Exercise. *Find the specific resistance of the given electrolyte.*

Hints. $\text{Specific resistance} = \dfrac{\text{Resistance} \times \text{area of cross-section}}{\text{Length of electrolyte}}$

Measure the length of the electrolyte between the two electrodes from the scale behind the electrolytic cell and the internal diameter by a vernier callipers.

Oral Questions

1. Electrolysis. (*i*) Define electro-chemical equivalent. (*ii*) At which electrode hydrogen and metals are liberated? (*iii*) Why should the plates be cleaned before use? (*iv*) What is the strength of current used for best deposit in a copper voltameter? (*v*) Why do you add a few drops of sulphuric acid to the copper sulphate solution? (*vi*) Why the plate having a deposit should be rinsed in dilute sulphuric acid after its removal? (*vii*) What is the rate at which hydrogen should be collected? (*viii*) Why should the tube be kept slightly raised above the base of the electrode? (*ix*) Can you use alternating current in electrolysis?

2. Reduction factor. (*i*) What is the principle of a tangent galvanometer ? (*ii*) What do you mean by galvanometer constant and reduction factor of a tangent galvanometer? (*iii*) Why do you set the coil of the tangent galvanometer in the magnetic meridian? (*iv*) Why do you use a reversing key in the circuit ? (*v*) Why do we use the two turn coil of the tangent galvanometer to find reduction factor?

3. Faraday and Avogadro's number. (*i*) Define Faraday. (*ii*) What is the standard value of one Faraday? (*iii*) Define Avogadro's number. (*iv*) What is the value of Avogadro's number? (*v*) What is the relation between Avogadro's number and Faraday? (*vi*) What is the relation between Faraday and gram equivalent of a substance?

4. Resistance of an electrolyte. (*i*) Define specific conductivity. (*ii*) What is the relation between the specific conductivity and specific resistance? (*iii*) Why do you replace the battery by a source of high frequency *A.C.* and the galvanometer by a headphone? (*iv*) Can you get a sharp balance point with no sound? If not, why? (*v*) Why don't you use a post office box?

30

Thermocouples

30.1. Thermocouples. If a thick iron wire is welded at its ends A and B to two pieces of a thick copper wire and the free ends of the copper wires are connected to a galvanometer as shown in Fig. 30.1 the galvanometer shows a deflection when one of the junctions is kept cold and the other is heated. It is found that a *current flows from copper to iron* at the hot junction. This current is called **thermo-electric current** and the arrangement is known as a *thermo-couple.* The *e.m.f.* generated by a thermo-couple is very small being of the order of a few microvolts.

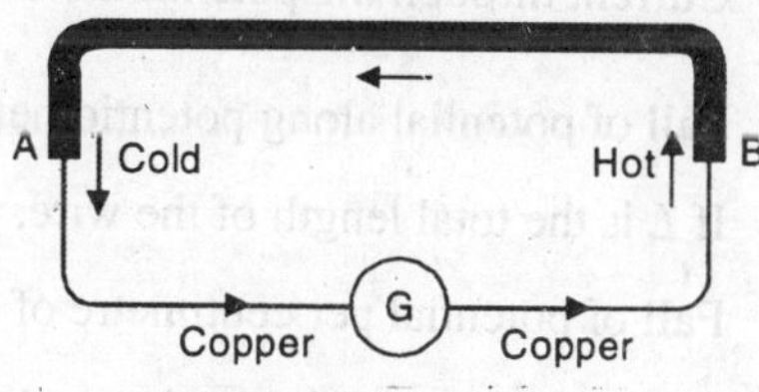

Fig. 30.1

If the temperature of the hot junction is raised keeping the temperature of the cold junction constant, the *e.m.f.* increases and becomes maximum at a certain temperature of the hot junction. This temperature is called the **neutral temperature.** Its value depends upon the pair of metals chosen and is independent of the temperature of the cold junction. The neutral temperature for a *copper-iron* couple is 270 °C.

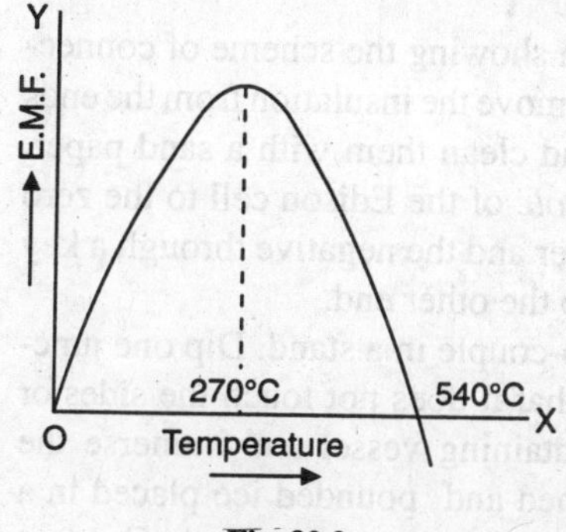

Fig. 30.2

If the temperature of the hot junction is increased beyond the *neutral temperature* the *e.m.f.* begins to decrease and becomes zero at a certain temperature known as the **temperature of inversion.** The temperature of inversion is not constant for the given thermo-couple but depends upon the temperature of the cold junction. *It is as much above the neutral temperature as the neutral temperature is above that of the cold junction.*

The value of thermo-*e.m.f.* for the various couples is given in the following table.

Couple	*e.m.f/deg. C*	*Direction of current at hot junction*
Cu – Fe	13.9×10^{-6} volt	Copper to iron
Cu – Constantan	39×10^{-5} volt	Constant an to copper
Bi – Sb		Bismuth to antimony

A thermo-couple is generally used to measure temperature. For this purpose a graph is plotted between temperature and deflection produced in a moving coil galvanometer. The deflection being proportional to the current which is in turn proportional to thermo-*e.m.f.*, the graph between deflection and temperature is of the same shape as the graph between *e.m.f.* and temperature. The *e.m.f.* generated in a thermo-couple can be measured with a micro-volt meter or a potentiometer calibrated with a standard cell.

Experiment 30.1. To plot a graph between the difference of temperature of the two junctions and the thermo–e.m.f. for a copper constantan thermo–couple using a potentiometer.

Apparatus. A copper constantan thermocouple, a clamp stand, a water bath, heating arrangement, a funnel, funnel stand, a beaker, ice, a thermometer, a potentiometer, a sensitive galvanometer, a resistance box, a post office box, an Edison cell, a one-way key, jockey, a voltmeter, connecting wires and a piece of sand paper.

Theory. The thermo-electric *e.m.f.* can be balanced against the *e.m.f.* of a cell with the help of a potentiometer and the actual value of the thermo-*e.m.f.* can be calculated. To achieve this the fall of potential per centimetre of the potentiometer wire must be known. This is calculated as follows:

Let E be the *e.m.f.* of the auxiliary battery, r the resistance of the potentiometer wire and R the resistance in series with the battery, then

$$\text{Current through the potentiometer wire} = \frac{E}{R+r}$$

$$\text{Fall of potential along potentiometer wire} = \frac{Er}{R+r}$$

If L is the total length of the wire, then

$$\text{Fall of potential per centimetre of the wire} = \frac{Er}{R+r} \times \frac{1}{L}$$

When a balance point is obtained at a distance l cm the *e.m.f.* of the thermocouple is given by

$$\frac{Er}{R+r} \times \frac{l}{L}$$

Procedure. 1. Find the resistance of the potentiometer wire either by a post office box or by an ammeter-voltmeter method.

2. Find the *e.m.f.* of the Edison cell.

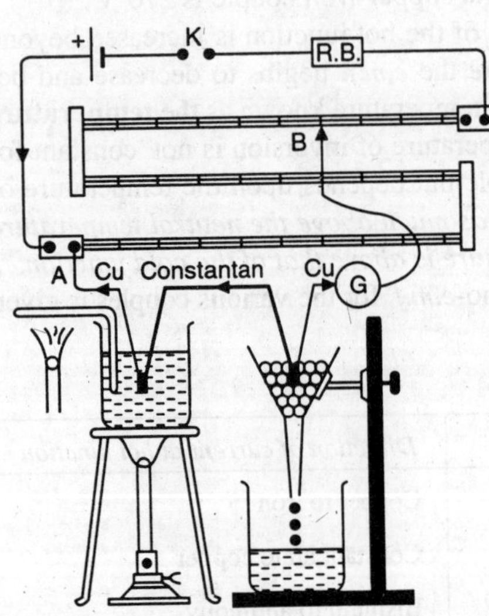

Fig. 30.3

3. Draw a diagram showing the scheme of connections as in Fig. 30.3. Remove the insulation from the ends of the copper wires and clean them with a sand paper. Connect the *positive pole* of the Edison cell to the zero end of the potentiometer and the negative through a key and a resistance box to the other end.

Clamp the thermo-couple in a stand. Dip one junction in water bath so that it does not touch the sides or the bottom of the containing vessel and immerse the other junction in washed and pounded ice placed in a funnel held in a stand with a beaker underneath. Connect the free end of the portion of copper wire at the hot junction to the zero end of the potentiometer and the portion of copper wire at the cold junction to a galvanometer. Connect the other terminal of galvanometer to a jockey.

Note. *The thermo-couple is connected to the potentiometer in such a way that the direction of the thermo-electric current in the potentiometer wire is the same as that due to the Edison cell.*

4. Set the galvanometer as already explained.

5. Heat the water bath so that the water begins to boil. Take out a resistance of about 2000 ohm from the resistance box *R.B.* Put the plug in the key *K* and find the balance point. If the balance point does not lie on the last wire adjust the resistance *R*.

6. Remove the burner and allow the temperature to fall. Keeping the temperature constant, at about 90°*C*, find the balance point. Similarly find the balance point at temperatures 80°*C* 70°*C*, and so on. Find the *e.m.f.* of the Edison cell again at the end of the experiment.

7. Plot a graph between temperature of the hot junction and thermo-electric *e.m.f.* in micro-volts.

Observations.

Resistance of potentiometer wire r = ohm
E.M.F. of Edison cell in the beginning =
E.M.F. of Edison cell at the end =
Mean *E.M.F.* E = volt
Resistance taken out from resistance box R = ohm
Length of the potentiometer wire L = cm

∴ Fall of potential per cm of the wire $e = \dfrac{Er}{(R+r)\,L}$ volt

= micro-volts

Sl. No.	*Temp.*	*Balance point l cm*	*E.M.F. e × l microvolts*
1			
2			
3			
4			
5			
6			
7			
8			
9			

Precautions. 1. The ends of the connecting wires should be cleaned and the connections should be tight, as the thermo.*e.m.f.* is usually small, any bad contact will lead to trouble.

2. Thermo-couple should be carefully examined before the experiment to see that the junctions are properly welded.

3. The Edison cell should be freshly charged so that its *e.m.f.* remains constant during the experiment.

4. The value of *R* should be so adjusted that the balance point at the highest temperature of the water bath lies on the last wire. The value of *R* should be such that the calculations are simple. For example, when the *e.m.f.* of the cell is 1.4 volt it is convenient to adjust $R + r$ equal to 1400 or 2800 ohm.

5. The positive of the thermo-couple should be connected to the point where the positive of the cell is connected. *The positive end of the thermo-couple is that end where the current leaves it.*

Exercise 1. *Find the melting point of wax with a thermocouple and a potentiometer.*

Hints. Plot a graph between the temperature of the hot junction and the balance point on the potentiometer. Dip the hot junction in melting wax. Find the balance point when the wax just begins to solidify and find the temperature corresponding to this balance point from the graph.

Exercise 2. *Find the neutral temperature for a copper constantan thermo-couple.*

Experiment 30.2. To plot a graph between the difference of temperature of the two junctions and the thermo–e.m.f. for a copper constantan thermocouple using a standard cell.

Apparatus. Same as in **Experiment 30.1.** An additional resistance box, a standard cell instead of the voltmeter, a two-way key etc.

Theory. Same as in **Experiment 30.1.**

Procedure. 1. Draw a diagram showing the scheme of connections as in Fig. 30.4.

2. Clean the ends of the connecting wires. Connect the positive pole of the Edison cell through two resistance boxes *R* and *S* to the zero end of the potentiometer and the *negative pole* through a key *K* to the other end. Connect the positive of the standard cell to the terminal of the resistance box *S* not connected to the potentiometer and the negative to the terminal *A* of the two-way key. Place the thermocouple in position and connect the free end of the copper wire at the hot junction to the zero end of the potentiometer and the free end of copper wire at the cold junction to the terminal *B* of the two-way key. Connect the common terminal *C* through the galvanometer to the jockey.

3. Take out a resistance of 1001 ohm from the resistance box *S*. Put the plugs in the key *K* between the points *A* and *C*.

Touch the jockey near the zero end of the potentiometer and adjust the resistance *R* till a change of one ohm causes a change in the direction of deflection. Now shift the jockey away from zero and find the balance point. Note the length l_1 up to this point.

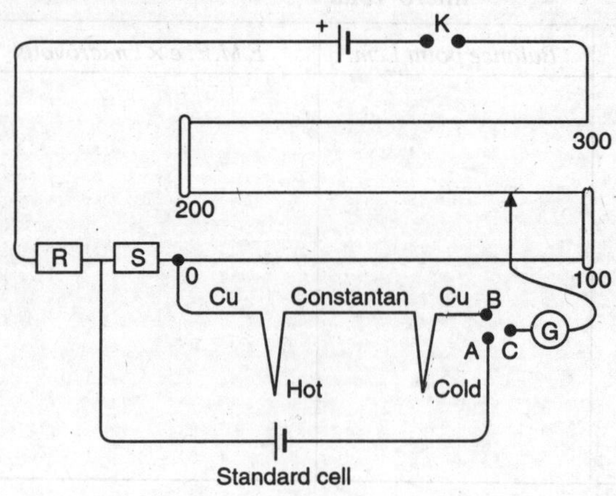

Fig. 30.4

Decrease S by one ohm and increase R by one ohm so that the current through the circuit remains constant. Find the balance point again and note the length l_2 up to this point. The difference $(l_2 - l_1)$ gives the length of the wire having a resistance of one ohm. Note the room temperature.

4. Remove the plug from *A* and *C* and insert the plug between *B* and *C* and find the balance point with the thermo-couple by keeping the temperature of the hot junction constant at 90°*C*, 80° *C* etc.

Observations. (*i*) *Readings with standard cell.*

Sl. No.	*S*	*R*	*Balance point*		
			(*i*)	(*ii*)	*Mean*
1	1001				l_1
2.	1000				l_2

Room temperature $t =$ °*C*

E.M.F. of the standard cell at room temp. $E =$ Volt

∴ Length of wire having 1 ohm resistance $= (l_2 - l_1)$ cm

Resistance of length l_2 of potentiometer wire $r = \dfrac{l_2}{l_2 - l_1}$ ohm

∴ Fall of potential per centimetre $e = \dfrac{E}{S+r} \cdot \dfrac{1}{l_2}$ volt

$=$ microvolt.

Record other observations as done in **Experiment 30.1.**

Oral Questions

1. Thermocouple. (*i*) What is a thermocouple? (*ii*) What is the direction of current in a (*a*) Cu-Fe (*b*) Cu-Constantan and (*c*) Bi-Sb thermocouples? (*iii*) How does the thermo.e.m.f. vary with the temperature of the hot junction? (*iv*) What do you understand by neutral temperature and temperature of inversion? (*v*) What are their values for the thermocouple that you are using? (*iv*) What is the order of e.m.f. generated?

2. Potentiometer method. (*i*) What should be the order of fall of potential along the potentiometer wire so that the readings extend over the whole of the length of wire? (*ii*) What is the use of a standard cell? (*iii*) With a copper constantan thermo-couple which wire will you connect at the zero end? (*iv*) What are the practical applications of a thermo- couple?

31

Ballistic Galvanometer and Magnetic Field

31.1. Ballistic galvanometer. A ballistic galvanometer is used to measure the total quantity of electrical charge that passes through it as a sudden discharge. The moment of inertia of the moving system is very large and hence it is slow to begin its motion under the impulse of the charge so that the whole of the charge passes through the galvanometer before the moving system has appreciably moved from its position of rest.

In such a galvanometer there should be no damping, but as it cannot be absolutely removed, a correction must be applied.

There are two types of ballistic galvanometers:

(i) **Moving magnet type.** In this galvanometer the charge q is given by the relation

$$q = \frac{B_H T}{\pi G} \sin \frac{\theta}{2}$$

where B_H is the controlling field due to the horizontal component of Earth's magnetic field, T the time period of the magnet in the earth's field alone, G the galvanometer constant *i.e.* the magnetic field at the centre of the coil due to a current of 1 Amp.

or
$$G = \frac{\mu_0}{4\pi} \frac{2\pi n}{r}$$

where n is the number of turns, r the radius of the coil and $\frac{\mu_0}{4\pi} = 10^{-7}$ Wb/A-m.

(ii) **Moving coil type.** In this galvanometer the charge q is given by the relation

$$q = \frac{c\,T}{2\pi nAB}\,\theta$$

where c is the restoring couple per unit angular twist of the suspension wire, T the time period of the coil, n the number of turns, A the area of the coil, B the magnetic field in which the coil is suspended and θ the first throw.

For a given instrument $\frac{c\,T}{2\pi nAB} = k$

$$\therefore \quad q = k\theta$$

or
$$\frac{q}{\theta} = k$$

The ratio $\frac{q}{\theta}$ is known as ballistic reduction factor or ballistic constant or *charge sensitivity* of the ballistic galvanometer.

When the deflection is measured with a lamp and scale arrangement **charge sensitivity** *is defined as the charge in micro-coulomb required to produce a deflection of one mm on a scale placed at a distance of one metre from the galvanometer mirror.*

If d is the deflection in mm produced by a charge of q micro-coulomb, then

$$k = \frac{q}{d} \text{ micro coulomb/mm}$$

In a moving coil type ballistic galvanometer the damping is reduced by winding the coil on wood, ivory, or bone *i.e.,* a non-conducting material. No soft iron core is used.

31.2. Damping correction. It is found in the case of a ballistic galvanometer that the amplitude of the successive swings goes on decreasing. If θ_1 is the swing on the right θ_2 on the left, θ_3 on the right and θ_4 on the left and so on, then

$$\frac{\theta_1}{\theta_2} = \frac{\theta_2}{\theta_3} = \frac{\theta_3}{\theta_4} \ldots\ldots = e^{\lambda} = \text{(a constant)}$$

where λ is a constant known as **logarithmic decrement.**

The damping correction to be applied to the very first throw is $e^{\lambda/2}$ because the coil simply makes half the swing as it travels from the mean position to the extreme position and not from one extreme to the other.

Hence if θ_1 is the first throw, then the corrected throw θ_0 is given by

$$\theta_0 = \theta_1 e^{\lambda/2}$$

$$= \theta_1 [1 + \lambda/2]$$

neglecting squares and higher powers of λ in the expansion of $e^{\lambda/2}$. Thus the first throw is corrected for damping by multiplying the observed throw with the factor $\left(1 + \frac{\lambda}{2}\right)$.

$$\therefore \qquad q = k\theta = k\theta_1 \left(1 + \frac{\lambda}{2}\right)$$

To determine the value of λ, the logarithmic decrement a capacitor is charged and then discharged through a ballistic galvanometer. If θ_1 be the first throw and θ_{11} the 11th throw counting successively, then

$$\frac{\theta_1}{\theta_{11}} = e^{10\lambda}$$

or

$$\lambda = \frac{1}{10} \log_e \frac{\theta_1}{\theta_{11}} = \frac{1}{10} \times 2.3026 \log_{10} \frac{\theta_1}{\theta_{11}}$$

In actual practice the damping is corrected by observing the first throw θ_1 to one side and again the next throw θ_3 to the same side of zero. The decrease $(\theta_1 - \theta_3)$ has been caused due to the motion of the coil from one extreme to the other and back again to the same position. Hence the decrease caused due to the motion of the coil from the mean position to the first extreme position is $\frac{\theta_1 - \theta_3}{4}$.

$$\therefore \qquad \text{Corrected deflection} = \theta_1 + \frac{\theta_1 - \theta_3}{4}$$

31.3. Lamp and scale arrangement. The deflection in ballistic galvanometer is generally measured with the help of a lamp and scale arrangement. It consists of a translucent ground glass or plastic scale having zero graduation at its centre and 25 cm division mark on either side of it. It is mounted horizontally on a stand as shown in Fig. 31.1. On the same stand is mounted a metal tube containing a 6 volt lamp fixed at one end and a convex lens with a crossline marked on it at the other end. The lens can be moved in and out by means of a rack-and-pinion arrangement *R*. The bulb is lighted either with a 6 volt battery or a suitable step down transformer connected to the *A.C.* mains.

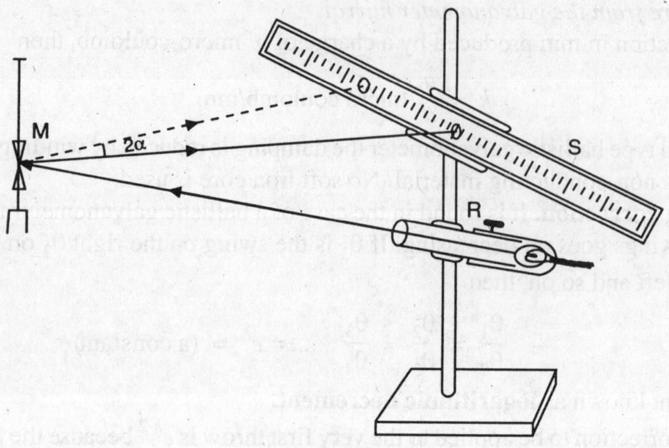

Fig. 31.1

The height of the lamp is so adjusted that the light from it falls on the mirror *M* of the galvanometer. The position of the scale is also adjusted so that light after reflection falls nearly in the centre of the scale. A sharp image of the cross-line in the spot of light is obtained on the scale by adjusting the position of the lens with the rack-and-pinion arrangement *R*. When the galvanometer coil is deflected, the mirror turns through a small angle α, the spot of light moves through twice the angle *i.e.*, 2α. Hence the distance through which the spot of light moves on the scale is proportional to the angle of deflection.

While using a lamp and scale arrangement, the ballistic galvanometer should be properly levelled so that the coil does not touch any part of the galvanometer and the spot of light moves freely on the scale.

Experiment 31.1. To calibrate the given moving coil ballistic galvanometer or to determine the ballistic reduction factor (ballistic constant) by steady deflection method.

Apparatus. A ballistic galvanometer, a lamp and scale arrangement, a cell of constant *e.m.f.*, three resistance boxes, a one-way key, a tap key, a reversing key, connecting wires and a metre rod.

Theory. If θ is the ballistic throw, then the charge *q* flowing through the moving coil ballistic galvanometer

$$q = k\theta$$

where *k* is the constant of ballistic galvanometer. For a given instrument

$$q = \frac{cT}{2\pi nAB}\theta$$

$$= \frac{T}{2\pi}\frac{c}{nAB}\theta$$

Now $\frac{c}{nAB}$ is the current sensitivity η for a steady current

$$\therefore \quad q = \frac{T}{2\pi}\eta\theta \qquad ...(i)$$

$$\therefore \quad \text{Ballistic constant} \quad k = \frac{q}{\theta} = \frac{T}{2\pi}\cdot\eta$$

$$\text{where } \eta = \frac{\text{steady current}}{\text{steady deflection}} = \frac{i}{\theta}$$

Thus to find the ballistic constant k, the figure of merit η and T the time period of the galvanometer are required.

Current sensitivity or figure of merit η. To determine η a cell of constant *e.m.f.* E and negligible internal resistance is connected in series with two resistance boxes P and Q. The ballistic galvanometer G in series with a resistance box R is connected across P through a reversing key K_1 as shown in Fig. 31.2.

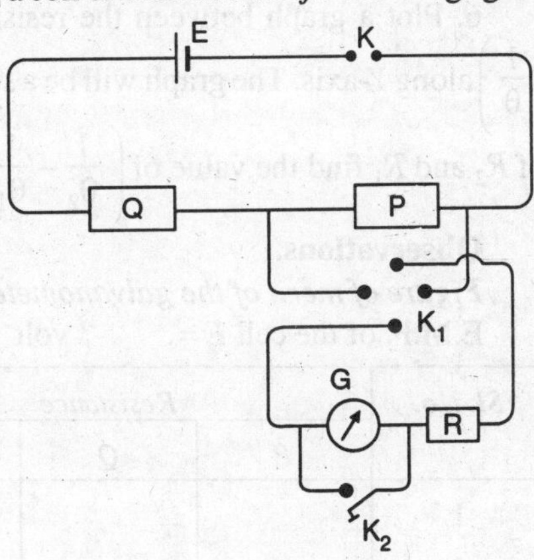

Fig. 31.2

$$E.M.F. \text{ across } P = \frac{EP}{P+Q}$$

If G is the resistance of the galvanometer and resistance P is negligle as compared to $(R + G)$, then the current i in the galvanometer circuit is given by

$$i = \frac{\frac{EP}{P+Q}}{(R+G)} = \frac{EP}{(P+Q)(R+G)}$$

If θ is the deflection produced, then

$$\eta = \frac{i}{\theta} = \frac{EP}{(P+Q)(R+G)} \frac{1}{\theta} \quad ...(ii)$$

The resistance G of the galvanometer can be eliminated by taking 2 observations for different values of R. Thus from equation *(ii)*, we have

$$R_1 + G = \frac{EP}{(P+Q)\eta} \frac{1}{\theta_1}$$

$$R_2 + G = \frac{EP}{(P+Q)\eta} \frac{1}{\theta_2}$$

$$\therefore \quad R_2 - R_1 = \frac{EP}{(P+Q)\eta}\left(\frac{1}{\theta_2} - \frac{1}{\theta_1}\right)$$

or

$$\eta = \frac{EP}{(P+Q)(R_2 - R_1)}\left(\frac{1}{\theta_2} - \frac{1}{\theta_1}\right) \quad ...(iii)$$

$\therefore$ Ballistic constant or charge sensitivity $k = \frac{q}{\theta} = \frac{T}{2\pi}\eta$

Procedure. 1. Adjust the lamp and scale arrangement so that the spot of light is obtained on the scale placed at a distance of *one metre* from the mirror of the galvanometer. See that the spot of light moves freely and its rest position is on the zero division mark.

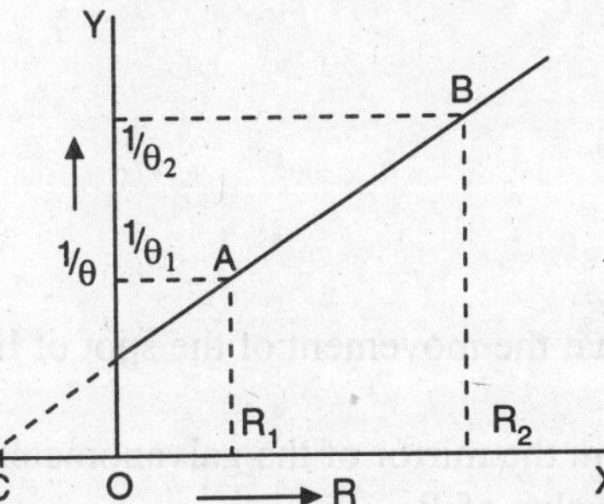

Fig. 31.3

2. Draw the scheme of connections as shown in Fig. 31.2. Make the connections and see that these are neat, clean and tight.

3. Take out a resistance from the resistance box R of the order of about 5000 ohm. Adjust the value of P to 2 ohm and Q to the desired value so that when the circuit is completed a good steady deflection on the scale is obtained. Note the deflection. Reverse the current and again note the deflection and find the mean value.

4. Keeping the values of P and Q the same increase the value of R in steps of 1000 ohm and determine the mean deflection. Repeat at least for six different values of R. Measure the *e.m.f.* of the cell with a voltmeter.

5. To measure the free time period of the galvanometer insert the key K momentarily. The sudden flow of charge through the galvanometer will give it a kick and the coil will oscillate in the open circuit. Measure the time of 20 vibrations thrice and find the mean time period T.

6. Plot a graph between the resistance R taken along the X-axis and *reciprocal of deflection* $\left(\frac{1}{\theta}\right)$ along Y-axis. The graph will be a straight line as shown in Fig. 31.3. Corresponding to two values of R_2 and R_1 find the value of $\left(\frac{1}{\theta_2} - \frac{1}{\theta_1}\right)$ and calculate the value of η from equation (*iii*).

Observations.

Figure of merit of the galvanometer.

E.M.F. of the cell E = volt

Sl. No.	*Resistance*			*Deflection*		*Mean* θ	$\frac{1}{\theta}$
	P	*Q*	*R*	*Direct*	*Reversed*		
1			5000				
2			6000				
3			7000				
4			8000				
5			9000				
6			10000				

Time period of the galvanometer

Sl. No.	*No. of vibrations*	*Time*	*Time period T*
1			
2			
3			

Mean time period T =

Calculations

Value of $\left\{\frac{1}{\theta_2} - \frac{1}{\theta_1}\right\}$ from graph =

Value of $(R_2 - R_1)$ = ohm

$\therefore$ Figure of merit $\eta = \frac{EP}{(P+Q)\,(R_2 - R_1)}\left\{\frac{1}{\theta_2} - \frac{1}{\theta_1}\right\}$

Hence ballistic constant $k = \frac{T}{2\pi}\eta =$

Precautions. 1. The galvanometer should be levelled so that the movement of the spot of light is free.

2. The scale should be set up at a distance of one metre from the mirror of the galvanometer.

3. The value of R should be very large as compared to the value of P.

4. A tap key should be placed across the galvanometer terminals and should be pressed to bring the spot of light to a stationary position.

5. A cell of constant *e.m.f.* and very low internal resistance should be used.

6. While finding the time-period of the galvanometer coil, the galvanometer circuit must be kept open.

Note. The resistance G of the ballistic galvanometer can also be found from the graph between the resistance R and $\frac{1}{\theta}$. For $\frac{1}{\theta} = 0$

$$R + G = 0$$

$$\therefore \quad G = -R$$

Hence the intercept OC on the X-axis in Fig. 31.3. gives the value of the resistance G of the galvanometer.

Note. When we use lamp and scale arrangement, the deflection θ is measured in mm on the scale.

Experiment 31.2. To determine the charge sensitivity of a moving coil ballistic galvanometer using a known capacitor.

Apparatus. A capacitor of a known capacitance (0.1 μF), a cell, a moving coil ballistic galvanometer, a morse key, a tap key, a lamp and scale arrangement, connecting wires, a piece of sand paper, a metre rod, voltmeter etc.

Theory. If a capacitor of capacitance C farad is charged to a potential difference V volt, then

Charge $q = CV$ coulomb

If this charge is allowed to pass through the coil of the ballistic galvanometer and θ is the corrected throw observed, then

$$q = CV = k\,\theta$$

Now charge sensitivity = $\frac{q}{\theta} = k$

$$\therefore \quad k = \frac{CV}{\theta} \times 10^{-6} \text{ micro-coulomb per mm}$$

If the capacitance is in micro-farad, then

$$k = \frac{CV}{\theta} \text{ micro-coulomb per mm}$$

Procedure. 1. Place the ballistic galvanometer on a rigid support and see that its coil rotates freely. If not, adjust the levelling screws. Set the lamp and scale arrangement so that a sharp image of the cross-line in the spot of light is obtained in the centre of the scale placed at a distance of one metre from the galvanometer mirror.

2. Draw the scheme of connections as shown in Fig. 31.4. and make connections accordingly.

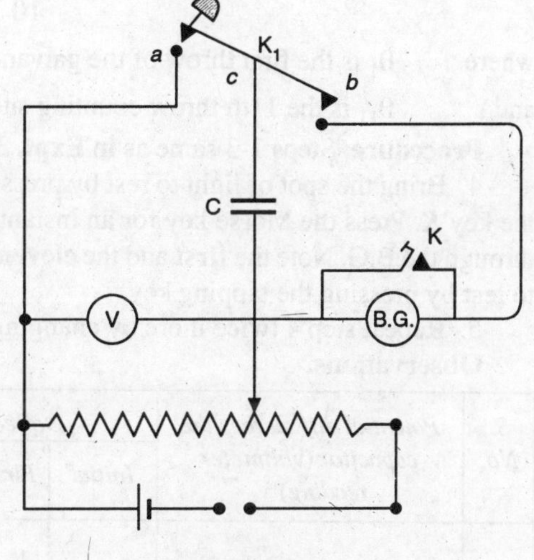

Fig. 31.4

3. Unlock the B.G. The spot of light shall start oscillating and bring the spot to rest at zero of the scale. Release the tap key. Press the Morse key K_1 for an instant. The capacitor C will get charged. Release the Morse key. The capacitor shall get discharged through the B.G. Observe the deflection of the spot of light. If it goes out of scale or is too small adjust the potential with the help of the rheostat (or the resistance boxes) of the potential dividing arrangement. Repeat it till you find that on pressing the Morse key for an instant and then releasing it the spot remains on the scale and gives a sufficiently large deflection.

4. Bring the spot of light to rest by pressing the key *K*. Note the initial reading. Release the key *K*, press the morse key, release it and note the **first** throw θ_1 of the spot of light on the scale and also the second throw θ_2 on the same side. Repeat the observation. Note the voltage across the cell (or the resistance box of the potential dividing arrangement). Find the mean corrected throw for the voltage used.

5. Change the voltage and repeat the experiment for two more observations.

Observations. Capacitance of the capacitor $C =$ μfd.

Sl. No.	*Pot. diff. V*	*Initial obs.*	*First throw* θ_1	*Second throw* θ_2	*Corrected throw* $\theta_1 + \frac{\theta_1 - \theta_2}{4}$	*Mean corrected throw* θ
1 (*i*)						
(*ii*)						
2						
3						

Calculations

Charge sensitivity $k = \frac{CV}{\theta} =$ micro-coulomb/mm

Precautions. 1. The connecting wires used should be short and straight.

2. Immediately after each observation bring the spot of light to rest by pressing the key *K* (electromagnetic damping).

3. The time of charging the capactor should be the same for each observation.

4. If the *e.m.f.* of the cell is large so that the ballistic throw goes out of the scale, then use potential dividing arrangement to decrease the applied charging potential difference.

Experiment 31.3. To find the logarithmic decrement for a ballistic galvanometer.

Apparatus. Same as in **Expt. 31.2.**

Theory. As explained in **Art. 31.2** the logarithmic decrement λ is given by

$$\lambda = \frac{1}{10} \times 2.3026 \log_{10} \frac{\theta_1}{\theta_{11}}$$

where θ_1 is the first throw of the galvanometer.

and θ_{11} is the 11th throw counting successively.

Procedure. Steps 1-3 same as in **Expt. 31.2.**

4. Bring the spot of light to rest by pressing the tapping key *K*. Note the initial reading. Release the key *K*. Press the Morse key for an instant and then release it. The capacitor shall get discharged through the B.G. Note the **first** and the **eleventh** throw of the spot of light on the scale. Bring the spot to rest by pressing the tapping key.

5. Repeat step 4 twice more by changing the voltmeter reading with the help of rheostat.

Observations.

S. No.	*Potential diff. across the capacitor (voltmeter reading) V*	*Deflection of spot of light*			*Logarithmic decrement* $\lambda = \frac{2.3026}{10} \log_{10} \frac{\theta_1}{\theta_{11}}$
		Initial	*First throw* θ_1	*11th throw* θ_{11}	

Result. Logarithmic decrement of the B.G =

Precautions. Same as in **Expt. 31.2.**

31.4. Earth inductor. The earth inductor as shown in Fig. 31.5 consists of a circular coil C of face area A having about 5000 turns of insulated copper wire. The coil is capable of rotation about an axis along its diameter. The rotation is restricted to 180° by means of two stops. When the coil is held between the two stops, it can be released by pushing the push-button P. The position of the coil and its axis of rotation can be adjusted. The ends of the coil are connected to the two segments of a commutator. Two small brass strips connected to the two terminals T_1 and T_2 press against the two halves of the commutator. This arrangement helps to draw current when the coil is rotated in the magnetic field.

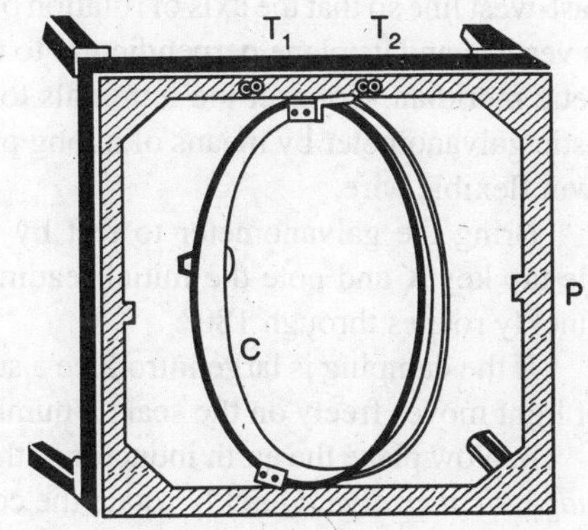

Fig. 31.5

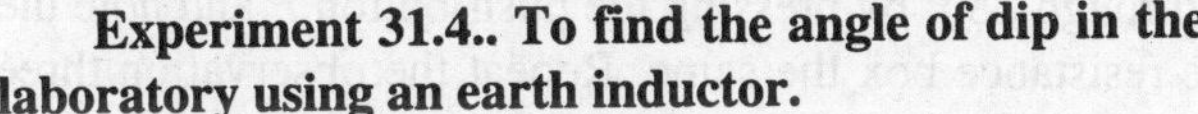

Experiment 31.4.. To find the angle of dip in the laboratory using an earth inductor.

Apparatus. An earth inductor, a moving coil ballistic galvanometer, a lamp and scale arrangement, a magnetic compass, a tap key, a set square, connecting wires and a sand paper.

Theory. The earth inductor is placed with the plane of the coil and axis of rotation in vertical position and at right angle to the magnetic meridian. When the coil is rotated rapidly through 180° it cuts the horizontal component B_H and a charge q_1 flows through the ballistic galvanometer producing a deflection θ_1. The charge q_1 that flows through the galvanometer is given by,

$$q_1 = \frac{2nAB_H}{R} = k\theta_1 \left(1 + \frac{\lambda}{2}\right) \qquad \text{...(i)}$$

where n is the number of turns, A the face area of the coil of the earth inductor, B_H the horizontal component of earth's magnetic field and R the resistance of the coil, leads and the galvanometer.

The earth inductor is now placed with its plane horizontal and its axis of rotation in the magnetic meridian. When the coil is rotated through 180° in this position, it cuts the vertical component B_V of the earth's field and a charge q_2 again flows through the galvanometer. If θ_2 is the deflection produced, then

$$q_2 = \frac{2nAB_V}{R} = k\,\theta_2 \left(1 + \frac{\lambda}{2}\right) \qquad \text{...(ii)}$$

Dividing (*ii*) by (*i*), we have

$$\frac{B_V}{B_H} = \frac{\theta_2}{\theta_1} = \tan\delta$$

where δ is the angle of dip.

If a lamp and scale arrangement is used and d_1 and d_2 are the distances through which the spot of light moves on the scale, then

$$\tan\delta = \frac{\theta_2}{\theta_1} = \frac{d_2}{d_1}$$

Procedure. 1. Set the ballistic galvanometer and the lamp and scale arrangement. Connect a tap key across the ballistic galvanometer.

2. Draw the scheme of connections as shown in Fig. 31.6. Connect the ballistic galvanometer, the earth inductor and the resistance box in series.

3. Draw the magnetic meridian on the table with the help of magnetic compass. Also draw a line at right angle to the magnetic meridian with a piece of chalk. Place the earth inductor on the east-west line so that the axis of rotation of the coil is vertical and its plane perpendicular to the magnetic meridian. Connect the terminals to the ballistic galvanometer by means of a long piece of a twin flexible wire.

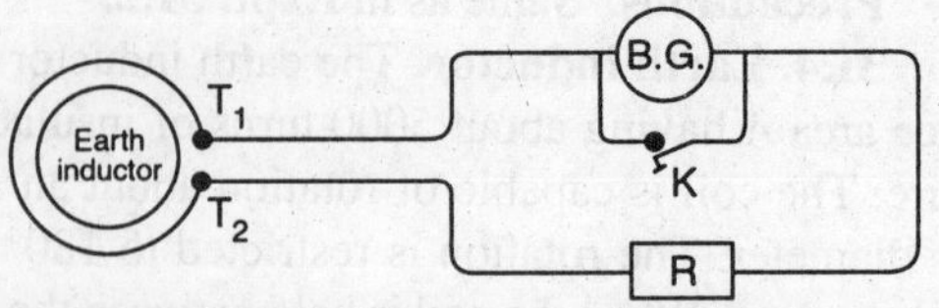

Fig. 31.6

Bring the galvanometer to rest by pressing the tap key K and note the initial reading. Press the push-button P, so that the coil is released and quickly rotates through 180°.

If the damping is large introduce a suitable high resistance from the resistance box R till the spot of light moves freely on the scale a number of times. Repeat the observation three times.

4. Now place the earth inductor with its coil in the horizontal plane and its axis of rotation in *the magnetic meridian.* Quickly rotate the coil through 180° by pressing the push-button P and note the first throw, keeping the resistance from the resistance box the same. Repeat the observation three times.

Observations.

Sl. No.	*Scale reading with coil in*							
	Vertical plane, rotation cutting horizontal component B_H				*Horizontal plane, rotation cutting vertical component* B_V			
	Initial R.P.	first throw	*Diff.*	*Mean* d_1	*Initial R.P.*	*First throw*	*Diff.*	*Mean* d_2
1								
2								
3								

Now $$\tan \delta = \left(\frac{d_2}{d_1}\right)$$

∴ Angle of dip $$\delta = \tan^{-1}\left(\frac{d_2}{d_1}\right)$$

Precautions. 1. The ballistic galvanometer should be at a large distance from the earth inductor and connections should be made with a twin flexible wire.

2. While observing the throw in the ballistic galvanometer, due to the cutting of earth's horizontal field by the earth inductor, the axis of rotation of the coil should be vertical and its plane exactly perpendicular to the magnetic meridian.

3. While observing the throw in the ballistic galvanometer due to the cutting of earth's vertical field by the earth inductor, the plane of the coil should be horizontal and its axis of rotation exactly, in the magnetic meridian.

31.5. Measurement of a strong magnetic field. To Measure a strong magnetic field as occurs between the pole-pieces of an electro-magnet or a strong magnet, a search coil is used.

Search coil. It is a small coil of about 50 turns of fine copper wire and about 1 sq. cm in area. When this coil, connected in series with a ballistic galvanometer is placed between the pole pieces, maximum flux passes through it. If it is removed quickly from the field, an induced current is set up which causes a throw in the ballistic galvanometer.

Experiment 31.5. To determine the magnetic induction field between the pole pieces of an electro-magnet with the help of a search coil and a ballistic galvanometer using an earth inductor for calibration of the ballistic galvanometer.

Apparatus. An electromagnet with a battery, an ammeter, a rheostat, a key for passing steady current through the field coils, a search coil, a ballistic galvanometer, a lamp and scale arrangement, an earth inductor, one way key, a tap key, connecting wires etc.

Theory. When a search coil is placed between the pole pieces of an electromagnet, with its face perpendicular to the magnetic lines of force maximum flux passes through it. This coil is also connected in series to the ballistic galvanometer and an earth inductor placed with its axis of rotation vertical and the plane of the coil perpendicular to the magnetic meridian. If a is the area of the face of the search coil and n the number of turns, then the change in magnetic flux N when the search coil is suddenly removed is given by

$$N = n\,a\,B$$

where B is the magnetic induction field and is defined as the magnetic flux per unit area.

The change in magnetic flux causes a charge q to flow through the galvanometer. If θ_1 is the first throw of the galvanometer, λ the logarithmic decrement and k the ballistic constant, then

$$q = k\,\theta_1 \left(1 + \frac{\lambda}{2}\right) \qquad (i)$$

If i is the current at any instant through the galvanometer and R the combined resistance of the galvanometer circuit, then the charge passing through the galvanometer at that instant is given by

$$q = \int i\,dt = \frac{1}{R}\int \frac{dN}{dt}\,dt$$

$$= \frac{N}{R} = \frac{naB}{R} \qquad ...(ii)$$

Equating (*i*) and (*ii*), we get

$$\frac{naB}{R} = k\,\theta_1\left(1 + \frac{\lambda}{2}\right) \qquad ...(iii)$$

If now the coil of earth inductor is rotated quickly through 180° and the throw θ_2 in the ballistic galvanometer is noted, then

$$\frac{2a'\,n'\,B_H}{R} = k\,\theta_2\left(1 + \frac{\lambda}{2}\right) \qquad ...(iv)$$

where B_H is the magnetic field due to horizontal component of earth's magnetic field, a' the face area of each turn of the coil and n' the number of turns of the earth inductor.

Dividing (*iii*) by (*iv*), we get

$$\frac{B}{B_H}\,\frac{na}{2a'n'} = \frac{\theta_1}{\theta_2}$$

or

$$B = \frac{2n'a'}{na}\,\frac{\theta_1}{\theta_2}\,B_H$$

Knowing the various quantities B can be calculated.

Procedure. 1. Level the ballistic galvanometer and set up the lamp and scale arrangement, so that the spot of light moves freely.

2. Draw a scheme of connections as shown in Fig. 31.7 and connect the ballistic galvanometer, the earth inductor and the search coil through a key K. Connect the tap key across the ballistic galvanometer. Set the plane of the coil of the earth inductor so that its axis is vertical and the plane of the coil is perpendicular to the magnetic meridian.

3. Now place the search coil in the pole-pieces of the electromagnet such that its plane is parallel to the pole pieces. Complete the circuit of the field coils through an ammeter, a rheostat, a battery and a key (not shown), and adjust a current of suitable value for a steady magnetic field.

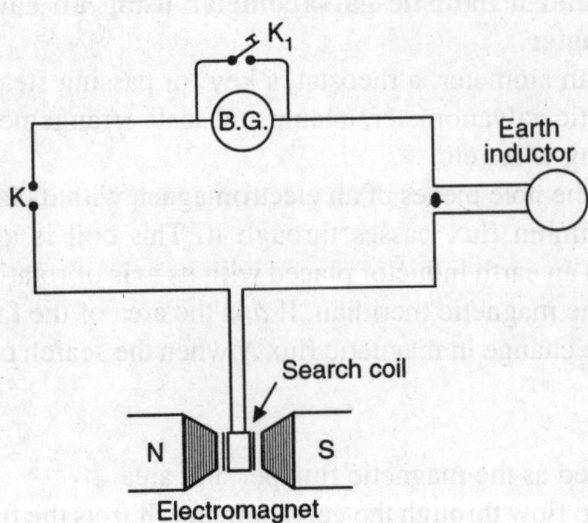

Fig. 31.7

4. Note the initial position of the spot on the scale. Close the key K and rapidly withdraw the search coil from the gap between the pole pieces of the electromagnet and note the first ballistic throw θ_1 of the spot of light.

5. Bring the galvanometer coil to rest by pressing the key K_1. Quickly rotate the coil of the earth inductor and observe the first ballistic throw θ_2 of the spot of light.

6. Change the current through the field coils and repeat the experiment for different values of the current i.

Observations.

Number of turns of search coil $n =$

Face area of each turn of search-coil $a =$ sq. m

Number of turns of earth inductor $n' =$

Face area of each turn $a' =$ sq. m

Horizontal component of earth's field $B_H =$ Tesla

Sl. No.	*Current in field coil*	*Initial position*	*First throw with*		*Corrected*		$\frac{\theta_1}{\theta_2}$
			S. coil	*E. inductor*	θ_1	θ_2	
1							
2							
3							
...							
...							

$$\text{Mean } \frac{\theta_1}{\theta_2} =$$

Calculations.

1st obs. $B = \frac{2n'a'}{na} \frac{\theta_1}{\theta_2} B_H =$ Tesla

Note. Make similar calculations for other observations. If a graph is plotted between B and i it will be a straight line.

Precautions. 1. The plane of the search coil should be parallel to the plane of the pole pieces.

2. The removal of search coil should be very quick.

3. The coil of the earth inductor should be set so that its axis of rotation is vertical and its plane perpendicular to the magnetic meridian.

4. A tap key should be connected across the ballistic galvanometer.

5. The earth inductor and the galvanometer should be at a considerable distance from each other and from the electromagnet.

6. The connections in the galvanometer circuit should be made with twin flexible wire.

31.6. Magnetic field due to a solenoid. Consider a closely wound solenoid *AB*. If *I* is the current through the solenoid in amperes, *N* the total number of terns and *l* the length of solenoid in metres, then total magnetic field at a point *P* on the axis of the solenoid is given by

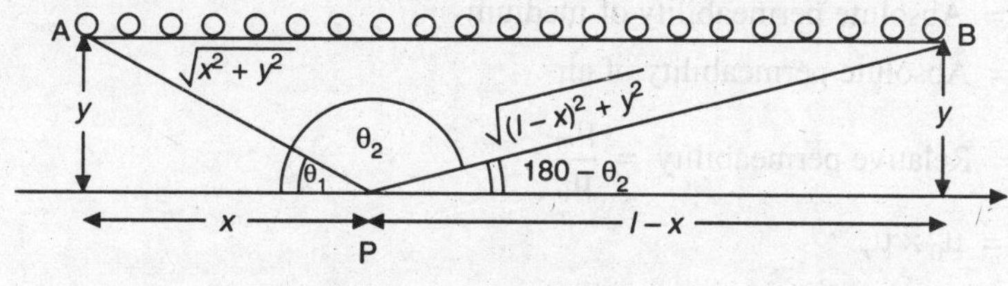

Fig. 31.8

$$B = \frac{\mu_0}{4\pi} \frac{2\pi NI}{l} [\cos \theta_1 - \cos \theta_2] \text{ Wb/sq. m} \quad ...(i)$$

$$= \frac{1}{2} \frac{\mu_0 NI}{l} [\cos \theta_1 - \cos \theta_2]$$

(*i*) ***When the solenoid is infinitely long***

$$\theta_2 = 180° \text{ and } \theta_1 = 0$$

Therefore, magnetic field at *P* which is in the middle of the solenoid

$$B = \frac{\mu_0}{4\pi} \frac{2\pi NI}{l} [1 + 1)$$

$$= \frac{\mu_0}{4\pi} \frac{4\pi NI}{l} = \frac{\mu_0 NI}{l}$$

(*ii*) ***At a point on the end of the solenoid***

$$\theta_2 = 0 \text{ and } \theta_1 = 90°$$

$$\therefore \quad B = \frac{\mu_0}{4\pi} \frac{2\pi NI}{l} = \frac{1}{2} \frac{\mu_0 NI}{l}$$

Thus the field at the end of *a long* solenoid is just one half of that at the centre. In other words, half the magnetic lines of force passing through the central section of a long solenoid pass out through the sides before reaching the ends. Within the solenoid the lines of force are directed along the length of the solenoid parallel to the axis. As the field lines come out of the solenoid these spread out widely before re-entering the solenoid at the other end.

(*iii*) If *y* is the radius of the solenoid, then for any point *P* *at a distance x* from one end and $(l - x)$ from the other end as shown in Fig. 31.8.

$$AP = (x^2 + y^2)^{1/2} \text{ and } BP = [(l - x)^2 + y^2]^{1/2}$$

$$\therefore \quad \cos \theta_1 = \frac{x}{(x^2 + y^2)^{1/2}} \text{ and } \cos (180° - \theta_2) = \frac{l - x}{[(l - x)^2 + y^2]^{1/2}}$$

$$\text{or} \quad \cos \theta_2 = \frac{l - x}{[(l - x)^2 + y^2]^{1/2}}$$

Substituting these values of $\cos \theta_1$ and $\cos \theta_2$ in relation (*i*) we can find the value of *B*.

31.7. Permeability. *The power of conduction of magnetic lines of induction by a medium is called its permeability.* In fact, permeability is a measure of the facility offered by the material to the flow of magnetic lines of force through it. The ferro-magnetic materials like iron, nickel and cobalt are better conductors of lines of magnetic induction and hence have greater permeability than air.

Generally, it is not possible to determine the absolute permeability of a material except for free space (vacuum) or air, which has permeability very close to the free space. Therefore, in actual practice we deal with *relative permeability which is defined as the ratio of absolute permeability of a medium to the absolute permeability of free space (vacuum) or air.*

If μ_m = Absolute permeability of medium

μ_0 = Absolute permeability of air

Than μ_r = Relative permeability $= \dfrac{\mu_m}{\mu_0}$

Hence $\mu_m = \mu_0 \times \mu_r$

For air or vacuum or free space $\mu_r = \dfrac{\mu_0}{\mu_0} = 1.$

Quantitatively permeability of a material is defined as *the ratio of magnetic induction B of the magnetised specimen to the intensity of magnetising field H*

or $$\mu_m = \frac{B}{H}$$

Experiment 31.6. To find the magnetic induction field along the axis of a long closely wound solenoid using a search coil and ballistic galvanometer and hence to find the value of permeability of air μ_0.

Apparatus. A long closely wound solenoid, a ballistic galvanometer with lamp and scale arrangement, a tap key, a search coil, a battery of a few cells, three resistance boxes, two one way keys, a reversing key, a rheostat an ammeter, connecting wires, sand paper etc.

Theory. (*i*) **Field at the centre.** If I is the current through the solenoid in amperes, N the total number of turns and l the length of the solenoid in metres, then

Magnetic induction field B at the centre of the solenoid $= \dfrac{\mu_0}{4\pi} \dfrac{4\pi NI}{l}$ Wb/sq.m $= \dfrac{\mu_0 NI}{l}$ Wb/sq.m.

If a search coil having n turns each of face area a sq. metre and resistance r is placed inside the solenoid along its axis with its face area perpendicular to the field, then

Total magnetic flux linked with the search coil = $B\ n\ a$ Wb

The search coil is connected to the ballistic galvanometer of resistance G. If the current in the solenoid is suddenly reversed total change in magnetic flux linked with the search coil

$$= 2\,Bn\,a \quad \text{Wb}$$

Due to this change in magnetic flux an induced *e.m.f.* or current is set up in the circuit containing the search coil and Ballistic galvanometer which sends a charge $q = \dfrac{2\,B\,n\,a}{G+r}$ and produces a deflection (or corrected throw) θ.

$$\therefore \qquad q = \frac{2Bna}{G+r} = k\,\theta$$

where k is the constant of ballistic galvanometer.

Hence $$B = \frac{(G+r)\,k}{2na}\,\theta \quad \text{Wb/Sq.m}$$

(*ii*) **Variation of field along the axis.** The search coil is moved from one end A of the solenoid to the other end B [Fig. 31.8] and the deflection θ is noted at various points by suddenly reversing.

The current in the solenoid. If now a graph is plotted between the distance from centre of the solenoid on either side and deflection (corrected throw) θ the shape of the graph is as shown in Fig. 31.9.

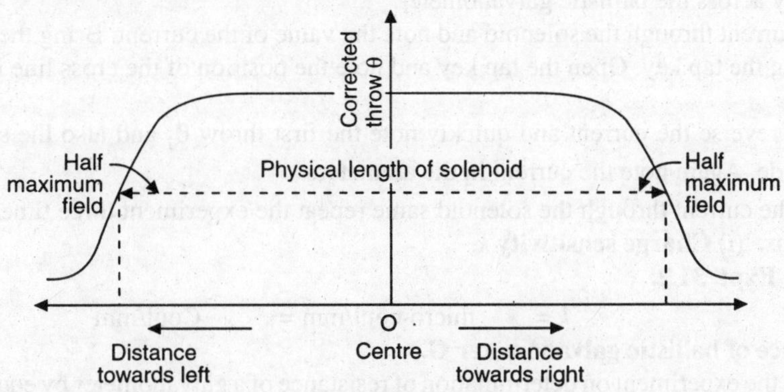

Fig. 31.9

(*iii*) **Permeability of air.** We have seen in (*i*) that magnetic induction field B at the centre of the solenoid is given by $B = \frac{\mu_0 NI}{l}$ Wb/sq.m. where I is the current through the solenoid in amperes, N the total number of turns, l the length of the solenoid in metres and μ_0 the permeability of air

$$\therefore \qquad \mu_0 = \frac{Bl}{NI} \text{ Wb/A-m}$$

We can, therefore, find the permeability of air by measuring the magnetic field B at the centre of a long closely wound solenoid having N turns per metre when a current I amperes passes through it.

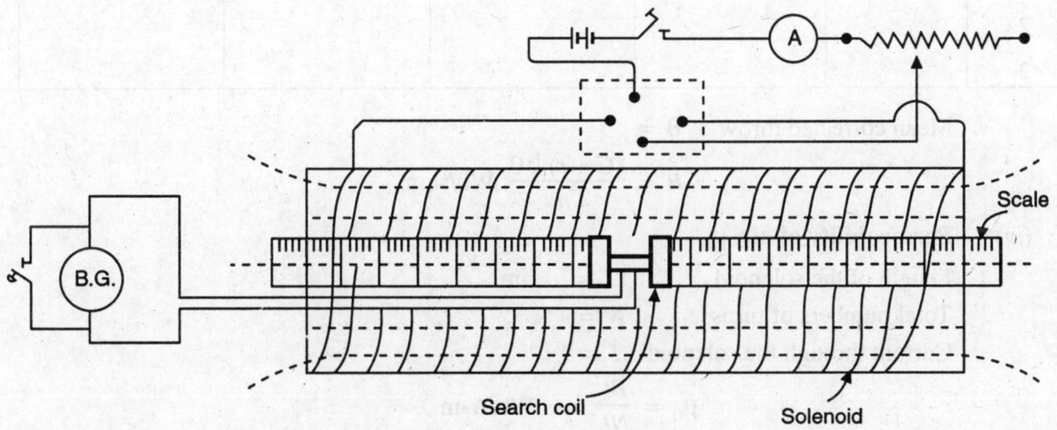

Fig. 31.10

Procedure. 1. Find the charge sensitivity k of the ballistic galvanometer, by steady deflection method, as explained in **Expt. 31.2.**

2. Find the resistance G of the galvanometer by equal deflection method.

3. Connect the two ends of the solenoid to a battery through a reversing key an ammeter and a rheostat. Adjust the current through the coil to a suitable value so that a strong magnetic field is set up along the axis of the coil.

4. Place the search coil in the middle of the solenoid along its axis with its face area perpendicular to the direction of the field. Connect the two ends of the search coil to a ballistic galvanometer. Connect a tap key across the ballistic galvanometer.

5. Pass the current through the solenoid and note the value of the current. Bring the spot of light to rest by pressing the tap key. Open the tap key and note the position of the cross line in the spot of light.

6. Suddenly reverse the current and quickly note the first throw θ_1 and also the second throw θ_2 on the same side. Again note the current in the ammeter.

7. Keeping the current through the solenoid same repeat the experiment three times.

Observations. (*i*) **Charge sensitivity** *k*.

Record as in **Expt. 31.2.**

$\therefore$ $k =$ micro-coul/mm = Coul/mm

(*ii*) **Resistance of ballistic galvanometer G.**

Record as in the experiment on determination of resistance of a galvanometer by equal deflection method.

$\therefore$ $G =$ ohm

(*iii*) **Magnetic induction field B.**

Number of turns of the search coil $n =$

Face area of each turn of search coil $a =$ sq. m

Resistance of the search coil $r =$ ohm

Sl. No.	*Current I*	*Initial Position*	*1st throw* θ_1	*2nd throw* θ_2	*corrected throw* $\theta = \theta_1 + \frac{\theta_1 - \theta_2}{4}$
1					
2					
3					

Mean corrected throw $\theta =$

$$B = \frac{(G+r)\,k\theta}{2\,na} \text{ Wb/sq.m}$$

(*iv*) **Permeability of air** μ_0

Length of the solenoid $L =$ m

Total numbers of turns $N =$

Current through the solenoid $I =$

$$\mu_0 = \frac{Bl}{NI} = \text{ Wb/A-m}$$

(*v*) **Variation of magnetic field along the axis of solenoid.**

8. To study the variation of magnetic field along the axis of the solenoid note the position of the search coil on the scale placed along the axis of the solenoid when placed in the centre (of the solenoid) and the corresponding deflection in the B.G on reversing the current.

9. Displace the search coil 2-3 cm along the axis *towards the left.* Again note the position of the search coil on the scale and corresponding deflection in B.G on reversing the current. Go on repeating the observations by displacing the search coil 2-3 cm at a time till we reach the left end.

10. Again bring the search coil in the centre of the solenoid and repeat the observation by displacing the search coil 2-3 cm at a time till we reach the right end.

Observations

S. No.	Distance along the axis of the solenoid measured from the centre of the solenoid d	Deflection in Ballistic galv. on reversing current			
		Initial position	1st throw θ_1	2nd throw θ_2	Corrected throw $\theta = \theta_1 + \frac{\theta_1 - \theta_2}{4}$
Left	..				
	..				
	..				
Centre	0				
Right	..				
	..				

11. Plot a graph between distance along the axis of the solenoid measured from the centre of the solenoid taken along the *X*-axis and corresponding corrected throw taken along the *Y*-axis with origin at the centre as shown in Fig. 31.9.

Conclusion. The graph shows that

(*i*) The magnetic field has its maximum value at the middle of the solenoid and remains fairly constant for a large distance on both sides of the centre.

(*ii*) The magnetic field at the ends of the solenoid is *half* that at the centre of the solenoid.

Precautions. 1. Same as in **Expt. 31.5**.

2. The plane of the search coil should be perpendicular to the magnetic field.

3. The current should be reversed quickly.

31.8. Alternative method. A very convenient method of studying the variation of magnetic field along the axis of a solenoid is by using a *cathode ray oscillograph* (C.R.O) and a low voltage A.C. source. The solenoid is connected to a low voltage A.C source through a regulating rheostat. The search coil is placed along the axis of the solenoid and its long flexible leads are connected to the *Y* plates of a C.R.O. When the solenoid is connected to the low voltage A.C. source the magnetic field at a point on the axis of the solenoid goes on continuously varying in accordance with the applied

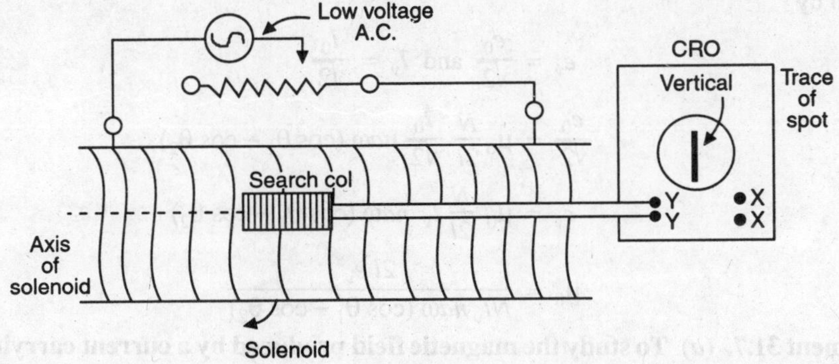

Fig. 31.11

A.C. voltage and an induced *e.m.f* is set up in the search coil which is proportional to the rate of change of solenoidal flux at that point. As the rate of change of flux remains constant because it is determined by the frequency of A.C. mains supply, the length of the trace in the C.R. tube will be proportional to the flux densily at the position of the search coil. Hence a graph between the trace length measured on the m.m scale provided on the cathode ray tube or an affixed transparent paper and the position of the search coil will give the variation of the magnetic flux along the axis of the solenoid.

31.9. Induced e.m.f. in the search coil. When the solenoid is connected to an A.C. source, the magnetic field at a point on its axis changes periodically. This gives rise to an induced *e.m.f.* in search coil which is placed in this varying magnetic field. The value of A.C through the solenoid at any instant is given by

$$I = I_0 \sin \omega t$$

where I = instantaneous value of current ...(*i*)

I_0 = Peak (maximum) value of current

$\omega = 2\pi\nu$ where ν is the frequency of A.C. supply.

The magnetic field at a point on the axis of the solenoid is given by [according to **Art. 31.5**]

$$B = \frac{\mu_0}{4\pi} \cdot \frac{2\pi NI}{l} (\cos \theta_1 - \cos \theta_2)$$

$$= \frac{\mu_0 N}{2l} I_0 \sin \omega t (\cos \theta_1 - \cos \theta_2)$$

When a search coil or a small pick-up coil having n turns and cross sectional area a is placed at this point with its face perpendicular to the magnetic lines of force of the solenoid an induced *e.m.f.* is set up in the coil due to changing flux through the coil given by

$$e = na \frac{dB}{dt} = \frac{\mu_0 N}{2l} I_0 na \frac{d}{dt} \sin \omega t (\cos \theta_1 - \cos \theta_2)$$

$$= \mu_0 \frac{N}{2l} I_0 n a \omega \cos \omega t (\cos \theta_1 - \cos \theta_2)$$

Substituting $\mu_0 \frac{N}{2l} I_0 n a \omega (\cos \theta_1 - \cos \theta_2) = e_0$ we have

$$e = e_0 \cos \omega t$$

$$\therefore \quad e_0 = \mu_0 \frac{N}{2l} I_0 n a \omega (\cos \theta_1 - \cos \theta_2)$$

The measured value of voltage e_v is the *r.m.s.* voltage and the measured value of current I_v is the *r.m.s.* current given by

$$e_v = \frac{e_0}{\sqrt{2}} \text{ and } I_v = \frac{I_0}{\sqrt{2}}$$

$$\therefore \quad \frac{e_0}{\sqrt{2}} = \mu_0 \frac{N}{2l} \frac{I_0}{\sqrt{2}} na\omega (\cos \theta_1 - \cos \theta_2)$$

or $$e_v = \mu_0 \frac{N}{2l} I_v \, na\omega (\cos \theta_1 - \cos \theta_2)$$

Hence $$\mu_0 = \frac{2l \, e_v}{NI_v \, na\omega (\cos \theta_1 - \cos \theta_2)}$$

Experiment 31.7. (*a*) **To study the magnetic field produced by a current carrying solenoid using a search coil and an A.C. source.**

(*b*) **To find the value of permeability of air (μ_0).**

Apparatus. A long closely wound solenoid with three layers of winding of different turns, a search coil (or pick-up coil) having few thousand turns, a step down transformer (24*V*/2*A*), a rheostat (2*A*), A.C. ammeter (2.5 A), A.C. millivoltmeter (0–100 mV), connecting wire, measuring scale etc.

Procedure 1. Draw a diagram showing the scheme of connections (Fig 31.12). Collect the required apparatus, arrange it on the table and make connections accordingly. Re-check the connections thoroughly and show it to the teacher/examiner before connecting the step-down transformer to the A.C. mains supply.

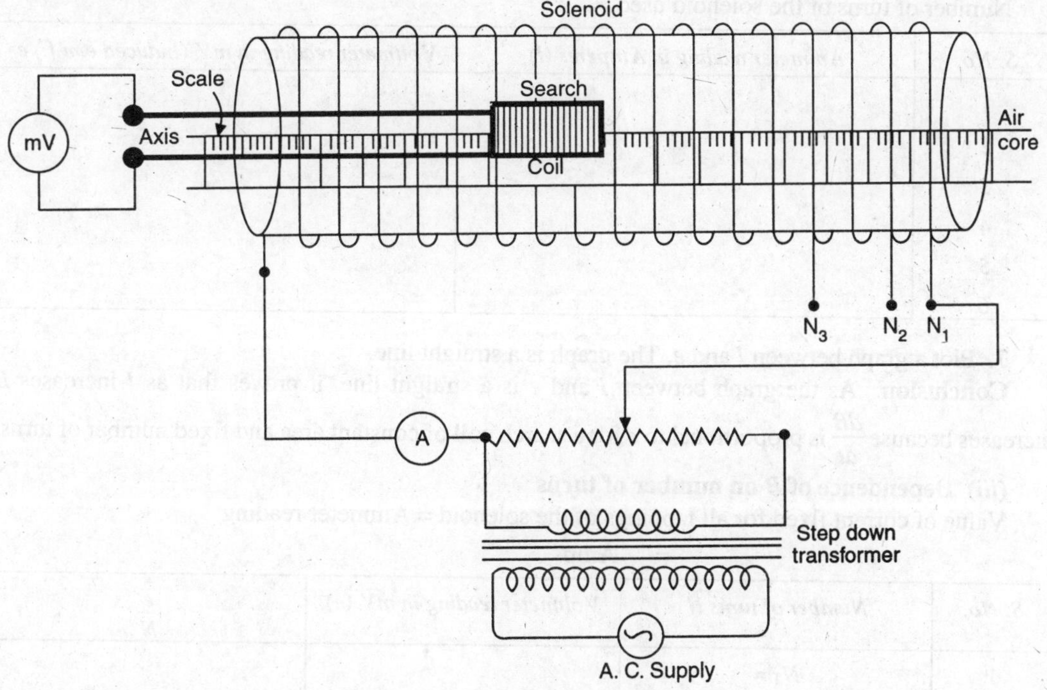

Fig. 31.12

2. Mount the search coil on the rod if it is not already mounted and see that it can be easily and conveniently moved along the axis of the solenoid.

3. Connect one of the tappings of the turns of the solenoid (say N_1) to the variable terminal of the rheostat. Adjust the zero reading of the A.C. ammeter and A.C. milli-voltmeter (if necessary).

4. To study the dependence of *B* the magnetic field on current, place the search coil at the centre of the solenoid. Switch on the transformer and note down the readings of the ammeter and milli-voltmeter. Without changing the number of turns in the solenoid and position of search coil change the value of the current with the help of the rheostat and record the readings of the ammeter and corresponding readings of the voltmeter. Repeat this process at least *six* times by changing the current with the help of the rheostat.

5. To study the effect of number of turns on the magnetic field keep the search coil within the solenoid at its centre and switch on the transformer. Record the readings of the ammeter and milli-voltmeter. Disconnect N_1 tapping and connect N_2 tapping to the rheostat-so that the number of turns changes. Adjust the value of the current with the helps of the rheostat so that *the current flowing with N_2 turns is the same as for N_1 turns.* Note the reading of the milli-voltmeter. Similarly disconnect N_2 and connect N_3. Adjusting the value of the current to be the same. Note down the reading of the millivoltmeter.

6. To study the variation of B with distance along the axis of the solenoid note down the reading of the voltmeter and the ammeter when the search coil is in the *middle* of the solenoid. Let the rheostat be connected to any terminal N_1, N_2 or N_3. This is not to be changed in one complete set of observations. Go on moving the search coil gradually along the axis of the solenoid (by 2-3 cms. at a time) and record the reading of the voltmeter for a **constant current** for various distances from the CENTRE of the solenoid on the left as well as right hand side. Also note the position and distance of the search coil from the centre.

Observations (*i*) Dependence of *B* on current

Number of turns of the solenoid used =

S. No.	*Ammeter reading in Amperes (I)*	*Voltmeter reading in mV (Induced e.m.f.) e*
1		
2		
3		
4		
5		
6		

7. Plot a graph between I and e. The graph is a straight line.

Conclusion. As the graph between I and e is a straight line, it proves that as I increases B increases because $\frac{dB}{dt}$ is proportional to e for a search coil of constant area and fixed number of turns.

(*iii*) Dependence of *B* on number of turns

Value of current fixed for all tappings of the solenoid = Ammeter reading

= Amp.

S. No.	*Number of turns N*	*Voltmeter reading in mV (e)*	$\frac{e}{N}$
1	N_1 =		
2	N_2 =		
3	N_3 =		

Conclusion. Within experimental error $\frac{e}{N}$ is a constant quantity.

(*iii*) Dependence of *B* on the position of search coil along the axis of the solenoid.

Number of turns used N =

Constant current passed = Ammeter reading I = Amp.

S. No.	*Distance of search coil from the centre of solenoid (x) (Left)*	*Voltmeter reading in mV (e)*	*Distance of search coil from the centre of solenoid (x) (Right)*	*Voltmeter reading in mV (e)*
1				
2				
3				
4				
5				
6				

8. Plot a graph between x and e (i.e. B) by taking x along the X-axis and e along the Y-axis. The graph is as shown in Fig. 31.9.

Conclusion. The graph shows that

(i) e and hence the magnetic field B is constant at its maximum value over a considerable length inside the solenoid.

(ii) The magnetic field at the ends of the solenoid is half that at the centre of the solenoid.

(*v*) **Value of permeability of air μ_0.** Frequency of A.C. $\nu = 50$ Hz

Angular frequency $= 2\pi\nu = 100\pi$

Number of turns in the solenoid (used) $N =$

Length of the solenoid $l =$ m

Diameter of the solenoid $= 2y =$ (*i*) .... (*ii*) .... (*iii*) ....

Mean diameter of the solenoid =

Radius of the solenoid $y =$

Number of turns in the search coil $n =$

Area of cross-section of the search coil $a =$

Current through the solenoid $= I_\nu =$ Amp.

S. No.	*Distance of the search coil from*		$\cos\theta_1$	$\cos\theta_2$	*Voltmeter reading* e_ν	
	Left end (x)	*Right end* $l-(x)$	$= \dfrac{x}{\sqrt{x^2+y^2}}$	$= \dfrac{-(l-x)}{\sqrt{(l-x)^2+y^2}}$	mV	Volt

Calculations

$$\mu_0 = \frac{2\,l\,e_\nu}{N I_\nu\, n\, a\,(\cos\theta_1 - \cos\theta_2)}$$

$$= \quad T\,m\,A^{-1}$$

Actual value (from tables) $= 4\pi \times 10^{-7}$ $T\,mA^{-1}$

% age error =

Oral Questions

1. Ballistic galvanometer. (*i*) What is a ballistic galvanometer? (*ii*) What does it measure? (*iii*) How does it differ from a dead beat galvanometer? (*iv*) Which type are you using a moving coil or a moving magnet type? (*v*) Why do you place a key in parallel with the galvanometer? (*iv*) What do you understand by the constant of a ballistic galvanometer? (*vii*) What is damping and on what factors does it depend? (*viii*) What is the use of a tap key connected across the ballistic galvanometer? (*ix*) Define current sensitivity and charge sensitivity. (*x*) What is the relation between the two?

2. Earth inductor. (*i*) What is an earth inductor? (*ii*) How is *e.m.f.* induced in it? (*iii*) What are the factors on which the induced *e.m.f.* depends? (*iv*) If the plane of the earth inductor is placed in the magnetic meridian and then rotated will you get any induced *e.m.f.*? (*v*) Does the charge produced in half a rotation depend upon time?

3. Search coil. (*i*) What is a search coil and what is its use? (*ii*) Why is it made small? (*iii*) What is the use of earth inductor in this experiment? (*vi*) Why should the plane of earth inductor be perpendicular to the magnetic meridian?

4. Magnetic field on the axis of a solenoid. (*i*) What is the value of magnetic induction field at any point on the axis of a closely wound solenoid? (*ii*) What is the value of magnetic induction field at the centre of a long closely wound solenoid? (*iii*) Is the magnetic field on the edges the same as on the central part. (*iv*) What do you mean by permeability of free space? (*v*) What is it value and units? (*vi*) What is relative permeability? (*vii*) What is its value for free space and what are its units? (*viii*) Is the value of magnetic field along the axis of a solenoid at a point the same if the air core is replaced by an iron core (*ix*) If not, what is the value with iron core of relative permeability. μ_r? (*x*) When we use alternating current method to find μ_0, state the formula used?

32

Capacitance

32.1. Capacitance. *Capacitance of a capacitor is defined as the quantity of charge required to set up a unit potential difference between its coatings.*

Let q be the quantity of charge required to set up a potential difference equal to V across a capacitor of capacitance C, then

$$q = CV$$

The unit of capacitance is a *farad* (F). A capacitor has a capacitance of one *farad* if a charge of one coulomb sets up a potential difference of one volt across its coatings. In practice a smaller unit called the *microfarad* is used. One microfarad $\mu F = 10^{-6}$ F.

32.2. Parallel plate capacitor. The capacitance of a parallel plate air capacitor in S.I. units is given by

$$C = \frac{\varepsilon_0 A}{d}$$

where A is the area of either plate (in sq. metres), d the distance between the two plates (in metres) and ε_0 the permittivity of air (or free space) the value of which is 8.85×10^{-12} coulomb^2/newton – m^2 [$C^2 N^{-1} m^{-2}$]

The capacitance of a parallel plate capacitor having a dielectric of dielectric constant k (or relative permittivity ε_r) between the two plates is given by

$$C = \frac{\varepsilon_0 \varepsilon_r A}{d}$$

or

$$C = k \frac{\varepsilon_0 A}{d}$$

Relative permittivity ε_r or dielectric constant k are mere numbers and have no dimensions or units.

32.3. Capacitances in series and parallel. When a number of capacitors are placed in series the total capacitance is given by

$$\frac{1}{C} = \frac{1}{C_1} + \frac{1}{C_2} + \frac{1}{C_3} + \quad \ldots\ldots$$

When a number of capacitors are placed in parallel, the total capacitance is given by

$$C = C_1 + C_2 + C_3$$

The capacitance of two capacitors may be compared by

(*i*) Deflection method and

(*ii*) De-Sauty's null method.

Experiment 32.1. To compare the capacitances of two capacitors by deflection method.

Apparatus. Two capacitors, a morse key, a cell, a moving coil ballistic galvanometer, a two-way key, a tap key, a lamp and scale arrangement, connecting wires, and a piece of sand paper.

Theory. If two capacitors of capacitances C_1 and C_2 are charged to the same potential V, then the charges q_1 and q_2 on the capacitors C_1 and C_2 respectively, are given by

$$q_1 = C_1 V \text{ and } q_2 = C_2 V$$

If the capacitors are allowed to discharge through a ballistic galvanometer one by one and θ_1 and θ_2 are the first throws produced by the two capacitors C_1 and C_2 respectively, then

$$q_1 = C_1 V = k\,\theta_1\left(1 + \frac{\lambda}{2}\right)$$

and

$$q_2 = C_2 V = k\,\theta_2\left(1 + \frac{\lambda}{2}\right)$$

$$\therefore \quad \frac{C_1}{C_2} = \frac{\theta_1}{\theta_2}$$

If a lamp and a scale arrangement is used and d_1 and d_2 are the distances through which the spot of light moves on the scale, then

$$\frac{C_1}{C_2} = \frac{\theta_1}{\theta_2} = \frac{d_1}{d_2}$$

Procedure. 1. Draw a diagram showing the scheme of connections as in Fig. 32.1. Mark the *central* terminal connected to the lever of the morse key as *c*, terminal making contact with the knob on *pressing* as *a* and the third terminal as *b*.

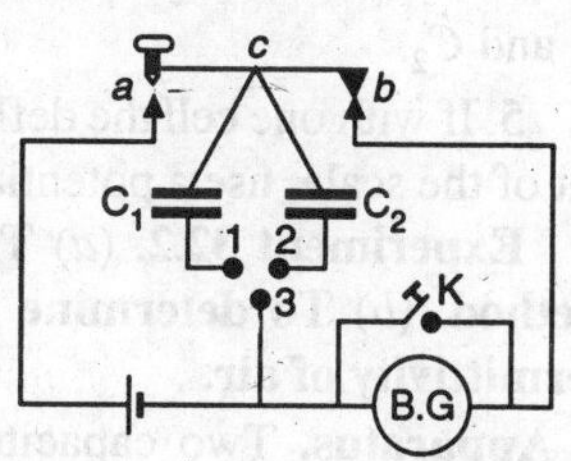

Fig. 32.1

2. Clean the ends of the connecting wires. Connect the terminal *c* of the morse key *through* a capacitor C_1 to the side terminal No. 1 of the two-way key and *through* another capacitor C_2 to the second side terminal No. 2 of the two way key. Connect the common terminal No. 3 of the two-way key through the *moving coil ballistic galvanometer* to the terminal *b* of the morse key. Connect a tap key *K* across the ballistic galvanometer. Connect a cell between the terminal No. 3 of the two-way key and the terminal *a* of the morse key.

3. Place the ballistic galvanometer on a rigid support say a stone table and adjust the levelling screws at the base so that the coil turns freely. Set the lamp and scale arrangement so that a sharp image of the cross-line in the spot of light is obtained nearly in the middle of the scale placed at a distance of one metre from the mirror of the galvanometer.

Note. Earlier capacitor was denoted as a condenser and capacitance as capacity. Some books may still use the older notation.

4. To test the connections put the plug between 1 and 3 in the two-way key. Press the Morse key for an instant and release it. If the spot of light goes beyond the scale, use *potential dividing arrangement with the cell.* Similarly, test by connecting the second capacitor in the circuit.

5. Bring the spot of light to rest *by pressing the tap key K just when the spot of light is at the resting point on the scale.* When the spot is completely at rest note the initial reading. Release the key *K* and put the plug between 1 and 3 in the two-way key. Press the morse key, release it and *note the first throw.* Bring the spot of light to rest and repeat three times.

Take out the plug from the terminals 1 and 3 and introduce between 2 and 3. Repeat the observation with the second capacitor. Take at least three different observations by changing the potential applied to the capacitors.

Note. *The coil of the ballistic galvanometer comes to rest on pressing the key K because an induced current is set up in it on short circuiting it. According to Lenz's law the induced current tends to stop the movement.*

Observations.

Sl. No.		*Scale reading with the capacitors*								
		C_1				C_2				$\frac{C_1}{C_2}=\frac{d_1}{d_2}$
		Initial resting point	*First throw*	*Difference*	*Mean* d_1	*Initial resting point*	*First throw*	*Difference*	*Mean* d_2	
1	(*i*) (*ii*) (*iii*)									
2										
3										

Precautions. 1. A moving coil type ballistic galvanometer should be used.

2. The connecting wires should be short and straight. These should not touch the table.

3. To bring the galvanometer to rest, the tap key placed across the galvanometer should be pressed before taking each observation.

4. The Morse key should be pressed for the same duration for each observation, with capacitors C_1 and C_2.

5. If with one cell the deflection of the ballistic galvanometer with any one of the capacitors goes out of the scale, use a potential dividing arrangement to reduce the charging potential.

Experiment 32.2. (*a*) **To compare the capacitances of two capacitors by de-Sauty's null method.** (*b*) **To determine the value of an air capacitor by de-Sauty's method and to find permittivity of air.**

Apparatus. Two capacitors, two high resistance boxes, a battery, a morse key, a ballistic galvanometer, lamp and scale arrangement and connecting wires.

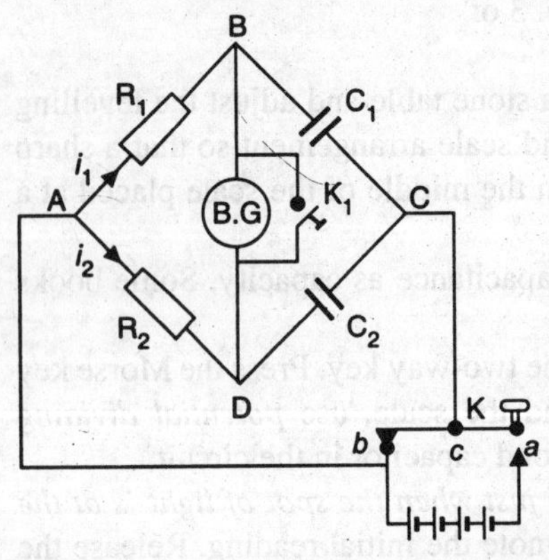

Fig. 32.2

Theory (*a*) The two capacitors of capacitances C_1 and C_2 and the two resistance boxes R_1 and R_2 are connected to form a Wheatstone bridge. A ballistic galvanometer is connected across the point *B* and *D*.

A battery of a few cells is connected across *A* and *C* through a Morse key as shown in Fig. 32.2.

On pressing the key *K* the capacitors C_1 and C_2 are charged and a momentary current flows through the ballistic galvanometer. Similarly on releasing the key the circuit is completed without the battery. The capacitors are discharged and again there is a momentary deflection in the ballistic galvanometer. The values of R_1 and R_2 are so adjusted that on *pressing the morse key or releasing it i.e., on charging or discharging the capacitors there is no deflection in the ballistic galvanometer.*

When a balance point is obtained, the points B and D are at the same potential. Hence the potential difference across the capacitor C_1 is equal to the potential difference across the capacitor C_2 and the potential difference across the resistance R_1 is equal to the potential difference across the resistance R_2. If i_1 and i_2 are the currents through the branches *ABC* and *ADC* respectively during the variable stage, then

$$i_1 R_1 = i_2 R_2$$

or $$\int i_1 R_1 \, dt = \int i_2 R_2 \, dt$$

or $$R_1 q_1 = R_2 q_2$$

or $$\frac{q_1}{q_2} = \frac{R_2}{R_1} \quad ...(i)$$

where q_1 and q_2 are the charges acquired by the capacitors of capacitances C_1 and C_2 respectively.

The potential difference across the capacitor C_2 is equal to the potential difference across the capacitor C_1.

$$\therefore \quad \frac{q_1}{C_1} = \frac{q_2}{C_2}$$

or $$\frac{C_1}{C_2} = \frac{q_1}{q_2} \quad ...(ii)$$

From (*i*) and (*ii*), we have

$$\frac{C_1}{C_2} = \frac{R_2}{R_1}$$

(*b*) If C_2 is a standard capacitor of known. capacitance and C_1 a parallel plate air capacitor, then

$$C_1 = C_2 \frac{R_2}{R_1}$$

Now if the area of either plate of air capacitor is A (sq. metre) and the distance between the plates d (metre), then

$$C_1 = \frac{\varepsilon_0 A}{d}$$

or $$\varepsilon_0 = \frac{C_1 d}{A}$$

Knowing A and d, ε_0 the permittivity of air can be determined.

If there are n parallel plate capacitors connected in *parallel* then

$$C_1 = nC$$

where C is the capacitance of each individual capacitor

$$\therefore \quad C = \frac{\varepsilon_0 A}{d}$$

or $$\frac{C_1}{n} = \frac{\varepsilon_0 A}{d}$$

$$\therefore \quad \varepsilon_0 = \frac{C_1 d}{n A}$$

In this case knowing A, d and n the permittivity of air can be determined.

Procedure 1. Draw a diagram showing the scheme of connections as in Fig. 32.2.

2. Connect the resistance boxes R_1 and R_2 as well as the two capacitors C_1 and C_2 in series. Connect the free terminals of C_1 and R_1 to one terminal and free terminals of C_2 and R_2 to the other terminal of the ballistic galvanometer. Connect the common terminals of R_1 and R_2 to the terminal b

and the common terminals of C_1 and C_2 to the terminal c of the Morse key as shown in Fig. 32.2. Connect the battery of 4 to 5 cells across the terminals a and b of the Morse key. Connect a tap key across the ballistic galvanometer so as to make the coil of galvanometer stationary as and when required.

3. Set the lamp and scale arrangement. Take out a large resistance of the order of 5000 ohm from R_1 and adjust the value of R_2 so that on pressing or releasing the morse key the ballistic galvanometer shows no deflection.

Note. If a perfect balance is not possible find two values of R_2 for which the ballistic galvanometer shows very small deflections of equal magnitude in opposite directions and take their mean.

4. Change the values of R_1 in steps of 2,000 ohm and find the corresponding values of R_2 for no deflection. Take at least five observations.

Observations.

Sl. No.	R_1	R_2 *lies between*		*Mean* R_2	$\frac{C_1}{C_2} = \frac{R_2}{R_1}$
		1	2		
1.					
2.					
3.					
4.					
5.					

$$\text{Mean } \frac{C_1}{C_2} =$$

(*b*) 5. Repeat the experiment by selecting a suitable parallel plate air capacitor for C_1 and a standard capacitor of known value for C_2. Hence calculate $C_1/C_2 = R_2/R_1$, then

Capacitance of standard capacitor $C_2 =$ Farad

$$\therefore \qquad C_1 = C_2 \frac{R_2}{R_1} = \text{Farad}$$

6. Find the number of capacitors connected in parallel forming the whole capacitor C_1. Measure the length and breadth of each capacitor plate with the help of a vernier callipers. Also measure the total distance between the extreme plates. Find the thickness of each capacitor plate with the help of a screw gauge.

Observations. Number of capacitors n =

Length of each capacitor plate = (*i*) ..... (*ii*) ..... (*iii*) ..... (*iv*) ....

Mean length of each plate l =

Breadth of each capacitor plate = (*i*) ..... (*ii*) ..... (*iii*) ..... (*iv*) ....

Mean breadth of each plate b = m

$\therefore$ Area of each capacitor plate $A = l \times b =$ m^2

Thickness of each capacitor plate = (*i*) ..... (*ii*) ..... (*iii*) ..... (*iv*) ....

(with screw gauge)

Mean thickness t = m

Total number of plates = $(n + 1)$ =

Total thickness of plates = $(n + 1)\, t$ = m

Total distance between extreme plates = x = m

$\therefore$ Total separation between n plates = $x - (n + 1)\, t$ = m

Hence average distance between two capacitor plates =

$$d = \frac{x-(n+1)\,t}{n} = \quad \text{m}$$

$\therefore$ Permittivity of air $\varepsilon_0 = \dfrac{C_1 d}{na} =$ Coulomb^2/Newton m^2 [$C^2N^{-1}\ m^{-2}$]

Actual value of $\varepsilon_0 = 8.85 \times 10^{-12}\ C^2\ N^{-1}\ m^{-2}$

% error =

Precautions. 1. The values of R_1 and R_2 should be very high.

2. The movement of ballistic galvanometer coil should be free.

3. The Morse key should be pressed only for a short time to reduce the effect of absorption of charge by the dielectric.

Sources of error. The method is not very accurate though simple. It is very difficult to obtain a perfect balance if the capacitors are not free from dielectric loss. It is only in the case of air capacitors that a perfect balance may be possible.

Alternative method. The method may be modified by using an alternating current. The battery is then replaced by a step-down transformer or an audio-frequency generator and the galvanometer by headphones. The balance is obtained when the sound in the headphones is a minimum. In such a case it is better to use a resistance box of the dial type for R_2 so that the resistance can be varied quickly and the effect on the intensity of sound is observed.

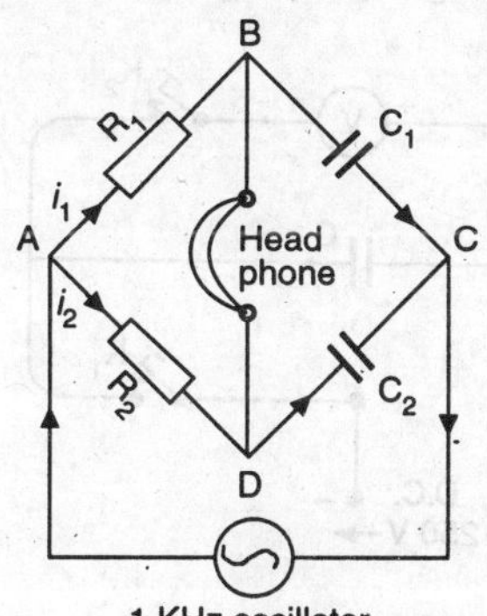

Fig. 32.3

32.4. Discharging of a capacitor. If a capacitor of capacitance C is charged to a potential V_0 and allowed to discharge through a resistance R, then after a time t, the voltage V across the capacitor plates is given by

$$V = V_0\, e^{-t/CR} \qquad ...(i)$$

After a time $t = CR$, the voltage

$$V = V_0\, e^{-1} = \frac{V_0}{e} = \frac{V_0}{2.738} = 0.368\ V_0$$

The time t during which the voltage across the capacitor plates drops to 1/e of its initial value is known as time constant of the circuit.

If the values of resistance R and the time constant t are known, then

$$C = t/R$$

Taking logarithms on both sides of equation (*i*), we get

$$\log_e V = \log_e V_0 - t/CR$$

or
$$\log_e V = \left(-\frac{1}{CR}\right) t + \log_e V_0$$

Thus a graph between $\log_e V$ and t is a *straight line* the slope of which is $-1/CR$.

$$\therefore \qquad -\frac{1}{CR} = \text{slope of the graph}$$

or $$C = -\frac{1}{R \times \text{slope of the graph}}$$

Experiment 32.3. To determine the capacitance of a capacitor by discharging it through a voltmeter.

Apparatus. A capacitor of large capacitance more than 64 μF, a high resistance voltmeter (sensitivity more than 1000 ohm per volt) and range 300 volt, two tap keys, D.C. regulated power supply capable of giving more than 250 volt and a stop-watch.

Procedure 1. Draw a diagram showing the scheme of connections as in Fig. 32.4. and make connections accordingly.

2. Press the tap key K_1 for some time so that the capacitor is charged to its peak value (say 250 V). Constant potential difference across the two plates of the capacitor C is shown by the voltmeter V when the tap key K_2 is also closed. Note the *constant* reading of the voltmeter.

3. Keeping the key K_2 closed open the key K_1 and *at the same time* start the stop watch. As the D.C. source is disconnected the capacitor begins to discharge through the voltmeter. Stop the stop–watch as soon as the voltmeter shows a drop of 25 volt (*i.e.*, when the voltmeter reads 225 volt). Note the time taken.

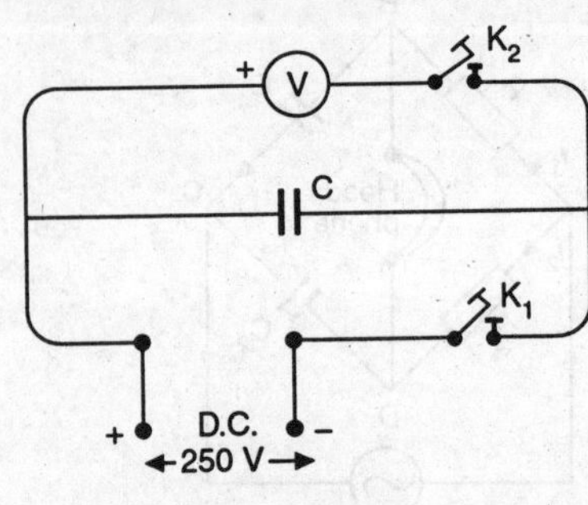

Fig. 32.4

4. Recharge the capacitor to the peak value by pressing again the tap key K_1. Now note the time taken for the voltage across the capacitor to drop to 200 volt (a fall of 50 volt).

5. Repeat the observation by noting the time taken for the peak voltage across the capacitor to drop to 175, 150, 125, 100 (steps of 25 volt), 80, 60 (steps of 20 volt), 50, 40, 30 (steps of 10 volt), 25, 20, 15, 10 and 5 (steps of 5 volt).

This is because due to the exponential nature of the discharge, voltage drop in the beginning is rapid but with fall in voltage becomes more and more slow. This is why at first the time taken for a voltage drop of 25 volt is noted and later on the time for 20, 10 and 5 volt drop respectively is noted.

6. Find the resistance of the voltmeter by half deflection method.

Note. Ordinarily the voltmeter has a resistance of 1000 ohm per volt and a range of 300 volt. Therefore, its resistance is about 3×10^5 ohm = 0.3 meg ohm. The shunt resistance box should be able to give a resistance of that order. A resistance box with resistance plugs of 100, 200, 200, 500, 1000, 2000, 2000 5000, 10,000, 20,000 20,000, 50,000 ohm and 0.1, 0.2, 0.2 and 0.5 meg ohm will serve the purpose.

Observations.

Sl. No.	*Voltage V*	$\log_e V$	*Time t*	*Sl. No.*	*Voltage V*	$\log_e V$	*Time t*
	250				50		
	...				...		
	...				30		
	125				25		
	100				...		
	...				...		
	60				5		

Resistance of the voltmeter.

Sl. No.	*Reading of voltmeter*		*Shunt resist.*	*Resistance of voltmeter*
	Without shunt	*With shunt*		

Mean resistance of the voltmeter $R =$ ohm

7. (*i*) Draw a graph between voltage across the capacitor (V) and time (t), (*ii*) Also draw a graph between ($\log_e V$) and time t.

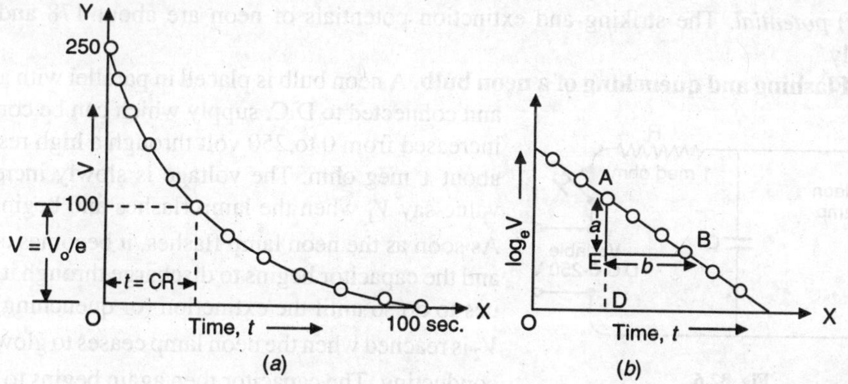

Fig. 32.5

Calculations. (*i*) $V_0 =$ volt

$\frac{V_0}{e} = .368\, V_0 =$ volt

From graph (*i*) $t =$ sec

Resistance of the voltmeter $R =$ ohm

$\therefore$ $C = \frac{t}{R} =$ Farad

$= \mu F$

Select two points A and B on the graph [Fig. 32.5 (*b*)] far apart. From A draw AD perpendicular on X-axis and from B draw BE perpendicular on AD. Measure $AE = a$ and $BE = b$

Slope of the graph $= -\frac{a}{b} =$

$\therefore$ $C = -\frac{1}{R \times \text{slope of the graph}}$

$= \frac{-1}{R \times \left(-\frac{a}{b}\right)} =$ Farad $=$ μF

Precautions. 1. Peak voltage should be noted when both the keys K_1 and K_2 are pressed.

2. The voltmeter used should have a high resistance of the order of 1000 ohm per volt.
3. The capacitance of the capacitor should be very large.
4. The resistance of the voltmeter should be found out by half deflection method.

32.5. Neon lamp. A neon lamp consists of a small glass bulb filled with neon gas at low pressure and fitted with two electrodes. When the electrodes are connected to a D.C. source stray electrons in the gas are attracted towards the positive electrode. As the voltage is increased, the speed of the electrons also increases and at a particular value of the applied voltage the speed of the electrons becomes sufficiently high to ionise the gas molecules by collision. The effect is cumulative and the lamp begins to conduct. This is indicated by the characteristic red glow of neon.

The potential at which the lamp starts to conduct (and begins to glow) is called striking (or flashing) potential. If the potential is now steadily reduced, the lamp does not stop glowing until the potential has dropped substantially (by about 30 to 40 volt) below the striking value. In other words de-ionisation takes place at a lower potential than ionisation potential.

The potential at which the lamp stops to conduct (and ceases to glow) is called the extinction (or quenching) potential. The striking and extinction potentials of neon are about 178 and 130 volt respectively.

32.6. Flashing and quenching of a neon bulb. A neon bulb is placed in parallel with a capacitor and connected to D.C. supply which can be continuously increased from 0 to 250 volt through a high resistance of about 1 meg ohm. The voltage is slowly increased to a value say V_1 when the lamp flashes and begins to glow. As soon as the neon lamp flashes, it becomes conducting and the capacitor begins to discharge through it. It continues to do so until the extinction (or quenching) potential V_2 is reached when the neon lamp ceases to glow and stops conducting. The capacitor then again begins to charge till the flashing potential V_1 is reached when again the lamp flashes and begins to glow. The process is repeated. During the time the capacitor is charging the neon lamp does not glow. In other words, the total time t between two consecutive flashes is equal to the time taken by the voltage first to fall from the flashing potential V_1 to quenching potential V_2 (discharge) and then to rise from V_2 to V_1 (charging). This flashing and quenching time can be determined by noting the time taken by the lamp to produce say 20 consecutive flashes and quenches.

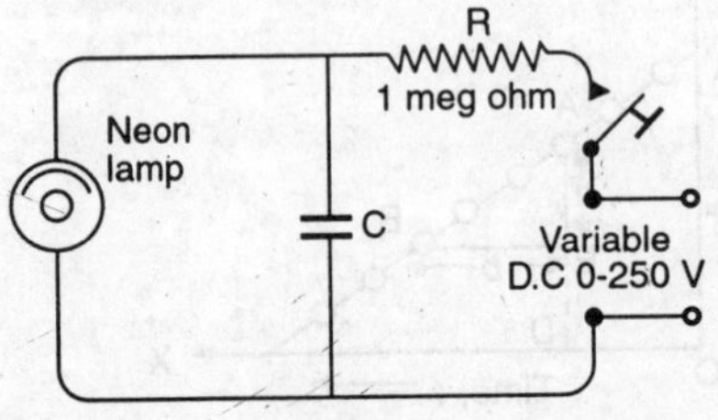

Fig. 32.6

If t_1 is the time taken by the capacitor voltage to fall from V_1 to V_2 and t_2 is the time taken by the voltage to rise from V_2 to V_1, then

$$V_2 = V_1 e^{-t_1/CR}$$

or

$$t_1 = -CR \log_e \frac{V_2}{V_1}$$

and

$$V_2 = V_1 (1 - e^{-t_2/CR})$$

or

$$t_2 = -CR \log_e \left(1 - \frac{V_2}{V_1}\right)$$

$$\therefore \quad t = t_1 + t_2 = C\left[-R \log_e \frac{V_2}{V_1} - R \log_e \left(1 - \frac{V_2}{V_1}\right)\right]$$

As R, V_2 and V_1 have constant fixed values, the function within brackets is a constant. Let it be $= k$

$$\therefore \quad t = kC$$

Thus the graph between t and C is a *straight line.* To plot the graph, the flashing and quenching time t is noted for a number of capacitors of *known* capacitances.

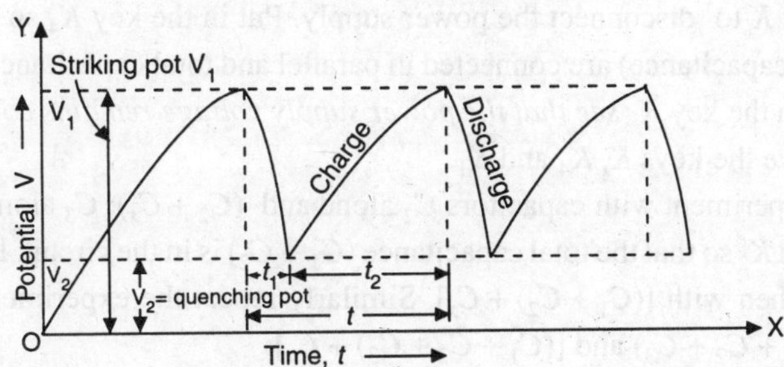

Fig. 32.7

32.7. Determination of capacitance. To find the value of an unknown capacitance a second graph is plotted by noting the flashing and quenching time for the sum of the unknown capacitance and each of the known capacitances. The two graphs are parallel straight lines as shown. To calculate the value of the unknown capacitance draw a straight line AB parallel to X-axis corresponding to any value of flashing and quenching time cutting the two graphs at A and B respectively. If C_1 and C_2 are the values of the known capacitances corresponding to the points A and B respectively C_x the unknown capacitance and C_s the stray capacitance of the circuit, then

Fig. 32.8

Total capacitance corresponding to point A

$$= C_s + C_1 + C_x$$

and total capacitance corresponding to point B

$$= C_s + C_2$$

Since the flashing and quenching time for both is the same

$$\therefore \quad k\,(C_s + C_1 + C_x) = k\,(C_s + C_2)$$

or

$$C_x = C_2 - C_1$$

Experiment 32.4. To find the capacitance of a capacitor using flashing and quenching of a neon lamp.

Apparatus. A capacitor of unknown capacitance, capacitors of known capacitance (say 0.1, 0.2, 0.5 μF), a neon flashing lamp, a resistance of about 1 meg ohm, regulated D.C. power supply capable of giving upto 250 volts and 5 keys.

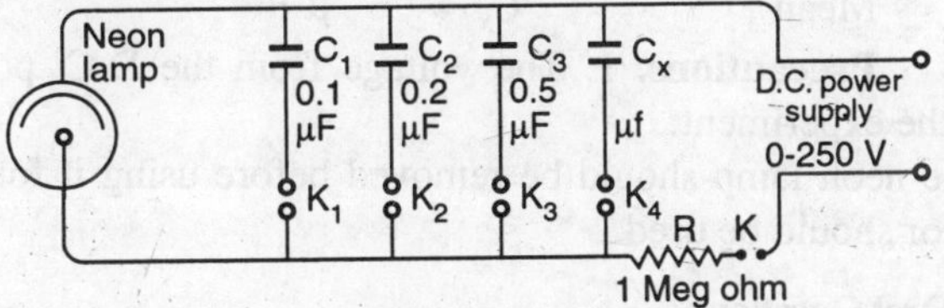

Fig. 32.9

Note. Neon lamps intended to be connected direct to H.T. have a ballast resistor of about 2000 ohm sealed into the cap to limit the current. This must be removed for this experiment. Alternatively small neon flash bulbs without any resistor sealed in are available from radio dealers.

Procedure. 1. Draw a diagram showing the scheme of connections as in Fig. 32.9 and make connections accordingly.

2. Connect the capacitance C_1 in the circuit by putting in the key K_1. Put in the key K and increase the power supply voltage *slowly* till the neon lamp *just begins to flash.* As it is connected in parallel

with capacitance it will start flashing and quenching. Note the flashing and quenching time for 20 flashes.

Remove the key K to disconnect the power supply. Put in the key K_4 so that the capacitors C_1 and C_x (of unknown capacitance) are connected in parallel and total capacitance is equal to their sum $C_1 + C_x$. Again put in the key K *(see that the power supply voltage remains constant).* Note the time of 20 flashes. Remove the keys K, K_4 and K_1.

3. Repeat the experiment with capacitors C_2 alone and $(C_2 + C_x)$; C_3 alone and $(C_3 + C_x)$. Now put in the keys K_1 and K_2 so that the total capacitance $(C_1 + C_2)$ is in the circuit. Repeat the experiment with $(C_1 + C_2)$ and then with $[(C_1 + C_2) + C_x]$. Similarly repeat the experiment with $(C_1 + C_3)$ and $[(C_1 + C_3) + C_x]$, $(C_1 + C_2 + C_3)$ and $[(C_1 + C_2 + C_3) + C_x]$.

Observations.

1 Sl. No.	*2* Known capacitance	*3* Time for 20 flashes without C_x	*4* Flashing and quenching time t	*5* Time for 20 flashes with C_x	*6* Flashing and quenching time t'
	$C_1 =$				
	$C_2 =$				
	$C_3 =$				
	$C_1 + C_2 =$				
	$C_1 + C_3 =$				
	$C_1 + C_2 + C_3$				

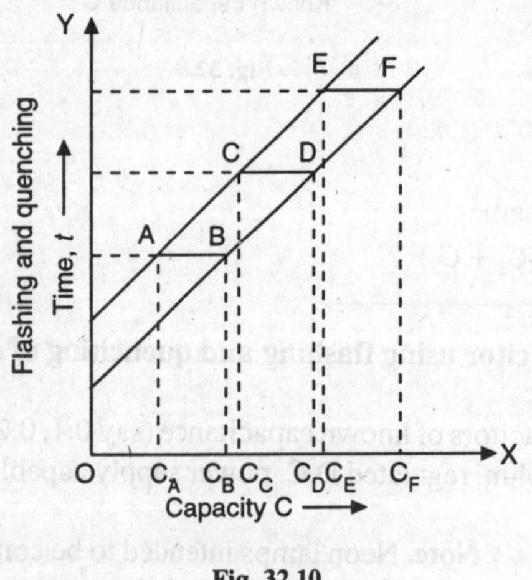

Fig. 32.10

4. Plot two graphs (*i*) between values of capacitance in column 2 taken along X-axis and flashing and quenching time t in column 4 (without unknown capacitance) and (*ii*) between values of capacitance in column 2 and quenching and flashing time t' in column 6 (with unknown capacitance) taken along the Y-axis. For *three* different values of flashing and quenching time draw three straight lines parallel to X-axis cutting the two graphs at A and B, C and D, and E and F respectively.

The unknown capacitance

$$C_x = AB = C_B - C_A$$
$$= CD = C_D - C_C$$
$$= EF = C_F - C_E$$

Mean $C_x =$ μ F

Precautions. 1. The voltage from the D.C. power supply should remain constant throughout the experiment.

2. The resistor sealed in the cap of the neon lamp should be removed before using it for the experiment, otherwise a bulb without resistor should be used.

Oral Questions

1. Capacitance (*i*) Define capacitance and state its units. (*ii*) What are the factors on which the capacitance of a parallel plate capacitor depend? (*iii*) What do you mean by an air capacitor? (*iv*) What other dielectrics are used in a capacitor? (*v*) What is the capacitance of a parallel plate air capacitor? (*vi*) What is the capacitance of a parallel plate capacitor having a dielectric of dielectric constance k (or relative permittivity ε_r)? (*vii*) What are the units of permittivity? (*viii*) What are the units of dielectric constant (or relative permittivity)? (*ix*) Two capacitors are placed in series. What will be the total capacitance? Will it increase or decrease? (*x*) Two capacitors are placed in parallel. What will be the total capacitance? Will it increase or decrease?

2. Comparison of capacitances. (*i*) In the deflection method can you use two separate keys instead of one morse key? (*ii*) Why is the key pressed for a short time only? (*iii*) Why it is not possible to get a perfect balance point in de-Sauty's method? (*iv*) In de-Sauty's method can you replace D.C. battery by A.C.? If so, how? (*v*) State the relation for comparison of capacitances in de-Sauty's method. (*vi*) What is the value of ε_0?

3. Neon lamp. (*i*) What is a neon lamp? (*ii*) What is striking potential? (*iii*) What is quenching potential.

4. Flashing and quenching of a neon bulb. (*i*) Define time constant of a circuit containing a resistance and capacitance. (*ii*) show that the graph between $\log_e V$ and t is a straight line. (*iii*) What do you mean by flashing and quenching time? (*iv*) Show that a graph between flashing and quenching time t and capacitance C is a straight line.

33

Electromagnetic Induction

33.1. Magnetic induction $(\vec{B})$. If a charge q moves with a velocity $\vec{v}$ in a magnetic induction field $\vec{B}$, the force $\vec{F}$ experienced by the moving charge is given by

$$\vec{F} = q\,(\vec{v} \times \vec{B}) = q\,v\,B \sin\theta$$

where θ is the angle between $\vec{v}$ and $\vec{B}$.

If $q = 1$, $v = 1$ and $\theta = 90°$, then $B = F$.

Thus magnetic induction field at a point is defined as the force experienced by a unit charge moving with unit velocity at right angles to the magnetic field. The S.I unit of magnetic induction is a *Tesla* (*T*).

The magnetic induction is said to be 1 Tesla if one coulomb of charge moving with a velocity of 1 ms^{-1} *at right angles to the magnetic field experiences a force of one Newton.*

33.2. Magnetic flux. The magnetic flux through any surface placed in a magnetic field is measured by the total number of magnetic lines of induction passing through the surface and is represented by the symbol ϕ.

Consider a plane surface of area A in a magnetic field. In vector notation this area can be represented by a vector $\vec{A}$ whose magnitude is A and direction is along the normal to the surface area as shown in Fig. 33.1(*a*). If the magnetic induction $\vec{B}$ makes an angle θ with the area vector $\vec{A}$, then the magnetic flux ϕ linked with the given surface is

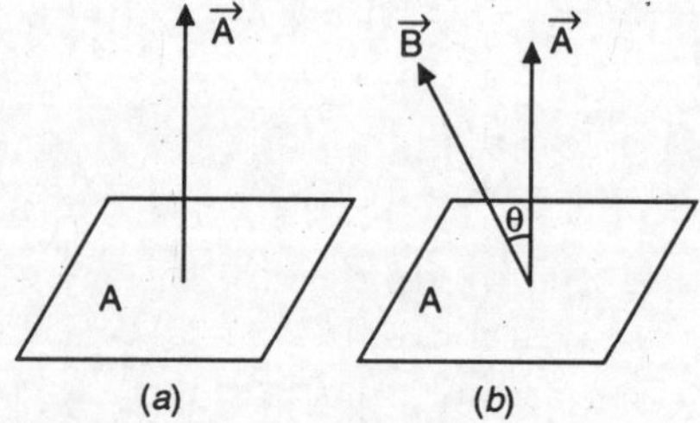

Fig. 33.1

$$\phi = \vec{B} \cdot \vec{A}$$

or $$\phi = BA \cos\theta$$

(*i*) When the magnetic field is perpendicular to the surface i.e it is along the normal to the surface, then $\theta = 0°$

$$\therefore \quad \phi = BA \cos 0° = BA$$

Thus we define magnetic flux linked with a given surface area as the product of the surface area and normal component of the magnetic field acting on that area.

(*ii*) When the magnetic field is parallel to the surface, then $\theta = 90°$

$$\therefore \quad \phi = BA \cos 90° = 0.$$

Positive and negative flux. Magnetic flux is taken to be *positive* if the direction of normal to the area points in the direction of the field *i.e.* $\theta = 0$ and *negative* if the direction of normal to the area points in a direction opposite to that of the magnetic field *i.e.* $\theta = 180°$.

Unit of magnetic flux. The S.I. unit of magnetic flux is a Weber (Wb).

Magnetic flux is said to be one Weber if a uniform magnetic flux of one Tesla acts normally over a plane area 1 m^2.

33.3. Electromagnetic induction. Whenever the magnetic flux linked with a circuit changes an induced *e.m.f.* or current is set up in it. This phenomenon is known as *electro-magnetic induction.* Faraday studied the phenomenon of electro magnetic induction and gave the following two laws :–

Faraday's laws of electro-magnetic induction (*i*) *Whenever the magnetic flux linked with a circuit changes an induced electro-motive force (or current) is set up in it which lasts only for the time the change is actually taking place.*

(*ii*) *The magnitude of induced e.m.f. (or current) is directly proportional to the rate of change of magnetic flux linked with the circuit.*

If $d\phi$ is the change in magnetic flux in a time dt, then

$$e = -\frac{d\phi}{dt}$$

The *negative* sign indicates that the induced *e.m.f.* opposes the change in magnetic flux.

If a circuit or coil has N number of turns, then

$$e = -N\frac{d\phi}{dt}$$

If the magnetic flux changes from ϕ_1 to ϕ_2 in time t, then *average e.m.f.* induced is given by

$$e = -N\frac{\phi_2 - \phi_1}{t}$$

In S.I. units e is in volts, magnetic flux ϕ in *Webers,* and time t in seconds.

If R is the resistance of the closed circuit, then

$$i = \frac{e}{R}.$$

33.4. Lenz's law. This law gives the direction of induced *e.m.f.* It states that '*The direction of induced e.m.f. (or current) is such that it always opposes the cause that produces it*'.

This is in accordance with the law of conservation of energy.

Consider a coil of a few turns to the ends of which a sensitive galvanometer has been connected. When we bring the *north* pole of a magnet quickly near the upper face A, the induced current is found to flow in the anti-clockwise direction. According to Ampere's rule, the upper face A should become a north pole. This will repel the north pole of the approaching magnet and thus oppose the motion of the magnet as predicted by Lenz's law. Work has to be done against this opposing force of repulsion to move the magnet forward. However, when the north pole of the magnet is pulled up, the induced current flows in the clock wise direction. Thus the face A should become a south pole and attract the north pole of the receding magnet again opposing the motion. Work is again done against this force of attraction in moving the magnet away. It is this mechanical work done on the magnet which is stored in the coil in the form of electrical energy.

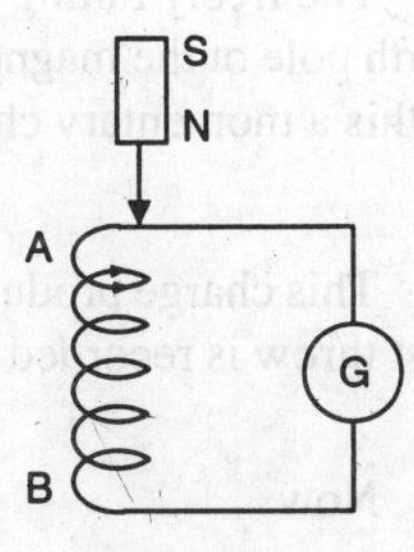

Fig. 33.2

This proves that Lenz's law is in accordance with the law of conservation of energy.

Experiment 33.1. To study the induced e.m.f. as a function of the velocity of the magnet (simple method).

Apparatus. A piece of soft iron and coil forming an electromagnet, A battery of six cells, a permanent bar magnet, two tapping keys, A wooden clamp stand with heavy base, a coil with large number of turns and wide cross-section and a ballistic galvanometer with lamp and scale arrangement.

Theory. When a small permanent bar magnet moves through a coil consisting of a large number of turns, then the magnetic flux linked with the coil due to the magnetic field produced by the magnet *increases* as the magnet approaches the coil Fig. 33.3(*a*) and the field *decreases* as the magnet moves away from the coil Fig. 33.3.(*b*). The rate of change of magnetic flux linked with the coil and hence the induced *e.m.f.* depends upon the velocity of the magnet.

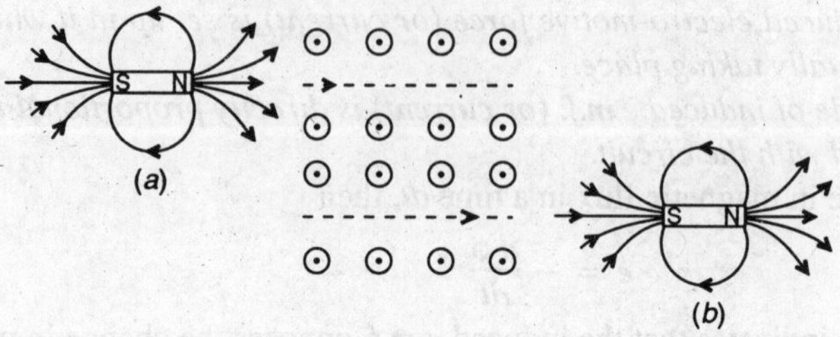

Fig. 33.3

According to Faraday's second law of electro-magnetic induction, the induced *e.m.f.* set up in a coil due to the motion of the magnet is given by

$$e = -\frac{d\phi}{dt}$$

But $\frac{a\phi}{dt}$, the rate of change of magnetic flux depends upon the relative motion between the coil and the magnet. If v is the relative velocity

$$\frac{d\phi}{dt} \propto v$$

$\therefore \qquad e \propto v$

If the magnet is allowed to fall through a height h, then velocity of free fall is given by

$$v = \sqrt{2gh} \text{ or } v \propto \sqrt{h}$$

The freely falling magnet is made to enter the coil placed just below the magnet. As soon as the north pole of the magnet enters the coil an induced *e.m.f.* and hence a current is set up in the coil. Due to this a momentary charge is induced, given by

$$q = \int I\,dt$$

This charge produces a throw in the ballistic galvanometer connected to the ends of the coil. The first throw is recorded on the scale using lamp and scale arrangement

Now $\qquad q = k\theta\left(1+\frac{\lambda}{2}\right)$

where k is the constant of the ballistic galvanometer and λ the logarithmic decrement

or $\qquad \int I\,dt = k\,\theta\left(1+\frac{\lambda}{2}\right)$

or $\qquad \int \frac{e}{R}\,dt = k\,\theta\left(1+\frac{\lambda}{2}\right)$

where e is the induced *e.m.f.* and R the resistance of the circuit.

Thus $\qquad e \propto \theta$

But $\qquad e \propto v \propto \sqrt{h}$

$\therefore \qquad \sqrt{h} \propto \theta$

Hence a graph between $\sqrt{h}$ and θ should be a straight line.

Procedure. 1. Draw a diagram showing the scheme of connections and set the apparatus as shown. The arrangement consists of a thick soft iron bar B held inside a coil C of thick insulated copper wire of large number of turns so that when the key K_1 is pressed it produces a strong magnetic field to hold the magnet NS vertically in position. This arrangement forming the electro magnet is held tightly in a wooden clamp with a heavy base for stability.

2. Connect the coil C to the battery through a tapping key K_1. Arrange a second coil S just below the magnet. Connect it to a ballistic galvanometer B.G. Connect a tap key K_2 in parallel with B.G. Set up a lamp and scale arrangement to get the deflection produced in B.G. and obtain the spot of light at the centre of the scale.

3. Press the key K_1 so that the electro-magnet is activated and the soft iron bar B is magnetised. It will now hold the magnet vertically in position.

4. Measure the height of the lower end of the pole of the magnet from the centre of the coil S placed below, with the help of a metre scale.

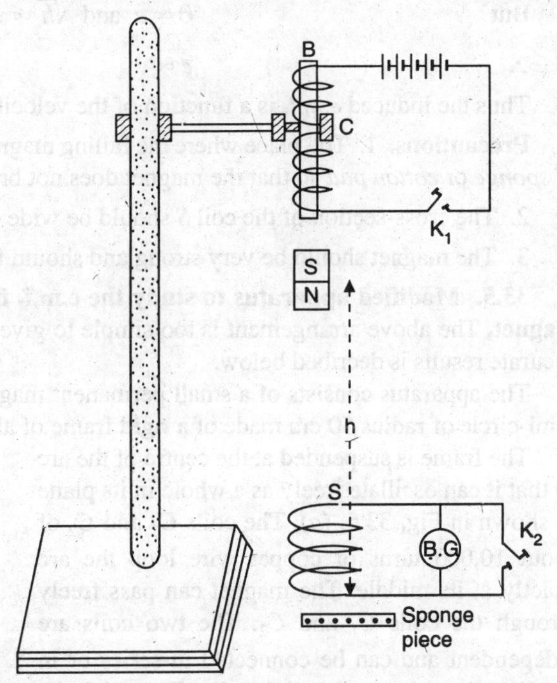

Fig. 33.4

5. Now release the key K_1 so that the magnet is released and falls freely into the coil S below. As soon the pole of the magnet approaches the coil an induced *e.m.f.* is set up in the coil which sends a momentary charge through the B.G. Note the *first* throw on the scale.

6. Repeat the experiment by changing the height of fall of the magnet as well as the polarity of the magnetic poles of the falling magnet.

Observations

S. No.	*Height of fall h*	$\sqrt{h}$	*Ballistic Galvanometer reading*						*Mean Deflection*
			North pole falling			*South pole falling*			$\theta = \frac{\theta_1 + \theta_2}{2}$
			(i)	*(ii)*	*mean* θ_1	*(i)*	*(ii)*	*mean* θ_2	
1									
2									
3									
4									
5									
6									

7. Draw a graph between $\sqrt{h}$ and θ. The graph is found to be a straight line.

Conclusion. As the graph is a straight line, therefore

$$\theta \propto \sqrt{h}$$

But $$\theta \propto e \text{ and } \sqrt{h} \propto v$$

$\therefore$ $$e \propto v$$

Thus the induced *e.m.f.* is a function of the velocity of the magnet.

Precautions. 1. The place where the falling magnet is to strike should be made *soft* with a piece of *sponge* or *cotton pad* so that the magnet does not break.

2. The cross-section of the coil *S* should be wide enough to let the magnet fall freely.

3. The magnet should be very strong and should fall vertically.

33.5. Modified apparatus to study the e.m.f. induced as a function of the velocity of the magnet. The above arrangement is too simple to give reliable results. A modified arrangement for accurate results is decribed below.

The apparatus consists of a small permanent magnet *SN* mounted at the middle of the arc of a semi-circle of radius 50 cm made of a rigid frame of aluminium.

The frame is suspended at the centre of the arc so that it can oscillate freely as a whole in its plane as shown in Fig. 33.5. (*a*). The coils C_1 and C_2 of about 10,000 turns of copper wire loop the arc exactly at its middle. The magnet can pass freely through the coils C_1 and C_2. The two coils are independent and can be connected in series or in parallel or can be used individually. The semi-circular arc is graduated in cm with zero in the middle. The diametric arm is also graduated in cm with zero in the centre. In the normal position of rest the zero mark on the semi-circular scale, the centre of the magnet and the centre of the coils C_1 C_2 all coincide. This is seen by a fixed reference mark. Two weights w_1 and w_2 are provided on the diametric arm. The time period of the aluminium frame can be changed by adjusting the position of the weights. The time period can be varied from about 1.5 to 3.0 second.

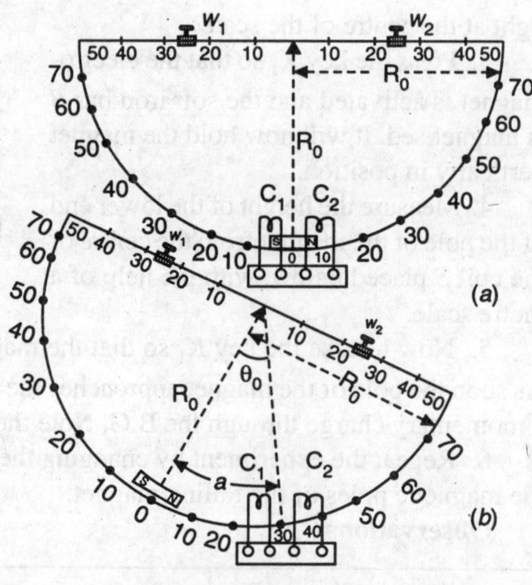

Fig. 33.5

The magnet is placed at the middle of the semi-circular arc. The arc is taken to one side from its mean position as shown in Fig. 33.5. (*b*) and released so that it begins to oscillate freely.

As the magnet starts from a point far away from the coil, moves through it and recedes, the magnetic field through the coil changes from a small value, increases to its maximum and becomes small again. The changes in magnetic flux thus produced induce *e.m.f.* in the coil. In fact, the field at the coil is substantial only when the magnet is very near the coil. The speed of the magnet is also the largest when it approaches very close to the coil as then it is approximately in the *mean position* of the oscillation. The speed again decreases as the magnet moves away from the coil and the magnetic field at the coil also becomes weaker and weaker, till the magnet begins its journey in the backward direction.

Thus the magnetic field changes very slowly with time when the magnet is far away and rapidly increases when it is close to the coil. Fig. 33.6. roughly shows the manner in which $\overrightarrow{B}$ at the coil changes with time. The centre of the hump refers to the time when the magnet is in the coil and the flat portion at the top corresponds to the finite length of the magnet.

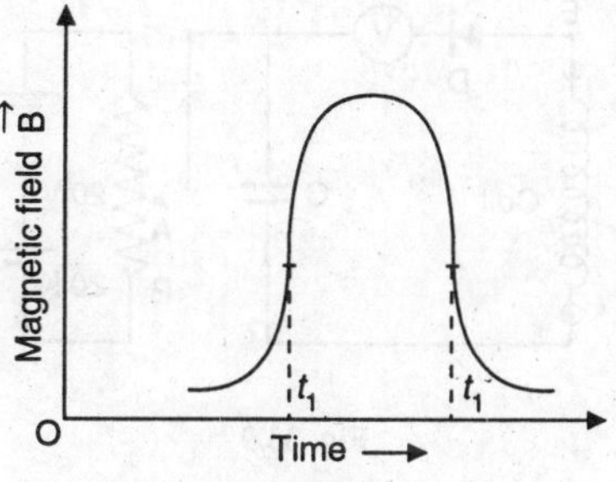

Fig. 33.6

The magnetic flux ϕ will depend upon magnetic field $\overrightarrow{B}$ and the dimensions of the coil *i.e.* the number of turns N and area of cross-section A of the coil. To a first approximation

$$\phi = BNA$$

As ϕ is proportional to $\overrightarrow{B}$, the variation of ϕ with time t is also represented by the graph in Fig. 33.7. Since the induced *e.m.f.* is proportional to $\frac{d\phi}{dt}$ and the value of $\frac{d\phi}{dt}$ is largest at times t_1 and t_2 when the slope is maximum the magnitude of induced *e.m.f.* e is the largest at these times. According to Lenz's law e is *negative* when ϕ is *increasing* at time t_1 and *positive* when ϕ is *decreasing* at time t_2. The sequence of these two pulses; one negative and the other positive occurs during just half a cycle when the magnet moves from one extreme to the other. On the return swing of the magnet it is repeated, being negative at t_2 and positive at t_1.

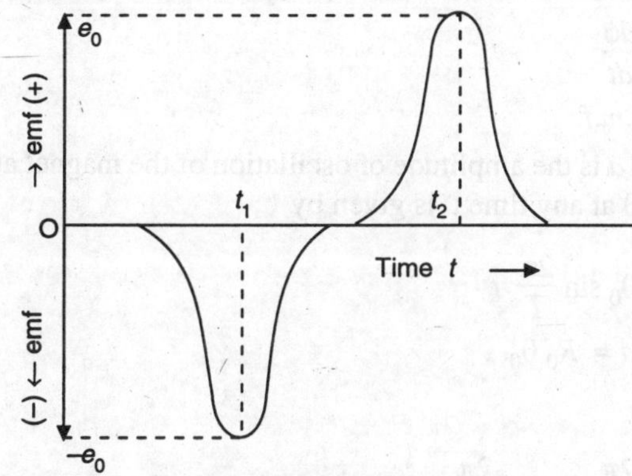

Fig. 33.7

33.6. Measurement of e.m.f. developed (*i*) In order to measure the alternating *e.m.f.* developed in the coil we use the method of charging a capacitor through a diode and measure the voltage developed cross the capacitor at a convenient time. The scheme of connections is shown in Fig. 33.8. The resistance R includes the resistance of the coil the forward resistance of the diode D and any additional resistance that may have to be introduced. Its value is of the order of 500 ohm. The capacitor C is of the range of 100 μf. The charging time RC is adjusted to about 40 milli-second. This time is somewhat larger than the time during which the *e.m.f.* in the coil is generated, so that the capacitor does not charge up to the peak value in a single swing and may take about 10 oscillations to do so. When the capacitor is fully charged to the peak value of the *e.m.f.* generated in the coil, the milli-ammeter (mA) in the circuit stops showing any deflection because the charging current ceases to flow.

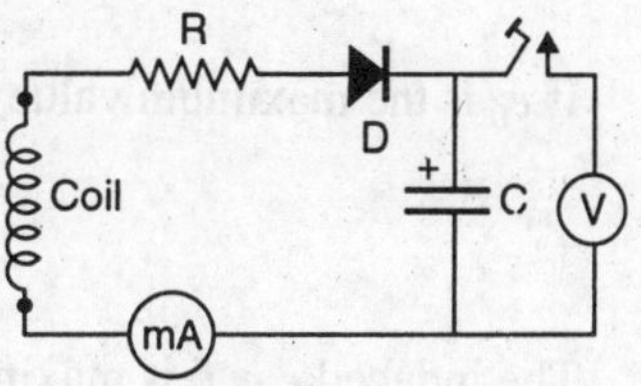

Fig. 33.8

(*ii*) The peak value of *e.m.f.* generated in the coil can also be measured by the null method as shown in Fig. 33.9.

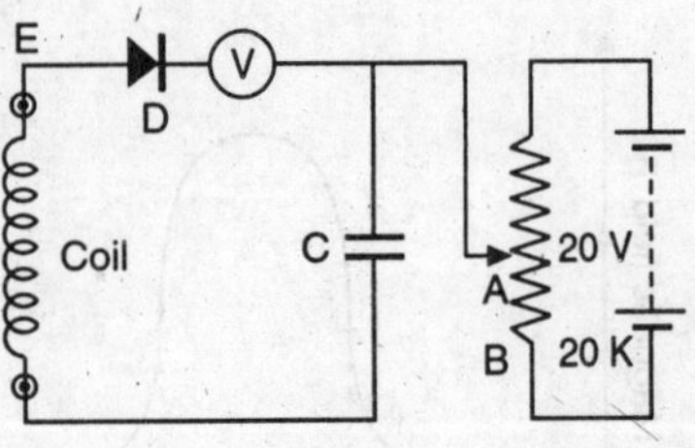

Fig. 33.9

The peak value of the alternating *e.m.f.* developed in the coil is compared with the D.C. voltage applied through a potentiometer arrangement. Suppose the potentionmeter is adjusted to give a D.C. voltage say x volts at A, then a current will flow through the voltmeter only when the point E is at a higher potential than A *i.e.* if the peak voltage developed in the coil is greater than x. The voltmeter will record a 'kick'. If the peak voltage is less than x there will be no 'kick'. If, therefore, the potential of the point A is so adjusted that the voltmeter just ceases to show any deflection (or kick), then

The peak voltage developed in the coil = Voltage across AB.

If we measure the voltage across AB at this stage, we get the peak voltage developed in the coil.

Here, the capacitor does not play a significant role.

Theory. When the magnet oscillates an induced *e.m.f.* is generated in the coil given by

$$|e| = \frac{d\phi}{dt} \qquad ...(i)$$

where $|e|$ is the magnitude of the induced *e.m.f.*

The oscillations are simple harmonic. If a is the amplitude of oscillation of the magnet and T the time period, then the angular displacement θ at any time t is given by

$$\theta = \theta_0 \sin \frac{2\pi}{T} t \qquad ...(ii)$$

where θ_0 is the angular amplitude given by $a = R_0 \theta_0$.

Differentiating, we get

$$\frac{d\theta}{dt} = \frac{2\pi}{T} \theta_0 \cos \frac{2\pi}{T} t \qquad ...(iii)$$

Relation (*i*) can be put in the form

$$|e| = \frac{d\phi}{dt} = \frac{d\phi}{d\theta} \frac{d\theta}{dt}$$

If e_0 is the maximum value of induced *e.m.f.* then

$$|e_0| = \left(\frac{d\phi}{d\theta}\right)_{max} \left(\frac{d\theta}{dt}\right)_{max}$$

The induced *e.m.f.* is maximum when the magnet is in the mean position *i.e.* $\theta = 0$

At $\theta = 0$ $\qquad \sin \frac{2\pi}{T} t = 0$

and $\qquad \cos \frac{2\pi}{T} t = 1$

Substituting $\cos \frac{2\pi}{T} t = 1$ in (*iii*), we have

$$\left(\frac{d\theta}{dt}\right)_{max} = \frac{2\pi}{T}\theta_0$$

$$\therefore \qquad |e_0| = \left(\frac{d\phi}{d\theta}\right)_{max} \frac{2\pi}{T}\theta_0 \qquad ...(iv)$$

From relation (*iv*), we find that for a given magnet and coil system

$$e_0 \propto \theta_0$$

and

$$e_0 \propto \frac{1}{T}$$

$\frac{2\pi}{T}\theta_0$ gives the maximum value of angular velocity. If R_0 is the radius of the semi-circular coil, then

Maximum linear velocity $\quad v_{max} = \frac{2\pi}{T}\theta_0 R_0$

or $\quad v_{max} = \frac{2\pi}{T}(R_0\theta_0)$

A graph between $|e_0|$ and v_{max} will be a straight line as shown.

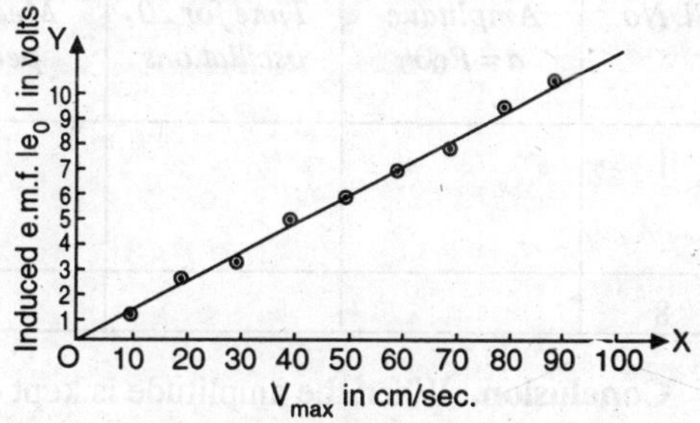

Fig. 33.10

Experiment 33.2. To study the induced *e.m.f.* as a function of velocity of the magnet.

Apparatus. Apparatus for the study of electromagnetic induction, a small strong permanent magnet, a stop-watch, circuit arrangement for measuring the peak value of induced *e.m.f.* etc.

Procedure 1. Mount the small permanent magnet at the middle point of the semi-circular arc and suspend the rigid aluminium frame containing the semi-circular arc from its centre so that the whole frame can oscillate freely through the coil. Level, if necessary.

2. Adjust the position of the two weights on the diameter arm of the arc to have minimum time period.

3. Connect the terminals of the coil to the diode circuit to note the peak voltage generated.

4. Take the magnet carrying arc to one side so that amplitude of vibration is about 20 cm (It can be noted from the graduation on the semi-circular arc) and release the arm. Note the time of 20 oscillations.

5. Repeat thrice keeping the amplitude the same and find the time period. Also note the peak voltage after 20 oscillations, every time.

6. Repeat the experiment after changing the amplitude and take at least 8 readings.

7. Now change the time period by adjusting the position of the weights on the diameter arm. Take at least three readings for each position keeping amplitude constant.

Take at least such 8 observations by changing the time period by adjusting the position of weights but keeping the amplitude the same for all the sets of observations.

(A) Time period constant, amplitude variable.

Mean position of the centre of the magnet =

Radius of the semi-circular arc R_0 = cm

Sl. No.	Amplitude $a = R_0 \theta_0$	Time for 20 oscillations	Mean time period (T)	e_0	$\frac{e_0}{a} = \frac{e_0}{R\theta_0}$	Linear vel. $v = \frac{2\pi}{T} R_0 \theta_0$
1	(i) (ii) (iii) Mean					
:						
8						

Conclusion. When the time period is kept constant

$$\frac{e_0}{a} = \frac{e_0}{R_0\theta_0} = \text{a constant}$$

or

$$e_0 \propto R_0 \theta_0$$

(B) Amplitude constant, time period variable

Sl. No.	Amplitdue $a = R_0\theta_0$	Time for 20 oscillations	Mean time period T	e_0	$e_0 T$	Linear vel $v = \frac{2\pi}{T} R_0 \theta_0$
1						
8						

Conclusion. When the amplitude is kept constant

$$e_0 T = \text{a constant}$$

or

$$e_0 \propto \frac{1}{T}$$

If we plot a graph between e_0 and $\frac{1}{T}$ it is a straight line.

8. To plot the graph between linear velocity v of the magnet and maximum induced *e.m.f.* e_0, take v along the X-axis and e_0 along Y-axis for all the values of e_0 and v in table A and B. The graph is a straight line as shown.

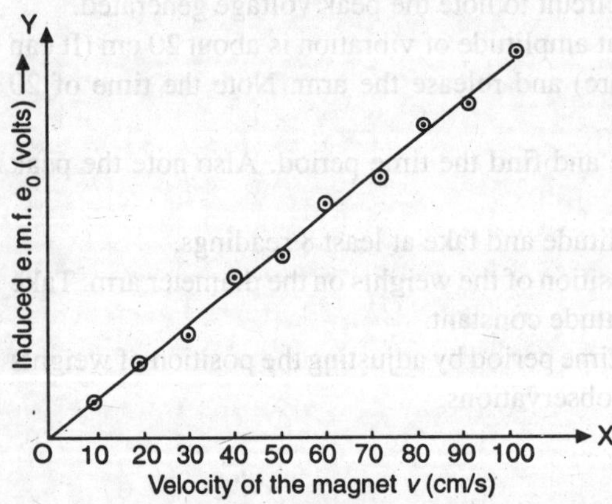

Fig. 33.11

This shows that the induced *e.m.f* is proportional to the linear velocity of the magnet.

Precautions. 1. The semi-circular frame should oscillate freely as a whole on the knife edge.

2. The magnet should pass freely through the coils C_1 and C_2.

3. The *e.m.f.* developed in the coil should be measured with the help of an electronic circuit.

4. The magnet should be small and should be mounted at the middle of the semi-circular arc.

33.7. Self-induction. When the current in a closed circuit is started, stopped or varied, the magnetic flux linked with the circuit changes and hence an *e.m.f.* or current is induced in the circuit. The phenomenon is known as self-induction.

Co-efficient of self-induction. When the permeability of the medium remains constant, the magnetic flux is proportional to the strength of current. If ϕ is the magnetic flux linked with a circuit through which a constant current i flows, then

$$\phi \propto i$$

or $$\phi = L\,i \qquad ...(i)$$

where L is a constant and is known as co-efficient of self-induction or inductance of the circuit. If $i = 1$, then $\phi = L$.

The co-efficient of self-induction of a circuit is defined as the magnetic flux linked with it when a unit current is passed through it.

Differentiating equation (*i*), we have

$$-\frac{d\phi}{dt} = -L\frac{di}{dt}$$

But $$-\frac{d\,\phi}{dt} = e$$

where e is the induced *e.m.f.* in the circuit when the magnetic flux is changing.

$\therefore$ $$e = -L\frac{di}{dt}$$

If $\frac{di}{dt} = 1$, then

$$L = -e$$

Thus co-efficient of self-induction of a circuit is numerically equal to the induced e.m.f. in the circuit due to a unit rate of change of current in it.

The practical (or *S.I.*) unit of self-inductance is a **henry.** *A circuit has a self-inductance of one henry if a magnetic flux of one weber is linked with it, when one ampere of current flows through it. Or if an e.m.f. of one volt is induced in it, when the current changes at the rate of one ampere per second.*

33.8. Co-efficient of self-induction of a solenoid. A loop of wire or a short straight conductor is practically non-inductive and has a negligible self-inductance. The induction effect is increased by having a large number of turns closely wound. The co-efficient of self-induction of a long closely wound solenoid of length l, total number of turns n and area of cross-section a having an air core is given by,

$$L = \frac{\mu_0 n^2 a}{l} \text{ henry}$$

where μ_0 is the permeability of air, the value of $\mu_0 = 4\pi \times 10^{-7}$ Web/amp-m.

If the core is made of a material of constant relative permeability μ_r and of the same cross-sectional area as that of the solenoid, then

$$L = \frac{\mu_0 \mu_r n^2 a}{l} \text{ henry}$$

33.9. Growth and decay of current in an L R circuit.

(*i*) **Growth.** Consider a circuit having an inductance L and resistance R placed in series with a battery of *e.m.f.* E. When the key K is pressed the current slowly increases from zero to a maximum steady value $I_0 = \frac{E}{R}$. During the time the current is growing, the value of the current I at the end of time t from the instant of closing the circuit, can be proved to be

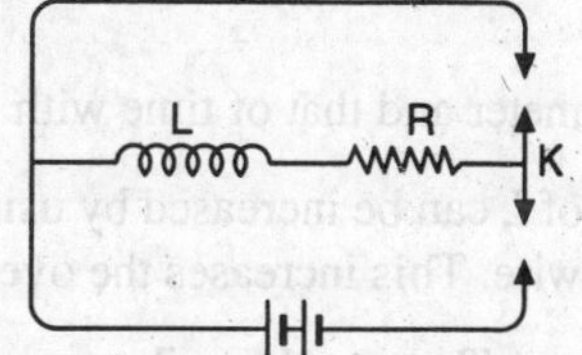

Fig. 33.12

$$I = I_0 \, [1 - e^{(-R/L)\,t}] \qquad ...(i)$$

From equation (*i*), we get

$$\log (I_0 - I) = \log I_0 - \frac{Rt}{L} \qquad ..(ii)$$

The fraction $\frac{L}{R}$ is called the *time constant* of the circuit and has the dimensions of time.

If $t = \frac{L}{R}$, then from equation (*i*), we have

$$I = I_0 \left(1 - \frac{1}{e} \right) = 0.6321 \, I_0$$

Thus time constant of an *LR* circuit is defined as the time during which the current grows from zero to $1 - \frac{1}{e} = 0.6321$ of the maximum constant value I_0.

(*ii*) **Decay.** As soon as the key *K* is lifted, contact is established on the upper side thereby again completing the *LR* circuit without the battery. The current slowly decreases and finally becomes zero. The value of the current at the end of time *t* from the instant of lifting the key can be proved to be

$$I = I_0 \, e^{\left(-\frac{R}{L}\right) t} \qquad ...(iii)$$

From equation (*iii*), we get

$$\log I = \log I_0 - \frac{R}{L} t \qquad ...(iv)$$

After a time $t = \frac{L}{R}$, from equation (*iii*).

$$I = I_0 \times \frac{1}{e} = 0.3679 I_0$$

The time constant can therefore, be also defined as the time during which the current falls from its maximum steady value I_0 to $\frac{1}{e} = 0.3679$ of the maximum value.

Equations (*i*) and (*iii*) clearly tell us that the growth and decay of current in an *LR* circuit is exponential in nature and depends upon the time constant $\frac{L}{R}$.

Also relation (*ii*) tells us that for growth, a graph between $\log (I_0 - I)$ and *t* will be a straight line.

Similarly relation (*iv*) tells us that for decay a graph between $\log I$ and *t* will also be a straight line.

(*iii*) **Study.** To study the growth and decay of current in an *LR* circuit it is desirable to have an inductor of large time constant $\frac{L}{R}$ so that measurements of current are made directly with a current meter and that of time with a metronome as the current grows or decays. To make $\frac{L}{R}$ large the value of *L* can be increased by using a large number of turns and *R* can be decreased by using thick copper wire. This increases the overall size of the inductor coil.

To keep the coil reasonably small an inductor with an iron core is used. Thus the value of $\frac{L}{R}$ can be made 10 second or so by using a coil of inductance 4,000 *H* and resistance about 400 ohm. Such inductors, however, cause some problems as their inductance varies considerably with current and growth and decay curves also differ due to hysterisis.

With an air core inductor $\frac{L}{R}$ can best be raised to a few mili-second. For measurement of time, therefore, we have to use an electronic device called the digital timer which can measure time with a least count of 0.1 milli-second.

Experiment 33.3. To study the growth and decay of current in a *LR* circuit using magnetic core inductor.

Apparatus. An iron core *inductor* of inductance nearly 4000 *H*, a resistance *R* of about 300 ohm, a milli-ammeter, a battery, two tap keys, a resistance $r << R$, a metronome and connecting wires.

Theory. The growth of current in a *LR* circuit is represented by the relation

$$I = I_0 \; 1 - e^{(-R/L)\,t} \qquad ...(i)$$

where *I* is the current at a time *t* from start and I_0 the maximum final steady value.

Similarly the decay current in such a circuit is given by

$$I = I_0 \, e^{(-R/L)\,t} \qquad ...(ii)$$

Relations (*i*) and (*ii*) clearly show that the growth and decay of current in an inductive circuit depends upon the value of $\frac{L}{R}$ and are exponential in nature.

Time constant. It is the time during which the current drops to $\frac{1}{e}$ of the initial steady value. It is denoted by $\frac{L}{R}$ and has the dimensions of time.

From relation (*i*), we have

$$\log (I_0 - I) = \log I_0 - \frac{R}{L} t \qquad ...(iii)$$

Hence for growth a graph between $\log (I_0 - I)$ and *t* is a straight line. The slope of the line gives the value of $\left(-\frac{L}{R}\right)$.

From relation (*ii*), we have

$$\log I = \log I_0 - \frac{R}{L} t \qquad ...(iv)$$

Hence for decay a graph between log *I* and *t* is a straight line. The slope of the line again gives the value of $\left(-\frac{L}{R}\right)$.

Procedure. 1. Draw a diagram showing the scheme of connections as in Fig. 33.13. Here *L* represents the pure inductance and *R* pure resistance including that of the inductor, the meter and actual resistance in the circuit. The battery *E* is shunted with a resistance $r << R$ to ensure that circuit resistance during decay remains close to the resistance *R*.

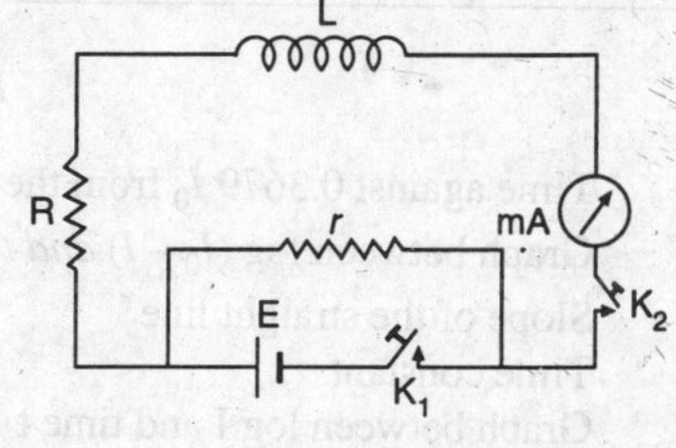

Fig. 33.13

In the absence of the shunt *r* the resistance at break would become infinite and sparking will occur when K_1 is switched off. The value of *r* should not be too small either because then, the battery would be short circuited. ($r \approx 20$ ohm).

2. Set the metronome to give '*ticks*' at intervals of $\frac{1}{2}$ sec. To study the growth of the current, close the key K_2 and then the key K_1 and start the metronome simultaneously.

3. Note the current in the meter at each regular 'tick' of the metronome till the current reaches a steady value. Stop the metronome.

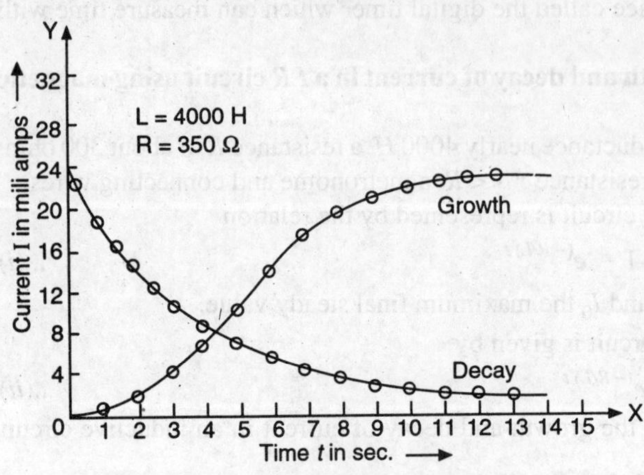

Fig. 33.14

4. When the current is maximum (steady), open the key K_1 (keeping K_2 pressed) and start the metronome simultaneously. Note the current in the meter at each regular 'tick' till the current reaches practically to zero value.

5. Taking time in seconds along the X-axis and current in milli- amps along the Y-axis plot the growth and decay curves. One set of such curves is shown in Fig. 33.14.

6. Also plot a graph between log $(I_0 - I)$ and time t for growth and log I and time t for decay by taking time along-X-axis and log $(I_0 - I)$ or log I_0 along Y-axis. The graphs are *very nearly* straight lines.

Measure the slope of each graph.

Growth

Sl. No.	*Time t*	*Current I*	*$(I_0 - I)$*	*log $(I_0 - I)$*
1	0			
2	0.5			
3	1			
.	1.5			
.	.			
.	.	$I_0 =$		
20	10			

Decay

Sl. No.	*Time t*	*Current I*	*Log I*

$$I = \frac{I_0}{e} = 0.3679\, I_0 =$$

Time against 0.3679 I_0 from the decay curve =

Graph between log $(I_0 - I)$ *and time t*

Slope of the straight line =

Time constant =

Graph between log I and time t

Slope of the straight line =

Time constant =

Precautions 1. An iron-core inductor should be used so that the time constant is large and measureable.

2. The battery should be shunted by a resistance very small as compared to the resistance of the *LR* circuit to ensure that circuit resistance during decay very nearly remains the same.

3. The ticks in the metronome should be regular and at small intervals.

33.10. Determination of self-inductance by Rayleigh's method. Lord Rayleigh used the principle of wheatstone bridge method for measuring the self-inductance of a coil. The arrangement of apparatus is shown in Fig. 33.15. Instead of an ordinary dead beat galvanometer a moving coil ballistic galvanometer is used. The arm *CD* contains an unknown inductance *L* in series with a small standard resistance *r* of the order of 0.1 or 0.01 ohm. This resistance is short-circuited by a key K_3.

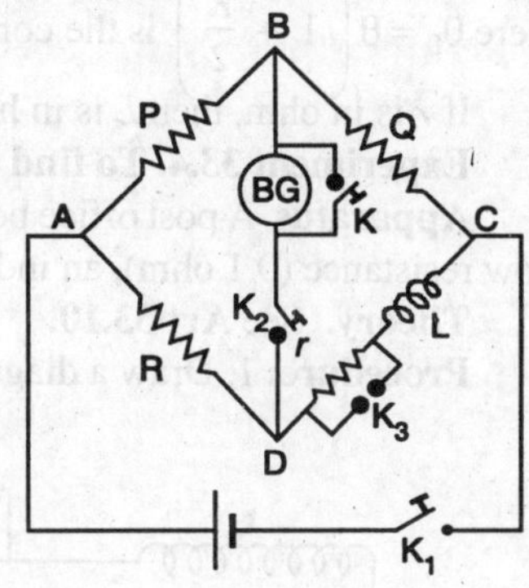

Fig. 33.15

The key K_3 is closed so that the resistance *r* is not in circuit. The battery key K_1 and then the galvanometer key K_2 are pressed and an exact balance point is obtained for a *steady current* by adjusting the value of *R* in the usual manner. If now the galvanometer key K_2 is pressed first and then battery key the ballistic galvanometer gives a sudden kick. It is because when the battery key is pressed (galvanometer key already being closed) the current slowly *grows* through the inductance and an induced *e.m.f.* $L \frac{di}{dt}$ is set up in the arm *CD*, where *L* is the co-efficient of self-induction and $\frac{di}{dt}$ the rate of change of current.

The presence of the induced *e.m.f.* makes a proportionate current to flow through the ballistic galvanometer. Hence the instantaneous current through the galvanometer is given by

$$kL \frac{di}{dt}$$

where *k* is the constant of proportionality.

∴ Charge *dq* through the ballistic galvanometer in time *dt*

$$= k L \frac{di}{dt} dt$$

or $$dq = kL\, di$$

If the current takes a time *t* to attain its constant value *I* in the arm *CD* the charge *q* passing through the ballistic galvanometer is given by

$$q = \int_0^I kL\, di$$

$$= k L I = \frac{cT}{2\pi nAB} \theta \left(1 + \frac{\lambda}{2}\right) \quad ...(i)$$

where *c, T, n* and *B* have their usual meaning for a moving coil ballistic galvanometer, θ the first throw and λ the logarithmic decrement. The key K_3 is now opened so that an additional small resistance *r* is introduced in the arm *CD*. This brings about a small change in *e.m.f.* = *Ir*.

If the key K_1 is now pressed first and then the key K_2 a proportionate steady current *kIr* flows through the ballistic galvanometer and produces a steady deflection θ′ given by

$$k I r = \frac{c}{nAB} \times \theta' \quad ...(ii)$$

Dividing (*i*) by (*ii*), we have

$$\frac{L}{r} = \frac{T}{2\pi} \frac{\theta}{\theta'} \left(1 + \frac{\lambda}{2}\right)$$

$$\therefore \qquad L = \frac{rT}{2\pi}\frac{\theta}{\theta'}\left(1+\frac{\lambda}{2}\right)$$

$$= \frac{rT}{2\pi}\frac{\theta_0}{\theta'}$$

here $\theta_0 = \theta\left(1+\frac{\lambda}{2}\right)$ is the corrected value of throw.

If r is in ohm, then L is in henry.

Experiment 33.4. To find the co-efficient of self-inductance by Rayleigh's method.

Apparatus. A post office box, a ballistic galvanometer, a lamp and scale arrangement, a standard low resistance (0.1 ohm), an inductance coil, a one-way key, a tap key, a cell and connecting wires.

Theory. See **Art 33.10.**

Procedure. 1. Draw a diagram showing the scheme of connections as in Fig. 33.16.

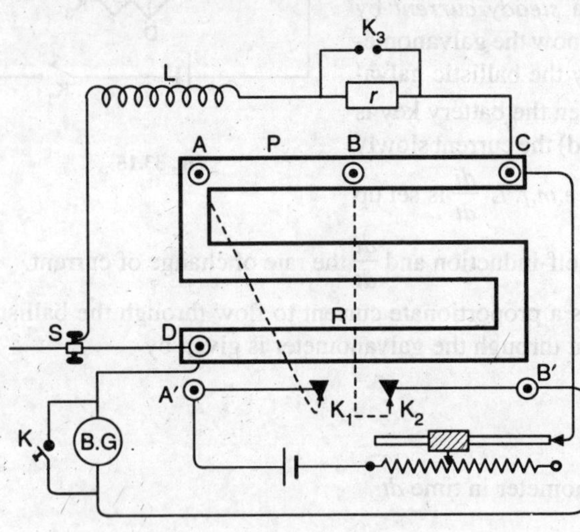

Fig. 33.16

2. Level the ballistic galvanometer and set up the lamp and scale arrangement so that the spot of light moves freely on the scale.

3. Connect one end of the inductance coil through a low resistance r of value 0.1 ohm to the terminal C of the post office box and the other end to a binding screw S which can be moved along a platinoid wire one end of which is connected to the terminal D of the post office box. Place a key K_3 across the low resistance r. Connect the ballistic galvanometer, the battery and the rheostat as shown.

4. Put in the plug in key K_3. Take out the 10 ohm plugs from the ratio arms P and Q.

Close the battery key K_1 first and then the galvanometer key K_2. If the deflection goes out of scale, adjust the value of the current with the rheostat in the battery circuit so that the deflection is within scale.

Now adjust the value of the resistance R till an increase of one ohm changes the direction. Adjust the position of binding screw S on the platinoid wire until a perfect balance of the bridge for a steady current is obtained.

Having obtained a perfect balance for a steady current, first release the battery key K_1 and then immediately after it the galvanometer key K_2. Note the first throw θ_1 of the spot of light on the scale and also the second throw θ_2 in the same direction. Repeat three times.

5. Bring the galvanometer coil to rest with the help of the tap key K. Test the bridge again for balance of a steady current. If the balance has been disturbed adjust the position of the binding screw S till a perfect balance is again obtained. Take out key K_3 so as to introduce the small resistance r in the arm CD. Press the battery key K_1 and then the galvanometer key K_2 and note the steady deflection of the spot of light. Similarly take two more observations.

6. Find the time period of the coil of the galvanometer by noting the time taken by the spot of light on the scale for 20 vibrations, keeping the galvanometer circuit open. Repeat twice and find mean value of time period.

Observations.

Sl. No.	First throw θ_1	Second throw θ_2	Corrected throw $\theta_0 = \theta_1 + \frac{\theta_1 - \theta_2}{4}$
1.			
2.			
3.			

Mean θ_0 =

Value of the low resistance r = ohm

Deflection for steady current θ' = 1. 2. 3.

Mean θ' =

Time period.

Sl. No.	Time for 20 vibrations	Time Period T
1		
2		
3		

Mean time period T =

Calculations.

$$\text{Co-efficient of self-inductance } L = \frac{rT}{2\pi}\frac{\theta_0}{\theta'} \text{ henry}$$

Precautions. 1. The ballistic galvanometer should be properly levelled so that the coil moves freely.

2. Short and straight connecting wires should be used to avoid spurious inductance effects and all resistances used in the experiment should be non-inductive.

3. The *e.m.f.* of the cell should be constant.

4. The current should be allowed to pass through the bridge for as small a time as possible to avoid the undue heating of the inductance coil of low resistance.

5. The balance for steady current should be perfect.

6. While finding the time period of the galvanometer coil, the galvanometer circuit must be kept open.

7. The low resistance r introduced in the arm CD in series with the inductance to produce a steady deflection θ' should be very small so that it does not appreciably affect the value of the steady current in that branch. Moreover it is better if its value is such that the steady deflection θ' produced by it is of the same order as θ.

33.11. Double key. the resistances P, Q, R and r are small and in a way shunt the galvanometer. A measurable and a reliable throw can be obtained by opening the galvanometer circuit immediately after the discharge has passed through it with the help of a special **double key** as shown in Fig. 33.17.

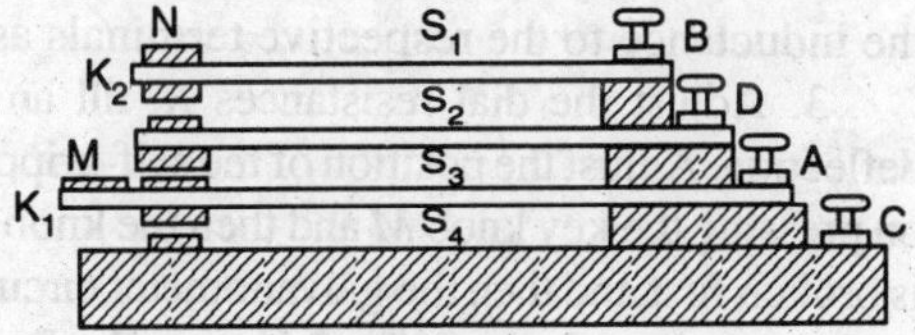

Fig. 33.17

It consists of four brass strips, S_1, S_2, S_3 and S_4 mounted horizontally on an ebonite base and provided with separate terminals B, D, A and C as shown. The battery is connected between A and C and the galvanometer between B and D. When the knob M is pressed contact is established between

A and C and the battery circuit is closed. If now the knob N is pressed contact is established between B and D and the galvanometer circuit is also closed. The battery key K_1 and the galvanometer key K_2 remain insulated from each other by the ebonite strips fixed to the brass strips S_2 and S_3.

Keeping N pressed if M is released, none of the two circuits will break. Now if N is released the battery circuit is broken first and the galvanometer circuit immediately after the discharge has passed through the galvanometer. If the time interval between the break of the battery and galvanometer circuit is small compared with the time period T of the galvanometer coil the first throw will be almost independent of electromagnetic damping due to low resistance circuit.

33.12. Rayleigh's self-inductance bridge. It contains in it a Wheatstone bridge network having fixed ratio arms P and Q, the ratio being 1 : 1. In series with this are terminals to connect the inductance coil. The arm containing the inductance coil is in series with a slide wire of manganin of resistance 0.12 ohm, calibrated to read 0.001 ohm. A self-gripping jockey is provided which can slide along a square section brass rod. This arrangement keeps the jockey in firm contact and thus can be moved to either side on lightly pressing the button. On releasing the pressure on the button, the jockey grips firmly on the wire. The fourth arm contains three dial resistance boxes of 10 × 0.1, 10 × 1 and 10 × 10, thus a total resistance of 111 ohm can be introduced. Two more terminals marked. '*bridge arm series*' are also provided for connecting additional resistance in the 4th arm, if necessary. Ordinarily these two terminals are connected by a thick brass strip. **A double key** as shown in Fig. 33.17. is also provided on the bridge. This key is known as Rayleigh's inductance key and has separate knobs for making battery and the galvanometer connections. When the pressure is released from the knob N, the battery circuit is first broken and then the galvanometer circuit immediately after. This provides elimination of excessive damping.

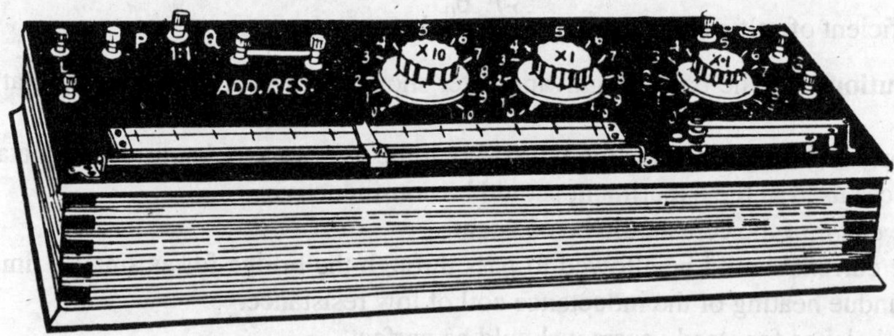

Fig. 33.18

Experiment 33.5. To find the co-efficient of self-inductance using Rayleigh's self-inductance bridge.

Apparatus. Rayleigh's self-inductance bridge, a ballistic galvanometer, a cell, a tap key, connecting wires, a stop watch, lamp and scale arrangement and an unknown self-inductance coil.

Procedure. 1. Level the ballistic galvanometer and set the lamp and scale arrangement so that the spot of light moves freely about the middle of the scale.

2. Draw the scheme of connections as in Fig. 33.19 and connect the battery, galvanometer and the inductance to the respective terminals as shown.

3. Adjust the dial resistances R, till an increase of 0.1 ohm resistance changes the direction of deflection. Adjust the position of the self-gripping jockey on the slide wire till a perfect balance is obtained on pressing the key knob M and then the knob N. Release both M and N together so that the battery circuit is broken first and then the galvanometer circuit. As M is released the flux decreases and an inductive kick is given to the galvanometer. Measure the first throw θ_1 and then θ_2 in the same direction.

4. Bring the galvanometer coil to rest with the help of the tap key placed across the galvanometer and check the balance point again.

Change the position of the self-gripping jockey by a known length (each division being 0.001 ohm) to create imbalance. On pressing the battery key M and then the galvanometer key N a steady deflection $'$ is produced. Note the resistance introduced in the circuit and the corresponding steady deflection produced.

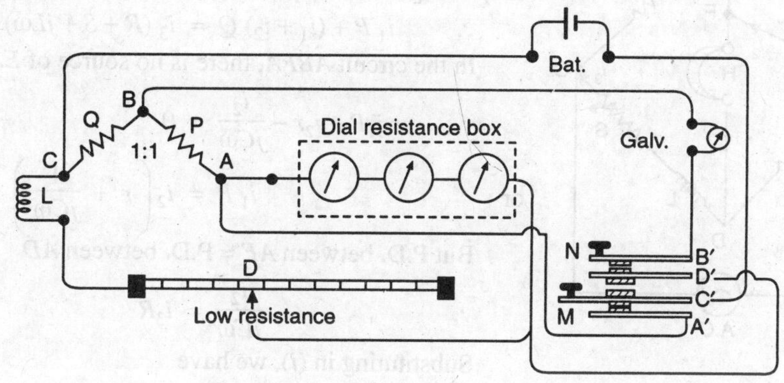

Fig. 33.19

5. Release the battery key M and then the galvanometer key N. The spot of light moves on the scale. Find the time taken for 20 vibrations with a stop watch. Repeat the observation twice.

Observations.

Record observations for θ as in **Experiment 33.4**

Resistance per division of manganin wire = 0.001 ohm

Sl. No.	*No. of division n*	*Resistance r = n × 0.001*	*Steady deflection θ′*	$\frac{r}{\theta'}$ =
1				
2				
3				

Mean $\frac{r}{\theta'}$ =

Record observations for time period T as given in **Experiment 33.4.**

Calculations.

Coefficient of self-induction $L = \frac{T}{2\pi} \cdot \frac{r}{\theta'} \theta_0$

= henry

Precautions. Same as in **Expt. 33.4.**

Note. 1. While noting the deflection θ, the damping sometimes is very large. The same can be reduced by putting a resistance in series with the galvanometer from the resistance box placed in series with it.

2. Different readings of θ' can be taken by introducing different resistances in the arm by changing the position of self-gripping jockey.

33.13. Anderson's bridge. Anderson's bridge is a modification of Wheatstone bridge. It enables us to measure the inductance of a coil in terms of capacitance and resistance and does not require repeated balancing of the bridge. The connections are shown in Fig. 33.20. The balance of the bridge is first obtained for a steady current by replacing the head phone H by a moving coil galvanometer and A.C. source by a battery. This is done by adjusting the variable resistance S.

After a steady balance has been obtained inductive balance is obtained by using the *A.C.* source and head phone. This is done by adjusting the variable resistance r in the arm BF without disturbing other resistances.

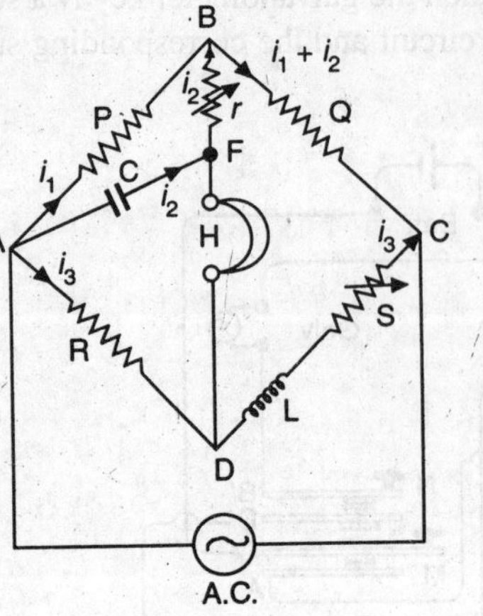

When the current in the head phone is zero, the potential drop along ABC is equal to that along ADC

$$\therefore \quad i_1 P + (i_1 + i_2) Q = i_3 (R + S + jL\omega) \quad ...(i)$$

In the circuit $ABFA$, there is no source of *E.M.F.*

$$\therefore \quad i_1 P - i_2 r - \frac{i_2}{jC\omega} = 0$$

or

$$i_1 P = i_2 \left(r + \frac{1}{jC\omega} \right) \quad ...(ii)$$

But P.D. between AF = P.D. between AD

$$\therefore \quad \frac{i_2}{jC\omega} = i_3 R$$

Substituting in (*i*), we have

$$i_1 (P + Q) = i_2 \left(\frac{R + S + j L \omega}{R j C \omega} - Q \right) ...(iii)$$

Fig. 33.20

Eliminating i_1 and i_2 from equations (*ii*) and (*iii*), we have by equating imaginary parts

$$S = \frac{RQ}{P}$$

This is the relation satisfied by the bridge for steady current. By equating real parts, we have

$$L = CR \frac{(Pr + Qr + PQ)}{P}$$

Thus we can find L in terms of resistances P, Q, R and r and capacitance C. The resistance S of the arm containing the inductance is not involved, hence ohmic resistance of the inductance is not required.

Further, we can make $P = Q$

$$\therefore \quad L = CR (P + 2r)$$

Caution. If the product CRP is greater than L, then r will be a negative quantity. It will, therefore, not be possible to obtain a balance point under such conditions. It is thus necessary to know the approximate value of L by a preliminary experiment. The value of P and R should then be adjusted so that r is a positive quantity and a balance can be obtained by varying the value of r.

To obtain the balance point a cathode ray oscilloscope can be used in place of head phones.

Experiment 33.6. To find the co-efficient of self-induction of a coil by Anderson's method using (*i*) a head phone and (*ii*) a cathode ray oscilloscope.

Apparatus. An air core coil of large inductance, 4 resistance boxes, a fractional resistance box, a capacitance of known capacity, a galvanometer, a battery, an audio oscillator (frequency 1000 Hz), a head phone, a cathode ray oscilloscope.

Procedure. 1. Draw a diagram showing the scheme of connections as in Fig. 33.20. Using a battery in place of the A.C. source and a galvanometer in place of the headphone make the connections accordingly.

Make the resistance P = resistance Q.

Taking a suitable value of R adjust the value of S so as to get a null point *i.e.*, balance for steady current. Note the values of the resistances P and R.

2. (*i*) **Using a head phone.** Replace the battery by the *A.C.* source say a 1000 Hz audio-oscillator and the galvanometer by a head phone.

(**Note.** If an audio oscillator is not available 50 cycle A.C. mains may be used with the help of a step down transformer.) Keeping the values of P, Q, R and S constant adjust the value of the variable resistance r till the sound in the head phone is reduced to a *minimum* value. This is the balance point for alternating (oscillating) current. Note the value of resistance r. Note the value of capacitance marked on it. Repeat three times.

(*ii*) **Using a cathode ray oscilloscope.** 3. To find the balance point for oscillating current connect the leads going to the head phones to the plates Y_1 and Y_2 of the C.R.O.

Set the C.R.O, so that a sharp, bright spot of light is obtained in the centre of the screen. Adjust the frequency of the time base circuit and vertical and horizontal gains so as to get a number of wave traces of large amptitude on a major part of the screen. 'Lock' the pattern.

(**Note.** The vertical amplitude of the wave trace may also be increased with the help of the amplifier in the oscillator itself by adjusting the oscillator gain).

Keeping the values of P, Q, R and S constant, now slowly adjust the value of r so that the vertical amplitude of the wave pattern decreases. Finally adjust the value of r for which the amplitude of the wave pattern is zero and the trace is almost a horizontal straight line. This is the balance point for oscillatory current. Note the value of r. Also note the marked value of the capacitance C. Repeat three times.

Note. The construction and use of a cathode ray oscilloscope is described later in the book.

Observations.

Resistance	$P =$	$Q =$	ohm
Resistance	$R =$		ohm
Resistance	$r =$	(*i*)	ohm
		(*ii*)	ohm
		(*iii*)	ohm
Mean	$r =$		ohm
Capacity	$C =$	$\mu Fd =$	Fd
Inductance	$L = CR\ (P + 2r) =$		Henry

Precautions. 1. To avoid inductive effect short straight wires should be used.

2. The resistances used should be non-inductive.

3. The product of CPR should be less than the inductance L, so that r is positive.

4. While using a cathode ray oscilloscope all precautions for its use must be observed.

33.14. Coefficient of mutual induction. When two circuits lie close to each other and the current in any one of them is started, stopped or varied, there is a change in magnetic flux passing through the second circuit and an *e.m.f.* is induced in the second circuit. This effect is known as **mutual induction.** The circuit in which the current is varied is known as the *primary* circuit and the one in which the *e.m.f.* is induced is known as the *secondary* circuit.

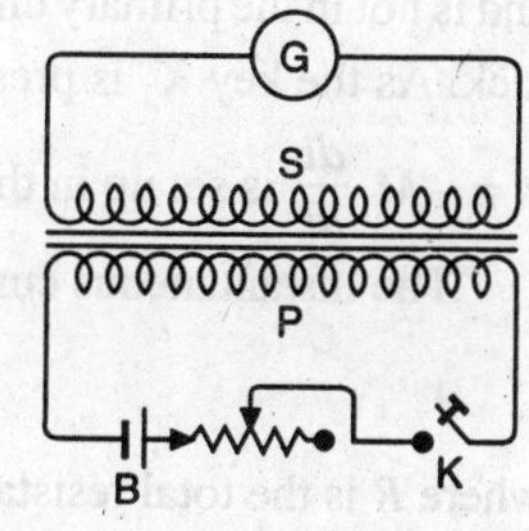

Fig. 33.21

Consider two coils P and S as shown in Fig. 33.21. Let ϕ be the magnetic flux linked with the coil S when a current i flows through the coil P, then

$$\phi \propto i$$

or

$$\phi = Mi \qquad ...(i)$$

where M is a constant and is called the coefficient of mutual inductance between the two coils,

If $i = 1$, then $\phi = M$

Thus the co-efficient of mutual induction between two circuits is defined as the magnetic flux linked with one due to a unit current flowing through the other.

Differentiating (*i*), we have

$$-\frac{d\phi}{dt} = -M\frac{di}{dt}$$

But
$$-\frac{d\phi}{dt} = e$$

where e is the induced *e.m.f.* in one when the magnetic flux is changing in the other.

$$\therefore \qquad e = -M\frac{di}{dt}$$

If $\frac{di}{dt} = 1$, then, $M = -e$.

Thus the co-efficient of mutual induction between two circuits is numerically equal to the induced e.m.f. in one circuit due to a unit rate of change of current in the other.

The co-efficient of mutual inductance of two circuits is one **henry** if a magnetic flux of one Weber is linked with one when a current of one ampere flows in the other, or an *e.m.f.* of one volt is induced in one when current in the other changes at the rate of one ampere per second.

A mutual inductance coil in general consists of two coils wound over the same air or iron core.

The co-efficient of mutual inductance between a secondary coil S of n_2 turns wound over the central part of a primary coil P of length l, area of cross-section a and having n_1 turns is given by

$$\frac{\mu_0 n_1 n_2 a}{l} \text{ henry}$$

where μ_0 is the permeability of air.

If the air core is replaced by an iron core of constant relative permeability μ_r and of the same area of cross-section, then

$$M = \frac{\mu_0 \mu_r n_1 n_2 a}{l} \text{ henry}$$

33.15. Determination of mutual inductance. The principle of the method of measuring the mutual induction of two coils P and S is shown in Fig. 33.22.

The points A and B are connected together so that the secondary circuit is completed through the ballistic galvanometer. The points C and D are connected so that the resistance r is short circuited and is not in the primary circuit. When the key K_1 is pressed, the ballistic galvanometer gives a sudden kick. As the key K_1 is pressed the current in the primary circuit slowly grows and an induced *e.m.f.* $e = -M\frac{di}{dt}$ is set up in the secondary circuit.

The instantaneous current through the secondary is given by

$$i = \frac{e}{R} = \frac{M}{R}\frac{di}{dt}$$

where R is the total resistance of the secondary circuit.

$\therefore$ The charge dq passing through the ballistic galvanometer in time dt is given by

$$dq = idt = \frac{M}{R}\frac{di}{dt}dt$$

or
$$dq = \frac{M}{R}di$$

If the current takes a time t to attain its steady value I in the primary circuit the charge q passing through the ballistic galvanometer is given by

$$q = \frac{M}{R}\int_0^I di = \frac{MI}{R} = \frac{cT}{2\pi nAB}\,\theta\left(1+\frac{\lambda}{2}\right) \qquad (i)$$

where c, T, n, A and B have their usual meanings for a moving coil ballistic galvanometer, θ is the first throw and λ the logarithmic decrement.

The contact between A and B and between C and D is now broken and contact between A and C and B and D is established. The resistance r is thus introduced in the primary circuit but its value is so small that the steady current I is not altered. The potential difference Ir across the resistance r sends a steady current $\frac{Ir}{R}$ through the ballistic galvanometer and produces a steady deflection θ' given by

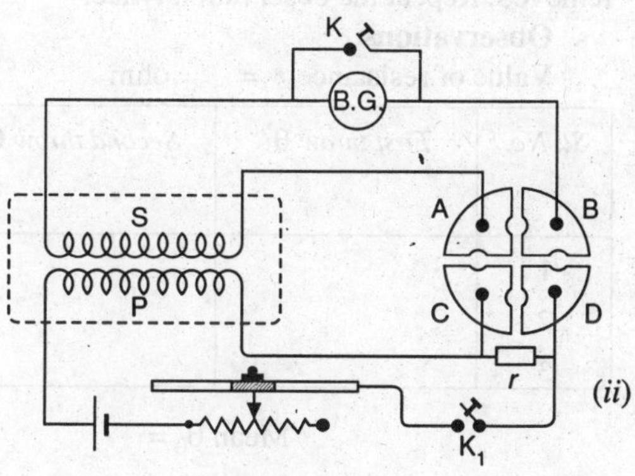

Fig. 33.22

$$\frac{Ir}{R} = \frac{c}{nAB}\theta' \quad \text{...(ii)}$$

Dividing (*i*) and (*ii*), we have

$$\frac{M}{r} = \frac{T}{2\pi}\frac{\theta}{\theta'}\left(1 + \frac{\lambda}{2}\right)$$

or
$$M = \frac{rT}{2\pi}\frac{\theta}{\theta'}\left(1 + \frac{\lambda}{2}\right)$$

$$= \frac{rT}{2\pi}\frac{\theta_0}{\theta'}$$

where $\theta_0 = \theta\left(1 + \frac{\lambda}{2}\right)$ is the corrected throw. If r is in ohm M is in henry.

Experiment 33.7. To find the co-efficient of mutual inductance of two coils.

Apparatus. A mutual inductance coil, a ballistic galvanometer, two tap keys, or a four-way key, rheostat, a standard resistance of 0.1 or 0.01 ohm, a battery, connecting wires, stop watch etc.

Theory. See Art. 33.15.

Procedure. 1. Draw a diagram showing the scheme of connections as in Fig. 33.22.

2. Level the ballistic galvanometer and set the lamp and scale arrangement so that the spot of light moves freely on the scale.

3. Connect the primary coil P to the terminals C and D of the four-way key through a cell, a rheostat and a tap key K_1, Connect the low resistance of 0.01 ohm across C and D. Connect the secondary coil to the terminals A and B of the four-way key through the ballistic galvanometer.

4. Put a plug between A and B so that the ballistic galvanometer is directly in the secondary circuit. Put a plug also between C and D to short circuit the resistance r. Press the key K_1 and if the kick of the ballistic galvanometer goes out of scale, adjust the value of resistance in the circuit with the rheostat.

Bring the spot of light to rest with the help of the key K connected across the ballistic galvanometer. Note the position of the cross line in the spot of light on the scale. Press the key K_1 and observe the first and the second throw on the same side. Repeat the observation twice.

5. Remove the plug keys from C and D and from A and B. Put in the plug key between A and C and between B and D. Press the key K_1 and observe the steady deflection in the ballistic galvanometer. Repeat this observation twice.

6. Find the time period of the ballistic galvanometer coil by noting the time of 20 vibrations with a stop watch in an open circuit *i.e.* when the plugs between *A* and *C*, *B* and *D* and *C* and *D* are removed. Repeat the observation twice.

Observations.

Value of resistance r = ohm

Sl. No.	*First throw* θ_1	*Second throw* θ_2	*Corrected throw* $\theta_0 = \theta_1 + \dfrac{\theta_1 - \theta_2}{4}$	*Steady deflection* θ'
1				
2				
3				

Mean θ_0 =

Mean θ' =

Time period

Sl. No.	*Time for 20 vibrations*	*Time period T*
1		
2		
3		

Mean time period T = sec

Calculations.

$$M = \frac{rT}{2\pi}\frac{\theta_0}{\theta'} \text{ henry.}$$

Precautions. Same as in **Experiment 33.4.**

Note. 1. In case the spot of light on pressing the key becomes stationary due to damping, then put a high resistance of the order of 0.3 to 0.5 meg-ohm in series with the galvanometer and increase the number of cells in the circuit.

Note 2. The above method is not applicable for the determination of mutual induction between two coils having an iron core. It is because in that case, the magnetic flux linked with the secondary is not proportional to the current flowing in the primary.

Experiment 33.8. To determine the magnetic induction field between the pole pieces of an electromagnet with the help of a search coil and ballistic galvanometer using a mutual inductance for calibration of ballistic galvanometer.

Apparatus. Electromagnet with battery and key, a ballistic galvanometer with lamp and scale arrangement, a tap key, A mutual inductance, a reversing key, an ammeter, a battery, a rheostat and connecting wires.

Theory. The arrangement to measure the magnetic field between the pole pieces of an electromagnet is shown in Fig. 33.23. A search coil is set between the pole pieces *NS* of the electromagnet so that its axis coincides with the axis of the electromagnet. A ballistic galvanometer B.G is connected in the secondary coil(*S*) circuit of the mutual inductance the primary coil (*P*) of which is connected to a battery through a reversing key *K* an ammeter *A* and a rheostat as shown.

If N_P is the number of turns *per unit length* of the primary coil *P* and *I* the current through it in Amperes, then

Magnetic field along the axis of the solenoid *P* (in S.I units)

$$B' = \frac{\mu_0}{4\pi} \cdot 4\,\pi\, N_P\, I.$$

If N_S is the total number of turns in secondary coil S and A its area of cross-section, then magnetic flux linked with the secondary is given by

$$\phi_S = N_S\,(B'A) = \frac{\mu_0}{4\pi} \cdot 4\,\pi\, N_P\, N_S\, IA$$

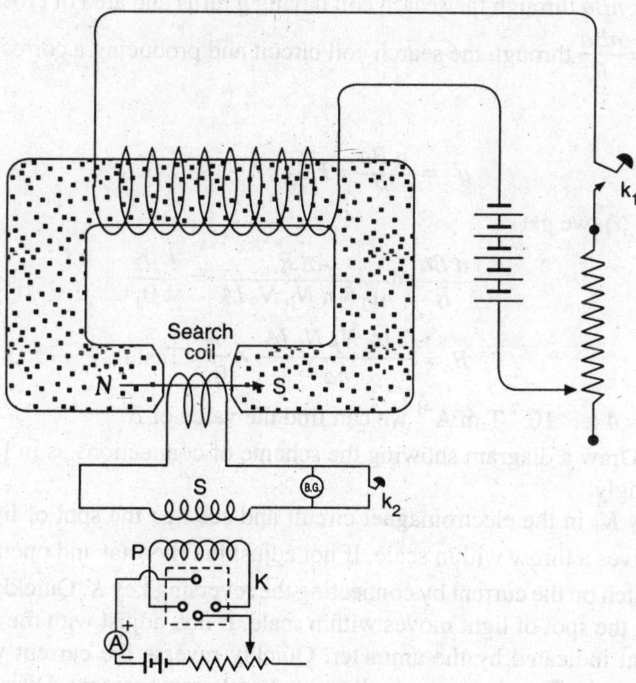

Fig. 33.23

The current through the primary coil is quickly reversed by turning the reversing key. Then change in magnetic flux

$$d\,\phi_S = \phi_s - (-\phi_s) = 2\,\phi_s$$

The change in magnetic flux sets up an induced *e.m.f.* in secondary given by

$$e = -\frac{d\,\phi_s}{dt}$$

An instantaneous current i flows in the secondary circuit given by

$$i = -\frac{e}{R} = \frac{1}{R}\,\frac{d\,\phi_s}{dt}$$

where R is the resistance of the secondary coil circuit. This current sends an instantaneous charge through the ballistic galvanometer given by

$$q = i\,dt = \frac{d\phi_s}{R}$$

If θ_1 is the first throw in the B.G and k the ballistic constant, then

$$q = \frac{d\,\phi_s}{R} = k\,\theta_1$$

or $$2\frac{\mu_0}{4\pi}\,\frac{4\pi\, N_P\, N_S\, IA}{R} = k\,\theta_1 \qquad ...(i)$$

The current through the primary coil P is switched off. The spot on the B.G scale is brought to rest and the key K_2 is opened.

Now the key K_1 is switched on, the current in the electromagnetic winding suddenly starts flowing. The magnetic field between the pole pieces NS quickly rises from 0 to B which immediately sends a magnetic flux nBa through the search coil having n turns and area of cross-section a, thereby sending a charge $q' = \frac{nBa}{R}$ through the search coil circuit and producing a corresponding first throw θ_2 in the B.G

$$\therefore \qquad q' = \frac{n\,Ba}{R} = k\,\theta_2 \qquad ...(ii)$$

Dividing (*ii*) by (*i*), we get

$$\frac{n\,Ba}{R} \times \frac{4\pi\,R}{\mu_0\, 8\,\pi\, N_P\, N_S\, IA} = \frac{k\,\theta_2}{k\,\theta_1}$$

or $$B = \frac{2\,\mu_0\, N_P\, N_S\, IA}{na}\cdot\frac{\theta_2}{\theta_1}\ \text{Tesla}$$

Substituting $\mu_0 = 4\,\pi \times 10^{-7}$ T m A^{-1} we can find the value of B.

Procedure 1. Draw a diagram showing the scheme of connections as in Fig. 33.23 and make connections accordingly.

2. Press the key K_1 in the electromagnet circuit and see that the spot of light of the lamp and scale arrangement gives a throw within scale. If not adjust the rheostat and open the key K_1.

3. Similarly switch on the current by connecting the reversing key K. Quickly reverse the current and again check that the spot of light moves within scale. If not, adjust with the help of the rheostat. Now note the current indicated by the ammeter. Quickly reverse the current with the help of the reversing key and note the first throw on the lamp and scale arrangement. Disconnect the reversing key.

4. Quickly press the key K_1 of the electromagnet and note the first throw on the lamp and scale arrangement.

5. Repeat the above observation a number of times.

Observations. Number of turns per metre in the Primary coil N_P = Total number of turns in the secondary coil N_S =

Total number of turns in the search coil n =

Diameter of the secondary coil

(*i*) (*ii*) (*iii*)

Mean diameter D = metre

Area of cross-section of the secondary coil (of mutual inductance) $A = \frac{\pi\, D^2}{4} = \quad m^2$

Diameter of the search coil

(*i*) (*ii*) (*iii*) (*iv*)

Mean diameter d = m

Area of search coil $a = \frac{\pi\, d^2}{4} = \quad m^2$

[The number of turns and the area of the search coil is generally marked on it by the makers]

Sl. No.	Current in Primary coil I (Amp.)	First throw in ballistic Galv					
		Current reversed in Primary coil (θ_1)			Current started in electromagnet (θ_2)		
		(i)	(ii)	mean	(i)	(ii)	mean

Calculations

$$B = \frac{2\mu_0 N_P N_S IA}{na} \frac{\theta_2}{\theta_1}$$

$$= 2 \times 4\pi \times 10^{-7} \frac{N_P N_S IA}{na} \frac{\theta_2}{\theta_1} = \quad \text{Tesla} \qquad [\mu_0 = 4\pi \times 10^{-7} \text{ T m A}^{-1}]$$

(i)
(ii)
(iii)
(iv)

Mean value of B = Tesla.

Precautions. 1. The ballistic galvanometer should be perfectly levelled so that the coil moves freely.

2. The search coil should be placed in the centre of the pole pieces *NS* of the electromagnet with its plane perpendicular to the lines of force.

3. The reversing of the current should be quick.

4. The mutual inductance coil should be at a distance from the electromagnet.

Oral Questions

1. Electromagnetic induction. (*i*) Define magnetic induction and magnetic flux. What are their units? (*ii*) What is electro-magnetic induction? (*iii*) State Faraday's laws of electro-magnetic induction. (*iv*) What is Lenz's law? (*v*) How does the induced *e.m.f.* change with the velocity of the magnet? (*vi*) Why should the semi-circular frame be made of a non magnetic material. (*vii*) What is the function of diode in the circuit? (*viii*) Why do we require rectification of voltage? (*ix*) What is the type of graph between induced *e.m.f* and maximum velocity? (*x*) What is the type of graph between mean time period and maximum velocity?

2. Electromagnetic induction (*i*) Define electromagnetic induction. (*ii*) State Farady's laws of electromagnetic induction. (*iii*) What is Lenz's law ? (*iv*) How does the induced *e.m.f.* change with the velocity of the magnet ? (*v*) Why should the semi-circular frame be made of a non-magnetic material.

3. Growth and decay of current (*i*) What is time constant of an *LR* circuit? (*iii*) What is the value of time constant in terms of *L* and *R*. ? (*ii*) What is the relation between current at any instant and the maximum saturation value in terms of *L* and *R* at growth and at decay ? (*iv*) What time should be taken theoretically for the current to grow from zero to maximum value ?

4. Self-induction. (*i*) Define co-efficient of self-induction. (*ii*) What is the absolute and practical unit of self-induction? (*iii*) What is the relation between a henry and an electromagnetic unit? (*iv*) What is the use of platinoid wire in Rayleight's method? (*v*) Why should the balance for steady current be perfect? (*vi*) Why should the resistance *r* be small? (*vii*) While finding the time-period of the galvanometer why the circuit should be kept open ? (*viii*) What is the use of a double key? (*ix*) State any other method of finding the coefficient of self-induction. (*x*) What is the condition that inductive balance point may be obtained in Anderson's method? (*xi*) In Anderson's method why the capacitance in the arm *AF* has no effect while obtaining a D.C. balance. (*xii*) In what way Anderson's method is better than Rayleigh's method?

5. Mutual inductance. (*i*) Define co-efficient of mutual inductance. (*ii*) Can we interchange the primary and secondary coils? (*iii*) What is the absolute and the practical unit of mutual inductance? (*iv*) Why should the resistance *r* be small? (*v*) Is the method suitable for finding mutual inductance with iron core? (*vi*) State some practical applications of mutual inductance. (*vii*) Why should the mutual inductance coil be placed at a large distance from the ballistic galvanometer?

34

Alternating Currents

34.1. Alternating currents. The current from a battery flows continuously in one direction and its strength remains constant with time. Such a current is known as direct current or *D.C.* There is another type of current which is induced in a coil when it is rotated in a magnetic field with uniform angular velocity. This current is known as an alternating current or *A.C.* and is represented in general by the relation

$$I = I_{max} \sin \omega t$$

where ω is the angular velocity of the coil. The strength of the current varies with time as shown in Fig. 34.1. The current increases from zero to a maximum value (I_{max}), then decreases to zero, changes direction, again increases to the maximum value in the opposite direction and finally attains zero value. This cycle of changes is again repeated. *The time taken by the alternating current to go from zero to maximum, maximum to zero, zero to maximum in the opposite direction and back again to zero is known as its* **time period.** *The number of such cycles made by the alternating current in one second is called its* **frequency.**

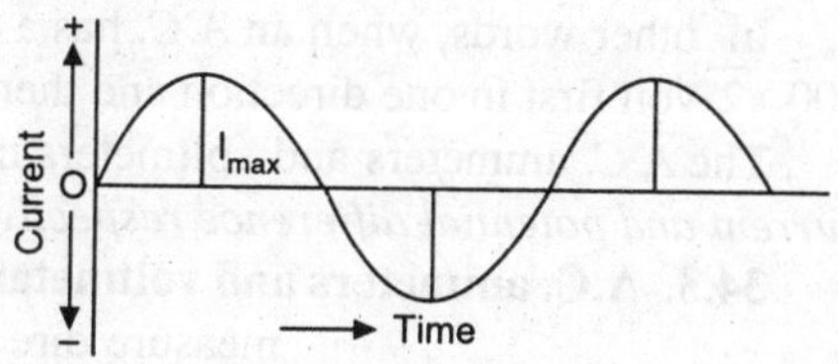

Fig. 34.1

The frequency commonly used for household lighting and power supply is about 50 cycles per second. If *n* is the frequency of the *A.C.* supply, then

$$\omega = 2\pi n$$

The current *I* is thus represented by

$$I = I_{max} \sin \omega t = I_{max} \sin 2\pi nt$$

and the corresponding voltage is represented by

$$E = E_{max} \sin \omega t = E_{max} \sin 2\pi nt$$

This relation shows that the *A.C.* voltage also varies with time and has a maximum value of E_{max}.

34.2. Measurement of alternating current. The strength of the alternating current is not constant. It is measured by that steady current which produces the same heating effect in a given time in a resistance as is done by the alternating current in the same time in the same resistance. This value of the current is known as the **effective** value or **virtual** value or **root mean square** (*R.M.S.*) value of the *A.C.*

Hence an alternating current has a strength of **one virtual ampere** *if it produces the same heating effect in a resistance as is done by a steady current of one ampere in the same time.*

The *R.M.S.* or the virtual value of current is represented by I_v and bears a constant ratio to the maximum value of the current. It is given by

$$I_v = \frac{I_{max}}{\sqrt{2}}$$

or

$$I_{max} = \sqrt{2}\, I_v = 1.414\, I_v$$

or

$$I_v = \frac{I_{max}}{\sqrt{2}} = \frac{I_{max}}{1.414} = 0.707\, I_{max}$$

It is evident from this relation that by an alternating current of *one ampere* we mean the root mean square or virtual value of current. In fact the current is changing its value from zero to a maximum value of $\sqrt{2}$ ampere in one direction and then in the opposite direction.

Similar remarks apply to the measurement of the alternating potential difference or *e.m.f.*

An alternating potential difference has a value of **one virtual volt** *if it produces the same heating effect when applied to the ends of a resistance as is done by a steady potential difference of one volt applied to the same resistance in the same time.*

The root mean square or the virtual value is represented by E_v and bears a constant ratio to the maximum value E_{max} and is given by

$$E_v = \frac{E_{max}}{\sqrt{2}} = 0.707\ E_{max}$$

In other words, when an *A.C.* has a virtual value of 100 volt, the potential varies from zero to $100\sqrt{2}$ volt first in one direction and then in the opposite direction.

The *A.C.* ammeters and voltmeters measure *the root mean square value or virtual value of the current and potential difference respectively.*

34.3. A.C. ammeters and voltmeters. The ordinary moving coil type ammeters and voltmeters measure direct currents and potential differences respectively. As such these are not suitable for the measurement of alternating currents and potential differences. These can be modified for use on *A.C.* with the help of small copper oxide metal rectifiers or semi-conductor diode rectifiers.

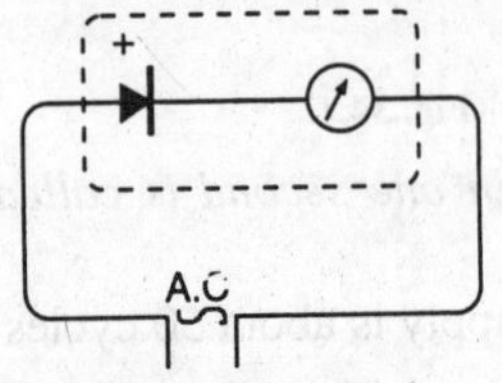

Fig. 34.2

(*i*) **Rectifier type instruments.** The metal rectifier consists of a copper plate of which the surface is oxidised to the reddish brown cuprous oxide state. If a point contact is made to the cuprous oxide surface by using soft lead, then the arrangement allows current to pass readily when the cuprous oxide is positive with respect to copper and does not allow the current to flow in the opposite direction. The rectifier thus converts the *A.C.* into a uni-directional current which is recorded by the moving coil meter. A simple arrangement of this type is shown in Fig. 34.2. Such an arrangement acts as a half-wave rectifier. To get a full wave rectification *four* metal rectifiers are arranged in the form of a bridge as shown in Fig. 34.3. When the end *A* is at a positive potential current flows from *A* to *B*, from *B* to *D* through the galvanometer and then from *D* to *C*. After half the cycle *C* is positive and the current flows from *C* to *B*, *B* to *D* through the galvanometer and then from *D* to *A*. Thus in both halves of the *A.C.* cycle current through the galvanometer flows in the same direction. Rectifier instruments can be adopted for use as ammeters or voltmeters with the help of suitable shunts or series resistance.

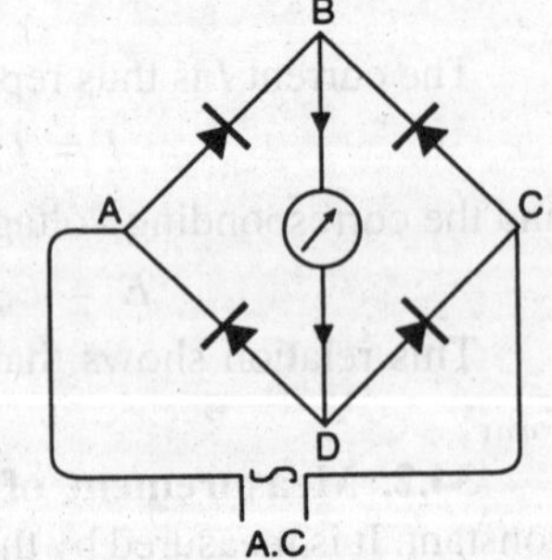

Fig. 34.3

For priniciple and working of semi-conductor dioide rectificers see **Chapter 50.**

(*ii*) **Thermo-couple type instruments.** Alternating current can be conveniently measured by thermo-couple meters. In these instruments the current to be measured is passed through a piece of resistance wire. The heating effect which is proportional to the square of the current raises the temperature of the wire. A copper constantan thermo-couple is fitted adjacent to or attached to the heater with the help of a small **ceramic bead** and connected to a moving coil galvanometer as shown in Fig. 34.4. The heat produced in the heater generates a thermo-electric current in the thermo–couple which is, therefore, proportional to the root mean square value of the alternating current. Such instruments measure both *A.C.* and *D.C.* and are also suitable for high frequency currents.

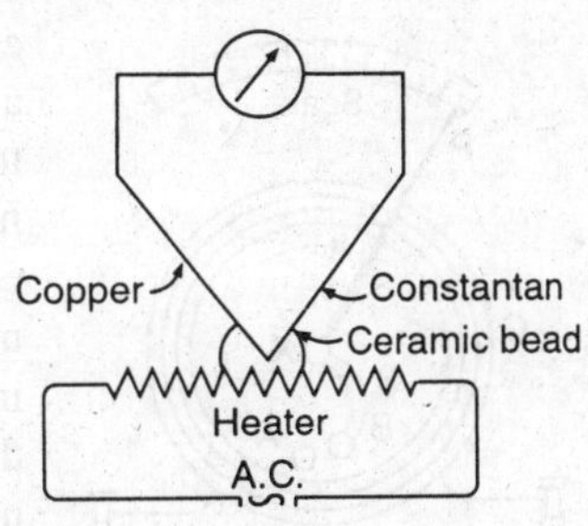

Fig. 34.4

(*iii*) **Hot-wire instruments.** The simple form of a hot-wire instrument, consists of a platinum-iridium wire *AB*, through which is passed the current to be measured. The middle of this wire is attached to a fine phosphor-bronze wire, which passes over a pulley *p* and is then fastened to a spring *S*. The axle of the pulley carries a pointer *P*, which moves over a scale. When the current passes through the wire, it is heated and expands. The 'slack' is taken up by the spring *S*, which pulls the phosphor-bronze wire, rotates the pulley and moves the pointer over the scale. The scale may be graduated to read the current in amperes or potential difference in volts.

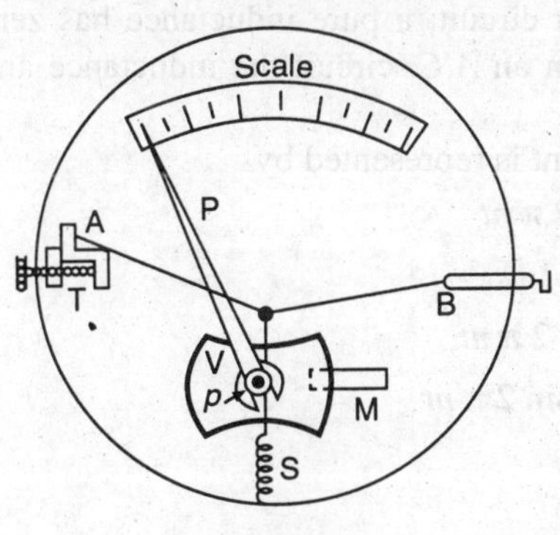

Fig. 34.5

An aluminium vane *V* is also mounted on the axle, which, when the pointer is deflected, moves between the poles of a small damping magnet *M*. The induced *e.m.f.* set up in the vane makes the instrument dead beat.

(*iv*) **Moving iron instruments.** These instruments are of two types:

(*a*) Attracted iron type, and

(*b*) Repulsion iron type.

(*a*) **The attracted iron type.** The instrument, in its simple form, consists of a solenoid *M* through which is passed the current to be measured. A soft iron rod *I* is inserted into solenoid *M* and held in position by a wire attached to a spring *S*, so that in the normal position, the rod remains partly in the solenoid, and partly outside it. The wire passes round a pulley *p*, mounted on a horizontal axle. When the current passes through the solenoid, the iron rod *I* is magnetised by induction and is attracted into the solenoid *M*. This causes *p* to rotate and the pointer *P* moves over the scale. The deflecting force due to the current is balanced by the controlling force due to the tension of the spring.

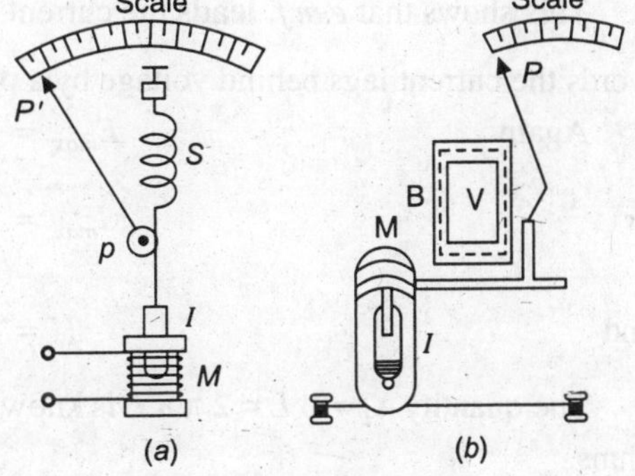

Fig. 34.6

In another form of the instrument, a soft iron disc *I* is pivoted eccentrically just outside a solenoid *M* of oval section, as shown in Fig. 34.6 (*b*). The working principle is the same as described above. To make the instrument dead beat, a vane *V* is attached to the moving system. It moves within a box *B* and damping is produced by the resistance to the motion of the vane by the air in the box.

(*b*) **Repulsion iron type.** In this type of the instrument, a coil C is arranged with its axis horizontal. A soft iron rod A is *fixed* along the coil almost from one end to the other. Another rod B is placed parallel to A. It is attached by a spring S to the horizontal axis X which carries the pointer. The current to be measured is passed through the coil. The rods A and B are magnetised with their near ends having the same polarity and repel each other. Hence B moves away from A, thereby rotating the axle and making the pointer to move over the scale. It is clear that in these instruments the deflection will be in the same direction, whatever be the direction of the current in the coil, so that the instruments can be used to measure alternating currents. These are further provided with iron shield or fitted in iron cases to screen the parts from external magnetic fields. The graduations on the scale are very close in the beginning and at the end.

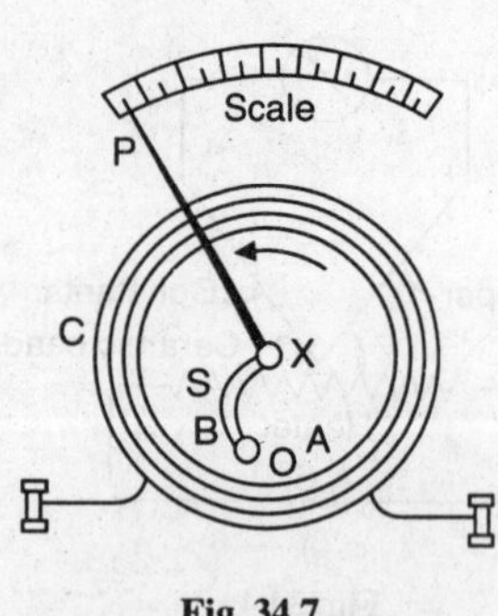

Fig. 34.7

34.4. Impedance of an A.C. circuit. In a direct current circuit, a pure inductance has zero resistance and an ideal capacitor has an infinite resistance. In an *A.C.* circuit the inductance and capacitor behave like ordinary resistances.

In an *A.C.* circuit containing only a resistance R, the current is represented by

$$I = I_{max} \sin \omega t = I_{max} \sin 2 \pi nt$$

and $$E_{max} = RI_{max}$$

Hence $$E = E_{max} \sin \omega t = E_{max} \sin 2 \pi nt$$

$$= RI_{max} \sin \omega t = RI_{max} \sin 2 \pi nt$$

$$\therefore \quad I_{max} = \frac{E_{max}}{R} \quad \text{or} \quad I_v = \frac{E_v}{R}$$

The current and the voltage are in phase.

When the *A.C.* circuit contains only a pure inductance L, the current and voltage are represented by

$$I = I_{max} \sin \omega t = I_{max} \sin 2\pi nt$$

and $$E = E_{max} \sin \left[\omega t + \frac{\pi}{2} \right] = E_{max} \sin \left(2 \pi nt + \frac{\pi}{2} \right)$$

This shows that *e.m.f.* **leads** the current by a phase angle $\frac{\pi}{2}$ or 90° *i.e.* the current lags. In other words the current lags behind voltage by a phase difference $\lambda/_2$ or 90°

Again $$E_{max} = \omega L \times I_{max} = 2\pi nL \, I_{max}$$

or $$I_{max} = \frac{E_{max}}{\omega L} = \frac{E_{max}}{2\pi nL} = \frac{E_{max}}{X_l}$$

and $$I_v = \frac{E_v}{X_l}$$

The quantity $X_l = \omega L = 2 \pi n L$ is known as the **reactance** due to **inductance.** It is measured in ohms.

When the *A.C.* circuit contains a pure capacitance C, the current is represented by

$$I = I_{max} \sin \omega t = I_{max} \sin 2 \pi nt$$

and *e.m.f.* by $$E = E_{max} \sin \left(\omega t - \frac{\pi}{2} \right) = E_{max} \sin \left(2 \pi nt - \frac{\pi}{2} \right)$$

This shows that the *e.m.f.* **lags** behind the current by a phase angle 90° $\left(\text{or } \frac{\pi}{2}\right)$ or the current leads. Moreover.

$$E_{max} = \frac{1}{\omega C} I_{max} = \frac{1}{2\pi n C} I_{max}$$

or
$$I_{max} = E_{max}\,\omega C = E_{max}\,2\pi nC$$

or
$$I_v = \frac{E_v}{X_c}$$

The quantity $X_c = \frac{1}{\omega C} = \frac{1}{2\pi n C}$ is known as the **reactance** due to **capacitance** and is also measured in ohms.

When the *A.C.* circuit contains resistance *R* ohm, an inductance *L* henry and a capacitance *C* farad in series the current is

$$I_v = \frac{E_v}{\sqrt{R^2 + (X_l - X_c)^2}} = \frac{E_v}{Z}$$

The quantity *Z* is known as **impedance** or the effective resistance of the *A.C.* circuit.

$$\therefore \quad Z = \sqrt{R^2 + (X_l - X_c)^2} = \sqrt{R^2 + (\omega L - 1/\omega C)^2}$$

The *E.M.F.* and current are not in phase. There is a phase difference θ between *E.M.F.* and current given by

$$\tan\theta = \frac{X_l - X_c}{R} = \frac{\omega L - \frac{1}{C\omega}}{R}$$

The *E.M.F.* leads the current if $X_l > X_c$ and *E.M.F.* lags behind the current if $X_l < X_c$

To obtain the value of *Z* graphically or by the vector diagram method, the resistance *R* is represented along the *X*-axis, the reactance due to inductance along the positive direction of the *Y*-axis and the reactance due to capacitance along the negative direction of the *Y*-axis. The diagonal *OC* of the rectangle *OACB* gives the value of the impedance *Z*, where

$$OA = R$$

and
$$OB = (X_l - X_c)$$

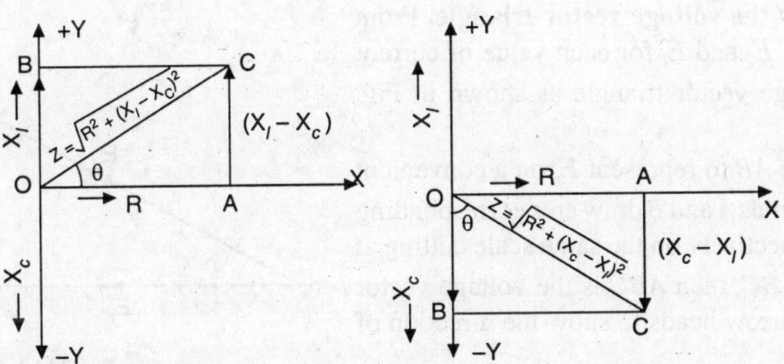

Fig. 34.8

When $X_l > X_c$ $OB = X_l - X_c$ shown in Fig 34.8 (*a*). The *E.M.F.* leads the current and the phase angle θ is *positive*. If $X_c > X_l$ then $OB = X_c - X_l$ as shown in Fig 34.8 (*b*). The *E.M.F.* lags the current and phase angle θ is *negative*.

To find the impedance of an *A.C.* circuit experimentally the current in the circuit is measured with an *A.C.* ammeter and the potential difference across the impedance with an *A.C.* voltmeter.

$$\text{The impedance } Z = \frac{E_v}{I_v} = \frac{\text{Voltmeter reading}}{\text{Ammeter reading}}$$

Experiment 34.1. to study an R L and an R C circuit.

Apparatus. A step down transformer (25 *V*) with variable adjustment, an inductance coil (L = 1 *H*) a capacitance (C = 15 μ*F*), resistance (R = 300 ohm), *A.C.* voltmeter (0 – 25 *V*), *AC* milliammeter (0–100 mA), connecting wires etc.

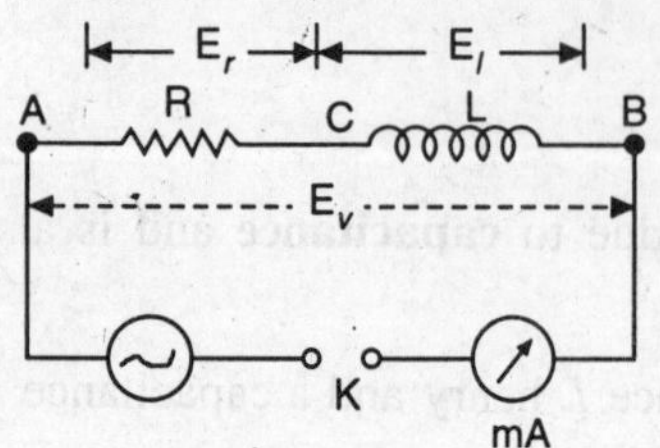

Fig. 34.9

Procedure. (*a*) **R.L circuit.** 1. Draw a diagram showing the scheme of connections as in Fig. 34.9 and connect up accordingly.

2. Connect the voltmeter across the points *A* and *B*, *switch on the current and adjust the applied voltage from the secondary of the step-down transformer so that a current of about 50–60 m. A flows in the circuit.* Note the milli-ammeter reading I_v and the voltmeter reading E_v .

3. Switch off and disconnect the voltmeter from the points *A* and *B*. *Connect it* (*across R*) between the points *A* and *C*. Switch on the current. See if the reading of the *mA* remains constant. If not adjust the voltage to keep *mA* reading the same. Note the reading of the voltmeter. Let it be E_r.

4. Switch off and disconnect the voltmeter from the points *A* and *C*. Connect it (across *L*) between the points *B* and *C*. Switch on the current. See if the reading of the *mA* remains constant. If not adjust the voltage to keep *mA* reading the same. Note the reading of the voltmeter. Let it be E_l.

5. Repeat for three different values of current.

Observations.

Sl. No.	*Current in*		*Voltage across R and L* (E_v)	*Voltage across*		*Resistance*
	mA	*Amp* I_v		*R* (E_r)	*L* (E_l)	$R = \frac{E_r}{I_v}$
1						
2						
3						

6. **To draw the voltage vector triangle.** From the values of E_v, E_r and E_l for each value of current I_v draw a voltage vector triangle as shown in Fig. 34.10 :-

Draw a line *AB* to represent E_r on a convenient scale. From the ends *A* and *B* draw arcs corresponding to E_v and E_l respectively on the same scale cutting at *C*. Join *AC* and *BC*, then *ABC* is the voltage vector triangle. Draw arrow heads to show the direction of each voltage.

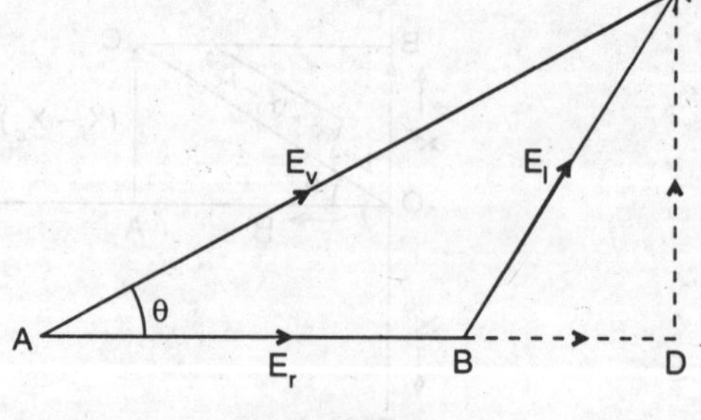

Fig. 34.10

The current I_v is in phase with the voltage E_r applied across *R*. Therefore, the vector *AB* also represents the phase of the current in the circuit.

Measure the angle θ. This gives the *phase* angle by which the voltage *leads* the current.

From the point *C*, draw *CD* perpendicular to *AB* meeting it at *D*. The triangle *BCD* represents the voltage vector triangle for the inductance coil *L*. *BD* represents the voltage across the (ohmic) resistance of the coil and DC the voltage across the pure inductance. The two combine to give the net voltage $BC = E_l$ across the coil.

If I_v is the current in the circuit, then

$$E_r = AB = R\,I_v$$

and $$BD = r\,I_v$$

where r is the (ohmic) resistance of the inductance coil

$$\therefore \quad \frac{r}{R} = \frac{BD}{AB}$$

or Ohmic resistance of the inductance coil $r = \dfrac{BD}{AB} R$

Measure BD and AB and calculate r. CD represents the voltage across the pure inductance L

$$\therefore \quad CD = X_l I_v = \omega L I_v = 2\pi nL I_v$$

$$\therefore \quad \frac{CD}{AB} = \frac{X_l}{R} = \frac{2\pi nL}{R}$$

or $$X_l = R\frac{CD}{AB}$$

and $$L = \frac{R}{2\pi n}\frac{CD}{AB} \text{ henry}$$

Calculations from voltage vector triangle

No.	θ	AB	AC	BC	BD	CD	$r = \dfrac{BD}{AB}R$	$X_l = R\dfrac{CD}{AB}$	L
1									
2									
3									

(*b*) **R.C. circuit** 7. Draw a diagram showing the scheme of connections as in Fig. 34.11 and connect accordingly.

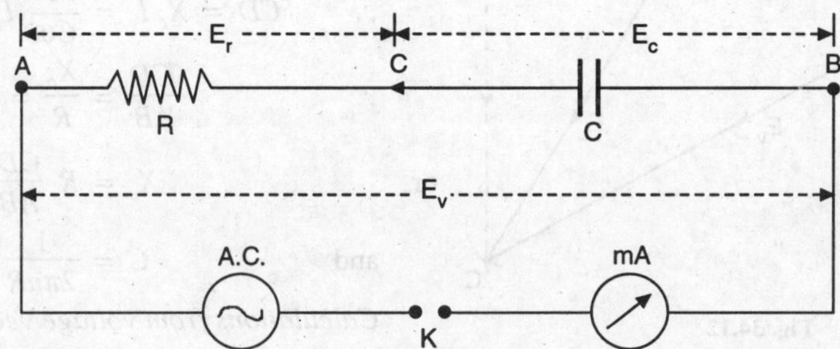

Fig. 34.11

8. Proceeding as in (*a*) note the *m A* reading and the values of voltage E_v between the points A and B, E_r (across R) between A and C and E_c (across the capacitor C) between C and B keeping the current constant as in steps 2–5 in (*a*). Repeat for three different values of current.

Observations.

No.	Current in		Voltage across R and C (E_v)	Voltage across		Resistance
	mA	Amp (I_v)		R (E_r)	C (E_c)	$R = \frac{E_r}{I_v}$
1						
2						
3						

9. **To draw the voltage vector triangle.** From the values of E_v, E_r and E_c for each value of current I_v draw a voltage vector triangle as explained in step 6 .

Note. As in an *RC* circuit the voltage lags behind the current the voltage vector triangle has been drawn in such a way that θ lies in the fourth quadrant and has a *negative value*. Also *DC* the voltage across the pure capacitance lies along the –*Y* direction *i.e.* lags behind the current by a phase angle π/2.

Measure the angle θ. This gives the phase angle by which the voltage *lags* the current. From the point *C* draw *CD* perpendicular to *AB* meeting it at *D*. The triangle *BCD* represents the voltage vector triangle for the capacitance *C*. *BD* represents the voltage across the *loss resistance* of the capacitor and *DC* the voltage across pure capacitance. The two combine to give the net voltage $BC = E_c$. If I_v is the current in the circuit, then

$$E_r = AB = R\, I_v$$

and $$BD = r\, I_v$$

where *r* is the *loss resistance* of the capacitor.

$$\therefore \quad \frac{r}{R} = \frac{BD}{AB}$$

or Loss resistance of the capacitor $r = \frac{BD}{AB} R$

Measure *BD* and *AB* and calculate *r*.

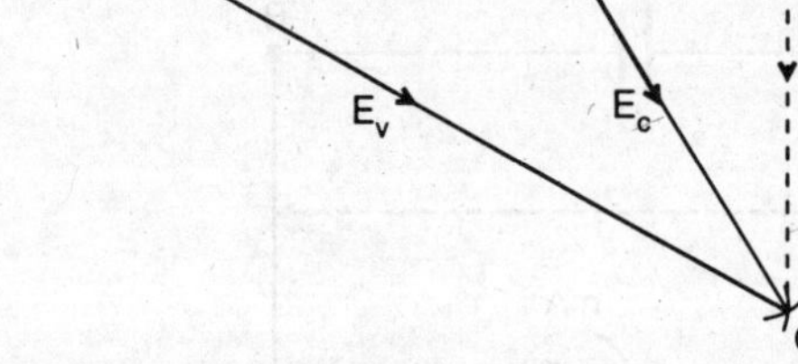

Fig. 34.12

CD represents the voltage across the pure capacitor *C*

$$\therefore \quad CD = X_c\, I_v = \frac{1}{C\omega} I_v = \frac{1}{2\pi nC} I_v$$

$$\therefore \quad \frac{CD}{AB} = \frac{X_c}{R} = \frac{1}{2\pi n\, CR}$$

or $$X_c = R\,\frac{CD}{AB}$$

and $$C = \frac{1}{2\pi nR}\,\frac{AB}{CD}$$

Calculations from voltage vector triangle.

No.	θ	AB	AC	BC	BD	CD	$r = \frac{BD}{AB} R$	$X_c = R\frac{CD}{AB}$	C
1									
2									
3									

Precautions. 1. Only well insulated pieces of flexible wire should be used.

2. Naked parts of the circuit should not be touched when the current is passing.

3. Suitable values of *R, L* and *C* should be chosen.

4. The current in the circuit should remain constant while measuring the voltage across *R* and *L* or across *R* and *C*.

Experiment 34.2. (*a*) To find the impedance of an A.C. circuit containing R, L and C in series.

(*b*) To draw the voltage vector triangle and to find the phase angle between the voltage and current.

Apparatus. A variable transformer, an *A.C.* milliammeter of range 0–500 milliamp, an *A.C.* voltmeter of range 0–250 volt, an A.C. voltmeter of range 0–25 volt, an inductance coil, a capacitor, lamp resistance, (60 watt) and connecting flexible wires.

Theory. The impedance *Z* of an *A.C.* circuit containing a resistance *R*, an inductor *L* and capacitor *C* is given by

$$Z = \sqrt{R^2 + (X_l - X_c)^2}$$

where $X_l = \omega L = 2\pi nL$ is the reactance due to inductance and

$$X_c = \frac{1}{\omega C} = \frac{1}{2\pi nC}$$

is the reactance due to capacitance, *n* being the frequency of the alternating current and $\omega = 2\pi n$ the *angular* frequency.

Procedure. 1. Draw a diagram showing the scheme of connections as shown in Fig. 34.13 and use a 60 watt lamp for the resistance *R*.

2. Connect the output terminals of the variable transformer to the lamp resistance *R*, the capacitor *C* and an inductance coil *L* through the milliammeter in series. Connect the voltmeter of 0–250 volt range across all these between the points *A* and *B*. Note the zero errors in the milliammeter and both the voltmeters, if any.

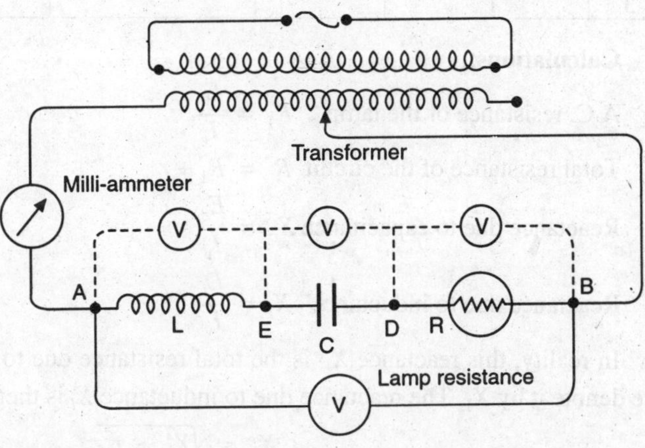

Fig. 34.13

3. Connect the input terminals of the transformer to the mains supply plug and adjust the voltage so that a current of about 150–200 milliamp flows through the circuit. Note the reading of the voltmeter and the milli-ammeter. Let the voltmeter reading be E_v and milliammeter reading I_v.

4. Switch off the current, disconnect the voltmeter and connect it between *A* and *D i.e.* across the inductance coil *L* and capacitor *C* combined. If the current changes adjust it to the previous value and note the voltmeter reading. Let it be E_{lc}.

5. Switch off the current, disconnect the voltmeter and connect it across the resistor *R*, between the points *B* and *D*. If the current changes adjust it to the previous value and note the voltmeter reading. Let it be E_r.

6. Switch off the current and now connect the voltmeter across the capacitor C between the terminals E and D and adjust the value of current if necessary to the previous value. Note the voltmeter reading and let it be E_c.

7. Switch off the current and connect the voltmeter of range 0–25 volt across the inductor L between the points A and E and adjust the value of the current to the previous value if necessary. Note the voltmeter reading. Let it be E_l.

8. Repeat the above observations by changing the current in the circuit to another convenient value.

9. Using a $D.C.$ ammeter, a $D.C.$ voltmeter and a battery of about 4–5 volt find the $D.C.$ resistance r of the inductance coil L as well, if not marked on it.

Observations.

Zero error of the milliammeter = milliamp
Zero error of the voltmeter (0–250 range) = volt
Zero error of the voltmeter (0–25 range) = volt
Resistance of inductance coil r = ohm

No.	*Current* I_v	*Voltage across*					*Reactance due to*			*Impedance*
		AB (E_v)	AD (E_{lc})	R (E_r)	C (E_c)	L (E_l)	R	C (X_c)	L (X_l)	$Z = \frac{E_v}{I_v}$
1										
2										
3										

Calculations.

A.C. resistance of the lamp $R_1 = \frac{E_r}{I_v}$

Total resistance of the circuit $R = R_1 + r$

Reactance due to capacitance $X_c = \frac{E_c}{I_v}$

Reactance due to inductance $X_l = \frac{E_l}{I_v}$

In reality, this reactance X_l is the total resistance due to inductance and resistance of the coil. We denote it by X_1. The reactance due to inductance X_l is then given by

$$X_1 = \sqrt{X_l^2 + r^2}$$

or

$$X_l = \sqrt{X_1^2 - r^2}$$

Impedance of the circuit

$$Z = \sqrt{R^2 + (X_l - X_c)^2}$$

To draw the voltage vector triangle. From the values of E_v, E_r and E_c *for each value of current* I_v draw a voltage vector triangle as shown. Measure the angle θ. This gives the phase angle. If $X_l > X_c$, θ is positive and voltage leads the current. If $X_l < X_c$, θ is negative and voltage lags the current. Also calculate $E_l - E_c$ and compare its value with E_{lc}.

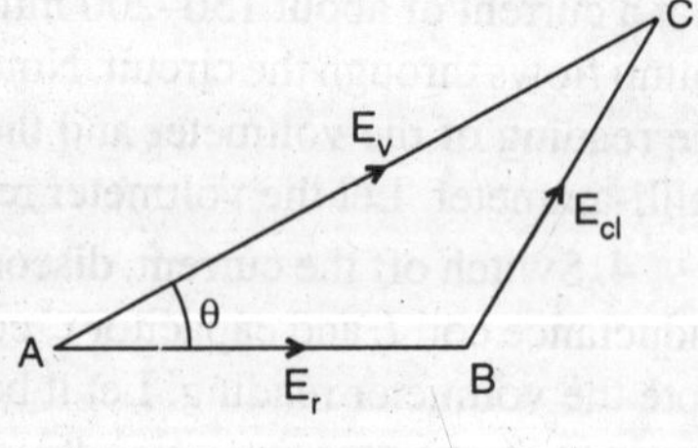

Fig. 34.14

Precautions. Same as in **Expt. 34.1.**

34.5. Series resonant circuit. We have seen that when an alternating voltage E_v is applied to an A.C. circuit containing a resistance R ohm, an inductance L henry and a capacitance C farad, then the current in the circuit

$$I_v = \frac{E_v}{\sqrt{R^2 + (X_l - X_c)^2}} = \frac{E_v}{Z}$$

where $X_l = \omega L = 2\pi nL$ is the inductive reactance $X_c = \frac{1}{\omega C} = \frac{1}{2\pi nC}$ the capacitative reactance and Z the impedance of the circuit, ω being the angular frequency $= 2\pi n$ where n is the frequency of the A.C.

$\therefore$ Impedance $Z = \sqrt{R^2 + (X_l - X_c)^2} = \sqrt{R^2 + (L\omega - 1/C\omega)^2}$

and the phase angle θ between the current and the applied voltage is given by

$$\tan\theta = \frac{X_l - X_c}{R}$$

Thus the impedance of an *A.C.* circuit depends upon the frequency of the applied *E.M.F.* and hence the current depends upon frequency.

When $n = 0$ $\quad X_l = 0$ $\quad$ and $X_c = \infty$

and $\quad$ When $n = \infty$ $\quad X_l = \infty$ $\quad$ and $X_c = 0$.

As the frequency of the applied voltage is gradually increased from zero to ∞, X_l *increases from zero to* ∞ *and* X_c *decreases from* ∞ *to zero. For a certain value of the frequency* $n = n_r$

$$X_l = X_c$$

or
$$2\pi n_r L = \frac{1}{2\pi n_r C}$$

$\therefore$
$$n_r = \frac{1}{2\pi\sqrt{LC}}$$

The frequency n_r is called the *resonant frequency.* $\frac{1}{2\pi\sqrt{LC}}$ *gives the natural frequency of an LC circuit and when the frequency of the applied E.M.F equals the natural frequency of the circuit resonance is said to take place.*

At resonant frequency $X_l = X_c$ and the total impedance of the *A.C.* circuit.

$$Z = \sqrt{R^2 + (X_l - X_c)^2} = R$$

$\therefore$ At resonance, current $I_v = I_r = \frac{E_v}{Z} = \frac{E_v}{R}$ has the *maximum value.*

Again $\tan\theta = \frac{X_l - X_c}{R} = \frac{0}{R} = 0$ or $\theta = 0$

i.e. the *current and the e.m.f. are in phase.* Also the power factor $\cos\theta = 1$

34.6. Quality factor Q. *The ratio of the potential drop across the inductance or capacitance to the potential drop across the resistance (or the applied voltage) at resonance is called the quality factor or Q-value of the circuit.*

If I_r is the current at resonance, then for an *inductor*

$$Q_l = \frac{E_l}{E_R} = \frac{I_r X_l}{I_r R} = \frac{X_l}{R} = \frac{L\omega}{R}$$

Similarly for a *capacitor*

$$Q_c = \frac{E_c}{E_R} = \frac{I_r X_c}{I_r R} = \frac{X_c}{R} = \frac{1/C\omega}{R} = \frac{1}{C\,\omega\,R}$$

At resonance $X_l = X_c$

$$\therefore \qquad Q_l = Q_c = Q$$

i.e. the quality factor for inductance is *exactly* equal to the quality factor for capacitance.

It can be proved that quality factor

$$Q = 2\pi \frac{n_r}{n_2 - n_1}$$

where n_r is the resonant frequency and n_1 and n_2 are the values of frequency above and below the resonant frequency for which the current in the circuit is $\frac{I_r}{\sqrt{2}} = 0.707\, I_r$, where I_r is the current at resonance.

The difference in frequencies n_1 and n_2 is called the band width β.

$$\therefore \qquad \beta = n_2 - n_1$$

Experiment 34.3. (*a*) To plot a graph between current and frequency in a series *LCR* circuit and to find the resonant frequency.

(*b*) To find the quality factor and band width of a series resonant circuit

Apparatus. An audio–frequency oscillator (range 10 *Hz* to 10 *KHz*), an inductance coil, a capacitor, a non-inductive resistance box, an electronic or vacuum tube voltmeter (*V.T.V.M.*) or a cathode ray oscilloscope (*C.R.O.*), a milliammeter connecting wires etc.

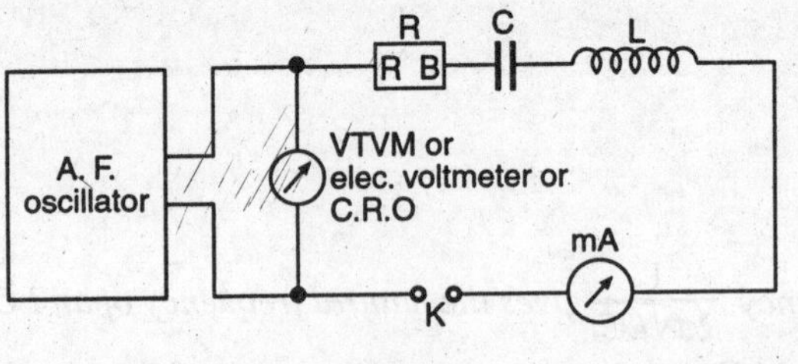

Fig. 34.15

Note The value of inductance *L* and capacitance *C* should be properly selected so that the natural frequency of the circuit given by $n = \frac{1}{2\pi\sqrt{LC}}$ lies in the range 1000–2000 *Hz*.

Procedure 1. Draw a diagram showing the scheme of connections as shown in Fig. 34.15 and connect accordingly.

2. From the given values of *L* and *C* calculate the natural frequency of the circuit by the relation

$$n_r = \frac{1}{2\pi\sqrt{LC}}$$

and adjust the oscillator to the frequency n_r for series resonance. This is indicated when the *mA* records the *maximum* current. Any increase or decrease from this value will decrease the current and the *mA* reading will go down.

3. Adjust the output voltage of the oscillator and the resistance *R* to get maximum current within the range of the *mA*. *The value of output voltage of the oscillator and R must be kept constant throughout the experiment.*

4. Now decrease the frequency of the *A.F.* oscillator in steps of 100 *Hz* and measure the corresponding value of current in the milli-ammeter for each observation.

Similarly *increase* the frequency above the resonance frequency n_r in steps of 100 *Hz* and note the value of current in each case.

5. Repeat the experiment for two more different values of *R*.

Note. The experiment may also be repeated by changing the value of L and C, in which case the resonance (or natural frequency) will also change.

Observations

Resistance R = ohm

Inductance L = H

Capacitance C = μF = F

Output voltage of the audio oscillator = Input voltage for LCR circuit

= E_v volts.

Sl. No.	*Frequency (Hz)*	*Current in the circuit in mA for*		
		R_1	R_2	R_3

6. Plot a graph between current in mA taken along the Y- axis and frequency in Hz taken along the X-axis. A typical graph is shown in Fig 34.16 where $R_3 > R_2 > R_1$

Calculations. Current at resonance from the graph for

(*i*) R_1 = mA = A

(*ii*) R_2 = mA = A

(*iii*) R_3 = mA = A

Calculated value of current at resonance for

(*i*) $R_1 = \frac{E_v}{R_1}$ = A

(*ii*) $R_2 = \frac{E_v}{R_2}$ = A

(*iii*) $R_3 = \frac{E_v}{R_3}$ = A

Difference for (*i*) (*ii*) (*iii*)

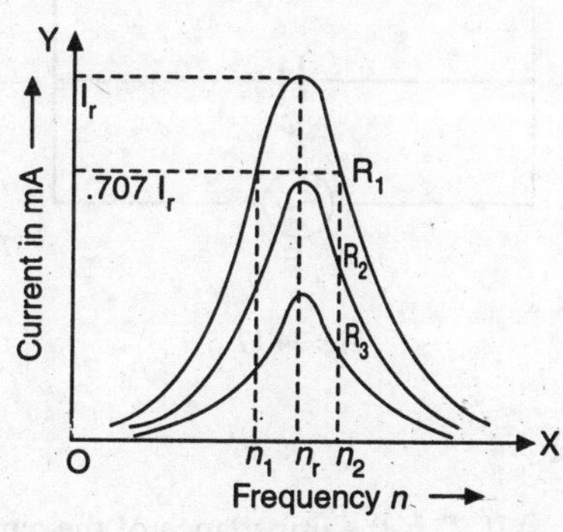

Fig. 34.16

Resonant frequency n_r = Hz

Calculated value of n_r = $\frac{1}{2\pi\sqrt{LC}}$ = Hz

Difference = Hz

Quality factor (For R_1)

Maximum value of current at resonance I_r = mA

Corresponding frequency n_r = Hz

$0.707\, I_r$ = mA

Corresponding value of frequency

below $n_r = n_1$ = Hz

Above $n_r = n_2$ = Hz

Band width $\beta = n_2 - n_1$ = Hz

$\therefore$ Quality factor $Q = 2\pi \left(\frac{n_r}{n_2 - n_1} \right)$

(*i*) Calculated value of Q from inductance $L = \frac{L\,\omega_r}{R} = \frac{2\pi n_r L}{R}$

(*ii*) Calculated value of Q from capacitance C

$$= \frac{1/C\omega_r}{R} = \frac{1}{2\,\pi\, n_r\, CR} =$$

Difference (*i*) (*ii*)

Precautions. 1. The value of inductance L and capacitance C should be properly selected so that the natural frequency of the circuit lies between 1000 and 2000 *Hz*.

2. Adjust the frequency of the *AF* oscillator when the milliammeter in the *RLC* circuit gives maximum current.

3. The value of output voltage of oscillator and resistance R must be kept constant throughout the experiment.

4. Resonance position must be obtained fist by decreasing the frequency and then by increasing the frequency.

5. The *VTVM* or the *CRO* must be used with its proper range.

34.7. Parallel resonant circuit. Now consider a circuit containing an inductor L and a capacitor C connected in *parallel* to an *A.C.* source. The resistance R shown in series with the inductor is the *load resistance* which also includes the resistance of the inductor coil.

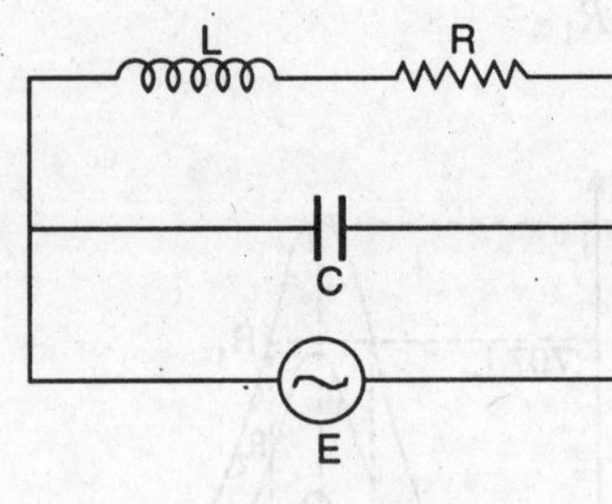

Fig. 34.17

Reactance due to capacitance

$$X_c = \frac{1}{C\omega}$$

As the presence of the capacitor produces a *phase lag* of $\pi/2$, we can represent it in the j notation as

$$X_c = -\frac{j}{C\,\omega}$$

or
$$\frac{1}{X_c} = j\,C\,\omega$$

If Z_1 is the impedance of the capacitor branch, then

$$\frac{1}{Z_1} = j\,C\,\omega$$

Similarly reactance due to inductance $X_l = L\,\omega$. As the presence of the inductance produces a *phase lead* of $\pi/2$, we can represent it in the j notation

as
$$X_l = j\,L\,\omega$$

$\therefore$ Impedance of the inductive branch

$$Z_2 = R + j\,L\,\omega$$

$$\therefore \quad \frac{1}{Z_2} = \frac{1}{R + jL\omega} = \frac{R - jL\omega}{(R - j\,L\,\omega)\,(R + j\,L\,\omega)}$$

$$= \frac{R - j\,L\,\omega}{R^2 + L^2\,\omega^2}$$

As Z_1 and Z_2 are in parallel, the net impedance is given by

$$\frac{1}{Z} = \frac{1}{Z_1} + \frac{1}{Z_2}$$

$$= \frac{R}{R^2 + L^2\omega^2} + j\left(C\omega - \frac{L\omega}{R^2 + L^2\omega^2}\right)$$

The reciprocal of impedance Z *i.e.* $\frac{1}{Z}$ is known as the **admittance** of the *A.C.* circuit.

Under conditions of resonance, current and voltage are in phase. In other words the co-efficient of j *i.e.* the reactive term which brings about a phase change is zero at the resonant frequency n_r. As phase difference θ is zero, the power factor $\cos\theta = 1$.

Note. It should be noted that at parallel resonance the term containing j being zero, the *admittance* has the *minimum value*, which means the *impedance* is *maximum* and the *current* is *minimum*. In a series resonance circuit the current at resonance is maximum (impedance Z being minimum) This is why parallel resonance is known as anti-resonance. Hence, at anti-resonance (or parallel resonance) for angular frequency ω_r, we have

$$C\omega_r - \frac{L\omega_r}{R^2 + L^2\omega_r^2} = 0$$

$$\therefore \quad \frac{L\omega_r}{R^2 + L^2\omega_r^2} = C\omega_r$$

or $$R^2 + L^2\omega_r^2 = \frac{L}{C}$$

or $$\omega_r = \sqrt{\frac{1}{LC} - \frac{R^2}{L^2}}$$

$$\therefore \text{ Anti resonant frequency } n_r = \frac{1}{2\pi}\sqrt{\frac{1}{LC} - \frac{R^2}{L^2}}$$

$$\text{Impedance at resonance} = \frac{R^2 + L^2\omega_r^2}{R} = R + \frac{L^2}{R}\omega_r^2$$

$$= R + \frac{L^2}{R}\left[\frac{1}{LC} - \frac{R^2}{L^2}\right]$$

$$= \frac{L}{RC}$$

The quantity $\frac{L}{RC}$ is also known as *dynamic resistance* of the parallel circuit.

To sum up, as the frequency of the applied voltage is gradually increased from zero to ∞, the inductive reactance X_l increases from zero to ∞ and capacitative reactance X_c decreases from ∞ to zero. The current through the inductive branch decreases and the current through the capacitative branch increases. At (anti) resonance frequency $X_l = X_c$, the current through the two branches is equal in value but differs in phase by 180° (π) so that the resultant current is zero.

When the inductive branch also has a resistance, the resultant current is minimum and not zero. Thus at (anti) resonance frequency

Net impedence $Z = \frac{L}{RC}$ = a maximum

and net current $I_r = \frac{E_v}{Z}$ = a minimum.

Experiment 34.4. To plot a graph between current and frequency in a parallel LCR circuit and to find the (anti) resonant frequency.

Apparatus. Same as in **Expt. 34.3.**

Procedure 1. Draw a diagram showing the scheme of connections and connect accordingly.

Note. A suitable set of L and C values for this experiment is $L = 900\ mH$ and $C = 0.03\ \mu$ F.

2. From the given values of L and C calculate the natural frequency of the circuit by the relation

$$n = \frac{1}{2\pi\sqrt{LC}}$$

Now tune the oscillator to this frequency and adjust the value so that the minimum current is indicated by the milli-ammeter.

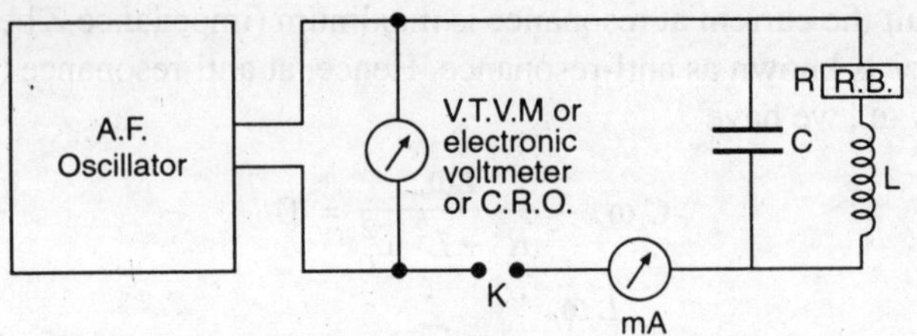

Fig. 34.18

3. Adjust the output voltage of the oscillator and the resistance R to get the minimum current (not zero). Adjust the frequency also, if necessary. *The value of out put voltage of the oscillator and R must be kept constant throughout the experiment.*

4. Now *decrease* the frequency of the *A.F.*oscillator. in steps of 100 *Hz* and measure the corresponding value of current in the milli-ammeter for each observation. Similarly increase the frequency above the (anti) resonance frequency in steps of 100 *Hz* and note the value of current in each case.

5. Repeat for one more value of R after adjusting the frequency of the oscillator for minimum current, if necessary.

Observations.

Inductance L = H

Capacitance C = μ F = F

Resistance R = ohm

Output voltage of the audio-oscillator =

Input voltage of parallel resonant circuit E_v = Volts

Sl. No.	*Frequency in Hz*	*Current in the circuit in mA*	
		R_1	R_2

6. Plot a graph between current in *mA* taken along *Y*-axis and frequency in *Hz* taken along the *X*-axis.

A typical graph in shown is Fig. 34.19. From the graph find n_r *i.e.* (anti) resonant frequency at which the current is minimum. Also find the value of current at (anti) resonance for R_1 as well as for R_2.

Current at (anti) resonance for

(*i*) R_1 = mA = A

(*ii*) R_2 = mA = A

Impedance at resonance $Z = \dfrac{L}{RC}$

(*i*) Z for $R_1 = \dfrac{L}{R_1 C} =$ ohm

(*ii*) Z for $R_2 = \dfrac{L}{R_2 C} =$ ohm

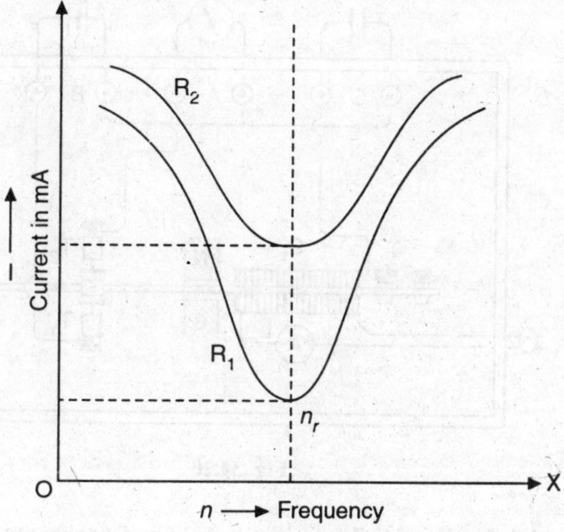

Fig. 34.19

Calculated value of current at (anti) resonance

for (*i*) $R_1 = \dfrac{E_v}{Z} = \dfrac{E_v R_1 C}{L} =$ A

(*ii*) $R_2 = \dfrac{E_v}{Z} = \dfrac{E_v R_2 C}{L} =$ A

Difference for (*i*) (*ii*)

(i) (Anti) resonant frequency from the graph for $R_1 = n_r =$ Hz

Calculated value for $R_1 = \dfrac{1}{2\pi}\sqrt{\dfrac{1}{LC} - \dfrac{R_1^2}{L^2}} =$ Hz

difference = Hz

(*ii*) Anti resonant frequency from the graph for $R_2 = n_r =$ Hz

Calculated value for $R_2 = \dfrac{1}{2\pi}\sqrt{\dfrac{1}{LC} - \dfrac{R^2}{L^2}} =$

Difference = Hz

Quality factor for $R_1, Q = \dfrac{L\omega}{R_1} = \dfrac{2\pi L n_r}{R_1}$

(From the graph)

Quality factor for $R_2, Q = \dfrac{L\omega}{R_2} = \dfrac{2\pi L n_r}{R_2} =$

(From the graph)

Precautions. Same as in **Expt. 34.3.**

34.8. Electrical vibrator. It consists of a solenoid S_1 through which passes a thin steel rod *R* which can be clamped at one end. The rod is slightly flattened at the free end and a hole is drilled in it so that a string can be tied to the rod through this hole. The solenoid can be connected to the *A.C.* mains through a suitable lamp resistance *L. A* permanent horse shoe magnet *NS* is mounted on the base board and the steel rod passes between the poles.

When alternating current is passed through the solenoid, the steel rod placed along its axis gets magnetised longitudinally. The polarity of the rod reverses after every half cycle of the current. If the rod is attracted towards the north pole of the permanent magnet in one half of the $A.C.$ cycle, it will be attracted towards the south pole of the magnet in the next half cycle. The rod thus begins to vibrate with the frequency of the $A.C.$ supply. The length of the rod can be adjusted to get resonance which is indicated by a large amplitude of vibration of the free end. The arrangement produces continuous vibrations and the energy is derived from the $A.C.$ supply.

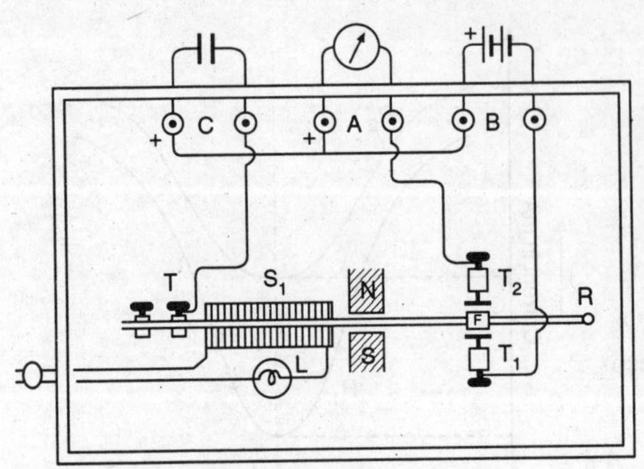

Fig. 34.20

To find the frequency of $A.C.$ supply one end of a thread about 2 metre long is tied to the free end of the rod. The other end passes over a pulley and is tied to a pan. Suitable weights are placed in the pan to keep the string taut. The vibrations of the rod are communicated to the string which begins to vibrate in a number of segments, the frequency of the string being the same as that of the rod or $A.C.$ supply.

Experiment 34.5. To find the frequency of A.C. supply using an electrical vibrator.

Apparatus. Electrical vibrator, a frictionless pulley, a string of uniform thickness, a light weight pan, a weight box, a balance and a metre rod.

Theory. If a string of mass m per unit length is connected to the vibrating rod of an electrical vibrator and stretched with a tension T, the string vibrates in a number of segments as in Melde's experiment. If the length of string is so adjusted that the nodes are clearly marked, the frequency of the stretched string is the same as that of the vibrating rod which is vibrating with the frequency of the $A.C.$ mains.

If l is the length of one loop of the string, T the stretching force and m the mass per unit length of the string, then the frequency of the vibrator is given by

$$n = \frac{1}{2\,l}\sqrt{\frac{T}{m}}$$

This is also the frequency of the $A.C.$ supply.

Procedure, 1. Switch on the current and adjust the length of the steel rod till the free end attains the maximum amplitude.

2. Switch off the current and tie one end of a string about 2 metre long to the free end of the steel rod. Pass the other end over a frictionless pulley fixed on the table and attach a light weight pan to the other end. Add some weights in the pan to make the string taut.

3. Switch on the current, the string will be found to vibrate in a number of loops. Adjust the load, if necessary. Change the length of the vibrating string by shifting the vibrator forward or backward so that the loops are sharply defined and the nodes are clearly marked.

4. Mark the position of the extreme nodes leaving out the first and the last loops as their position cannot be ascertained correctly. Measure the distance between the two marks and divide it by the number of loops contained to find the value of l. Repeat three times for the same stretching force by changing the length of the string so as to change the number of loops. Note the weights added in the pan. Repeat for three different loads.

5. Find the weight of the pan with a balance and also find the weight of an exact length, say 200 cm of string.

Observations.

Weight of the pan = gm

Mass of 200 cm of string = gm

Mass per unit length m = gm

Sl. No.	*No. of loops*	*Distance between extreme nodes*	*Length of each loop (l)*	*Wt. in the pan*	*Tension (T)*	$n=\frac{1}{2t}\sqrt{\frac{T}{m}}$
1						
2						
3						

Mean frequeny n = cycles/sec

Precautions. 1. The string should be of uniform thickness so that it has a fairly constant mass per unit length. A fishing cord fulfils this condition and is most suitable for this experiment.

2. The length of the steel rod should be adjusted before attaching the string so that it vibrates with a large amplitude in resonance with the frequency of the *A.C.* supply.

3. The nodes and anti-nodes should be sharply defined.

4. There should be no friction in the pulley.

34.9. Determination of capacitance. The electrical vibrator can also be used to find the capacitance of a capacitor. For this purpose three pairs of terminals are provided on the base board as shown at *A, B* and *C* in Fig. 34.20. Across the terminals marked *A* is connected a milliammeter or a micro-ammeter, across those marked *B* is connected a battery; and across those marked *C* is connected the capaçitor whose capacitance is to be determined The internal connections are as indicated in Fig. 34.20. The steel rod carries near its end a small free metal piece *F* with flat sides. On either side of the metal piece are two small metal discs passing through the two terminals T_1 and T_2. The distance between the two discs is adjustable.

The rod vibrates when the alternating current is passed through the solenoid. The distance between the metal discs is adjusted so that the metal piece carried by the steel rod just touches them during its vibration. When the metal piece touches the disc on one side (T_1) the capacitor is charged and when it touches the other side (T_2) the capacitor is discharged through the micro-ammeter.

Experiment 34.6. To find the capacitance of a capacitor using an electrical vibrator given that the frequency of A.C. supply is 50 cycles per second.

Apparatus. An electrical vibrator, a battery of 6 volts, a capacitor, a micrommeter, a high resistance voltmeter and connecting wires.

Theory. The capacitor is connected across the terminals marked *C*, the battery across the terminals marked *B* and a milliammeter across *A*. If n is the frequency of the *A.C.* supply the rod of the electrical vibrator will vibrate n times per second and the capacitor will also be charged and discharged n times in one second. If E is the *e.m.f.* of the battery and C the capacitance of the capacitor then the charge passing through the micro-ammeter in each discharge

$$= CE$$

Hence charge passing per sec = nCE. As current is the rate of flow of charge the current i indicated by the mircro-ammeter is given by

$$i = nCE$$

$$\therefore \quad C = \frac{i}{nE} = \frac{\text{current}}{\text{frequency} \times e.m.f. \text{ of battery}}$$

Procedure. 1. Draw a diagram showing the scheme of connections. Connect the battery, the micro-ammeter and the capacitor as shown in Fig. 34.20.

2. Switch on the *A.C.* supply, adjust the length of the steel rod so that it has a maximum amplitude. Now adjust the distance between the metal discs so that the metal piece attached to the rod just touches them during its vibration. This will happen when the micro-ammeter shows a steady deflection.

3. Note the reading of the micro-ammeter. Switch off the *A.C.* supply and find the *e.m.f.* of the battery with a high resistance voltmeter.

4. Repeat the observation three times by using different *e.m.fs.* to charge the capacitor.

Observations. Frequency of the *A.C.* mains $n =$

Zero reading of voltmeter =

Zero reading of micro-ammeter =

Sl. No.	*Reading of Micro-ammeter*		*Current in amperes i*	*E.M.F. of battery in volts*		$\frac{i}{E}$
	observed	*corrected*		*observed*	*corrected (E)*	
1						
2						
3						

$$\text{Mean value of } \frac{i}{E} = \quad \text{ampere/volt}$$

$$\text{Capacitance } C = \frac{i}{E} \times \frac{1}{n} \text{ farad}$$

$$= \quad \text{micro-farad}$$

Precautions. 1. Zero-error of the micro-ammeter and voltmeter if any should be noted.

2. Suitable values of the applied voltage should be used to get appreciable deflection in the micro-ammeter.

3. The contact of the metal piece on the vibrating rod with the flat discs should be so adjusted that the microammeter shows a steady deflection when the vibrator is working.

Exercise. *Find the capacitance of the given capacitor using an electrical vibrator when the frequency of A.C. mains is not given.*

Hints. Find the frequency of *A.C.* mains as explained in **Experiment 34.5** and use this value of frequency to calculate the capacitance as explained in **Experiment 34.6.**

Oral Questions

1. Alternating current. (*i*) What is an alternating current? How does it differ from direct current? (*ii*) What is meant by the frequency of A.C.? (*iii*) What is the frequency of the A.C. supply that you are using? (*iv*) How is A.C. measured? (*v*) What do you mean by the root mean square value of A.C.? What is its relation with the maximum value of the current? (*vi*) Define a virtual volt and a virtual ampere. (*vii*) What do you understand by the reactance due to inductance and reactance due to capacitance? (*viii*) Define impedance, of an A.C. circuit. (*ix*) State the relation that exists between ohmic resistance, reactance and impedance. (*x*) What is the relation between potential difference and current in an *A.C.* circuit ? (*xi*) What is the principle of A.C. ammeter and voltmeter that you are using? Can you use these instruments on direct current?

2. LCR circuit. (*i*) In an *RL* circuit does the voltage lead the current or lags? (*ii*) What is the angle of lag or lead ? (*iii*) In an *RC* circuit does the voltage lead the current or lags ? (*iv*) What is the value of the angle of lag or lead ? (*v*) What determines the lag or lead in an *LCR* circuit? (*vi*) What do you mean by voltage vector triangle ?

3. Series and parallel resonance (*i*) What is the natural frequency of an LC circuit ? (*ii*) What do you mean by series resonance ? (*iii*) What is the phase difference between voltage and current at resonance? (*iv*) What is the relation between inductive reactance and capacitative reactance at series resonance ? (*v*) Does the current at series resonance depend upon inductance or capacitance? (*vi*) What is quality factor? (*vii*) What is the relation between quality factor for inductance and quality factor for capacitance, at resonance? (*viii*) What do you understand by band width? (*x*) What do you mean by sharpness of resonance? (*x*) What are the factors on which sharpness of resonance depend ? (*xi*) Is impedance minimum or maximum at parallel resonance ? (*xii*) Why do we call the parallel resonance as anti resonance ? (*xiii*) What is the impedance at anti-resonance? (*xiv*) What do you mean by dynamic resistance of a parallel LCR circuit? (*xv*) A series resonant circuit is known as an acceptor circuit and a parallel resonant circuit as a rejecter circuit Why ?

4. Measurement of frequency (*i*) Explain the principle of the vibrator. (*ii*) Why do you use a lamp in the main circuit? (*iii*) State the formula for calculating the frequency. (*iv*) When do you get the maximum amplitude of the vibrations of the string? (*v*) While measuring the length why do you leave the first and the last loop ?

5. Capacitance of a capacitor (*i*) State the formula for calculating the capacitance of a capacitor with a vibrator. (*ii*) Is the micro-ammeter that you are using A.C. or D.C.? (*iii*) Explain how the capacitor is charged and discharged as the steel rod vibrates?

35

Multimeter and Energy Meter

35.1. Multimeter. A multimeter is an electrical measuring instrument which can directly measure current, potential difference and resistance. In other words, it is an ammeter, a voltmeter and an ohmmeter all combined into one. For this reason a multimeter is generally known as an **AVO** (Ampere, Volt, Ohm) **meter**. Some multimeters can also measure capacitances. The multimeters used in the laboratory may measure only *D.C.* voltages and currents or both *A.C.* and *D.C.*

Construction. A multimeter essentially consists of a sensitive moving coil galvanometer. It is provided with several scales on its dial reading the current in amperes, potential difference in volts and resistance in ohms.

To use it as an ammeter suitable values of **shunt** resistances have been provided in the instrument and connected in parallel to the galvanometer coil so as to give the appropriate values of current in micro-amps, milli-amps or amperes.

To use it as a voltmeter suitable values of resistances have been provided in the instrument and connected in **series** with the galvanometer coil, so as to give appropriate values of potential difference in millivolts or volts.

To make the instrument work on *A.C.* also, a solid state crystal diode rectifier is connected in the galvanometer circuit. The rectifier converts the *A.C.* (current or voltage to be measured) into *D.C.* which causes a suitable deflection in the galvanometer.

For the measurement of resistance a dry cell of 1.5 volt is fitted inside the multimeter. Two **testing leads** one **black** and the other **red** in colour with uninsulated metallic pins are provided to give input to the multimeter.

A number of switches are provided on the front panel of the multimeter.

(*i*) **Range switch.** The range switch *S* in the centre of the lower part of the front panel can be adjusted to several positions marked round it. A simplified diagram of the front panel is shown in Fig. 35.1 in which the range switch has been shown to have 12 positions indicating different ranges of current, potential difference and resistance. The range switch can be turned in either direction *i.e.* to the left or right and set on the required range.

(*ii*) **Function switch.** The function switch *F* is located towards the *left* of the range switch *S*. The function of the switch is to set the multimeter for measurement of *D.C.* or *A.C.* For measuring *D.C.* the knob is set on *D.C.* mark and for measuring *A.C.* it is set on *A.C.* mark.

(*iii*) **Zero ohm switch.** The zero ohm switch *Z* is located towards the *right* of the range switch *S*. To use this switch one end of the black coloured test lead is inserted in the socket marked '*common*' or (–) whereas one end of the red lead is inserted in the socket marked (+). The test leads are now short circuited. Ordinarily the pointer of the multimeter should now indicate '*zero*' at the extreme *right* side of the scale. If the pointer is not exactly at zero, the adjustable knob of the zero-ohm switch is slowly rotated in the clockwise direction till the pointer is exactly on the zero mark of the scale.

Note. The zero of the resistance scale starts from the *extreme right hand side* of the front panel whereas those of current and voltage scales start from the *extreme left hand side* of the front panel.

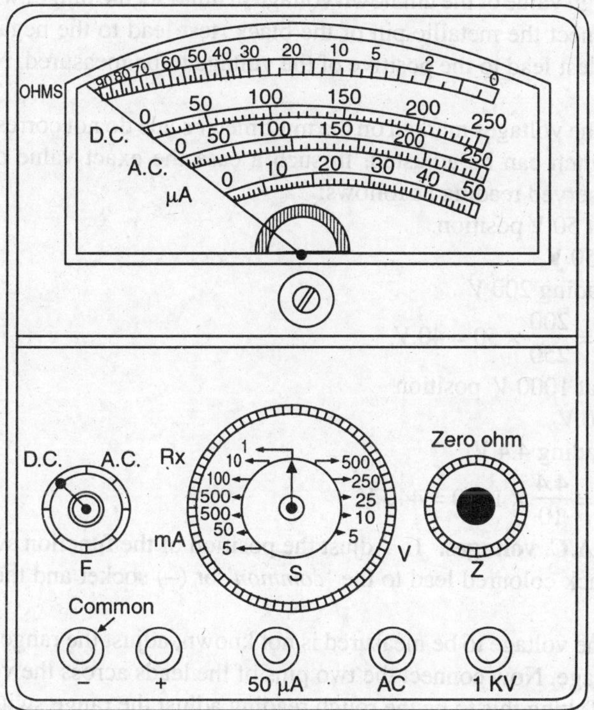

Fig. 35.1

This is because when there is zero ohm resistance in the multimeter circuit, current in the coil is maximum and hence the deflection in the coil is maximum so that the pointer moves to the extreme right hand side where the zero of the resistance scale is situated. As higher and higher resistance is introduced in the multimeter circuit the current goes on becoming less and less, therefore, smaller and smaller deflection is produced. Thus the resistance scale increases from right to left and voltage and current scales from left to right. In other words, the two scales increase in opposite directions.

(*iv*) **Sockets.** There are a number of sockets (metallic holes) provided on the front panel in which banana plugs can be introduced to use the instrument for various purposes. Some of these sockets are shown in Fig. 35.1.

Experiment 35.1. To use a multimeter for measuring

(*a*) **Voltage** (*A.C./D.C.*)

(*b*) **Current** (*A.C./D.C*)

(*c*) **Resistance.**

Apparatus. A multimeter with a dry cell fitted inside it. Two unknown resistances, a *D.C.* source (say a 6 volt battery) *A.C.* mains, connecting wires etc.

Procedure. (*a*) Measurement of *D.C.* voltage.

1. Set the position of the function switch *F* to *D.C.*

2. Connect the black coloured lead to '*common*' or (–) socket and the red coloured lead to the (+) socket.

3. Generally the order of the voltage to be measured is not known. In such a case adjust the range switch to the position of maximum voltage range. Now connect the metallic pin of the *black* test lead to the negative terminal of the voltage to be measured and the pin of the *red* test lead to the *positive*

terminal and note the scale reading. This gives an approximate rough value of the voltage. Now disconnect the pins from the negative and positive terminal of the voltage to be measured.

4. Keeping this rough value of the unknown voltage in mind set the range switch to the appropriate position and again connect the metallic pin of the black test lead to the negative terminal and the metallic pin of the red test lead to the positive of the voltage to be measured. Note the scale reading accurately.

Note. Sometimes the voltages marked on the multimeter scale do not correspond to all the ranges to which the range switch can be adjusted. In such a case the exact value of the voltage can be calculated from the observed reading as follows:

(*i*) Range switch at 50 *V* position.
Scale marked 0–250 *V*
Observed scale reading 200 *V*

$$\therefore \text{ Actual reading} = \frac{200}{250} \times 50 = 40 \ V.$$

(*ii*) Range switch at 1000 *V* position
Scale marked 0–10 *V*
Observed scale reading 4.4 *V*

$$\therefore \text{ Actual reading} = \frac{4.4}{10} \times 1000 = 440 \ V.$$

Measurement of *A.C.* voltages. 1. Adjust the position of the function switch *F* to *A.C.*

2. Connect the black coloured lead to the '*common*' or (–) socket and the red coloured lead to the *A.C.* socket.

3. If the order of the voltage to be measured is not known, adjust the range switch to the position of maximum *A.C.* voltage. Now connect the two pins of the leads across the voltage to be measured and note the reading. Taking this to be the rough reading adjust the range switch to the proper value and note the reading accurately.

4. If the *A.C.* voltage from the mains is to be measured set the range switch to 250 *V*. Connect the two pins of the leads across the voltage to be measured and note the reading on the suitable scale.

Observations. *D.C.* potential difference =
A.C. potential difference =

Precautions. 1. For measuring *D.C.* voltage set the function switch at *D.C.* and for measuring *A.C.* voltage set it at *A.C.* point.

2. For accurate measurement of voltage set the range switch at the appropriate range.

3. Connect the black coloured test lead to common (–) socket.

4. Connect the red coloured test lead to the (+) socket for *D.C.* voltage and to *A.C.* socket for measurement of *A.C.* voltage.

5. Do not touch the meter or the test leads when the voltage to be measured is on and switch off the current before disconnecting the leads.

(*b*) **Measurement of current.**

Direct current 0–50 μA

1. Set the positive of the function switch *F* to *D.C.*

2. Connect the black coloured lead to '*common*' or (–) socket and the red coloured lead to 50 μA socket.

3. Turn the range switch to 50 μA position (generally common with 50 *V*).

Now *open* the circuit in which the current is to be measured. Connect the multimeter in series with the circuit by connecting the metallic pin of the red test lead to the positive and that of the black test lead to the negative end of the circuit. Now switch on the current and note the reading in the multimeter.

If the pointer is deflected to the left the current polarity has been connected in the opposite direction. Turn off the power and reverse the lead connections of the circuit. Again switch on the current and note the reading directly on the 0–50 scale. This current is directly in μA. Turn off the power before disconnecting test leads.

Direct current.

1. Set the position of the function switch to *D.C.*

2. Connect the black coloured test lead to '*common*' or (–) socket and the red coloured test lead to (+) socket.

3. Adjust the range switch *S* to the position of maximum current range.

Now *open* the circuit in which the current is to be measured and connect the multimeter in series with the circuit by connecting the metallic pin of the red test lead to the positive and that of the black test lead to the negative end of the circuit. Switch on the current and note the reading. If the pointer is deflected towards the left the current polarity has been connected in the opposite direction. Switch off the current, and reverse the lead connections of the circuit. Again switch on the current and note the reading. Turn off the power.

4. If the current reading is too small take it as the rough value and adjust the range switch to suitable current range. Again switch on the current and note the reading accurately.

Measurement of alternating current (*A.C*).

1. Set the position of the function switch *F* to *A.C.*

2. Connect the black coloured lead to the '*common*' or (–) socket and the red coloured lead to '*A.C*'. socket.

3. Adjust the range switch *S* to the position of appropriate range. Now *open* the circuit in which the current is to be measured and connect the multimeter in series by connecting the metallic pins of the two leads to be open ends. Note the reading accurately.

Observations. Value of direct current *D.C.* = μA

Value of direct current *D.C.* = mA

Value of alternating current *A.C.* = A

Precautions. 1. For measuring direct current set the functions switch at *D.C.* position and for measuring alternating current set it at *A.C.* position.

2. For accurate measurement of current set the range switch at the appropriate range.

3. Connect the black colured test lead to the '*common*' or (–) socket.

4. Connect the red coloured lead to 50 μ A socket for measurement of current upto 50 μ A and to the (+) socket for measurement of direct current. For the measurement of alternating current connect the red coloured lead to the A.C. socket.

(*c*) **Measurement of resistance.** 1. Set the position of the function switch to *D.C.*

2. Connect the black coloured test lead to the '*common*' (–) socket and the red coloured test lead to the (+) socket. Connect the metallic pins of the test leads with each other. If the pointer of the multimeter does not stand at zero ohm reading on the resistance scale adjust the 'zero-ohm switch' till the pointer stands at zero.

Note. The adjustment has to be made each time when different ranges of resistance scale are used.

3. Adjust the position of the range switch to ($R \times 1$) Connect the pins of the two test leads to the ends of the unknown resistance and note the reading of the pointer on the scale.

If the reading is too small, the value of the resistance is too high. In such a case adjust the range switch to $R \times 10$ position, again adjust zero and note the reading as explained above.

If the reading is still small, adjust the range switch to $R \times 100$ and if necessary to $R \times 1000$ position, adjusting zero each time.

When the range switch is at 10, the multiplying, factor is 10. Similarly for 100 and 1000 position of the range switch the multiplying factor is 100 and 1000 respectively. Measure three different values of resistance say, R_1 R_2 and R_3.

Observations.

Sl. No.	*Resistance*	*Multimeter reading*	*Multiplying factor*	*Resistance*
1	R_1			
2	R_2			
3	R_3			

Precautions. 1. The function switch should be adjusted to the *D.C.* position.

2. Check the zero error of the multimeter before making any measurement of resistance.

3. The observed scale reading must be multiplied by appropriate multiplying factor depending upon the position of the range switch to get the correct value of resistance.

35.2. Energy meter or Watt-hour meter. A watt-hour meter is a device which measures the energy consumed in an electric circuit. Its scale is calibrated to give the reading directly in kilowatt-hours or UNITS. The device may be mechanical or digital. Fig. 35.2 represents the sketch of such a meter.

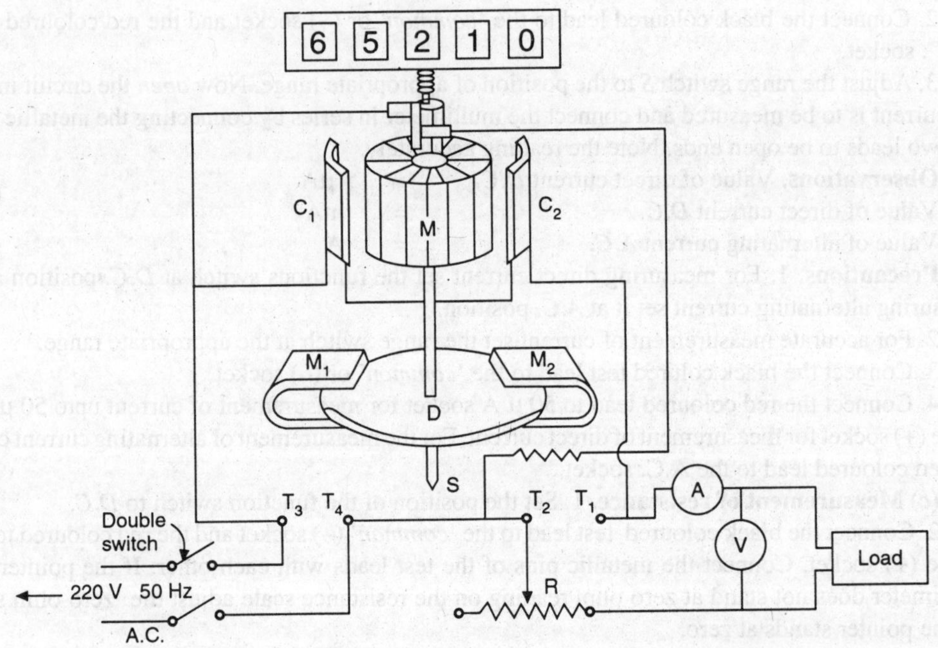

Fig. 35.2

It consists of an armature *M* which can rotate between the field coils C_1 C_2 surrounding it. The field coils are connected in series with the load and thus *produce a magnetic field proportional to the current in the load circuit.* The armature coil is connected in parallel to the load. Thus *its magnetic field is proportional to the voltage applied to the load.* The combined effect of these two fields, one proportional to *V* and the other proportional to *I* is to produce a resulting torque proportional to their product *VI*. Due to this the armature *M* rotates at a speed which is proportional to *VI i.e.* the power in Watts consumed in the load circuit.

The lower end of the armature spindle S carries an aluminium disc D which rotates between the pole pieces of two magnets M_1 and M_2. The eddy currents induced in the disc control the speed of rotation. At the upper end of the spindle S a gear arrangement (or a clock-work mechanism) operates the mechanical counter which records the UNITS, the number of rotations made by the disc being proportional to the number of units (kilo-watt hours) consumed.

Experiment 35.2. To study the working of household energy meter or Watt-hour meter and to determine the power consumption of some electric applicances.

Apparatus. Energy meter fitted on a board, A.C. ammeter (0–10A), A.C. voltmeter (0–250 V), Variable resistance or rheostat (250V – 10A), plastic or rubber insulation covered connecting wires, extension board and appliances like heater, bulb or electric iron (press).

Procedure 1. Set up the energy meter (fitted on board) on the experimental table. There are four wires leading from the base of the meter from terminals marked T_1, T_2, T_3 and T_4. The terminals T_1 and T_2 are meant for the output load or electric appliance consuming energy and the terminals T_3 and T_4 are meant for the input from the voltage supply source or the *line*.

2. Note the energy meter constant *i.e.* the energy in Watt-hour per rotation of the aluminium disc. It is generally stamped near the disc D or marked under the plate covering the lead-connection.

3. Draw a diagram showing the scheme of connections as in Fig. 35.2 and make connections accordingly.

4. Now connect the two input terminals to the extension board and make the value of the resistance in the rheostat maximum. Switch on the supply voltage and see that the voltmeter and ammeter reading is well within scale. Adjust the rheostat and make its resistance value zero as it is only a protective measure.

5. As the disc of the watt-hour meter begins to rotate count the number of rotations for a known time say 10 minutes. (This can be done with the help of a reference *red* mark on the edge of the disc) measured by an accurate stop watch. Record the time and number of rotations. Also note the ammeter reading, voltmeter reading and the reading as shown by the energy meter scale. Switch off the mains switch for some time. Again switch on the mains and repeat the observation to find the number of rotations of the energy meter for the same time. Take three observations.

6. Put off the switch. Disconnect the first appliance and connect the second appliance and repeat the whole of the procedure again taking three observations.

Observations. Energy meter constant = k = Wh/rotation

Appliance	*P.D in volts (V)*	*Current in Amp (I)*	*Number of rotations of disc (n)*	*Time of rotation of disc (t)*	*Electric energy recorded by energy meter* nk *(Wh)* $= \frac{nk}{1000}$ *(kWh)*	*Electric power recorded* $\frac{nk}{t}$ *watt,*	*Electric power calculated VI watt.*
A (*i*)							
(*ii*)							
(*iii*)							
B (*i*)							
(*ii*)							
(*iii*)							
C (*i*)							
(*ii*)							
(*iii*)							

Result

Appliance A Electric energy recorded by energymeter = $\frac{nk}{1000}$ kWh

Electric energy calculated = $\frac{VIt}{1000}$ kWh =

Difference =

Electric power from energy meter = $\frac{nk}{t}$ = Watt

Electric power calculated VI = Watt

Difference

Similarly record for appliances B and C.

Precautions. 1. While performing the experiments with mains supply do not touch anything with naked hands. Switch off the mains before handling.

2. Use ammeter and voltmeter of proper range.

3. Use maximum resistance of rheostat to begin with and make it zero when performing the experiment.

Oral Questions

1. Multimeter. (*i*) What is a multimeter? (*ii*) What are other names of a multimeter? (*iii*) What is the construction of a multimeter? (*iv*) How does the multimeter measure *A.C.* as well as *D.C.* current? (*v*) Why does the zero of the resistance scale start from the extreme right hand side whereas those of current and voltage scales start from extreme left? (*vi*) The division of the current and voltage scales are of equal size where as those of resistance scale go on decreasing in size for higher values of resistance. Why ? (*vii*) Why is the adjustment of zero ohm necessary before measuring resistance? (*viii*) While measuring resistance why does the pointer move even when no external source of e.m.f. is connected to the multimeter? (*ix*) Why is it not some time possible to make zero adjustment completely ? (*x*) If the deflection of the meter while measuring current is to the left, what is the reason and how do you get the correct deflection?

2. Kilowatt hour meter. (*i*) Define a kilowatt-hour. (*ii*) How many Joules of energy make one kilowatt hour. (*iii*) What is the unit of power? (*iv*) An electric iron is marked 750W–250V. What does it mean? (*v*) Define energy meter constant. (*vi*) In an energy meter 600 revolutions of the disc are equivalent to 1 k Wh, what is the energy meter constant?

PART VII

ATOMIC, MOLECULAR AND NUCLEAR PHYSICS

36

Measurement of e/m and Electronic Charge

36.1. e/m of an electron. *e/m* of an electron is the ratio of the charge on an electron to its rest mass. It is also known as specific charge. The value of *e/m* $= 1.76 \times 10^{11}$ coul/kg.

$= 1.76 \times 10^{7}$ e.m.u./gm. $= 5.28 \times 10^{17}$ e.s.u/gm.

36.2. Determination of e/m. The value of *e/m* for an electron can be determined by Busch method or Helical method or long solenoid method by making use of a *cathode ray tube*. The electrons emitted by the *cathode* and collimated by the *grid* are accelerated by applying a suitable positive potential to the *anode* system. These are brought to a sharp point focus at the centre of the *screen.*

If V_0 is the accelerating voltage, then the velocity v acquired by the electrons is given by

$$\frac{1}{2}\, mv^2 = e\, V_0$$

or

$$v = \sqrt{\frac{2\, e\, V_0}{m}}$$

If l is the distance between the centre of the *deflecting plates* of the cathode ray tube and the centre of the screen, Fig 36.3, then

Time taken by the electron to travel from the centre of the deflecting plates to the centre of the screen

$$t = \frac{l}{v} = \frac{l}{\sqrt{\dfrac{2\, e\, V_0}{m}}}$$

In S.I. units l is in metres, m in kg, V_0 in volts and e in coulomb.

If an alternating potential difference is applied either across the *X* or *Y* deflecting plates, then due to the alternating electric field, a force acts on the electron beam at right angles to the direction of the beam. Under the action of two perpendicular forces, one due to the accelerating voltage and the other due to the alternating potential difference the electron beam has a parabolic path. The electrons travel along the parabolic path with increased transverse velocity as they pass through the deflecting plates after which they travel along straight lines so that the beam on reaching the screen sweeps back and forth tracing out a straight line. This line is horizontal if the alternating potential difference is applied across the *X*-plates and vertical if the potential difference is applied across the *Y*-plates. The length of the line depends upon the magnitude of the alternating potential difference.

Finally a magnetic field is applied with the help of a long closely wound solenoid in such a way that the lines of magnetic force are parallel to the axis of the cathode ray tube. This field, therefore, does not affect the forward motion of the electrons *i.e.* their longitudinal velocity but acts on the transverse motion and the electrons describe a *circular path* due to this *transverse velocity.* The resultant of the combined uniform circular and uniform linear motion is that the electrons travel along a *helical* path in the tube.

If B is the strength of the magnetic field and v' the transverse velocity of an electron, then the equilibrium equation is given by

$$B\,e\,v' = \frac{mv'^2}{r}$$

where r is the radius of the circular path along which the electron moves.

$$\therefore \qquad \frac{v'}{r} = \omega = B\frac{e}{m}$$

Now $\frac{v'}{r} = \omega$, the angular velocity depends upon B only and is a constant for a given value of B.

$\therefore$ Time period to describe the circle

$$T = \frac{2\pi}{\omega} = \frac{2\,\pi\,m}{B\,e} \qquad \text{...(ii)}$$

The value of B is adjusted to make the time T of electron to go once around a circular path equal to the time t to travel forward a distance l from the centre of deflecting plates to the screen as given in relation (*i*). As this condition is independent of the transverse velocity, it holds true for all the electrons and therefore all the electrons again collect on the axis and focused at a tiny point on the screen *i.e.* the line again reduces to a point focus.

Such a condition may be obtained again for higher values of B when the time period $T = 1/2$ or $1/3$ etc of the time t. Thus we may obtain a number of foci by varying B. If B_0 is the minimum value of B for which this condition is obtained, then

$$\frac{2\pi m}{B_0\,e} = \frac{l}{\sqrt{2\frac{e}{m}V_0}}$$

$$\therefore \qquad \frac{e}{m} = \frac{8\,\pi^2\,V_0}{B_0^2\,l^2} \qquad \text{...(iii)}$$

The magnetic field B_0 along the axis of a long closely wound solenoid in S.I. units (Tesla) is given by

$$B_0 = \frac{\mu_0}{2}\frac{NI}{L}(\cos\theta_1 - \cos\theta_2)$$

where θ_1 and θ_2 are the angles subtended at the centre by the extremities of the solenoid and

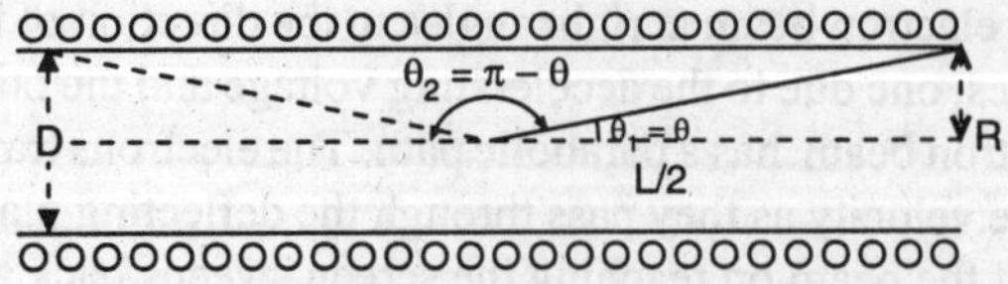

Fig. 36.1

$\mu_0 = 4\,\pi \times 10^{-7}$ Weber/Amp–metre

N = Total number of turns

I = current in Amps.

L = Length of the solenoid in metre

When the centre of the solenoid coincides with the centre of the deflecting plates

$$\theta_1 = \theta \text{ and } \theta_2 = \pi - \theta$$

$$\therefore \quad B_0 = \frac{\mu_0 NI}{2L} 2\cos\theta = \frac{\mu_0 NI}{L}\cos\theta \quad ...(iv)$$

If R is the radius of the solenoid, and D its diameter then

$$\cos\theta = \frac{L/2}{\sqrt{R^2 + (L/2)^2}} = \frac{L/2}{\sqrt{(D/2)^2 + (L/2)^2}} = \frac{L}{\sqrt{D^2 + L^2}}$$

Substituting the value of B_0 from (iv) in equation (iii), we have

$$\frac{e}{m} = \frac{V_0}{I^2} \cdot \frac{1}{2}\left[\frac{4\pi}{\mu_0} \frac{L}{N\,l\cos\theta}\right]^2$$

Substituting the value of $\frac{\mu_0}{4\pi} = 10^{-7}$, we have

$$\frac{e}{m} = \frac{V_0}{I^2}\left[\frac{L}{N\,l\cos\theta}\right]^2 \times 5\times 10^{13}$$

$$= K\frac{V_0}{I^2}$$

where
$$K = 5\times 10^{13}\left(\frac{L}{N\,l\cos\theta}\right)$$

If l_x is the distance from the centre of X-deflection plates to the screen and l_y the corresponding value for Y-plates, then

For X-plates

$$K_x = 5\times 10^{13}\left(\frac{L}{N\,l_x\cos\theta}\right)$$

and for Y-plates

$$K_y = 5\times 10^{13}\left(\frac{L}{N\,l_y\cos\theta}\right)$$

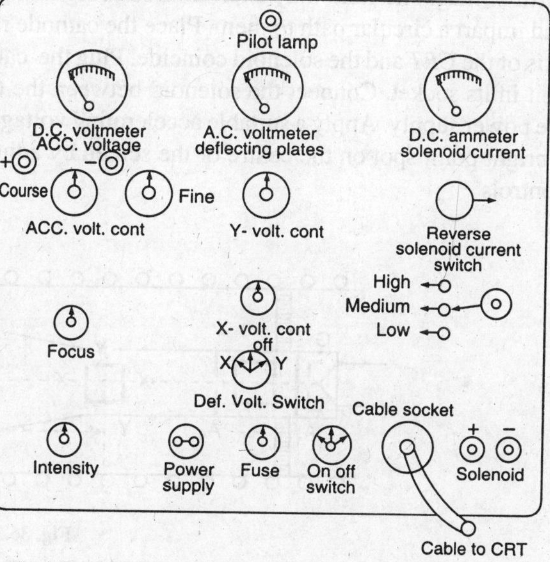

Fig. 36.2

36.3. Apparatus. The apparatus for determination of *e/m* consists of three parts.

(*i*) A cathode ray tube

(*ii*) A power supply unit

(*iii*) A long closely wound solenoid in which the cathode ray tube can be placed so that the axis of the cathode ray tube and that of the solenoid coincides.

The cathode ray tube is connected to the power supply unit through a cable. The cable encloses the leads originating from the filament, grid, accelerating anodes and deflecting plates. The other ends of the leads are connected internally to suitable points in the power supply unit.

The front panel of the power supply unit is shown in Fig. 36.2. When the ON-OFF SWITCH is put in the ON position the supply is on and the filament of the cathode ray tube gets suitable heating. The cathode then starts emitting electrons. Accelerating voltage can be applied to the accelerating anodes by adjusting the accelerating voltage control. Separate controls are provided for coarse and fine adjustment. The value of accelerating voltage can be measured by the accelerating voltage (*D.C.*) voltmeter. The beam can be focused with the FOCUS control and its brightness adjusted with the INTENSITY control.

When a sharp bright point focus is obtained in the middle of the screen, the DEFLECTING VOLTAGE SWITCH is turned to the *X*-plates and with the help of *X*-voltage control regulator the *A.C.* voltage applied to the *X*-plates is adjusted to obtain a line 2-3 cm long.

Similarly the *Y*-deflection switch and control can be used. The value of *AC* voltage is measured by the *A.C.* voltmeter.

To apply the magnetic field the cathode ray tube is placed axially in a suitable position within the long closely wound solenoid and the solenoid is connected to the power supply unit between the two terminals marked SOLENOID. The solenoid current can be adjusted from low to high value with a regulating control. The current can also be reversed with a switch marked REVERSE. The value of the current is read with a D.C. ammeter.

Experiment 36.1. To determine the value of e/m for electron by long solenoid (Helical) method

Apparatus. A power supply unit fitted with suitable high resistance *DC* and *AC* voltmeters and an ammeter, a cathode ray tube, a long closely wound solenoid.

Procedure. 1. Place the solenoid in the East-West direction so that the horizontal component of earth's magnetic field may exert a force at right angles to the direction of motion of the electrons and impart a circular path to them. Place the cathode ray tube centrally inside the solenoid so that the axis of the *CRT* and the solenoid coincide. Plug the cable leading from the *CRT* into the power supply unit in its socket. Connect the solenoid between the two terminals marked SOLENOID. Switch on the power supply. Apply a suitable accelerating voltage by adjusting the ACC. VOLT. CONT. Obtain a bright point spot on the centre of the screen by adjusting accelerating voltage, focus and intensity controls.

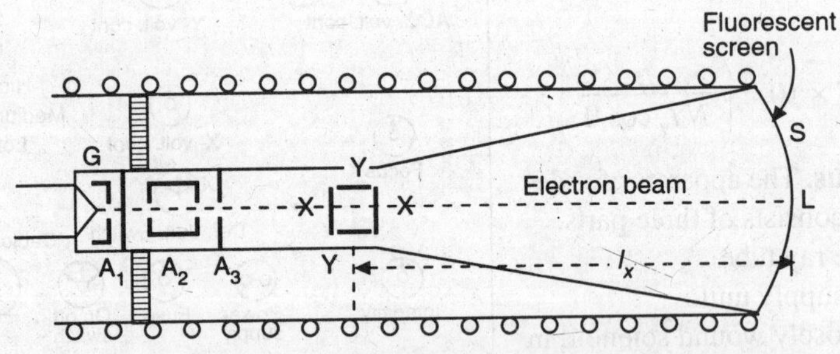

Fig. 36.3

2. Put the DEFLECTING PLATE VOLTAGE switch to the *X*-plate position and adjust the *X*-VOLT CONT to apply a suitable *A.C.* voltage to produce a horizontal line 2-3 cm long on the screen.

3. Now switch on the current in the solenoid so as to produce a magnetic field along its axis. Adjust the value of the current till a point focus is again obtained on the screen. Note the value of the current. Also note the value of accelerating voltage. Reverse the current in the solenoid by the REVERSE switch and again adjust its value, if necessary till a point focus is obtained.

4. Take a number of observations by changing the accelerating voltage and adjusting the solenoid current corresponding to each voltage.

5. Now put the DEFLECTING PLATE VOLTAGE SWITCH to the Y-plate position and adjust the Y-VOLT CONT to apply a suitable $A.C.$ voltage to produce a vertical line 2-3 cm long on the screen and again take a number of observations.

Observations. Length of the solenoid $L =$

Outer radius of the solenoid tube $R_1 =$

Inner radius of the solenoid tube $R_2 =$

Mean effective radius of the solendoid $R = \dfrac{R_1 + R_2}{2} =$

Mean effective diameter of the solenoid $= 2R = D$

Total Number of turns in the solenoid $N =$

Mean distance of the X-plates from the screen $l_x =$

Mean distance of the Y-plates from the screen $l_y =$

[All the above data for the solenoid and the *CRT* is generally provided by the manufacturers]

$$\therefore \qquad \cos\theta = \frac{L/2}{\sqrt{R^2 + (L/2)^2}} = \frac{L}{\sqrt{D^2 + L^2}} =$$

$$K_x = 5 \times 10^{13}\left(\frac{L}{N\, l_x \cos\theta}\right) =$$

$$K_y = 5 \times 10^{13}\left(\frac{L}{N\, l_y \cos\theta}\right)$$

Sl. No.	*Accelerating Voltage in Volts* (V_0)	*Horizontal X-deflection current in amperes I*				$\frac{V_0}{I_0^2}$	$\frac{e}{m} = K_x \frac{V_0}{I_0^2}$
		Forward	*Reverse*	*Mean* (I_0)	I_0^2		
1							
2							
3							
4							
5							
6							
7							

Sl. No.	*Accelerating Voltage* (V_0)	*Vertical Y-deflection current in Amp. I*				$\frac{V_0}{I_0^2}$	$\frac{e}{m} = K_y \frac{V_0}{I_0^2}$
		Forward	*Reverse*	*Mean* (I_0)	I_0^2		
1							
2							
3							
4							
5							
6							
7							

Mean *e/m* for horizontal X-deflection = coulomb/kg

Mean e/m for vertical Y-deflection = coulomb/kg

Actual value of e/m = coulomb/kg

$$\text{Percentage error} = \frac{\text{Observed value–Actual value}}{\text{Actual value}} \times 100$$

$$= \quad \%$$

6. **Graph.** We can also calculate the mean value of $\frac{V_0}{I_0^2}$ by plotting a graph. To plot the graph take '*Accelerating voltage*' V_0 along the X-axis and '*square of current through the solenoid*' I_0^2 along Y-axis. Draw two separate graphs for '*X-deflection*' and '*Y-deflection*' currents. The graph is a straight line as shown in Fig. 36.4. Take two points A and C on the graph and draw a right angled triangle ABC as shown, then

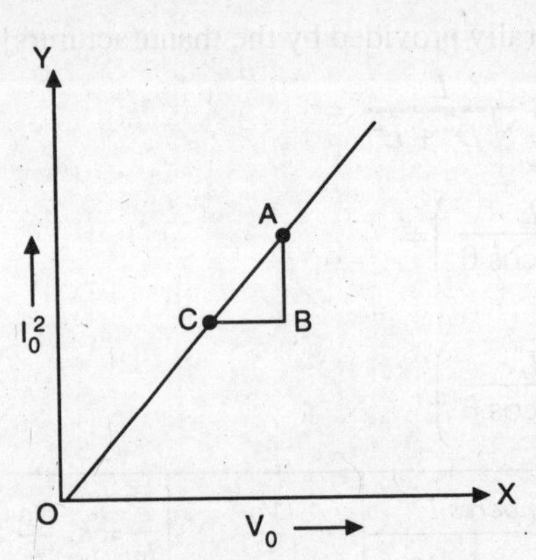

Fig. 36.4

$$\text{Slope of the graph} = \frac{AB}{BC} = \frac{I_0^2}{V_0}$$

$$\therefore \quad \frac{V_0}{I_0^2} = \frac{BC}{AB}$$

Record (*i*) $\frac{V_0}{I_0^2}$ from X-deflection current graph

$$= \frac{BC}{AB} =$$

$$\frac{e}{m} = K_x \frac{V_0}{I_0^2} = \quad \text{coul/kg (C Kg}^{-1}\text{)}$$

(*ii*) $\frac{V_0}{I_0^2}$ from Y-deflection current graph $= \frac{BC}{AB} =$

$$\frac{e}{m} = K_y \frac{V_0}{I_0^2} = \quad \text{coul/kg} \quad \text{(C Kg}^{-1}\text{)}$$

Precautions. 1. The solenoid should be placed in the East-West direction.

2. The cathode ray tube must be placed centrally inside the solenoid.

3. The electron beam must trace a line 2 to 3 cm long on the screen before the magnetic field is applied.

4. The line trace of electronic beam must be brought to a point focus by applying the magnetic field in one direction and then by reversing the current.

5. Reading must be taken by applying the electric field between X-plates as well as Y-plates.

36.4. Thomson's method for e/m. The value of e/m can be determined by the method originally used by Prof. J.J. Thomson.

The apparatus consists of a cathode ray tube which has three essential parts.

(*i*) *The electron gun*. This part consists of the cathode K and the anode A. It produces, accelerates and focuses the electrons into a fine beam so as to produce a small bright spot on the fluorescent screen S.

(*ii*) *The screen*. The screen S is coated with some fluorescent material like zinc sulphide etc so that a greenish blue spot of light is observed where the electron beam impinges upon the screen. A transparent mm graph is attached to the screen to note the exact position of the fluorescent spot.

(iii) The electric and magnetic deflecting system. Two plates P and Q are fitted in the tube symmetrically on either side of the electron beam so that an electric field *perpendicular to the plane of the paper* can be applied between them.

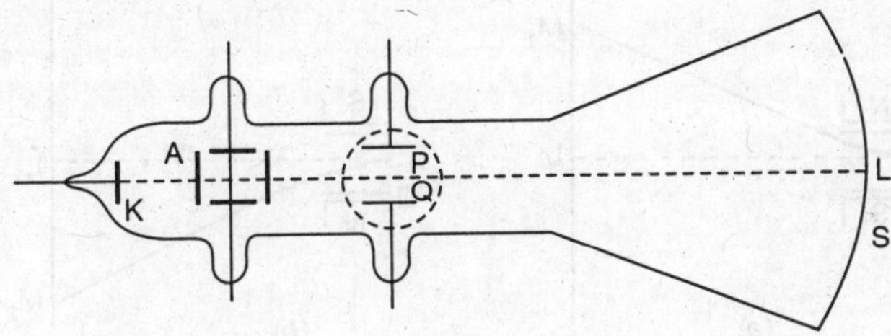

Fig. 36.5

A magnetic field can also be applied at the same place in a direction perpendicular to the direction of the electric field *i.e.* in the plane of the paper. Thus the deflection of the cathode rays due to the electric as well as the magnetic field takes place in a direction perpendicular to the plane of the paper.

To apply the magnetic field the cathode ray tube is fitted on a wooden stand provided with two arms, one on either side perpendicular to the direction of the electron beam. A scale is marked on each arm which gives the distance from the axis of the cathode ray tube. When two bar magnets are placed with their lengths parallel to the arms with their opposite poles towards the cathode ray tube, these produce a *uniform* magnetic field exactly at the same place where an electric field is produced when a potential difference is applied between the plates P and Q.

Fig. 36.6

Theory. Let E be the intensity of the electric field applied between the plates P and Q in a direction perpendicular to the plane of the paper then the force acting on the electron of charge e

$$= Ee$$

This force acts at right angles to the direction of motion of the electron. The electron, therefore, moves along a circular path within the electric field and on leaving the field flies off tangntially to the path meeting the screen at M_1 or M_2 depending upon the direction of the electric field as shown in Fig. 36.7. (*a*) and (*b*). When a magnetic field B is applied in the plane of the paper perpendicular to the direction of the beam as shown in Fig. 36.8 (*a*) then according to Fleming's left hand rule the beam is deflected upward out of the paper in a direction perpendicular to both, the direction of motion of the electron and that of the field.

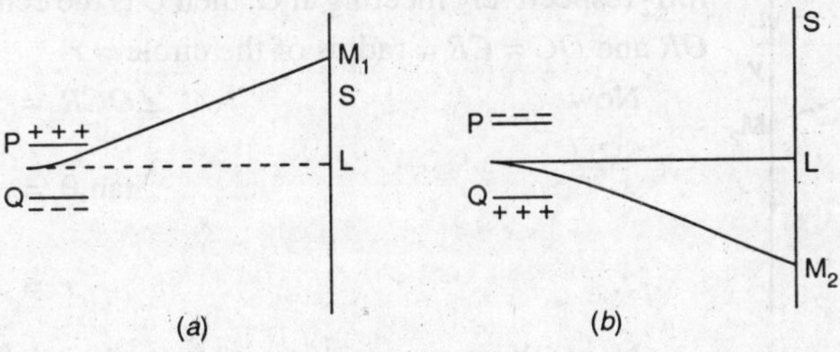

Fig. 36.7

When the magnetic field is applied as shown in Fig. 36.8 (*b*) the electron beam is deflected, downward into the paper. If v is the velocity of the electrons in the beam, then magnitude of the force acting on each electron = $B e v$. As this force also acts at right angles to the direction of motion of the electron, the electron moves along a circular path in the magnetic field and on leaving the field flies off tangentially meeting the screen at M_1 or M_2 as explained above.

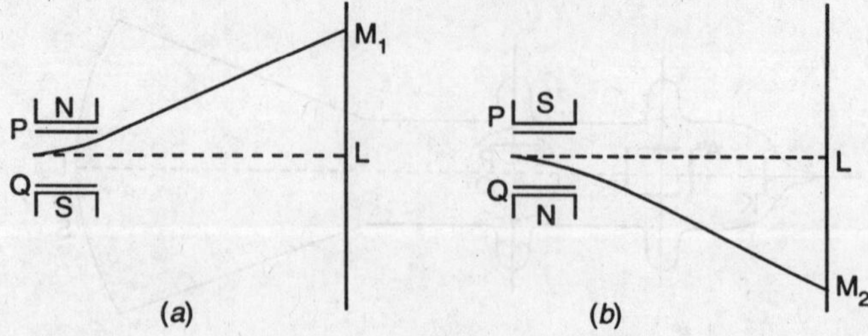

Fig. 36.8

If r is the radius of the circular path along which the electron moves in the magnetic field, then

$$B e v = \frac{m v^2}{r}$$

or $$\frac{e}{m} = \frac{v}{r B} \quad ...(i)$$

To find the velocity of elctrron v: To find the value of v, one of the fields say the magnetic field is first applied. The spot of light moves from its initial position to the position say M_1. The electric field is now applied simultaneonsly so as to produce deflection of the electron beam in the opposite direction. The value of the electric field is so adjusted that the spot of light comes back to its initial position. In such a case the forces acting on the electron due to the electric and magnetic fields are equal and opposite.

$\therefore$ $$Ee = B e v$$

or $$v = \frac{E}{B}$$

Substituting in (*i*), we have

$$\frac{e}{m} = \frac{E}{rB^2} \quad ...(ii)$$

To find the radius r of circular path. To find the value of r, the radius of the circular path in the magnetic field, note the position of the luminous spot on the screen when no field is applied and again when a field B is applied. Let the deflection of the spot be $y = LM_2$ as shown. The circular path of the beam OR in the magnetic field is as shown in Fig 36.8. The beam leaves the magnetic field at R tangenially to the path and meets the screen at M_2 so that $M_2 R$ produced meets OX in T. Draw OC and RC perpendiculars to OL and RM_2 respectively meeting at C, then C is the centre of the circular arc OR and $OC = CR$ = radius of the circle = r

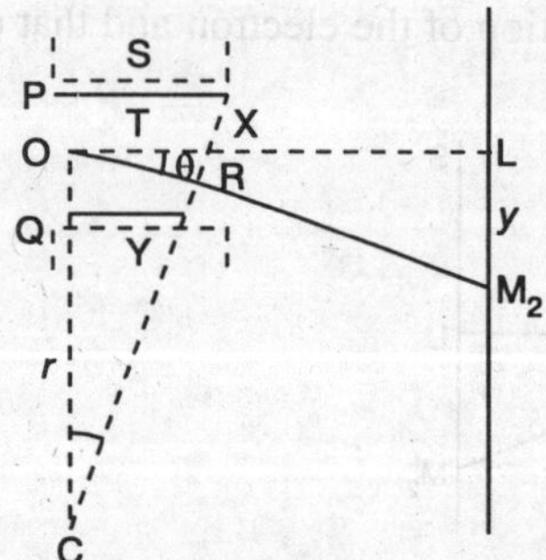

Fig. 36.9

Now $$\angle OCR = \angle LT M_2 = \theta$$

$\therefore$ $$\tan\theta = \frac{LM_2}{LT} = \frac{OX}{OC}$$

$\therefore$ $$r = OC = \frac{LT \times OX}{LM_2}$$

Now OX is very nearly equal to the length of the region in which the magnetic field is applied. This is also equal to the length of the region in which the electric field is applied *i.e.* the length of the plates P and Q. Let it be = l (This is generally given by the makers).

LT is the distance of the centre of the screen from the centre of the region of the magnetic field (or electric field). Let it be = L

$$\therefore \qquad r = \frac{L\,l}{y}$$

Substituting in (*ii*), we have

$$\frac{e}{m} = \frac{y\,E}{L\,l\,B^2} \qquad \text{...}(iii)$$

The electric field $E = \frac{V}{d}$ where V is the potential difference applied between the plates P and Q and d the distance between the plates (d is also given by the makers). Substituting in (*iii*), we have

$$\frac{e}{m} = \frac{yV}{L\,l\,d\,B^2}$$

When V is in volts, B in Tesla and y, L, l, and d in metres *e/m* is in Coulomb/kg.

Experiment 36.2. To find the value of e/m for an electron by Thomson's method using bar magnets.

Apparatus. A cathode ray tube fitted on a wooden stand provided with two arms perpendicular to the axis of the tube and having scales with zero at the point where the central line of the arms cuts the axis of the tube; a power supply unit capable of giving the accelerating *D.C* voltage for the cathode ray tube and the deflecting *D.C.* voltage for the electric field through a reversing key with suitable voltmeters, two bar magnets, a compass box, a compass needle, etc.

Procedure 1. Draw the North South line using a compass needle. Also draw the East-West line. Place the cathode ray tube fitted in the wooden frame with its axis along the North South line so that the arms of the frame lie along the East West line.

2. Connect the cathode ray tube to the power supply unit. Switch on the current and wait till a luminous bright spot appears on the screen. Adjust the brightness and focus controls so as to get a sharp bright point spot in the middle of the screen. Note the initial position of the spot on the scale fitted on the screen.

3. Now apply a suitable deflecting voltage so that the luminous spot is deflected by about 0.5 to 1.0 cm. Note the deflecting voltage V and the position of the spot. Measure the distance through which the spot has moved and let it be y.

4. Place the bar magnets symmetrically on either side of the cathode ray tube along the arms of the wooden stand on which the tube is fitted such that their opposite poles face each other and their common axis is exactly at right angles to the axis of the cathode ray tube. Adjust the polarity as well as the distance of the magnets so that the luminous spot comes back to its initial position. When the adjustment is perfect note the distance of the poles of the magnets on the side nearer to the cathode ray tube. Let the distances be r_1 and r_2.

5. Remove the bar magnet, switch off the electric field applied to the deflecting plates and again note the initial position of the luminous spot. Reverse the polarity of the potential difference applied to the electric deflecting plates with the help of the reversing switch fitted in the power supply unit thereby reversing the electric field. Again note the final position of the luminous spot and calculate y.

Again place the bar magnets on the arms of the wooden stand as in the previous step and adjust their polarity as well as the distance so that the spot comes back to its initial position. When the adjustment is perfect again note the distances of the poles of the magnets on the side nearer to the cathode ray tube. Let the distances be r'_1 and r'_2. Switch off the power supply.

6. To find the value of the magnetic field B, carefully remove the magnets and the cathode ray tube from the wooden stand. Place the compass box (of a deflection magnetometer or tangent galvanometer) such that its centre lies exactly on the point where the common axis of the bar magnets

and the axis of the cathode ray tube intersect. Rotate the compass box about its vertical axis so that the pointer lies along the 0–0 line.

Place the magnets exactly in the same positions as in step 4 at distances r_1 and r_2. This produces a deflection in the magnetometer compass box and the two ends of the pointer give the deflection. Let the readings be θ_1 and θ_2.

Now place the magnets exactly, in the same positions as in step 5 at distances r'_1 and r'_2 and again note the deflections θ'_1 and θ'_2 from the two ends of the pointer of the compass box. The mean of these four deflections θ_1, θ_2, θ'_2 and θ'_2 gives the mean deflection θ. If B_H is the horizontal component of earth's magnetic field, then

$$B = B_H \tan \theta$$

7. Take two more sets of observations by changing the value of V and hence that of the electric field.

Observations. (Data provided on the cathode ray tube by the manufacturers).

Length of the deflecting plates P or $Q = l$ = m

Distance of the screen from the centre of the plates L = m

Distance between the deflecting plates P and Q d = m

Sl. No.	Applied Voltage V	DIRECT FIELD							REVERSED FIELD						
		Position of Spot			Magnetic pole and Distance				Position of Spot			Magnetic Pole and Distance			
		Initial	Final	Deflection y in met.	Pole	r_1	Pole	r_2	Initial	Final	Deflection y in met.	Pole	r'_1	Pole	r'_2
1	V_1 =										y_1 =				
2	V_2 =										y_2 =				
3	V_3 =										y_3 =				

Horizontal component of earth's field B_H = Wb/sq.m (Tesla)

For determination of B

Sl. No.	Applied Voltage V	Reading of two ends of pointer when mag. poles are at r_1 and r_2		Reading of two ends of pointer when mag. poles are at r'_1 and r'_2		Mean θ	$B = B_H \tan \theta$
		θ_1	θ_2	θ'_1	θ'_2		
1	V_1						B_1
2	V_2						B_2
3	V_3						B_3

Calculations

(i) $e/m = \dfrac{y_1 V_1}{L\, l\, d\, B_1^2} =$ C kg^{-1}

(ii) $e/m = \dfrac{y_2 V_2}{L\, l\, d\, B_2^2} =$ C kg^{-1}

(iii) $e/m = \dfrac{y_3 V_3}{L\, l\, d\, B_3^2} =$ C Kg^{-1}

Mean *e/m* = C kg^{-1}

Precautions. 1. The cathode ray tube must be set in the north south direction.

2. The luminous spot on the screen should be sharp, bright and in the middle.

3. The deflecting voltage must produce a deflection of the order of 0.5 to 1.0 cm.

4. The axis of the bar magnets and the axis of the cathode ray tube must be exactly at right angles and in a horizontal plane.

5. The deflecting voltage must be reversed to produce the deflection of the spot of light in the opposite direction and the magnetic field must also be reversed to bring it back to initial position.

6. The cathode ray tube must be handled carefully.

36.5. *e/m* by magnetron method. A magnetron is a diode valve in which the electron can move in a magnetic field as well as in an electrostatic field between the cathode and the anode (or plate). The diode consists of a cathode in the form of a straight wire and a cylindrical plate co-axial with the cathode. In the absence of any magnetic field the electron normally leaves the cathode along a straight path in a direction perpendicular to the anode due to the electrostatic field between the cathode and the anode.

If V is the potential of the anode (plate) with respect to the cathode, the velocity gained by the electron on reaching the plate is given by

$$eV = \frac{1}{2} mv^2$$

or

$$v = \sqrt{\frac{2\, eV}{m}}$$

where e is the charge on the electron and m its mass. A longitudinal magnetic field along the axis of the wire filament (cathode) is applied by placing the cylindrical diode valve inside a straight solenoid consisting of a large number of layers of thick insulated copper wires able to carry a current of 10-15 amperes. It is long enough to produce a uniform magnetic field along its axis at its middle point. The diode valve can easily slide in or out of the solenoid. When a current is passed through the solenoid the electron experiences a force given by

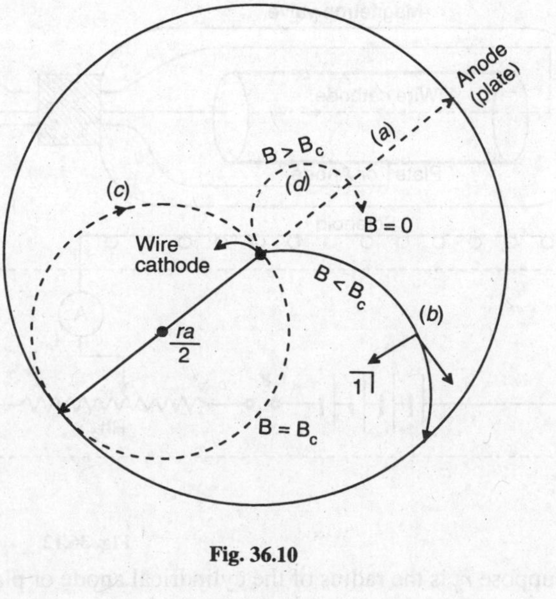

Fig. 36.10

$$F = Bev$$

at right angles to its direction of motion. B is the intensity of the magnetic field at a point on the axis of the solenoid at its middle. Because of this force the electron is made to move in circular path. If the applied magnetic field is weak, this force will give some curvature to the electron path which is straight in the absence of the field. As the field is increased still further, the electron path is more and more curved till for a value of the field $= B_c$ known as *critical field* the electron just reaches the plate or anode i.e. the electron just grazes along the anode. (or plate) For values of the field above B_c the path is highly curved and the electron does not reach the anode or plate. At this stage the anode or plate current is zero.

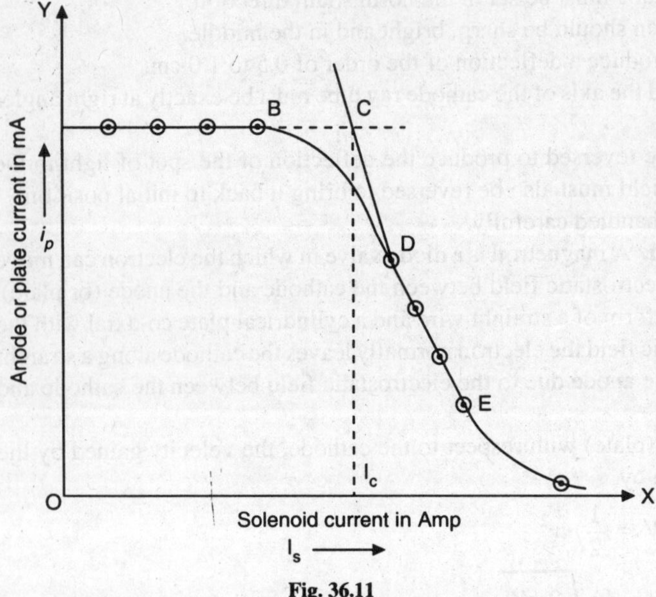

Fig. 36.11

In other words, as the magnetic field is increased gradually, the anode current remains almost constant but above a particular critical value B_c the anode current drops suddenly as shown in Fig. 36.11. The cut off is by no means sharp, the curve showing a distinct knee and foot.

It is clear from Fig. 36.10 that when $B = 0$ the electron goes straight from the cathode to anode (plate) [line (*a*)]. When $B < B_c$ it follows a curved path [curve (*b*)]. When $B = B_c$ the electron just grazes along the anode [curve (*c*)]. When $B > B_c$ the electron moves in a circular path and misses the anode [curve(*d*)].

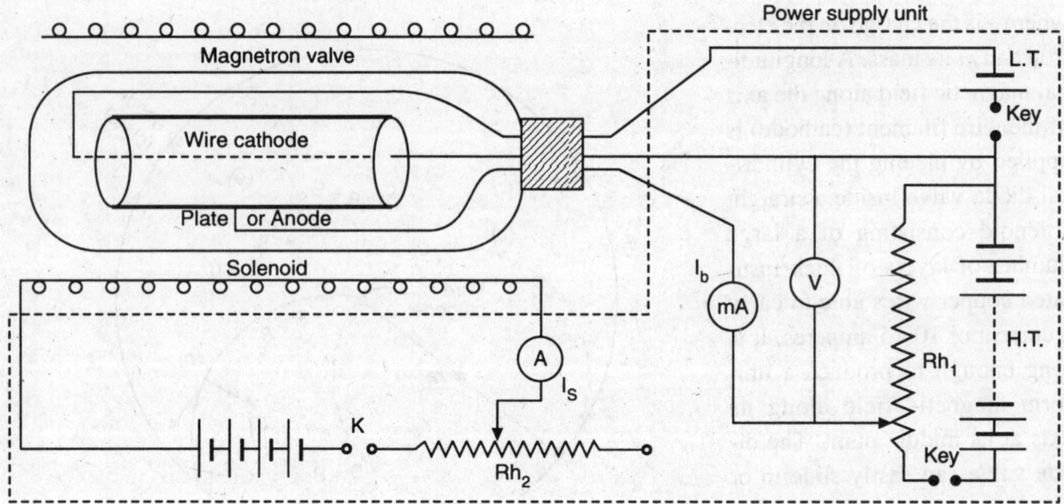

Fig. 36.12

Suppose r_a is the radius of the cylindrical anode or plate of the diode valve and B_c is the critical magnetic field, then

$$B_c ev = \frac{mv^2}{r_a/2}$$

or
$$mv = \frac{B_c\, e\, r_a}{2}$$

But
$$v = \sqrt{\frac{2eV}{m}}$$

Substituting in (*i*), we get
$$m\sqrt{\frac{2eV}{m}} = \frac{B_c\, e\, r_a}{2}$$

Squaring both sides, we have
$$2e\,Vm = \frac{B_c^2\, e^2\, r_a^2}{4}$$

or
$$\frac{e}{m} = \frac{8V}{B_c^2\, r_a^2} \quad \text{coulomb/kg (C Kg}^{-1}) \quad ...(ii)$$

The critical magnetic field is produced by a solenoid of length L, having n turns per metre. If I_c is the current through the solenoid for critical field (i.e. for cut off plate current), then

$$B_c = \mu_0\, n\, I_c \cos\theta \quad \text{Weber}$$

where
$$\mu_0 = 4\pi \times 10^{-7} \quad \text{Wb/A–m}$$

and
$$\cos\theta = \frac{L/2}{\left[R^2 + \left(\frac{L}{2}\right)^2\right]^{1/2}} = \frac{L/2}{\sqrt{(D/2)^2 + (L/2)^2}} = \frac{L}{\sqrt{D^2 + L^2}}$$

R being the average radius of the solenoid and D its corresponding diameter.

Experiment 36.3. To determine e/m by magnetron method or small solenoid method.

Apparatus. Magnetron (diode valve) R.C.A 41 or 42, *E* 1189, standard long axial solenoid in which the diode can be mounted at the centre, Power supply unit consisting of a L.T. supply (For heating the cathode of diode), a supply of about 30V (to supply magnetising current to the solenoid) and an H.T supply for the plate of diode, two suitable rheostats, a volt meter, an ammeter, a milli-ammeter, keys and connecting wires.

Procedure. 1. Place the solenoid in the *E-W* direction. Slide the magnetron tube gently inside the solenoid so that it is at its centre.

2. Draw a diagram showing the scheme of connections. Make necessary connections from the power supply unit as shown in Fig. 36.12.

3. Switch on the L.T. supply and let the cathode get heated up.

4. Switch on the H.T. supply. With the help of the rheostat Rh_1 apply a voltage say 60V to the plate of the diode. The milli-ammeter will show some deflection indicating the presence of plate current I_p.

5. Now insert the key K and send a current through the solenoid using the rheostat Rh_2. With the help of this rheostat increase the value of current I_s through the solenoid from zero to such a value that the plate current in the magnetron diode shows a sharp fall. This sharp fall occurs when the magnetic field produced by the solenoid just exceeds the critical value B_c. Record all the values of the solenoid current I_s (in Amps) by increasing it from zero upward in small steps and corresponding values of plate current I_p (in mA).

6. Plot a graph between '*current through the solenoid*' I_s taken along *X*-axis and corresponding '*Diode plate current*' I_p taken along *Y*-axis. Draw two tangents to the curve at two points *B* and *D*

7. Repeat the experiment for plate voltages 70, 80, 90 and 100 volts respectively. Calculate the value of I_c in each case by plotting a separate graph.

Observations. Length of the solenoid $L =$ m

Number of turns $N =$

Number of turns per unit length (per metre) $n = \frac{N}{L} =$

Diameter of solenoid $D =$ m

Radius of solenoid $R =$ m.

[For a standard solenoid L, N and R is generally given by the makers]

$$\cos\theta = \frac{L/2}{[R^2 + (L/2)^2]^{1/2}} = \frac{L}{[D^2 + L^2]^{1/2}} =$$

Radius of the cylindrical anode (of diode) $r_a =$ m (Also given by the manufacturer of Magnetron diode)

S. No.	Solenoid current in Amp I_s	Plate current I_p in m A				
		Plate voltage (V_a) $V_1 = 60V$	Plate voltage (V_a) $V_2 = 70$ V	Plate voltage (V_a) $V_3 = 80V$	Plate voltage (V_a) $V_4 = 90V$	Plate voltage (V_a) $V_5 = 100V$
1						
2						
3						
4						
5						
6						
7						

Calculations

S. No.	Plate voltage V_a	Critical solenoid current I_c (from graph)	Critical Field $B_c = \mu_0 n I_c \cos\theta$	$\frac{e}{m} = \frac{8 V_a}{B_c^2 r_a^2}$ Coul/kgm
	$V_1 = 60$ V			
	$V_2 = 70$ V			
	$V_3 = 80$ V			
	$V_4 = 90$ V			
	$V_5 = 100$ V			

$$\text{Mean } \frac{e}{m} = \qquad \text{Coul/kg m}$$

Precautions. 1. The diode valve (magnetron) should be placed in the middle of the solenoid with its axis coincident with the axis of the solenoid.

2. At the critical value the current should not be allowed to flow in the solenoid for a long time as it would heat up the wires.

3. The voltage applied to the diode plate should not exceed the maximum permissible value as recommended by the manufacturers.

4. The current in the filament circuit should remains constant.

36.6. Millikan's apparatus. The apparatus consists of an earth connected metallic chamber C having two windows: one for light to illuminate the field of view and the other for X- rays to ionise the gas inside, if necessary. Two optically flat brass plates P and Q are fitted inside the chamber. The lower plate Q is earth connected while the upper plate P can be given a suitable positive potential from a high tension battery. A constant and very accurately measurable field thus can be applied across the central space between the two plates.

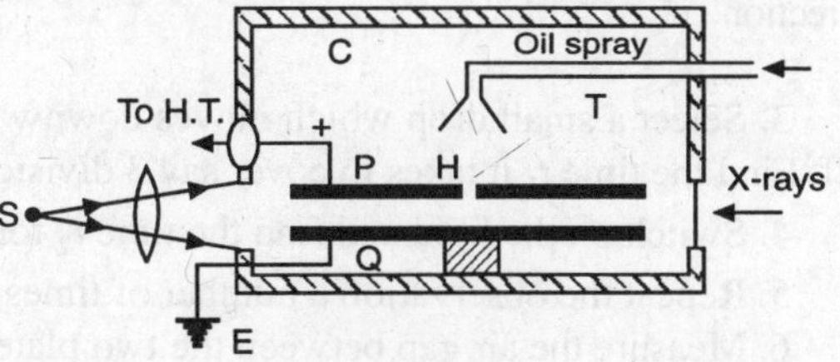

Fig. 36.13

The upper plate P has a hole H at its centre through which enter a few oil drops produced by an atomiser in the chamber C through the tube T ending in a funnel.

Experiment 36.4. To determine the electronic charge by Millikan's method.

Apparatus. Millikan's apparatus, oil sprayer containing thin oil of low vapour pressure (clock oil or light lubricating oil), spirit level, stop watch, vernier callipers, screw gauge, a dip-stick wire etc.

Theory. If a is the radius of the drop of oil of density ρ, then

Weight of the drop $= 4/3\, \pi a^3 \rho g$

As the drop is lying in a gas of density d, it experiences an upthurst

$$= 4/3\, \pi a^3 d g$$

$\therefore$ Resultant downward force

$$= 4/3\, \pi a^3 (\rho - d)g$$

If the drop acquires a terminal velocity v, while moving downwards then

$$4/3\, \pi a^3 (\rho - d)g = 6 \pi \eta a v \quad [\eta \text{ is the viscosity of air}]$$

or

$$a = \left(\frac{9\eta v}{2(\rho - d)\, g} \right)^{\frac{1}{2}}$$

If q is the charge on the drop and E is the electric field applied between the plates so that the drop begins to move upward with a uniform velocity v_1, then

Resultant upward force $= Eq - 4/3\, \pi a^3 (\rho - d)g$

$$\therefore \quad Eq - 4/3\, \pi a^3 (\rho - d)g = 6 \pi \eta a v_1$$

or

$$Eq = 6 \pi \eta a (v + v_1)$$

$$\therefore \quad q = \frac{6 \pi \eta a (v + v_1)}{E}$$

$$= \frac{6 \pi \eta}{E} \left[\frac{9 v \eta}{2 (\rho - d)\, g} \right]^{\frac{1}{2}} (v + v_1)$$

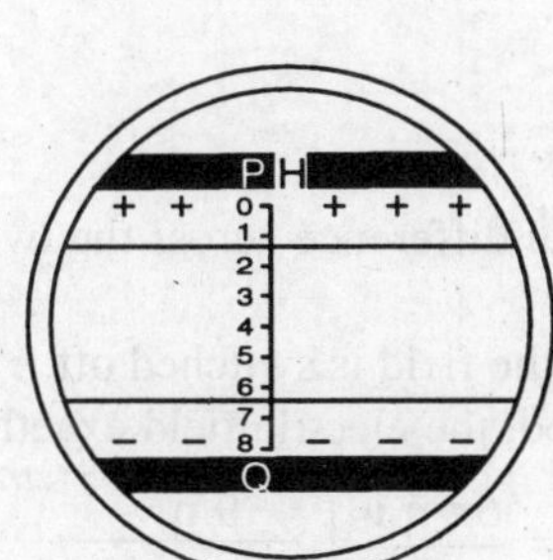

Fig. 36.14

Procedure. 1. Insert the dip-stick wire in the hole H and focus the microscope on the wire. Find the number of divisions of the microscope scale equivalent to the thickness of the wire.

2. Remove the dip-stick and spray oil in the chamber by means of the atomiser. After a few seconds the oil drops will appear as tiny spots moving upward in the field of view of the microscope. Switch on the $H.T.$ and it will be noticed that a majority of the oil drops are unaffected. Some of the drops will appear to accelerate in the upward direction. (These are positively charged

drops and should be ignored). A few drops will appear to reverse direction and move in the downward direction.

3. Select a small drop which moves downwards and note its motion when the field is switched off. Find the time t_1 it takes to cover say 3 divisions of the microscope scale.

4. Switch on the field and find the time t_2 for its passage in the reverse direction.

5. Repeat the observation a number of times for different drops.

6. Measure the air gap between the two plates with a vernier callipers.

Observations.

Diameter of dip-stick with screw gauge R = 1. 2 3. 4

Mean = cm

No. of divisions of microscope scale equivalent to wire thickness n =

∴ Magnitude of each microscope scale div. $\beta = \frac{R}{n}$ cm = m

No. of div. (y_1) metres	Time t_1 1	2	mean	$v = \beta \frac{y_1}{t_1}\ ms^{-1}$	No. of div. (y_2) metres	Time t_2 1	2	mean	$v_1 = \beta \frac{y_2}{t_2}\ ms^{-1}$	Pot. diff. V (volts)

Distance between plates x = cm = m

Density of oil ρ = gm/cc = kg/ m^3

Density of air d = gm/cc = kg/ m^3

Viscosity of air at $t\,^\circ C$ η = poise = Deca poise

Electric field $E = \frac{V}{x}$ = volt/metre

Charge on the drop $q = \frac{6\pi\eta}{E}\left[\frac{9\eta v}{2(\rho - d)g}\right]^{1/2}(v + v_1)$ = Coulomb

Note. 1. It will be seen that the charge q on drops are all multiples of a quantiy of charge (e) which is the electronic charge. Find the value of e by taking a large number of observations.

Precautions. 1. The apparatus should be levelled carefully.

2. The plates should be cleaned before spraying.

3. Low vapour pressure oil should be used.

4. The positive of $H.T.$ should be connected to the top plate P.

5. The microscope scale should be calibrated.

6. High resistance voltmeter should be used to measure the potential difference across the two plates.

7. Smallest drop should be selected which moves downward when the field is switched off.

Note. Balanced oil drop method. In the balanced oil drop method, the electric field exactly balances the gravitational field so that $v_1 = 0$ and charge on the drop $q = \frac{6\pi\eta v}{E}\left(\frac{9\eta v}{2(\rho - d)g}\right)^{1/2}$

$$= \frac{18\pi\eta^{3/2} v^{3/2}}{E\,[2(\rho - d)g]^{1/2}}$$

Oral Questions

1. e/m by Helical method (*i*) What do you mean by e/m of an electron? (*ii*) What is its value in S.I. units? (*iii*) State the formula for determining e/m by Helical (or long solenoid method). (*iv*) State the value of magnetic induction field along the axis of a long closely wound solenoid. (*v*) Why do we obtain a number of foci by using higher values of magnetic induction field *B*.

2. e/m by Thomson's method (*i*) What are cathode rays? (*ii*) Why do we call the tube as cathode ray tube? (*ii*) State the formula for e/m in Thomson's method in terms of *v*, and *B*. (*iv*) What is the relation between the velocity of the electrons, electric field *E* and magnetic induction field *B* when both the fields applied simultaneously produce no deflection ? (*vi*) What is the direction of the electric and magnetic field in Thomson's method? (*vii*) How do you find the radius of the circular trace of the electron beam?

3. $\frac{e}{m}$ by Magnetron method. (*i*) What is a Magnetron valve? (*i*) What happens when a magnetic field is applied to the Magnetron valve? (*iii*) What is the direction of magnetic field? (*iv*) How is the Magnetron placed in the solenoid. (*v*) What is the critical magnetic field? (*vi*) What is the magnitude of the force on an electron moving with a velocity *v* at right angles to the magnetic field *B*? (*vii*) What is the formula used to calculate *e/m* by magnetron method? (*viii*) If we plot a graph between B_c^2 and *V* what is the shape of the graph?

4. Electronic charge by Millikan's method (*i*) What is the value of electronic charge ? (*ii*) Knowing *e/m* and *e* can you find the mass of the electron? (*iii*) Why does the charge carrying droplet acquire a downward terminal velocity when no electric field is applied? (*iv*) When the electric field is applied and the drop moves upward, what is the direction of electric field? (*v*) Does the droplet carry a unit electronic charge? (*vi*) What is balanced oil drop method?

37

Ionisation Potential of Mercury

37.1. Ionisation potential. *The least energy in electron–volts required to remove the most loosely bound electron from an isolated free neutral atom is known as the ionisation potential of the atom.*

Ionisation potential of mercury. The ionisation potential of mercury can be determined by introducing mercury vapour at a low pressure of 10 mm to 50 mm of mercury column in an evacuated tube fitted with a cathode and an anode. A mercury vapour filled gas diode is the most suitable for the purpose. The cathode of the gas diode may be directly or indirectly heated type. A hot cathode gas filled diode is known as a *phanotron* [Incidentally a gas filled triode is known as a *Thyratron*] A gas filled diode is symbolically represented as shown in Fig. 37.1. The **dot** in the tube shows the presence of the gas (or vapour).

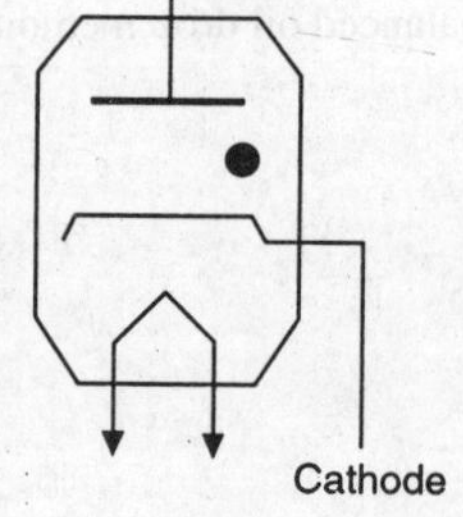

Fig. 37.1

When the anode or plate of the gas filled diode is at a positive potential with respect to the cathode, electrons move across the tube from the cathode to the anode. This electronic current depends upon two factors :

(*i*) The number of electrons emitted per unit area from the cathode and its temperature.

(*ii*) *The effect of space-charge* —the negatively charged region containing the electron cloud due to the accumulation of electrons emitted by the cathode.

As the plate potential is increased the plate current slowly increases. But when the plate potential is increased beyond a particular value the plate current increases much more rapidly than it does below that critical value. This is because when the plate potential approaches this critical value the electrons arriving at the anode gain enough energy to knock out the electrons from the atoms of the gas close to the anode. These electrons are also attracted by the anode causing an increase in plate current and the positive ions neutralise some of the space charge, which further helps to increase the kinetic energy of the thermo-electrons. This potential is equal to the ionisation potential of the gas (or vapour) and for this value of plate potential there is a marked increase in plate current.

If a graph is plotted between plate potential and plate current (for a constant value of filament current) the plate current at first increases slowly for a given increase in plate voltage and when the plate potential is equal to or greater than the ionisation potential there is a greater increase in plate current for the same increase in plate potential.

The change in slope is, however, not very abrupt but there is a short curved portion within which the change in plate current goes on becoming more and more rapid.

To find the value of ionisation potential the two straight portions *AB* and *CD* of the graph are produced to meet at a point *E*. If we draw a perpendicular *EF* on the *X*-axis, then *OF* represents the ionisation potential.

Experiment. 37.1. To find the ionisation potential of mercury using a gas filled diode.

Apparatus. A gas filled (mercury vapour) diode, a *D.C.* power supply, a voltmeter, a milliammeter, a rheostat, connecting wires etc.

Procedure 1. Draw a diagram showing the scheme of connections as in Fig. 37.3. and make the connections accordingly taking care that the positive marked terminal of the voltmeter and that of the milliammeter are connected to the positive of the *D.C.* power supply.

2. Apply a suitable potential to the heating filament *FF* of the gas diode as specified by the manufacturers and switch on the power supply. The filament is heated in a short time to become red hot.

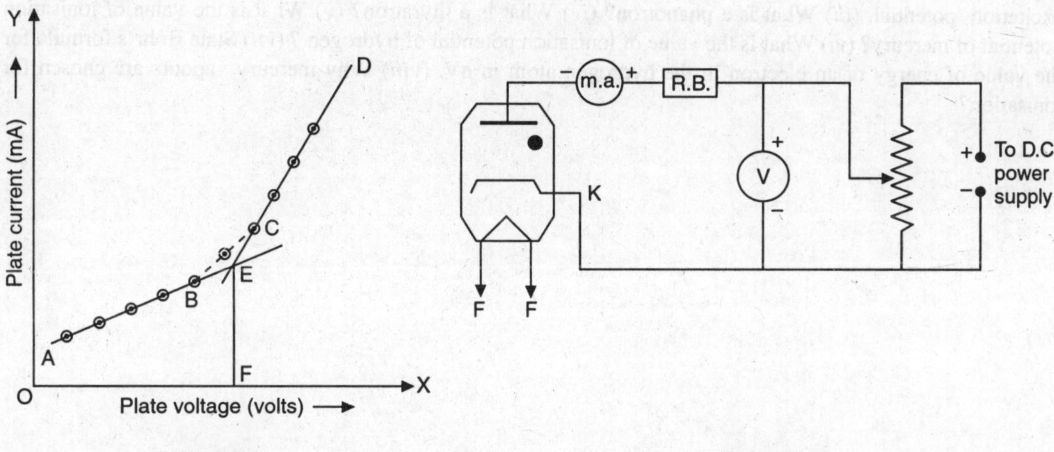

Fig. 37.2 Fig. 37.3

3. Adjust the voltmeter reading to 1 volt and note the corresponding value of the current in the milli-ammeter.

Increase the plate potential by one volt and note the voltmeter reading as well as the milliammeter reading. Proceed till the plate potential is about 15 volts.

4. Taking the plate voltage along the *X*-axis and plate current along the *Y*-axis plot a graph between plate current and plate voltage as shown in Fig. 37.2. Draw the straight line *AB* between the first few points and the straight line *CD* between the last few points and produce *AB* and *DC* to meet at *E*. Draw *EF* perpendicular to the *X*-axis, then *OF* gives the value of ionisation potential of mercury.

Observations.

No.	*Plate voltage in volts*	*Plate current in m A*

Ionisation potential of mercury from the graph = Volts

Standard value of ionisation potential = Volts

$$\text{Percentage error} = \frac{\text{Obs. value–stand. value}}{\text{stand. value}} \times 100 =$$

Precautions 1. A gas filled mercury vapour diode must be used.

2. The positive of the voltmeter as also that of the milliammeter must be connected to the positive of the *D.C.* supply.

3. The plate potential should not exceed 15 volts

4. To find the exact position of ionisation potential two straight lines joining the first few points and the last few points should be produced to meet. A smooth curve joining all the points should not be drawn.

5. For accurate measurement of voltage a *V.T.V.M* may be used in place of an ordinary voltmeter.

Oral Questions

Ionisation potential. (*i*) Define ionisation potential. (*ii*) Distinguish between ionisation potential and excitation potential. (*iii*) What is a phanotron? (*iv*) What is a thyratron? (*v*) What is the value of ionisation potential of mercury? (*vi*) What is the value of ionisation potential of hydrogen ? (*vii*) State Bohr's formula for the value of energy of an electron in the hydrogen atom in eV. (*viii*) Why mercury vapours are chosen for ionisation?

38

Photoelectric Effect

38.1. Photo-electric effect. It is the phenomenon of emission of electrons from a metal surface when light of suitable wavelength falls on it. The electrons emitted are known as *photo electrons.* Alkalie metals like sodium, potassium and caesium emit electrons even when visible light falls on them whereas zinc, cadmium etc. are sensitive only to ultra-violet light. Light of shorter wavelength is more effective in producing photo-electrons than light of longer wave length. The effect is also observed in non-metals but as these are highly electro-negative light of very short wavelength is required for the purpose.

38.2. Laws of photo-electric emission. 1. The emission of photo-electrons takes place only when the frequency of incident radiation is above a certain critical value characteristic of that metal. The critical value is known as **Threshold frequency** for the metal.

2. The emission of photo-electrons takes place as soon as light falls on the metal surface i.e. there is *no time-lag* between the incidence of photon and emission of electron.

3. The maximum kinetic energy with which an electron is emitted from a metal surface is independent of the intensity of incident light but depends only upon its frequency.

4. The number of photo-electrons emitted i.e photo-electric current is independent of frequency of the incident light but only depends upon its intensity.

38.3. Photoelectric cell. A photo-electric cell is an arrangement to convert light energy into electrical energy.

According to their construction the photo-electric cells are of three types

(i) *Photo-emissive cell*, (*ii*) *Photo-conductive cell* and (*iii*) *Photo voltaic cell.*

Photo emissive cell. There are two types of photo emissive cells

(*i*) Vacuum type and

(*ii*) Gasfilled type.

A vacuum type photo electric (emissive) cell consists of a glass or a quartz bulb according as it is to be used with visible or ultra violet light. It consists of a cathode C in the form of a semi cylindrical plate having a large surface area and an anode A in the form of a straight wire or a wire frame as shown in Fig. 38.1. When the cell is to be used with visible light a sensitive material like sodium, potassium or caesium is deposited on the cathode. For use with ultra violet light the material used is cadmium.

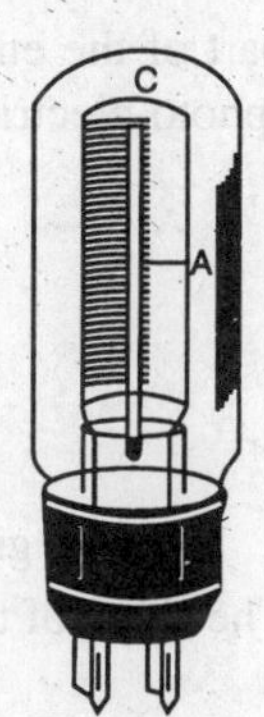

Fig. 38.1

To have a large number of electrons per photon of light, composite materials such as caesium on silver oxide ($Ag - O - Cs$) or antimony caesium alloy ($Sb - Cs$) or a recently developed combination of bismuth, silver, oxygen and caesium ($Bi - Ag - O - Cs$) are used as photo-cathode materials.

The current in a vacuum type (emissive) cell is small. To increase the current the cell is filled with an inert gas like neon or argon at a pressure of a few millimetre of mercury. Such a cell is known as a *gas-filled cell.*

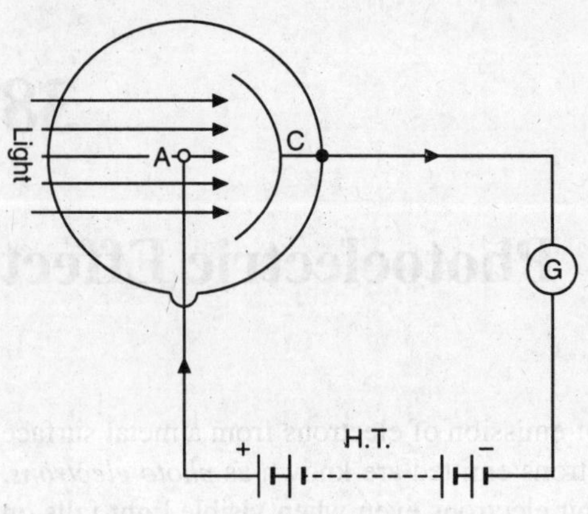

Fig. 38.2

A simple circuit showing the working of a cell is shown in Fig. 38.2. If a light of suitable wavelength is allowed to fall on the cathode C it gives out electrons. When the anode A is maintained at a positive potential the electrons are attracted towards it and a conventional current flows in the external circuit as shown. If A is maintained at a *negative* potential, the electrons are repelled by it and a smaller number reaches it. The current, therefore, decreases and at a certain negative potential no photo electrons reach A. This potential is called *stopping potential or cut off voltage*.

If V_s is the stopping potential, m, e and v_m the mass, charge and maximum velocity of the ejected electrons respectively, then

$$V_s e = \frac{1}{2} m v_m^2$$

or
$$v_m = \sqrt{\frac{2 V_s e}{m}}$$

The velocity of ejected electrons increases with increase in frequency of the incident light. When the frequency is below a certain critical value called **threshold frequency** no electrons are ejected. If ν_o is the threshold frequency the energy required to eject the electrons *just* out of the surface is given by

$$\omega_0 = h\nu_0$$

where ω_0 is known as *photo electric work function.*

In photo electric effect **all** or **none** of the energy of the incident photon is transferred to the electron in the metal. If, therefore, light of frequency $\nu > \nu_0$ falls on the photo-sensitive material a part of the energy of the incident photon is used by the electron to come out of the metal surface (photo-electric work function) and the rest is stored in it as kinetic energy.

$$\therefore \quad h\nu = \frac{1}{2} m v_m^2 + \omega_0$$

$$= V_s e + h\nu_0$$

or
$$V_s = \frac{h}{e}\nu - \frac{h}{e}\nu_0$$

Thus a graph between stopping potential V_s and frequency of incident light ν is a straight line. The slope of the straight line

$$\tan\theta = \frac{h}{e}$$

or
$$h = e\tan\theta$$

$$= e\frac{\Delta V_s}{\Delta \nu}$$

$$= e\,\frac{V_2 - V_1}{\nu_2 - \nu_1}$$

For $\quad V_s = 0,\ \nu = \nu_0$

For $\quad \nu = 0$

$$V_s = \frac{h\nu_0}{e} = \frac{\omega_0}{e}$$

$$= OB$$

∴ Photo electric work function

$$\omega_0 = e.\,OB$$

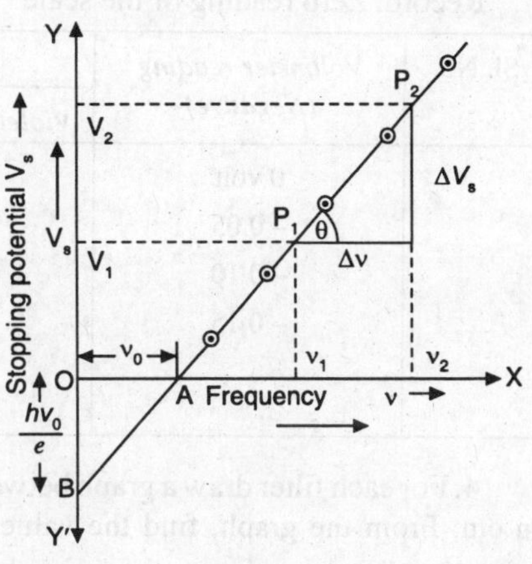

Fig. 38.3

Experiment 38.1. To find the value of Planck's constant and photo electric work function of the material of the cathode using a photo-electric cell.

Apparatus. A vacuum type photo-cell enclosed in a light-tight box, (R.C.A. 935 or its equivalent), 6 volt *D.C.* supply, a voltmeter (0-3 volt capable of reading up to 0.05 volt), a rheostat, a moving coil reflecting type galvanometer with lamp and scale arrangement, a one way key, a mercury lamp, light filters (red, yellow, green, blue, violet).

Procedure. 1. Draw a diagram showing the scheme of connections as in Fig. 38.4 and connect accordingly. Arrange the mercury lamp and the photo-cell as shown and set up the lamp and scale arrangement of the galvanometer so that the spot of light moves freely on the scale.

With the box closed (no light falling on the photo-cell) adjust the sliding contact to the extreme point *P* so that when the key *K* is introduced the voltmeter reads zero. Adjust the position of the galvanometer scale so that the spot of light with its central cross–wire is at zero graduation. Place the violet filter (λ= 4050Å) in front of the photo cell and switch on the mercury vapour lamp. Adjust the distance of the lamp so as to get a reasonable deflection on the galvanometer scale. Note the reading of spot of light on the scale and also that of the voltmeter.

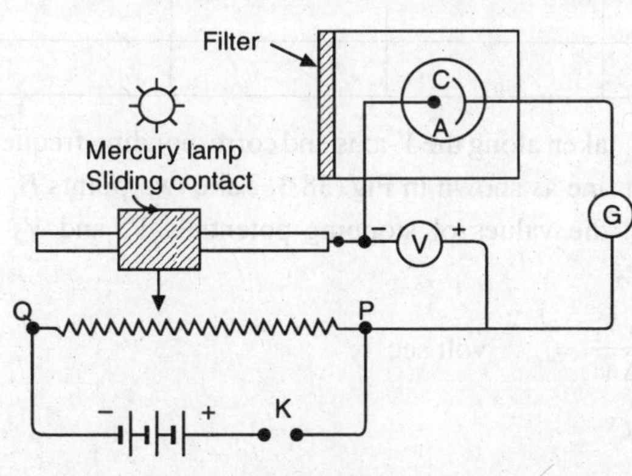

Fig. 38.4

2. Slowly move the sliding contact toward *Q* (negative end of the battery) so that a negative potential of 0.05 *V* is applied to the anode *A* of the photo-electric cell. The deflection will slightly decrease. Note the scale reading and voltmeter reading. Go on increasing the negative potential applied to the anode *A* in steps of 0.05 volt and go on noting the voltmeter reading and corresponding scale reading until the spot of light returns to zero position. Take at least *three* more readings by further increasing the negative potential in steps of 0.05 volt.

3. Repeat the experiment with blue (λ = 4360Å), green (λ = 5460Å) yellow (λ = 5780Å) and red (λ = 6910Å) filters.

Record. Zero reading of the scale =

Sl. No.	*Voltmeter reading (Negative)*	*Scale deflection with*				
		Violet filter	*Blue filter*	*Green filter*	*Yellow filter*	*Red filter*
	0 volt					
	– 0.05					
	– 0.10					
	– 0.15					
	...					
	...					

4. For each filter draw a graph between voltmeter reading (negative anode voltage) and deflection in cm. From the graph, find the value of stopping potential V_s for which scale deflection is zero corresponding to each wavelength and record.

Note. Alternatively the value of stopping potential corresponding to each filter may be found directly without plotting the graph. For this purpose the negative potential applied to the anode of the photo-cell is increased *very slowly* so that the deflection on the galvanometer scale *just* becomes zero. The corresponding voltmeter reading gives the stopping potential for each filter.

Filter	*Violet*	*Blue*	*Green*	*Yellow*	*Red*
Wavelength λ	4050×10^{-10} m	4360×10^{-10} m	5460×10^{-10} m	5780×10^{-10} m	6910×10^{-10} m
Frequency $\nu = \dfrac{c}{\lambda}$					
Stopping pot V_s					

5. Plot a graph between stopping potential V_s taken along the Y-axis and corresponding frequency ν taken along the X-axis. The graph is a straight line as shown in Fig. 38.3. Take two points P_1 and P_2 sufficiently apart on the straight line. Find the values of stopping potentials V_1 and V_2 and frequencies ν_1 and ν_2 corresponding to P_1 and P_2.

$$\text{Slope of the graph} = \tan\theta = \frac{V_2 - V_1}{\nu_2 - \nu_1} = \frac{\Delta V_s}{\Delta \nu} = \quad \text{volt sec}$$

Now $\qquad e = 1.59 \times 10^{-19}\ C$

$\therefore$ Planck's constant $h = e \tan\theta$

$$= e\frac{\Delta V_s}{\Delta \nu} = 1.59 \times 10^{-19} \times \frac{\Delta V_s}{\Delta \nu}$$

$$= \quad \text{Joule sec.}$$

Threshold frequency (corresponding to point A) $\nu_0 =$

Photo-electric work function $\omega_0 = h\nu_0 =$ Joule

Photo-electric work function (corresponding to point B)

$$= OB \text{ electron volt}$$

$$= e\,OB = 1.59 \times 10^{-19}\ OB \text{ Joule}$$

Precautions. 1. While adjusting the galvanometer reading at zero mark and voltmeter reading at 0 volt, no light should fall on the photo-cell.

2. The distance between the mercury lamp and the photo-cell should be kept constant during the experiment.

3. When the value of negative potential for which the deflection is zero has been reached, at least 3 more readings must be taken by further increasing negative potential in very small steps.

Experiment 38.2. To verify inverse square law of radiation using a photo-electric cell.

Apparatus. Same as in **Expt. 38.1** and an optical bench.

Theory. If L is the luminous intensity of an electric lamp and E the illuminascence (or intensity of illumination) at a point distant r from it, then according to inverse square law.

$$E = \frac{L}{r^2}$$

If this light is allowed to fall on the cathode of a photo-electric cell, then the photo-electric current given out is proportional to E and if θ is the corresponding deflection in a sensitive galvanometer connected to the photo-electric cell,then

$$E = \frac{L}{r^2} = k\theta$$

Hence a graph between $\frac{1}{r^2}$ and θ is a straight line, which verifies inverse square law of radiation.

Also $r^2\,\theta$ = a constant, which again verifies inverse square law.

Procedure. 1. Set the mercury vapour lamp and the photo-electric cell enclosed in a box as described in **Expt. 38.1.** on an optical bench. The connections of the photo-cell are made as shown with the difference that the positive terminal of the battery is connected to the anode A and the negative terminal to the cathode C.

2. Place the violet filter λ = 4050Å in front of the mercury lamp and obtain a deflection in the galvanometer with anode at zero potential. Adjust the distance of the mercury lamp from the photocell and the positive voltage applied to the anode so that there is saturation value of photo-electric current and the galvanometer gives the maximum deflection within scale. Note the positive potential applied to the photo-electric cell.

3. Note the position of the mercury lamp and the photo-electric cell on the optical bench. Also note the deflection of the galvanometer.

4. Keeping the voltage constant and position of the photo-cell fixed, increase the distance of the mercury lamp from the photo-cell in small steps. In each case note the position of the lamp on the optical bench and the deflection in the galvanometer.

5. Repeat using blue (4360Å) and green (5780Å) filters. In each case starting with the minimum distance adjust the positive voltage of the anode for saturation current for deflection of the galvanometer within scale. Keeping the voltage constant take different observations by varying the distance.

Observations. Filter violet λ = 4050Å

Positive potential applied to the photo-electric cell = V

Reading of photo-electric cell on optical bench =

Sl. No.	*Scale reading of lamp*	*Distance between lamp and photo-cell (r)*	r^2	$\frac{1}{r^2}$	*Deflection* θ	$r^2\,\theta$
1						
2						
.						

Similarly record observations for blue and green filters.

As $r^2\,\theta$ is constant, the inverse square law is verified.

6. Plot a graph between $\frac{1}{r^2}$ taken along the X-axis and θ taken along the Y-axis. The graph (for each filter) is a straight line proving the inverse square law of radiation.

Precautions same as in **Expt 38.1.**

38.4. Photo conductive cell. A photo conductive cell is based on the principle that the electrical resistance of a semi-conductor material like Selenium, Silicon, Germanium, Cadmium sulphide, Lead sulphide etc decreases with the increase in the intensity of radiation falling upon the material. In other words, the conductivity increases with the intensity of radiation falling upon a semi-conductor material.

A photo-conductive cell consists of a sheet of selenium or one of the semi-conducting materials given above. It is connected to an external source of *e.m.f* say a battery with the help of two wires of gold or some other conducting material. A current flows through the external circuit and when light falls on the cell the resistance considerably decreases due to which the current in the external circuit rises appreciably. However, the change in current is not proportional to the change in the intensity of light. There is also sufficient time lag between the change of light intensity and change in current.

38.5. Photo-voltaic cell. A photo–voltaic cell is the only true type of cell which generates its own electromotive force. *It requires no external battery to operate* it.

It consists of a *layer of semi-conductor material* spread over a metallic surface by heat treatment. The metallic surface is known as the *base*. On the other side of the semi-conductor there is a very thin layer of *translucent* deposit of gold, silver or platinum in the form of a *surface film* which allows the surface of the semi-conductor to be illuminated by radiation.

38.6. Back wall photo voltaic cell. In this type of photo-voltaic cell the *base* metal plate is made of *copper* the *semi-conductor material* is cuprous oxide (Cu_2O) and the *surface film* is of *gold.* During the preparation of the semi conducting oxide layer a very thin insulating layer is formed between the copper metal and the oxide. This layer is known as *barrier* layer. When light falls on such a cell, it passes through translucent surface film, liberates electrons from the transparent semi-conducting layer of Cu_2O which over come the barrier and pass out of the cell through the copper plate at the back. For this reason such a cell is known as **back-wall cell.** *In the external circuit electrons flow from the back plate to front surface film.*

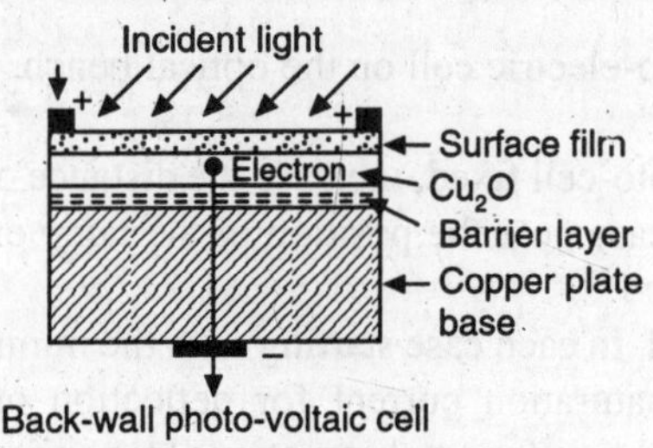

Fig. 38.5

Front wall photo voltaic cell. In this type the *barrier layer* is formed between the *translucent metal film* and the semi-conducting material which is **opaque**. The copper or metal plate (*base*) is pressed to the semi-conducting layer to form a perfect electrical contact. In such a cell the electrons emitted by the semi conductor say (Cu_2O) due to exposure to radiation over come the *barrier layer* and pass out of the cell through the translucent surface film at the front. Such a cell is, therefore, known as a **front wall cell.** *The electrons in the external circuit flow from the surface film to the back plate.*

38.7. Selenium cell. In this type the base is nickel plated iron over which is spread a thin layer of selenium (semi-conducting material). The surface film is a translucent layer of silver or platinum. The barrier layer is formed when silver is deposited by electrical sputtering process. A copper ring in contact with the iron base serves as one electrode and another copper ring in contact with surface silver film acts as the second electrode. The cell is enclosed in a *case* fitted with a *protective window.*

When light falls on the cell it passes through the window the translucent surface film of silver and the transparent barrier layer. On reaching the selenium layer it sets free some electrons which are not ejected into the free space but only produce an *e.m.f.* As selenium is opaque to light, this cell is also *front wall type i.e.* the electrons flow from the surface film to the nickel plated iron base in the external circuit (Fig. 38.6).

38.8. Characteristics of a photo-voltaic cell. A photo voltaic cell is much more sensitive than even a gas filled *photo-emissive cell and is used as a solar cell.*

In such a cell the current due to the generated photo-electric *e.m.f.* is proportional to the intensity of incident light only when the electrodes are short-circuited *i.e.* there is no external resistance.

The *e.m.f.* generated by the photo-voltaic cell in open circuit, *i.e.* when no current is drawn from it is denoted by V_{OC} (*V*-open circuit)

This is the maximum value of *e.m.f.* When a high resistance is introduced in the external circuit a *small* current flows through it and the voltage decreases. The voltage goes on falling and the current goes on increasing as the resistance in the external circuit is reduced. When the resistance is reduced to zero the current rises to its *maximum* value known as *saturation current and is denoted as* I_{SC}, the voltage becomes zero. An *I–V* characteristic of a photo-voltaic cell is shown in Fig. 38.7.

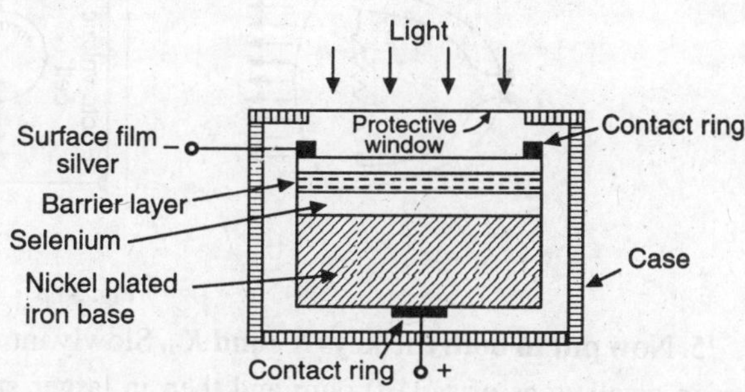

Fig. 38.6

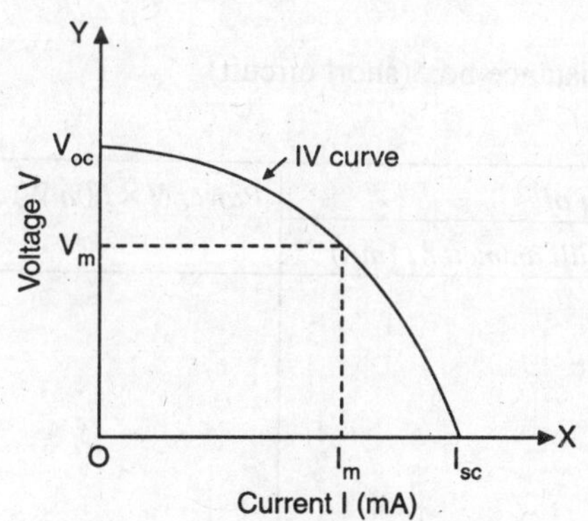

Fig. 38.7

The product of open circuit voltage V_{OC} and short circuit current I_{SC} is known as *ideal power.*

$$\therefore \quad \text{Ideal power} = V_{OC} \times I_{SC}$$

The maximum useful power is the area of the largest rectangle that can be formed under the *I-V* curve. If V_m and I_m are the values of voltage and current for this condition, then

Maximum useful power = $V_m \times I_m$. The ratio of maximum useful power to ideal power is called the *Fill factor*

$$\therefore \text{ Fill factor} = \frac{V_m \times I_m}{V_{OC} \times I_{SC}}$$

Experiment 38.3. To study the characteristics of a photo-voltaic cell (solar cell)

Apparatus. A solar cell (photo voltaic cell), a *D.C.* voltmeter, (or millivoltmeter) a *D.C.* milliammeter (or micro-ammeter) a dial type resistance box, 2 one way keys, a 100 W lamp, connecting wires etc.

Note. A solar module 6 V, 100 m A is suitable as a solar cell.

Procedure 1. Keep the solar module (solar cell) in the sunlight for 15-20 minutes with the light rays falling normally over it, so that it gets properly activated.

2. Draw a diagram showing the scheme of connections of the photo-voltaic cell as shown in Fig. 38.8 and connect accordingly. The experiment is preferably performed in a dark room to avoid the effect of any stray light falling on the photo voltale cell.

3. Place the 100 W lamp close to the cell so that light from the lamp falls normally on the cell. Put in the key K_1 (keeping K_2 open) and note the reading of the voltmeter for open circuit V_{OC}. Take out the key K_1.

4. Now put the key K_2 (keeping K_1 open) and make resistance R from the resistance box *zero* and note the reading of milliammeter under short circuit conditions I_{SC}.

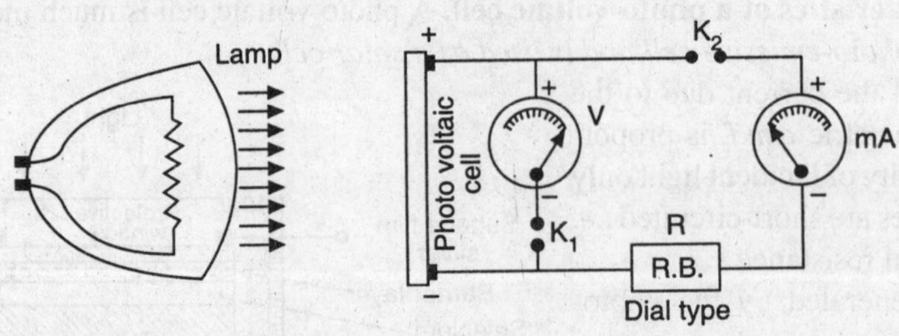

Fig. 38.8

5. **Now put in both the keys K_1 and K_2.** Slowly introduce the resistance from the resistance box first in small steps of say 20 ohm and then in larger steps of say 50 ohm. For each value of the resistance note the resistance, the voltmeter reading and the millammeter reading. Take out the keys K_1 and K_2 and switch off the lamp. Repeat the experiment with a lamp of a higher or lower power (wattage).

Observations. Voltmeter reading for open circuit

V_{OC} = volts

Milliammeter reading with zero resistance from the resistance box (short circuit)

I_{SC} = m A

Sl. No.	*R ohms*	*Reading of*		*Power V × I (m W)*
		Voltmeter V (volt)	*Milli ammeter I (mA)*	
1				
2				
3				
.				
.				
.				
.				
11				
12				

6. Draw the following graphs :-

(*i*) *Voltage current or VI graph.* Taking I along the X-axis and V along the Y-axis draw the *V-I* graph. Draw a rectangle having the *maximum area* under the *V I* curve and note the values of V_m and I_m.

Value of V_m from the graph = Volts

Value of I_m from the graph = m A

Maximum useful power $V_m \times I_m$ = m W

Ideal power $V_{OC} \times I_{SC}$ = m W

Fill factor $= \dfrac{V_m \times I_m}{V_{OC} \times I_{SC}} =$

(*ii*) *Voltage resistance or V-R. graph.* Taking R along the X-axis and V along the Y-axis draw the V–R graph.

(iii) Power-resistance or W-R graph. Taking R along the X-axis and power $W = V \times I$ along the Y-axis draw W-R graph. From the graph find maximum useful power. This is also known as **power load graph.**

Precautions 1. The solar cell should be exposed to sun light before using it in the experiment.

2. Light from the lamp should fall normally on the cell.

3. A resistance in the cell circuit should be introduced so that the current does not exceed the safe operating limit.

Experiment 38.4. To study the voltage current (*V-I*) Power load (*VI-R*), areal and azimuthal characteristiccs of a photo-voltaic cell.

Apparatus. Same as in **Expt. 38.3**, an adjustable slit to change the area, (with mounting arrangement) a circular frame dial system to change the angle (azimuthal position) of the photo-voltaic cell with proper mounting arrangement, an optical bench with three uprights, connecting wires etc.

Procedure *(i) Voltage current (V-I) characteristics.* See **Expt. 38.3.**

(ii) Power load (VI-R) characteristics See **Expt 38.3.**

(iii) Areal characteristics. 1. Set the apparatus as shown in Fig. 38.9 making proper connections of photo-voltaic cell with voltmeter, dial type resistance box (R.B), milli-ammeter (or micro-ammeter) and keys K_1 and K_2.

2. Keep the key K_2 closed and K_1 open. Mount the adjustable slit on an upright. Adjust the slit with the screws S_1 and S_2 on it so that the length and width of the slit are minimum. Switch on the lamp and adjust its distance from the photo-voltaic cell so that the micro-ammeter (or milli-ammeter) gives some reading. Note the length and breadth of the slit. Also note the reading of the micro-ammeter. A scale is generally provided along the length as well as along the breadth of the slit from which readings can be taken.

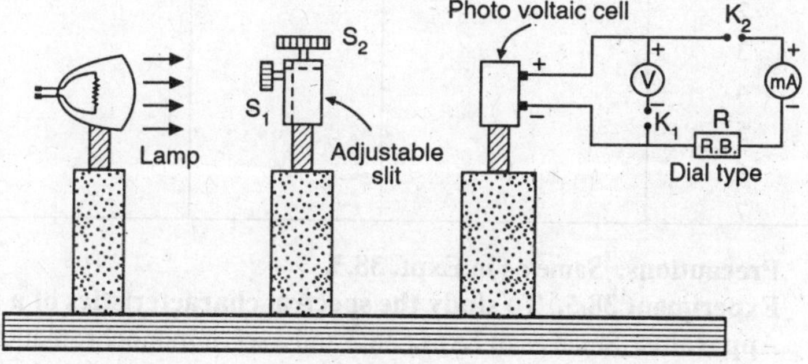

Fig. 38.9

3. Gradually start increasing the area of the slit by increasing the length and breadth by 1 mm at a time. Note the length, breadth and the deflection i.e. current in each case. Take about 6-7 readings.

Observations. Distance between the lamp and photo-voltaic cell =

S. No.	*Length the slit (l) in m.m.*	*Breadth of slit (b) in mm*	*Area of slit* $A = l \times b$ *mm*^{2}	*Current in mA/μ A (I)*
1				
2				
3				
4				
5				
6				
7				

4. **Graph.** Plot a graph between area A taken along the X-axis and the corresponding values of current I along Y-axis.

(iv) **Azimuthal angle characteristics.** 5. Set the apparatus as in step 1. Mount the photo-voltaic cell on the circular frame with dial system to change the angle of the cell on an upright. Set the area of the slit to a suitable convenient value. Adjust the photo-voltaic cell in the vertical position so that the reading of the angle on the circular dial is zero.

6. Switch on the lamp and adjust the distance between the lamp and the photo-voltaic cell so that the micro-ameter gives a proper large deflection.

7. Gradually start changing the angle of the photo-voltaic cell with the help of the graduated dial by 5°–10° at a time and go on recording the value of the angle turned and corresponding reading of the micro-ammeter keeping the intensity and distance of the lamp constant. Go up to 90° first to one side and then to the other side.

8. Repeat the observations by changing the distance.

9. Plot a graph between *'angle turned'* taken along X-axis and *'current'* along Y-axis.

Observations

S.No.	*Distance between lamp and photo-voltaic cell (d)*	*Angle turned θ (towards right)*	*Micro-ammeter reading I*	*Angle turned θ (towards left)*	*Micro ammeter reading I*
1					
2					
3					
4					
5					
6					
7					

Precautions. Same as in **Expt. 38.3.**

Experiment 38.5. To study the spectral characteristics of a photo-voltaic cell.

Apparatus. Same as in **Expt. 38.3** and a set of monochromatic filters of different wavelengths. A tin or card board box cover which completely fits over the photo-voltaic cell and has a window in which the filter can be mounted, A potential dividing arrangement, a battery of a few cells, a key and a reversing key (or commutator).

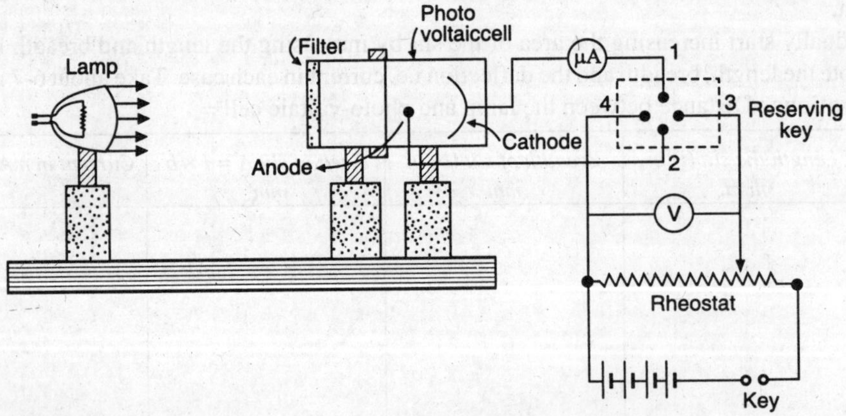

Fig. 38.10

Procedure 1. Set the lamp and photo voltaic cell with filter cap on proper uprights on an optical bench. Draw a diagram showing the scheme of connections as in Fig. 38.10 and connect accordingly using a reflecting type galvanometer or a micro-ammeter. Connect the top terminals 1 and 2 of the reversing key to the photo-voltaic cell and bottom terminals 3 and 4 to the potential dividing arrangement as shown.

2. Switch on the lamp. Adjust its position and distance so that the micro-ammeter gives a large deflection.

3. Bring the moveable contact of the potential dividing arrangement to the extreme left and rotate the reversing key so that terminal 1 is connected to terminal 3 and 2 to 4 and thus the cathode of the photo-voltaic cell is at a negative potential with respect to the anode. Insert one filter in the filter cabin or case and note the reading of the micro-ammeter for voltmeter reading $V = 0$. Slowly increase the value of V by sliding the moveable contact of the rheostat in very small steps and note the voltmeter as well as micro-ammeter reading.

4. Now reverse the current by rotating the reversing key so that terminal 1 is connected to terminal 4 and 2 to 3. In this case the cathode of the photo-voltaic cell is at a positive potential with respect to the anode. Again note the reading of the micro-ameter for $V = 0$. Slowly increase the value of V in small steps and note down the micro-ammeter as well as corresponding voltmeter reading. For a particular value of V the micro-ammeter reading will become zero. This is *stopping potential.* Note this value.

5. Change the filter. Repeat steps 3 and 4 making the cathode once negative and once positive with respect to the anode of the photo-voltaic cell. Repeat the experiment with other filters one by one.

Observations. Position of the lamp = cm.

Position of the photo-voltaic cell = cm.

Sl. No.	*Filter*	*Wavelength* λ	*Frequency* ν	*Cathode positive*		*Cathode negative*	
				V volts	*I μA*	*V volts*	*I μA.*
1 2 3 4 5 6	1	$\lambda_1 =$ A°	$\nu_1 = \frac{c}{\lambda_1} =$				
1 2 3 4 5 6	2	$\lambda_2 =$ A°	$\nu_2 = \frac{c}{\lambda_2} =$				
1 2 3 4 5 6	3	$\lambda_3 =$ A°	$\nu_3 = \frac{c}{\lambda_3} =$				
1 2 3 4 5 6	4	$\lambda_4 =$ A°	$\nu_4 = \frac{c}{\lambda_4} =$				

6. **Graph.** For each wavelength plot a separate graph taking V(in volts) along + X direction for cathode at *negative* potential and along $-X$ direction for cathode at *positive* potential and current I(in μ A) along Y-axis. From each graph find the value of stopping potential as shown in Fig. 38.11.

7. Now plot a graph between stopping potential V in volts taken along Y-axis and frequency υ taken along X-axis as shown in Fig. 38.12.

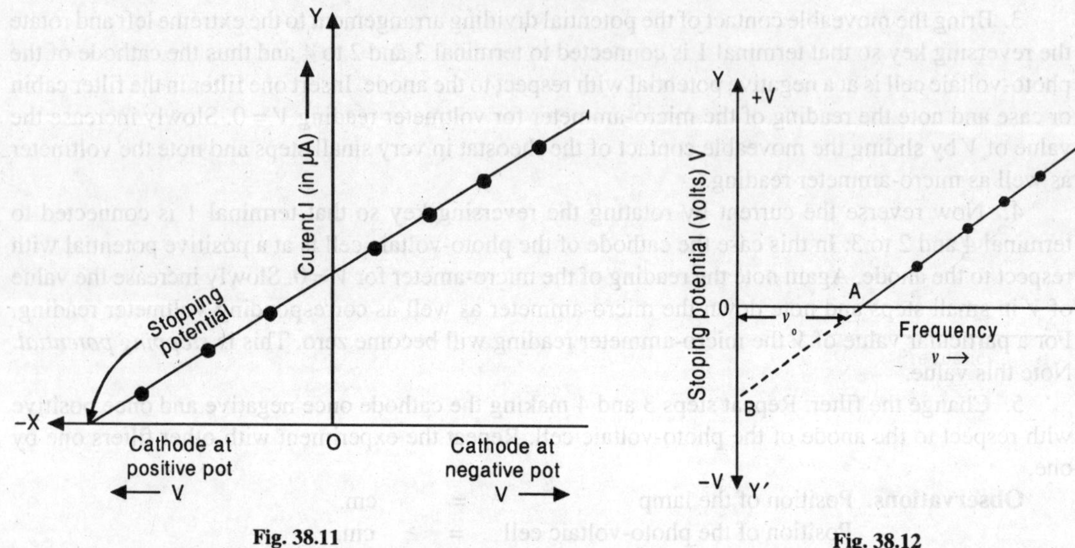

Fig. 38.11 **Fig. 38.12**

The graph cuts the X-axis at A. Extra- polate the graph so that it cuts the Y-axis at B.

Threshold frequency ν_0 = OA = Hz

Work function ω_0 = OB = eV

(For theory and details see **Expt. 38.1**)

Precautions. Same as in **Expt. 38.3.**

Experiment 38.6. To verify inverse square law of light using a photo voltaic cell.

Apparatus. A photo voltaic cell (barrier layer type) with mounting arrangement, A source of light with mounting arrangement, a sensitive galvanometer or a micro-ammeter, optical bench with two uprights.

Theory. Same as in **Expt. 38.2.**

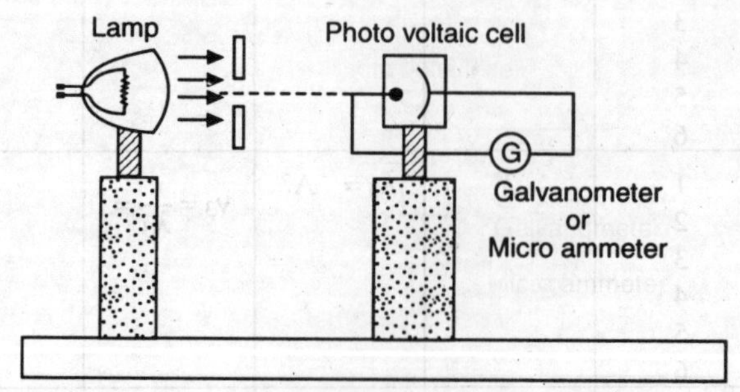

Fig. 38.13

Procedure. 1. Fit the photo voltaic cell in an upright and place it on the optical bench. Connect the two terminals of the cell to a sensitive galvanometer. On the other upright fit the light source and place it on the optical bench in such a way that light from the source falls on the photo-voltaic cell. Adjust the height and position of the lamp and the photovoltaic cell so that they lie in the same horizontal line and light from the lamp (light source) falls normally on the cell. Now adjust the distance of the light source so that the galvanometer gives a large deflection within scale. Note the position of the uprights carrying the cell and the lamp on the optical bench. Also note the deflection θ in the galvanometer.

2. Slowly change the distance of the lamp from the photo voltaic cell and take a number of observations noting the distance and corresponding deflection in the galvanometer.

Observations. Scale reading of photo-voltale cell on the optical bench = (*a*) =

Sl. No.	*Scale reading of lamp (b)*	*Distance between lamp and photo-voltaic cell r = (a – b)*	r^2	$\frac{1}{r^2}$	*Galvanometer deflection* θ	$r^2 \theta$
1						
2						
3						
.						
.						
7						

Result. As $r^2\theta$ is a constant, the inverse square law is verified.

3. Plot a graph between $\frac{1}{r^2}$ taken along *X*-axis and galvanometer deflection θ taken along the *Y*-axis. The graph is a straight line proving that the inverse square law is verified.

Precautions. 1. Light should fall normally on the photo-voltaic cell.

2. The galvanometer connected to the photo voltaic cell should have a low resistance.

3. A cover should be placed on the cell which should be removed only when the light is to fall on it. The photo-voltaic cell should not be exposed to light for a long time continuously.

38.9. To compare the luminous intensities (Illuminating powers) of two light sources. A photo voltaic cell can be used to compare the luminous intensities of two light sources.

If L_1 is the luminous intensity of a light source and it is kept at a distance r_1 from a photovoltaic cell so that light from the source falls normally on the cell, then according to the inverse square law, the illuminascence (or intensity of illumination) at the cell.

$$E = \frac{L_1}{r_1^2}$$

If the photo-electric current produced in the cell circuit is proportional to *E* and if it produces a deflection θ in a galvanometer, then

$$E = \frac{L_1}{r_1^2} = k\,\theta \qquad ...(i)$$

where *k* is constant for a given pair of photo voltaic cell and galvanometer.

If a second source of luminous intensity L_2 produces the same deflection θ, when placed at a distance r_2 from the photo-voltaic cell, then

$$E = \frac{L_2}{r_2^2} = k\,\theta \qquad ...(ii)$$

From (*i*) and (*ii*), we have

$$\frac{L_1}{r_1^2} = \frac{L_2}{r_2^2}$$

or

$$\frac{L_1}{L_2} = \frac{r_1^2}{r_2^2} = \left(\frac{r_1}{r_2}\right)^2$$

Experiment 38.7. To compare the luminous intensities (illuminating powers) of two light sources using a photo-voltaic cell.

Apparatus. A photo voltaic cell with mounting arrangement, two sources of light with mounting arrangement, an optical bench with two uprights, a sensitive galvanometer, connecting wires etc.

Procedure. 1. Same as in **Expt. 38.6.**

2. Remove the first light source from the upright and in its place fit the second light source. Adjust its height and angle so that again the light source and the cell are in the same horizontal line and light falls normally on the cell.

Adjust the distance of the light source by bringing it closer to the cell or taking it away from it till the galvanometer again shows the same deflection θ. Note the position of the uprights carrying the cell and the lamp on the optical bench. Also note the deflection θ in the galvanometer.

3. Repeat the experiment 3 times for different values of the deflection θ.

Observations.

No.	FIRST LIGHT SOURCE				SECOND LIGHT SOURCE				$\frac{L_1}{L_2} =$
	Position of photo voltaic cell (*a*)	Position of light source (*b*)	$a - b = r_1$	Deflection θ	Position of photo voltaic cell (*c*)	Position of light source (*d*)	$d - c = r_2$	Deflection θ	$\left(\frac{r_1}{r_2}\right)^2$

Mean value of $\frac{L_1}{L_2} =$

Note. No index correction is applied to find r_1 and r_2 because the lamp filamant as well as the photo-sensitive surface of the photo–voltaic cell lie inside the glass bulbs and index correction can not be determined.

Oral Questions

1. Photo-electric cell (*i*) What is photo-electric effect? (*ii*) What is a photo electric cell? (*iii*) What is a vacuum type photo cell ? (*iv*) What sensitive material is used when the cell is to be used for a visible light ? (*v*) What material is to be used when the cell is to be used with ultraviolet light? (*vi*) What is stopping potential or cut off voltage? (*vii*) What is threshhold frequency? (*viii*) What is photo electric work function? (*ix*) How will you determine the stopping potential? (*x*) What is Planck's constant? What are its units? (*xi*) What is the relation between energy of photon and its wavelength? (*xii*) If the wavelength of light increases will the energy increase or decrease? (*xiii*) Does the violet light have more energy or the red light? (*xiv*) State the laws of photoelectric emission. (*xv*) What is the order of current in a photo-electric cell?.

2. Photo-voltaic cell. (*i*) Explain the construction and working principle of the photo-voltaic cell: (*ii*) What are the various types of the photo-voltaic cell in use? (*ii*) What is solar cell? (*iv*) Define fill factor of a photo-voltaic cell. (*v*) What is the function of barrier layer in a photo-voltaic cell? (*vi*) What do you mean by band-gap energy as referred to a photo voltaic cell?

39

Geiger Mueller Counter

39.1. Geiger Mueller counter. A Geiger Mueller counter tube consists of a glass cylinder *G* with a metal coating *M* on the inside and a co-axial wire anode *A* (usually of tungsten) suspended along its axis. The cylinder is filled with a gas such as argon mixed with halogen vapour at a reduced pressure of about 10 cm of mercury. The tube may have a very thin mica end window *W* shielded by a protective gauze *P*. The metal tube is maintained at a *negative* potential of about 1000 volt with respect to the wire. The potential difference between *M* and *A* should be greater than certain *threshold* value but less than the voltage which would cause a continuous discharge of electricity through the gas.

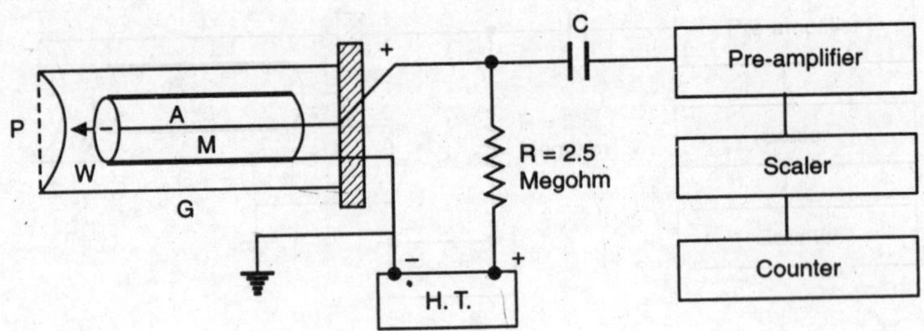

Fig. 39.1

Under this condition a single ionising particle like a β-particle or a γ-ray photon entering the counter through the end window will produce ionisation of the gas molecules. The ions are accelerated by the electric field and produce more ions by collision causing ionisation current to build up rapidly which gives rise to an *avalanche*. The avalanche in turn produces a current in the external resistance *R*. The resulting drop in potential across *R* reduces the potential across the tube below the threshold value and the current decays rapidly as the circuit has a small time constant. There is, therefore, a momentary surge of current which is referred to as an *electric pulse*. Each pulse corresponds to one particle entering the counter.

It takes the counter some time to recover and be ready for the next particle. This period is known as recovery time or *dead time* and is of the order of a few hundred micro-seconds for most counters. If a second particle enters the counter before it has recovered the particle will not be counted. Thus the *dead time* of the tube sets up an upper limit to the *counting rate* (*CR*) of the tube. The successive current pulses generated in the resistance *R* corresponding to the number of ionising particles passing into the tube are *detected* and *counted* by a suitable arrangement.

The wire-cylinder potential difference *V* plotted against the counting rate (*CR*) owing to a radio active source of fixed intensity placed near the tube is known as a *plateau curve*. A typical plateau curve is shown in Fig. 39.2.

Corresponding to voltage at point *A* pulses begin to appear. This voltage at *A* corresponds to *Threshold voltage or Geiger Mueller Threshold potential.* As the voltage across the tube is increased the counting rate increases rapidly upto a point *B* (corresponding to voltage V_1) and then levels off to an almost constant value until voltage V_2 corresponding to point *C* is reached. The portion *BC* is known as *Geiger Mueller region.* The voltage from *B* to *C* is known as the *length of the plateau. The slope of the plateau is the average percentage change in the counting rate per unit increment of applied voltage and is known as relative plateau slope. (R.P.S).* It is, therefore, given by

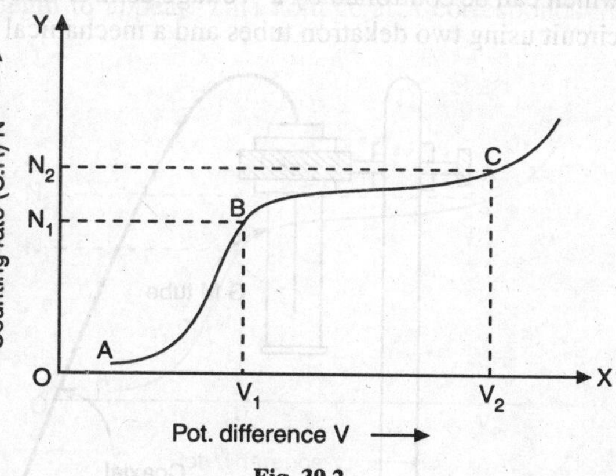

Fig. 39.2

$$R.P.S = \frac{(N_2 - N_1)/N}{V_2 - V_1} \times 100$$

where N_1 is the counting rate at V_1 and N_2 that at voltage V_2. Also

$$N = \frac{N_1 + N_2}{2}$$

As the voltage is increased beyond V_2 a continuous discharge takes place. The operating voltage used in the experiment, therefore, should be some where between V_1 and V_2 in the region *BC* because in this region the counting rate is almost independent of the voltage. The voltage interval between *B* and *C* $(V_2 - V_1)$ is commonly of the order of 200 – 400 volt.

The Geiger counter tube is housed in a massive iron box which shields it from stray radio-active particles. The tube itself is made of very thin metal or glass with metal coating inside and is, therefore, very fragile. It should be handled very carefully. The Geiger tube is mounted on a stand provided with a *source holder* on which the radio active source can be conveniently placed as shown in Fig. 39.3.

Scaler unit. The electrical pulses from the Geiger tube after being suitably amplified must be counted in some way so that the rate can be measured. For very slow counting rates *i.e.* less than 10 counts/sec the pulses can be directly made to operate an electro-mechanical recording device. For fast rates the mechanical counters may either jam completely or miss many counts. We, therefore, use a *scaling unit* or *scaler* which is an electronic instrument that selects only certain pulses and passes them on to the mechanical recorder. A scalar with a scaling ratio of 64 selects every 64th pulse and passes it on to the recorder. In order to have a simple electronic circuit scaling ratio is a power of 2 *i.e.*, (32, 64, 128 etc.).

A power of 10, 100, 1000 etc. is used in *Dekatron counter*. A dekatron counter consists of two or more dekatron tubes each containing a glow or discharge which can move round a circular scale graduated from 0 to 9 together with a mechanical counter. Each pulse causes the discharge in the first tube which counts *units* to advance by one digit. The circuit is so designed that on the *tenth* pulse which returns the first counter to zero a pulse is sent to the second tube. The glow in it moves to *one.* The second tube thus counts the number of tens of pulses. After 10 pulses are sent to the second tube corresponding to a count of 100, the output pulse from the second tube is fed to the mechanical counter. This, therefore, registers the hundreds, thousands and so on. The front panel of a common form of scaler unit is shown in Fig. 39.4. It contains electronically regulated extra high voltage supply unit which can be controlled by a 'voltage control' having an accurately calibrated voltmeter, the scaling circuit using two dekatron tubes and a mechanical counter.

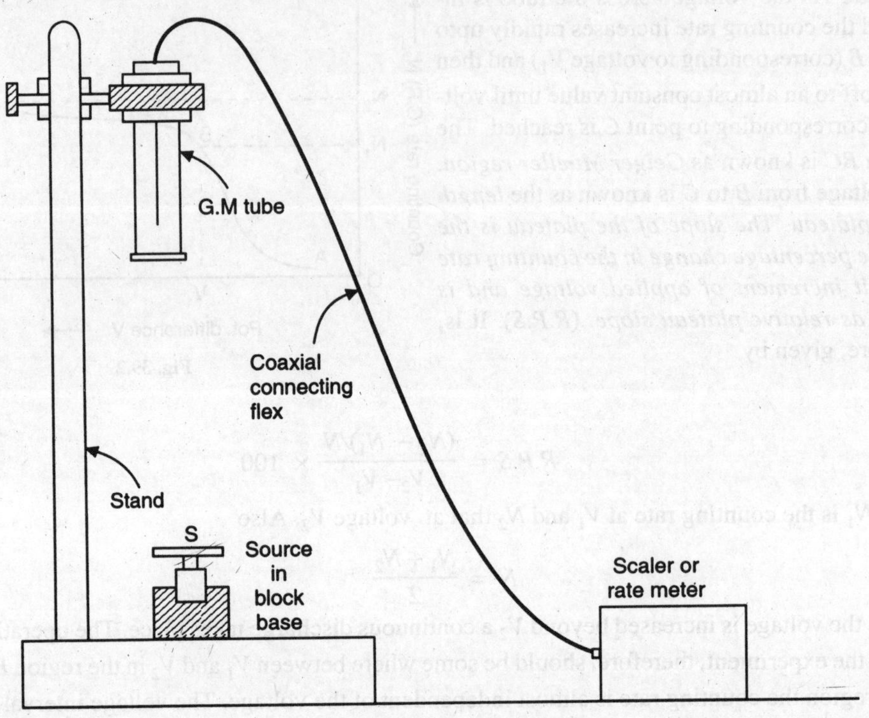

Fig. 39.3

Some scaler units are also fitted with a time recording device. Such a unit is known as a *rate-meter.* The scaler unit is connected to the G.M. tube or detector through a *co-axial connecting flex.*

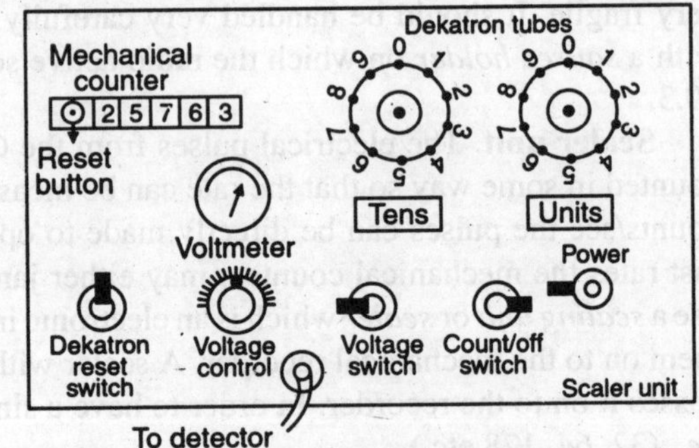

Fig. 39.4

Experiment 39.1. To draw the plateau curve for a Geiger Mueller counter.

Apparatus. A Geiger Mueller counter tube with stand, scalar unit having arrangement for extra high voltage *D.C.* supply and fitted with a voltmeter. A stop watch (if no time recording device is fitted in the scaler unit) and a radioactive source.

Note. A commonly used β-source for laboratory purpose is 1 μc Sr^{90}. Suitable radioactive sources for laboratory use are available from Bhaba Atomic Research Centre, Bombay. A suitable G. M tube is No. 1030.

Procedure. 1. Draw a diagram showing the scheme of connections and connect accordingly as in Fig. 39.1. Here *R* is a high resistance of 2–5 megaohm. Note that the scaler unit is not connected to the power supply. Connect the Geiger tube securely to the scaler unit. Turn the '*voltage control*', knob of the scaler unit to its lowest position (counter-clockwise) and switch on the '*voltage switch*' and the '*count switch*'. Now connect the scaler unit to the mains supply and switch on '*Power*'.

2. Using a forceps, place the *radioactive source* on the source holder just below the Geiger tube as shown in Fig. 39.3 and increase the high voltage *slowly* until the scaler unit begins to indicate pulses. Note the reading of the voltmeter (fitted on the panel of the scaler unit). Pulses generally start appearing at about 800 volt or less. Determine the counting rate (*CR*) by noting the number of counts for one minute or for a smaller time say 10 sec or 20 sec or 30 sec depending upon the counting rate). Repeat three times.

3. From this *threshold* value increase the voltage in steps of 20-25 volt upto 1000 volt or to a voltage for which the counting, rate increases to 10% above the plateau value, whichever is less. Note counting rate corresponding to each step by taking a one minute reading and repeat each reading thrice.

Caution (*i*) The counter switch should always be kept *on* to avoid excessive counting rates.

(*ii*) Do not exceed the high voltage beyond a value that raises the counting rate 10% above the plateau value. *Excessive voltage applied to the counter will damage it.*

4. Reduce the applied voltage to zero and switch off the mains supply. Remove the radio-active source from the source holder.

Again switch on the mains supply and note the background counting rate (without the radio active source) corresponding to each voltage used in step 3. Again reduce the applied voltage to zero and switch off the mains supply. Also disconnect the Geiger tube from the scaler unit.

5. Plot a graph by taking voltage along the *X*-axis and the corrected counting rate N [No. of pulses per minute with the radio-active source N' – No. of pulses per minute without the radio active source *i.e.* background only (b)] = $N' - b$. taken along the *Y*-axis. The graph is of the form shown in Fig. 39.2.

Draw perpendiculars from the points *B* and C of the *almost* horizontal part meeting the *X*-axis at V_1 and V_2 and *Y*-axis at N_1 and N_2 respectively. Note the corresponding voltage V_1 and V_2. The voltage interval $V_2 - V_1$ gives the plateau region. The operating voltage of the counter is given by

$$V = V_1 + \frac{1}{3}(V_2 - V_1)$$

It is approximately 50 volt above the lower end of the plateau region *i.e.* V_1.

Observations.

Sl. No.	*Voltage*	*Observed counting rate*					*Corrected counting rate (Pulses/minute)* $N = N' - b$
		With radioactive source				*Only background (b)*	
		(i)	*(ii)*	*(iii)*	*mean N'*		

Note. A more accurate method of applying background correction is to take $N' - b = N$. Find $\sqrt{N}$ and calculate $N + \sqrt{N}$ and $N - \sqrt{N}$. While plotting the graph, points are plotted for $N + \sqrt{N}$ (shown by small square □). and $N - \sqrt{N}$. (shown by small circles.) corresponding to the respective voltage. The curve passes through the mean position corresponding to points for $N + \sqrt{N}$ and $N - \sqrt{N}$.

From the graph

Threshold voltage = voltage at *A* = volt

Voltage at which plateau begins

= voltage at $B = V_1$ = volt

Voltage at which plateau ends

= voltage at $C = V_2$ = volt

Plateau region = volt to volt

Operating voltage from the graph

$$V = V_1 + \frac{1}{3}(V_2 - V_1) \quad = \quad \text{volt}$$

Relative plateau slope

Mean corrected counting rate at $V_1 = N_1$ = pulses per minute

Mean corrected counting rate at $V_2 = N_2$ = pulses per minute

$$N = \frac{N_1 + N_2}{2}$$

$$\therefore \qquad R.P.S. = \frac{(N_2 - N_1)/N}{V_2 - V_1} \times 100$$

$$= \%$$

Precautions. 1. The radioactive material should be handled very carefully by a pair of tongs or forceps.

2. The counting switch should always be kept on so that the counting rate does not exceed 10% above the plateau voltage. Excessive voltage will damage the G.M. counter.

3. The radio active source should be placed in the source holder just below the Geiger tube before increasing the high voltage slowly.

39.2. Dead time of a G.M. counter. The time interval after recording a count, for which the counter is completely insensitive and is unable to detect and record other ionising events inside the counter, is called the *dead time* of the counter.

Suppose a β-source gives N counts per second in a counter with *zero* dead time. Let it give n counts per second in a counter with dead time T sec, then

$$T = \frac{1}{n} - \frac{1}{N}$$

or

$$N - n = N n T$$

In other words, the counter fails to record $N - n$ or NnT counts per second.

To find the *dead time* of a counter we use two different β-sources, say Sr^{90} and Tl^{204}. Let the two sources give n_1 and n_2 counts per second separately and n_3 counts per second when placed together, then

$$N_1 - n_1 = N_1 n_1 T \qquad ...(i)$$

$$N_2 - n_2 = N_2 n_2 T \qquad ...(ii)$$

and

$$(N_1 + N_2) - n_3 = (N_1 + N_2) n_3 T \qquad ...(iii)$$

where N_1, N_2 are counts per second as recorded by a counter with *zero* dead time for the two sources separately and $(N_1 + N_2)$ counts per second when placed together.

From (*i*), we have

$$N_1(1 - n_1 T) = n_1 \text{ or } N_1 = \frac{n_1}{1 - n_1 T}$$

Similarly from (*ii*), we have

$$N_2 = \frac{n_2}{1 - n_2 T}$$

Substituting in (*iii*), we have

$$\frac{n_1}{1-n_1 T}+\frac{n_2}{1-n_2 T}-n_3=\left(\frac{n_1}{1-n_1 T}+\frac{n_2}{1-n_2 T}\right) n_3 T$$

or $$n_1+n_2-n_3-2 n_1 n_2 T+n_2 n_3 T+n_1 n_3 T-n_1 n_2 n_3 T^2$$

$$= n_1 n_3 T - n_1 n_2 n_3 T^2 + n_2 n_3 T - n_1 n_2 n_3 T^2$$

or $$n_1+n_2-n_3-2 n_1 n_2 T=0 \quad \text{...}(iv)$$

Neglecting terms containing T^2

From (*iv*), we have

$$T=\frac{n_1+n_2-n_3}{2 n_1 n_2}$$

Experiment 39.2. To find the dead time of a G.M. counter.

Apparatus. Same as in **Expt. 39.1** and a double source holder, two radio active sources Sr^{90} and Tl^{204}.

Procedure 1. Find the value of operating voltage as in **Expt. 39.1.** Adjust the power supply at the *operating voltage*. Find the background counting rate *i.e.* number of pulses per minute corresponding to the operating voltage. Using a forceps place the first radioactive source S_1 say Sr^{90} on the double source holder just below the Geiger tube as shown in Fig. 39.5 and determine the counting rate (n_1). Also note the operating *voltage*.

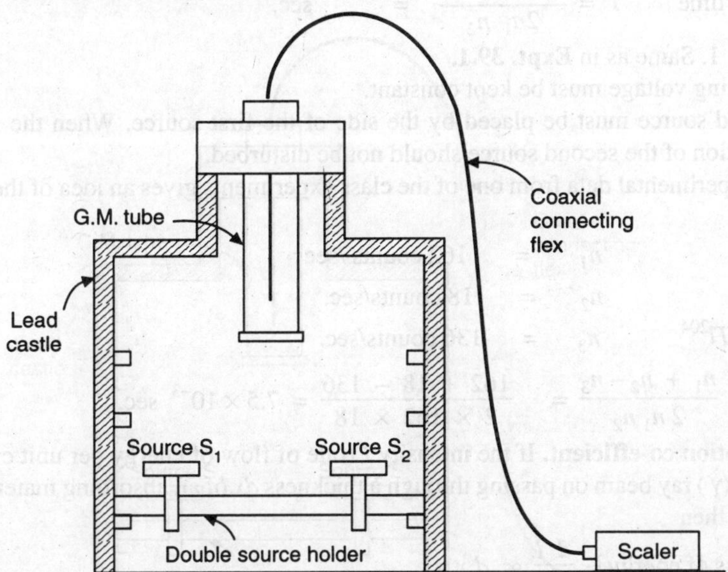

Fig. 39.5

2. Fix the *second* radioactive source S_2 say Tl^{204} in the double source holder by the side of the *first* source and determine the counting rate of the two sources together keeping the operating voltage the same. Let it be n_3.

3. Now remove the *first* radioactive source S_1 without disturbing S_2 and determine the counting rate for the second source S_2 (Tl^{204}) without disturbing its position, and keeping the operating voltage the same.

4. Remove the *second* source S_2 and again find the background counting rate corresponding to the operating voltage.

5. Repeat all the observations three times keeping the operating voltage the same.

Observations.

Operating voltage applied to G.M. counter = volts

Background counting rate in the beginning b_1 = pulses/min

Background counting rate in the end b_2 = puless/min

Average background counting rate = $\dfrac{b_1 + b_2}{2}$ = pulses/min

Sl. No.	*Source*	*Counting rate (Pulses/minute)*				*Observed counts per second*	*Corrected counts per second*
		(i)	*(ii)*	*(iii)*	*mean*		
1	$(S_1)\ Sr^{90}$						$n_1 =$
2	$Sr^{90} + Tl^{204}$ $(S_1 + S_2)$						$n_3 =$
3	$(S_2)\ Tl^{204}$						$n_2 =$

Dead time $T = \dfrac{n_1 + n_2 - n_3}{2 n_1 n_2}$ = sec.

Precautions. 1. Same as in **Expt. 39.1.**

2. The operating voltage must be kept constant.

3. The second source must be placed by the side of the first source. When the first source is removed the position of the second source should not be disturbed.

Note. The experimental data from one of the class experiments gives an idea of the dead time of the counter.

For Sr^{90} n_1 = 162 counts /sec

For Tl^{204} n_2 = 18 counts/sec.

For $Sr^{90} + Tl^{204}$ n_3 = 136 counts/sec.

$$\therefore \quad T = \frac{n_1 + n_2 - n_3}{2\, n_1\, n_2} = \frac{162 + 18 - 136}{2 \times 162 \times 18} = 7.5 \times 10^{-3} \text{ sec.}$$

39.3. Absorption co-efficient. If the intensity I (rate of flow of energy per unit cross-sectional area) of a gamma (γ) ray beam on passing through a thickness dx of an absorbing material is reduced by an amount dI, then

$$\text{Fractional loss of energy} - \frac{dI}{I} \propto dx$$

or
$$\frac{-dI}{I} = \mu dx$$

where μ is the *linear attenuation co- efficient* or *absorption co-efficient* for the material. Integrating, we get

$$\Big[\log_e I\Big]_{I_0}^{I} = \Big[-\mu x\Big]_0^x$$

or
$$\log_e I - \log_e I_0 = -\mu x \qquad \text{...}(i)$$

or
$$\log_e \frac{I}{I_0} = -\mu x$$

or $$I = I_0 e^{-\mu x}$$

where I_0 is the intensity at $x = 0$ and I, the intensity of the beam at $x = x$

Thus the intensity of the beam decreases exponentially with thickness

From relation (*i*), we have

$$\log_e I = \log_e I_0 - \mu x$$

Therefore a graph between $\log_e I$ and x is a straight line and the slope of the line gives the value of μ the absorption co-efficient of the material with a negative sign.

The β-particles are charged particles; they interact with the absorbing material, but their range being pretty large for materials like aluminium, their intensity decreases exponentially with thickness for small thicknesses. Hence the relation

$$I = I_0 e^{-\mu x}$$

or $$\log_e I = \log_e I_0 - \mu x$$

also applies to absorption of β-particles by thin foils of aluminium.

The value of I_0 and that of I for different values of thickens x is determined with the help of a G.M. counter. The value of I is proportional to the number of β-particles in the beam which in turn is proportional to the number of counts per second given by the G.M. counter. If N_0 and N are the number of counts corresponding to I_0 and I, we have

$$N = N_0 e^{-\mu x}$$

or $$\log_e N = \log_e N_0 - \mu x$$

Hence a graph between $\log_e N$ and x is a straight line, the slope of which gives the absorption co-efficient with a negative sign.

$$\therefore \quad \text{Slope of the straight line} = \mu = \frac{\log_e N}{x}$$

or $$\mu = \frac{2.3026 \log_{10} N}{x}$$

Hence if a graph is plotted between $\log_{10} N$ and x, then

$$\mu = 2.3026 \times \text{slope of the graph}$$

$$= 2.3026 \times \frac{\log_{10} N}{x}$$

If x is measured in metres, then μ is in m^{-1}. Similarly if x is measured in cm, μ is in cm^{-1}. However, x is usually measured in cm^2/gm or cm^2/mg. Sometimes x is measured in atoms/cm^2 or electrons/cm^2, then μ is measured in cm^2/atom or cm^2/ electron.

Experiment 39.3. To find the absorption co-efficient of β-particles in aluminium using a G.M. counter.

Apparatus. Same as in **Expt. 39.1** and at least six aluminium foils having different values of mass per unit area.

Procedure 1. Select suitable square/rectangular pieces of aluminium foil. Find the area and mass of each piece and calculate the value of mass per unit area in mg/cm^2. Arrange the foils in increasing order of mass per unit area.

Note. Standard Aluminium foils of suitable size are available. The values of their mass per unit area are marked on each foil in mg/cm^2.

2. Proceed as in **Expt. 39.2** and adjust the power supply at the operating voltage and find the background counting rate corresponding to this voltage. Using forceps place the β-source on the

source holder just below Geiger tube and determine the counting rate without placing any aluminium foil piece *i.e.* with *zero* thickness of the absorbing material. Again note the operating voltage.

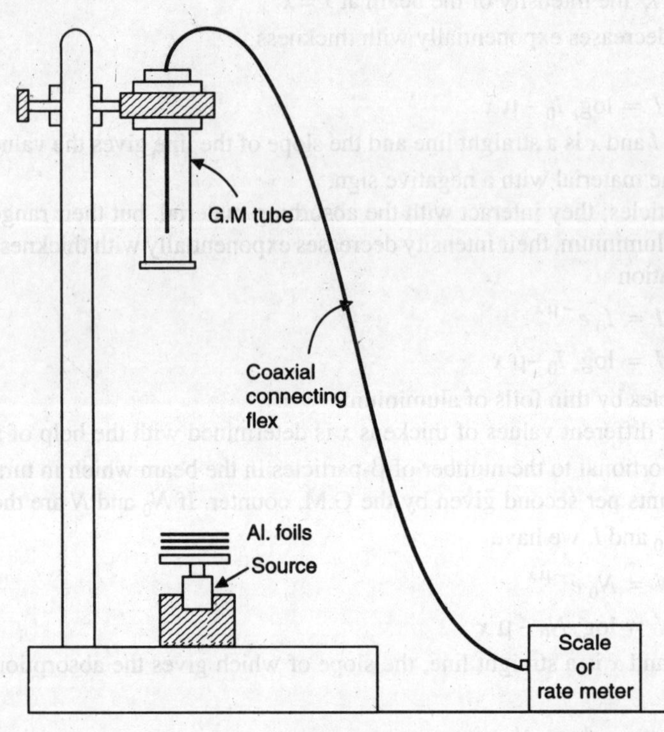

Fig. 39.6

3. Now place the Aluminium foil piece (fitted in a proper holder or frame) having the *smallest* value of mass per unit area, *above* the source and *below* the Geiger tube at a *fixed point* as shown in Fig. 39.6 and find the counting rate keeping the operating voltage the same.

4. Repeat the observation with all the pieces of Aluminium foil placing them always at the same fixed point and keeping the same operating voltage. Also take readings by placing two or three *Al* foils together at a time.

5. Note the operating voltage and find the background counting rate at the end.

Observations.

Applied operating voltage to G.M counter = Volt

Background counting rate in the beginning b_1 = pulses/min

Background counting rate in the end b_2 = pulses/min

Average background counting rate $b = \dfrac{b_1 + b_2}{2}$ = pulses/min

Sl. No.	*Thickness of Al. foil in mg/cm² x*	*Counting rate pulses/min*	*No. of counts per sec*		*log₁₀ N*
			Observed N′	*Corrected N = N′ − b*	
1					
2					
3					
.					
.					
7					

6. Plot a graph taking x the thickness of Aluminium foil in mg/cm^2 along the *X*-axis. and corresponding value of $\log_{10} N$ along *Y*-axis. The graph is as shown in Fig. 39.7.

5. Select two suitable points *A* and *B* on the graph. From the points *A* and *B* draw perpendiculars on the *X*-axis and *Y*-axis. Let the corresponding values be x_1, x_2 and $\log_{10} N_1$ and $\log_{10} N_2$, then

$$\text{Slope of the graph} = \frac{AC}{BC}$$

$$= \frac{\log_{10} N_1 - \log_{10} N_2}{x_1 - x_2}$$

or $\mu = 2.3026\,(-\text{slope}) = \qquad \text{cm}^2/\text{mg}$

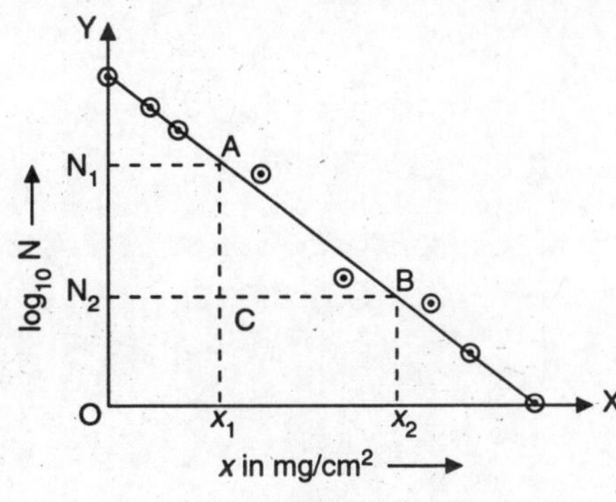

Fig. 39.7

Note. 1. The slope of the graph is a negative quantity but the absorption co-efficient μ is a positive quantity.

2. While using two or three *Al* foils together take the total thickness of the foils for calculation.

Precautions. 1 to 3 same as in **Expt. 39.1.**

4. Readings should start with aluminium foil having the smallest value of mass per unit area.

5. The aluminium foil should be placed above the source and below the Geiger tube always at the same place.

Oral Questions

1. G.M. counter (*i*) What is G.M. counter? (*ii*) What is a plateau curve? (*iii*) Why the G.M. counter is shielded by a massive iron box ? (*iv*) How is the rate of amplified pulses measured, so that no pulse is missed? (*v*) What is a dekatron counter? (*vi*) If you count the pulses for 1 sec and then for 2 sec, will the number be exactly double? (*vii*) What is plateau voltage? (*viii*) What is G.M. threshold potential? (*ix*) Define relative slope of the plateau for a G.M. tube (*x*) Why do pulses appear even when no radio-active source is placed near the Geiger tube? (*xi*) What do you mean by the dead time of a G.M. counter?

2. Absorption coefficient. (*i*) Define absorption coefficient of a material for radio active rays. (*ii*) What do you mean by the intensity of a γ-ray beam? (*iii*) When a γ-ray beam passes through the aluminium foil, in what manner does the intensity change? (*vi*) Do β-rays behave like γ-rays for this purpose? (*v*) What are the units in which the absorption coefficient is measured?

40

Radioactivity

40.1. Radioactivity. Radioactivity is the phenomenon of spontaneous and continuous emission of high energy radiations due to the disintegration of atomic nuclei.

The substances whose nuclei give out radioactive radiations are known as radio-active substances.

The radiations emitted by radioactive substances are of three types. These are known as α – particles, β–particles and γ–rays. An α-particle is a nucleus of helium atom. It has a mass 4 units and a positive charge twice the charge on an electron. A β–particle is a fast moving electron. β–emission includes electron emission, positron emission and electron capture. γ – rays are high energy waves of very short wavelength.

The phenomenon of natural radioactivity is exhibited by heavier atomic nuclei (atomic number above 82). However, lighter atomic nuclei can also be made artificially radioactive by bombarding them with fast moving particles. Most of these artificially produced isotopes are unstable and undergo disintegration of their own accord.

It is not necessary that a radioactive substance may emit all the three types of radiation. It may only emit an α– particle, a β–particle or a γ–ray or an α-particle or β– particle along with energy in the form of γ–ray.

40.2. Laws of radioactive disintegration. The laws connected with radioactive disintegration can be explained on the basis of following hypothesis due to Rutherford and Soddy.

(*i*) *Atoms of every radioactive substance are continuously breaking into fresh radio-active products with the emission of* α*– particles,* β*–particles and* γ*–rays.*

(*ii*) *The rate of breaking is not effected by external factors (temperature, pressure. chemical combination, etc.) but depends entirely on the law of chance, i.e., the number of atoms breaking per second at any instant is proportional to the number present.*

Let N_0 represent the number of radioactive atoms of a certain element at time $t = 0$ when the disintegration just starts. If N is the number of atoms left after time t, the rate of disintegration at any instant depends upon the total number of the atoms present at that instant and is, therefore, directly proportional to N. Let dN be the number of atoms which disintegrate in a small interval dt, then

$$\text{Rate of disintegration} = \frac{dN}{dt}$$

and $$-\frac{dN}{dt} \propto N \text{ or } \frac{dN}{dt} = -\lambda N \qquad ...(i)$$

where λ is a constant called the **disintegration or decay constant** of the given radioactive element. Negative sign indicates that N is decreasing with time.

Disintegration constant λ can, therefore, be defined as the ratio of the amount of the substance which decays in a unit time to the amount of the substance present.

From relation (*i*), we have

$$dN = -\lambda N dt \text{ or } \frac{dN}{N} = -\lambda\, dt$$

Integrating, we have

$$\log_e N = -\lambda t + C$$

where C is a constant of integration. When $t = 0,\ N = N_0$

$$\therefore \qquad \log_e N_0 = C$$

Hence $$\log_e N = -\lambda t + \log_e N_0$$

or $$\log_e \frac{N}{N_0} = -\lambda t$$

and $$\frac{N}{N_0} = e^{-\lambda t} \text{ or } N = N_0\, e^{-\lambda t} \qquad ...(ii)$$

This equation shows that the number of atoms of a given radioactive element decreases exponentially with time and greater the value of decay constant λ, more rapid will be the decrease in radioactivity of the substance. From relation (*ii*), we have after a time

$$t = \frac{1}{\lambda},$$

the number of atoms left behind is given by

$$N = N_0\ e^{-\lambda t} = N_0\, e^{-\lambda \times \frac{1}{\lambda}}$$

or $$= N_0\, e^{-1} = \frac{N_0}{e}$$

or $$\frac{N}{N_0} = \frac{1}{e}$$

Hence the radioactive constant is also defined as the reciprocal of the time during which the number of atoms of a radioactive substance falls to $\frac{1}{e}$ *of its original value.*

40.3. Half-life period. *The half-life period of a radioactive substance is the time in which half of the radioactive substance will disintegrate*. To determine the half-life period we have to find the time T_H in which the number of atoms N left behind becomes equal to $N_0 / 2$ where N_0 is the number of atoms present in the beginning.

Now $$N = N_0\ e^{-\lambda t}$$

$$\therefore \qquad \frac{N_0}{2} = N_0\, e^{-\lambda t}$$

or $$\frac{1}{2} = e^{-\lambda t}$$

or $$2 = e^{+\lambda t}$$

Denoting $t = T_H$ we have

$$\lambda T_H = \log_e 2$$

$$T_H = \frac{\log_e 2}{\lambda} = \frac{0.6931}{\lambda}.$$

40.4. Average life of a radioactive substance. The atoms of a radio-active substance are constantly disintegrating and thus the life of every atom is different. The average life of a radioactive substance

$$T_A = \frac{\text{The sum of the lives of all the atoms}}{\text{Total number of atoms}}$$

It can be shown that $T_A = \dfrac{1}{\lambda}$

Half life $T_H = \dfrac{0.6931}{\lambda} = 0.6931\ T_A.$

40.5. Determination of half life period with a G.M. counter. Let n_1 be the number of counts per minute in a G.M. counter at a time t_1 from start and n_2 the number of counts per minute at time t_2. If N_0 is the number of atoms present at time $t = 0$, N_1 at $t = t_1$, N_2 at $t = t_2$, then the rate of disintegration at time t_1;

$$\frac{dN_1}{dt} = -\lambda N_1 = \lambda N_0 e^{-\lambda t_1}$$

Similarly the rate of disintegration at time t_2;

$$\frac{dN_2}{dt} = -\lambda N_2 = \lambda N_0 e^{\lambda t_2}$$

The number of counts per minute in the G.M. counter is equal to the number of atoms disintegrating per minute or the rate of disintegration.

$$\therefore \quad n_1 = \lambda N_0 e^{-\lambda t_1}$$

and

$$n_2 = \lambda N_0 e^{-\lambda t_2}$$

$$\therefore \quad \frac{n_1}{n_2} = e^{\lambda (t_2 - t_1)}$$

or

$$\log_e \left(\frac{n_1}{n_2}\right) = \lambda \left(t_2 - t_1\right)$$

$$\therefore \quad \lambda = \frac{\log_e \left(\dfrac{n_1}{n_2}\right)}{t_2 - t_1} = \frac{2.303 \log_{10} \left(\dfrac{n_1}{n_2}\right)}{(t_2 - t_1)}$$

or Half life

$$T_H = \frac{0.6931}{\lambda} = \frac{0.6931}{2.303} \frac{(t_2 - t_1)}{\log_{10} \left(\dfrac{n_1}{n_2}\right)}$$

or

$$T_H = \frac{0.301\,(t_2 - t_1)}{\log_{10} \left(\dfrac{n_1}{n_2}\right)}$$

where n_1 = Count rate of G.M. counter per minute at time t_1 from start and n_2 = Count rate of G.M. counter per minute at time t_2 from start.

Experiment 40.1. To find the half life period of a given radioactive substance using a G.M. counter.

Apparatus. Same as in **Expt 39.1** and given suitable radioactive source.

Procedure 1. Proceed as in **Expt. 39.1** and find the value of the operating voltage for the G.M. counter. Adjust the power supply at the operating voltage and find the background counting rate.

2. Place the *radioactive source* on the source holder just below the Gieger tube and reset the scaler unit to zero. Start the scaler unit and simultaneously start the stop watch. (If a time recording device is provided in the scaler unit, then simultaneously switch on the device). Exactly after 1 minute stop the scaler, unit but allow the stop watch or the time recording device to continue working).

3. When the stop watch (or the time recording device) records exact 5 minutes, restart the scaler unit and record the number of counts in exact *one* minute *i.e.* from the end of 5th minute or beginning of 6th minute to the end of 6th minute. Again stop the scaler unit but allow the stop watch or the time recording device to continue working.

4. Again record the number of counts in one minute from the beginning to the end of 11th minute. Similarly record the number of counts for exact one minute after every 5 minutes for about 2 hours.

5. Remove the radioactive source and again record the background counting rate.

Observations

Record observations for plateau curve to find the operating voltage as in **Expt. 39.1.**

Operating voltage of the G.M. counter = Volts

Background counting rate

in the beginning $b_1 =$ pulses/min.

in the end $b_2 =$ pulses/min.

Average background counting rate $b = \dfrac{b_1 + b_2}{2} =$ pulses/min.

Sl. No.	*Time of start of scalar unit from the beginning in minutes t*	*Counting rate per minute due to radioactive sample*	
		Observed n′	*Corrected n = n′ − b*
1	0		
2	5		
3	10		
4	15		
5	20		

6. Plot a graph between time *t* taken along the *X*-axis and the corrected counting rate per minute *n* due to the radioactive sample taken along the *Y*-axis. Take two points *A* and *B* in the central portion of the graph and find the corresponding values of counting rate per minute n_1 and n_2 and the time t_1 and t_2.

From the graph.

Counting rate at time $t_1 =$ min.

$n_1 =$ pulses/min.

Counting rate at time $t_2 =$ min

$n_2 =$ pulses/min.

$\therefore$ Half life period $T_H = \dfrac{0.301\,(t_2 - t_1)}{\log_{10}\left(\dfrac{n_1}{n_2}\right)} =$ min.

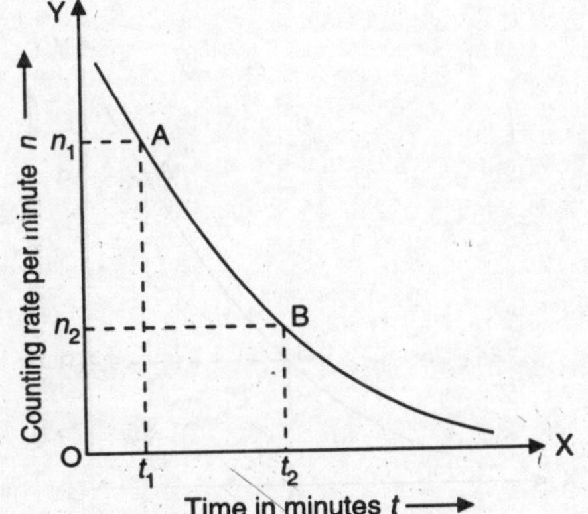

Fig. 40.1

Precautions. Observe precautions 1 to 3 of **Expt. 39.1.**

4. The radio-active source selected should have a half life reasonably longer than the time taken for noting the radio-active decay.

5. Start the scaler unit and the time recording device simultaneously.

6. Record the number of counts in one minute after every 5 minutes.
7. The radio-active source should not be disturbed during the experiment.

Oral Questions

Radioactivity (*i*) What is radio-activity ? (*ii*) What are the various radioactive radiations? (*iii*) State the relation between rate of decay and radioactive constant of a radioactive substance. (*iv*) What is the relation between radioactive constant and half life of a radioactive substance. (*v*) Name one substance each which is α-active, β-active and γ-active. (*vi*) What do you mean by average life of a radioactive substance? (*vii*) What is the relation between the average life and radioactive constant as also between average life and half life?

41

Laser

41.1. Laser. A Laser is a device which produces a highly concentrated, monochromatic, coherent and unidirectional beam of light. The term LASER stands for Light Amplification by Stimulated Emission of Radiation. The typical operating frequency of laser is 10^{15} Hz in the visible region. Laser, therefore, produces a line spectrum of extreme purity and coherence.

41.2. Emission of light. Whenever an atom is excited it absorbs energy and moves from the ground state to one of the excited states. Normally an atom in the excited state comes back to the original or ground state within a very short time of about 10^{-8} sec by emission of a photon.

41.3. Spontaneous emission. When an atom in an excited state E_2 goes to the ground state E_1 by spontaneously emitting a photon, the process in known as *spontaneous emission* and the frequency of photon is given by

$$\nu = \frac{E_2 - E_1}{h}$$

In such a case the emitted photon can move in any random direction. Photons emitted from various atoms in the assembly have *no phase relationship* between themselves and thus the radiation given out is *incoherent*. Ordinary light is of this type.

41.4. Stimulated or induced emission. When an atom in an excited state E_2 interacts with an incident photon of *right* frequency ν, and is thereby induced to move to the ground state E_1 by emitting the differentce of energy $E_2 - E_1$ as a photon of the *same* frequency ν, the process is known as *stimulated or induced emission*. In such a case

(*i*) The emitted photon travels in the same direction in which the incident photon is moving.

(*ii*) The two photons are *in phase* with each other.

Since the emitted photon has the *same frequency* and is *in phase* with the incident photon, the two further interact with other atoms and thus the process builds up and a sort of *chain emission* starts in which all the photons are of the *same frequency, in phase* and *travel in the same direction*. The number of identical photons goes on increasing with each emission and a single photon thus produces a very large number of *identical* photons so that the radiation beam is now highly amplified. *The beam of electro–magnetic radiation coming from such a source is highly intense, monochromatic, coherent and unidirectional, i.e. it is a laser beam.*

To achieve a higher probability of stimulated emission, we make use of the phenomenon of **population inversion.**

The establishment of a situation in which the number of atoms in the higher energy level is greater than that in the lower energy level is called population inversion.

The procedure adopted to achieve population inversion is called *pumping*.

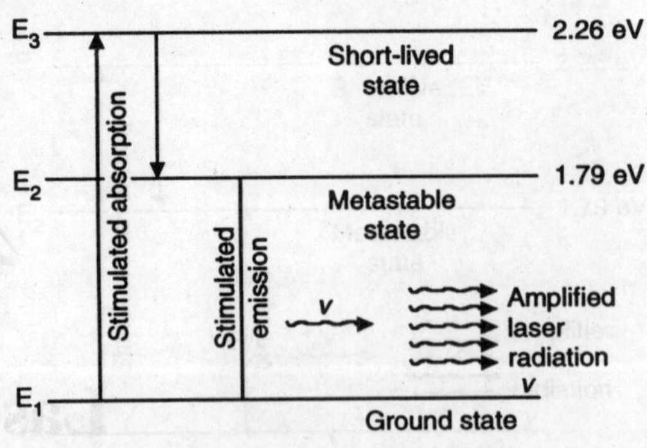

Fig. 41.1

In optical pumping the active material is illuminated with light of suitable frequency $v = \frac{E_2 - E_1}{h}$. As a result, an atom in the lower energy state E_1 absorbs the incident photon of energy hv and is raised to the higher energy level E_2. As the excited atoms lose their energy by spontaneous emission and drop to the lower energy level in a very short time, the process fails to produce necessary population inversion.

In practice, therefore, population inversion is brought about by a *three level scheme*. Consider an atom with energy levels E_1, E_2 and E_3 with increasing values of energy. E_1 is the ground state, E_3 is a short lived state and E_2 is an intermediate meta-stable state. *i.e. an energy state having mean life of more than* 10^{-3} *sec*. Further the transition from E_3 to E_1 is forbidden but transition from E_3 to E_2 is allowed. When these atoms are irradiated with an exciting radiation of the *right* frequency $v_0 = \frac{E_3 - E_1}{h}$, the atoms are excited to the E_3 state by the process of *stimulated absorption*. Some excited atoms quickly drop to the intermediate energy level E_2 by spontaneous emission or by a *non-radiative* process thereby converting their excess energy into vibrational kinetic energy of the atoms forming the substance. As E_2 is a meta-stable state the atoms remain in this excited state for a comparatively longer time of 10^{-3} sec as compared to 10^{-8} sec for the short lived state E_3 and for this time the population of the state E_2 is more than that of the state E_1 thus resulting in population inversion of the collection of atoms.

If an atom in the state E_2 decays by spontaneous emission or stimulated emission it emits a radiation of frequency $v = \frac{E_2 - E_1}{h}$. The photon may produce stimulated emission from another atom thereby giving two coherent photons moving in the same direction. These two photons produce two more photons and so on producing an amplified beam. The system explains the *principle of laser*.

41.5. Laser. Various types of lasers in use are:

(*i*) *Solid lasers*. The most commonly used solid laser is a *Ruby laser*. The other solid laser used is $Y_3\ A_5\ O_{12}$ doped with Nd^{+3} ions in place of Y (Ytterium). Its most useful operating wavelength is the infra red light $\lambda = 1.064\ \mu$m.

(*ii*) *Gas lasers*. Helium-Neon gas lasers (consisting of about 10 part of *Ne* and 1 part of *He*) is the best known gas laser. It operates on three wavelengths 6328Å in the red and 1.15 and 3.39 μ m in the infra-red region.

(*iii*) Other important types include semi-conductor lasers, chemical lasers and dye lasers.

Ruby Laser. It is one of the first lasers to be constructed. It makes use of the *three level scheme* of population inversion with a ruby as a solid state laser element. For this purpose the ruby crystals are grown in special furnaces and then shaped into cylindrical rod 2 to 30 cm in length and 0.5 to 2 cm in diameter. Both the end faces of the crystal are made *optically-flat* and *silvered*. One face *A* is *thickly coated* to become *fully reflecting* while the other face *B* is *partially coated* so that an intense beam can emerge out of it. The curved surface is also coated to make it reflecting. The ruby cylinder prepared in this manner is called *resonant cell* or *cavity*. The rod is surrounded by a cylindrical glass tube through which liquid Nitrogen is circulated to keep the rod cool. A Xenon flash tube is wound round the glass cylinder so that the ruby rod lies along its axis. It is connected to a suitable source of power supply and made to flash for a few milli-seconds.

Ruby consists of a crystal of Al_2O_3 (Aluminium oxide) doped with 0.03 to 0.05% of Cr_2O_3 (Chromium oxide) so that Cr^{+3} ions replace some of the Al atoms. The crystal field splits up the energy levels of Cr^{+3} ions in such a way that it has a short lived energy state $E_3 = 2.26\ eV$ and an intermediate metastable energy state $E_2 = 1.79\ eV$ above the ground state E_1.

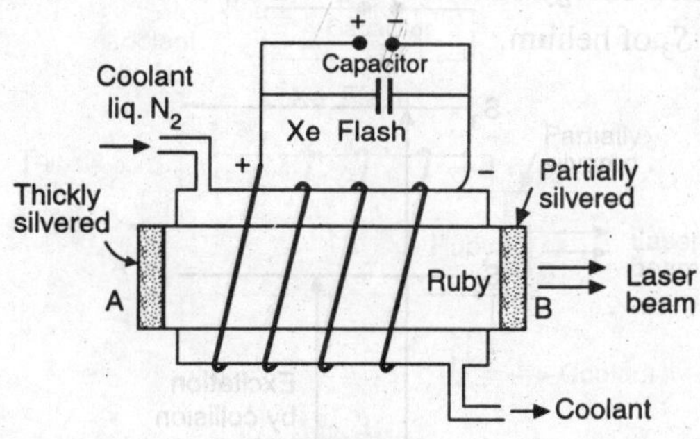

Fig. 41.2

[The energy state E_3 for Cr^{+3} consists of two wide energy bands and the state E_2 consists of a closely spaced double level, but most of the contribution to laser action comes from the lower energy level 1.79 eV.]

The Xe flash tube acts as an optical pump and only a small fraction of the energy emitted by it in the form of blue green radiation of $\lambda = 5500$ Å corresponding to energy level $E_3 - E_1$ is absorbed by the ruby and used to excite the Cr^{+3} ions. The rest of the energy is used up in heating the apparatus, which therefore has to be kept cold by circulating liquid nitrogen through the glass tube surrounding the ruby rod. The Cr^{+3} ions excited to the short lived 2.26 eV level jump to the 1.79 eV meta-stable level (life time 4×10^{-3} sec) in the first step by losing 0.47 eV of energy. As a result the level E_2 becomes more populated than the level E_1.

Such a system with population inversion is very unstable. Once the stimulated emission is started, photons of energy 1.79 eV corresponding to red light of $\lambda = 6943$ Å are continuously produced. These photons undergo multiple reflections from the silvered ends of the crystal till the beam becomes sufficiently intense to emerge out of the partially silvered end. Thus a highly intense, monochromatic, coherent and unidirectional beam is obtained.

The minimum energy required to be pumped for laser action to start is of the order of 50–100 joules of absorbed energy and its typical output is 10–20 kilowatts in milli-second pulses.

Helium Neon Laser. It consist of a long and narrow discharge tube about 80 cm long and 1 cm in diameter filled with a mixture of helium at a pressure of 1 mm and neon at a pressure of 0.1 mm of mercury column, which forms the *laser medium.* Two electrodes E_1 and E_2 are fitted to the discharge tube as shown. M_1 and M_2 are the two mirrors which form a resonant cavity. The mirror M_1 is fully reflecting whereas the mirror M_2 is partially reflecting so as to allow the laser beam to pass out of it.

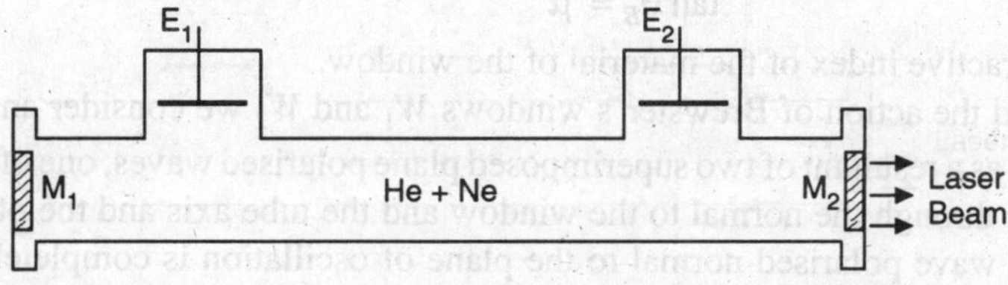

Fig. 41.3

Working. As soon as the electric discharge passes through the mixture of He and Ne gases, electrons in the tube are accelerated. These collide with helium atoms and excite them to higher energy levels S_2 and S_3 as shown in Fig. 41.4. These levels of helium are meta-stable and the excited helium atoms remain in these levels for a 'long' time before being de-excited. On the other hand some of the excited energy states say E_4 and E_6 of neon correspond very approximately to the energy states S_2 and S_3 of helium.

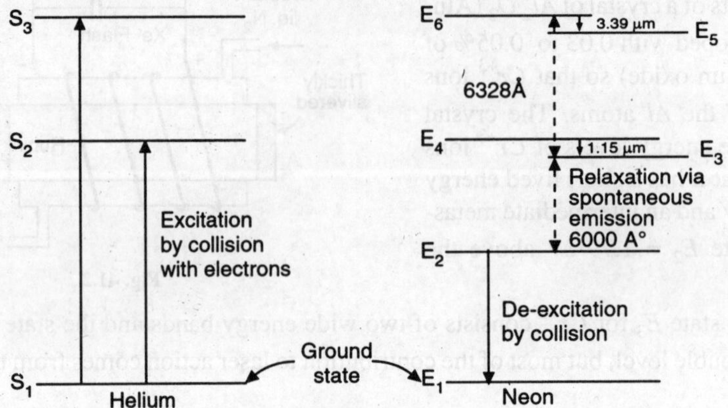

Fig. 41.4

Because of this when helium atoms in the energy states S_2 and S_3 collide with neon atoms in the ground state E_1, *Ne* atoms absorb energy and are excited to the energy states E_4 and E_6, whereas *He* atoms lose energy and are de-excited to the ground state S_1. This process continuously transfers more and more neon atoms from the ground state to the excited energy states E_4 and E_6.

There are two more energy states E_3 and E_5 slightly below the states E_4 and E_6 and another energy state E_2 in *Ne*. Since the energy state E_4 and E_6 are *highly populated*, there is a *population inversion* between the states E_4 and E_6 and the lower energy states E_3 and E_5. As shown transition between E_6 and E_5, E_4 and E_3 and E_6 and E_3 respectively lead to emission of wavelengths 3.39 μ m, 1.15 μ m and 6328 Å. The first two lie in the infra red region and the last corresponds to the red light from the *He – Ne* laser in the visible region which is generally used as a laser beam for experimental and other purposes. Selection of a suitable frequency is made by using mirrors which reflect only a small range of frequency band.

In an improved form of *He–Ne* laser the resonator mirrors M_1 and M_2 are used *externally* to the laser cavity because the mirrors sealed inside the discharge tube are eroded by the gas discharge and have to be replaced. These are spherical in shape and the radius of curvature of each mirror is equal to the cavity length.

Further to minimise the loss due to reflection at the ends of the discharge tube Brewster windows W_1 and W_2 are used as shown. These are fitted at an angle known as Brewster angle θ_B given by

$$\tan \theta_B = \mu$$

where μ is the refractive index of the material of the window.

To understand the action of Brewster's windows W_1 and W_2 we consider an unpolarised wave incident on a plate as a resultant of two superimposed plane polarised waves, one of which is polarised in a plane passing through the normal to the window and the tube axis and the other at right angles to this plane. The wave polarised normal to the plane of oscillation is completely reflected by the window plate and therefore does not play any part. The other wave polarised in the plane of oscillation is transmitted in the same direction and is completely reflected by the mirrors M_1 and M_2 again and again thus producing a strong *plane polarised* laser beam.

The power output of *He-Ne* Laser lies between 1 mW to 50 mW of continuous wave for inputs of about 5 W to 10 W.

Experiment 41.1. To study double slit interference by Helium Neon laser.

Apparatus. *He–Ne* laser source, a double slit composed of two slits each of width 0.0075 mm separated by a distance 0.0541 mm fitted in a frame with screws for vertical and horizontal motion, a higher power microscope, a metre scale etc.

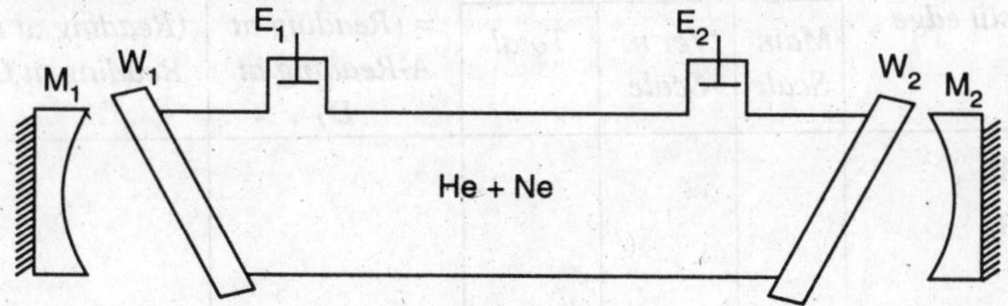

Fig. 41.5

Procedure. 1. Place the double slit frame in front of the *He–Ne* laser source such that the pair of slits lies across a diameter of the laser source close to the output window. Adjust the position of the *He–Ne* laser source so that the axis of the *He–Ne* tube is horizontal.

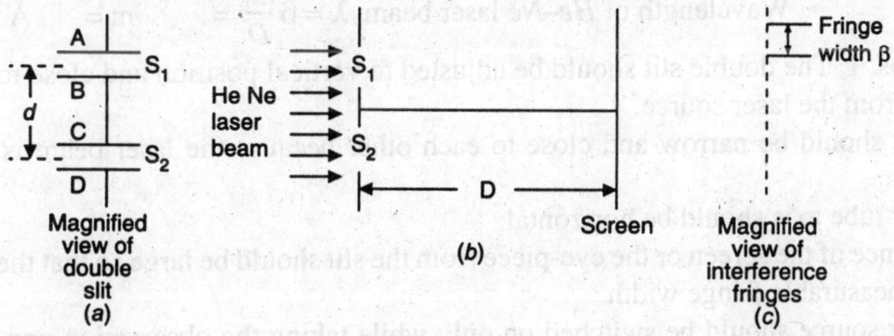

Fig. 41.6

2. Rotate the double slit frame about a horizontal axis and in the vertical plane by means of a tangential screw so that the slits are vertical. Place a screen fitted with a mm scale at a small distance from the double slit so that it is parallel to the double slit frame.

3. Switch on the power supply to the *He–Ne* laser source and let the laser beam fall on the double slit. Rotate the double slit frame about a vertical axis so that the laser beam falls normally on it. At this stage bright and dark interference fringes of equal width (of small height) will be seen on the screen. Rotate the plane of the screen about a vertical axis, if necessary so that the fringes are sharp and show maximum contrast.

4. Note the position of the centres of successive bright fringes on the mm scale fitted on the screen or with the help of a micrometer eye piece. Measure the width of each slit and the distance between the centres of the two slits with the help of a microscope.

5. Note the distance between the centres of the double slit frame and the screen.

Observations. *For fringe width* β

Sl. No.	*Order of fringe*	*Reading on mm scale*	*Width of 3 fringes (mm)*	*Fringe width* β *(mm)*
0				
1			(3) – (0) =	
2			(4) – (1) =	
3			(5) – (2) =	
4				
5				

Mean fringe width $\beta =$ mm = m

Measurement of distance between the centres of two slits (d).

Least count of microscope = mm

Sl. No.	*Cross wire on double slit edge*	*Microscope reading*			*Distance AD = (Reading at A-Reading at D)*	*Distance BC = (Reading at B-Reading at C)*	$d = \frac{AD + BC}{2}$
		Main Scale	*Ver n. Scale*	*Total*			
	A						
	B						
	C						
	D						

$\therefore \quad d = \frac{AD + BC}{2} =$ mm = m

Distance between double slit frame and screen D = cm = m

$\therefore$ Wavelength of *He–Ne* laser beam $\lambda = \beta \frac{d}{D} =$ m = Å

Precautions. 1. The double slit should be adjusted to vertical position and close to the outlet of the laser beam from the laser source.

2. The slits should be narrow and close to each other because the laser beam is always very narrow.

3. The laser tube axis should be horizontal.

4. The distance of the screen or the eye-piece from the slit should be large so that the interference fringes have a measurable fringe width.

5. The laser source should be switched on only while taking the observation and immediately switched off thereafter.

Oral Questions

Helium–Neon Laser. *(i)* What does the word LASER represent? *(ii)* Distinguish between spontaneous emission and stimulated or induced emission of radiation. *(iii)* What do you mean by population inversion? *(iv)* What is a meta-stable state? *(v)* What are various types of Lasers in common use? *(vi)* What is a resonant cavity? *(vii)* What is the function of Brewster's windows in *He–Ne* laser source? *(viii)* What is the distance between two mirrors in this Laser source ? *(ix)* What is optical pumping? *(x)* What is the wavelength of light emitted by a *He–Ne* laser source? *(xi)* What is the advantage of a *He–Ne* laser source over a ruby laser? *(xii)* What are various practical uses of laser?

42

Cathode Ray Oscilloscope

42.1. Cathode ray oscilloscope. The cathode ray oscilloscope is an instrument which can record instantaneous values of rapidly varying voltages. It is used to observe various alternating current wave forms. There is no mechanical moving part in the oscilloscope but the '*moving*' is done by a beam of electrons. There is no inertia and the electron beam serves as an ideal indicator for any rapidly changing voltage. The electron beam is focused on a fluorescent screen where it produces a fine, bright spot of light. By applying an electrostatic (or magnetic) field in the vicinity of the beam it can be deflected and the spot of light made to move over the face of the screen tracing out a visible pattern which is determined by the potential producing the electrostatic field (or the current producing the magnetic field).

A cathode ray oscilloscope (C.R.O.) essentially consists of a cathode ray tube (C.R.T), a sweep circuit or time base circuit, a synchronisation circuit, high and low voltage supplies, and horizontal and vertical amplifiers.

1. **Cathode ray tube.** The cathode ray tube is an evacuated tube of the shape shown in Fig. 42.1 (*a*). Its main parts are an *electron emitter* called the *cathode* (*K*), a device to control the intensity called the *grid* (*G*); accelerating and focusing electrodes (A_1, A_2, A_3) known as *anodes; deflecting plates* vertical (Y_1, Y_2) and horizontal (X_1 X_2) and a *screen* which fluoresces when electrons impinge on it.

(*i*) *Electron emitter*. The method of obtaining electrons is similar to that used in a thermionic valve. The *cathode K* is indirectly heated by passing a suitable current through the *filament FF*. It is electrically insulated but thermally connected to the filament. The cathode is oxide coated type (with a material like *MgO*) and when heated gives out a copious supply of electrons.

(*ii*) *Intensity control*. The cathode is surrounded by a metal shield in the form of a cylinder with a hole at the end farther from the cathode. The cylinder is given a negative potential with respect to the cathode so that the electrons are repelled from it and any off axis electrons join the axial stream to form a fine beam which passes through the hole. This electrode is called the *grid* (or shield or *modulator*). By varying the *negative* potential of the grid, the emission of electrons from the cathode can be controlled and thus the *brightness* of the spot on the screen can be varied. This control is usually labelled '*brightness*.' '*brilliance*' or '*intensity*' on the C.R.O. panel board.

(*iii*) *Focus*. The beam of electrons from the grid passes through a series of electrodes known as the first, second or third anode (A_1, A_2 and A_3) all of which are at positive potentials with respect to the cathode. These electrodes act like an electron lens system and bring about a sharp focusing of the electron beam on the fluorescent screen. Generally, the anodes A_1 and A_3 are connected together and have the *highest positive potential* with respect to the cathode while the anode A_2 has a potential lower than that of A_1 and A_3 but higher than that of the cathode. The potential of the anode A_2 can be varied and thus the spot of light brought to a fine focus on the screen. The potentiometer knob controlling the potential of A_2 is usually labelled *'focus'*.

The whole system consisting of the cathode, the grid and the anode shoots out a fine pencil of fast moving electrons and is therefore, called an **electron gun.**

(ii) Deflection of electron beam. The common form of a cathode ray tube uses electrostatic deflection arrangement. For this purpose two sets of plates X_1 X_2 and Y_1 Y_2 are mounted between the final anode A_3 and the screen. As shown in Fig. 42.1. *(a)* the plates Y_1 Y_2 are horizontal and represent a parallel plate capacitor in which an electric field in the *vertical* (*Y*) direction is set up when a potential difference is applied to the plates thereby causing the electron beam to move *up* or *down* according to the polarity of the plates. The plates X_1 and X_2 are vertical and an electric field in the' *horizontal* (*X*) direction is set up when the potential difference is applied to these plates thereby causing the electron beam to move from left to right or right to left according to the polarity of the plates. If electric fields are simultaneously set up between the vertical and the horizontal deflecting plates, the resultant displacement of the electron will be in accordance with the vector sum of the vertical and horizontal displacements.

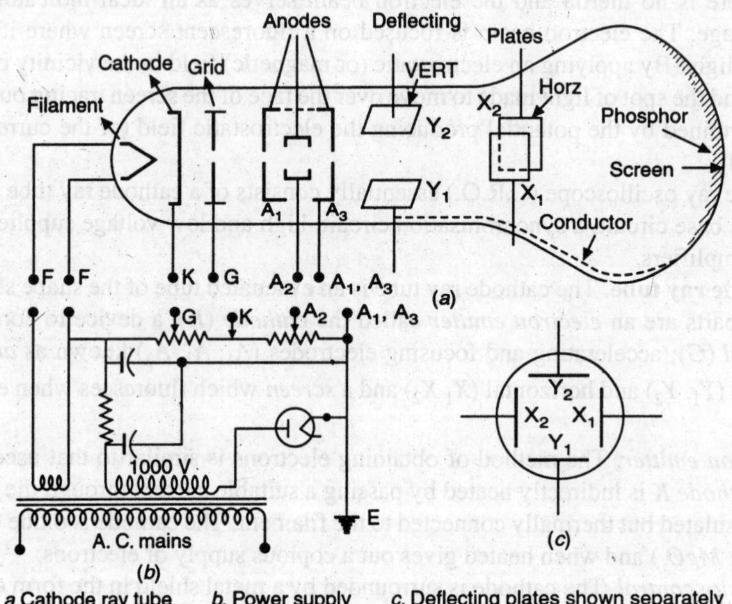

a Cathode ray tube *b.* Power supply *c.* Deflecting plates shown separately

Fig. 42.1

The vertical and horizontal plates though shown separately are at the same place.

The position of the spot on the screen can, therefore, be adjusted to a pre-determined position by applying suitable *D.C.* potentials to each of the two sets of deflecting plates. The variable *D.C.* potential for this purpose is regulated by control knobs labelled *'Horizontal positioning* (*Hor. Pos*) and '*Vertical positioning* (*Vert. Pos.*)

One plate of each pair (X_1 and Y_1) is connected to the anode A_3 so that there is no electric field between the plates and the electron gun. Finally the anode A_3 is earth connected so that the deflector plates are not in the vicinity of high potentials which would cause the spot to behave in an erratic manner and may also be dangerous. Thus the cathode and the other electrodes are at a negative potential with respect to the earth.

Caution. *It is dangerous to handle the cathode, the filament (heater) and the focusing anode A_2 when the tube is working.*

(v) *Screen.* The screen is coated with a fluorescent material such as zinc orthosilicate and a brilliant spot is visible on the screen due to fluorescence where the electron beam strikes. The inside of the neck of the tube is coated with *aquadag* or some other conducting material thereby connecting the screen to the final plate A_3 which keeps the electron beam from A_3 to the screen in free flight as the space through which the electron beam passes is free from all electrostatic (or magnetic) fields except those set up between the deflecting plates. As the electrons strike the screen these *'leak off'* to the anode A_3 and finally through the power supply to the cathode.

2. **Time base.** One of the great advantages of a cathode ray oscillograph is its ability to indicate very short time intervals and to record variations of voltage with time. If an alternating potential difference is applied to the plates Y_1 Y_2, the spot on the screen will trace out a vertical line proportional to the peak value of the applied voltage. If the wave form of the alternating voltage is to be obtained the spot must be made to move horizontally along the *X*-direction as well as move up and down along the *Y*-axis so that the pattern may spread out on the screen. This is done by connecting the horizontal deflecting plates X_1 X_2 to a source of voltage that rises gradually at a constant rate to a maximum value and then suddenly drops back to zero. Such a voltage is said to have a *saw tooth* shape. It causes the beam to move horizontally across the screen at a uniform speed and then snaps back to its starting point. The electronic circuit producing the saw tooth voltage is called the '*sweep circuit*' or '*time base circuit*'. A time base is said to be *linear* if it produces a uniform movement of the spot. The sweep circuit or time base enables us to obtain a pattern on the screen which is exactly the same as a curve of varying voltage applied across the vertical deflecting plates Y_1 Y_2 as a function of time. The only condition that has to be fulfilled is that the period of the saw tooth voltage must be equal to or an exact multiple of the period of the applied voltage to be studied. For example, suppose the frequency of the alternating voltage applied to the *Y*-plates is 50 c.p.s., then if the time period of the saw tooth voltage is also 1/50 sec (frequency 50 c.p.s.), then *one* complete cycle of the applied voltage will appear on the screen. If the time period of the saw tooth voltage is double *i.e.* 1/25 sec (frequency 25 c.p.s.), then *two* complete cycles of the applied voltage will appear and so on.

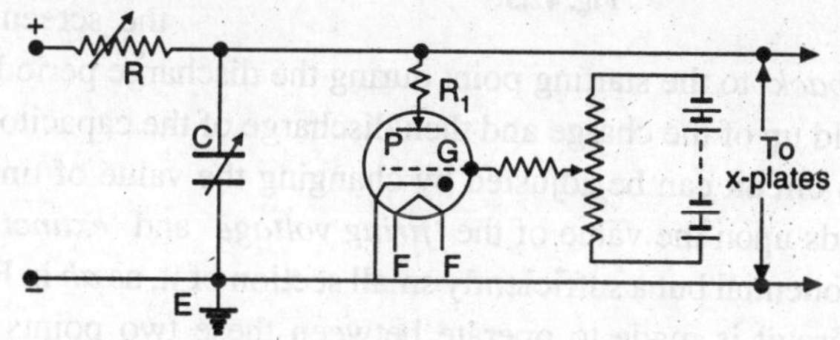

Fig. 42.2

A simple form of '*time base*' makes use of a *gas filled triode* known as *thyratron*. The circuit is shown in Fig. 42.2. The capacitor *C* is charged through a high resistance *R* by connecting it to a *D.C.* source of constant *e.m.f.* V_0. With the thyratron connected in parallel with the capacitor, the voltage V_c builds up from zero to V_c according to the equation.

$$V_c = V_0 \left(1 - e^{-t/RC}\right)$$

The curve for the voltage V_c plotted against time is shown in Fig. 42.3. The quantity *RC* is called the *time constant* and is defined as the time during which the voltage rises to

$$\left(1 - \frac{1}{e}\right) = \left(1 - \frac{1}{2.7}\right) = 0.63$$

of the maximum value. When the voltage across C is sufficient to '*fire*' the thyratron the capacitor discharges rapidly through the valve. (The current being limited to a safe value by the resistance R_1) till the capacitor potential falls to '*extinction value*' at which it is unable to maintain the discharge. At this stage the discharge stops and the capacitor commences to recharge. The process repeats itself and the resulting potential variation across C is shown in Fig. 42.4. The pattern is similar to the edge of a saw and hence the name '*saw tooth*'. If the X-plates are connected across the capacitor, then during the charging period the spot moves across the screen at a nearly uniform rate and suddenly '*flies back*' to the starting point during the discharge period. The time required for a complete cycle of build up of the charge and then discharge of the capacitor *i.e.*, the frequency of oscillations of the sweep circuit can be adjusted by changing the value of time constant RC. The length of the sweep depends upon the value of the '*firing voltage*' and '*extinction voltage*'. The capacitor charge curve is exponential but a sufficiently small section of it, as *ab* in Fig. 42.3 can be regarded as almost linear. The circuit is made to operate between these two points by adjusting the *firing* and '*extinction*' voltages corresponding to V_a and V_b respectively. For greater linearity the resistance R is replaced by a pentode valve.

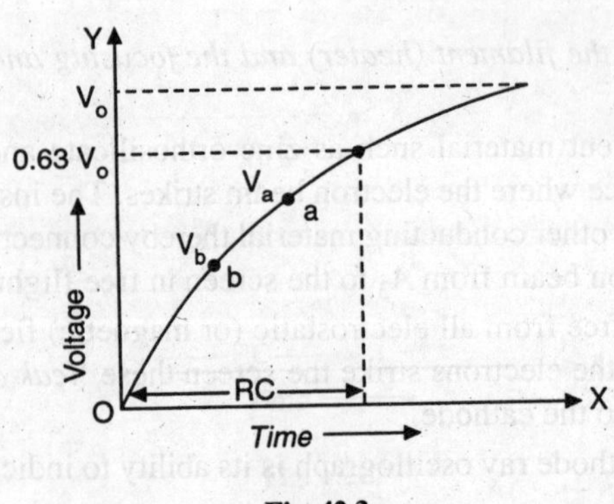

Fig. 42.3

In order to study the wave forms of different frequencies, the sweep frequency can be suitably adjusted with the help of two controls–a '*Range switch*' giving different ranges of frequencies in coarse steps by changing the capacitance C and marked '*frequency selector*' and a '*frequency control switch*' marked '*frequency vernier*' which by gradually varying the resistance R permits fine adjustment of sweep frequency within the limits of any one setting on the range switch.

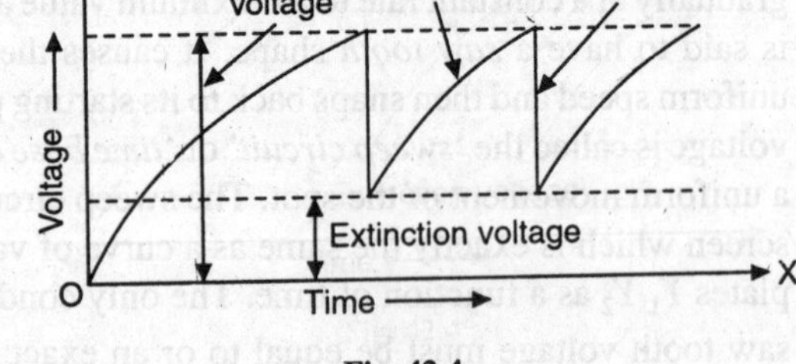

Fig. 42.4

3. **Synchronisation.** When two events occur simulatneously they are said to be synchronised. The input signal applied to the Y-plates, if it is periodic is synchronised with the sweep signal and will appear *stationary* or '*locked in step*' when the beginning of a pulse of the signal wave train is caused to appear at the beginning of the sweep trace. If there is a slight lack of synchronisation the wave pattern will move across the screen either left to right or right to left. To keep the trace steady without frequent recourse to fine frequency control a portion of the incoming signal is fed to the saw tooth generator and serves to '*lock*' the generator in step with the vertical input frequency. The '*synch*' control varies the signal fed to the sweep oscillator and a slight adjustment of fine '*synch*' control knob helps to '*freeze*' (or *lock*') the pattern on the screen. When synchronisation has been achieved the trace is repeated again and again in the same position and thus appears stationary on the screen.

4. **Power supply.** The power supply provides the high voltage for the cathode ray tube (1–15 *KV*), filament voltages for vacuum tubes and low voltages (1–150*V*) for other valves in the circuit. A simple form is shown in Fig. 42.1 (*b*).

5. **Amplifiers.** To amplify weak signals to increase the deflection in the vertical and horizontal directions without distortion of the incoming signals, the X-and Y-plates producing the horizontal and vertical deflections are connected to suitable amplifiers. Regulating knobs provided on the C.R.O.

panel are respectively labelled '*Horizontal gain*' (HOR. GAIN) and '*Vertical gain*' (VERT. GAIN). With the help of these the gain or amplification of the signal may be varied for the horizontal and vertical plates. The horizontal amplifier can also amplify the sweep circuit oscillations and is provided with a switch so that it can be connected to the internal sweep circuit or the external signal.

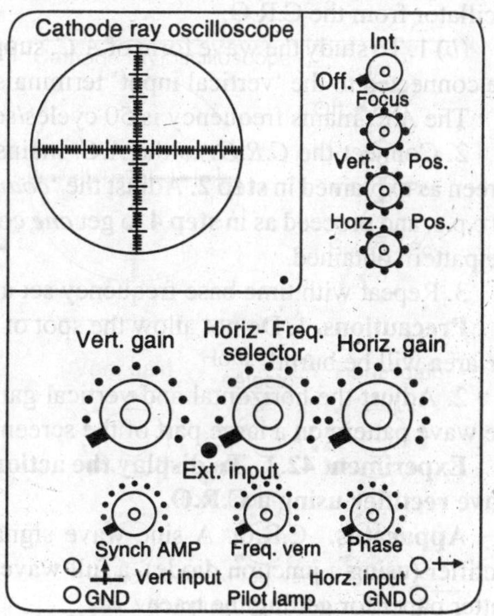

Fig. 42.5

Experiment 42.1. (*a*) To obtain the wave form of a given oscillator using a cathode ray oscilloscope.

(*b*) To obtain the wave form of A.C. mains supply using a cathode ray oscilloscope.

Apparatus. Cathode ray oscilloscope, audio oscillator (signal generator), a step down transformer.

Procedure (*a*) 1. Connect the output terminal marked '*ground*' (GND) or '*earth*' of the oscillator to the terminal marked '*ground* or '*earth*' of the vertical input of the C.R.O.

2. Connect the C.R.O. to the A.C. mains and switch on the '*intensity*' control (or *power switch*). Note the position of the '*spot*' as it appears on the screen and adjust the '*intensity*' and '*focus*' controls to get a sharp and fine spot. Now adjust the bias on the *X* and *Y* deflecting plates by using the '*horizontal*' and *vertical positioning*' (centering) controls so that the spot is in the centre of the screen.

CAUTION. *See that the bright spot does not remain on the screen for any length of time otherwise that area of the tube will be burnt.*

3. Switch on the '*time base*' or '*sweep circuit*' so that the spot sweeps across the screen in rapid successions. See that the time axis is straight horizontal and clearly defined.

4. Connect the oscillator to A.C. mains and switch it on. Set the oscillator frequency at 1000 cycles/sec. Now set the '*coarse*' frequency control of the sweep circuit of the C.R.O. with the help of '*frequency selector*' control to a step with 1000 c.p.s. and adjust the '*vernier frequency*' control carefully to '*stop*' the wave pattern obtained. Adjust the oscillator gain as well as the vertical and horizontal gains of the C.R.O. to get a proper amplitude of the wave form on a large part of the screen. When the hand is taken off the '*vernier frequency*' control the pattern will, probably, begin to *walk* along the sweep. If it does, adjust the '*synch. amp*' control to '*freeze*' the pattern. You will get *one* complete wave form of the oscillator with this adjustment. Trace the wave form.

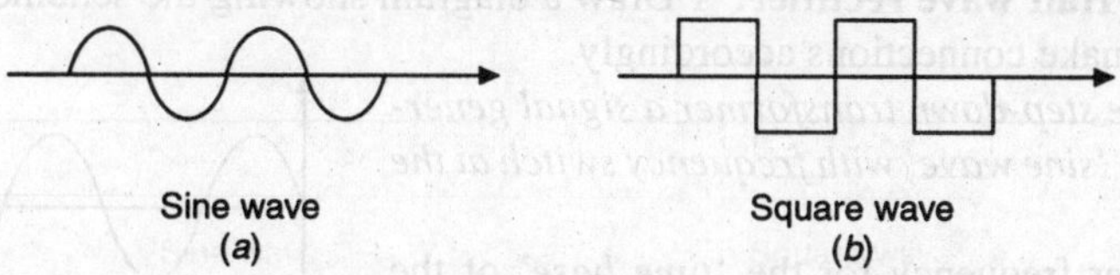

Fig. 42.6

5. Now set the 'COARSE' FREQUENCY CONTROL to a step with 500 c.p.s. and again adjust the '*vernier frequency*' control to '*stop*' the pattern. '*Freeze*' the pattern with the help of '*synch amp*' control, if necessary. You should now get *two* similar patterns. Similarly get 4, 5, and 10 patterns. Note the type of wave pattern *i.e.* sinusoidal, square wave, triangular wave or any other form. Trace the wave form an a tracing paper. A sine wave and a square wave is shown in Fig. 42.6 (*a*) and (*b*).

6. Switch off the oscillator and the C.R.O. and disconnect from A.C. mains. Disconnect the oscillator from the C.R.O.

(*b*) 1. To study the wave form of *A.C.* supply, the secondary terminals of a step-down transformer are connected to the 'vertical input' terminals of the *C.R.O.*

The *A.C.* mains frequency is 50 cycles/sec and is usually held to a very small tolerance.

2. Connect the *C.R.O.* to the *A.C.* mains and obtain a sharp and fine spot on the centre of the screen as explained in **step 2.** Adjust the '*coarse frequency*' control of the sweep circuit to a step with 50 c.p.s. and proceed as in **step 4** to get *one* complete *A.C.* wave form '*locked in*' on the screen. Trace the pattern obtained.

3. Repeat with time base frequency set at 25 c.p.s. and 100 c.p.s.

Precautions. 1. Do not allow the spot of light to remain on the screen for a long time, otherwise the area will be burnt.

2. Adjust the horizontal and vertical gain controls of the *C.R.O.* to obtain proper amplitude of the wave pattern on a large part of the screen.

Experiment 42.2. To display the action of a junction diode as (*a*) half wave rectifier (*b*) full wave rectifier using a C.R.O.

Apparatus. C.R.O. A sine wave signal generator or a step down transformer, a half wave rectifier (using a junction diode), a full wave rectifier (using two junction diodes), a resistance box, butter paper for getting the trace.

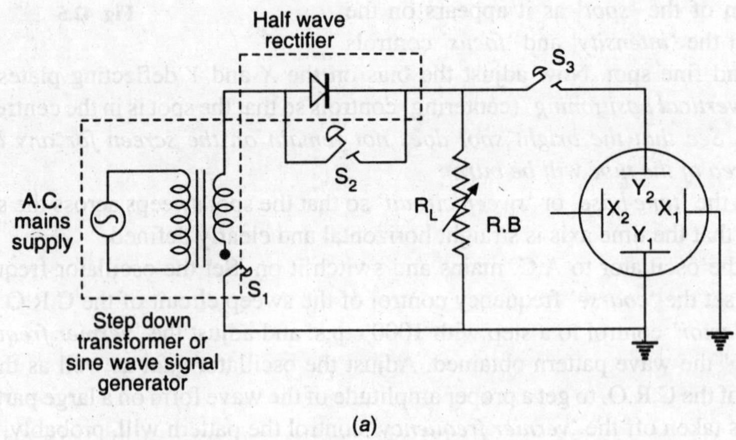

Fig. 42.7(*a*)

Procedure (*a*) Half wave rectifier. 1 Draw a diagram showing the scheme of connections as in Fig. 42.7(*a*) and make connections accordingly.

[*If instead of the step-down transformer a signal generator is used, set it on 'sine wave' with frequency switch at the proper range.*]

2. Select proper frequency for the '*time base*' of the C.R.O. suitable for viewing the wave form of the given frequency.

3. Close all the three switches S_1, S_2 and S_3 so that half wave rectifier is bye-passed. The C.R.O. will show the wave form of the A.C. supply from the voltage developed across R_L. Synchronise the signal being fed to C.R.O and trace it. It is shown in Fig. 42.7 (*b*) (*i*).

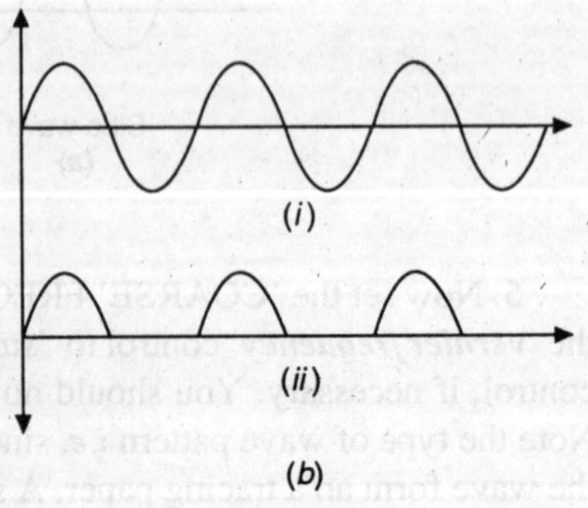

Fig. 42.7(*b*)

4. Now open switch S_2 and observe the new wave form of the trace on C.R.O from the voltage across R_L as shown in Fig. 42.7 (*b*) (*ii*). This is half wave rectification. The diode conducts during positive half cycle and we get the output wave. The diode does not conduct during the negative half cycle.

5. Trace the rectified signal for different values of load resistance R_L from the resistance box R.B.

6. Reverse the diode and again trace the signal. What is the change?

(*b*) **Full wave rectifier.** 7. Draw a diagram showing the scheme of connections as in Fig. 42.8 (*a*) using a centre tapped transformer and make connections accordingly.

8. Select proper frequency for the time base of the C.R.O suitable for viewing the wave form of the given frequency.

9. Trace the rectified signal for different values of R_L from the resistance box R.B. The trace is shown in Fig. 42.8 (*b*). For one half cycle the end *A* of the secondary of the transformer is positive with respect to the centre tap *C* and the diode D_1 conducts. For the second half cycle the end *B* of the secondary of the transformer is positive with respect to the centre tap *C* and the diode D_2 conducts. We, therefore get both the halves of the wave form.

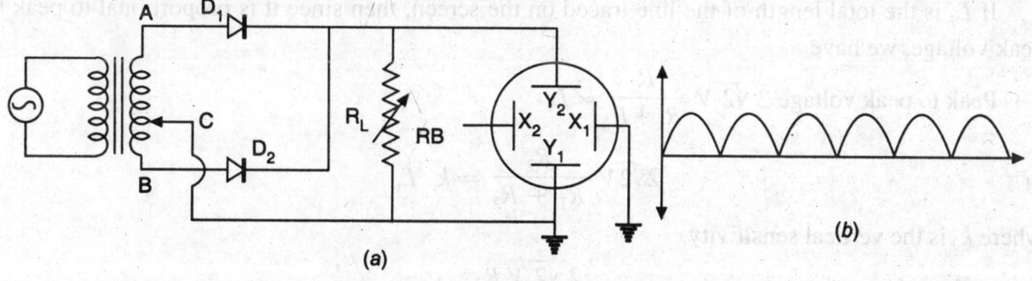

Fig. 42.8

10. Trace the rectified signal for different values of load R_L from the resistance box R.B.

Precautions. Same as in **Expt. 42.1.**

42.2. Calibration of an oscilloscope. If one plate of a pair of deflecting plates of a C.R.O. is maintained at a positive potential with respect to the other, then electron beam will be attracted towards the positive plate and the spot of light will move across the screen.

If *D.C.* potential is applied to the *X* -plates the spot will move from the '*centre*' position to a point either on the left or right of the centre depending upon the polarity of the plates. If the potential is applied to the *Y*-plates, the spot will move up or down for the same reason. The deflection of the spot will be proportional to the difference of potential between the plates.

If alternating voltages are applied, *the spot will be drawn out into a line*. The length of the line is a measure of the *peak to peak* voltage applied. The *D.C.* voltage corresponding to the length of the line *i.e.* the sum of the deflections on either side of the centre will be equal to $2\sqrt{2}$ times the r.m.s. voltage.

The calibration of an oscilloscope is generally done with *A.C.* voltage to avoid the possibility of burning of the screen due to the electron spot having remained in one particular position for a long time. For calibration purposes some oscilloscopes are provided with an in built standard 1 volt *A.C.* signal. The usual circuit for calibration is shown in Fig. 42.9. (*a*). *T* is a step-down transformer, the voltage across its secondary is measured by an *A.C.* voltmeter *V*. This voltage is applied to a potential dividing arrangement consisting of two resistance boxes R_1 and R_2. To find the vertical sensitivity the potential difference across R_2 is applied to the plate Y_2 of the vertical deflecting plates and the plate X_2 of the horizontal deflecting plates is earthed by connecting it to the terminal marked '*ground*'. The deflecting plates X_1 X_2 and Y_1 Y_2 are shown separately in Fig. 42.9 (*b*). To find the horizontal sensitivity the potential difference across R_2 is applied to the plate X_2 and plate Y_2 is earthed by connecting it to the terminal marked '*ground*'. If V is the voltmeter reading, then

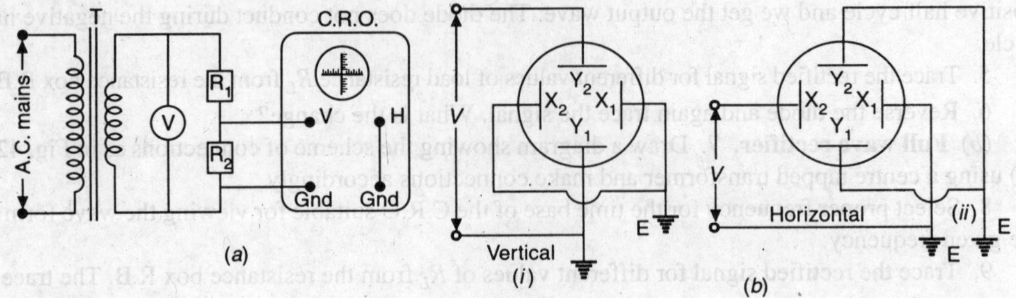

Fig. 42.9

$$P.D. \text{ across } \quad R_2 = V \frac{R_2}{R_1 + R_2}$$

If L_v is the total length of the line traced on the screen, then since it is proportional to peak to peak voltage, we have

$$\text{Peak to peak voltage } 2\sqrt{2}\, V \frac{R_2}{R_1 + R_2} \propto L_v$$

or

$$2\sqrt{2}\, V \frac{R_2}{R_1 + R_2} = k_v\, L_v$$

where k_v is the vertical sensitivity.

$$\therefore \quad k_v = \frac{2\sqrt{2}\, V R_2}{L_v (R_1 + R_2)}$$

Similarly If L_h is the total length of the line traced on the screen in the horizontal direction, then horizontal sensitivity

$$k_h = \frac{2\sqrt{2}\, V R_2}{L_h (R_1 + R_2)}$$

The vertical and horizontal sensitivity of a cathode ray oscilloscope depends upon the gain from the respective amplifiers. The sensitivity is *minimum* for *no gain* position, *maximum* for *full* gain position and has an intermediate value for any other gain position. For example, the sensitivity of a C.R.O. may be 1 volt per cm for no gain position and 0.1 volt per cm for full gain. The vertical and horizontal sensitivities may also be different. In general, the sensitivity of a C.R.O. is required to be determined in a *particular* gain position which may have been used in an *experiment*. In such a case the gain position should not be disturbed after the experiment. In other words, the vertical and horizontal gain control knobs *should be kept fixed in position.*

Experiment 42.3. To find the vertical and horizontal sensitivity of a cathode ray oscilloscope (*i*) with no amplifier gain and (*ii*) with full amplifier gain.

Apparatus. A cathode ray oscilloscope, two resistance boxes, an *A.C.* voltmeter, a step down transformer etc.

Procedure. 1. Draw a diagram showing the scheme of connections as in Fig. 42.9 (*a*) and connect accordingly

2. (*a*) **Vertical sensitivity.** Connect the '*hor. input*' terminal to the '*ground*' terminal. Set the '*hor. freq. select*' control knob to the '*ext. input*' position. Set the '*vert. gain*' control knob to the *no* (zero) gain position.

3. Connect the C.R.O. to the A.C. mains. Put off the time base circuit and obtain a sharp, bright spot on the screen. Switch on the A.C. mains supply to the step down transformer T and adjust the value of R_1 and R_2 so that a vertical trace of a suitable length is obtained on the C.R.O. screen. Note the reading of the voltmeter, values of R_1 and R_2 and trace the line displayed on the screen on a *ransparent paper*. Measure the length L_v of the line

Take six such readings by changing the values of R_1 and R_2 and getting the line trace from a minimum to the maximum length.

4. Now set the '*vert. gain*' control knob to the *maximum* (full) gain position and repeat the experiment.

5. (*b*) **Horizontal sensitivity.** Connect the '*vert. input*' terminal to its '*Ground*' terminal. Set the '*hor. freq. select*' control knob to the *no* (zero) gain position. Take at least six observations as explained in step 3 by tracing the horizontal line displayed on the screen.

6. Now set the '*hor. gain*' control knob to the *maximum* (full) gain position and repeat the experiment.

Record. (*a*) Vertical sensitivity

No gain					*Full gain*				
Voltmeter-Reading (V)	R_1	R_2	L_v	$k_v = \frac{2\sqrt{2}\,VR_2}{(R_1+R_2)\,L_v}$	*Voltmeter-Reading (V)*	R_1	R_2	L_v	$k_v = \frac{2\sqrt{2}\,VR_2}{(R_1+R_2)\,L_v}$

Mean sensitivity = volt/cm Mean sensitivity = volt/cm

(*b*) Horizontal sensitivity

No gain					*Full gain*				
Voltmeter-Reading (V)	R_1	R_2	L_h	$k_h = \frac{2\sqrt{2}\,VR_2}{(R_1+R_2)\,L_h}$	*Voltmeter-Reading (V)*	R_1	R_2	L_h	$k_h = \frac{2\sqrt{2}\,VR_2}{(R_1+R_2)\,L_h}$

Mean sensitivity = volt/cm Mean sensitivity = volt/cm

Precautions. 1. While finding the vertical sensitivity both the horizontal deflecting plates must be earthed and vice versa.

2. The sensitivity depends upon the amplifier gain. The gain control knob should not be disturbed during the experiment.

3. The A.C. voltmeter must be connected in parallel with the secondary of the step down transformer.

Exercise. *Find the vertical and horizontal sensitivity of a C.R.O. at half gain.*

Hint. In each case set the '*gain control*' knob midway between the *zero* and *full* gain markings and proceed as in **Expt. 42.3.**

Experiment 42.4. To measure A.C. voltages using a C.R.O. and to calculate the deflection sensitivity in m.m. per r.m.s. volt.

Apparatus. Same as in **Expt. 42.3.**

Theory. When an A.C. voltage is applied to the *Y-Y* plates of a C.R.O. keeping *X-X* plates earthed, the spot remains moving up and down due to the variation of potential. However, due to *persistence of vision* a *line* is traced on the screen. The length of the line corresponds to peak to peak value of the A.C. voltage applied and thus measures $2\sqrt{2}$ *r.m.s.* voltage. If V *is the r.m.s.* voltage, then

$$V \propto L$$

where L is the length of the line traced.

Procedure 1. Draw a diagram showing the scheme of connections as shown in Fig. 42.10 and make connections accordingly. The variable resistance R is a wire potentiometer type with a variable knob having resistance in the range 2-5 KΩ.

2. Switch on the C.R.O putting off the time base circuit and get a sharp and bright, stationary spot at O the centre of the screen. Using the *Brightens control* make the brightness of the spot minimum.

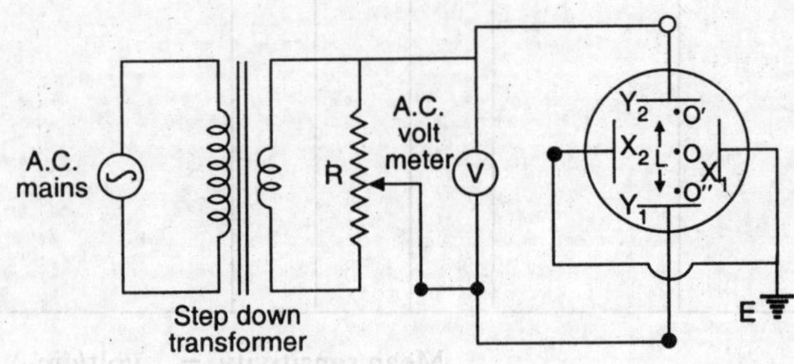

Fig. 42.10

3. Now apply a suitable A.C. voltage to the *Y-Y* plates by adjusting the variable knob of the resistance R. Note the voltmeter reading. Measure the length of the line $O'O''$ traced by the spot of light on the screen by tracing it on a butter paper.

4. Go on changing the voltage and note the voltage as well as the corresponding length of the line traced by the spot of light on the screen. Take 6-7 readings first by increasing the voltage and then decreasing it.

5. Now apply the unknown A.C. voltage E_v to the *Y-Y* plates of the C.R.O and note the length of the line on the screen traced by the spot of light.

Observations.

S. No.	*Applied voltage (V)*	*Length of the line*		
		Voltage increasing *Length* $L_1 = O'O''$	*Voltage decreasing* *Length* $L_2 = O'O''$	Mean $L = \frac{L_1 + L_2}{2}$
1				
2				
3				
4				
5.				
6				
7				

6. Plot a graph taking the voltage V (*r.m.s.* value) along the X-axis and the mean length L of the line traced by the spot of light (in m.m.) along Y-axis. The graph is a straight line. [NOTE: The deflection $O'O''$ measures the peak to peak voltage whereas the voltage measured by the voltmeter is the *r.m.s.* value].

Deflection sensitivity v = slope of the graph = $\frac{AC}{BC}$ = mm per *r.m.s.* volt.

Fig. 42.11

Unknown voltage. Length of the line for the unknown A.C. voltage = l = mm

Unknown voltage $E_v = v \times l$ = volts (*r.m.s.*)

Preacations. Same as in **Expt. 42.3.**

Experiment 42.5. To measure a d.c. voltage with the help of a C.R.O.

Apparatus. A *d.c.* source, A variable resistance of the order of 20-50 kΩ, a *d.c.* voltmeter (of suitable range) a key, connecting wires.

Theory. When the X-X plates of a C.R.O are earthed, its time base-circuit switched off and no voltage is applied to the Y-Y plates we get a sharp, bright spot of light at O the centre of the screen. If a *d.c.* voltage is now applied by connecting Y_2 to the positive and Y_1 to the negative of the *d.c.* source the spot of light is displaced towards the positive plate Y_2 and appears at O' instead of O. The displacement of the spot OO' is directly proportional to the applied voltage as indicated by the voltmeter V connected in parallel with Y-Y plates.

Procedure. 1. Draw a diagram showing the scheme of connections as shown in Fig. 42.12 and

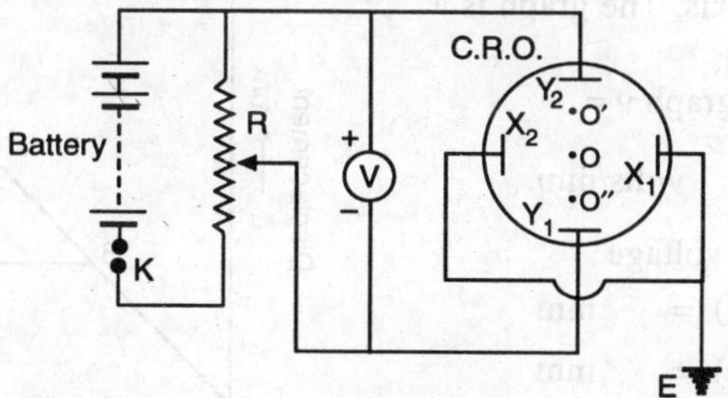

Fig. 42.12

make connections accordingly. The resistance R is of the order of 20-50 kΩ. It is a wire potentiometer type with a variable knob.

2. Switch on the C.R.O. putting off the time base circuit and get a sharp, bright and stationary spot at O the centre of the screen. Rotate the brightness control to the *left* to reduce the brightness of the spot temporarily.

3. Now apply a known suitable *d.c.* voltage to the Y_1 Y_2 plates by inserting the key K in the battery circuit. Rotate the brightness control to the *right* so as to increase the brightness. The spot of light is now displaced to the point O' towards the Y_2 plate. Note the displacement OO' in m.m. either by tracing the position of the spot on a tracing paper fixed to the screen or on the screen itself if it is already graduated.

Note the corresponding voltage from the voltmeter connected in parallel with Y-Y plates.

4. Now change the value of the voltage applied to the Y-Y plates and go on noting the voltage as well as the corresponding displacement of the spot of light on the screen.

NOTE The spot should not be allowed to stay at one point for a long time. Always bring the brightness control to minimum before taking the next reading.

5. Now interchange the polarities of Y_1 and Y_2 plates connecting Y_1 to the *positive* and Y_2 to the *negative* of the battery and repeat the observation. The displacement of O is now in the downward direction from O to O'' towards the plate Y_1. Note the reading of the voltmeter and corresponding displacement of the spot of light.

6. Now apply the unknown *d.c.* voltage to the Y_1 Y_2 plates of the C.R.O. and note the displacement first with Y_2 positive and then with Y_1 positive.

Observations

No.	*Applied voltage in volts*	*Displacement OO′ in mm. Y2 (+ve)*	*Displacement OO″ in mm. Y1 (+ve)*	*Mean displacement d*
1				
2				
3				
4				
5				
6				
7				

Graph. 7. Plot a graph taking voltage v in volts along the X-axis and the mean displacement of the spot of light d along the Y-axis. The graph is a straight line.

Voltage per m.m. from the graph $v =$

$$\frac{AC}{BC} = \quad \text{volts/mm.}$$

Displacement for unknown voltage

Y_2 (+ve) = mm

Y_1 (+ve) = mm

Mean displacement $d =$ mm.

Unknown *d.c.* voltage $V = v \times d =$ volts.

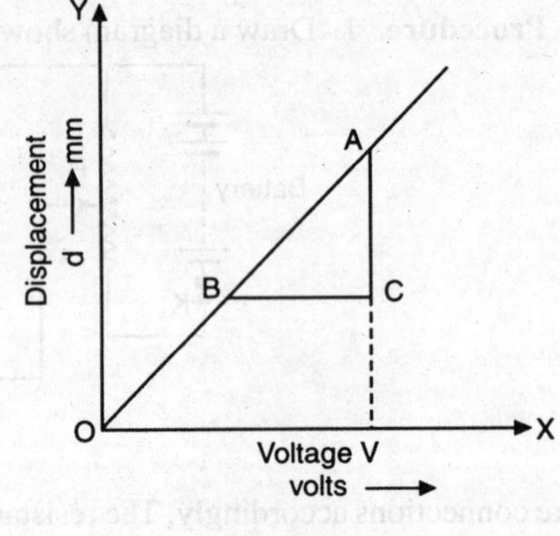

Fig. 42.13

Precautions. 1. The spot of light should not be allowed to stay at one point to avoid damage to the screen.

2. The variable resistance R should have a suitable value.

3. Always decrease the brightness of the spot of light to minimum before proceeding to the next observation.

4. The C.R.O should be handled very carefully.

42.3. Phase difference across resistor, inductor and capacitor. In an A.C. circuit containing only a resistance the current I and *e.m.f.* (voltage) V_R are in phase as shown in Fig. 42.14 (*a*). When the A.C. circuit contains a pure inductance current *lags* behind the *e.m.f.* (voltage) by a phase angle $\pi/2$ or 90° *i.e.* the *e.m.f.* (voltage) V_L *leads* the current I by 90° as shown in Fig. 42.14 (*b*). When the A.C. circuit contains a pure capacitor the current *leads* the *e.m.f.* (voltage) by 90° *i.e.* the *e.m.f.* (voltage) V_c *lags* behind the current I by 90° as shown in Fig. 42.14 (*c*).

In an A.C. circuit containing L and R in series the voltage across inductor L (V_L) leads the voltage across R (V_R) by a phase angle $\pi/2$ (90°) so that the resultant voltage across LR (V_{LR}) *leads* the voltage

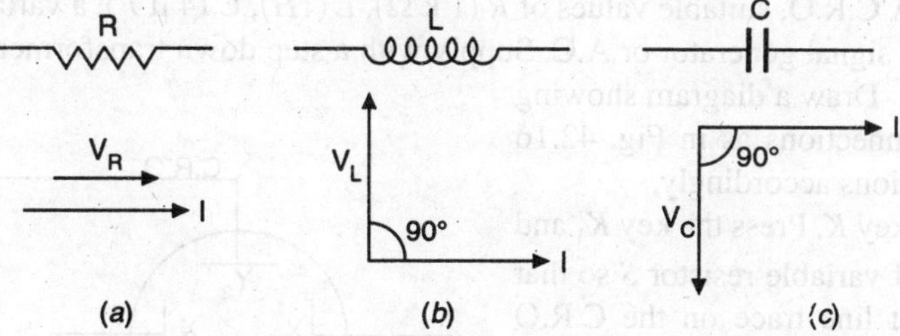

Fig. 42.14

across R (or the current I) by a phase angle ϕ given by $\tan\phi = \dfrac{L\omega}{R}$ where $L\omega = X_L$ = reactance due to inductance and $\omega = 2\pi f$, f being the frequency of A.C. [Fig. 42.15(a)].

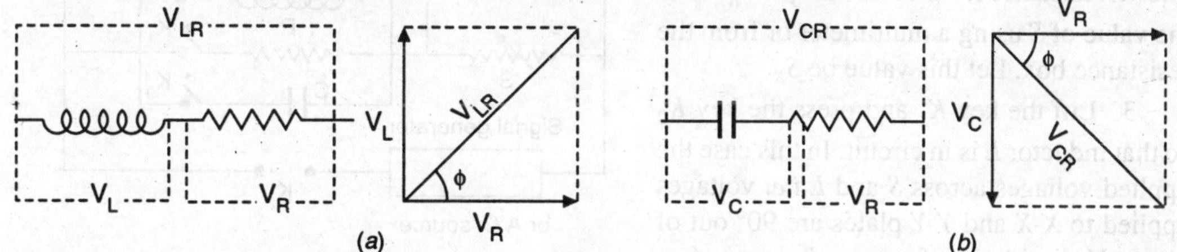

Fig. 42.15

In an A.C. circuit containing C and R in series the voltage across the capacitor C (V_c) lags behind the voltage across R (V_R) by a phase angle $\pi/2$ (90°) so that the resultant voltage across CR (V_{CR}) *lags* behind the voltage across R (or current I) by a phase angle ϕ given by $\tan\phi = \dfrac{1/C\omega}{R}$ where $\dfrac{1}{C\omega} = X_c$ = reactance due to capacitance.

When the A.C. voltage applied to the X-X plates (V_x) and Y-Y plates (V_y) of a C.R.O have only a resistance in the circuit the phase difference between V_x and V_y is zero. In such a case the trace on the C.R.O. screen is a *straight line*. If out of the two we put an inductor or a capacitor in one of the circuits there is a phase difference of 90° between V_x and V_y. In such a case the trace on the screen is an ellipse. If one circuit contains an inductor and the other a capacitor the phase difference between the two voltages V_x and V_y is 180° and the trace on the screen is again a straight line. If one circuit contains a pure resistance and the other a combination of inductance and resistance (L and R) or a combination of capacitance and resistance (C and R), then there is a phase difference $\phi \neq 90°$ between them. In such a case the trace on the screen in an inclined ellipse.

[For proof see **Art. 42.4.**]

The above results are shown in a tabular form below.

	X-X plates	*Y-Y plates*	*Phase diff.*	*Trace*
1.	V_R	V_R	0	straight line
2.	V_R	V_L	90°	ellipse
3.	V_R	V_C	90°	ellipse
4.	V_L	V_C	180°	straight line
5.	V_R	V_{LR}	$\phi \neq 90°$	inclined ellipse
6.	V_R	V_{CR}	$\phi \neq 90°$	inclined ellipse

Experiment 42.6. To demonstrate the phase difference in the case of resistance, inductance and capacitance and to measure their values using a C.R.O.

Apparatus A C.R.O, suitable values of R (1 k Ω), L (1H), C (4 μ F), a variable resistance (0–1 K Ω), a sine wave signal generator or A.C. Supply with a step down transformer, 4 keys etc.

Procedure 1. Draw a diagram showing the scheme of connections as in Fig. 42.16 and make connections accordingly.

2. Put in the key K. Press the key K_1 and adjust the value of variable resistor S so that you get a straight line trace on the C.R.O screen. The straight line is inclined to the horizontal at an angle of 45°. In this case the voltages applied to the X-X and Y-Y plates have a phase difference of 0° and the unknown resistance $R = S$ because $V_S = V_R$. Note the value of S using a multimeter or from the resistance box. Let this value be S_R.

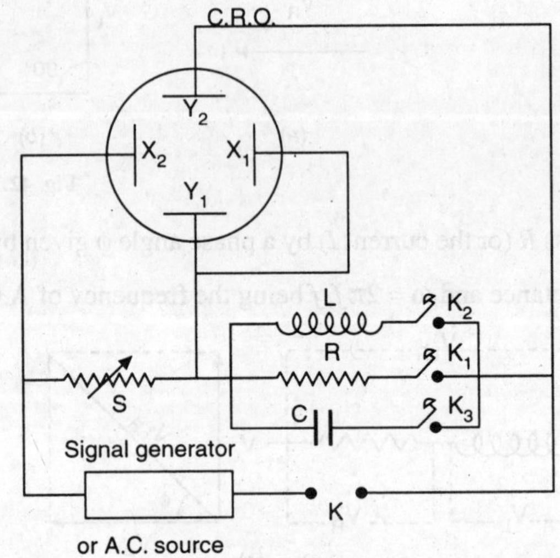

Fig. 42.16

3. Lift the key K_1 and press the key K_2 so that inductor L is in circuit. In this case the applied voltages across S and L *i.e.* voltages applied to X-X and Y-Y plates are 90° out of phase. You will, therefore, get the trace of an ellipse on the C.R.O screen. Adjust the value of variable resistor S in such a way that you get the trace of a circle on the C.R.O screen. In such a case the inductive reactance $X_L = S$. Again note the value of S using a multimeter or from the resistance box. Let this value be S_L.

4. Lift the key K_2 and press the key K_3 so that the capacitor C is in circuit. In this case the applied voltages across S and C *i.e.* the voltages applied to X-X and Y-Y plates are again 90° out of phase. You will, therefore, again get the trace of an ellipse on the C.R.O. screen. Adjust the value of variable resistor S in such a way that you again get the trace of a circle on the C.R.O screen. In such a case the capacitative reactance $X_C = S$. Again note the value of S using a multimeter or from the resistance box. Let this value be S_C.

5. Repeat the experiment three times.

Observations Observed value of S_R = (*i*) (*ii*) (*iii*)

Mean observed value of S_R = Ω

Observed value of S_L = (*i*) (*ii*) (*iii*)

Mean observed value of S_L = Ω

Observed value of S_C = (*i*) (*ii*) (*iii*)

Mean observed value of S_C = Ω

Calculations *Resistance* $R = S_R =$

Inductance L. $S_L = X_L$

but $X_L = \omega L = 2\pi fL$

where f is the frequency of A.C. signal and $\omega = 2\pi f$ is the angular frequency

$\therefore$ Inductance $L = \dfrac{S_L}{2\pi f} =$ henry

Capacitance C. $S_C = X_C$

but $X_C = \dfrac{1}{\omega C} = \dfrac{1}{2\pi f C}$

$\therefore$ Capacitance $C = \dfrac{1}{2\pi f S_C} = \quad$ Farad.

Precautions. Same as in **Expt. 42.5.**

Exercise. *To calculate the frequency of A.C. mains using a C.R.O.*

Hint. Use either an inductor L or a capacitor C of known value. Adjust the value of variable resistor S in such a way that you get the trace of a circle on C.R.O. screen using circuit connection of **Expt. 42.6.** In such a case

(*i*) If an inductor is used $\quad X_L = L\omega = 2\pi f L = S$

or $\quad f = \dfrac{S}{2\pi L}$

(*ii*) If a capacitor is used $\quad X_C = \dfrac{1}{C\omega} = \dfrac{1}{2\pi f C} = S$

$\therefore \quad f = \dfrac{1}{2\pi CS}$

Experiment 42.7. To measure the phase difference between current and voltage for C-R and L-R A.C. circuit using a C.R.O.

Apparatus. A C.R.O, variable step down transformer, a known variable resistance or resistance box, a capacitor of known capacitance, an inductor of known inductance, two keys etc.

Theory. When the voltage applied to the *X-X* or *Y-Y* plates of a C.R.O has a resistor and inductor or a resistor and capacitor in the two circuits as shown n Fig. 42.17(*a*) the trace obtained on the C.R.O. screen is an inclined ellipse. It can be shown that the phase difference is given by

$$\sin\phi = \frac{b}{a} \quad \text{or} \quad \phi = \sin^{-1}\frac{b}{a}$$

where the parameter a and b are shown in Fig. 42.17(*b*).

Procedure 1. Draw a diagram showing the scheme of connections as in Fig. 42.17 (*a*) and make connections accordingly.

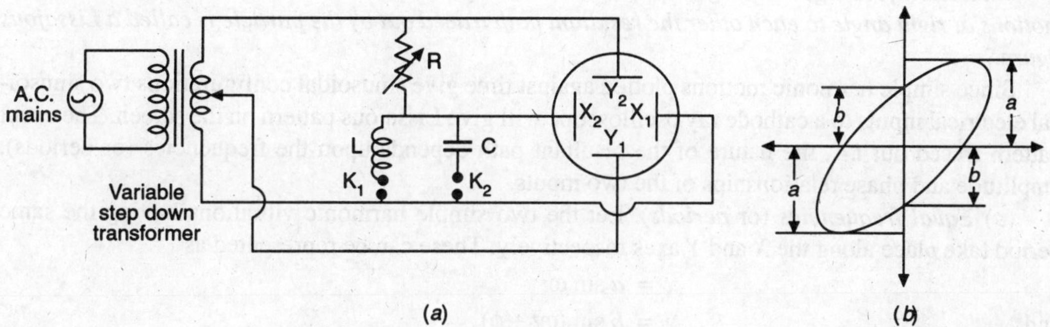

Fig. 42.17

2. Put in the key K_1 so that the inductor L is in circuit. In this case the A.C. signals given to the *X-X* and *Y-Y* plates are of the same frequency but differ in phase due to the presence of passive elements like L and R. Adjust the value of variable resistance R so that we get the trace of an inclined ellipse on the C.R.O screen as shown in Fig. 42.17(*b*).

3. Trace the ellipse on a tracing paper and measure the lengths *a* and *b*. Also note the value of *R* from the resistance box or measure it with a multimeter. Take a number of readings with different values of *R*.

4. Take out the key K_1 so as to disconnect the inductor and put in the key K_2 so that the capacitor *C* is in circuit along with resistance *R*. Repeat the experiment with different values of *R* and note the value of *R* and that of *a* and *b* in each case.

Observations

Value of fixed inductance L = Henry

Value of fixed capacitance C = Farad

Frequency of A.C. voltage f = Hz

Angular frequency $\omega = 2\pi f$ = sec^{-1}

No.	*Component in circuit*	*Resistance R (Ω)*	*Length a (cm)*	*Length b (cm)*	*Observed phase difference* $\phi = \sin^{-1} b/a$
1	Inductor *L*				
2					
3					
1	Capacitor *C*				
2					
3					

Observed mean value of ϕ =
(From *LR* circuit)

Theoretical calculated value of $\phi = \tan^{-1} \dfrac{L\omega}{R} =$

Observed mean value of ϕ =
(From *CR* circuit)

Theoretical calculated value of $\phi = \tan^{-1} \dfrac{1}{C\omega R} =$

Precautions. Same as in **Expt. 42.5.**

42.4. Lissajous figures. *When a particle is acted upon simultaneously by two simple harmonic motions at right angle to each other the resultant path traced out by the particle is called a Lissajous figure.*

Since simple harmonic motions plotted against time give sinusoidal configurations two sinusoidal electrical inputs to a cathode ray oscilloscope will give Lissajous pattern on the screen. The exact pattern traced out *i.e.*, the nature of the resultant path depends upon the frequencies (or periods), amplitude and phase relationships of the two inputs.

(*a*) *Equal frequencies* (or *periods*). Let the two simple harmonic vibrations having the same period take place along the *X* and *Y* axes respectively. These can be represented as

$$x = a \sin \omega t$$

and
$$y = b \sin (\omega t + \phi)$$

where *a* and *b* are the respective amplitudes and the *Y*-vibrations is *ahead* of the *X*-vibration by a phase angle ϕ.

If $\phi = 0$

$$x = a \sin \omega t$$

$$y = b \sin \omega t$$

or
$$y = \frac{b}{a} x$$

This represents a straight line through the origin as shown in Fig. 42.18. (*i*). The slope is given by

$$\tan\theta = \frac{b}{a}$$

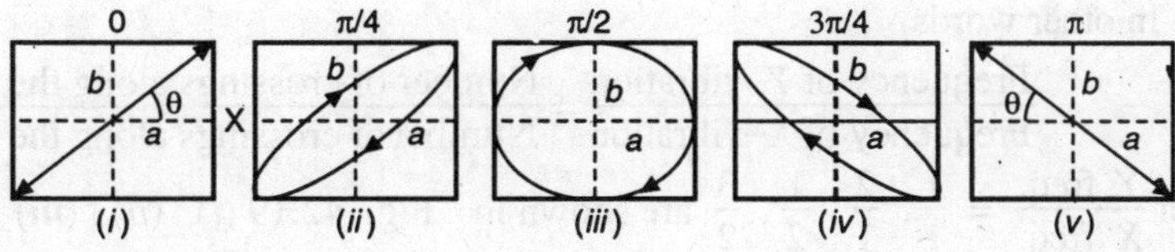

Fig. 42.18

If $\phi = \pi/2$ $\quad x = a \sin \omega t$

$$y = b \sin(\omega t + \pi/2) = b \cos \omega t$$

$$\therefore \quad \frac{x^2}{a^2} + \frac{y_2}{b^2} = 1$$

This represents a *symmetrical* ellipse whose major and minor axes coincide with *X* and *Y* axes respectively as shown in Fig. 42.18 (*iii*).

If $a = b$ the resultant curve is a circle given by

$$x^2 + y^2 = a^2$$

If $\phi = \pi$ $\quad x = a \sin \omega t$

$$y = b \sin(\omega t + \pi) = -b \sin \omega t$$

$$\therefore \quad y = -\frac{b}{a} x \cdot$$

This again represents a straight line through the origin as shown is Fig. 42.18. (*v*). The slope is given by

$$\tan\theta = -\frac{b}{a}$$

For $\phi = \pi/4$ and $\phi = 3\pi/4$ we get oblique ellipses as shown in Fig. 42.18. (*ii*) and (*iv*)

As ϕ exceeds π the whole cycle is repeated in the *reverse* order.

(*b*) *Frequencies in the rations of 1 : 2.* When the frequency of the *Y*-vibration is double that of the *X*-vibration but the two are in phase ($\phi = 0$), the curve traced out is shown in Fig. 42.19 (*ii*).

If, however, the frequency of the *X*-vibration is double that of *Y*-vibration, then the pattern obtained is as shown in Fig. 42.19. (*iii*).

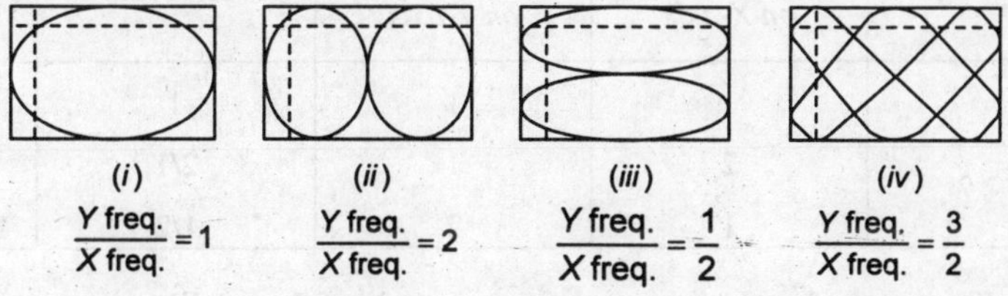

Fig. 42.19

In fact, the frequency ratio of the two inputs may be determined from an analyses of the Lissajous figures produced. A simple method is to enclose the Lissajous figures in a rectangle whose sides are parallel to the formation axes of the figure and note the *number* of *tangency points* along the *vertical*

(Y-axis) and the *horizontal* (X-axis). The ratio of the number of tangency points is in the *inverse* ratio of the two frequencies *i.e.*

$$\frac{\text{Number of tangency points along } X\text{–axis}}{\text{Number of tangency points along } Y\text{–axis}} = \frac{\text{Frequency of } Y\text{–vibration}}{\text{Frequency of } X\text{–vibration}}$$

An alternate method is to count the number of points at which the vertical and horizontal lines cross the figure. In other words

$$\frac{\text{Frequency of } Y\text{–vibration}}{\text{Frequency of } X\text{–vibration}} = \frac{\text{Number of crossings along the } X\text{–axis}}{\text{Number of crossings along the } Y\text{–axis}}$$

The cases of $\frac{Y \text{ freq.}}{X \text{ freq.}} = \frac{1}{1}, \frac{2}{1}, \frac{1}{2}, \frac{3}{2}$ are shown in Fig. 42.19 (*i*), (*ii*) (*iii*) and (*iv*).

Experiment 42.8. To compare the frequencies of oscillations produced by two audio-oscillators using Lissajous figures.

OR

Given a standard oscillator of frequency 1000 Hz to calibrate an audio oscillator.

Apparatus. A standard 1000 Hz audio-oscillator, a variable frequency audio-oscillator and cathode ray oscilloscope.

Procedure. 1. Connect the standard frequency oscillator [1000 Hz] to the ''*vertical input*' terminals of an oscilloscope. Connect the audio-oscillator whose frequencies are to be compared with the standard oscillator to the '*horizontal input*' terminals. Connect together the '*ground*' terminals of both the oscillators.

2. Set the C.R.O. as already explained so that a sharp, bright spot is obtained in the centre of the screen. Turn the '*sweep control*' to *off* position by setting it to '*ext. input*' position. Set the audio-oscillator frequency to the marked value of 1000 cycles/-sec.

3. Switch on both the oscillators and adjust the gain controls of the two oscillators as well as the '*horizontal*' and '*Vertical*' gains of the oscilloscope so that a good size ellipse (a 1:1 Lissajous figure) appears on the screen. Adjust the necessary controls to '*stop*' the ellipse. The actual oscillator frequency is now 1000 Hz. Record the dial reading. (It will, in general, have a slightly different value, say 990 Hz).

By switching off the audio-oscillator allow the ellipse to change phase and note various shapes that appear on the screen. By phase change and amplitude adjustment try to obtain a '*circular*' trace.

4. Set the oscillator frequency to the marked value of 500 and adjust slowly so that a 1 : 2 Lissajous figure as shown in Fig. 42.19 (*ii*) is obtained. If possible, adjust the controls to '*lock*' the figure. Record the dial reading.

Similarly obtain (1 : 3, 3 : 1), (2 : 3, 3 : 2) Lissajous figures and so on upto (1 : 5, 5 : 1).

Observations. Vertical input standard frequency = 1000 Hz.

Hor. input Marked dial Freq.	*Shape of Fig.*	*No. of tangency points*		$\frac{\text{Vert Freq.}}{\text{Hor. Freq.}}$	*Actual Hor. Freq.*
		on X-axis	*on Y-axis*		
990		1	1	1	1000/1 = 1000
495		2	1	2/1	1000/2 = 500
1980		1	2	1/2	1000× 2 = 2000

Precautions. 1. The vertical and horizontal gain controls of the oscilloscope should be adjusted to obtain a proper size of lissajous figures.

2. The frequency of the audio-oscillator should be slowly adjusted so as to lock the pattern.

Exercise. *Find the phase difference between two sinusoidal voltages of the same frequency and amplilude.*

Hints : The two sinusoidal voltages are simultaneously fed to the *horizontal* and *vertical* plates of a C.R.O. From the shape of Lissajours figures obtained the phase difference can be calculated in accordance with **Art** 42.3.

Experiment 42.9. To determine the frequency of an electrically maintained tuning fork using a C.R.O.

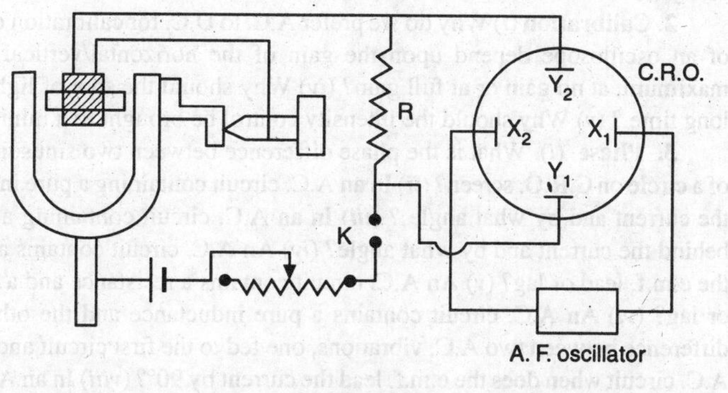

Apparatus. Electrically maintained tuning fork, a battery, a C.R.O, an audio frequency oscillator, a rheostat, resistor etc.

Fig. 42.20

Procedure. 1. Draw a diagram showing the scheme of connections as in Fig. 42.20 and make connections accordingly.

2. Switch on the power supply to the C.R.O and obtain a bright, sharp and fine spot of light at the centre of the C.R.O. screen by adjusting the intensity, focussing and positioning controls of the C.R.O.

3. Set the tuning fork into vibrations by inserting the key *K*. As the tuning fork vibrates an A.C. voltage is developed across the resistance *R*, which will trace a vertical line on the screen. Keep this line within the scale by adjusting the *Y*-amplifier gain.

4. Switch on the audio frequency (A.F.) oscillator and adjust its frequency so as to obtain the trace of a circle on the C.R.O screen. Trace the circle. Note the frequency produced by the audio-oscillator. This is the frequency of the tuning fork.

5. Now adjust the frequency of the audio-oscillator so that a figure of eight (8) appears in the vertical direction on the screen. Trace the figure and note the frequency produced by the audio oscillator. The frequency of the tuning fork will be *one half* the frequency produced by the audio-osicillator.

6. Again adjust the frequency of the audio-oscillator such that a figure of eight (∞) in the horizontal direction appears on the screen. Trace the figure and note the frequency produced by the audio-osicllator. The frequency of the tuning fork will be *double* the frequency produced by the audio-oscillator.

Observations

No.	*Horizontal input Frequency from audio-oscillator(f)*	*Shape of pattern*	*No. of tangency points*		*Vert. freq. / Hor. freq.*	*Freq. of tuning fork.*
			on X-axis	*on Y-axis*		
1		O circle				
2		8				
3		∞				

Mean frequency of tuning fork =

Precautions. Same as in **Expt. 42.8.**

Oral Questions

1. Cathode ray oscilloscope. (*i*) Name the various parts of a cathode ray tube. (*ii*) How does the cathode emit electrons? (*iii*) What is the function of the grid? (*iv*) How do the focusing electrodes work? (*v*) How do you adjust the position of the fluorescent spot in the centre of the screen? (*vi*) Why is the last anode and one plate of each of the *X* and *Y* pairs earth connected? (*vii*) What is the material used on the fluorescent screen? (*viii*) How is the fluorescent screen connected to the final plate and why? (*ix*) What is the function of time base circuit? (*x*)

Draw a time base circuit using a gas filled triode (*xi*) What is a thyratron? (*xii*) Can a neon lamp be used in place of thyratron? What are its draw backs?

2. Calibration (*i*) Why do we prefer A.C. to D.C. for calibration of an oscilloscope? (*ii*) Does the sensitivity of an oscillosope depend upon the gain of the horizontal/vertical amplifier? (*iii*) When is the sensitivity maximum; at no gain or at full gain? (*iv*) Why should the spot of light not be allowed to stay at one point for a long time ? (*v*) Why should the intensity control be brought to a minimum before taking the next observation?

3. Phase (*i*) What is the phase difference between two sinusoidal oscillations which produce the pattern of a circle on C.R.O. screen? (*ii*) In an A.C. circuit containing a pure inductance does the e.m.f. lead or lag behind the current and by what angle ? (*iii*) In an A.C. circuit containing a pure capacitor does the e.m.f lead or lag behind the current and by what angle? (*iv*) An A.C. circuit contains a resistance and inductance in series. Does the e.m.f. lead or lag? (*v*) An A.C. circuit contains a resistance and a capacitance in series. Does the e.m.f. lead or lag? (*vi*) An A.C. circuit contains a pure inductance and the other a pure capacitance, what is the phase difference between two A.C. vibrations, one fed to the first circuit and other fed to the second circuit? (*vii*) In an A.C. circuit when does the e.m.f. lead the current by 90°? (*viii*) In an A.C. circuit when does the e.m.f. lag behind the current by 90°?

4. Lissajous figures. (*i*) What is a Lissajous figure? (*ii*) Draw the Lissajous figures for frequency ratio 1 : 1, 2: 1, 1: 2 for a phase difference $\pi/2$. (*iii*) Draw diagrams to show how the shape of the Lissajous figure varies when the phase changes from 0 to 2π and the frequency ratio is 1 : 1 (*iv*) What is the condition that the figure may be circle? (*v*) How will you calculate the frequency ratio from the shape of Lissajous figure?

43

R.C. Circuit

43.1. A.C. Wheatstone bridge. The principle of wheatstone bridge is applicable to A.C. circuits as well. Four impedances Z_1, Z_2, Z_3, Z_4 forming a wheatstone bridge arrangement are shown in Fig. 43.1. The A.C. source is a signal generator having a frequency range 200-2000 Hz. The detecting arrangement can be a head phone (as shown) or alternatively a C.R.O.

When the points B and D are at the same potential no current flows in the branch BD and the bridge is balanced. At this stage the head phone indicates zero or minimum sound. In this case, according to Wheatstone principle

$$\frac{Z_1}{Z_2} = \frac{Z_3}{Z_4}$$

The bridge is, however, most sensitive when all the four impedances are equal and the source and detector have the same frequency.

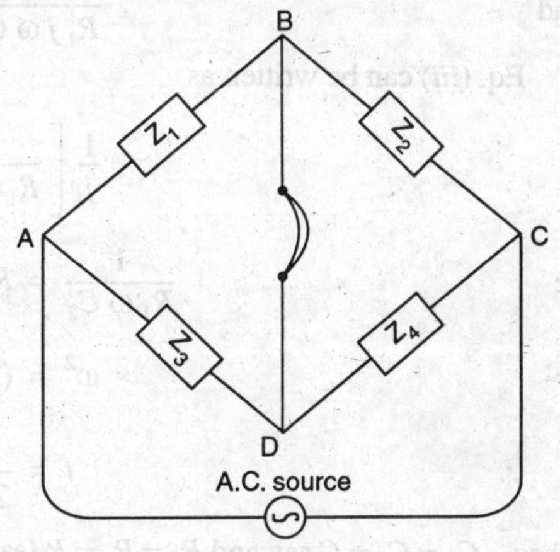

Fig. 43.1

Experiment 43.1. To study the behaviour of an RC circuit at various frequencies.

Apparatus. Two capacitors having the same capacitance, two resistance boxes (plug type), two resistance boxes (dial type), Audio frequency oscillator, head phone (or C.R.O) and connecting wires.

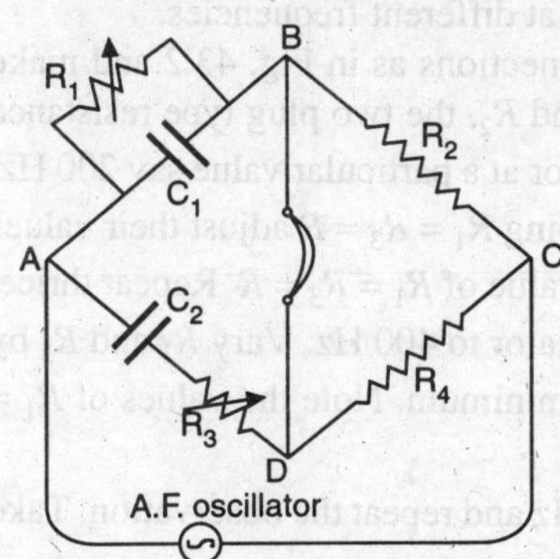

Fig. 43.2

Theory. A Wien's bridge arrangement is shown in Fig. 43.2. The impedance Z_1 consists of a resistance R_1 and capacitance C_1 in *parallel.* The value of Z_1 is, therefore, given by

$$\frac{1}{Z_1} = \frac{1}{R_1} + j\,\omega\,C_1$$

where $\omega = 2\pi f$, f being the frequency of the A.F. oscillator the impedance Z_3 consists of a resistance R_3 and capacitance C_2 in series. The value of Z_3 is, therefore given by

$$Z_3 = R_3 + \frac{1}{j\,\omega\,C_2}.$$

Impedance $Z_2 = R_2$ and $Z_4 = R_4$. When the bridge is balanced

$$\frac{Z_1}{Z_2} = \frac{Z_3}{Z_4}$$

or $$\frac{Z_4}{Z_2} = \frac{Z_3}{Z_1}$$

or $$\frac{R_4}{R_2} = \left(R_3 + \frac{1}{j\,\omega\,C_2}\right)\left(\frac{1}{R_1} + j\,\omega\,C_1\right)$$

$\therefore$ $$\frac{R_4}{R_2} = \frac{R_3}{R_1} + \frac{1}{R_1\,j\,\omega\,C_2} + R_3\,j\,\omega\,C_1 + \frac{C_1}{C_2} \qquad ...(i)$$

Equating the real and imaginary parts on both sides of Eq. (*i*), we get

$$\frac{R_4}{R_2} = \frac{R_3}{R_1} + \frac{C_1}{C_2} \qquad ...(ii)$$

and $$\frac{1}{R_1\,j\,\omega\,C_2} + R_3\,j\,\omega\,C_1 = 0 \qquad ...(iii)$$

Eq. (*iii*) can be written as

$$\frac{1}{j}\left[\frac{1}{R_1\,\omega\,C_2} - R_3\,\omega\,C_1\right] = 0$$

or $$\frac{1}{R_1\,\omega\,C_2} = R_3\,\omega\,C_1$$

or $$\omega^2 = (2\,\pi f)^2 = \frac{1}{R_1\,R_3\,C_1\,C_2}$$

$\therefore$ $$f = \frac{1}{2\,\pi\,(R_1\,R_3\,C_1\,C_2)^{1/2}}$$

when $C_1 = C_2 = C$ say and $R_1 = R_3 = R$ (say), we have

$$f = \frac{1}{2\,\pi\,RC} \qquad ...(iv)$$

Eq. (*ii*), now becomes $\frac{R_4}{R_2} = 2$ or $R_4 = 2\,R_2$

Using Eq. (*iv*), we can study the behaviour of an *R.C.* circuit at different frequencies.

Procedure 1. Draw a diagram showing the scheme of connections as in Fig. 43.2 and make connections accordingly. Take $C_1 = C_2 = C = 0.2\ \mu f$. From R_4 and R_2, the two plug type resistance boxes take out $R_4 = 2\,R_2$. Adjust the frequency of the A.F. oscillator at a particular value say 200 Hz.

2. R_1 and R_3 are two dial type resistance boxes. Always keeping $R_1 = R_3 = R$ adjust their values so that the sound heard in the head phone is minimum. Note the value of $R_1 = R_3 = R$ Repeat thrice.

3. Keeping $R_4 = 2\,R_2$ change the frequency of the A.F. oscillator to 400 Hz. Vary R_1 and R_3 by equal values so that the sound heard in the head phone is again minimum. Note the values of $R_1 = R_3 = R$. Repeat thrice.

4. Again change the frequency of the A.F. oscillator by 200 Hz and repeat the observation. Take such six readings with the same value of $C_1 = C_2 = C = 0.2\ \mu F$.

5. Change the value of $C_1 = C_2 = 0.3\ \mu f$ and then $C_1 = C_2 = 0.4\ \mu f$ and repeat all the steps.

Observations. $R_4 = 2\,R_2 =$ ohms.

Sl. No.	Freq. of A.F. oscillator in Hz	$C_1 = C_2 = C$ in Farads	$R_1 = R_3 = R$ in ohms	$\frac{1}{RC}$	Calculated frequency $f = \frac{1}{2\pi RC}$	Diff. between marked freqn. of oscillator and f.
1	200	0.2×10^{-6} 0.3×10^{-6} 0.4×10^{-6}				
2	400	0.2×10^{-6} 0.3×10^{-6} 0.4×10^{-6}				
3	600	0.2×10^{-6} 0.3×10^{-6} 0.4×10^{-6}				
4	800	0.2×10^{-6} 0.3×10^{-6} 0.4×10^{-6}				
5	1000	0.2×10^{-6} 0.3×10^{-6} 0.4×10^{-6}				
6	1200	0.2×10^{-6} 0.3×10^{-6} 0.4×10^{-6}				

6. Graph. Taking frequency in Hz along the X-axis and $\frac{1}{RC}$ along the Y-axis plot a graph. The graph is a straight line.

Conclusion. As the graph is a straight line $f \propto \frac{1}{RC}$.

Precautions. 1. Standard capacitors having the same capacitance should be used.

2. While calculating $\frac{1}{RC}$, C must be taken in farads.

3. R_4 should always be equal to $2R_2$.

4. R_1 and R_3 should be equal and should always be varied together in equal steps.

5. For getting the null point the sound in the head phone should be almost inaudible. Preferably a C.R.O should be used for the purpose.

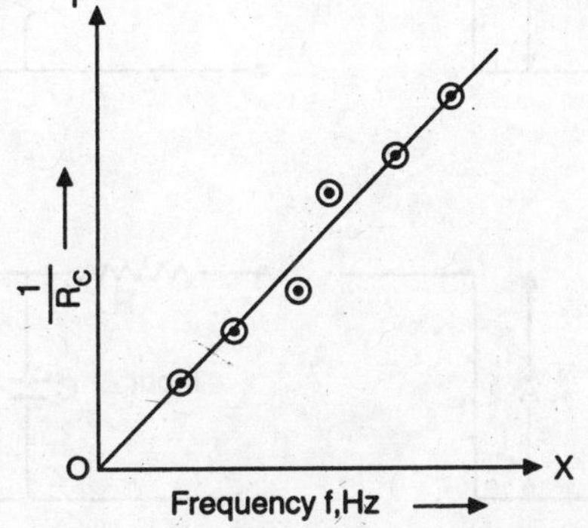

Fig. 43.3

43.2. Differentiating circuit. *A circuit in which output voltage is directly proportional to the derivative of the input is known as a differentiating circuit.*

i.e. (output) $\propto \frac{d}{dt}$ (Input)

A differentiating circuit is a simple *RC* series circuit in which output is taken across the resistance *R*. A typical differentiating circuit is shown in Fig. 43.4. If a *d.c.* or constant input is applied to such a circuit, the output will be zero, because the derivative of a constant is zero.

If e_i is the input alternating voltage, i the alternating current due to it, e_c the voltage drop across the capacitor C at any instant, then the charge on the capacitor is given by

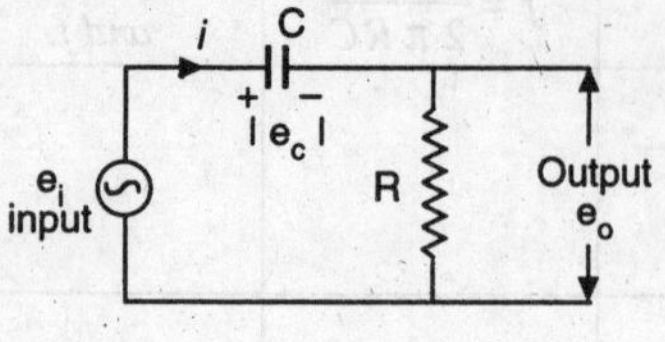

Fig. 43.4

$$q = C e_c$$

$$\therefore \quad i = \frac{dq}{dt} = C\frac{de_c}{dt}$$

As the capacitative reactance is very much greater than R, the input voltage e_i is almost equal to the voltage drop across the capacitor e_c

$$\therefore \quad i = C\frac{de_i}{dt}$$

Output voltage $\quad e_0 = iR = CR\dfrac{de_i}{dt} \propto \dfrac{d}{dt}e_i$

Hence Output voltage ∝ Derivative of input voltage.

Thus a *sinewave* input becomes a *cosine wave* and a cosine wave input becomes an *inverted sine wave* at the output. When a *saw tooth* wave is fed to a differentiating circuit the output is a succession of *rectangular* waves having equal or unequal intervals depending upon the shape of the saw tooth input.

The response of a *differentiator* (circuit) to a *step function* (or a *square pulse*) for $CR << T$, $CR = T$ and $CR >> T$ where T is the time period of the *input wave* is shown in Fig. 43.5 (*a*).

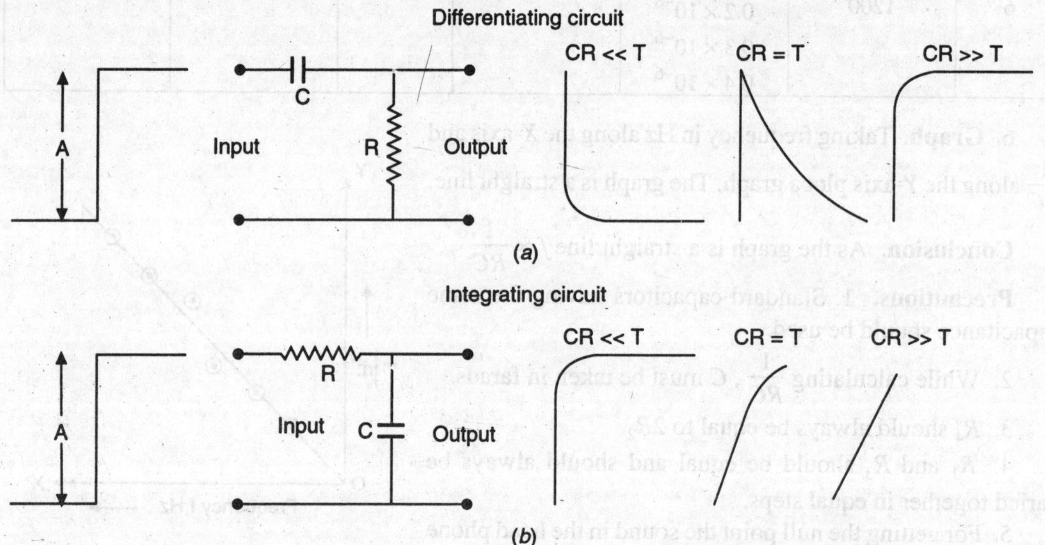

Fig. 43.5

Experiment 43.2. To study the characteristics of a differentiating circuit (or differentiator) [CR circuit].

Apparatus. A square wave generator, Resistors 20 K Ω, 200 K Ω; capacitors 0.001 μ*f*, 0.01 μ*f*; a cathode ray oscilloscope.

Procedure 1. A block diagram of the connections is shown in Fig. 43.6.

2. Switch on the square wave generator and adjust the output frequency to 1000 cycles per second. Switch on the C.R.O and connect it directly to the square wave generator. Observe the pattern

of the oscillations. Adjust so that the pulse width is 0.5 mm. Measure the peak to peak voltage of the C.R.O input.

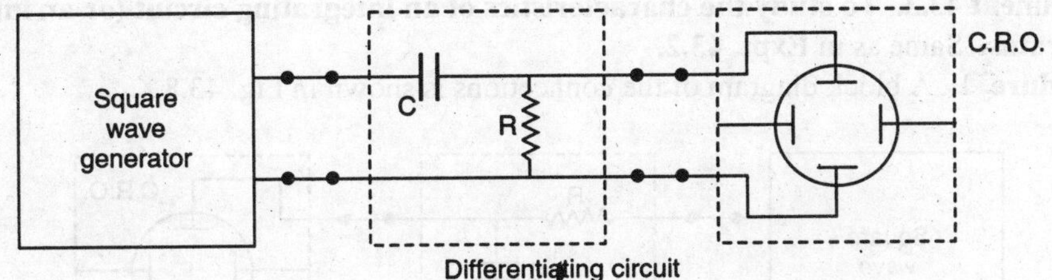

Fig. 43.6

3. Connect the pulse generator to the C.R.O through the differentiating circuit as shown. Using $C = .001\ \mu F$ and $R = 20\ K\ \Omega$ trace the wave form which appears on the C.R.O across R. Also measure peak to peak voltage. Repeat by changing R to $200\ K\ \Omega$.

4. Repeat step 3 with $C = .01\ \mu F$ and $R = 20\ K\ \Omega$ and $200\ K\ \Omega$.

5. Change the frequency of square wave pulse generator to 100 c/s and 1000 c/s and again trace the wave form and measure peak to peak voltage as in step 3 and 4.

Conclusion. It is observed that the shape of the output pulses depends upon the relation between time constant of the differentiating circuit CR and the time period of the input wave T.

In order to obtain good differentiation following two conditions are necessary:—

(*i*) The time constant RC of the circuit should be much smaller than the time period of the input wave.

(*ii*) The value of X_C should be greater than $10\ R$ at the operating frequency.

43.3. Integrating circuit. *A circuit in which output voltage is directly proportional to the integral of the input is known as an integrating circuit.*

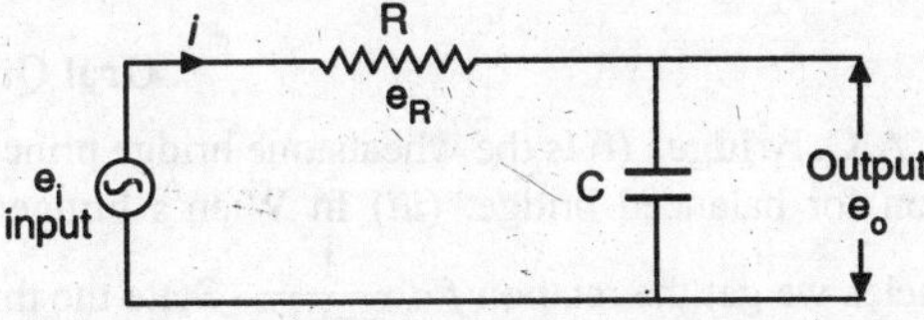

Fig. 43.7

i.e. (output) $\propto \int$ (Input)

An integrating circuit is a simple RC series circuit in which output is taken across the capacitance C. Let e_i be the input alternating voltage and i the alternating current due to it. The voltage drop across the resistor R is e_R. If R is very large as compared to capacitative reactance X_C, then

$$e_R = e_i$$

and
$$i = \frac{e_R}{R} = \frac{e_i}{R}$$

The charge on the capacitor at any instant

$$q = \int i\,dt$$

and output voltage
$$e_0 = \frac{q}{C} = \frac{\int i\,dt}{C} = \frac{\int \frac{e_i}{R}\,dt}{C}$$

$$= \frac{1}{RC}\int e_i\,dt$$

or (Output voltge) $e_0 \propto$ (Integral of input voltage).

The response of an integrator (circuit) to a step function (or a square pulse) for $CR << T$, $CR = T$ and $CR >> T$ where T is the time period of the *input wave* is shown in Fig. 43.5(*b*).

Experiment 43.3. To study the characteristics of an integrating circuit (or an integrator)

Apparatus. Same as in **Expt. 43.2.**

Procedure 1. A block diagram of the connections is shown in Fig. 43.8.

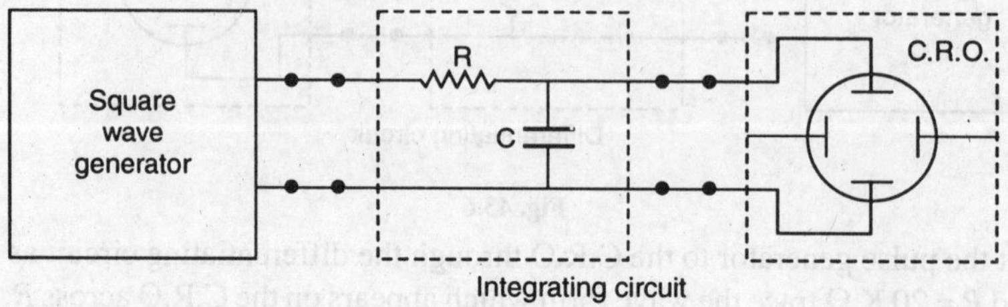

Fig. 43.8

2. Same as in **Expt. 43.2.**

3. Connect the square wave generator to the C.R.O. through the integrating circuit as shown and repeat all the steps given in **Expt. 43.2.**

Conclusion. It is observed that the shape of the output pulse depends upon the relation between the time constant CR of the integrating circuit and the time period T of the input wave. The shape of the output wave is made up of exponential arc which approaches a rectangular form as the value of time constant is decreased.

In order to obtain good integration the following two conditions are necessary.

(*i*) The time constant RC of the circuit should be much greater than the time period of the input wave.

(*ii*) The value of R should be greater than 10 X_C at the operating frequency.

Oral Questions

1. A.C. Bridge. (*i*) Is the wheatstone bridge principle applicable to alternating current? (*ii*) If so, state the condition for balanced bridge. (*iii*) In Wien's bridge to study the behaviour of an R.C. circuit at various frequencies we get the relation $f = \frac{1}{2\pi RC}$ · State the three conditions under which this relation holds. (*iv*) In a Wien's bridge if $R_1 = R_3 = R$, $C_1 = C_2 = C$ and $R_4 = 2R_2$, what is the relation for frequency when the bridge is balanced. (*v*) Can you find the value of an unknown capacitance by Wien's bridge?

2. Differentiating and integrating circuits. (*i*) What do you mean by time constant of an *R.C.* circuit? (*ii*) Mark the correct answer.

(*a*) In a differentiating circuit output is drawn from

1. Resistor 2. Capacitor

(*b*) In an integrating circuit output is drawn from

1. Resistor 2. Capacitor

(*iii*) What should be the relation between X_C and R for (*a*) good differentiation (*b*) good integration

(*iv*) What is a picofarad? (*v*) State one application of an integrating circuit.

44

Magnetic Properties of Materials

44.1. Magnetic materials. All substances are more or less effected by a magnetic field. Some of them are very strongly attracted, others are only feebly attracted while some others are even repelled. As such substances are divided into three classes

(*i*) Paramagnetic substances (*ii*) Ferromagnetic substances and (*iii*) Diamagnetic substances.

(*i*) **Paramagnetic substances** (*Paramagnetics*). Paramagnetic substances are those which are feebly attracted by magnets and when placed in a non-uniform magnetic field move from weaker to stronger parts of the field. In a magnetising field paramagnetics acquire a feeble magnetism in the same sense as the external magnetising field. Some familiar examples are:– Aluminium, Platinum, Oxygen, Manganese; Chromium and solution of salts of Nickel.

(*ii*) **Ferro magnetic substances** (*Ferromagnetics*). Ferromagnetic substances are those which are strongly attracted by magnets. When placed in a magnetising field ferromagnetics acquire a strong magnetism in the same sense as the external magnetising field. Familiar examples are, Iron, Nickel, Cobalt and their alloys like perm, mumetal alnico etc.

(*iii*) **Diamagnetic substances** (*Diamagnetics*) Diamagnetic substances are those which are repelled by magnets and when placed in a non-uniform magnetic field move from stronger to weaker parts of the field. When placed in a magnetising field diamagnetics acquire magnetism in a sense opposite to that of external magnetising field. Familiar examples are, Bismuth, Phosphorus, Antimony, Copper, Water, Alcohol and Hydrogen.

44.2. Magnetising field H. *The magnetic field which magnetises a substance when placed in it is called a magnetising field.*

The S.I. unit of magnetising field H is a *Tesla* and the C.G.S. unit is an *Oersted.*

44.3. Intensity of magnetisation. The magnetisation acquired by a substance when placed in a magnetising field is called *intensity of magnetisation I*

It is defined as magnetic moment per unit volume

$$\therefore \qquad I = \frac{P_m}{V}$$

where P_m is the magnetic moment acquired by the specimen.

44.4. Magnetic susceptibility. *Magnetic susceptibility is defined as the ratio of intensity of magnetisation I developed in a specimen to the intensity of the magnetising field H in which it is placed.*

$$\therefore \qquad \chi = \frac{I}{H}.$$

As intensity of magnetisation is defined as magnetic moment per unit volume, χ denotes the *volume susceptibility* of the specimen.

The mass susceptibility is defined as the ratio of volume susceptibility to mass per unit volume or density of the material.

$$\therefore \quad \text{Mass susceptibility} \quad k = \frac{\text{volume susceptibility}}{\text{density}} = \frac{\chi}{\rho}$$

44.5. Magnetic Induction B. When a material is placed in a magnetising field it gets magnetised. The force experienced by a unit north pole placed inside the material is due to the combined effect of the magnetising field and the induced magnetism acquired by the specimen.

The total force experienced by a unit north pole placed at any point inside the specimen is known as magnetic induction or magnetic flux density.

The S.I. unit of magnetic induction is a *Tesla.*

The C.G.S. unit of magnetic induction is a *Gauss.*

NOTE In our treatment in this chapter we shall use C.G.S. system for convenience. Also

$$1 \text{ Tesla} = 10^4 \text{ Gauss.}$$

44.6. Permeability. *Permeability of a material is defined as the ratio of magnetic induction B of the magnetised specimen of the material to the magnetising field H.*

$$\therefore \qquad \mu = \frac{B}{H}.$$

44.7. Hysterisis. Consider an unmagnetised bar of a ferromagnetic material placed in a magnetising solenoid. If an electric current is passed through the solenoid the bar gets magnetised. If the value of the current is gradually increased the magnetising field *H* increases and the magnitude of the intensity of magnetisation *I* developed in the ferro magnetic specimen also increases. A graph between *I* and *H* is shown in Fig. 44.1.

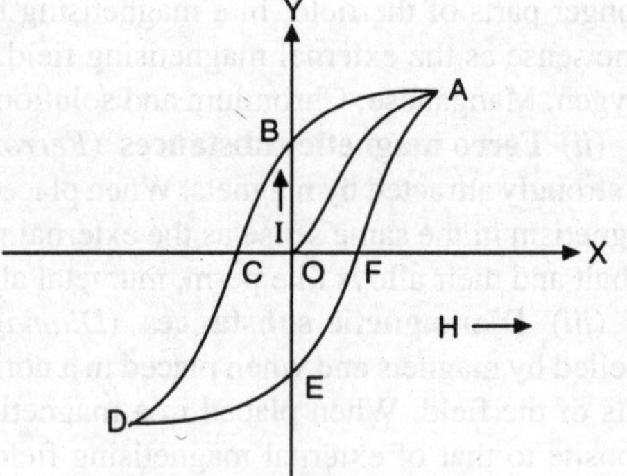

Fig. 44.1

(*i*) At *O* when *H* is zero, *I* is also zero. When *H* increases *I* also increases but not very uniformly till we reach *A*. Beyond *A*, if *H* is further increased *I* remains constant. The specimen in this condition is said to be *saturated.*

(*ii*) If *H* is now decreased the value of *I* decreases but at a much slower rate and when *H* is zero the value of *I* is not zero but has a finite value represented by *OB*.

(*iii*) If the value of *H* is now increased in the opposite direction the value of *I* further decreases and becomes zero when the value of magnetising field *H* = *OC*.

(*iv*) If the value of *H* is further increased again the saturation stage is reached in the opposite direction and the curve *CD* is obtained.

(*v*) On decreasing the value of *H* to zero again we get the portion *DE* of the curve. It is clear from the curve that *I* has a value given by *OE* even when *H* = 0.

(*vi*) On increasing the value of *H* in the positive direction again the part *EFA* of the cycle is repeated till we again reach the saturated stage at *A*. [The portion *OA* is not repeated unless the specimen is completely demagnetised]. This completes the cycle of magnetistation.

The closed curve *ABCDEFA* is called *hysterisis curve.* The word hysterises means '*lagging behind*'.

The graph between magnetic induction *B* and the magnetising field *H* is almost similar in shape as that between *I* and *H* but is slightly bigger in size. Since, at the saturation stage, increase in *B* does not stop but is equal to increase in H, this part of the B-H curve slopes very slightly upwards and unlike the I-H curve is not parallel to H-axis.

The lagging of magnetic induction B or intensity of magnetisation I behind the magnetising force H producing it when a specimen of a ferromagnetic material is taken through a cycle of magnetisation is called hysterisis.

44.8. Hysterisis loop. *The graph showing how I or B increases with H from zero to a maximum in one direction and then back through zero to a maximum in the opposite direction and finally back again through zero to the first maximum is called a hysterisis loop.*

44.9. Remanence. *The value of magnetic induction B which remains after the material has been magnetised and the magnetising force H has been reduced to zero is known as remanence or remanent induction or retentivity.*

44.10. Coercivily. *The magnitude of the reverse magnetising field required to reduce the remanent induction to zero is known as coercive force or coercivity.*

Magnetic materials having a *small* coercive force are known as **soft** whereas those possessing *large* coercive force are known a **hard.**

44.11. Loss of energy due to hysterisis. When a specimen is taken through a complete cycle of magnetisation more work is done **on** the material than that recovered **from it**. Thus there is a net loss of energy in each cycle of hysterisis. *The loss of energy per unit volume of the material per cycle of hysterisis is equal to the area of I - H loop* $= \frac{1}{4\pi} \times$ *Area of B-H loop* [In C.G.S. units].

44.12. Hysterisis loop with a cathode ray oscilloscope. To obtain a B-H hysterisis loop with a C.R.O. the circuit is shown in Fig. 44.2.

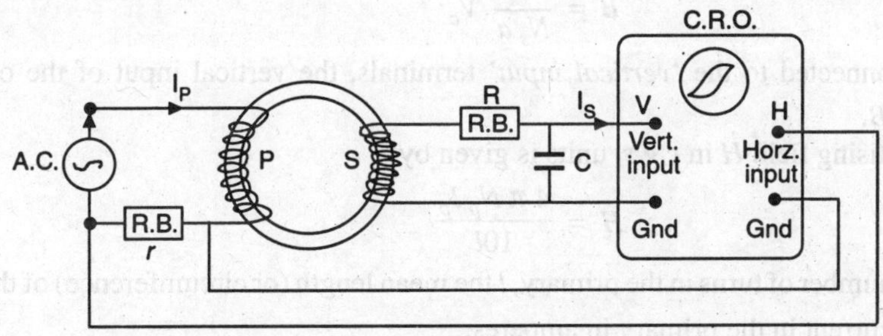

Fig. 44.2

The *A.C.* voltage drop across the resistance r in the primary circuit of a transformer is fed to the '*horizontal input*' terminals of *C.R.O.* and the *A.C.* voltage drop across the capacitance C in the secondary circuit of the transformer is fed to the '*vertical input*' terminals of the *C.R.O.* As shown below, the vertical input voltage is proportional to the magnetising field H. As the *A.C.* voltage supplied to the primary of the transformer completes one cycle, the material of the transformer core is taken through a complete cycle of magnetisation. The fluorescent spot on the *C.R.O.* moves in response to the B-H values. The frequency of the *A.C.* mains source being 50 cycles per second, the spot traces the loop 50 times in one second and due to persistance of vision we observe a stationary hysteresis loop on the *C.R.O.* screen as shown.

Let I_p be the current in the primary circuit of the transformer and ϕ_s the magnetic flux linked with each turn of the secondary due to it. If a is the area of cross-section of the transformer core and B the magnetic induction (Flux per unit area), then

$$\phi_s = Ba$$

As the current I_p is continuously changing, the flux in the secondary also goes on changing and an induced voltage V_s is set up in the secondary the *magnitude* of which is given by

$$V_s = N_s \frac{d\phi_s}{dt} = N_s\, a \frac{dB}{dt} \qquad ...(i)$$

where N_s is the total number of turns in the secondary. If I_s is the current in the secondary having a resistance R and capacitance C and V_c is the voltage drop across the capacitor, then

$$V_s = V_c + RI_s$$

The charge on the capacitance $Q = CV_c$

$$\therefore \qquad I_s = \frac{dQ}{dt} = C\frac{dV_c}{dt}$$

Hence $$V_s = \frac{Q}{C} + RC\frac{dV_c}{dt}$$

In general Q/C is very small as compared to the second term and may be neglected. Hence

$$V_s = RC\frac{dV_c}{dt} \qquad ...(ii)$$

Comparing (*i*) and (*ii*), we have

$$N_s\, a \frac{dB}{dt} = RC\frac{dV_c}{dt}$$

or $$dB = \frac{RC}{N_s\, a}\, dV_c \qquad (iii)$$

Integrating relation (*iii*), we get

$$B = \frac{RC}{N_s\, a}\, V_c$$

As V_c is connected to the '*vertical input*' terminals, the vertical input of the oscilloscope is proportional to B.

The magnetising field H in *c.g.s.* units is given by

$$H = \frac{4\pi N_p I_p}{10 l} \qquad ...(iv)$$

where N_p is the number of turns in the primary, l the mean length (or circumference) of the transformer core and I_p the current in the primary in amperes.

If V_x is the voltage across the resistance r in the primary, then

$$I_p = \frac{V_x}{r}$$

$$\therefore \qquad H = \frac{4\pi}{10}\frac{N_p}{rl}V_x \qquad ...(v)$$

As V_x is connected to the '*horizontal input*' terminals of the oscilloscope the horizontal input is proportional to H.

The energy loss per unit volume per cycle of hysteresis in C.G.S. units is given by

$$u = \frac{1}{4\pi}\int H dB$$

From (*iii*) and (*v*), we get

$$u = \frac{1}{4\pi}\frac{RC}{N_s\, a}\frac{4\pi N_p}{10\, rl}\int V_x\, dV_c$$

The total loss of energy in the transformer core

$U = u \times$ volume of the core $= ual$

$$\therefore \qquad U = \frac{R}{r}\frac{N_p}{N_s}\frac{C}{10}\int V_x\, dV_c$$

The integral $\int V_x\, dV_c$ gives the area enclosed by the loop obtained on the oscilloscope screen. If C is in Farad, V_x and V_s in volts, then U is in Joules.

$$\therefore \qquad U = \frac{R}{r}\frac{N_p}{N_s}\frac{C}{10}\int V_x\, dV_c \text{ Joule} \qquad ...(vi)$$

$$= \frac{R}{r}\frac{N_p}{N_s}\frac{C}{10}\int V_x\, dV_c \times 10^7 \qquad \text{erg}$$

If k_v and k_h are the vertical and the horizontal sensitivity of the *C.R.O.* in volt/cm and A the area of the hysteresis loop in sq. cm, then

$$\int V_x\, dV_c = k_h k_v\, A$$

Hence
$$U = \frac{R}{r}\frac{N_p}{N_s}\frac{C}{10}\, k_h\, k_v\, A \times 10^7 \qquad \text{erg}$$

$\therefore$ Loss of energy per unit volume per cycle of hysteresis

$$u = \frac{R}{r}\frac{N_p}{N_s}\frac{A}{al}\, C\, k_h\, k_v \times 10^6 \text{ erg/c.c./cycle} \qquad ...(vii)$$

Experiment 44.1. To find the loss of energy per cycle due to hysteresis for the material of the given metal ring using a C.R.O.

Apparatus. A cathode ray oscilloscope, a metal ring having a primary and a secondary coil (of known number of turns) wound on it, a resistance box, a fractional resistance box, a capacitor of known capacitance, connecting wires etc.

Procedure 1. Draw a diagram showing the scheme of connections as in Fig. 44.2, and make connections accordingly. The resistance box r is a fractional resistance box.

2. Connect the *C.R.O.* to the *A.C.* mains and obtain a fine, bright spot in the centre of the screen as already explained. Turn the '*sweep control*' to '*off*' position by setting it to '*ext. input*' position.

3. Take out a resistance of about 2000 ohm from the resistance box R and about 2 ohm from r. Switch on the *A.C.* mains supply to the transformer. A hysteresis loop will be obtained on the screen. Adjust the vertical and horizontal gains so as to get a loop of the proper size. If necessary, adjust the values of R and r. Trace the loop obtained on a transparent paper.

4. Keeping the position of the vertical and the horizontal gain controls *fixed* find the vertical and horizontal sensitivity for *this* gain position as already explained in **Expt. 42.3**. Take six observations by varying R and r.

Observations.

Horizontal sensitivity					*Vertical sensitivity*				
Voltmeter reading	R_1	R_2	L_h	k_h	*Voltmeter reading*	R_1	R_2	L_v	k_v

Number of turns in the primary N_p =
Number of turns in the secondary N_s =
Capacitance of the capacitor C =

Sl. No.	*Resistance*		*Area of the loop A*	*Loss of energy per cycle in ergs*
	R	*r*		$U = \frac{R}{r}\frac{N_p}{N_s}\frac{C}{10} \times k_v\, k_h A \times 10^7$

Precautions. 1. The number of turns in the secondary should be about 1/10 of the number of turns in the primary so that it acts as a step down transformer in the ratio 10 : 1.

Exercise 1. *Find the loss of energy per unit volume per cycle of hysteresis of the above specimen.*

Hint. Find the length l (or circumference $l = 2\,\pi\,R$ where R is the radius of the ring and the area of cross-section $a = \frac{\pi d^2}{4}$ where d is the diameter of cross-section of the ring. Loss of energy per unit volume per cycle = U/al = erg/c.c./cycle.

Exercise 2. (*i*) *Find the maximum value of magnetic induction B and corresponding values of magnetising force H and permeability from one of the hysteresis loops obtained in Expt.* 44.1.

(*ii*) *Also find the value of remanent magnetic induction and coercive force.*

or

From a B-H hysterisis loop find the value of magnetic parameters.

Hints. A typical *B-H* hysteresis loop is shown in Fig. 44.3 Maximum vertical deflection $(L_v)_{max} = \frac{1}{2} MN$.

$$\therefore \quad B_{\max} = \frac{RC}{N_s a} V_c = \frac{RC}{N_s a} k_v (L_v)_{\max}$$

Corresponding horizontal deflection $(L_h)_{\max} = \frac{1}{2} QP$

$$\therefore \quad H_{\max} = \frac{4\pi}{10}\frac{N_p}{rl} V_x$$

$$= \frac{4\pi}{10}\frac{N_p}{rl} k_h (L_h)_{\max}$$

Permeability $\quad \mu = \frac{B_{\max}}{H_{\max}} =$

Vertical deflection $(L_v)_{rem}$ corresponding to remanent magnetic induction $= \frac{1}{2} ST$.

$$\therefore \quad B_{rem} = \frac{RC}{N_s a} (V_c)_{rem} = \frac{RC}{N_s a} k_v (L_v)_{rem}$$

Horizontal deflection $(L_h)_{cor}$ corresponding to coercive force $= \frac{1}{2}$ EF

$$\therefore \quad H_{cor} = \frac{4\pi}{10}\frac{N_p}{rl} (V_x)_{cor} = \frac{4\pi}{10}\frac{N_p}{rl} k_h (L_h)_{cor}$$

Fig. 44.3

NOTE. 1. In **Art 44.12,** if we use S.I. units, then relation (*iv*) is given by

$$H = \mu_0 \frac{N_p I_p}{l}$$

where I_p is in Amp and l in metres. Similarly relation (*v*) becomes $H = \mu_o \frac{N_p}{rl} V_x$ and relation (*vi*) becomes $\frac{\mu_0}{4\pi} \frac{R}{r} \frac{N_p}{N_s} C \int V_x \, dV_c$ Joule. Finally relation (*vii*) becomes $u = \frac{\mu_0}{4\pi} \frac{R}{r} \frac{N_p}{N_s} \frac{A}{al} C k_h k_v$ where u is in Joules, k_h and k_v are the horizontal and vertical sensitivity of *C.R.O.* in volt/metre, *A* the area of the loop in sq. metre, *a* the area of cross section of the ring in sq. metre and *l* the length (circumference) of the iron ring in metre.

NOTE 2. In the last experiment to draw the hysterisis loop the specimen has been used in the form of an anchor ring. However, it is more convenient to use the material in the form of a rod. We shall describe the alternative method using a rod of ferro-magnetic material in the next experiment.

Experiment 44.2. To trace the B-H curve for a ferromagnetic material using C.R.O and to find the magnetic parameters from the B-H hysterisis loop.

Apparatus. A C.R.O, step down transformer (6-12V, 2-3A), A.C. ammeter (0-3A), a rheostat, 2 resistances (1 Ω each) carbon resistance (47 K Ω), variable carbon resistances (20 K Ω and 5 K Ω) two capacitors (1 μ *f* and 2 μ *f*). A solenoid, a specimen rod.

Description of the apparatus (*i*) *SOLENOID.* The solenoid consists of a primary coil and a secondary coil on the central part of the primary.

(*ii*) *PRIMARY.* The primary coil consists of a non metallic, non-magnetic frame of approximate diameter 2-3 cm and length 30 cm with two or three layers of winding of enamelled copper wire S.W.G 22 on it.

(*iii*) *SECONDARY.* The secondary is wound on a spool of about 10 cm length and 4.5 cm diameter and consists of 10,000 turns of enamelled copper wire SWG 48.

(*iv*) *SPECIMEN ROD.* The specimen rod consists of 40-50 thin iron wires grouped to form a rod. The cycle spokes can serve the purpose. The rod is inserted completely in the solenoid.

(*v*) *INTEGRATING CIRCUIT.* The 47 K Ω resistance in series with a 0.2 μ *f* capacitor where the output is drawn from the capacitor forms an integrating circuit. For principle and theory see **Chapter 43 Art 43.3.**

Theory. The specimen is placed inside the primary of the solenoid which produces a varying magnetic field due to the alternating current passing through it. The value of the varying magnetic field (in C.G.S units) is given by

$$H = \frac{4 \pi n \sqrt{2} \, I_{r.m.s.} \cos \omega t}{10}$$

where n = number of turns per unit length of the primary of the solenoid.

$I_{r.m.s.}$ = *r.m.s* value of current as measured by *a.c.* ammeter. This current also develops an *a.c.* voltage across the resistance $R_1 = 1\ \Omega$, which is applied to the $X_1 - X_2$ plates of the C.R.O. This *a.c.* voltage produces an *X*-deflection on the screen of the C.R.O. As the voltage across R_1 as well as the magnetic field *H* produced by the solenoid are both proportional to $I_{r.m.s.}$ the *X*-deflection is proportional to the varying magnetic field *H*.

The magnetising field *H* in the primary produces a magnetic induction *B* in the specimen. As *H* is a varying magnetic field, *B* also changes at the rate $\frac{dB}{dt}$ and induces an e.m.f. in the secondary proportional to $\frac{dB}{dt}$

The circuit consisting of 47 KΩ resistance connected in series with 2 μ *f* capacitor acts as an integrating circuit, which therefore produces a potential difference across 2 μ *f* capacitor proportional to magnetic induction *B*. This potential difference is applied to the $Y_1 - Y_2$ plates of the C.R.O. due to which there is a *Y*-deflection proportional to *B*.

Thus we find that the *X*-deflection in the C.R.O is proportional to the magnetising field *H* and *Y* deflection to the magnetic induction *B*, which results in producing the hysterisis (*B-H*) loop on the C.R.O screen.

Procedure 1. Draw a diagram showing the scheme of connections as in Fig. 44.4 and make all the connections accordingly.

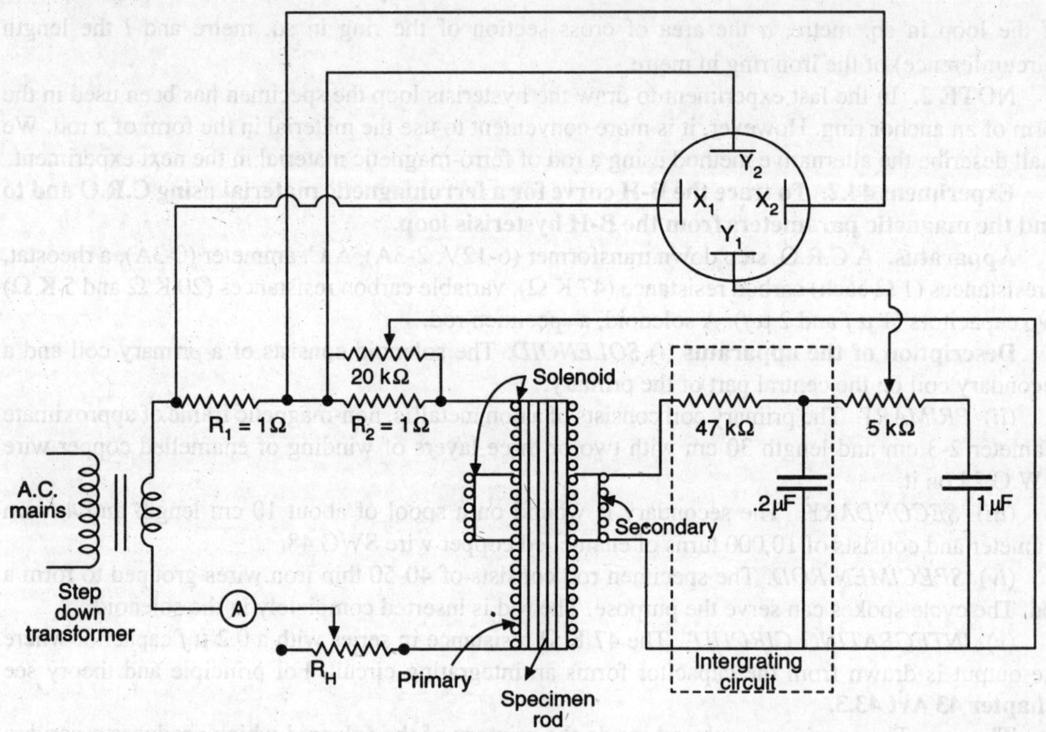

Fig. 44.4

2. Switch on the A.C. mains supply and adjust the value of maximum current in the primary circuit (of the solenoid) with the help of rheostat R_H.

3. Now adjust the *X* and *Y* amplifier, gain so that we get a pattern within the screen. The pattern obtained should be as shown in Fig. 44.5.(*a*).

Usually we do not get this type of pattern.

(*i*) If the pattern obtained is as shown in Fig. 44.5 (*b*), then the secondary leads to the C.R.O should be interchanged.

(*ii*) If the pattern obtained is as shown in Fig. 44.5 (*c*), then 5 K Ω potentiometer should be adjusted in such a way that the loops formed at the tips of the pattern disappear and flatness is restored.

(*iii*) If the pattern obtained is as shown in Fig. 44.5(*d*), then 20 K Ω potentiometer should be adjusted so that the magnetic saturation stage becomes parallel to *H*-axis. If, however, the slope of the ends of the curve increases instead of decreasing then the leads attached to the *Y*-plates should be interchanged.

NOTE. For fine adjustment the values of various resistors and capacitors specified in the figure may be slightly changed, if required.

4. After getting the proper hysterisis pattern on the screen, mark the X-axis (H) and Y-axis (B) as explained.

(i) *To mark H-axis.* Set Y-amplifier gain to zero without disturbing the X-amplifier gain. You will get a horizontal straight line. This is H-axis. Mark this line on the screen. Bring the Y-amplifier gain to its original value.

(i) *To mark B-axis.* Now set X-amplifier gain to zero without disturbing the Y-amplifier gain. You will now get a vertical straight line. This is B-axis. Mark this line also on the screen.

After marking the H-axis and B-axis adjust the X and Y-amplifier again to get proper pattern on the screen. Place a tracing paper on the screen and trace the B-H hysterisis loop (pattern) on it alongwith H-axis and Y-axis.

Note the *r.m.s.* value of current in the solenoid as indicated by the ammeter A.

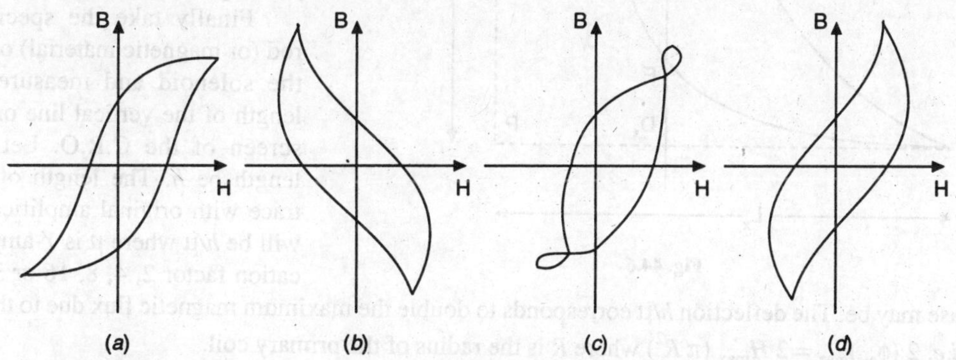

Fig. 44.5

5. **Calibration of H-axis.** The B-H hysterisis loop in shown in Fig. 44.6. Paste the tracing paper on a graph paper. From the tip A draw AF_1 P parallel to Y-axis and from D draw DD_1P parallel to X-axis. Measure the length $DP = L_x$ on the graph paper. This value corresponds to twice the maximum value of magnetising field *i.e.* H_{max}.

$$\therefore \qquad k_x L_x = 2 H_{max}$$

where k_x is the *calibration constant* for H-axis and

$$H_{max} = \frac{4\pi n \sqrt{2} I_{r.m.s}}{10} \text{ Oersted.}$$

∴ Calibration constant for H-axis

$$k_x = \frac{2 H_{max}}{L_x} = \frac{2 \times 4\pi n \sqrt{2} I_{r.m.s}}{10 L_x} \text{ Oersted/cm.}$$

Calibration of B-axis. Reduce X-amplifier gain and resistance of 20 K Ω potentiometer to zero. You will get a vertical straight line (along the Y-axis). Measure the length of the line $AP = A_1D_1 = L_y$. Slowly move the specimen rod out of the solenoid until the length L_y of the vertical line is reduced to $\frac{1}{2} L_y$. Keeping the specimen rod in this position adjust the Y-amplifier gain of the C.R.O to increase the length of the line to its original value L_y. This implies that Y-amplifier gain is now double of its original value.

Now move the specimen rod further out of the solenoid till the length L_y again becomes $\frac{1}{2} L_y$.

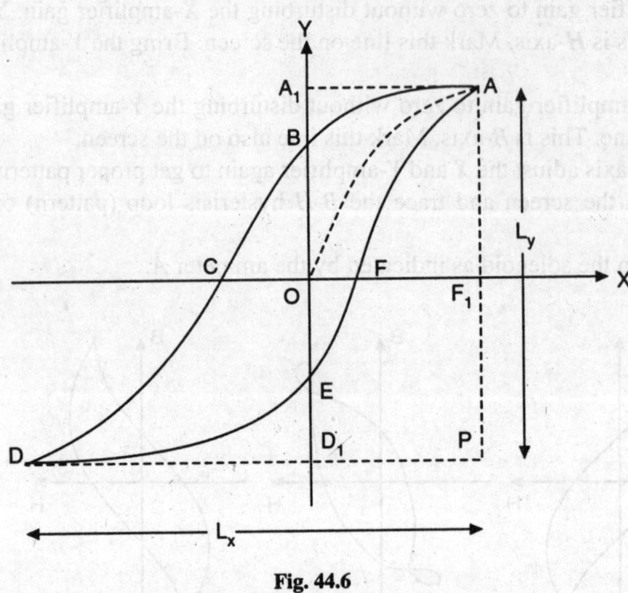

Fig. 44.6

Keeping the specimen rod in this position increase the Y-amplifier gain till the length is once again restored to L_y. This means that the Y-amplifier gain is now four times the original value.

Go on moving the specimen rod out of the solenoid and repeat the above procedure to get Y-amplifier gain 8, 16 or 32 times.

Finally take the specimen rod (or magnetic material) out of the solenoid and measure the length of the vertical line on the screen of the C.R.O. Let this length be h. The length of this trace with original amplification will be h/μ where μ is Y-amplification factor 2, 4, 8, 16 or 32 as the case may be. The deflection h/μ corresponds to double the maximum magnetic flux due to the air core *i.e.* $2\,(\phi_{air})_{max} = 2\,H_{max}\,(\pi R^2)$ where R is the radius of the primary coil.

$\therefore$ Calibration constant for magnetic flux

$$C = \frac{2\,H_{max}\,(\pi R^2)}{h/\mu} \quad \text{Maxwell/cm}$$

If B is the magnetic induction for the specimen rod and A its total area of cross section, then

$$B = \frac{\phi_{max}}{A}$$

$$\text{Hence calibration constant for } B = k_y = \frac{C}{A} = \frac{2\,H_{max}\,\pi R^2}{h/\mu\, A}$$

$$= \frac{2\,H_{max}\,\pi R^2\,\mu}{h\,A} \quad \text{gauss/cm.}$$

Observations. Diameter of the primary of the solenoid $D =$ cm

Radius of the primary of the solenoid $R = D/2 =$

Diameter of one specimen wire $d =$ cm

Radius of one specimen wire $r = \dfrac{d}{2} =$ cm.

Total number of wires (spokes) used $= n_1 =$

$\therefore$ Total area of cors-section of the specimen rod

$$A = \pi r^2 n_1 = \quad \text{cm}^2$$

Number of turns per cm of the primary of the solenoid $= n =$

Y-amplification factor, $\mu =$

Length of the vertical line $h =$ cm

Current in the primary circuit, $I_{r.m.s} =$ Amp.

Calculations

$$H_{max} = \frac{4\pi\, n \sqrt{2}\, I_{r.m.s.}}{10} = \quad \text{Oersted}$$

From B-H loop. Breadth of *B-H* loop, $DP = L_x =$ cm.

∴ Calibration constant for *H*-axis $k_x = \frac{2\, H_{max}}{L_x} = \quad$ Oersted/cm

Height of *BH* loop, $A\, F_1\, P = L_y =$ cm.

∴ Calibration constant for *B*-axis

$$k_y = \frac{2\, H_{max}\, \pi R^2\, \mu}{hA} = \quad \text{gauss/cm.}$$

Saturation value of magnetic induction B_{max}

$$= \frac{1}{2} k_y\, L_y = \quad \text{gauss.}$$

Remanence or remanent induction or retenivity (from *B-H* loop) $= OB$ cm

$= OB \times k_y =$ Gauss

Coercivity or coercive force (from *B-H* loop)

$= OC$ cm

$= OC \times k_x =$ Oersted.

Loss of energy per unit volume due to hysterisis.

6. To find the loss of energy per unit volume per cycle of hysterisis calculate the area of *B-H* loop from the graph paper. Find the area of one small square on the graph. Count the total number of squares on the graph within the *B-H* loop taking each square more than half as one, ignoring each square less than half and taking two half squares equal to one. Then

Hysterisis loss per unit volume per cycle $= \frac{1}{4\pi} \times$ (Area of *B-H* loop)

$=$ ergs/cycle/cm^2.

Precautions. 1. The magnetising current in the primary coil of the solenoid should be large enough to properly magnetise the given material.

2. The *B-H* loop should be stable.

3. The A.C. supply should be switched on only while taking the observation and switched off immediately thereafter.

4. The C.R.O controls should be operated carefully and within specified limits.

Oral Questions

1. Magnetic properties (*i*) What are paramagnetic, ferromagnetic and diamagnetic substances? (*ii*) Define intensity of magnetisation and magnetic induction. (*iii*) What is the relation between magnetic induction and intensity of magnetisation in the C.G.S. system? (*iv*) Define susceptibility? (*v*) What is volume susceptibility and mass susceptibility? (*vi*) Define permeability. (*vii*) What is the relation between permeabilty and susceptibility in C.G.S. system.

2. Hysteresis loop. (*i*) What is hysteresis? (*ii*) What is the relation between the area of a *B-H* loop and the loss of energy per cycle per unit volume of the material? (*iii*) Out of soft iron and hard steel, which has a greater hysteresis loss? (*iv*) Define magnetic flux and magnetic induction field. State the relation between the two and give their C.G.S. and S.I. units (*v*) Define retentivity and coercivity. How do you find their values from a *B–H* loop?

PART VIII

SOLID STATE PHYSICS AND ELECTRONICS

45

Thermionic Emission

45.1. Thermionic emission. The phenomenon of emission of electrons from the surfaces of certain materials, when they are heated up suitably, is called thermionic emission. The electrons so emitted are called **thermions.** If these electrons are made to flow in closed circuit and cause an electric current to flow in the circuit, the current is known as **thermionic current.**

The thermionic electrons are emitted from the outermost orbit of the atoms of the material. As these electrons are far away from the nucleus, they are loosely bound and are called, free electrons. When the material is heated, the electrons get thermal energy from the external source and acquire kinetic energy, which is sufficient to overcome the binding forces and get detached from the atom as thermions.

The phenomenon of emission of electrons from the surface of a metal when thermal energy is supplied to it is called thermionic emission.

45.2. Thermionic work function. *The minimum amount of thermal energy that must be supplied to the metal to eject an electron from it is called thermionic work function.*

It is denoted by ϕ and depends upon the nature of the material, condition of the surface and impurities.

Thermionic emitters. The thermionic emitters generally used in vacuum tubes are of three types:

(*i*) *Pure tungsten emitter.* Pure tungsten is an ideal material to be used as emitter in high vacuum tubes. It has a high melting point (3655 K), a low vapour pressure and a work function 4.52 eV. It can, therefore, be operated at a high temperature and yields an appreciable emission current. Its emission efficiency is 8.5 mA per watt at operating temperature of 2600 K. The efficiency, however, decreases with absorption of gases. This is why it can be used only in high vacuum tubes. The surface of a tungsten emitter is not damaged by bombardment of positive ions.

(*ii*) *Thoriated tungsten emitter.* The thoriated tungsten emitter consists of tungsten containing 1 to 2% of thorium oxide and carbon which acts as a reducing agent. A thin layer of thorium is thus formed on the surface of tungsten. This layer being electro-positive with respect to tungsten, lowers the work function of the material. It, therefore, gives a copius supply of electrons at about 1900 K and has an emission efficiency of 100 mA per watt. Thoriated tungsten emitters are likely to be damaged due to bombardment by positive ions and can, therefore, be used only under conditions of high vacuum and low anode voltage.

(*iii*) *Oxide coated emitter.* An oxide coated emitter consists of a nickel or nickel alloy cylinder the outer surface of which is coated with a mixture of barium and strontium oxide. The cylinder is indirectly heated by placing a heating filament inside it and well insulated from it. It gives a copius supply of electrons at about 1000 K and its emission efficiency ranges from 100 to 10,000 mA per watt.

The oxide coated emitters are now widely used. They, however, have a tendency to lose their activity due to bombardment by positive ions and are, therefore, used in tubes with low anode voltage. Moreover, these emitters give out some gases during operation. These cannot, therefore, be used in high vacuum tubes.

Some typical values of thermionic work function are

Material	*Work function in (eV)*	*Material*	*Work function in (eV)*
Platinum	5.32	Molybdenum	4.30
Nickel	5.00	Thorium	3.40
Carbon	4.70	Cesium	1.18
Tungsten	4.52	Oxide coated metals	1.00

45.3. Richardson Dushman equation. The amount of thermionic emission increases rapidly with temperature of the emitter. The emission current density is given by Richardson Dushman equation *i.e.*

$$J = AT^2 \exp\left[-(\phi - E_0)/kT\right] \qquad ...(i)$$

where J is the emission current density *i.e.* the electronic charge per second per unit area of the emitter or the current per unit area, A a constant depending upon the nature of the material of the emitter.

According to Fermi Dirac statistics

$$A = 4\pi \frac{mek^2}{h^3} \qquad ...(ii)$$

where m is the mass of the electron $(9.1 \times 10^{-31}$ kg) e the charge on it $(1.6 \times 10^{-19}$ C), k Boltzmann's constant $(1.38 \times 10^{-23}$ J/degree) and h Planck's constant $(6.62 \times 10^{-34}$ Joule sec)

Substituting the above values in relation (*ii*), we get

$$A = 12.04 \times 10^5 \text{ Am}^{-2} K^{-2}$$

ϕ is the thermionic work function of the metal and E_0 the zero point energy *i.e.* the maximum energy possesed by an electron at zero K.

On the basis of quantum theory, the value of

$$E_0 = \frac{h^2}{2m}\left(\frac{3n}{8\pi}\right)^{2/3}$$

where n is the number of electrons per unit volume. The classical value of E_0 is zero and even the quantum mechanical value of E_0 is very small being of the order of 10^{-10} and may be neglected as compared to ϕ particularly at high temperatures.

$$\therefore \quad J = AT^2 \exp\left[-\phi/kT\right] \qquad ...(iii)$$

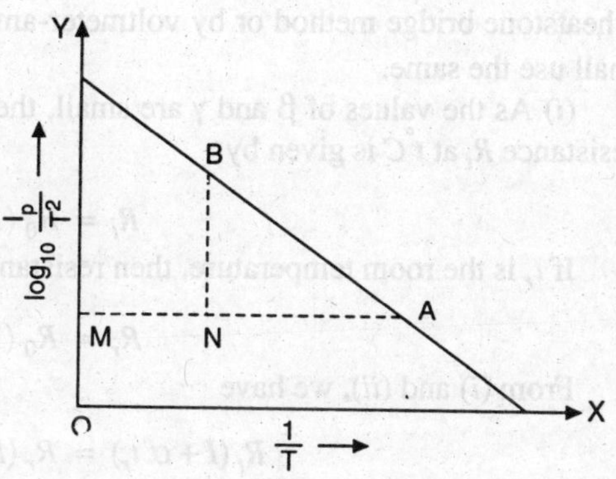

Fig. 45.1

45.4. To find the work function using a directly heated diode. The plate (or anode) current I_p and plate (or anode) potential E_p characteristics of a diode for a constant filament current shows that the plate current attains a constant (saturation) value for a certain value of plate potential. At this stage all the thermoelectrons emitted by the filament reach the plate. If a is the is the area of cross section of the filament surface, then saturation value of the plate current is given by

$$I_p = A\,a\,T^2 \exp(-\phi/kT)$$
$$= A_0\,T^2 \exp(-\phi/kT)$$

or
$$\frac{I_p}{T^2} = A_0 \exp(-\phi/kT)$$

Taking logarithms on both sides, we have

$$\log_e \frac{I_p}{T^2} = -\frac{\phi}{k}\frac{1}{T} + \log_e A_0$$

or
$$2.30\,3 \log_{10} \frac{I_p}{T^2} = -\frac{\phi}{k}\frac{1}{T} + 2.303 \log_{10} A_0$$

or
$$\log_{10} \frac{I_p}{T^2} = -\frac{1}{2.303}\frac{\phi}{k}\frac{1}{T} + \log_{10} A_0$$

This is the equation of a straight line of the type $y = mx + c$. Hence a graph between $\log_{10} \frac{I_p}{T^2}$ taken along the Y-axis and $\frac{1}{T}$ taken along the X-axis gives a straight line as shown in Fig. 45.1 having a negative slope given by $\frac{\phi}{2.303\,k}$. The intercept on the Y-axis gives the value of A_0 from which, if necessary the area of cross-section of the diode filament (a) can be calculated as $A_0 = aA$ and $A = 12.04 \times 10^5\ A\,m^{-2}\,K^{-2}$.

45.5. Determination of temperature of filament. The temperature of the glowing filament can be determined by one of the following methods.

(a) Pyrometric method *(b) Resistance method*

Resistance method. The variation of resistance of pure metals with temperature is given by the relation

$$R_t = R_0 (1 + \alpha t + \beta t^2 + \gamma t^3)$$

where R_0, R_t are the resistances at 0°C and t°C, the constant $\alpha = 5.24 \times 10^{-3}\ C^{-1}$, $\beta = 7.00 \times 10^{-7}\ C^{-2}$ and $\gamma = 6.00 \times 10^{-11}\ C^{-3}$ for tungsten. The value of R_t and R_0 can either be determined by wheatstone bridge method or by voltmeter-ammeter method. The latter is more convenient and we shall use the same.

(i) As the values of β and γ are small, these may be neglected. Using the simpler relation, the resistance R_t at t°C is given by

$$R_t = R_0 (1 + \alpha t) \quad ...(i)$$

If t_r is the room temperature, then resistance at room temperature

$$R_r = R_0 (1 + \alpha t_r) \quad ...(ii)$$

From *(i)* and *(ii)*, we have

$$R_t (I + \alpha t_r) = R_r (I + \alpha t)$$

as in this experiment t is very large, $(1 + \alpha t)$ is very nearly equal to αt

$$\therefore \quad R_t (1 + \alpha t_r) = R_r \alpha t$$

or
$$t = \frac{R_t (1 + \alpha t_r)}{R_r \alpha}$$

(*ii*) A more accurate method in which β and γ are not neglected, is explained below : -

Suppose R_g is the resistance of the filament when it is *just* glowing. If V_g is the potential difference across it, as measured by the voltmeter V_F and I_g the current flowing through it, as measured by the ammeter A_F, then

$$R_g = \frac{V_g}{I_g} \text{ ohm}$$

If β and γ are not neglected, an empirical relation betweeen R_g and R_0 (R_{273}), the resistance at 0°C (273 K) is given by

$$\frac{R_g}{R_0} = f$$

The value of *f* varies from 3.9 to 5, depending upon the nature of material of the filament. Taking $f = 5$ in this experiment, the value of R_{273} can be calculated. If R_T is the resistance at any higher temperature T K, then the ratio $\frac{R_T}{R_{273}}$ depends upon the temperature of the filament and is as given below, in a tabular form, for temperatures on and above 1273 K.

Table I

T in degrees K	$\frac{R_T}{R_{273}}$	*T in degrees K*	$\frac{R_T}{R_{273}}$
1273	7.00	2073	13.05
1373	7.69	2173	13.89
1473	8.40	2273	14.76
1573	9.13	2373	15.64
1673	9.87	2473	16.55
1773	10.64	2573	17.48
1873	11.42	2673	18.43
1973	12.23	2773	19.41

From a graph between T and $\frac{R_T}{R_{273}}$, the temperature of the filament can be calculated for different values of filament current.

Experiment 45.1. To determine the thermionic work-function of tungsten, using a directly heated diode.

Apparatus. A directly heated type diode valve, a *D.C.* power supply, an ammeter, a milliammeter, two voltmeters (all meters of suitable range according to diode data supplied), a rheostat, two one way keys, a battery of 4 cells, connecting wires, etc.

Procedure 1. Draw a diagram showing the scheme of connections, as in Fig. 45.2. and make connections accordingly.

2. Put-in the key K_1 to switch-on the filament current, keeping the plate (or anode) circuit off. Adjust the position of the variable contact of the rheostat R_h so that the voltmeter V_F in the filament circuit reads almost zero. Now increase the current in the filament circuit (as indicated by the ammeter A_F) *till the filament FF* just glows. Note the reading of the filament current I_g and filament voltage V_g at glow and calculate the filament resistance R_g.

$$R_g = \frac{V_g}{I_g}$$

Then taking $\frac{R_g}{R_0} = 5$, we have

$$R_0 \text{ (or } R_{273}) = \frac{R_g}{5}$$

Repeat three times and find the mean value of R_{273}.

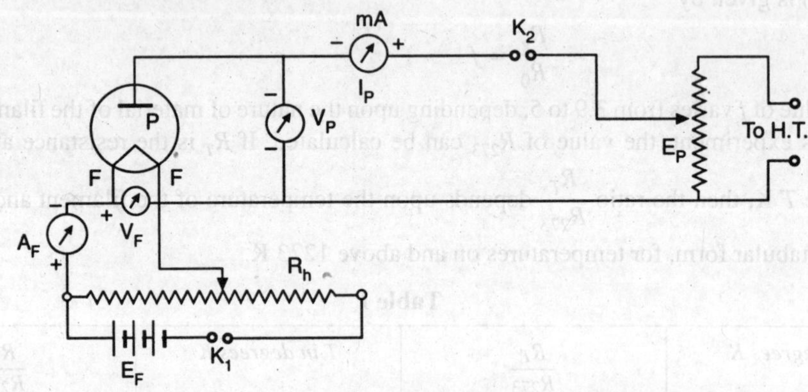

Fig. 45.2

3. Now put in the key K_2 to switch on the current in the plate circuit. Adjust the current in the filament circuit so as to obtain a suitable measurable current in the plate circuit at minimum plate voltage. Note the value of filament current I_T and filament voltage V_T corresponding to the temperature T of the filament (in degrees K). Calculate $R_T = V_T/I_T$. Repeat three times keeping filament current constant.

4. Plot a graph between $\frac{R_T}{R_{273}}$ and T from the data given in Table 1 taking $\frac{R_T}{R_{273}}$ along the Y-axis and T along the X-axis. From the graph find the value of temperature of the filament (in degrees K) corresponding to the calculated value of R_T in the last step.

5. Keeping the filament current constant, go on increasing the value of plate voltage V_p in small steps and note the values of V_p and corresponding values of plate current I_p till the saturation stage is reached *i.e.* there is no increase in plate current with increase in plate voltage. Note the maximum or saturation value of the current I_p (*sat*).

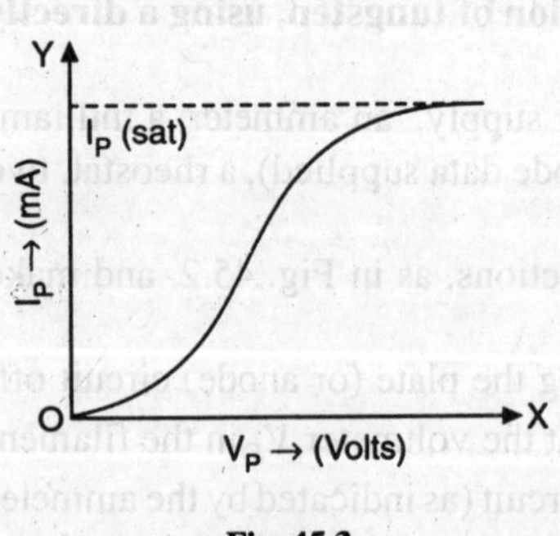

Fig. 45.3

6. Also plot a graph between V_p and I_p taking I_p along the Y-axis and V_p along the X-axis. The graph is of the type shown in Fig. 45.3. Also find I_p (*sat*) from the graph.

7. Repeat steps 3, 4, 5 and 6 at least 8 times by increasing the values of filament current *i.e* at increasing filament temperatures and find the corresponding values of saturation plate current I_p (*sat*) directly.

Observations. To find the value of R_0 (R_{273}), given $f = 5$

Sl. No.	Filament voltage at glow V_g Volts	Filament current at glow I_g Amps.	R_g Ohms	$R_0 = \frac{R_g}{5}$ ($f = 5$) or R_{273} Ohms
1				
2				
3				

Mean R_0 = ohm

To find the temperature in degrees Kelvin.

Sl. No.	Filament resistance at T K (R_T)			$\frac{R_T}{R_{273}}$	T in deg. K	1/T	T^2	I_p (sat) (Direct)	$\frac{I_p (sat)}{T^2}$	$\log_{10} \frac{I_p(sat)}{T^2}$
	V_T Volts	I_T Amps	$R_T = \frac{V_T}{I_T}$ Ohms							
1										
2										
3										
4										
5										
6										
7										
8										

To find the saturation plate current I_p *(sat)* from the graph.

Sl. No.	Plate Current I_p	Plate Voltage (V_p)	Sl. No.	Plate Current (I_p)	Plate Voltage (V_p)
1			5		
2			6		
3			7		
4			8		

I_p (sat) from the graph = mA = amp

Repeat the table for all values of filament current

8. Taking $\frac{1}{T}$ along the *X*-axis and $\log_{10} \frac{I_p}{T^2}$ *(sat)* along the *Y*-axis, plot a graph. The graph is a straight line as shown in Fig. 45.1. and has a negative slope. To find the magnitude of the slope take two points *A* and *B* on the graph. From *A* draw *A M* perpendicular to *Y*-axis and from *B* draw perpendicular on *X*-axis meeting *A M* at *N*.

Measure *B N* and *A N*, then

$$\text{Magnitude of the slope} = \frac{BN}{AN} =$$

Calculations.

Boltzmann's constant $k = 1.38 \times 10^{-23}$ J/degree

$$\frac{\phi}{2.303k} = \frac{BN}{AN}$$

$$\therefore \quad \phi = 2.303\,k\,\frac{BN}{AN}$$

$$= \text{Joules.}$$

Precautions. 1. The values of voltage and current must be noted when the filament just glows and no plate voltage is applied.

2. Suitable value of $f = R_g / R_0$ must be taken for the material of the filament of the directly heated diode.

3. Filament current must be kept constant for each set of observations.

4. The value of saturation current must be determined directly as well as from the graph.

5. Plate voltage must be increased in small steps.

Oral Questions

1. Thermionic emission (*i*) What is thermionic emission? (*ii*) Define thermionic work function. (*iii*) In what units is it generally expressed? (*iv*) State Richardson Dushman equation. (*v*) Distiguish between emission current density and emission current. (*vi*) What is the value of constant *A* in Richardson Dushman equation according to Fermi-Dirac statistics? (*viii*) The value of E_0 in the above equation has a small value on the basis of quantum theory. What is the value on the basis of classical theory? (*viii*) How do you find the resistance of the filament of the diode at 0 °C, the resistance when it just glows is known? (*ix*) How do you find the temperature of the filament of the valve? (*x*) What is the value of thermionic work function ϕ for tungsten ?

46

Thermionic Valves

46.1. Diode valve. It is the simplest type of a thermionic valve. It is a two electrode tube containing an emitting electrode called the *cathode K* and a current collecting electrode called the *anode* or *plate P*. The emitting electrode may be directly or indirectly heated oxide coated cathode. In a directly heated type shown in Fig. 46.1 (*a*) the cathode is an oxide coated filament which becomes the source of electrons when current is passed through it. The diode commonly in use are indirectly heated type as shown in Fig. 46.1. (*b*). The cathode *K* is coated with barium or strontium oxides and lies very close to the filament *FF* which is heated by a low tension battery or 6.3 volt *A.C.*

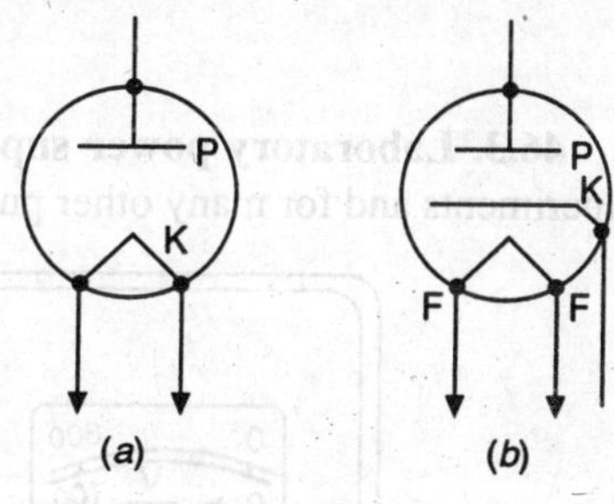

Fig. 46.1

The electrons emitted by the cathode *K* collect in its immediate vicinity. This gives rise to a negative *space charge* and repels more electrons back into the filament. If the anode or plate is at a positive potential the electrons are attracted towards it. An electronic current thus flows in the plate circuit. The value of the plate current depends upon the plate potential. As the plate potential is raised the plate current increases, initially rapidly, then slowly and finally acquires a constant saturation value. The value of saturation current, however, depends upon the filament current *i.e.* the temperature of the cathode.

When the plate is negative with respect to the cathode, the electrons are not attracted by it and hence no current flows in the plate circuit.

A diode is, therefore, a perfect rectifier and is used for rectification of alternating current. A single diode brings about only a half wave rectification and thus utilises only half of the alternating voltage, the other half going waste. Full wave rectification in which both sides of the alternating voltage wave are obtained after rectification is brought about by using two single diodes or one double diode. A double diode has two plates and one filament in the same enclosure and its action is equivalent to two diodes.

46.2. Rectifier. The circuit diagram of a simple laboratory rectifier used for converting *A.C.* supply into *D.C.* supply for valve experiments is shown in Fig. 46.2.

The mains supply is fed to the rectifier through the primary of the transformer *T*. The connections for 250, 230 and 210 volt supply are also indicated. The secondary contains two windings S_1 and S_2. The winding S_1 supplies the low tension for heating the filament and the winding S_2 supplies the high tension to be applied to the plates of the double diode. When the plate P_1 is positive with respect to the filament, electrons are attracted towards it and a conventional current flows from P_1 to *K* and from *A* to *B* through the load resistance *R*. After half a cycle when the plate P_2 is positive electrons are attracted by it and conventional current again flows from A to B in the resistance R. Thus in both halves of the *A.C.* cycle, current through R flows in the same direction *i.e.* from A to B. In other words A is the **positive** and B the **negative**. The capacitors C_1, C_2 and the inductor L form a π–type filter circuit which smoothens the rectified *A.C.* into a constant direct current. When a variable *D.C.* is required one contact is made with the negative end B and the other with sliding contact *S.C.*.

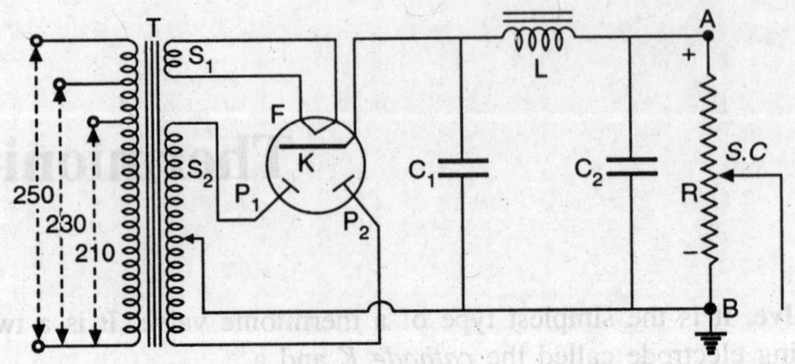

Fig. 46.2

46.3. Laboratory power supplies. Commercially built power supplies, for use in electronic experiments and for many other purposes, are now available.

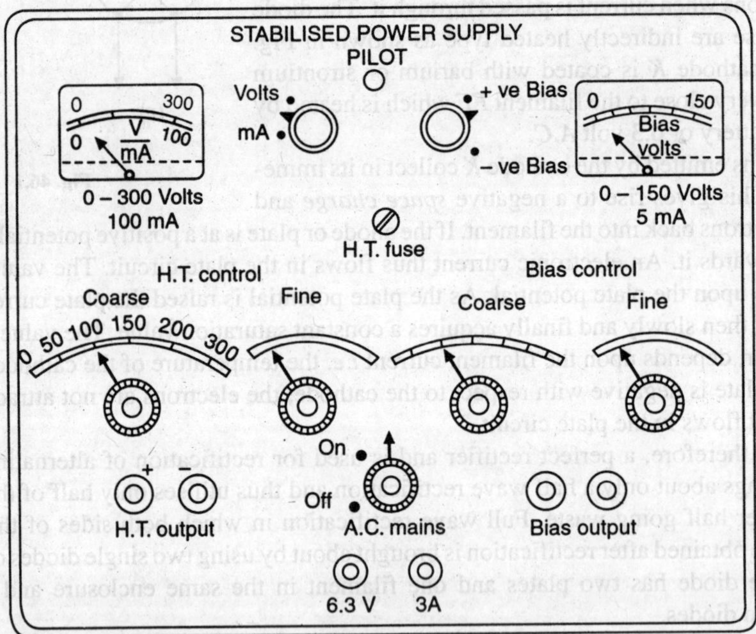

Fig. 46.3

Simple power supply. A simple form of *D.C.* power supply consists of a full wave rectifier circuit with suitable filters. The input is 220-250 volts *A.C.* mains supply. The output can be obtained from three pairs of terminals. The first pair supplies *D.C.* voltage continuously variable from 0-300 volts for H.T. supply, to the plate circuit of a valve. The variation in voltage is brought about by a built-in potential dividing arrangement and is measured by a high resistance voltmeter, fixed on the panel. The corresponding current is also measured by a suitable milli-ammeter.

The second pair of terminals, marked *G. B* (or Bias input) supplies *D.C.* voltage, which can be varied form + 30 V to – 30 V by another potential dividing arrangement. This is also measured by a voltmeter of suitable range, fixed on the panel.

The third pair of terminals supplies 6.3. volt *A.C.*, which is used for heating the filament of the valve.

Note. As most of the college laboratories have only simple power supply, as described above, the experiments given in this chapter are designed for work on such a power supply. However, it will be advisable to use *regulated* power supply, as described below.

Regulated power supply. The *D.C.* output obtained in a simple power supply is *unregulated* and is suitable only for certain limited purposes. The output voltage is kept *constant* by electronic regulating circuits. The ability of a power supply to maintain a constant voltage against load variation is indicated on it.

There are two types of *regulated* power supplies generally used in the laboratories.

(*i*) **Tube power supply.** It provides a variable *D.C.* voltage 0 to 300 volt at current up to 100 mA and is used to supply *D.C.* voltages to the plate and grid circuits of vacuum tubes. It is highly stabilised against load and voltage variations and gives a continuously variable, ripple free output between 0-300 volt upto a maximum load current of 100 m A.

The front panel of such a power supply is shown in Fig. 46.3. The *D.C.* output terminals are provided on the panel and are labelled 'H.T. OUTPUT'. The meter V/mA reads the H.T. output voltage as well as current in mA. For this purpose a separate double switch is provided which can be adjusted to 'VOLTS' or 'mA' for measurement of voltage or current. The H.T. voltage can be adjusted to the desired value with the help of two regulating knobs labelled 'H.T. CONTROL'. The knob marked 'COARSE' enables the H.T. voltage to be adjusted in steps of 50 volt and the knob marked 'FINE' enables a fine adjustment of voltage for each step.

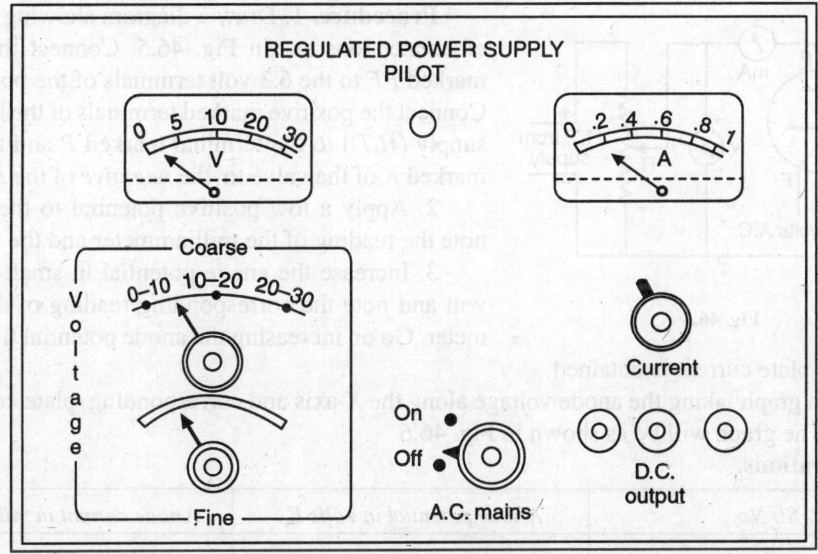

Fig. 46.4

A *D.C.* bias voltage (to be applied to the grid of a vacuum tube 0-150 volt at a maximum current of 5 mA can be obtained from the terminals marked 'BIAS OUTPUT' and its value can be adjusted with the help of two knob controls labelled 'BIAS CONTROL'. The knob marked 'COARSE' enables to select the suitable range and the knob marked 'FINE' enables a fine adjustment of voltage for each step. The bias voltage is given by the meter marked 'BIAS VOLTS'. A double switch enables the application of a 'POSITIVE' as well as a 'NEGATIVE' bias.

The A.C. 6.3 volt supply at a maximum current of 3 amp. for heating the filament of the vacuum tube can be obtained from the terminals marked 6.3 V/3 A.

To operate the power supply, the H.T. knob labelled 'COARSE' is set to the extreme *left* position. (In this position H.T. is 'OFF') and the A.C. mains is switched on. The H.T. is switched on after a short time during which the circuit warms up and the 'COARSE' knob is adjusted to a suitable step. Exact required value of H.T. is obtained by adjusting the 'FINE' knob. *To turn off the power supply the H.T. is first switched off and then the A.C. mains.*

(*ii*) **Transistor power supply.** It provides a variable *D.C.* voltage 0–30 volt in steps of 0–10, 10–20, 20–30 volt at currents up to 500 mA/1A and is used to supply *D.C.* voltage to solid state devices and transistor circuits. The front panel of such a regulated power supply is shown in Fig. 46.4. The D.C. output is obtained from the terminals labelled '*D.C.* OUTPUT'. The voltage is adjusted in steps of 10 volt by the knob marked 'COARSE'. Fine adjustment is made with the knob marked 'FINE'. The voltmeter *V* reads the voltage applied and the ammeter *A* reads the load current.

Experiment 46.1. To draw the anode current anode voltage characteristic curve of a diode and hence to find the internal resistance of the valve.

Apparatus. A diode valve (6H6) mounted on a panel, a power supply capable of giving high tension voltage with steps of 10 volt and a tapping of 6.3 volt *A.C.* for heating the filament.

Theory. The internal resistance of a diode valve is defined as the ratio of a small change in anode potential to the corresponding change in the anode current. If δE_p is the change in plate potential and δI_p, the corresponding change in plate current, then

$$\text{Internal (or slope) resistance } R_p = \frac{\delta E_p}{\delta I_p}$$

This resistance is not constant and must be measured at a specified mean anode potential *i.e.*, in the straight part of the curve as shown in Fig. 46.6.

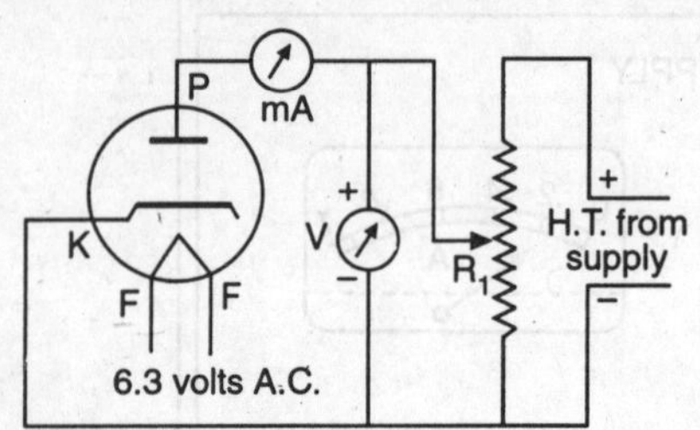

Fig. 46.5

Procedure. 1. Draw a diagram showing the scheme of connections as in Fig. 46.5. Connect the terminals marked *FF* to the 6.3 volt terminals of the power supply. Connect the positive marked terminals of the high tension supply (*H.T.*) to the terminal marked *P* and the terminal marked *K* of the valve to the *negative* of the *H.T.* supply.

2. Apply a low positive potential to the anode and note the reading of the milliammeter and the voltmeter.

3. Increase the anode potential in small steps of 10 volt and note the corresponding reading of the milliammeter. Go on increasing the anode potential till saturation value of the plate current is obtained.

4. Plot a graph taking the anode voltage along the *X*-axis and corresponding plate current along the *Y*-axis. The graph will be as shown in Fig. 46.6.

Observations.

Sl. No.	*Anode potential in volts E_p*	*Anode current in milliamps I_p*
1		
2		
3		
...		
...		
...		

Calculations. Select a point P in the straight portion AB of the curve and draw a tangent QR to the curve at P. From Q and R draw straight lines parallel to Y-axis and X-axis respectively. Find the values of RS and QS according to the scales selected for anode voltage and anode current. From graph, we have

Current increase $QS = \delta I_p$

$=$ milli-amp $=$ amp

Voltage increase $RS = \delta E_p =$ volt

$$\therefore \text{ Resistance } R = \frac{\delta E_p}{\delta I_p} = \frac{RS}{QS} = \text{ ohm}$$

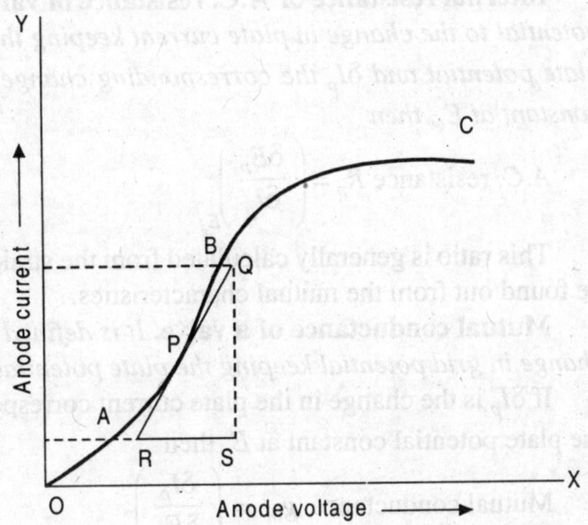

Fig. 46.6

Precautions. 1. Zero correction if any should be applied to the milliammeter reading as also to the voltmeter reading.

2. The positive marked terminal of $H.T.$ should be connected to the plate terminal P.

3. Straight portion of the curve should be selected to find the resistance of the valve.

46.4. Triode valve. In a triode valve in addition to the plate and cathode a third element called the **grid** G is provided as shown. The grid consists of a wire gauze cylinder surrounding the cathode and surrounding it is the cylindrical plate or anode. The grid is nearer to the cathode than the plate.

Normally the plate of a triode is connected to a high positive voltage and grid is kept at a negative potential with respect to the cathode K, so that negatively charged electrons emitted by the cathode are repelled by the grid. The grid being closer to the cathode than the plate, has a greater effect in controlling the tube current than does the positive plate. If the grid becomes less negative, more electrons will get through to the plate and if the grid becomes more negative, fewer electrons will get through to the plate. This is what happens when the bias voltage is varied. If the grid is made positive with respect to the cathode, still more current flows. A point is reached when the grid is so positive that the electrons flow from the cathode as fast as the cathode can emit them. This is the saturation point and a still further increase in grid voltage will not cause further increase in the plate current.

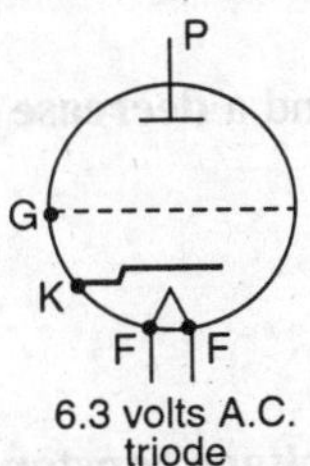

Fig. 46.7

Base connection of a triode. The valve fits on the base in a definite position. The various holes into which the leg terminals fit are marked 1, 2 .....8 with respect to the reference mark as shown in Fig. 46.8. The terminals No. 2 and 7 pertain to the filament, terminal No. 8 is the cathode, terminal No. 3 the anode or plate and the terminal No. 5 is connected to the grid.

46.5. Triode characteristic. The current in the plate circuit of a triode varies with the variations in the plate potential as well as in the grid potential. A graph between plate current and plate potential at a constant grid potential is known as the **plate characteristic** or **the anode characteristic** of the valve.

A graph between the plate current and grid potential at constant plate potential is called the **grid characteristic** or **mutual characteristic** of the valve.

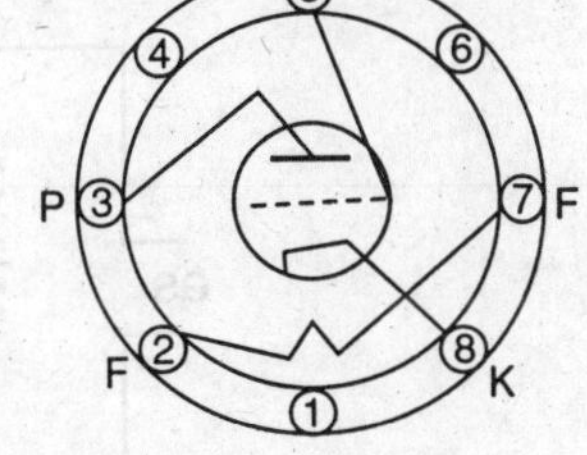

Fig. 46.8

46.6. Valve parameters. The current in the plate circuit can be increased

(*i*) By keeping the grid potential constant and by increasing the plate potential.

(*ii*) By keeping the plate potential constant and by increasing the grid potential.

Internal resistance or A.C. resistance of valve. *It is defined as the ratio of the change in plate potential to the change in plate current keeping the grid potential constant. If δE_p is the change in plate potential and δI_p the corresponding change in plate current when the grid potential is kept constant at E_g, then*

$$\text{A.C. resistance } R_p = \left(\frac{\delta E_p}{\delta I_p}\right)_{E_g}$$

This ratio is generally calculated from the straight part of the anode characteristics, but may also be found out from the mutual characteristics.

Mutual conductance of a valve. *It is defined as the ratio of the change in plate current to the change in grid potential keeping the plate potential constant.*

If δI_p is the change in the plate current corresponding to a change in grid potential δE_g, keeping the plate potential constant at E_p then

$$\text{Mutual conductance } g_m = \left(\frac{\delta I_p}{\delta E_g}\right)_{E_p}$$

Amplification factor. *It is defined as the ratio of the change in plate potential keeping grid potential constant to the change in grid potential keeping plate potential constant, required to bring about the same change in plate current.*

If δE_p is the change in the plate potential and δE_g *the change in grid potential required to bring about the same change in plate current, then*

$$\text{Amplification factor } \mu = \left(\frac{\delta E_p}{\delta E_g}\right)_{I_p}$$

In other words it means that an **increase** in the plate potential equal to δE_p and a **decrease** in grid potential equal to δE_g will keep the plate current constant at the value I_p.

The three constants or the co-efficients of a triode are related by the equation

$$\mu = R_p \times g_m$$

Experiment 46.2. To study for a triode

(*a*) Variation of plate current with change in plate voltage keeping grid voltage constant.

(*b*) Variation of plate current with change in grid voltage keeping plate voltage constant.

(*c*) Variation of plate voltage with change in grid voltage keeping plate current constant.

Apparatus. Triode valve 6 J 5 with base connections, a power supply with *HT* terminals, *GB* terminals and 6.3 volts A.C. terminals for heating the filament, pieces of flexible wire etc. (The power supply is fitted with milliammeter and high resistance voltmeters to measure *HT* as also *GB*.)

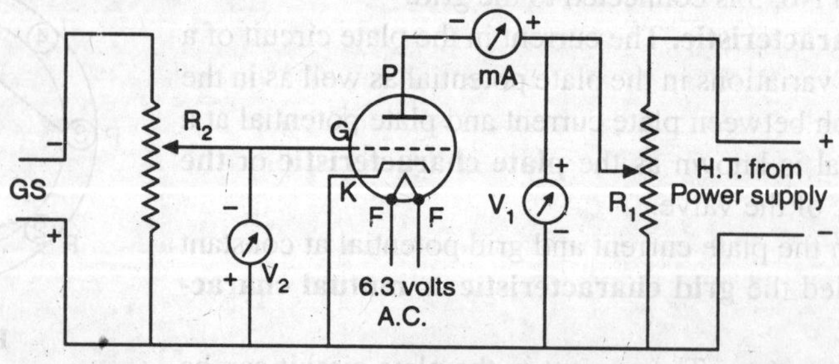

Fig. 46.9

Procedure. 1. Draw the scheme of connections and make connections circuit by circuit as shown in Fig. 46.9.

(*i*) Connect the filament terminals marked *FF* to the 6.3 volt terminals of the power supply.

(*ii*) Connect the positive marked terminal of the *HT* to the plate *P* and the negative terminal to the cathode *K*.

(*iii*) Connect the negative terminal of *GB* to the grid terminal *G* of the valve and positive terminal to the cathode *K*

(*a*) To study variation of plate current with change in plate voltage keeping grid voltage constant.

2. Adjust the grid voltage to –2volt and keeping it *constant* adjust the plate voltage to 50 volt. Note the reading of the voltmeter as well as the current in the milliammeter in the plate circuit. Increase the plate voltage in steps of 20 volt keeping the grid voltage constant and note the reading of plate potential as well as the plate current. Take at least 8 observations.

3. Repeat the experiment with two more constant values of grid potential say – 4 volt and –6 volt.

4. **Anode characteristics.** A graph between plate current I_p and plate potential E_p at constant grid potential is known as *Anode characteristics.*.

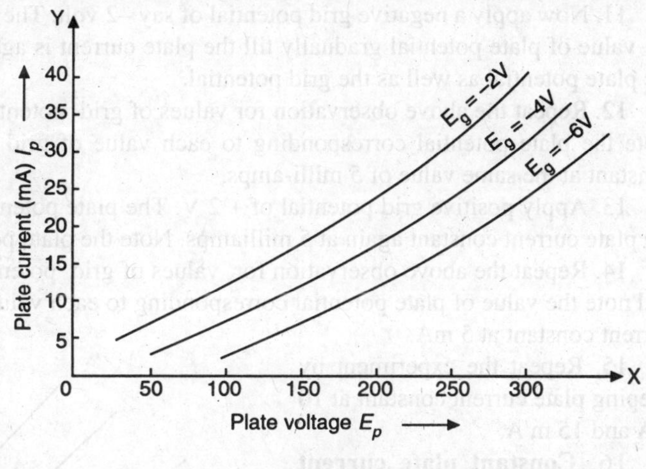

Fig. 46.10

To plot the graph take the values of E_p along the *X*-axis and corresponding values of I_p along the *Y*-axis. The anode characteristics for E_g = –2V, –4V, and –6V (kept constant) are shown in Fig. 46.10.

(*b*) To study variation of plate current with change in grid voltage keeping plate voltage constant.

5. Apply a positive potential of about 80 volt to the plate *P* and apply about –10 volts to the grid by adjusting the grid voltage from the *GB* tapping. Adjust the plate voltage to exactly 80 volts and note the current indicated by the milliammeter in the plate circuit.

6. Decrease the grid voltage in steps of 1.5 to 2.0 volt keeping the plate voltage constant at 80 volt, and note the corresponding reading of the milliammeter. Repeat observations till zero *GB* voltage is reached.

7. Now connect the positive of the grid bias to the grid and negative to the cathode. Increase the grid potential in steps of 2 volt and note the milliammeter reading keeping the plate voltage constant at 80 volt till saturation value of current is indicated by the milliammeter.

8. Repeat the above procedure for a plate voltage of 100 volt and 120 volt.

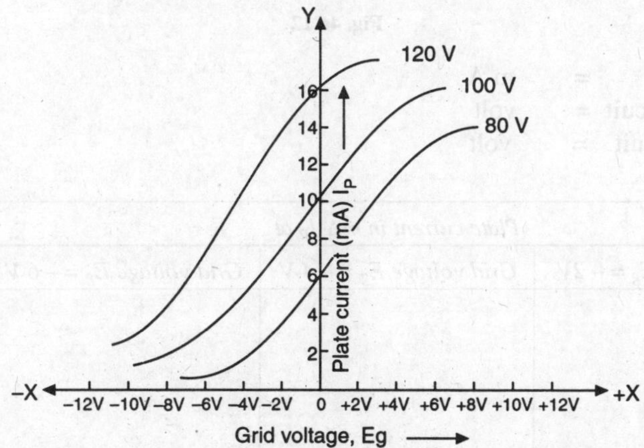

Fig. 46.11

9. **Mutual characteristics.** A graph between plate current I_p and grid potential E_g at constant plate potential is known as *mutual characteristics*.

To plot the graph take values of $+ E_g$ along positive *X*-axis and $-E_g$ along negative *X*-axis and corresponding values of I_p along *Y*-axis for $E_p = +80$ V, + 100 V and + 120 V (kept constant) as shown in Fig. 46.11.

(*c*) **To study variation of plate voltage with change in grid voltage keeping plate current constant**

10. Make the grid potential zero and adjust the plate potential so that we get a certain value of plate current say 5 milli-amps.

11. Now apply a negative grid potential of say –2 volt. The plate current will decrease. Increase the value of plate potential gradually till the plate current is again the same *i.e.* 5 milli-amps. Note the plate potential as well as the grid potential.

12. Repeat the above observation for values of grid potential of –4V, - 6 V, – 8 V and so on. Note the plate potential corresponding to each value of grid potential keeping the plate current constant at the same value of 5 milli-amps.

13. Apply positive grid potential of + 2 V. The plate potential will have to be reduced to bring the plate current constant again at 5 milliamps. Note the plate potential.

14. Repeat the above observation for values of grid potential of + 4V, + 6V, + 8 V and so on and note the value of plate potential corresponding to each value of grid potential keeping the plate current constant at 5 mA.

15. Repeat the experiment by keeping plate current constant at 10 mA and 15 m A.

16. **Constant plate current characteristics.** A graph between plate potential E_p and grid potential E_g at constant plate current is known as *constant plate current characteristic*.

To plot the graph take positive grid voltage along + *X*-axis, the negative grid voltage along –*X*-axis and plate voltage along *Y*-axis. The graph is shown in Fig. 46.12.

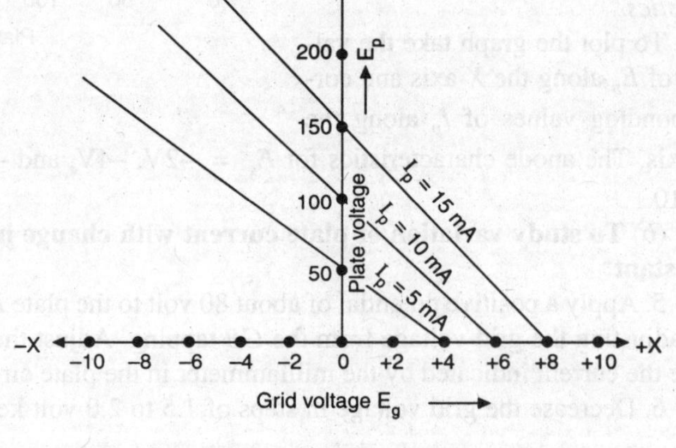

Fig. 46.12

Observations.

Least count of the milli-ammeter = m A
Least count of voltmeter in plate circuit = volt
Least count of voltmeter in grid circuit = volt

(*i*) **For anode characteristics**

Sl. No.	*Plate voltage E_p*	*Plate current in mA I_p at*		
		Grid voltage E_g = – 2V	*Grid voltage E_g = – 4 V*	*Grid voltage E_g = – 6 V*
1	80			
2	100			
3				
...				
...				
...				

(*ii*) **For mutual characteristics**

Sl. No.	Grid voltage E_g	Plate current in mA I_p at		
		Plate potential $E_p = 80$ V	Plate potential $E_p = 100$ V	Plate potential $E_p = 120$ V
	–10			
	–8			
	–6			
	...			
	...			
	...			
	+2			
	+4			
	+6			
	...			

(*iii*) **Constant plate current characteristics**

Sl. No.	Grid voltage E_g	Plate Voltage E_p for		
		Plate current $I_p = 5$ mA	Plate current $I_p = 10$ mA	Plate current $I_p = 15$ mA
	–10			
	–8			
	–6			
	...			
	...			
	...			
	+2			
	+4			
	+6			
	...			

Precautions. 1. Zero error, if any in the milli-ammeter and voltmeters in plate and grid circuit must be noted. It is desirable to adjust the pointer to zero mark when no current is passing.

2. Specifications of the given valve about maximum plate voltage, heating voltage and grid voltage prescribed by the makers must be strictly adhered to.

3. The positive marked terminal of the power supply must be connected to the plate *P* of the triode.

4. (*i*) For study of anode characteristics the grid voltage must be kept constant for each set of observations.

(*ii*) For study of mutual characteristics the plate voltage must be kept constant for each set of observations.

(*iii*) For study of constant plate current characteristics the plate current must be kept constant for each set of observations.

Experiment 46.3. To plot a graph between grid potential and the corresponding plate current for a triode valve and hence to determine the amplification factor, internal resistance and the mutual conductance for the valve

Apparatus. Same as in **Expt. 46.2.**

Theory. Internal resistance $R_p = \left(\frac{\delta E_p}{\delta I_p}\right)_{E_g}$

Mutual conductance $g_m = \left(\frac{\delta I_p}{\delta E_g}\right)_{E_p}$

Amplification factor $\mu = \left(\frac{\delta E_p}{\delta E_g}\right)_{I_p}$

Procedure. Same as in **Expt. 46.2** for **Mutual characteristics**

Observations. Same as in **Expt. 46.2** for **Mutual characteristics.**

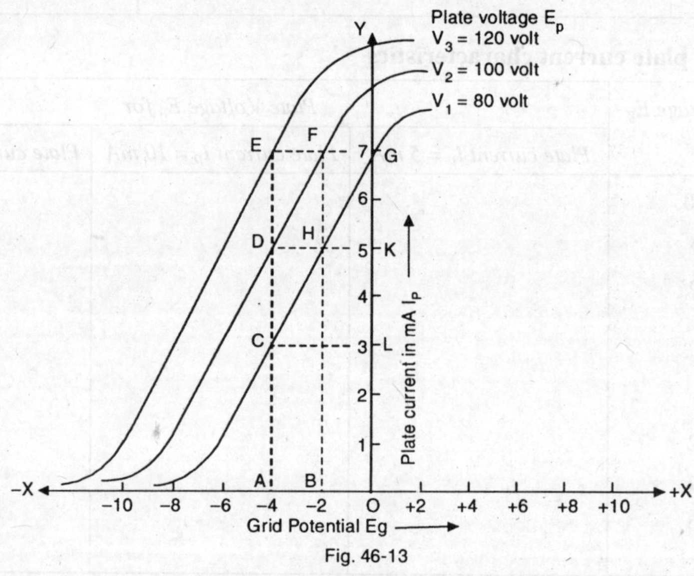

Fig. 46.13

Calculations. Plot a graph between plate current I_p and grid potential E_g for three constant values of plate potential $E_P = V_1 = 80$ volt, $E_p = V_2 = 100$ volt and $E_p = V_3 = 120$ volt as shown in Fig. 46.13.

Take a point A on the X-axis representing a grid potential $(-OA)$ and draw a line $ACDE$ parallel to Y-axis cutting the characteristics for plate potential V_1, V_2 and V_3 at C, D and E respectively. From C draw CL parallel to X-axis meeting Y-axis at L, from D draw DHK cutting the curve for plate potential V_1 at H and meeting Y-axis at K. From E draw EFG parallel to X-axis cutting the curve for plate potential V_2 at F and meeting Y-axisat G. From F draw FHB parallel to Y-axis cutting the curve for plate potential V_1 at H. and meeting X-axis at B. From E draw $EDCA$ parallel to Y-axis cutting the curve for plate potential V_2 at D, the curve for plate potential V_1 at C and meeting X-axis at A.

(*i*) **Internal resistance**

Sl. No.	Grid pot. E_g	Plate pot. E_p	Plate Current I_p	δE_p	δI_p	$R_p = \left(\frac{\delta E_p}{\delta I_p}\right)_{E_g}$
1	$-OA$ = volts $-OA$ = volts	V_1 = volt V_2 = volt	OL = mA OK = mA	$V_2 - V_1$ volt	$OK - OL$ = mA	$R_p = \frac{V_2 - V_1}{OK - OL}$ $\times 1000$ = ohm
2	$-OB$ = volts $-OB$ = Volts	V_1 = volt V_2 = volt	OK = mA OG = mA	$V_2 - V_1$ volt	$OG - OK$ = mA	$R_p = \frac{V_2 - V_1}{OG - OK}$ $\times 1000$ = ohm
3	$-OA$ = volts $-OA$ = Volts	V_2 = volt V_3 = volt	OK = mA OG = mA	$V_3 - V_2$ volt	$OG - OK$ = mA	$R_p = \frac{V_3 - V_2}{OG - OK}$ $\times 1000$ = ohm

Mean internal resistance R_p = ohm

(*ii*) **Mutual conductance**

Sl. No.	Plate pot. E_p	Grid pot. E_g	Plate Current I_p	δE_g	δI_p	$g_m = \left(\frac{\delta I_p}{\delta E_p}\right)_{E_p}$
1	V_1 = volt V_1 = volt	$-OA$ = volts $-OB$ = volts	OL = mA OK = mA	$-OB-$ $(-OA)$ = volt	$OK - OL$ = mA	$\frac{OK - OL}{OA - OB}$ = mA/volt
2	V_2 = volt V_2 = volt	$-OA$ = volts $-OB$ = volts	OK = mA OG = mA	$-OB-$ $(-OA)$ = volt	$OG - OK$ = mA	$\frac{OG - OK}{OA - OB}$ = mA/volt

Mean mutual conductance g_m = m A/volt.

(*iii*) **Amplification factor.** 1. From the graph we find that at grid potential $E_g = (-OA)$ and plate potential $E_p = V_1$ the plate current $I_p = OL$. If the grid potential is increased from $(-OB)$ to $(-OB)$ keeping plate potential constant at V_1 the plate current increases to OK. Alternatively if the grid potential is kept constant at $-OA$ and plate potential is increased from V_1 to V_2 the plate current is again OK.

$\therefore$ Amplification factor = $\frac{\text{Change in plate potential}}{\text{Change in grid potential}}$ to bring about the same change in plate current is given by

$$\mu = \frac{V_2 - V_1}{OA - (-OB)} =$$

2. Similarly we find that the plate current increases from OK to OG if either the plate potential is increased from V_2 to V_3 at constant grid potential $(-OA)$ or the grid potential is increased from $(-OA)$ to $(-OB)$ at constant plate potential V_2

$$\therefore \quad \mu = \frac{V_3 - V_2}{-OA - (-OB)} =$$

Mean amplification factor =

Precautions. Same as in **Expt 46.2.**

46.7. Tetrode. In a thermionic tube each pair of electrodes acts as if it were a small capacitor. In a triode valve there are three such capacitors consisting of (*i*) grid and plate (*ii*) grid and cathode and (*iii*) the plate and the cathode. In a triode the grid plate capacitor has a capacitance of 10 micro-micro farad. It feeds back energy from plate-circuit to the grid-circuit and thus amplification falls off at high frequencies of the order of 10 kc/sec. This causes whistling noise.

To reduce the inter-electrode capacitances in a tetrode an additional grid G_2 called the **screen grid** or simply screen is provided. This arrangement creates two small capacitors in series as shown in Fig. 46.14, so that the total capacity is reduced. The screen grid acts as a shield between the control grid G_1 and the plate and thus enables the tetrode to work at high frequencies. The control grid governs the flow of plate current in the same manner as in a triode as it is not shielded from the space charge by the screen grid. Thus a small change in the control grid voltage will produce a large change in the plate current. The tetrode requires a very large change in plate voltage to produce a small change in plate current. Hence amplification factor of a tetrode is **far higher than that of a comparable triode.** It is 400 to 800 as compared to the values of about 5 to 50 for a triode.

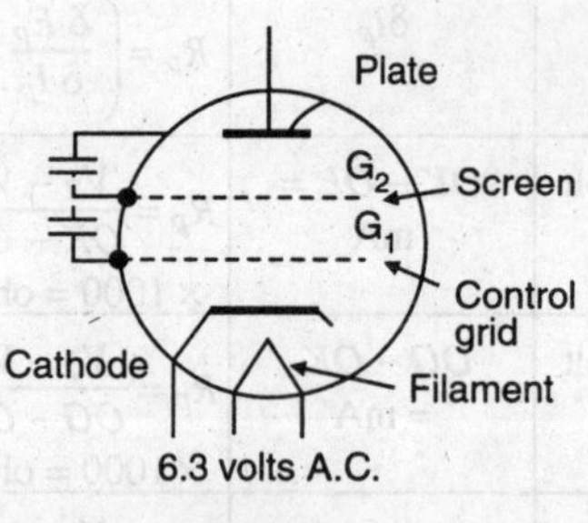

Fig. 46.14

Experiment 46.4. To plot a static characteristic of a tetrode by varying plate voltage keeping the grid and screen voltages constant.

Apparatus. A tetrode with base connections, a power supply with *H.T.* and *G.B.* terminals, source of constant voltage for screen grid, a milliammeter, connecting wires etc.

Procedure. 1. Draw the scheme of connections as shown in Fig. 46.15 and make connections accordingly.

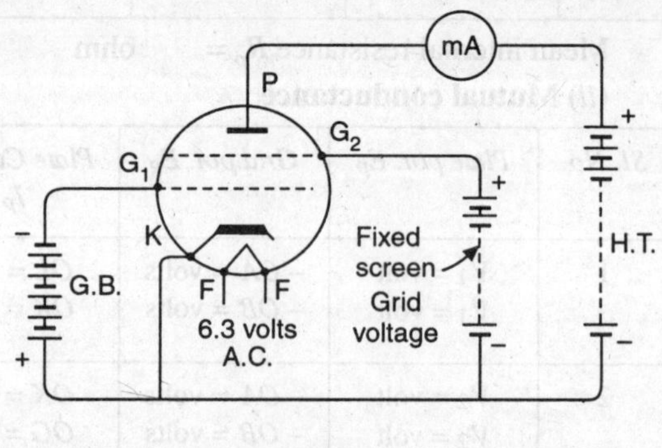

Fig. 46.15

(*i*) Connect the filament *FF to 6.3 volt A.C.* supply.

(*ii*) Connect the positive of *H.T.* to the plate *P* and the negative terminal to the cathode.

(*iii*) Connect the negative terminal of the *GB* to the control grid G_1 of the valve and the positive to the cathode.

(*iv*) Connect the positive of the second *HT* to the screen grid terminal G_2 and the negative to the cathode.

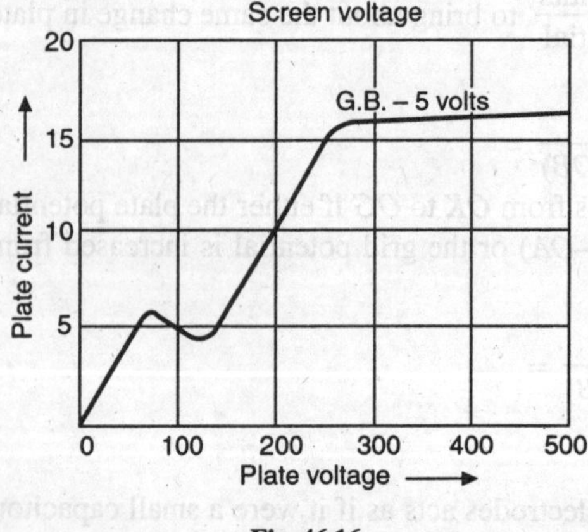

Fig. 46.16

2. Adjust the screen grid voltage to about 200 volt and the negative grid bias applied to the control grid to about –5 volt.

3. Vary the plate voltage in steps of about 30 volt and note the corresponding current in the milliammeter.

4. Taking plate voltage along *X*-axis and plate current along the *Y*-axis, plot the curve, The curve obtained will be as shown in Fig. 46.16. At about 100 volt the increase in plate voltage causes a decrease in plate current due to secondary emission. As long as the plate voltage is below the screen voltage almost all the secondary electrons go to the screen. Beyond 300 volt the plate current becomes constant as the screen shields the space charge at the cathode. This causes the screen to exert a greater control over plate current than the plate itself.

5. Repeat for a different control grid voltage and screen voltage.

Observations.

Control grid voltage = (*i*).... (*ii*)....

Screen voltage = (*i*).... (*ii*)....

Sl. No.	*Plate voltage*	*Plate current*	
		Observation (i)	*Observation (ii)*
1.	0		
2.	30		
3.	60		
:	:		
:	450		

Precautions. Same as for triode experiment.

46.8. Pentode. The main disadvantage of the tetrode is the need for a high plate voltage to prevent distortion in its output due to the effect of secondary emission. The insertion of an additional grid G_3 called the **suppressor grid** between the plate and the screen grid G_2 of the tetrode overcomes the undesirable feature by eliminating the **flow** of secondary electrons between the plate and the screen grid. By adding this electrode the tube becomes a five electrode unit and is known as the **pentode.** In addition to overcoming of distortion the pentode retains all the advantages of the tetrode such as low grid-plate capacitance, high amplification factor and high power output.

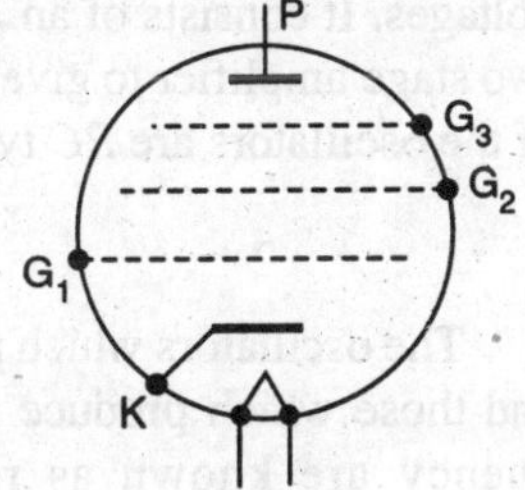

Fig. 46.17

Construction. A typical pentode contains an emitter *K* and three grids G_1, G_2 and G_3 and a plate as shown in Fig. 46.17. The grid G_1 closest to the cathode is called the *control grid*, next is the *screen grid* G_2 and third is the *suppressor grid* G_3 located between the screen grid G_2 and the plate.

Action of suppressor grid. The suppressor grid G_3 is connected to the cathode and thus is at a substantial negative potential with respect to the plate. The secondary electrons emitted by the plate when high speed electrons strike it are repelled by G_3 and are attracted by the plate and thus it eliminates the effect of secondary emission. As a result the plate current rises smoothly as the plate voltage is increased from zero to maximum for each value of the control grid voltage. The presence of the suppressor grid further increases the shielding action between the plate and the control grid. This further decreases the grid-plate capacitance. For this reason the plate current becomes more independent of the plate voltage than in tetrode.

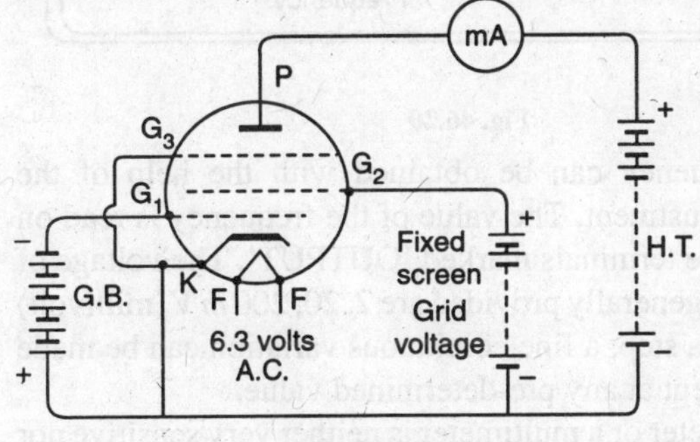

Fig. 46.18

Experiment 46.5. To plot a static characteristic of a pentode by varying the plate voltage keeping the grid and screen voltage constant.

Apparatus. Same as in **Expt. 46.4.** and a pentode with base connections instead of tetrode.

Procedure. 1. Draw the circuit diagram as shown in Fig. 46.18. Make connections accordingly. Connect the supressor grid G_3 to the cathode so that it is at a substantial negative potential with respect to the plate.

2. Adjust the screen voltage to about 200 volt and apply negative grid bias of –3 volt to the control grid.

3. Vary the plate voltage in small steps upto 50 volt and then in steps of about 50 volt and note the corresponding plate current values.

4. Plot the graph between the plate voltage and the plate current. The curve obtained will be as shown in Fig. 46.19.

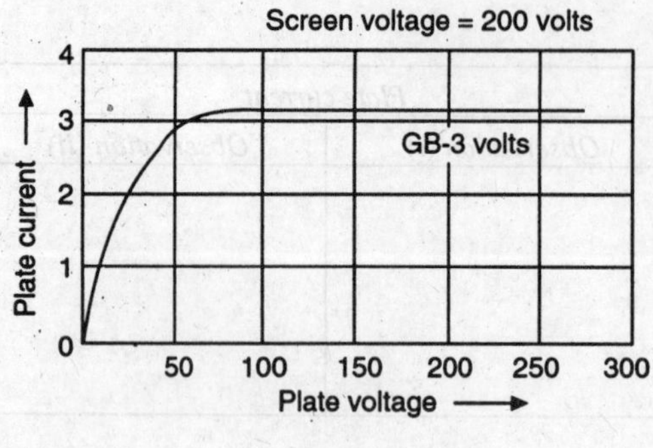

Fig. 46.19

5. From the graph it is seen that the plate current first rises and then becomes almost constant, being little affected by the increase of the plate voltage. Due to high screening, the amplification factor, as also the plate resistance, becomes extremely high. The amplification factor is of the order of 1000–1500 and plate resistance of the order of one megohm.

6. Repeat the experiment for a different value of *G.B* and screen voltage.

Observations. Record as in **Expt. 46.4.**

Precautions. Same as for triode experiment.

46.9. Signal generator. A signal generator is an electronic oscillator which produces *A.C* signal voltages. It consists of an *LC* or *RC* oscillating circuit (known as *tank circuit*) provided with one or two stage amplifier to give suitable output power. As *LC* circuit requires large size components most of the oscillators are *RC* type. The frequency generated by *LC* circuit is given by

$$f = \frac{1}{2\pi \sqrt{LC}}$$

The oscillators which produce audio frequencies are known as audio frequency (A.F.) oscillators and those which produce radio frequency are known as radio frequency (R.F.) oscillators. The wave form produced by the oscillator may be sinusoidal or non-sinusoidal *i.e.* square wave or any other shape.

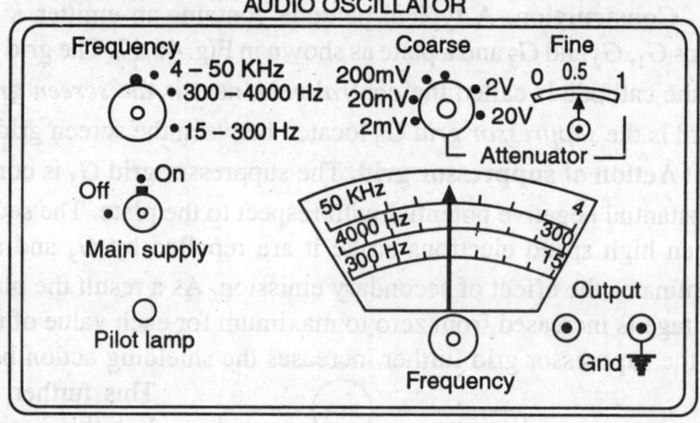

Fig. 46.20

Audio frequency signal generator. The front panel of an A.F. signal generator is shown in Fig. 46.20.

To operate the oscillator it is connected to the *AC* mains. When the current is switched on the 'PILOT LAMP' lights up. The frequency range switch can be set at one of the ranges 15-300 *Hz*, 300 to 4000 *Hz* or 4 – 50 *KHz*. A desired value of the frequency can be obtained with the help of the 'FREQUENCY' knob which enables continuous adjustment. The value of the frequency is read on the corresponding scale. The output is drawn from the terminals marked 'OUTPUT'. The voltage of the output signal can be varied from 0-20 *V*. The steps generally provided are 2, 20, 200 *m V* (millivolt) 2 *V* and 20*V*. After setting the 'COARSE' switch to a step, a fine continuous variation can be made with the control labelled 'FINE' and the voltage is kept at any pre-determined value.

46.10. Electronic voltmeter. An ordinary voltmeter or a multimeter is neither very sensitive nor it has a very high resistance. For electronic experiments we need a voltmeter which has a very high input resistance of the order of 10-100 mega – ohm so that it produces a very small load on the circuit under test. For such measurements, we use an *electronic voltmeter*.

The electronic voltmeter is of two main types :

(*i*) *Vacuum tube voltmeter* (*V.T.V.M*). It uses vacuum tubes and a moving coil meter.

(ii) *Transistor voltmeter*. A transistor voltmeter uses ordinary bi-polar junction transistors (*BJT*) and is known as *T.V.M* or it uses field effect transistors and is known as *FETVM*. The *FETVM* is most commonly used because of its various advantages and we shall describe only its working principle. Electronic voltmeters may be *digital* or *analog* type.

46.11. Vacuum tube voltmeter. A vacuum tube voltmeter (*V.T.V.M.*) is an instrument that can measure *D.C.* and *A.C.* voltages and resistances. Some *V.T.V.M's* are also capable of measuring currents.

The basic circuit of a balanced bridge type *V.T.V.M.* is shown in Fig. 46.21.

It consists of two similar triodes V_1 and V_2, a sensitive meter *M* having a number of scales for different ranges and a power supply.

Fig. 46.21

The meter *M* is connected between the plates of V_1 and V_2. The voltage to be measured is applied at the *input* points *A* and *B* of the *grid-cathode circuit* of the valve V_1 in such a way that the grid is always negative with respect to the cathode. Triode V_2 acts as a neutralising valve.

When no voltage is applied at the input terminals *A* and *B*, the plate current flowing in both the triodes is *equal* because the valves are *similar*. Due to this the plates of the two valves are at the *same potential* and the current through the meter *M* is *zero i.e.* the meter reads zero. If, however, there is any constructional difference in the two valves the current may not be exactly zero and the meter may given some deflection. In such a case the current is brought to zero value, so that the meter also reads zero by adjusting the variable resistance *r* in the circuit of valve V_2.

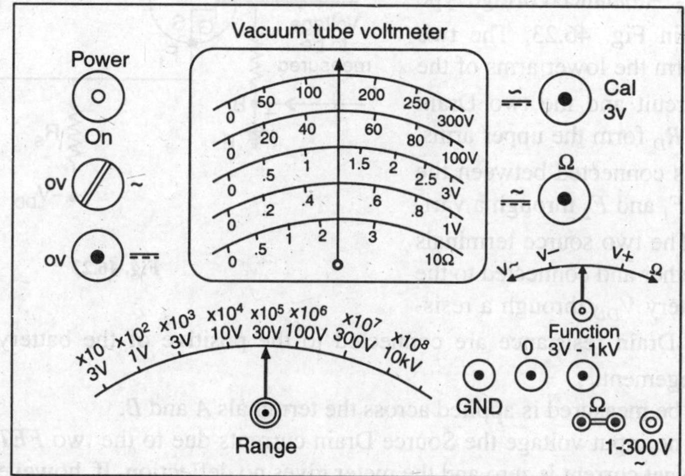

Fig. 46.22

The voltage to be measured is applied at the grid of the triode V_1 making the grid negative with respect to the cathode. This changes the plate current of valve V_1 with the result that the plates of the valves V_1 and V_2 are not at the same potential. The difference in the potential of the two plates causes a small current to flow through the meter proportional to the potential difference. The meter is calibrated to read the voltage directly.

The front–panel of a typical *V.T.V.M.* is shown in Fig. 46.22.

D.C. voltages can be directly measured by applying between the sockets marked 0 and 3 *V-1KV*. The 'FUNCTION' switch is set to + *V* for measuring positive voltages and to –*V* for measuring negative voltages. The range switch is kept at the highest value to begin with and then brought to the proper step depending upon the voltage to be measured. For *A.C.* voltage measurement the 'FUNCTION' switch is set at *A.C.* volts (*V* ~~). The measurement is made through a 'probe' consisting of rectifier diodes. One lead from the probe is inserted into the socket labelled (1-300 *V* ~) and the tip of the probe is connected to the 'GND' terminal. The voltage to be measured is connected to the probe terminals.

For resistance measurement the *V.T.V.M.* is calibrated to directly indicate the resistance values from the voltage drop across the unknown resistance. The constant *D.C.* voltage for resistance measurement is provided internally from a separate winding of the mains transformer after rectification. For resistance measurement the FUNCTION switch is set on ohms (Ω). The unknown resistance is connected across the two terminals labelled (Ω). The 'RANGE' switch is adjusted to maximum and the resistance value is read on the scale. For accurate reading the 'RANGE' switch is adjusted to a suitable position.

The *V.T.V.M.* shown has *D.C.* voltage measuring ranges of 1, 3, 10, 100, 300 and 1000 *V*, *A.C.* voltages measuring ranges 1, 3, 10, 100 and 300 *V*, and resistance measuring ranges 10 ohm to 10^8 ohm. The voltage and resistance ranges are indicated on the 'RANGE' switch.

The input resistance of a *V.T.V. M.* is very high being of the order of 10 mega ohm. Thus the measurements with a *V.T.V.M.* have very little effect on the current under test.

Field effect transistor voltmeter (*F.E.T.V.M.*). Just like the *V.T.V.M.* a *FETVM* also has two identical field effect transistors F_1 and F_2. A balanced bridge type *FETVM* is shown in Fig. 46.23. The two *FET*'s F_1 and F_2 form the lower arms of the balanced bridge circuit and the two Drain resistances R_D and R_D form the upper arms. A sensitive meter is connected between the Drain terminals of F_1 and F_2 through a variable resistance *R*. The two source terminals are also joined together and connected to the negative of the battery V_{DD} through a resistance R_s. The two Drain resistance are connected to the positive of the battery V_{DD} through a potentiometer arrangement.

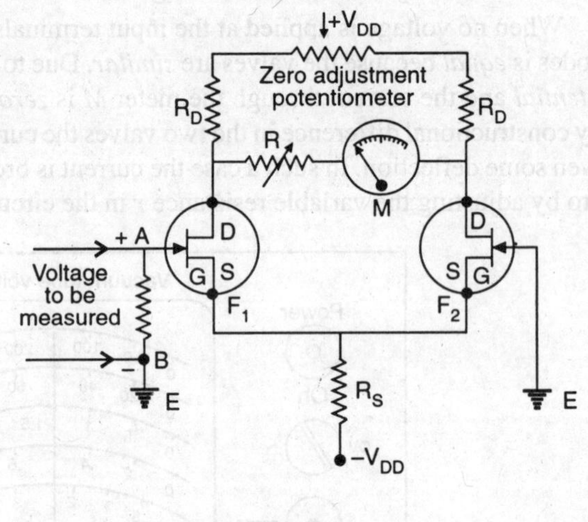

Fig. 46.23

The voltage to be measured is applied across the terminals *A* and *B*.

When there is no input voltage the Source Drain currents due to the two *FET's* are equal and opposite so that the net current is zero and the meter gives no deflection. If, however, the two *FET's* are not exactly identical, a small current flows through the meter. The current is adjusted to *zero* value with the help of zero adjustment potentiometer arrangement.

The voltage to be measured is applied between the input terminals *A* and *B* such that the gate *G* of F_1 is *positive*. The current through F_1 changes, the balance is disturbed and a current flows through the meter *M*. The magnitude of the current is proportional to the voltage applied to the Gate of F_1.

The scale of the meter is calibrated to read the applied voltage directly. Full scale calibration of the meter is adjusted with the help of variable resistance *R*.

Experiment 46.6. To measure the peak, average and r.m.s. value of a signal using an electronic voltmeter.

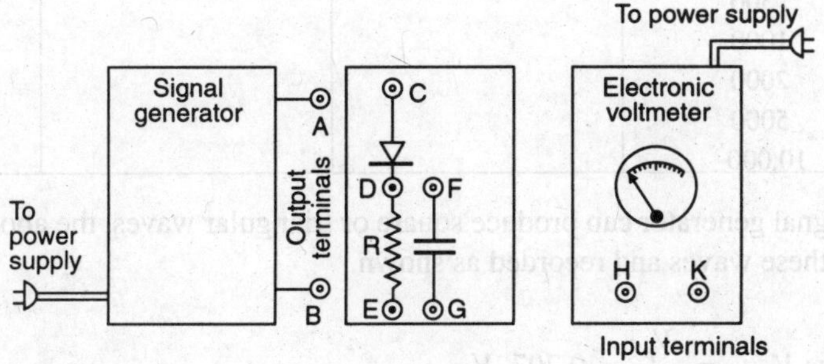

Fig. 46.24

Apparatus. Signal generator, electronic voltmeter [*V.T.V.M.* or *F.E.T.V.M.*], a diode, a resistance (about 10 $K\ \Omega$), a capacitor (32 μF) all fitted on a board with suitable terminals.

Procedure 1. Arrange the signal generator, the board on which a diode, a resistor and a capacitor are fitted and an electronic voltmeter as shown.

2. *r.m.s. value of a. c. signal*. Turn the FUNCTION SWITCH of the electronic voltmeter to *A.C.* Select the scale for measuring highest *A.C.* voltage range. Connect the input terminals *H* and *K* of the *E.V.M.* to the output terminals *A* and *B* of the signal generator. Set the frequency switch to say 200 *Hz* and the output to a suitable value. Connect the electronic voltmeter and the signal generator to their *proper* power supplies. Note the reading of the meter in the *EVM*. If the scale chosen is too large or two small, select the suitable scale and note the final reading. This gives the *r.m.s.* value of the *A.C.* signal produced by the signal generator. Disconnect the signal generator and the electronic voltmeter from their power supplies and also disconnect the *EVM* from the signal generator.

3. *Average Value of the A.C. signal*. Now connect the output terminals *A* and *B* of the signal generator to the terminals *C* and *E* of the diode and resistance *R* placed in series. Connect the terminals *H* and *K* of the electronic voltmeter across the resistor *R* to the terminals *D* and *E*. Switch on the power supply to the *EVM* and signal generator. The output across the resistor *R* has under gone half wave rectification on passing through the diode. The reading of the *EVM*, therefore, gives the average value of the *A.C. signal produced by the signal generator. Note the reading. Switch off the power supplies.*

4. *Peak value of A.C. signal*. Now connect the terminal *D* to *F* and *E* to *G* so that the capacitor is connected in parallel with the resistor *R*. Connect the output terminals *A* and *B* of the signal generator to the terminals *C* and *E* of the diode and resistance capacitance combination. Connect the terminals *H* and *K* of the electronic voltmeter across the capacitor to the terminals *F* and *G*. Switch on the power supply to the *EVM* and signal generator. As the capacitor is charged to the peak value of the signal the reading in the electronic voltmeter corresponds to the of the peak value of the *A.C.* signal produced by the signal generator.

7. Repeat the observations for *r.m.s.* value, average value, and peak value of *A.C.* for different frequencies produced by the signal generator.

Observations.

Sl. No.	*Frequency produced by signal generator*	*r.m.s. value V_{rms} volts*	*Average value V_a volts*	*Peak value V_p volts*
	200			
	500			
	1000			
	2000			
	5000			
	10,000			

Note. If the signal generator can produce square or triangular waves, the above values may also be determined for these waves and recorded as shown.

Calculations.

For a sine wave $V_{r.m.s.} = \dfrac{V_p}{\sqrt{2}} = 0.707 \ V_p$

$$V_a = \frac{2V_p}{\pi} = 0.9 \ V_{r.m.s.}$$

Verify the above relations.

46.12. Frequency response of an R.C. coupled amplifier. No amplifier can satisfactorily amplify signals of all frequencies. The limitations are due to the amplifying devices themselves.

The variation of relative voltage gain with frequency is shown in Fig 46.25. The curve is plotted on a semi-log-graph paper with frequency on logarithmic scale so as to accomodate large frequency range. The curve is horizontal in mid frequency range showing that in this region the gain is constant and does not change with frequency. At low frequency the gain decreases. Lower the frequency, smaller is the gain. This is due to the fact that the reactance of the coupling capacitor $\dfrac{1}{2\,\pi f C}$ at low frequencies is high and hence the gain is low.

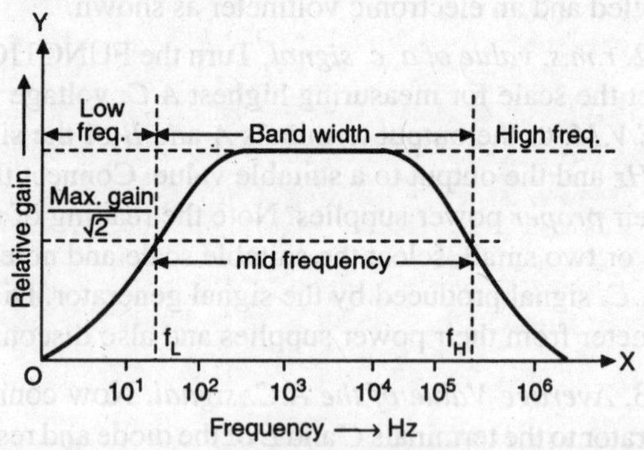

Fig 46.25

At high frequencies the gain again decreases. Higher the frequency, the smaller is the gain. This is due to the fact that the shunt capacitances become effective though the reactance of the coupling capacitor decreases. The frequency response of an amplifier is measured by the *band width* β. *It is defined as the frequency band in which the amplifier gain remains above 0.707 of the mid frequency gain.* The low frequency f_1 at which the gain falls to $1/\sqrt{2}$ *i.e.* (0.707) of its mid-frequency value is know as *lower cut off frequency*. The high frequency f_h at which the gain falls to $1/\sqrt{2}$ *i.e.* (0.707) of its mid frequency value is known as *higher cut off frequency*.

$\therefore$ Bandwidth $\beta = f_h - f_l$

The bandwidth measurement is done by first measuring the amplifier gain in mid-frequency region (1 *KHz*). Keeping the input signal constant, the frequency is varied on the low as well as high frequency side till the gain falls to 0.707 of the gain at 1 *KHz*. To obtain the frequency response curve the amplifier gain is plotted as a function of input signal frequency.

Experiment 46.7. To set up R.C. coupled amplifier using a triode valve (6J5 or 6C5) and to plot its frequency response curve.

Apparatus. A triode valve 6J5, with base socket provided with terminals, a power supply with *HT* and 6.3 volt *A.C.* supply, signal generator, vacuum tube voltmeter (*V.T.V.M.*), resistance of 1 *meg. ohm* 470 *k* Ω,100 *k* Ω, 50 *k* Ω, electrolytic capacitors of 2.5 μ *fd* and 0.05 *μf.d.*

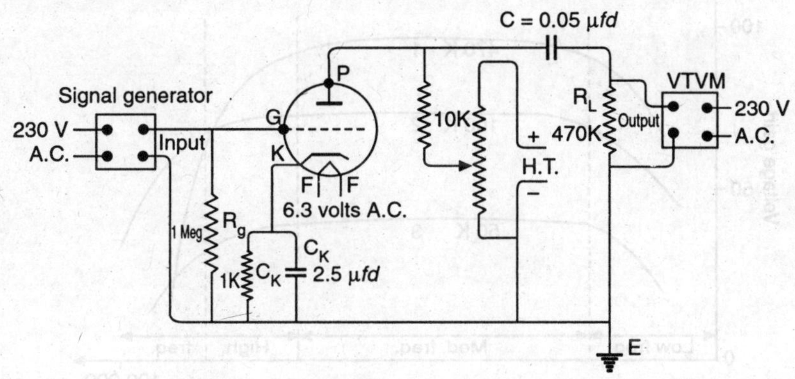

Fig. 46.26

Procedure. 1. Draw a diagram showing the scheme of connections as in Fig. 46.26. and make connections accordingly.

2. Connect the signal generator across the input circuit and *V.T.V.M.* across the output circuit as shown.

3. Connect the plate *P* to the positive of the *H.T.* through a 10*k* Ω resistance and negative to the earth. Connect the filament of the valve to the 6.3 volt *A.C.* terminals of the power supply.

4. Connect the plate through a capacitor of capacitance 0.05 *μfd* to one terminal of the resistance of 470 *k* Ω the other terminal of which is connected to the negative of the *H.T.* Connect the vacuum tube voltmeter across the resistance 470 *k* Ω the grid *G* to the signal generator and the cathode *K* as shown in diagram.

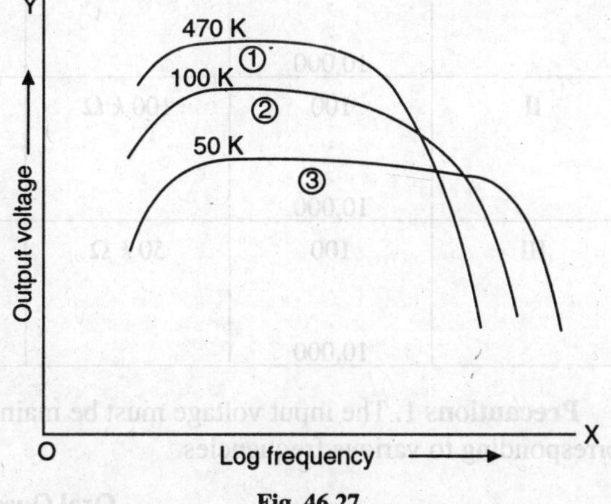

Fig. 46.27

5. Connect the oscillator (signal generator) to 230 volt *AC* mains and set it to supply a sinusoidal input of 1000 *Hz* at a suitable voltage say about 0.1 *V*. Measure the output voltage across 470 *k* Ω resistance with a *V.T.V.M.* Keeping the voltage across the signal generator constant first change the frequency to the lower side say to 100 *Hz* and note the output voltage at intervals of 100 *Hz* Keeping the voltage across the signal generator constant change the frequency to the higher side say 10,000 *Hz* at intervals of 200–300 *Hz* and note the output voltage of the signal generator with *V.T.V.M.* This is the input voltage for the amplifier circuit.

Note. The actual range of frequencies to be selected depends upon the amplifier circuit and the range provided in the signal generator.

6. Replace the resistance 470 *k* Ω by 100 *k* Ω and then by 50 *k* Ω and record observations of the output voltage in a similar manner. If necessary adjust the output voltage of the signal generator (which is the input for the amplifier circuit), when the load resistance is changed.

7. Plot a graph between the log of frequency and the output voltage. The graph will be of the form as shown in Fig. 46.27.

8. Find the voltage gain which is the ratio of output voltage and input voltage and plot a graph between voltage gain and the frequency. The graph will be as shown in Fig. 46.28.

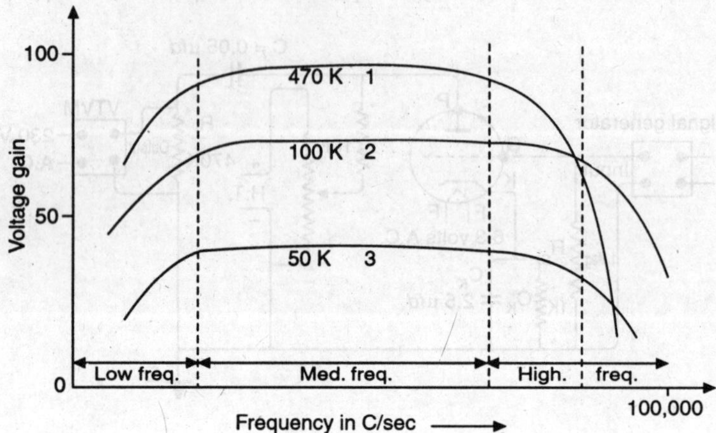

Fig. 46.28

Observations.

Sl. No.	*Frequency Hz*	*Resistance R*	*Input voltage*	*Output voltage*	*Voltage gain*
I	100 . . 10,000	470 $k\ \Omega$			
II	100 . . 10,000	100 $k\ \Omega$			
III	100 . . 10,000	50 $k\ \Omega$			

Precautions 1. The input voltage must be maintained constant when the observations are made corresponding to various frequencies.

Oral Questions

1. **Diode.** (*i*) What is a diode valve? Explain its function as a rectifier. (*ii*) Draw a diagram showing the connections of a full-wave rectifier. (*iii*) What is the function of the choke and the two capacitors? (*iv*) Which is the positive of the H.T. supply?

2. **Triode valve.** (*i*) What is a triode valve? (*ii*) What is the difference between a triode and a diode valve? (*iii*) On what factors does the plate current depend? (*iv*) How is the plate current affected when (*a*) plate is negative (*b*) grid is negative? (*v*) Define A.C. resistance, mutual conductance and amplification factor of a valve (*vi*) What is a characteristic curve and what is its shape? (*vii*) Does a triode valve itself oscillate? (*viii*) How do you tune a radio set? (*ix*) What changes take place when you work the band switch and the tuning knob? (*x*) What is the difference between R.F. and A.F. transformers?

3. **Tetrode and pentode** (*i*) What is the difference between a tetrode and a triode; a pentode and a tetrode? (*ii*) What is the function of screen grid in a tetrode? (*iii*) What do you mean by the negative resistance of a tetrode

and how do you explain it? (*iv*) Is the amplification factor of a tetrode greater than that of the triode? If so why? (*v*) What is the action of suppressor grid in a pentode? (*vi*) How does the plate current in a pentode become more independent of plate voltage? (*vii*) Does the pentode also have a region of negative plate resistance? If not, why?

4. **Electronic voltmeter.** (*i*) What is an electronic voltmeter? (*ii*) What is a *V.T.V.M.* ? (*iii*) What is the main difference between a multimeter an *EVM* and a *V T V M*? (*iv*) How can an *E V M* be used for A.C. measurements ? (*v*) What is the order of input resistance of an EVM? (*vi*) How is such a high resistance obtained? (*vii*) Why do you conne :t a diode and a resistance in the circuit for measuring the average value of signal voltage? (*viii*) Why do you connect a capacitor in the circuit while measuring the peak value of signal voltage? (*ix*) What should be the value of the capacitance and why? (*x*) What is a field effect transistor voltmeter (*F E T V M*)?

5. **R.C. coupled amplifier** (*i*) What are various types of amplifier coupling in use? (*ii*) What are the advantages and disadvantages of R.C. coupling ? (*iii*) How does an R.C. coupled single stage valve amplifier respond to various input frequencies? (*iv*) What do you mean by lower and higher cut off frequencies? (*v*) What is band width of an amplifier?

47

Semi-conductor Diode

47.1. Semi-conductors. Transistors and junction diodes are made from a class of crystals called *semi-conductors* whose conductivity lies in between the conductors and insulators. Familiar examples of semi-conductor elements are germanium (Ge) and silicon (Si) etc. Germanium is most widely used in the manufacture of transistors.

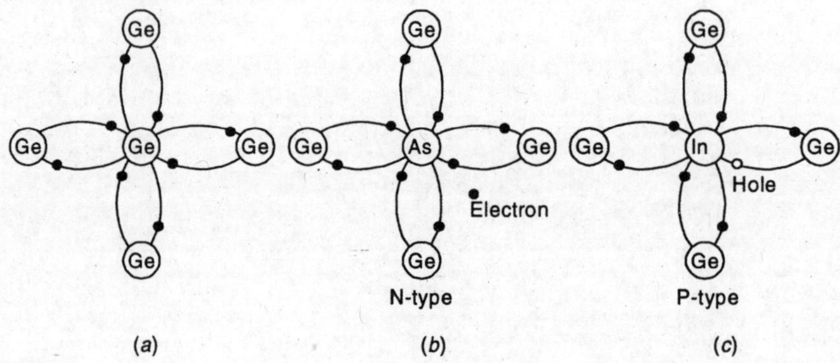

Fig. 47.1

The atomic number of germanium is 32 and it contains 2 electrons in the *K*-shell, 8 in the *L*-shell, 18 in the *M*-shell and 4 in the **outermost** *N*-shell. In Germanium which has a tetrahedral (cubical) structure, no free electrons exist at very low temperatures. Each atom of Germanium shares one electron each, with four neighbouring atoms by covalent bonds that involve all the valency electrons, the outermost orbit becomes saturated with eight electrons, four of its own and four shared between four other atoms as shown in Fig. 47.1. (*a*). At very low temperature germanium is, therefore, not a conductor of electricity.

The thermionic work function, *i.e.* the energy required to free an electron of Germanium is very small (being only 0.7 *eV*) and hence at room temperature an electron here and there may break away from its bond due to thermal agitation and become free. This electron drifts through the crystal lattice in a random manner. The atom from which the electron breaks away is left with a **'hole'** and has a tendency to catch an electron due to its *electropositive* character. An electron from the covalent bond of a neighbouring atom shifts to fill this *hole*, thereby creating a new hole in the atom from which it shifts and so on. Thus the *hole* shifts from atom to atom and behaves like a positive electronic charge. In the absence of an electric field the negative electrons and the positive *holes* move randomly at thermal velocities. When an electric field is applied positive carriers (holes) drift in the direction of the field from positive to negative and the negative carriers (electrons) in the opposite direction. Thus the material behaves as an intrinsic semi-conductor. In a semi-conductor both the electrons and the holes contribute towards conductivity.

N-type semi-conductor. The electrical conductivity of germanium can be very much increased by adding a suitable impurity in a very small proportion (say one in ten million parts) during crystal growth. This process is known as **doping**.

If arsenic (*As*) or phosporous with five valence electrons in the outermost orbit is added as an impurity to germanium during cystal growth, it will replace an equal number of germanium atoms in the crystal lattice. The atom of *As* will use only four of its valance electrons to form covalent bonds with the germanium atoms so that the *fifth* electron becomes free.

Thus presence of arsenic (*As*) as impurity donates free electrons to the germanium crystals. It is known as a *donor* element for germanium as it donates free electrons. It, however, does not add mobile holes because even though the arsenic (*As*) atom becomes positive ion, there is no vacant bond position into which a neighbouring electron can move as shown in Fig. 47.1. (*b*), The positive charge left at the location of *As* ion is fixed and not free to move. The crystal now has an excess of electrons and is called an *N-type* semi-conductor crystal as its conductivity is due to the negative charge carriers.

P-type semi-conductor. If Indium (*In*) or boron with three valance electrons in the outermost valence orbit is added as an impurity, then each atom of indium (*In*) will form covalent bonds with three of its neighbouring germanium atoms. Each *In* atom, therefore, results in a *hole* where one electron is missing from the four bonds Fig 47.1(*c*) Such a hole is mobile because it can accept an electron from a neighbouring atom and can move about as a positive carrier. Indium is called an *acceptor* for germanium. The presence of an acceptor forms a *P-type* semi-conductor crystal as it has an excess of positive charge carriers (holes) which are available for conduction.

47.2. P-N junction diode. When a layer of *P*-type crystal is joined to a layer of *N*-type crystal,

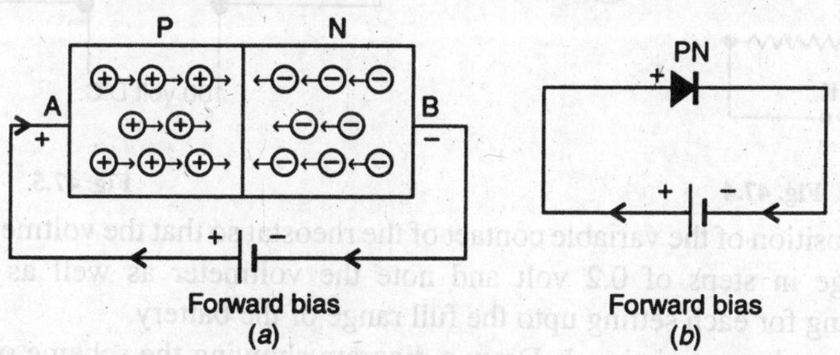

Fig. 47.2

in such a way that the atoms in *N*-type combine, across the surface of contact with the atoms in *P*-type and vice-versa, we get a *P-N junction diode*. A *P-N* junction diode along with its schemic diagram is shown in Fig. 47.2 (*a*) and (*b*). When the positive terminal of a battery is connected to the *P*-side of

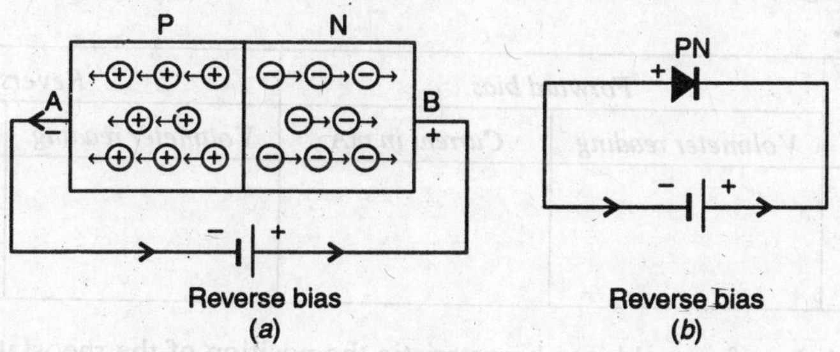

Fig. 47.3

the diode and the negative to the *N*-side, the diode is said to be *forward biased*. In such a case positive charges (holes) are urged across the *P-N* junction from *P* to *N* and negative charges (electrons) from *N* to *P* and an appreciable current is obtained. If the *P*-side is connected to the negative terminal of battery and the *N*- side to the positive, the diode is said to be *reverse biased*. In this case the free *electrons* in *N*-type rush towards the positive while the *holes* rush towards the negative. Very little current thus flows across the junction.

As a *P-N* junction permits current to flow only in one direction it acts as a *rectifier* and this is why it is known as a junction diode.

Experiment. 47.1. To draw the V-I characteristics of a forward and reverse bias P-N junction diode and to draw the load line.

Apparatus A junction diode [*BY* 127], milliammeter (0-500 mA), voltmeter (0 - 3 V), 4 volt battery, key, rheostat, micro-ammeter (0–100 μA), volt meter (0–100 V), a *D.C.* power supply capable of giving about 100 volt supply.

Procedure (*i*) *Forward bias characteristics*. 1. Draw a diagram showing the scheme of connection as in Fig. 47.4 and make connections accordingly.

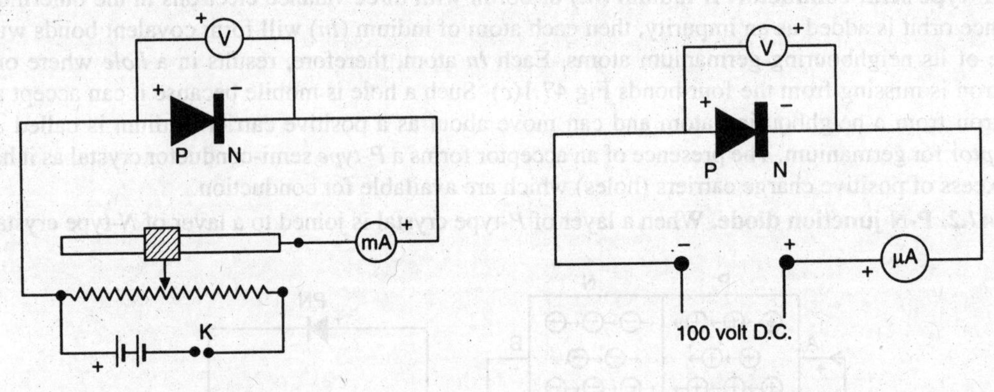

Fig. 47.4 **Fig. 47.5**

2. Adjust the position of the variable contact of the rheostat so that the voltmeter reads zero. Now increase the voltage in steps of 0.2 volt and note the voltmeter as well as the corresponding milliammeter reading for each setting upto the full range of the battery.

(*ii*) *Reverse bias characteristics*. 1. Draw a diagram showing the scheme of connections as in Fig. 47.5 and make connections accordingly. Starting from zero volt first increase the voltage in steps of one volt and then in steps of 5 volt. Note the corresponding current in micro-ammeter. Draw a graph between voltage and current taking the voltage for forward bias along the +*X*-axis and current along +*Y*-axis and for reverse bias the voltage along the –*X*-axis and current along –*Y*-axis. A typical graph is shown in Fig. 47.7.

Observations.

S.No.	*Forward bias*		*Reverse bias*	
	Voltmeter reading	*Current in mA*	*Voltmeter reading*	*Current in μA*

Precautions. 1. In a forward bias characteristic the position of the rheostat should be adjusted so that the voltmeter reading is zero. The voltage then should be increased in steps of 0.2 volt.

2. In reverse bias characteristics, the voltage should be increased in steps of 1 volt starting from 0 volt reading in the beginning and then in steps of 5 volt.

47.3. Load line of a diode circuit. Consider the circuit shown in Fig. 47.6. where the diode is forward biased by a source of voltage V. If the source has a negligible (zero) resistance $V = V_D + IR_L$ where V_D is the voltage drop across the diode, I the current through the circuit and R_L the load resistance.

$$\therefore \quad I = -\frac{1}{R_L}V_D + \frac{V}{R_L} \quad ...(i)$$

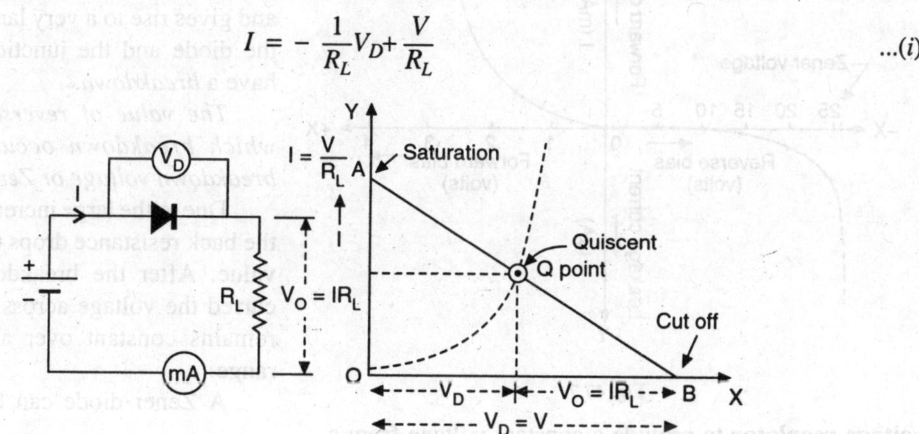

Fig. 47.6

This is an equation of the type $y = mx + c$ the equation of a straight line. Hence a graph between I taken along the Y-axis and V_D taken along the X-axis is a straight line.

If $V_D = 0$, then $I = \frac{V}{R_L}$

This gives the point A on the I or Y-axis and represents the maximum value of the current called saturation current $I = \frac{V}{R_L}$

If $I = 0 \quad V_D = V$

This gives the point B on the V_D or X-axis and represents the maximum value of V_D corresponding to zero current called the cut off voltage.

The straight line joining AB is called the *load-line*

If we draw a forward bias static characteristic of the diode also alongwith the load line, then the intersection of the static characteristic curve with the load-line gives the operating point or *Quiescent* point Q. The point Q represents the current through the load resistance R_L and the corresponding value of V_D, the voltage drop across the diode.

To draw the load line. To find the point A, *short circuit* the diode so the $I = \frac{V}{R_L}$ and measure the current. To find the point B measure the voltage across the source in *open circuit*. Join the point A and B, then AB is the load line.

47.4. Zener diode. A Zener diode is a silicon crystal diode which conducts in the reverse direction at a certain fixed voltage. It is a highly doped, P-N junction having a high back resistance.

As shown in Fig. 47.7, it is found that when a reverse bias is applied to a crystal diode a very small reverse current (known as leakage current) flows due to high reverse resistance. As the value of the reverse bias is increased, a small current due to *minority* carriers flows in the circuit. As the reverse bias is increased further, a stage reaches when the applied field is so high that a large number of co-valent bonds are broken with consequent abrupt increase in current. Further, the electrons are also accelerated to high velocities due to the large field and bring about ionisation by collision with atoms of the crystal. The electrons thus produced are again accelerated and produce more free electrons due to further ionisation (by collision). This process is known as *avalanche process* and gives rise to a very large current in the diode and the junction is said to have a *breakdown.*

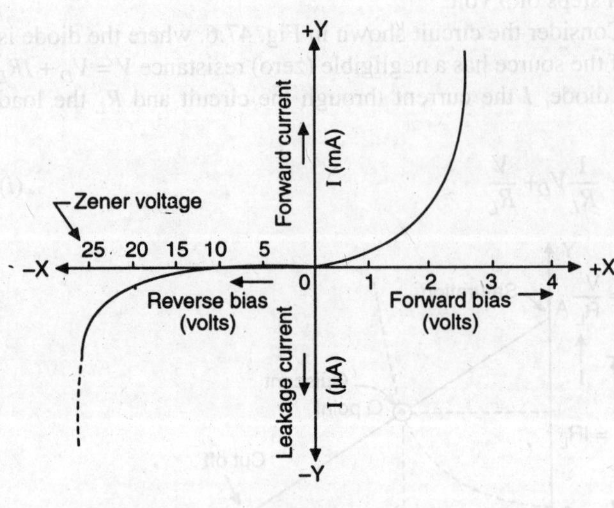

Fig. 47.7

The value of reverse voltage at which breakdown occurs is called breakdown voltage or Zener voltage.

Due to the large increase in current the back resistance drops to a very low value. After the breakdown has occurred the voltage across the junction remains constant over a fairly wide range.

A Zener diode can be used as a **voltage regulator** to provide a constant voltage from a source whose voltage may vary over a sufficient range. The Zener voltage depends upon the semi-conductor material, amount of doping and method of producing the *P-N* junction. Zener diodes can be obtained with breakdown voltage ranging from 2 volt to several hundred volts by adjusting the concentration of the impurity in the materials producing the *P-N* junction.

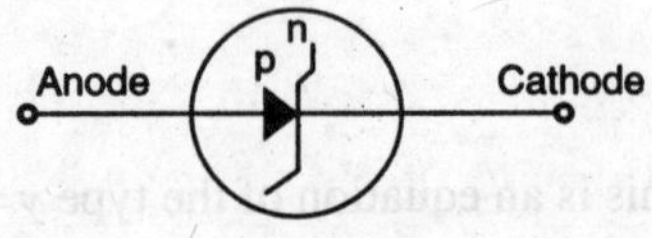

Fig. 47.8

A zener diode is symbolically represented as in Fig. 47.8.

47.5. Voltage regulation characteristics. The basic Zener diode D.C. voltage regulating circuit is shown in Fig 47.9.

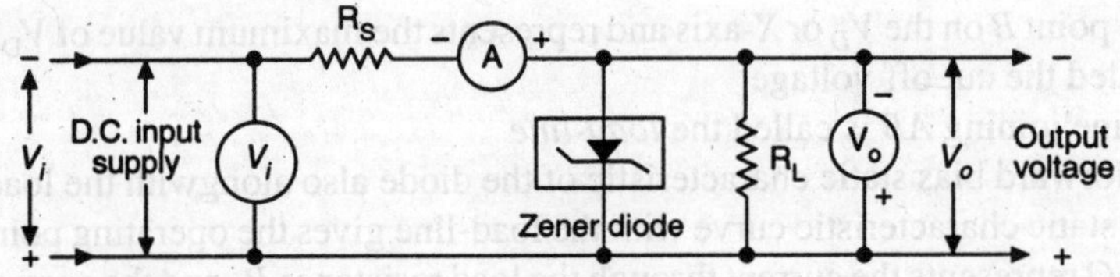

Fig. 47.9

The Zener diode with Zener voltage V_z is connected in parallel with the load resistance R_L across which a constant output voltage is desired. The Zener diode is reverse biased with output voltage whose variations are to be regulated. It will maintain a constant output voltage $V_0 = V_z$ across the load as long as the input voltage does not fall below V_z. When the input voltage V_i is greater than V_z, a large current I_z will flow through the Zener diode and the series resistance R_s and the voltage drop across R_s will increase maintaining the voltage drop across R_L at constant value $V_0 = V_z$.

When the input voltage is less than the breakdown (or zener voltage) very small current flows through the Zener diode and the whole input appears across the output terminals. Consequently within this range the output voltage increases linearly with the input voltage. The voltage regulation characteristic curve of Zener diode is shown in Fig 47.10 (*a*).

To study the voltage regulation property of Zener diode with load resistance R_L, the input is maintained at a value greater than the zener voltage for the given diode. On gradually increasing the

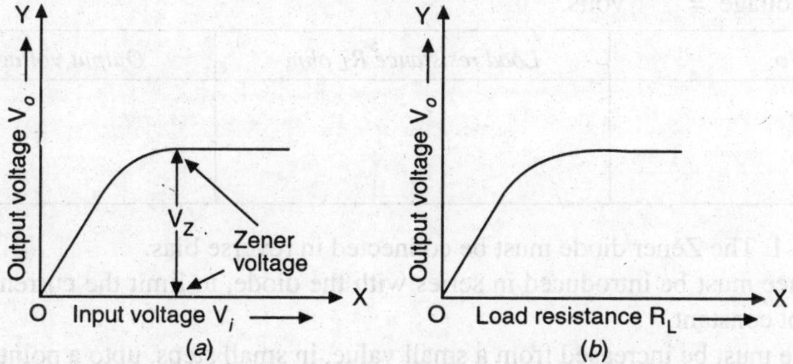

Fig. 47.10

load resistance it is found that for small values of load resistance the output voltage increases linearly with load. For large values of load resistance the output voltage remains constant. The variation of output voltage with load resistance is shown in Fig 47.10 (*b*)

Experiment 47.2. To study Zener diode voltage regulating characteristics.

Apparatus. A variable *D.C.* supply, a Zener diode, voltmeter and an ammeter with ranges according to zener diode supplied, two resistance boxes R_s and R_L, connecting wires etc.

Procedure. 1. Draw a diagram showing the scheme of connections as in Fig. 47.9 and make connections accordingly.

Note. The Zener diode must be connected in reverse bias.

2. Introduce a fixed resistance R_s (100 to 500 Ω) in series with the diode. Keep it constant throughout the experiment. Introduce a load resistance R_L of about 1000 ohm.

3. Starting from very low values of input voltage increase it in small steps well above the zener voltage for the given diode. For each observation note the input voltage V_i and corresponding output voltage V_0. Also note the current in the input circuit.

4. Plot a graph between input voltage V_i taken along the *X*-axis and output voltage V_0 taken along the *Y*-axis. The graph is as shown in Fig 47.10 (*a*).

5. **Voltage regulation with load.** Keep the input voltage at a value more than the zener voltage. Starting from a small value of about 50 ohm for R_L, increase it in steps of 50-100 ohm. Note the resistance and the corresponding values of output voltage.

6. Plot a graph between load resistance R_L taken along the *X*-axis and output voltage V_0 taken along *Y*-axis. The graph is as shown in Fig. 47.10. (*b*)

Observations.

Series resistance R_s = ohm

Load resistance R_L = ohm

Sl. No.	*Input voltage V_i*	*Current*	*Output voltage V_o*
	0 V		
	0.5 V		
	1.0 V		
	.		
	.		
	.		

Variation of output voltage with load resistance.

Series resistance R_s = ohm

Fixed input voltage = volts.

Sl. No.	*Load resistance R_L ohm*	*Output voltage V_0 volts*

Precautions. 1. The Zener diode must be connected in reverse bias.

2. A resistance must be introduced in series with the diode, to limit the current at breakdown and should be kept constant.

3. The voltage must be increased from a small value, in small steps, upto a point well above the zener voltage.

4. To study the variation of output voltage with resistance, the input voltage must be kept well above the zener voltage.

Conclusion. The graph between load resistance R_L and output voltage V_0 shows that the output voltage increases linearly with load but for large values of load resistance R_L, the output voltage remains constant, as shown in Fig 47.10(*b*)

Oral Questions

1. Semi-conductors (*i*) What is a semi-conductor? give examples (*ii*) What is doping? (*iii*) What is a donor and what is an acceptor element? (*iv*) What is a *P*- type and a *N*-type semi-conductor? (*v*) Which element makes Germanium *P–type* and which makes it *N–type* ? (*vi*) Name the mobile charge carrier in each type of semi-conductor ?

2. Diode. (*i*) What is a *P-N* junction diode? (*ii*) When is a junction diode forward biased and when is it reverse biased? (*iii*) Explain the rectifying action of a junction diode. (*iv*) Describe various uses of solid state diodes.

3. Zener diode (*i*) What is zener voltage? (*ii*) What do you mean by avalanche break down? (*iii*) On what factors does Zener voltage depend? (*iv*) What do you mean by voltage regulation characteristic of a Zener diode. (*v*) Can you use a Zener diode as an ordinary diode rectifier.

48

Thermistor

48.1. Thermistor. The resistance of metals generally increases with temperature. The variation of resistance with temperature is given by

$$R_t = R_0 (1 + \alpha t)$$

where α is the temperature co-efficient of resistance, R_t the resistance of the resistor at t °C and R_0 at 0 °C.

$$\therefore \qquad \alpha = \frac{R_t - R_0}{R_0 t}$$

In general if R_2 is the resistance at t_2 °C, R_1 at t_1 °C, then to a first approximation

$$\alpha = \frac{R_2 - R_1}{R_1 (t_2 - t_1)}$$

The word thermistor is derived from '*thermal resistor*'. A **thermistor** is a device whose resistance varies quite markedly with temperature. Depending upon their composition the thermistor can have either a *positive temperature co-efficient* (*p.t.c.*) or *a negative temperature co-efficient* (*n.t.c.*). The *n.t.c.* type consists of a mixture of oxides of iron, nickel and cobalt with small amounts of other substances and is used in electronic circuits to compensate for resistance increase in other components when the temperature rises and also as thermometer for temperature measurement. For example, it is possible that if the resistance of a thermistor at 0°C is 10,000 ohm, it may be 200 Ω at 100°C and only 10 Ω at 300°C. Thermistors are available in different shapes and sizes as shown

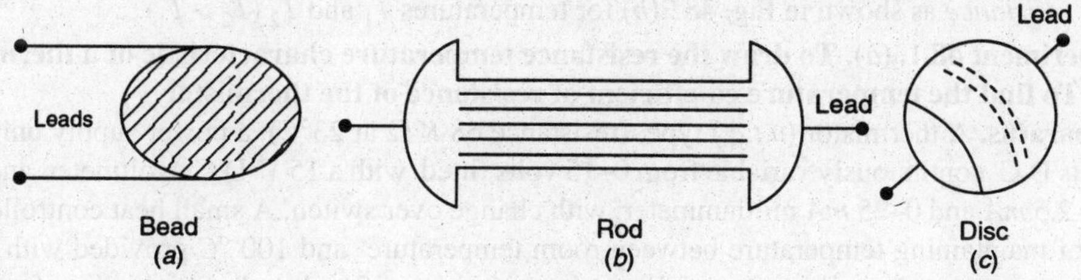

Fig. 48.1

The *p.t.c.* type is based on barium titanate and shows a resistance increase of 50 to 200 times for a temperature rise of a few degrees. It is useful as a temperature controlled switch. Such resistors are called **posistors** *Thermistors* and *posistors* are collectively known as *varistors i.e* their resistance changes markedly with temperature.

Metals and alloys have a positive temperature co-efficient *i.e.* they are *p.t.c.* materials. Graphite, semi-conductors and most non-metals have a negative temperature co-efficient *i.e.* they are *n. t. c.* materials.

48.2. Resistance temperature (*R-T*) characteristics. The variation of resistance with temperature of a thermistor is given by

$$R_2 = R_1 \exp\left[\beta\left(\frac{1}{T_1} - \frac{1}{T_2}\right)\right]$$

where β is a constant which depends upon the material of the thermistor.

R_1 is the resistance of thermistor at temperature T_1 *K* and R_2 is the resistance of thermistor at temperature T_2 *K*.

A graph between *R* and *T* shows a rapid fall in resistance with temperature [Fig. 48.2.(*a*)].

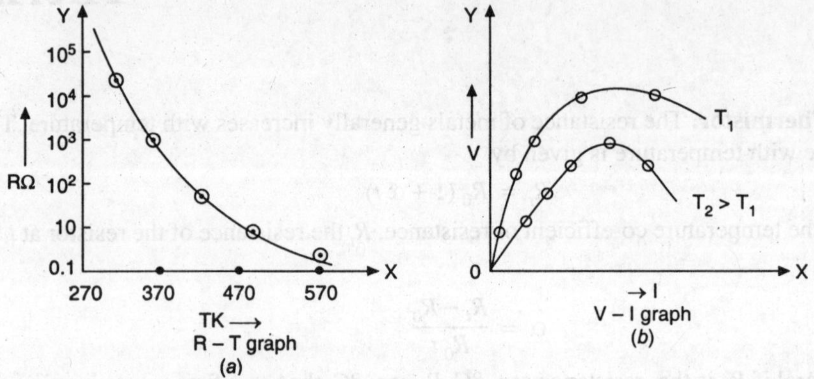

Fig. 48.2

48.3. Voltage current (*V-I*) characteristics. When a voltage is applied to a thermistor at constant temperature, at first the current increases slowly with increase in voltage *i.e.* Ohm's law is obeyed to some extent. As the applied voltage is further increased the temperature of the thermistor increases which decreases the resistivity. As a result there is an increase in current, which further increases the temperature of the thermistor causing a further fall in resistance and consequent increase in current. The process continues till the heat dissipated by the thermistor is equal to the electrical power supplied to the thermistor.

Thus to begin with current increases with increase in voltage till a maximum value of voltage is reached. Beyond this voltage the current increases with decrease in voltage i.e. the thermistor has a negative resistance as shown in Fig. 48.2(*b*) for temperatures T_1 and T_2 ($T_2 > T_1$).

Experiment 48.1. (*a*). To draw the resistance temperature characteristic of a thermistor (*b*). To find the temperature co-efficient of resistance of the thermistor.

Apparatus. A thermistor (*n.t.c.*) type, (resistance 68 *K* Ω at 23°*C*), a power supply unit giving 0–15 volts D.C. continuously variable from 0–15 volts fitted with a 15 *V* D.C. voltmeter and a dual range, 0–2.5 *mA* and 0–25 *mA* milliammeter, with change over switch, A small heat controlled oven capable of maintaining temperature between room temperature and 100 °*C* provided with a panel type thermometer or thermistor mounted on a base having a 12 volt bulb fitted below for heating purposes provided with two stage intensity adjustment switch at 'LOW' and 'HIGH' points and fitted with a clamp above in which a half degree thermometer can be inserted to read the temperature of the thermistor.

Procedure 1. Connect the thermistor (mounted on the base) fitted with a lamp below and a thermometer clamped above it, through a double range (2.5 m A and 25 m A) milli-ammeter provided with a range switch to 15 V power supply. Connect a 15 V voltmeter in parallel with the thermistor. Connect the lamp to 12 V supply and keep the intensity switch in the low position.

2. Keep the milli-ammeter at 2.5 mA range. Switch on the current in the thermistor circuit and adjust the voltage so that the voltmeter reads 12-13 volts. Note the reading of the voltmeter and milli-ammeter. Also note the temperature.

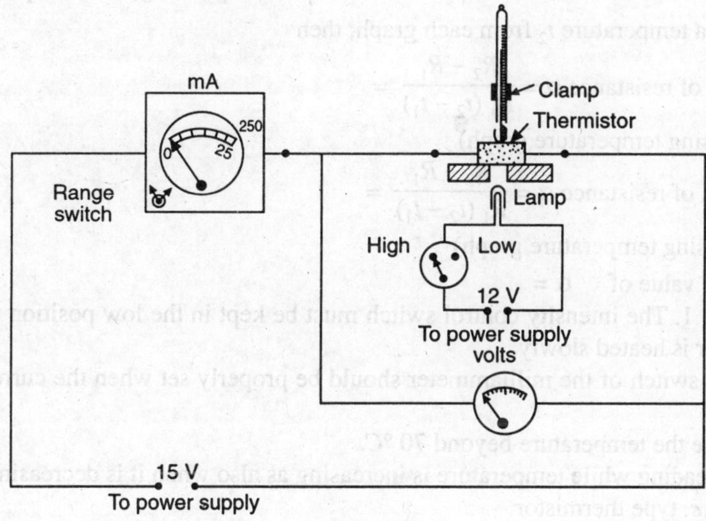

Fig. 48.3

3. Now switch on the power supply to the lamp (kept in the LOW intensity position). The temperature of the thermistor will slowly rise and its resistance will fall thereby increasing the current. Note the value of the current, voltage and temperature, for every increase of about 5°*C* till the current is about 2.5 m A.

4. Now set the range switch to 25 mA range and intensity switch to HIGH position (if necessary) and again note the value of current, voltage and temperature for every increase of about 5°*C* till the temperature reaches to about 65°*C*. *Do not increase the temperature beyond 70°C in any case.*

5. Now switch off the lamp. The temperature will slowly fall. Again note the value of the current, voltage and temperature for every decrease of about 5°*C* till the current is about 2.5 mA. Now set the milliammeter to the 2.5 mA range and go on noting the current, voltage and temperature for every 5 °*C* decrease of temperature till the room temperature is reached.

Observations.

Sl. No.	*Temp (rising)*	*Current mA*	*Voltage (Volts)*	*Resistance (K Ω)*
	Temp. (falling)			

6. (*a*) Taking temperature along *X*-axis and corresponding resistance in K Ω along *Y*-axis draw a graph between resistance and temperature (*i*) when the temperature is rising and (*ii*) when the temperature is falling.

7. (*b*) To find the temperature co-efficient of resistance we shall assume that the temperature co-efficient of resistance is given by

$$\alpha = \frac{R_2 - R_1}{R_1 (t_2 - t_1)}$$

as in the case of conductors. Note the resistance R_1 corresponding to a temperature t_1 and R_2 corresponding to a temperature t_2 from each graph, then

Temp. coeff. of resistance $\alpha = \frac{R_2 - R_1}{R_1 (t_2 - t_1)} =$

(from increasing temperature graph)

Temp. coeff. of resistance $\alpha = \frac{R_2 - R_1}{R_1 (t_2 - t_1)} =$

(from decreasing temperature graph)

Mean value of $\alpha =$

Precautions. 1. The intensity control switch must be kept in the low position to begin with so that the thermistor is heated slowly.

2. The range switch of the milliammeter should be properly set when the current goes beyond scale.

3. Do not take the temperature beyond 70 °C.

4. Take the reading while temperature is increasing as also when it is decreasing.

5. Use an *n.t.c.* type thermistor.

Experiment 48.2. To draw the voltage current (V-I) characteristics of a thermistor.

Apparatus. Same as in **Expt. 48.1.**

Procedure 1. Take a thermistor (mounted on the base, with a thermometer clamped above it and a lamp below it connected to a 12*V* supply through an ammeter, a rheostat and a thermostat to maintain the thermistor at a constant temperature during one set of observations. Alternatively the thermistor may be placed in a bath whose temperature can be maintained constant.

2. Draw a diagram showing the scheme of connections and make connections accordingly [Fig. 48.4].

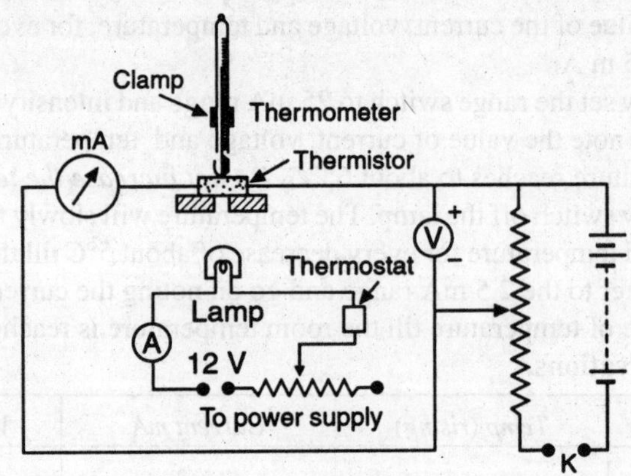

Fig. 48.4

3. Maintain the thermistor at a constant temperature T_1. Go on slowly increasing the potential applied to the thermistor. Go on noting the values of the potential from the voltmeter *V* and the corresponding values of current from the milli-ammeter (mA). A stage comes when with increase of current potential falls. When this transition stage is reached take only 2-3 readings more and stop.

4. Now maintain the thermistor at a constant temperature T_2. Again go on slowly increasing the potential and note the values of potential and corresponding values of current. A stage again comes when with increase of current potential falls. When this stage is reached stop after taking 2-3 readings.

Observations

S. No.	*Temperature of the thermostat or bath*	*Voltmeter reading (volts)*	*Ammeter reading (mA)*
1			
2			
3			
4	T_1 =		
5			
6			
7			
8			
1			
2			
3			
4	T_2 =		
5			
6			
7			
8			

5. Plot a graph taking current along *X*-axis and potential along *Y*-axis separately for temperature T_1 and T_2. The graph is shown in Fig. 48.2(*b*)

Conclusion. At first the current increases with increase of voltage till a maximum value of voltage is reached. Thereafter with increase in voltage current decreases. The higher the temperature, the lower is the maximum value of voltage.

Precautions. 1. Only 2-3 readings must be taken after the transition stage is reached.

2. The temperature of the bath or thermistor must be kept constant during one set of observations.

Oral Questions

1. Thermistor (*i*) What is a thermistor? (*ii*) What are various types of thermistors? (*iii*) What do you mean by *p.t.c.* and *n.t.c.* type thermistor? (*iv*) Give some examples of two types of thermistors, (*v*) What are the practical advantages of thermistors? (*vi*) What do you mean by negative resistance.

49

Band gap of Semi-conductors

49.1. Band gap of semi-conductor. The highest filled energy band which includes electrons shared in co-valent bonds or transferred in ionic bonds in a semi-conductor is known as *valence band.*

The energy level corresponding to the top of the valence band in an intrinsic semi-conductor is denoted as E_v.

The energy band which includes free electrons is known as the *conduction band.*

The energy level corresponding to the bottom of the conduction band in an *intrinsic semi-conductor* is denoted as E_c.

The energy gap between the top of the valence band and bottom of the conduction band is known as band gap.

$\therefore$ Band gap $E_g = E_c - E_v$

This region between the two energy bands is known as *forbidden band.* For a semi-conductor the energy gap for the forbidden band or band gap is of the order of $1 V$. For example, the energy gap for *Si* is $1.1 V$ and for *Ge* is $0.7 V$.

49.2. Fermi-level. In between the valence band energy level and conduction band energy level, there is another level known as *Fermi level.* It is denoted as E_F.

According to Fermi-Dirac distribution law an energy state corresponding to the Fermi-level has a probability of 1/2 of being occupied by an electron.

For every electron which moves from the valence band to the conduction band a *hole* is created in the valence band. A hole carries the same positive charge as the negative charge on an electron and behaves like a positive charge carrier in the same manner as the electron behaves as a negative charge carrier.

For an intrinsic semi-conductor the concentration of electrons in the conduction band is equal to the concentration of holes in the valence band and hence the Fermi level lies in the *middle* of the band gap as shown in Fig. 49.1 (*a*).

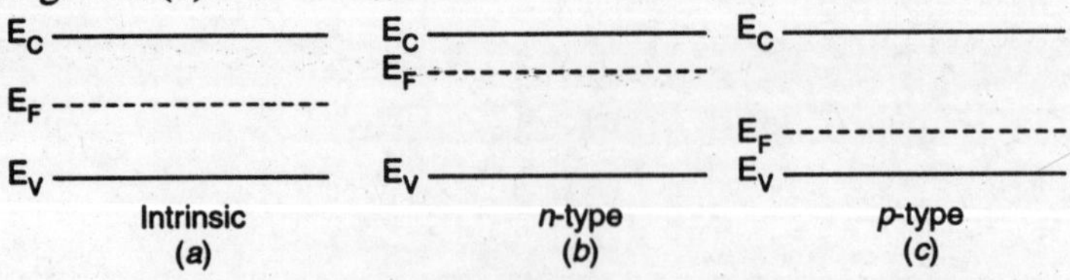

Fig. 49.1

When a semi-conductor is doped with a pentavalent impurity it has a very high concentration of electrons in the conduction band and it is then known as *n*-type extrinsic semi-conductor.

Similarly when a semi-conductor is doped with a trivalent impurity it has a very high concentration of holes in the valence band and it is then known as *p*-type extrinsic semi-conductor.

In an n-type semi-conductor there is a high concentration of electrons in the conduction band as compared to the hole concentration in valence band, the Fermi level, therefore, lies near the conduction band below the bottom of the band as shown in Fig. 49.1 (*b*).

In a p-type semi-conductor there is a high concentration of holes in the valence band as compared to the electron concentration in conduction band, the Fermi level, therefore, lies near the valence band above the top of the band as shown in Fig. 49.1 (*c*).

49.3. P-N junction diode. A p-n junction diode consists of two separate regions of p and n type semi-conductor materials brought together to form a junction with uniform p-type doping on one side and a uniform n-type doping on the other side.

At the junction the valence and conduction band energy levels on the p-side are higher than the corresponding energy levels on the n-side by an amount called the contact potential difference as shown in Fig. 49.2, where E_{cp}, E_{Fp} and E_{vp} denote the conduction band energy level, Fermi level and valence band energy level for the p-material and E_{cn}, E_{Fn} and E_{vn} corresponding values for the n-material.

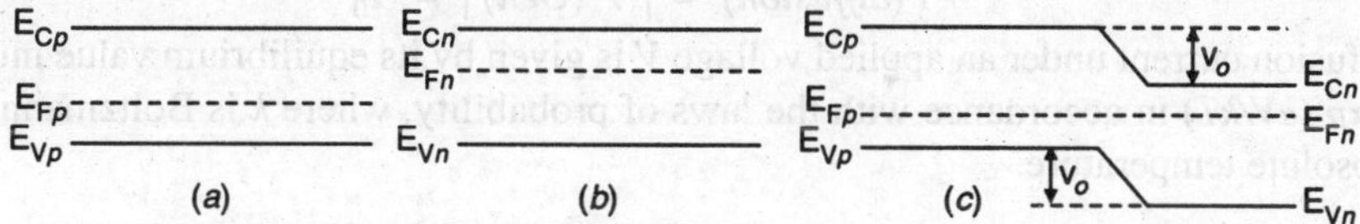

Fig. 49.2

This is due to the fact that Fermi level should remain constant through the whole device. Any gradient in the Fermi level will mean a net current which is not possible under equilibrium conditions.

The contact potential is given by

$$e\, V_0 = E_{cp} - E_{cn} = E_{vp} - E_{vn}$$

where e is the charge on the electron.

The n-material has a large concentration of electrons and a few holes whereas the p-material has a large concentration of holes and a few electrons. When a junction is formed the holes diffuse from the p-side into the n-side and electrons diffuse from n to p. The holes diffusing from p to n and electrons from n to p both give rise to a *conventional diffusion current* from p to n.

The diffusion of holes from p to n gives rise to a positive space charge at the boundary of the n-side of the junction. Similarly the diffusion of electrons from n to p gives rise to a negative space charge at the boundary of the p-side of the junction. Thus both produce an electric field from n to p.

This electric field allows the few holes in the the n- material to slowly shift from n to p and the few electrons in the p-material from p to n both giving rise to a drift current from n to p in a direction opposite to the diffusion current.

At equilibrium drift and diffusion currents for each type of carrier (holes as well as electrons) separately cancel each other and there is no net current.

When a forward bias is applied to the p-n junction *i.e.* the p-side is connected to the positive and the n-side to the negative of a battery the diffusion current composed of *majority carrier electrons* flowing from n to p-side and *majority carrier holes* from p to n-side rises rapidly due to the lowering of the potential barrier by the forward bias.

For reverse bias *i.e.* when p-side is connected to the negative and n-side to the positive of the battery the potential barrier becomes very large and virtually no electrons in the n-side conduction band and holes in the p-side valence band have enough energy to overcome the barrier. Hence, the diffusion current is almost negligible for reverse bias.

The drift current composed of minority carrier holes flowing from n to p-side and electrons from p-side to n-side is, however, independent of the applied voltage. This is because the minority carriers on each side of the junction which give rise to the drift component of the current are generated by thermal excitation of electron-hole pairs (EHP). The magnitude of the drift current depends entirely on the *rate of generation* of electron-hole pairs which depends only on temperature and not on the applied voltage. The drift current is also known as generation current *I* (*Gen.*).

The total current across the *p-n* junction is the sum of the diffusion and drift components. The net current at equilibrium is zero.

Under reverse bias the diffusion current due to both the types of carriers (electrons and holes) is negligible because of large potential barrier. The only current is, therefore, drift current or generation current *I* (*Gen*) due to carrier generation in the transition region or minority carriers which drift to the junction and are collected.

At equilibrium applied voltage $V = 0$

$$I = I\,(diffusion) - |\,I\,(Gen.)\,| = 0$$

In other words equilibrium diffusion current is equal in magnitude to the generation current *I* (Gen)

or
$$I\,(diffusion) = |\,I\,(Gen)\,| = I_0$$

The diffusion current under an applied voltage *V* is given by its equilibrium value multiplied by the factor *exp* (*eV/kT*) in accordance with the laws of probability, where *k* is Boltzmann's constant and *T* the absolute temperature.

$$\therefore \quad I\,(diffusion) = I_0 \; exp\left(\frac{eV}{kT}\right)$$

As
$$I = I\,(diff) - |\,I\,(Gen)\,|$$

$$\therefore \quad I = I_0 \; exp\left(\frac{eV}{kT}\right) - I_0$$

$$= I_0\left[exp\left(\frac{eV}{kT}\right) - 1\right]$$

When *V* is *positive* as in a *forward bias* and is greater than a few *e/kT*, the exponential term is much greater than unity and *I* increases exponentially with forward bias. When *V* is *negative* as in *reverse bias* the exponential term approaches zero and

$$I = -I_0$$

which is in the *negative n-p* direction. This current is known as *reverse saturation current* and denoted as I_s

$$\therefore \quad I = -I_s\left[exp\left(\frac{eV}{kT}\right) - 1\right] \qquad ...(i)$$

The reverse saturation current is given by

$$I_s = Ae\left(n_p\, v_n - p_n\, v_p\right) \qquad ...(ii)$$

where A = area of the junction

e = electronic charge

n_p = concentration of electrons in *p*-material

p_n = concentration of holes in *n*-material

v_n = mean velocity of electrons

v_p = mean velocity of holes

Now
$$n_p = N_n\, exp\left[-\left(E_{cp} - E_{Fp}\right) / kT\right]$$

$$p_p = N_p \, exp \left[-\left(E_{Fp} - E_{vp}\right) / kT \right]$$

$$n_n = N_n \, exp \left[-\left(E_{cn} - E_{Fn}\right) / kT \right]$$

$$p_n = N_p \, exp \left[-\left(E_{Fn} - E_{vn}\right) / kT \right]$$

where n_p , p_n , E_{cp} , E_{Fp} , E_{vp} , E_{cn} , E_{Fn} and E_{vn} have the same meaning as already explained.

Further $p_p =$ Concentration of holes in p-material

$n_n =$ Concentration of electrons in n-material

$$N_n = \frac{2\,(2\,\pi\, m_n\, kT)^{3/2}}{h^3}$$

and $$N_p = \frac{2\,(2\,\pi\, m_p\, kT)^{3/2}}{h^3}$$

where m_n is the effective mass of the electron, m_p, the effective mass of the hole and h the Planck's constant.

As already stated, at the junction Fermi levels in the n and p materials adjust to equalise so that

$$E_{Fp} = E_{Fn}$$

Further $$n_p\, p_p = n_n\, p_n = N_n\, N_p \; exp\,[-\Delta\, E \;/ kT]$$

where $$\Delta\, E = E_{cp} - E_{vp} = E_{cn} - E_{vn}$$

Hence $$n_p = \frac{N_n\; N_p}{p_p}\; exp\;[-\Delta\, E\; / kT]$$

and $$p_n = \frac{N_n\; N_p}{n_n}\; exp\;[-\Delta\; E / kT]$$

Substituting the values of n_p and p_n in (*ii*), we have

$$I_s = A\, e \left[N_n\, N_p \left(\frac{v_n}{p_p} + \frac{v_p}{n_n} \right) exp\,(-\,\Delta\, E / kT) \right]$$

At temperatures where *donor* and *acceptor* levels are freely activated, the density of electrons n_n in the n-type and density of holes p_p in the p-type material will be approximately equal to impurity concentration and will be constant for a small range of temperature. Further when the junction is appreciably reverse biased the exponential term in relation (*i*) approaches zero and

$$I = -\, I_s$$

Hence $$-\log_e I_s = \log_e \left[-\,Ae\, N_n\, N_p \left(\frac{v_n}{p_p} + \frac{v_p}{n_n} \right) \right] + \frac{-\Delta E}{kT} \quad ..(iii)$$

Over a small range of temperature I_s depends only on term containing $1/T$, as the other term is a constant being independent of temperature. Hence relation (*iii*) can be put as

$$2.303 \log_{10} I_s = \text{constant} - \frac{\Delta E}{kT} \quad ...(iv)$$

If I_s is in micro - amperes and ΔE in eV, then

$$2.303 \log_{10} (I_s \times 10^{-6}) = \text{constant} - \frac{\Delta\, E \times 1.6 \times 10^{-19}}{kT}$$

or $$\log_{10} I_s = \text{Another constant} - \frac{\Delta E \times 1.6 \times 10^{-19}}{2.303 \times 1.38 \times 10^{-23}\, T}$$

where $k = 1.38 \times 10^{-23}\ J/K$

or
$$\log_{10} I_s = \text{a constant} - 5.036 \times \Delta E \times \frac{10^3}{T} \qquad ...(v)$$

Hence a graph between $\log_{10} I_s$ and $\frac{10^3}{T}$ is a straight line of which the slope is $\frac{\log_{10} I_s}{10^3/T}$

$\therefore$ Slope of the line $= -5.036\,\Delta E$

or Band gap of the semi-conductor $\Delta E = \frac{\text{slope of the line}}{5.036}$

Experiment 49.1. (*a*) To draw the characteristics of a p-n junction diode for current under reverse bias and temperature.

(*b*) To find the band gap of the semi-conductor material.

Apparatus. A junction diode, a *D.C.* power supply, a small heat controlled oven, a thermometer, a micro-ammeter, connecting wires etc.

Procedure. (*a*) 1. A reverse biased *p-n* junction diode is shown in Fig. 49.3 (*a*). Insert the *p-n* junction diode in the oven and connect its two leads to the terminals provided for the purpose. Connect the terminals through a 50 μ *A D.C.* micro-ammeter to a *D.C.* supply so that the diode is reverse biased. Insert a thermometer in its socket.

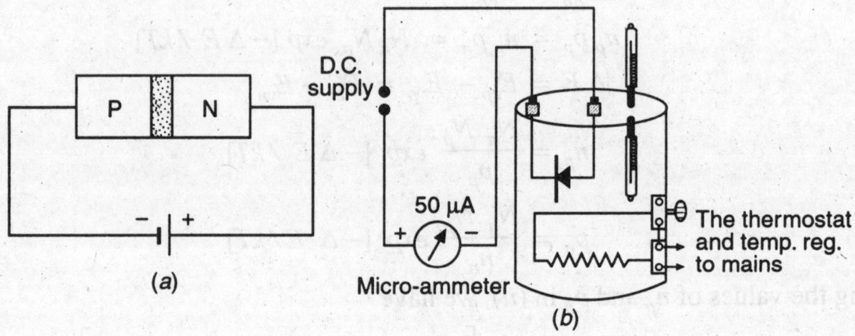

Fig. 49.3

2. Connect the heating element of the oven to *A.C.* main supply. Switch on the mains and allow the temperature of the oven to rise to about 70–75°*C*. Adjust the thermostat and temperature regulator so that the temperature remains steady at 70°*C*.

3. Note the temperature, Adjust the reverse bias so that the micro-ammeter reads about 30 μ *A*. Note the current reading.

4. Now switch off the mains supply, so that the oven cools slowly. As the temperature falls the micro-ammeter reading goes on decreasing. When the current decreases by 5 μ *A* note the current and the corresponding reading of temperature. Go on noting the current and corresponding temperature for every 5 μ *A* decrease in current in the diode circuit.

Observations.

No.	*Current I_s (in μA)*	*Temp. °C*	*Temp. T K*	$\frac{10^3}{T}$	$\log_{10} I_s$

(*b*) 5. Taking $10^3/T$ along the *X*-axis and $\log_{10} I_s$ along the *Y*-axis, plot a graph between $\log_{10} I_s$ and $10^3/T$. The graph is a straight line as shown in Fig. 49.4. Take two points *A* and *B* on the graph. From *A* draw a line parallel to *Y*-axis and from *B* a line parallel to *X*-axis meeting at *C*. Then slope of the line $= \dfrac{AC}{BC}$

$$\therefore \quad \text{Band gap } \Delta E = \frac{\text{Slope}}{5.036} = \quad eV.$$

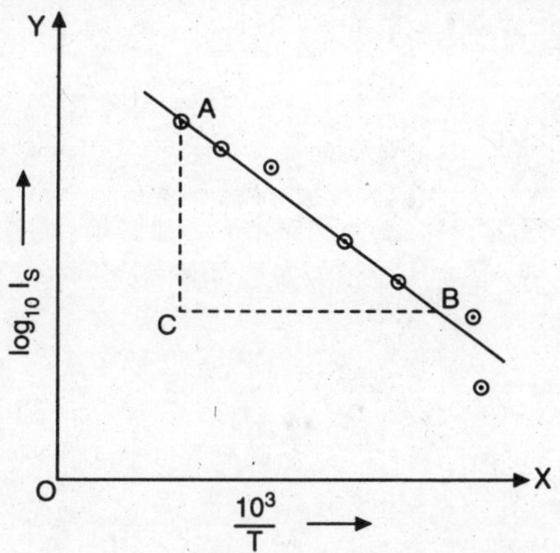

Fig. 49.4

Precautions. 1. A reverse biased *p-n* junction diode must be used.

2. Reading of current must be taken when the temperature is decreasing.

3. Reading of current and temperature must be taken simultaneously.

Oral Questions

1. Band gap of semi-conductor (*i*) What do you understand by band gap of a semi-conductor? (*ii*) What do you mean by valence band, conduction band and forbidden band? (*iii*) What is the value of band gap energy for silicon and for germanium? (*iv*) What do you mean by Fermi level? (*v*) What is the position of Fermi level in an intrinsic semi-conductor and in a *p*-type or *n*-type semi conductor with respect to the positions of conduction and valence bands? (*vi*) How is the Fermi level adjusted in a *p-n* junction diode? (*vii*) Distinguish between diffusion current and generation current in a semi-conductor. (*viii*) State the formula for reverse saturation current in a semi-conductor diode and state what various symbols represent. (*ix*) What is the shape of the graph between log I_s and $1/T$? (*x*) How do you find band gap energy from this graph?

50

Rectifiers and Filters

50.1. Rectification. *Rectification is the process of converting an alternating current (a.c.) into a unidirectional direct current* (*d.c.*)

A diode valve as well as a semiconductor *P–N* –junction diode are almost perfect rectifiers and are used for producting rectification.

When a single diode valve or only one *P–N*–junction diode is used as a rectifier, rectification of only one half of *a.c.* wave form takes place. This is known as **half wave rectification.**

Full wave rectification. In this both sides of the alternating current are obtained after rectification. Full wave rectification is brought about by using two single diodes (or a double diode) or two *P–N* junction diodes (or four *P–N* junction diodes forming a bridge).

50.2. Diode valve rectifiers. (*i*) *Half wave rectifier*. The circuit for a half wave diode valve rectifier is shown in Fig. 50.1.

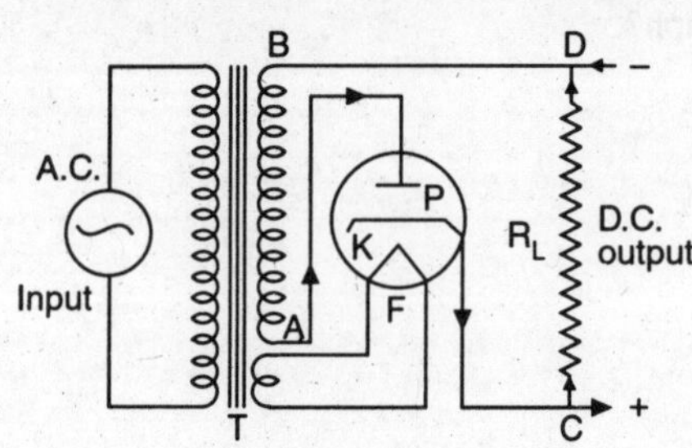

Fig. 50.1

The alternating current supply is fed to the primary of a transformer *T*. The alternating voltage developed in the secondary of the transformer *T* is applied to the plate of the diode. The filament of the diode is heated by the current from a separate secondary winding of a few turns on the transformer *T*. The direct current supply is taken across the load resistance R_L placed in series with the valve.

The wave form of the *A.C.* input voltage which appears across the secondary of the transformer is shown in Fig. 50.2. (*a*) (*i*).

When the direction of the alternating current is such that the end *A* of the secondary of the transformer is at a positive potential electrons flow from the cathode to the plate. Thus the conventional current flows from the plate to the cathode, as shown, in the whole circuit. The end *C* of the resistance *R* is at a positive potential with respect to the end *D* and the current flows from *C* to *D*. This continues for the (positive) half cycle of *A.C.* waveform and the current in the plate circuit varies exactly in the same manner as the voltage applied to it, as shown in Fig. 50.2. (*a*) (*ii*).

In the second (negative) half of the cycle, the alternating current is such that the end *A* is at a negative potential. The electrons are then repelled and no current flows in the plate circuit. Hence the current in the plate circuit flows in *one direction* but it only flows for one-half of the cycle and does not flow for the remaining half of the cycle. The output current is thus not continuous. It is also not constant but varies between zero and a certain maximum value.

In this case only one half of the alternating voltage wave is utilised and this rectification is called *half-wave rectification*. The output current of a half wave rectifier is *pulsating, intermittent* but *unidirectional*.

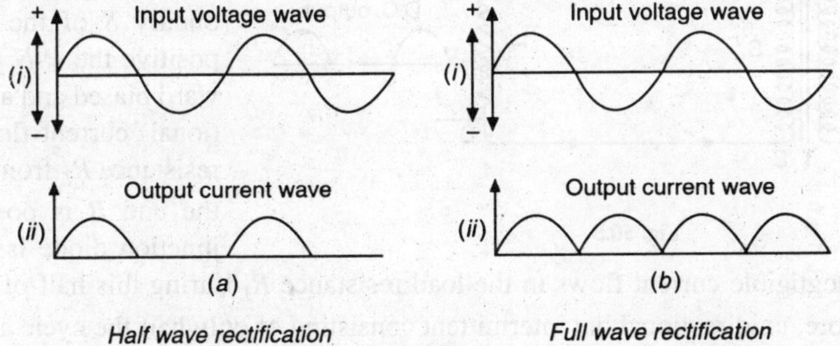

Fig. 50.2

(*ii*) **Full wave rectifier.** A full-wave rectifier using a double diode is shown in Fig. 50.3. A double diode has two plates and only one filament and in action it is equivalent to two diodes. In this case the centre E of the secondary of the transformer T, is connected to one end D of the load resistance R_L. The two ends A and B of the transformer secondary are connected to the plates P_1 and P_2 respectively. When the end A is at a positive potential with respect to the centre tap E, the plate P_1 is positive and electrons flow from the cathode to the plate P_1. This is equivalent to a conventional current from P_1 to the cathode in the direction indicated. The end C is at a positive potential whereas the end D is at a negative potential.

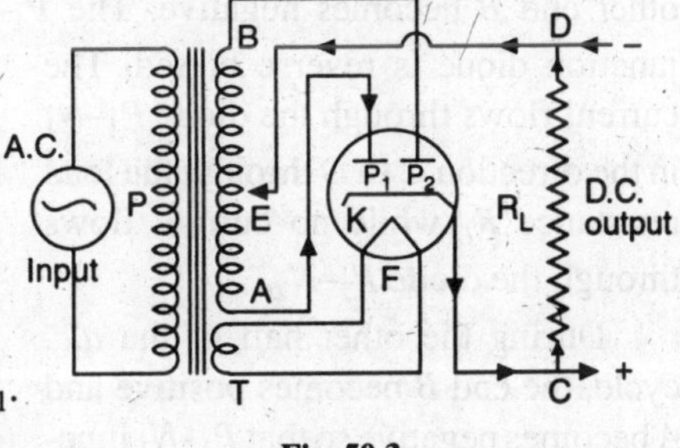

Fig. 50.3

After half the cycle, the end B is at a positive potential and the electrons flow from the cathode to the plate P_2. The (conventional) current in the outside circuit again flows in the same direction making C positive and D negative. Thus, the current in both the cases flows in the same direction, no matter which plate is positive.

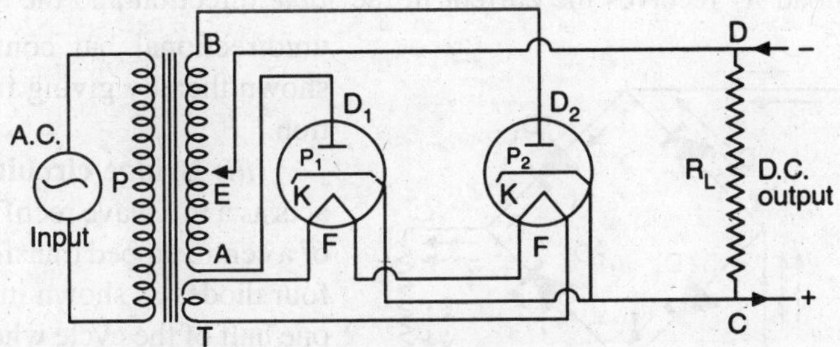

Fig. 50.4

(*iii*) The circuit of a full wave rectifier using **two separate single diodes** is shown in Fig 50.4. Full wave rectification is brought about in the same manner as by a double diode.

The *A.C.* applied voltage is shown in Fig 50.2. (*b*) (*i*) and the unidirectional current in the load resistance after full wave rectification is shown in Fig. 50.2 (*b*) (*ii*). The output current of a full wave rectifier is *unidirectional, pulsating* and *continuous*.

50.3. Semi conductor P-N junction diode rectifiers (*i*) **Half wave rectifier**. The basic circuit of a *P-N* junction diode as a half wave rectifier is shown in Fig. 50.5. When the end *A* of the secondary *S* of the transformer is positive the *P-N* junction is forward biased and a large (conventional) current flows in the load resistance R_L from *C* to *D*. When the end *B* is positive, the *P-N* junction diode is reverse biased and an almost negligible current flows in the load resistance R_L during this half of the cycle. The output is, therefore, unidirectional but intermittent consisting of only half the cycle as shown.

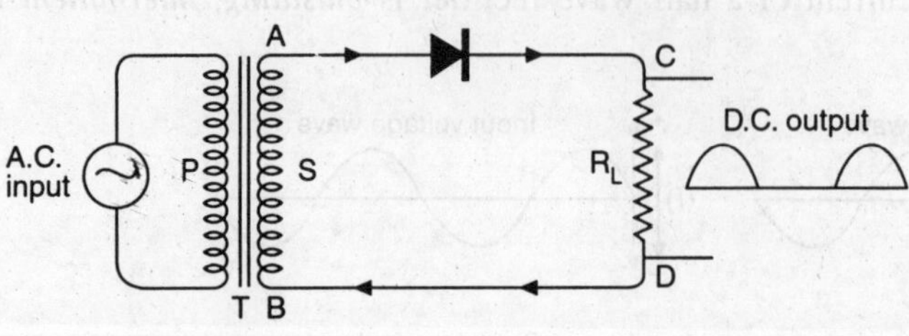

Fig. 50.5

(*ii*) **Full wave rectifier.** The basic circuit of a full wave rectifier using two *P-N* junction diodes is shown in Fig. 50.6.

During the first half of the *a.c.* cycle one end of the secondary say *A* becomes positive and the other end *B* becomes negative. The P_1–N_1 junction diode is forward biased, whereas the P_2–N_2 junction diode is reverse biased. The current flows through the diode P_1–N_1 in the direction *C* to *D* through the load resistance R_L while no current flows through the diode P_2–N_2.

During the other half of the *a.c.* cycle, the end *B* becomes positive and *A* becomes negative so that P_1–N_1 junction is reverse biased while the P_2–N_2 junction is forward biased. Thus the current flows through the junction diode P_2–N_2 in the direction *C* to *D* through the load resistance R_L while no current flows through P_1–N_1. Thus during both halves, the load R_L receives the current in the *same* direction and the output consists of unidirectional but continuous pulses as shown thereby giving full wave rectification.

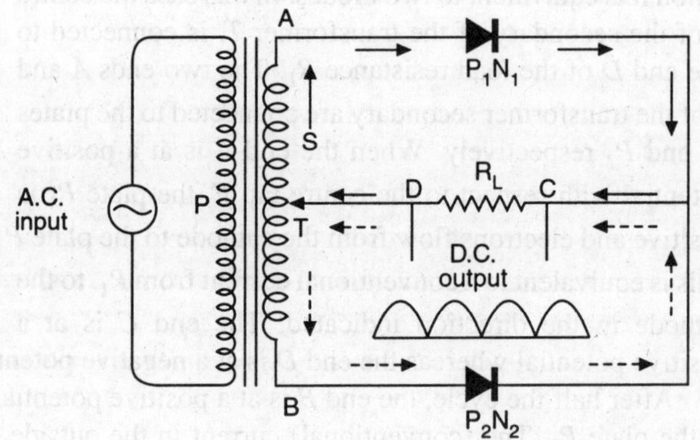

Fig. 50.6

(*iii*) **Bridge circuit.** A bridge circuit acts as a full wave rectifier without the use of a centre tapped transformer but requires four diodes as shown in Fig. 50.7. During one half of the cycle when the end *A* of the secondary of the transformer is positive the diodes D_1 conduct and current flows in the load resistance R_L from *C* to *D* as shown by full arrows. During the second half when the end *B* is positive the dioides D_2 conduct and the current through the load resistance R_L again flows from *C* to *D* as shown by dotted arrows.

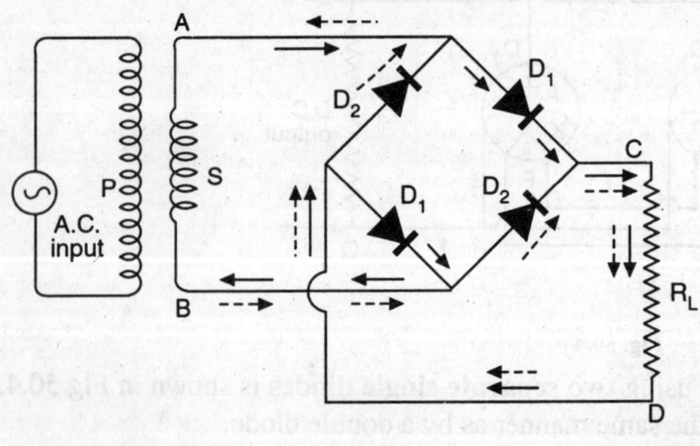

Fig. 50.7

The advantages of a bridge rectifier are :-

(*i*) No centre tap is required in the secondary of the transformer

(*ii*) Current in both, the primary and the secondary of the transformer flows for the entire cycle. Hence for a given power output, power transformer of smaller size can be used.

(*iii*) Since two rectifier elements are present in each conduction path, peak inverse voltage is shared equally. The bridge rectifier circuit is, therefore, suitable for high voltage applications.

50.4. Efficiency. *The efficiency of a rectifier is defined as the ratio of d.c. output power P(d.c) to the a.c. input power P(a.c) supplied to the rectifier.*

$$\eta = \frac{\text{d.c. power output } P\ (d.c)}{\text{a.c. power input } P\ (a.c.)}$$

Now $$P\ (d.c.) = I^2\ (d.c.)\ R_L$$

where $I\ (d.c.)$ is the *average value* of output current through the load resistance R_L

$$P\ (a.c.) = I^2_{r.m.s} \times (R_L + r)$$

where r is the diode resistance

For a half wave rectifier $$I\ (d.c.) = \frac{I_0}{\pi}$$

and $$I\ (r.m.s.) = \frac{I_0}{2}$$

$$\therefore \text{ Efficiency } \eta = \frac{P\ (d.c.)}{P\ (a.c.)} = \frac{\frac{I_0^2}{\pi^2} R_L}{\frac{I_0^2}{4}(R_L + r)} = \frac{4}{\pi^2} \frac{R_L}{R_L + r} = \frac{0.406}{1 + \frac{r}{R_L}}$$

If r is very small as compared to R_L $\eta = 0.406 = 40.6\%$

For a full wave rectifier $$I\ (d.c.) = \frac{2\ I_0}{\pi}$$

and $$I\ (r.m.s.) = \frac{I_0}{\sqrt{2}}$$

$$\therefore \text{ Efficiency } \eta = \frac{P\ (d.c.)}{P\ (a.c.)} = \frac{\frac{4 I_0^2}{\pi^2} R_L}{\frac{I_0^2}{2}(R_L + r)} = \frac{8}{\pi^2} \frac{R_L}{R_L + r} = \frac{0.812}{1 + \frac{r}{R_L}}$$

If r is very small as compared to R_L, $\eta = 0.812 = 81.2\%$

50.5. Ripple Factor. The current supplied by a rectifier and the voltage developed by it across a load (resistance) is unidirectional but its magnitude is constantly fluctuating *i.e.*, the output contains a *d.c.* as well as an *a.c.* component. The *a.c.* component contained in the output of a rectifier is known as *ripple*.

The ratio of the effective value of the alternating component of current (or voltage) to the average d.c. component of output current (or voltage) is known as ripple factor

$\therefore$ Voltage ripple factor $\gamma = \dfrac{E\ (a.c)}{E\ (d.c)}$

Current ripple factor $\gamma = \dfrac{I\ (a.c)}{I\ (d.c.)}$

Ripple factor for a half wave and a full wave rectifier.

The output current contains an *a.c.* component as well as a *d.c.* component. The root mean square value (*r.m.s.*) of the output current is given by

$$I^2_{r.m.s.} = I^2_{d.c.} + I^2_{a.c.}$$

$$\therefore \quad I_{(a.c.)} = \sqrt{I^2_{rms} - I^2_{(dc)}}$$

$$\therefore \quad \gamma = \frac{I_{(a.c.)}}{I_{(d.c.)}} = \frac{\sqrt{I^2_{rms} - I^2_{dc}}}{I_{dc}}$$

$$= \left[\left(\frac{I_{r.m.s.}}{I_{(dc)}}\right)^2 - 1\right]^{1/2}$$

$I_{(d.c.)}$ *is the value of the current measured by a d.c. ammeter* placed in series with the load resistance.

$I_{r.m.s.}$ *is the value of the current measured by an a.c. ammmeter* connected in series with the load resistance.

For a half wave rectifier

$$I_{r.m.s.} = \frac{I_o}{2} \text{ and } I_{dc} = \frac{I_o}{\pi}.$$

$$\therefore \text{ Ripple factor} \qquad \gamma = \left\{\left(\frac{I_0^2}{4} \times \frac{\pi^2}{I_o^2}\right) - 1\right\}^{1/2} = 1.21$$

For a full wave rectifier

$$I_{r.m.s.} = \frac{I_o}{\sqrt{2}} \text{ and } I_{d.c.} = \frac{2\,I_0}{\pi}$$

$$\therefore \text{ Ripple factor} \qquad \gamma = \left\{\left(\frac{I_o^2}{2} \times \frac{\pi^2}{4I_0^2}\right) - 1\right\}^{1/2} = 0.482.$$

Ripple factor in terms of voltage. In terms of voltage

$$\text{Ripple factor} \qquad \gamma = \left[\left(\frac{E_{r.m.s.}}{E_{d.c.}}\right)^2 - 1\right]^{1/2}$$

50.6. Filter circuits. The output of even a full wave rectifier is unidirectional and continuous but it is not steady and has a pulsating character. In most of the electrical and electronic circuits a *steady* direct current is required. It is, therefore, necessary to reduce the *ripple* of a rectifier output to a minimum by keeping away or smoothing out the *a.c.* component.

The process of reducing the ripple of a rectifier output is called *filteration*. A device which allows to pass *d.c.* component to the load and blocks the *a.c.* component of the rectifier output is called a *filter circuit*. A filter circuit is installed between the rectifier output and the load and consists of a suitable combination of inductors and capacitors.

(*a*) **Shunt capacitor filter.** The simplest filtering device is a capacitor C connected in parallel with the load resistance as shown in Fig. 50.8 (*a*). The output of a full wave rectifier is shown in Fig. 50.8 (*b*). As the output voltage across the load increases, the capacitor gets charged to the value of maximum voltage and stores energy in its electrostatic field. When the output voltage begins to fall, the capacitor discharges itself through the load resistance and in the process releases energy from its electrostatic field, thereby maintaining the voltage across the load resistance for a long time and reducing the ripple, as shown by thick curve in Fig. 50.8(*b*).

Fig. 50.8

The ripple factor is given by

For half wave rectifier $\gamma = \dfrac{\pi}{\sqrt{3}\,\omega\, C\, R_L}$

where $\omega = 2\pi f$, f being the frequency of A.C.

For a full wave rectifier

$$\gamma = \frac{\pi}{2\sqrt{3}\,\omega\, C\, R_L}$$

(*b*) **Series inductor filter.** An inductor L connected in series with the load, as shown in Fig. 50.9(*a*), is called a *series inductor filter*. The filtering action of an inductor is due to the fact that it opposes any *change* in current flowing through it.

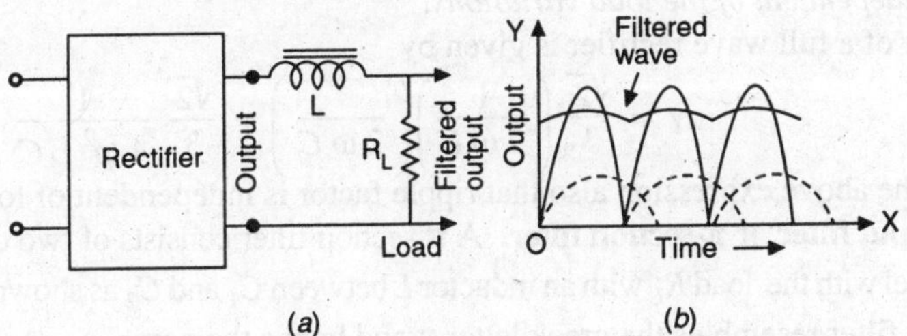

Fig. 50.9

When the current rises through an inductor, a self induced *e.m.f.* is set up in it, which according to Lenz's law, opposes the rise in current. The electrical energy is stored in the inductor as energy of the magnetic field. When the current starts falling through the inductor, the process is again opposed by supplying the stored energy from the magnetic field as electrical energy. Thus the period, for which the current continues to flow through the load, is increased due to the presence of the inductor. The period is proportional to the time constant of the circuit $L\omega/R_L$. Larger the value of time constant, larger is the time for which the current continues to flow. Hence smaller the value of R_L longer is the time taken and lesser is the ripple. The corresponding effect on a full wave rectifier is shown in Fig. 50.9 (*b*). The full line curve shows the output from a full wave rectifier without inductor, the dotted line curve shows the curve for each wave and the thick line, the resultant form with series inductor filter. The filtered wave shows a considerable reduction in ripple.

The ripple factor for a **full wave rectifier** is given by

$$\gamma = \frac{\sqrt{2}}{3}\,\frac{1}{\sqrt{1+\left[\dfrac{2\omega L}{R}\right]^2}}$$

If $R >> 2\,\omega L$ $\quad \gamma = \frac{\sqrt{2}}{3} = 0.47$

If $2\,\omega L >> R$ $\quad \gamma = \frac{\sqrt{2}}{3}\,\frac{R}{2\,\omega L} = 0.235\,\frac{R}{\omega L}$

(*c*) **Choke input filter or L-section filter.** A choke input filter is a combination of a series inductor L and a shunt capacitor C. The inductor is the first component connected in series with the rectifier output and shunt capacitor C is the second component, placed in parallel with the load resistance R_L as shown. The filter derives its name L-section from its shape formed by the combination of the inductor L and capacitor C.

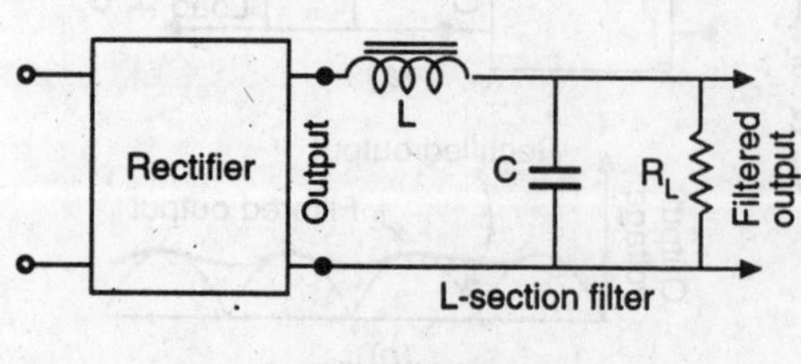

Fig. 50.10

The series inductor L offers a high impedance $X_L = L\,\omega$ to the *a.c.* component of the rectifier output and thus blocks the major part of the *a.c.* and allows the *d.c.* component to pass through it. The capacitance C offers a very low impedance $X_C = 1/C\omega$ to the *a.c.* component, as compared to the resistance R_L and thus offers a low impedance passage to the *a.c.* component which comes out of the inductor. Thus the combined action of both the inductor and the capacitor is to maintain a fairly constant voltage across the load resistance R_L.

A series inductor type filter produces low ripple for a *light* load and a shunt capacitor-type filter produces low ripple for a *heavy* load. Hence the L-section filter, which is a combination of both, gives low ripple for light as well as heavy loads. In other words, *the ripple factor and the output voltage can be kept almost independent of the load variations.*

The ripple factor of a full wave rectifier is given by

$$\gamma = \frac{\sqrt{2}}{3}\left(\frac{1}{2\,\omega L}\right)\left(\frac{1}{2\,\omega C}\right) = \frac{\sqrt{2}}{3}\,\frac{1}{4\,\omega^2 L C}$$

If is clear from the above expression also that ripple factor is independent of load resistance.

(*d*) **Capacitor input filter or π-section filter.** A π-section filter consists of two capacitors C_1 and C_2 connected in parallel with the load R_L with an inductor L between C_1 and C_2 as shown in Fig. 50.11(*a*). The shape of the filter resembles the greek letter π and hence the name.

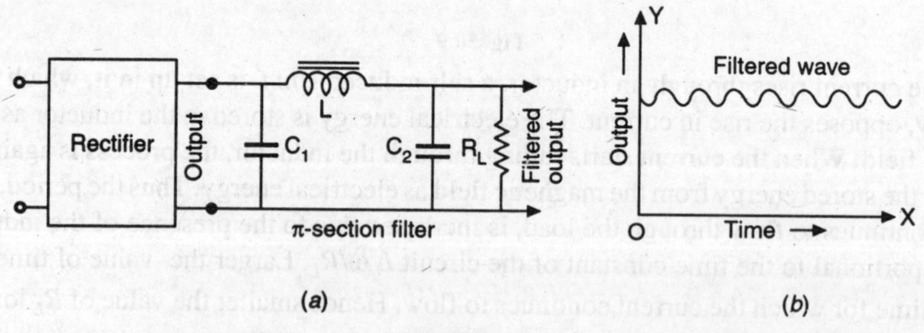

Fig. 50.11

A full wave rectifier consists of two diodes and each diode conducts only for a fraction of the half cycle. During each conducting interval, the capacitor C_1 is charged to the peak value of the output pulse. During the non-conducting period the capacitor discharges slowly through the inductor L and the load resistor R_L. Therefore, though the output voltage falls between successive pulses, it remains near the peak value. Thus the capacitor C_1 smoothens the output current by reducing the ripple and maintains the output voltage at a *higher* value. In other words a π-section filter provides a higher *d.c.* potential than the L-section filter.

Thereafter, the combination of inductor L and capacitor C_2 forming an L-section filter acts on the output of C_1 and further reduces the ripple. The voltage output from a π-section filter is thus highly filtered but is not as stable as for a L-section filter.

The load resistance R_L is also known as *bleeder resistance*. It is provided with taps to supply different voltages and hence besides providing the required bleeder current, the filter serves as a voltage divider.

A π-section filter is a combination of a shunt capacitor filter and an L-section filter. The ripple factor is, therefore given by the product of the shunt capacitor filter ripple factor γ_1 and L-section filter ripple factor γ_2

$$\therefore \quad \text{Ripple factor} \quad \gamma = \gamma_1 \times \gamma_2 = \frac{\pi}{2\sqrt{3}} \frac{1}{\omega C_1 R_L} \times \frac{\sqrt{2}}{3} \frac{1}{4\omega^2 L C_2}$$

$$= \frac{\pi}{12\sqrt{6}} \frac{1}{\omega^3 C_1 C_2 L R_l}$$

50.7. Voltage regulation. *The variation in output voltage of a rectifier with changes in load current is known as voltage regulation of the circuit. It is defined as the ratio of voltage at no load minus voltage at full load divided by voltage at full load*

It is given by

$$\therefore \% \text{ Voltage regulation} = \frac{E_{NL} - E_L}{E_L} \times 100$$

where E_{NL} is the no-load voltage and E_L is the full load voltage.

50.8. Current regulation. *The current regulation of the circuit is defined as the ratio of output current at no load minus current at full load divided by the current at full load*

$$\therefore \quad \% \text{ current regulation} \quad = \frac{I_{NL} - I_L}{I_L} \times 100$$

where I_{NL} is the no load current and I_L the full load current.

Experiment 50.1. To study the voltage regulation and ripple factor of a (*i*) Half wave rectifier and (*ii*) Full wave valve rectifier with π-type and L-type filter circuits.

Apparatus A full wave rectifier with change over switch to convert it into a half wave rectifier, π-type filter circuit consisting of an inductance L of 8 H, and two capacitors of 32 μF each with two off-on switches, a milliammeter (100 mA range), a vacuum tube voltmeter *(V.T.V.M.)*, an *A.C.* voltmeter (0-5 V) or a *V.T.V.M.* capable of measuring *A.C.* volts or a cathode ray oscilloscope (*C.R.O.*) and a blocking capacitor of 16 μF.

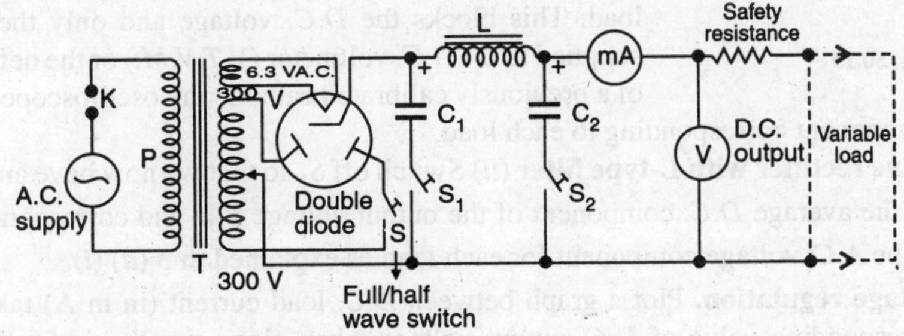

Fig. 50.12

Procedure 1. Draw a diagram showing the scheme of connections as in Fig. 50.12 and make connections accordingly.

Power transformer T. It is a transformer having a centre tapped secondary giving 300 V – 0 – 300 V at 100 mA rating and a 6.3 V winding for heating the filament of a double diode (normally EZ 80). The switch *S* allows it to be used as a full wave or a half wave rectifier according as it is closed or open. The filter system is a π-type which can be made into *L*-type when the switch S_1 is open and S_2 is closed. *V* is a high resistance D.C. voltmeter or a V.T.V.M. Other parts are clearly labelled.

2. Voltage regulation (*a*) Full wave rectifier with π-type filter

(*i*) Close the switch S to make the double diode to act as a full wave rectifier. Also close switches S_1 and S_2. Connect the primary of the transformer *T* to *A.C.* mains supply and switch on the key *K*. Measure the R.M.S. value of the *A.C.* voltage applied to each plate of the double diode with an *A.C.* voltmeter. Without applying the load to the output of the rectifier measure the *D.C.* output voltage with a high resistance voltmeter *V* or *V.T.V.M*. This gives *D.C.* voltage at no load E_{NL}. Also measure the *D.C.* output current at no load.

(*ii*) Now connect the *safety resistance* and a variable resistance having steps of 1 *K* Ω to serve as a *variable load.* Keeping the variable resistor at its maximum value (say 50 *K* Ω) *i.e.* load at its minimum, measure the *D.C.* output voltage and output current. Go on decreasing the variable resistance in steps of 1 *K* Ω *i.e.* increasing the load and measure the corresponding values of *D.C.* output voltage and current till the variable resistance is zero *i.e.* the rectifier is on **Full load** with only the safety resistance in circuit.

(*b*) **Full wave rectifier with *L*-type filter.** (*iii*) Switch off S_1 (keeping S_2 closed) so that the *L*-type filter consisting of choke coil *L* and a capacitor C_2 is only in circuit. Repeat all observations as in step 2 (*a*) (*i*). and (*ii*)

(*c*) **Half wave rectifier with π-type filter** (*iii*) Switch off *S* so that only one plate of double diode is in action and the circuit works as a half wave rectifier. Keeping both S_1 and S_2 on so that a π-type filter is in circuit, again. Repeat all observations as in step 2 (*a*) (*i*). and (*ii*)

(*d*) **Half wave rectifier with *L*-type filter.** (*iv*) Switch off S_1 so that only *L*-type filter is in circuit and repeat all observations as in 2 (*a*) (*i*). and (*ii*)

3. **Ripple factor. Note** *We shall study ripple factor only for a full wave rectifier. The study of ripple factor for a half wave rectifier is left to the students as an exercise.*

(*a*) **Full wave rectifier with π-type filter.** (*i*) Close the switches *S*, S_1 and S_2. Measure the *D.C.* output voltage and current at no load and also at variable loads as explained in step 2 (*a*) (*i*). This gives the average *D.C.* component of the output voltage E_{DC} and corresponding value of $I_{D.C.}$ (average *D.C.* output or load current).

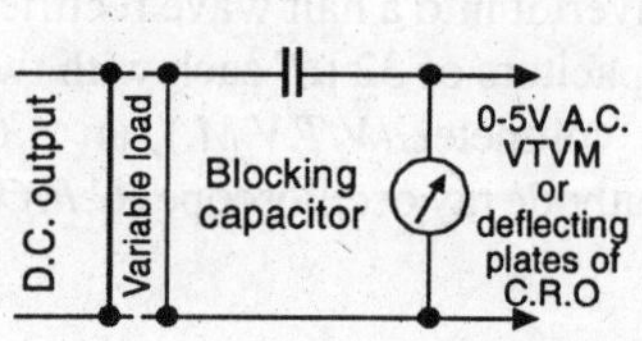

Fig. 50.13

To find the effective value of the *A.C.* component connect a blocking capacitor C_B (16 μ*F* capacitance) as shown across the load. This blocks the *D.C.* voltage and only the *A.C.* part is supplied to the *A.C.* voltmeter (*V.T.V.M.*) or the deflecting plates of a previously calibrated cathode ray oscilloscope. Measure the *A.C.* voltage component corresponding to each load.

(*b*) **Full wave rectifier with L-type filter** (*ii*) Switch off S_1 so that we now have an *L*-type filter. Again measure the average *D.C.* component of the output voltage E_{DC} and corresponding value of $I_{D.C.}$ as well as the *A.C.* voltage component for each load as explained in 3 (*a*) (*i*).

4. **For voltage regulation.** Plot a graph between *D.C.* load current (in m A) taken along the *X*-axis and corresponding value of *D.C.* output voltage taken along the *Y*-axis for (*a*) Full wave rectifier with (*i*) π-filter and (*ii*) *L*-filter and (*b*) half wave rectifier with (*i*) π-filter and (*ii*) *L*-filter.

For *ripple factor* plot a graph between *D.C.* load current in (m A) taken along the *X*-axis and corresponding values of % ripple factor taken along the *Y*-axis for (*i*) a π-type filter and (*ii*) *L*-type filter.

Observations. (*a*) **Voltage regulation**

A.C. voltage applied to each plate of a double diode

= volt

Full wave rectifier

π type filter				*L-type filter*			
No.	*Variable load K Ω*	*D.C. load current (mA)*	*D.C. output voltage (volts)*	*No.*	*Variable load K Ω*	*D.C. load current (mA)*	*D.C. output voltage (volts)*
1	NO LOAD		$E_{NL} =$	1	NO LOAD		$E_{NL} =$
2				2			
3				3			
.				.			
.				.			
.				.			
	FULL LOAD		$E_L =$		FULL LOAD		$E_L =$
Voltage regulation = $\frac{E_{NL} - E_L}{E_L} \times 100 = \%$				Voltage regulation = $\frac{E_{NL} - E_L}{E_L} \times 100 = \%$			

Half wave rectifier

π type filter				*L-type filter*			
No.	*Variable load K Ω*	*D.C. load current (mA)*	*D.C. output voltage (volts)*	*No.*	*Variable load K Ω*	*D.C. load current (mA)*	*D.C. output voltage (volts)*
1	NO LOAD		$E_{NL} =$	1	NO LOAD		$E_{NL} =$
2				2			
3				3			
.				.			
.				.			
.				.			
	FULL LOAD		$E_L =$		FULL LOAD		$E_L =$
Voltage regulation = $\frac{E_{NL} - E_L}{E_L} \times 100 = \%$				Voltage regulation = $\frac{E_{NL} - E_L}{E_L} \times 100 = \%$			

(*b*) **Ripple factor**

Full wave rectifier (with π-type filter)

No.	Variable Load K Ω	D.C. Load current (mA)	E_{ac} (volts)	E_{dc} (volts)	Ripple factor $\gamma = \frac{E_{a.c.}}{E_{d.c.}} \times 100$
1	NO LOAD				
2					
3					
..					
.					
.					
.	FULL LOAD				
			With L-type filter		
1	NO LOAD				
2					
3					
..					
.					
.					
.	FULL LOAD				

Precautions 1. A safety resistance must be connected in series with the load to avoid excessive current.

2. To find the effective value of *a.c.* component a blocking capacitor of 16 μF capacitance must be used.

3. The load in the output circuit must be varied by changing the resistance by 1 *K* Ω at a time.

Experiment 50.2. To study *L*-type and π-type filter in a (*i*) half wave and (*ii*) full wave semi-conductor diode rectifier and to calculate the percentage current regulation and current ripple factor.

Apparatus. A full wave rectifier using two semi-conductor diodes with a change over switch to convert it into half wave rectifier, π-type filter circuit consisting of a choke coil and two capacitors (2μF each) with two turn off switches, a *d.c.* milliammeter, an *a.c.* milliammeter with 1μ*F* capacitor in series, A cathode ray oscilloscope (C.R.O), a step down transformer with centre tapping.

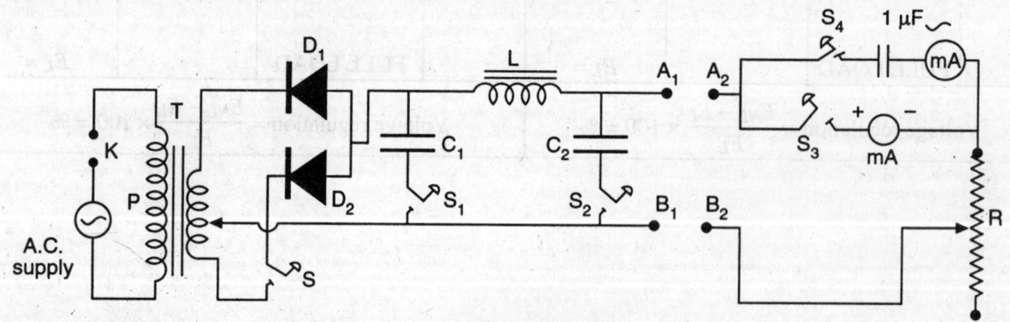

Fig. 50.14

Procedure 1. Draw a diagram showing the scheme of connections as in Fig. 50.14 and make connections accordingly.

2. **Full wave rectifier with π-tpe filter.** Close the switch S to bring both the semi-conductor diodes D_1 and D_2 in circuit so that the arrangement acts as a full wave rectifier. Also close switches S_1 and S_2 to get a π-type filter. Connect the terminals A_1 and B_1 to the Y-Y plates of a C.R.O. Connect the primary of the transformer T to A.C. mains supply and switch on the key K. Obtain the pattern of the full-wave rectified voltage through the π-type filter on the C.R.O. screen and trace it.

3. Now connect A_1 to A_2 and B_1 to B_2. Keeping R at its maximum value switch on S_3 and measure the output *d.c.* current. This may be taken as output *d.c.* current for no load. Switch off S_3 and switch on S_4 to measure the effective value of *a.c.* current. The capacitor blocks the *d.c.* current and only the *a.c.* part is measured by the a.c. milliammeter.

4. Now reduce the value of R by a small amount from its maximum value and again measure *d.c.* output current (by switching on S_4 with S_3 off) and then *a.c.* output current (by switching on S_4 and S_3 off). Go on decreasing the variable resistance R in small steps (*i.e.* increasing the load) and measure the *d.c.* output current and corresponding *a.c.* output current for various values of R till the variable resistance is zero *i.e.* the rectifier is on full load.

5. **Full wave rectifier with *L*-type filter.** Switch off S_1 keeping S_2 closed so that L-type filter consisting of choke coil L and capacitor C_2 is only in circuit. Repeat all observations as in step 2, 3 and 4.

6. **Half wave rectifier with π-type filter.** Switch off S so that only the semi conductor diode D_1 is in action and the circuit works as a half wave rectifier. Keeping both S_1 and S_2 closed so that a π-type filter is in circuit, again repeat all observations as in step 2, 3 and 4.

7. **Half wave rectifier with *L*-type filter.** Switch off S_1 (keeping S_2 closed) so that only L type filter is in circuit. Repeat all observations as in step 2, 3 and 4.

Observations

Full wave rectifier

No.	π-type filter				No.	L-type filter			
	Resistance	d.c. output current $I_{d.c.}$	a.c. output current $I_{a.c.}$	current ripple factor $\gamma = \frac{I_{a.c.}}{I_{d.c.}} \times 100$		Resistance	d.c. output current $I_{d.c.}$	a.c. output current $I_{a.c.}$	current ripple factor $\gamma = \frac{I_{a.c.}}{I_{d.c.}} \times 100$
1	No Load				1	No Load			
2					2				
3					3				
.					.				
.					.				
5					5				
6	Full Load				6	Full Load			
d.c. output current at no load, I_{NL} =					d.c. output current at no load, I_{NL} =				
d.c. output current at full load, I_L =					d.c. output current at full load, I_L =				
% current regulation $\frac{I_{NL} - I_L}{I_L} \times 100 =$					% current regulation $\frac{I_{NL} - I_L}{I_L} \times 100 =$				

Half wave rectifier

No.	π-type filter				No.	L-type filter			
	Resistance	d.c. output current $I_{d.c.}$	a.c. output current $I_{a.c.}$	current ripple factor $\gamma = \frac{I_{a.c.}}{I_{d.c.}} \times 100$		Resistance	d.c. output current $I_{d.c.}$	a.c. output current $I_{a.c.}$	current ripple factor $\gamma = \frac{I_{a.c.}}{I_{d.c.}} \times 100$
1	No Load				1	No Load			
2					2				
3					3				
.					.				
.					.				
.					.				
5					5				
6	Full Load				6	Full Load			
d.c. output current at no load, I_{NL} =					*d.c.* output current at no load, I_{NL} =				
d.c. output current at full load, I_L =					*d.c.* output current at full load, I_L =				
% current regulation $\frac{I_{NL} - I_L}{I_L} \times 100 =$					% current regulation $\frac{I_{NL} - I_L}{I_L} \times 100 =$				

Precautions. Same as in Expt. 50.1.

Experiment. 50.3. To fabricate and study a full wave bridge rectifier using a step down transformer with several output tappings.

Apparatus. A step down transformer with several output tappings, 4 *P–N* junction diodes. A π-section filter circuit consisting of a choke and two capacitors, a high resistance, a vacuum tube voltmeter, a cathode ray oscilloscope.

Procedure. 1. Draw a diagram showing the scheme of connections and connect the 4 *P–N* junction diodes to form a bridge and the choke and two capacitors to form a π-type filter element.

2. Connect the terminal A_1 of the transformer to the terminal C_3 of the load resistance and terminal B_1 to D_3. Connect the *V.T.V.M.* to the terminals C_4 and D_4 and set it to measure *A.C.* volts. Switch on the *A.C.* mains and measure the *A.C.* voltage between A_1 and B_1 across the load resistance R_L.

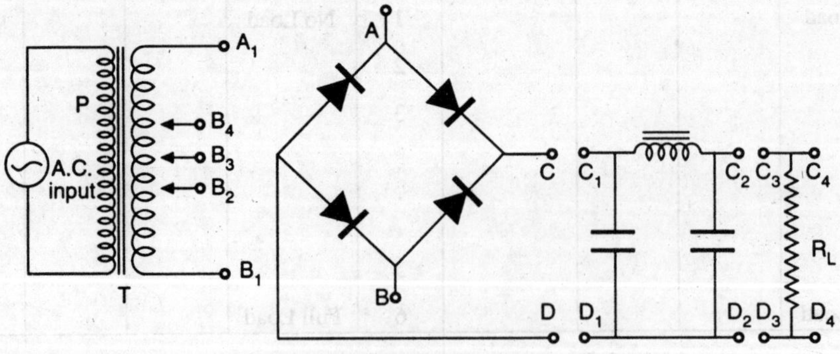

Fig. 50.15

Similarly measure the *A.C.* voltage between A_1 and B_2, A_1 and B_3, A_1 and B_4 across the load resistance R_L.

3. Now connect A_1 to A, B_1 to B, C to C_3 and D to D_3. Connect the terminals C_4 and D_4 to the 'vert. input' terminals of a *C.R.O.* Set the *C.R.O.* so that a sharp, fine spot is obtained in the centre of the screen. Set the 'Time base' frequency to obtain a stationary pattern consisting of 2–3 wave traces on the screen. Trace the pattern on a transparent paper.

Similarly trace the pattern by connecting B to B_2, B to B_3, and B to B_4, each time adjusting the '*vertical gain*' and '*synch amp*' if necessary.

4. Now connect A_1 to A, B_1 to B, C to C_1, D to D_1, C_2 to C_3 and D_2 to D_3 so as to introduce the filter circuit between the rectifier and the load resistance. Obtain the pattern on the screen of the *C.R.O.*

Repeat by connecting B to B_2, B to B_3 and B to B_4 respectively.

5. Disconnect the *C.R.O.* from the terminals C_4 and D_4 of the load resistance R_L. Set the *V.T.V.M.* to *D.C.* volts and measure with it the voltage across R_L between the points C_4 and D_4 by connecting B to B_1, B to B_2 and B to B_3 and B to B_4 respectively.

Observations.

Terminal B of bridge rectifier connected to	*Input A.C. voltage before rectification*	*Output D.C. voltage after rectification*
B_1		
B_2		
B_3		
B_4		

Precautions. 1. Always connect the primary of the transformer to the *A.C.* mains.

2. The *P–N* junction diodes must be properly checked before use and their positive ends must be connected in the right direction.

3. Connections must be secure and tight and wherever possible should be properly soldered.

50.9. Diode as a clipper (Limitor). A diode can be used as a rectifier. It can also be used as a *clipper* (or limitor). A clipper is used to clip off or remove a portion of *a.c.* signal. The half wave rectifier is, in fact, a clipper which eliminates either the upper or lower half of the *a.c.* signal.

A circuit with which the wave form is shaped by removing (or clipping) a portion of the applied wave is known as a clipping circuit.

Positive clipper. A positive clipper is that which removes the positive half cycles of the input voltage. A typical circuit of a positive clipper using a diode is shown in Fig. 50.16. It is seen that the output voltage has all the positive half cycles removed or clipped off.

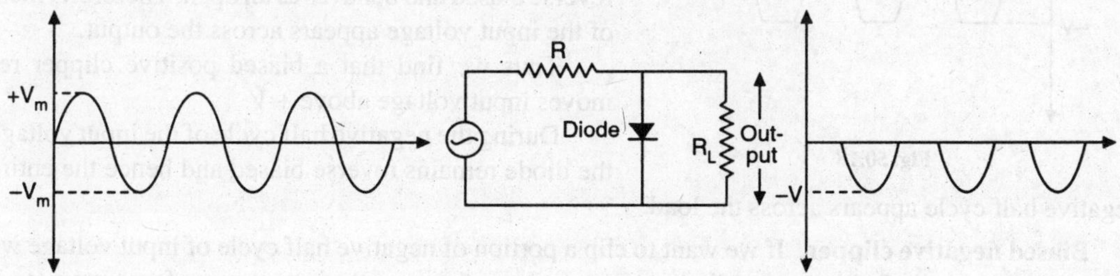

Fig. 50.16

Action. During the positive half cycle of the input voltage the diode is forward biased and conducts heavily *i.e.* behaves as a *short*. Therefore the voltage across the diode and hence across the load resistance R_L is zero.

During the negative half cycle of the input voltage, the diode is reverse biased and behaves as *open.*

$\therefore$ Output voltage across $R_L = -V = \dfrac{R_L}{R+R_L} V_m$

If $R_L >> R$, $V = -V_m$

Negative clipper. If it is desired to remove the negative half cycle of the input, we have to reverse the polarities of the diode in the above circuit. In such a case the circuit behaves as a *negative clipper.*

NOTE. In the above case the clipping is not perfect. If we are using a silicon diode, it drops approximately 0.7V. As the first 0.7V is used to overcome the barrier potential the output signal is clipped near +0.7V rather than 0*V*, in the case of positive clipper. For a negative clipper the clipping level is –0.7V.

50.10. Biased clipper. When it is required to remove a small portion of positive or negative half cycle of the signal voltage, a *biased clipper* is used.

The circuit of a biased clipper using a diode with a battery of *V* volts is shown in Fig. 50.17. With the polarities of the battery as shown a portion of each positive half cycle will be clipped whereas the negative half cycle will appear as such across the load. Such a clipper is, therefore, called **biased positive clipper.**

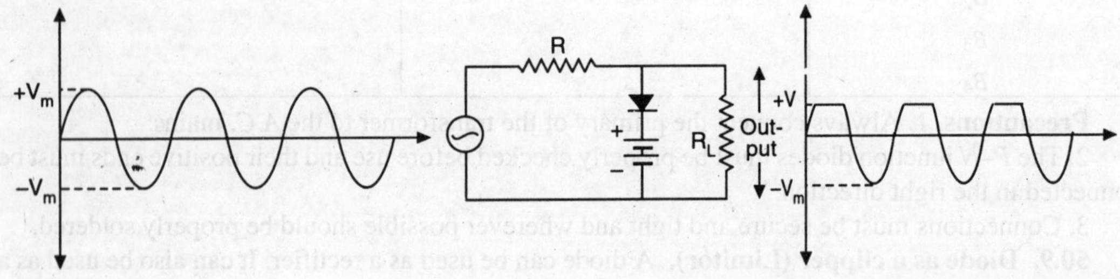

Fig. 50.17

When the input voltage is greater than +*V* [In fact + *V* + 0.7 for a silicon diode] the diode conducts heavily and behaves as a *short* so that the output is held at + *V* [or + *V* + 0.7 for silicon]. The output will stay at +*V* so long as the input is greater than +*V*. During the time the input voltage is less than +*V* the diode is reverse biased and behaves as an open. Therefore, most of the input voltage appears across the output.

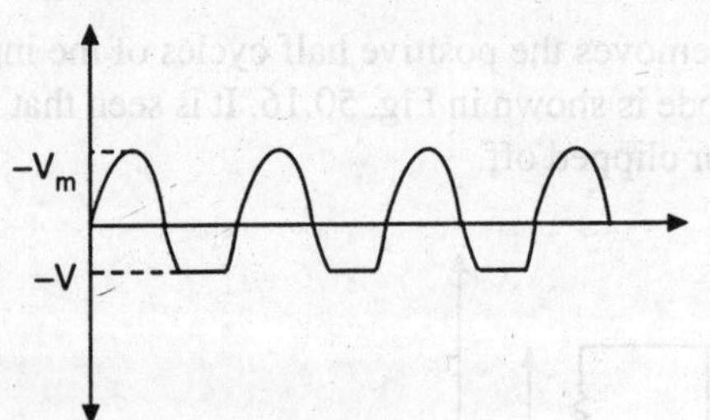

Fig. 50.18

Thus we find that a biased positive clipper removes input voltage above + *V*.

During the negative half cycle of the input voltage the diode remains reverse biased and hence the entire negative half cycle appears across the load.

Biased negative clipper. If we want to clip a portion of negative half cycle of input voltage we have to reverse the polarities of the diode or battery. In such a case the output wave form stays at –*V* (or –*V*–0.7 for silicon) as long as input voltage is greater than –*V*. However, the entire positive half cycle appears as such across the load.

Combination clipper. It is a combination of biased positive and negative clippers. In this a case a portion of both positive and negative half cycles of input voltage can be clipped off as shown in Fig. 50.19.

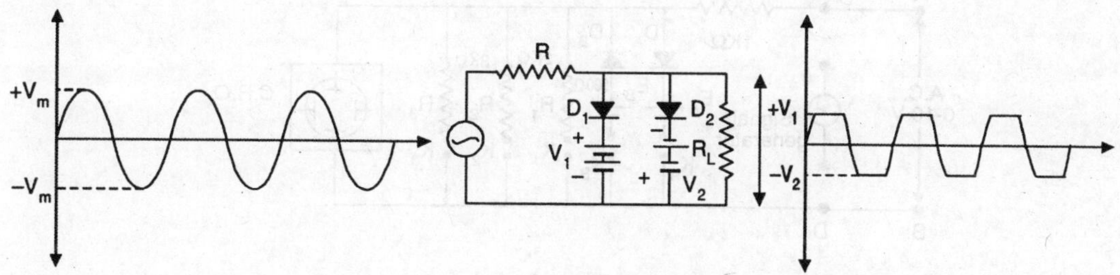

Fig. 50.19

When the positive input voltage is greater than $+V_1$ diode D_1 conducts heavily but the diode D_2 remains reverse biased. Therefore, a voltage $+ V_1$ appears across the load. This is the *positive clipping level.*

On the other hand during the negative half cycle the diode D_2 conducts heavily when the input is more than $- V_2$. The output stays at $- V_2$ so long as the input voltage is greater than $- V_2$. This is the *negative clipping level.*

Between $+ V_1$ and $- V_2$ none of the two diodes is on and in this condition most of the input voltage appears across the load. If V_m is much greater than the clipping levels, such a clipping circuit gives a *square wave.*

Experiment 50.4. To study the characteristics of a diode limitor (or clipper)

Apparatus. Sine wave oscillator, cathode ray oscilloscope (C.R.O) A diode, a resistance 1 K Ω and a resistance 50 K Ω.

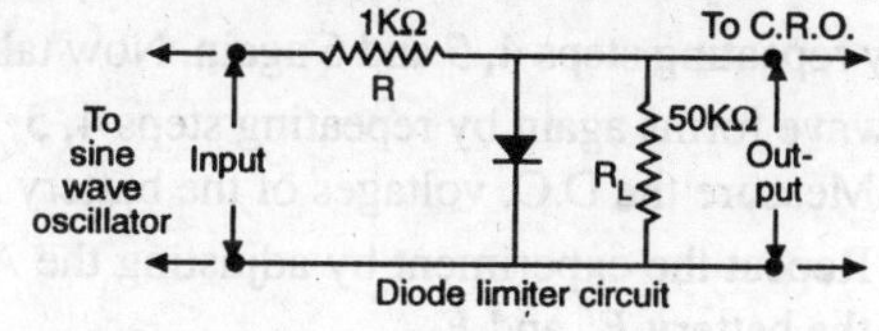

Fig. 50.20

Procedure 1. Draw a diagram showing the scheme of connections as in Fig. 50.20 and make connections accordingly.

2. Adjust the frequency of the sine wave oscillator to 1000 cycles per second. Connect it to the C.R.O and measure the peak to peak amplitude of the signal on C.R.O.

3. Now connect the signal generator to the diode limitor circuit. Connect the output side to the C.R.O. Observe the output wave form on C.R.O. and trace it.

4. Slowly increase the signal level and observe the output wave form in each case and trace the same.

5. Now reverse the polarities of the diode in the circuit and again observe the output wave form on the C.R.O for the same levels of the signal and trace the wave form in each case.

Conclusion. In the first case the positive half of the signal is clipped *i.e.* the circuit behaves as a *positive clipper* whereas in the second case the negative half of the signal is clipped *i.e.* the circuit behaves as a *negative clipper.*

Precautions. The value of R_L should be about 50 to 100 times the value of R.

Experiment 50.5. To study the function of diode as a clipping element.

Apparatus. A step down variable transformer or a signal generator, cathode ray oscilloscope (C.R.O), Resistances 500 Ω, 1 K Ω, 2 K Ω, 3 K Ω, two diodes, two batteries and keys.

Procedure. 1. Drawn a diagram showing the scheme of connections and make connections accordingly. (Generally the components are fixed on a board.)

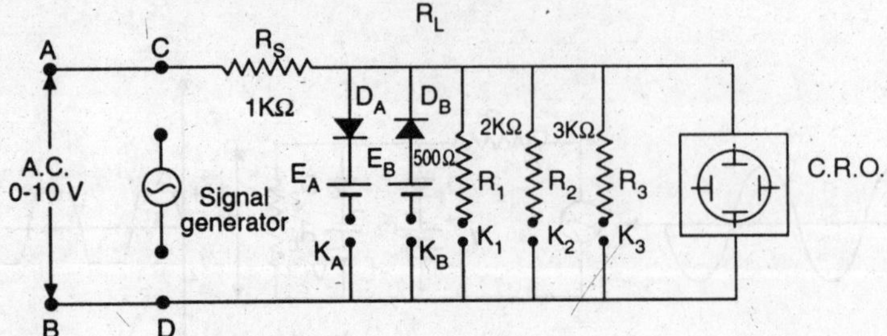

Fig. 50.21

2. Connect the sine wave signal generator between *C* and *D*. If, for some reason signal generator cannot be used or is not available connect the secondary of the step down variable transformer to the points *A* and *B*.

3. Put in the key K_1 and adjust the A.C. signal voltage to about $3V$. We shall get a sine wave on the C.R.O. Trace the sine wave.

4. Now insert the key K_A so that the diode D_A and the battery E_A are in circuit. A part of one portion of one half of the wave shall appear clipped. Trace the wave.

5. Take out the key K_A and insert the key K_B so that the diode D_B and the battery E_B are in circuit. Now a part of one portion of the second half of the wave shall appear clipped. Trace the wave.

6. Now insert both the keys K_A and K_B. A part of the output wave shall appear clipped from both the halves. Trace the wave.

7. To study the effect of load resistance on clipping take out key K_1 and insert key K_2. Trace the wave by repeating steps 4, 5 and 6 again. Now take out the key K_2 and insert the key K_3 and trace the output wave forms again by repeating steps 4, 5, and 6.

8. Measure the D.C. voltages of the battery E_A and E_B. Let it be V_A and V_B respectively.

9. Repeat the experiment by adjusting the A.C. signal voltage to about $5V$ and using two cells each in the battery E_A and E_B.

Results. When only the diode D_A is in the circuit the upper portion of the wave form is partly clipped and stays at V_A (more precisely at $V_A + 0.7$ volts for a silicon diode).

When only the diode D_B is in the circuit the lower portion of the wave form is partly clipped and stays at $- V_B$ (more precisely at $- V_B - 0.7$ volts for a silicon diode).

When both the diodes are in the circuit a part of upper as wells as lower portion of the wave form is clipped and stay at $+ V_A$ and $- V_B$ (more precisely at $V_A + 0.7$ and $- V_B - 0.7$ volts for silicon diodes).

Precautions 1. The input signal voltage should be greater than the bias voltage applied to the diode.

Oral Questions

1. Rectification. (*i*) What do you mean by rectification ? (*ii*) What is half wave rectification ? (*iii*) How can you bring about half wave rectification ? (*iv*) What is full wave rectification and how is it achieved ? (*v*) Is there any difference between rectification brought about by a diode valve and a semi-conductor *P–N* junction ? (*vi*) What is a double diode ? (*vii*) What is a bridge rectifier ? (*viii*) How many diodes does it use ? (*ix*) What is its advantages over other types of full wave rectifiers ? (*x*) What is a centre tapped transformer? (*xi*) What is the difference between step up and step down transformer?

2. Efficiency and ripple factor. (*i*) What do you understand by efficiency of a rectifier? (*ii*) What is voltage efficiency? (*iii*) What is current efficiency. (*iv*) What is maximum efficiency of a half wave rectifier? (*v*) What is maximum efficiency of a full wave rectifier. (*vi*) Define ripple factor in terms of voltage as well as current. (*vii*) What is the value of ripple factor for (*a*) half wave rectifier (*b*) full wave rectifier. (*viii*) What is percentage voltage regulation? (*ix*) What is percentage current regulation?

3. Filter circuit. (*i*) What is a filter circuit and what is its necessity ? (*ii*) What is a shunt capacitor filter. (*iii*) How does it reduce the ripple ? (*iv*) What is a series inductor filter and how does it reduce the ripple? (*v*) What is a choke input or L–section filter? (vi) What are its advantages? (viii) What is a capacitor input or π-section filter and what are its advantages? (viii) What is the relation for ripple factor in shunt capacitor filter for (a) half wave rectifier (b) full wave rectifier? (ix) What is the relation for ripple factor for a series inductor filter in a full wave rectifier when $R >> 2\,\omega L$ and when $2\,\omega L >> R$? (x) What is the ripple factor for a full wave rectifier using an L-type filter? (xi) What is the ripple factor for a full wave rectifier using a π-type filter. (xii) In a series inductor type filter the ripple factor is almost independent of load. Why?

4. Clipping circuits. (*i*) What is a clipping circuit? (ii) How does a semi-conductor diode act as a clipper? (iii) What is the difference between a positive clipper and negative clipper? (iv) What is the necessity of clipping? (v) What is a biased clipper? (vi) How does a biased clipping circuit change the wave form? (vii) Why should the input potential be more than the diode bias? (viii) What are clamping circuits?

51

Transistors

51.1. Transistors. A transistor is an extremely small but simple device that can perform the function of a thermionic valve in an electronic circuit. There are no filaments to produce electrons and transistors work on very low voltages thereby consuming much less power. These are, therefore, widely used in portable radio and television receivers.

51.2. Junction transistor. A junction transistor (or simply a transistor) is formed by growing either a very thin layer of *N*-type crystal between two very thick layers of *P*-type crystals or a thin layer of *P*-type crystal between two very thick layers of *N*-type crystals. The first type of transistor is known as a *P-N-P* transistor and the second type as *N-P-N* transistor.

In both these the middle thin layer is called the **base** of the transistor and can be compared roughly to the control grid of a therminoic vacuum tube. One of the outer thick layers is called the **emitter,** often compared to the cathode of a vacuum tube. The emitter is highly doped and is the source of majority charge carriers *i.e. holes* in case of *P-N-P transistor* and *electrons* in case of *N-P-N transistor.*

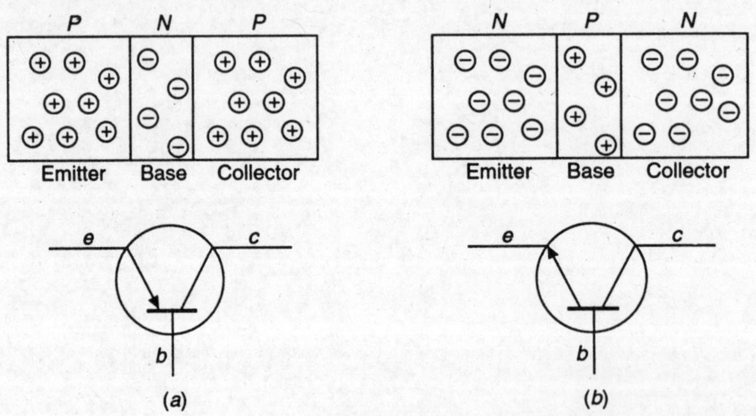

Fig. 51.1

The second thick layer is known as the **collector,** often compared to the plate of a vacuum tube. The collector is comparatively *less* doped and its function is to collect the charge carriers. More over the length of the collector is greater than that of the **emitter** and hence the emitter and the collector cannot be interchanged. Symbolically the two types of transistors are represented as shown in Fig. 51.1 (*a*) and (*b*). An arrow head is drawn on the emitter to distinguish it from the collector, pointing towards the direction of flow of **positive charge** *i.e. holes*. Thus the direction of the arrow head is from *P* to

N in the emitter base combination *i.e.* from emitter to base in a *P-N-P* transistor and from base to emitter in an *N-P-N* transistor.

51.3. Working of P-N-P type. To study the working of *P-N-P* type transistor, the emitter base circuit is forward biased with a small voltage V_{eb} while collector base circuit is reverse biased with a voltage V_{cb}. The emitter base circuit, being forward biased helps, the movement of the *holes* in *P*-type towards the junction between the emitter and base. On the other hand the collector base junction, being reverse biased does not allow the holes in *P*-type and electrons in *N*-type to move toward the collector base junction. The holes in the emitter and the electrons in the base are repelled by the positive and the negative terminals respectively of the battery V_{eb}. On reaching the emitter base junction a small fraction about 5% of the total number of holes combine with the electrons in the base to get neutralised. As the base region is very thin and the collector is given a high negative potential the holes will rush towards the collector and about 95% of them are thus collected by it. For each hole reaching the collector, an electron is released from the negative of the collector base battery V_{cb} and neutralises the hole. Also for each hole which is lost in the collector region a covalent bond near the emitter breaks down and an electron is liberated which enters the positive terminal of the emitter-base battery. Thus we see that the current is carried by the holes inside the crystal and by electrons in the external circuit. In the emitter base external circuit the flow of electrons is from *P* to *N* and, therefore, the conventional current I_e flows from *N* to *P* or base to emitter.

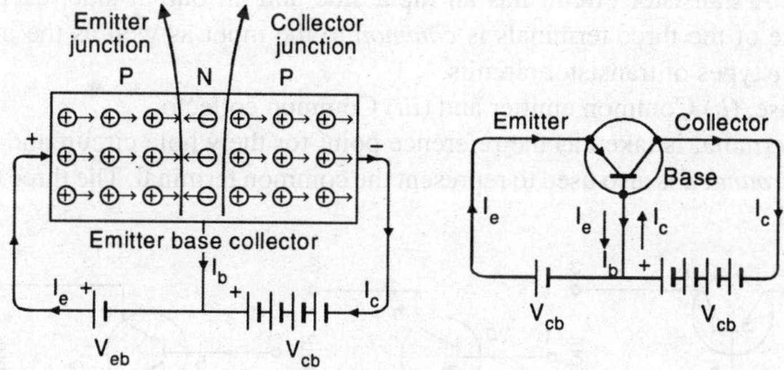

Fig. 51.2

Similarly in the collector base circuit the flow of electrons is from *N* to *P* in the external circuit and hence a current I_c flows in the collector base circuit from *P* to *N*. The difference between I_e and I_c is the base current I_b which flows in the direction of I_e. Thus

$$I_b = I_e - I_c$$

or

$$I_e = I_b + I_c$$

51.4. Working of N-P-N type. In this case also the emitter base junction is forward biased while the collector-base circuit is reverse biased. The holes in the base being repelled by positive terminal of the forward bias and the electrons in the emitter being repelled by the negative terminal move towards the emitter base junction. On reaching the junction a small fraction of about 5% of total number of electrons combine with the holes at the base to get neutralised. As the base layer is very thin and the collector is given a very high positive potential, the electrons will rush towards the collector and about 95% of them are collected by it. These electrons collected by the collector flow towards the positive of battery V_{bc}. The deficiency of the electrons in the emitter is made up by the negative of the emitter base battery V_{be}. Thus the current is carried by the electrons both inside the crystals as well as in the external circuit. In the emitter base external circuit the flow of electrons is from *P* to *N*. Hence the conventional current I_e is from *N* to *P*. In the collector base circuit the flow of electrons in the external circuit is from N to P resulting in the conventional current I_c from P to N. In the base circuit the current

$$I_b = I_e - I_c$$

or

$$I_e = I_b + I_c$$

Fig. 51.3

51.5. Transistor configurations. A transistor is three terminal or three electrode device, emitter, base and collector. A transistor circuit has an input side and an output side, each requiring two terminals. Thus, one of the three terminals is *common* to the input as well as the output side. We, therefore, have *three* types of transistor circuits.

(*i*) Common base, (*ii*) Common emitter and (*iii*) Common collector.

The common terminal is taken as the reference point for the whole circuit and is connected to *ground*. The word *grounded* is also used to represent the common terminal. The three types of circuits are, therefore called

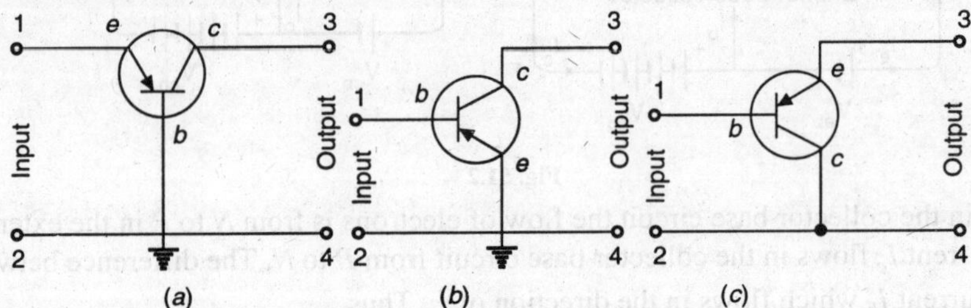

Fig. 51.4

(*i*) grounded base, (*ii*) grounded emitter and (*iii*) grounded collector.

These three configurations have been shown in Fig 51.4 (*a*), (*b*) and (*c*) for a *P-N-P* transistor. The input terminals have been marked 1 and 2 and the output terminals as 3 and 4. The terminals 2 and 4 are common.

Experiment 51.1. To study the common base characteristics of a P-N-P junction transistor.

Apparatus. A *P-N-P* transistor mounted on a board with three terminals marked *e* (emitter), *b* (base) and *c* (collector), a voltmeter of range 0-3 volt, a voltmeter of range 0-10 volt, two milli-ammeters of range 0-25 milliamp, a potentiometer of total resistance of the order of one megohm, a potentiometer of total resistance of the order of 25000 onm, 3-volt and 9-volt batteries, keys, connecting wires etc.

Object. The aim of the experiment is to study the relation.

(*i*) between I_e (emitter current) and V_{eb} (potential difference between emitter and the base) at constant values of the collector voltage. This is known as **input characteristic.**

(*ii*) between I_c (collector current) and V_{cb} (potential difference between the collector and the base) at constant value of emitter current. This is known as **output characteristic.**

Procedure. 1. Draw a diagram showing the scheme of connections as in Fig. 51.5 and make the connections accordingly.

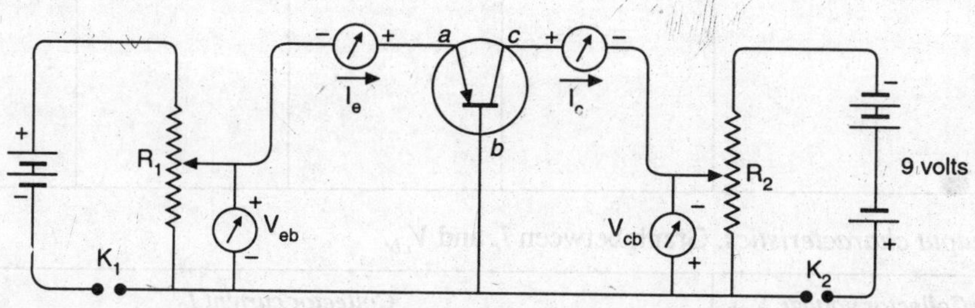

Fig. 51.5

2. (*i*) To draw the **input characteristics,** adjust the value of $V_{cb} = 0$. Keeping the value of $V_{eb} = 0$ note the value of I_e.

(*ii*) Increase the value of V_{eb} in very small steps and note the corresponding values of I_e.

(*iii*) Now adjust the value of $V_{cb} = 1$ volt and keeping the value of $V_{eb} = 0$ note the value of I_e. Again increase the value of V_{eb} in very small steps and note the corresponding values of I_e keeping V_{cb} always equal to one volt.

(*iv*) Similarly repeat the experiment corresponding to the values of $V_{cb} = 2$ volt, 3 volt, etc.

(*v*) Taking V_{eb} along the *X*-axis and I_e along the *Y*-axis plot the various curves. Some typical curves are as shown in Fig. 51.6.

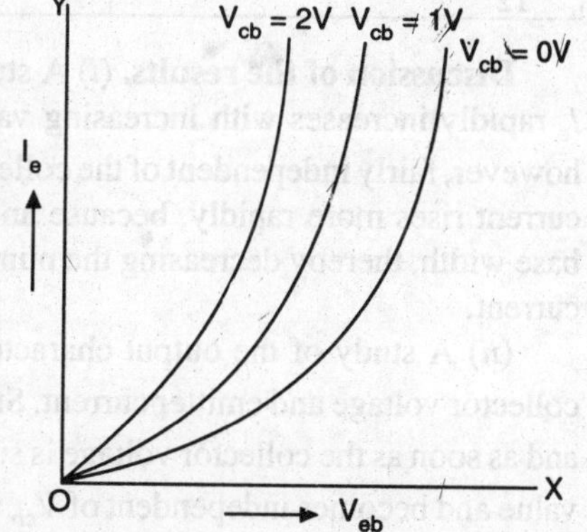

Fig. 51.6

3. (*i*) To draw the **output characteristics**; Adjust the value of I_e at some fixed value say 1 milli-ampere with the help of the potentiometer R_1. Keeping the value of $V_{cb} = 0$ note the value of I_c.

(*ii*) Increase the value of V_{cb} in small steps and note the corresponding values of I_c keeping I_e always constant at 1 mA.

(*iii*) Now adjust the value of $I_e = 2$ *mA* and keeping the value of $V_{cb} = 0$ note the value of I_c. Again increase the value of V_{cb} in small steps and note the corresponding values of I_c, keeping I_e constant at 2 *mA*.

(*iv*) Similarly repeat the experiment corresponding to I_e = 3*mA*, 4 *mA* etc.

(*v*) Taking V_{cb} along the *X*-axis, and I_c along the *Y*-axis, plot the various curves as shown in Fig. 51.7.

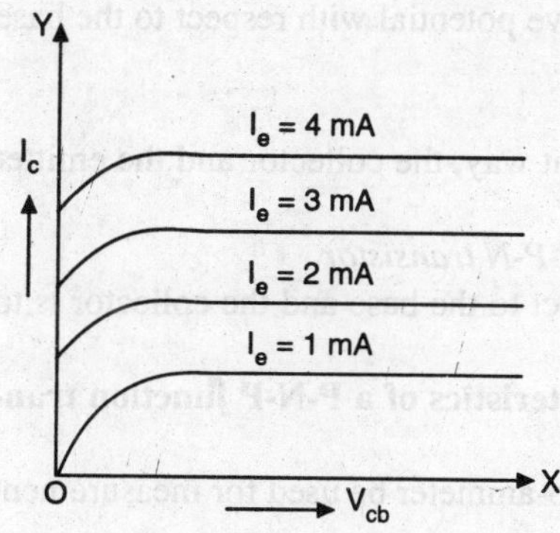

Fig. 51.7

Observations.

(*i*) *Input characteristics.* Graph between I_e and V_{eb}.

No.	Emitter voltage V_{eb}	Emitter current I_e			
		$V_{cb} = 0V$	$V_{cb} = 1\ V$	$V_{cb} = 2\ V$	$V_{cb} = 3\ V$
1					
2					
...					
...					
...					
12					

(*ii*) *Output characteristics*. Graph between I_c and V_{cb}.

No.	Collector voltage V_{cb}	Collector current I_c			
		$I_e = 1\ mA$	$I_e = 2\ mA$	$I_e = 3\ mA$	$I_e = 4\ mA$
1					
2					
...					
...					
...					
12					

Discussion of the results. (*i*) A study of the input characteristics shows that the emitter current I_e rapidly increases with increasing values of the emitter-base voltage V_{eb}. The emitter current is, however, fairly independent of the collector voltage. With the increase in collector voltage, the emitter current rises more rapidly, because an increase in the collector voltage results in a decrease in the base width, thereby decreasing the number of holes that diffuse back and thus increasing the emitter current.

(*ii*) A study of the output characteristics shows that the collector current I_c depends upon the collector voltage and emitter current. Starting from $V_{cb} = 0$ volt I_c at first increases as the voltage rises and as soon as the collector voltage is sufficient to collect all the majority carriers, I_c attains a constant value and becomes independent of V_{cb} as shown by the straight line portion of the curve, the collector current at this stage being only slightly less than the emitter current.

Precautions. 1. The collector voltage must not exceed the breakdown voltage for the junction.

2. For a *P-N-P* transistor the emitter must be given a positive potential with respect to the base and the collector a negative voltage with respect to the base.

3. Correct values of biasing potential must be employed.

4. The leads of the transistor must be connected in the right way, the collector and the emitter junctions should not be interchanged.

Exercise. *To study the common base characteristics of a N-P-N transistor.*

Note. The emitter is to be given a negative bias with respect to the base and the collector is to be given a positive bias with respect to the base.

Experiment 51.2. To study the common emitter characteristics of a P-N-P junction transistor.

Apparatus. Same as in **Experiment 51.1** except that micro-ammeter be used for measurement of I_b.

Object. The aim of the experiment is to study the relation.

(*i*) Between I_b (base current) and V_{be} (potential difference between the base and the emitter) at constant value of the collector voltage. This is known as **input characteristics**.

(*ii*) Between I_c (collector current) and V_{ce} (potential difference between the collector and the emitter) at constant value of the base current. This is known as **output characteristics.**

Procedure. 1. Draw a diagram showing the scheme of connections as in Fig. 51.8 and make the connections accordingly.

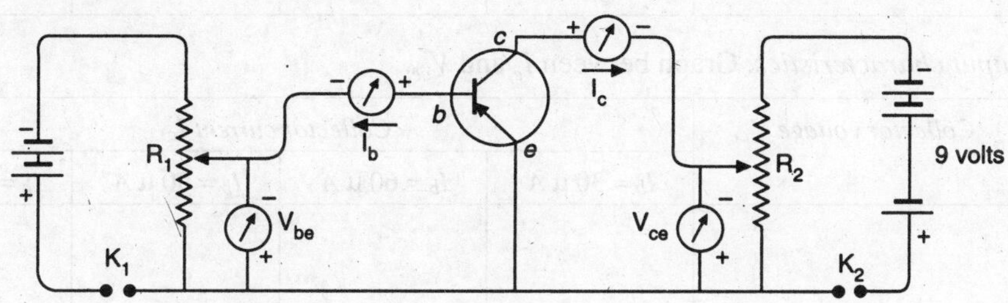

Fig. 51.8

2. (*i*) To draw the **input characteristics** adjust the value of $V_{ce} = 0$. Keeping the value of $V_{be} = 0$ note the value of I_b.

(*ii*) Increase the value of V_{be} in very small steps and note the corresponding values of I_b.

(*iii*) Now adjust the value of $V_{ce} = 1$ volt and keeping the value of $V_{be} = 0$ note the value of I_b. Again increase the value of V_{be} in very small steps and note the corresponding values of I_b keeping V_{ce} always at one volt.

(*iv*) Similarly repeat the experiment corresponding to values of $V_{ce} = 2$ volt, 3 volt, etc.

(*v*) Taking V_{be} along the *X*-axis and I_b along the *Y*-axis plot the various curves. Some typical curves are shown in Fig. 51.9.

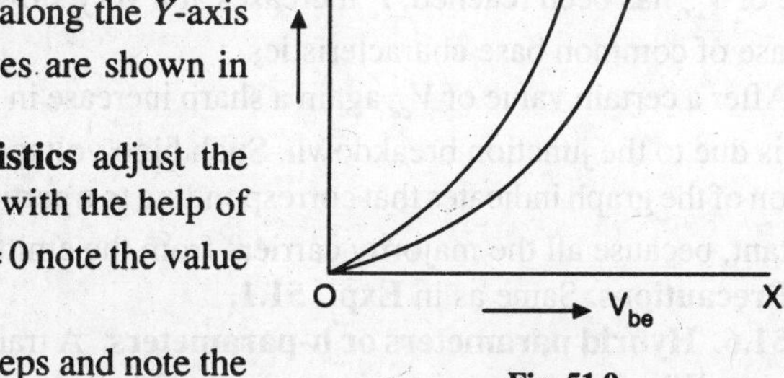

Fig. 51.9

3. (*i*) To draw the **output characteristics** adjust the value of I_b at some fixed value say 30 μ *A* with the help of potentiometer R_1. Keeping the value of $V_{ce} = 0$ note the value of I_c.

(*ii*) Increase the value of V_{ce} in small steps and note the corresponding values of I_c, keeping I_b always constant at 30 μ *A*.

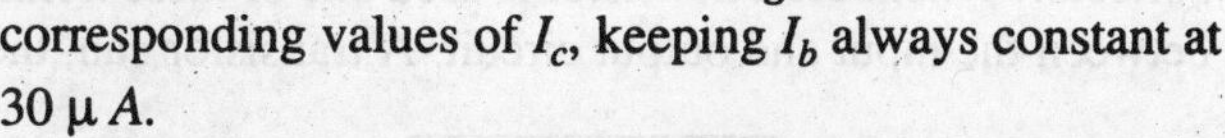

(*iii*) Now adjust the value of I_b at 60 μ *A* and keeping the value of $V_{ce} = 0$ note the value of I_c. Again increase the value of V_{ce} in small steps and note the corresponding value of I_c, keeping I_b constant at 60 μ *A*.

(*iv*) Similarly repeat the experiment corresponding to $I_b = 90$ μ *A* and 120 μ *A*.

(*v*) Taking V_{ce} along the *X*-axis and I_c along the *Y*-axis, plot the various curves as shown in Fig. 51.10.

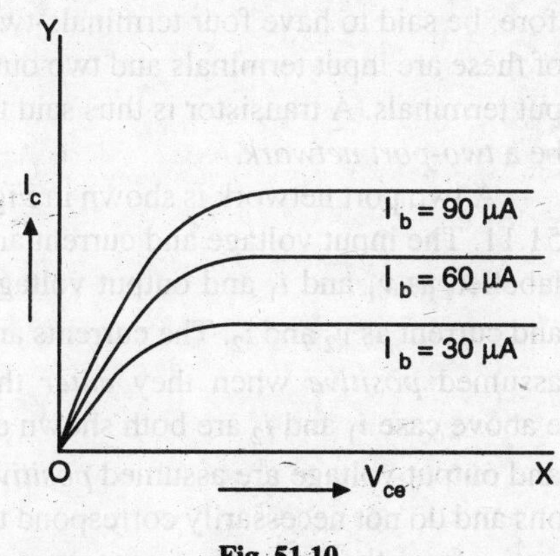

Fig. 51.10

Observations. (*i*) *Input characteristics*. Graph between I_b and V_{be}

No.	Base voltage V_{be}	Base current I_b			
		$V_{ce} = 0$	$V_{ce} = 1$ V	$V_{ce} = 2$ V	$V_{ce} = 3$ V
1					
2					
...					
12					

(*ii*) *Output characteristics*. Graph between I_c and V_{ce}.

Sl. No.	Collector voltage V_{ce}	Collector current I_c			
		$I_b = 30\ \mu A$	$I_b = 60\ \mu A$	$I_b = 90\ \mu A$	$I_b = 120\ \mu A$
1					
2					
...					
...					
...					
...					

Discussion of results. (*i*) A study of the input characteristics shows that the base current I_b rapidly increases as the value of V_{be} is increased from zero to the positive values. The collector voltage V_{ce} has very little effect on the values of I_b.

(*ii*) A study of the output characteristics shows that the collector current I_c depends upon the collector voltage and the base current. Starting from $V_{ce} = 0$, I_c rises at first steeply and after a particular value of V_{ce} has been reached, I_c increases at a very slow rate. The linearity is not as much as it is in the case of common base characteristics.

After a certain value of V_{ce} again a sharp increase in I_c can be observed (not shown in the graph). This is due to the junction breakdown. Such high voltages are, therefore, always avoided. The linear portion of the graph indicates that corresponding to a particular value of V_{ce}, the base current I_b remains constant, because all the majority carriers from the emitter are collected by the collector.

Precautions. Same as in **Expt. 51.1.**

51.6. Hybrid parameters or h-parameters. A transistor has three terminals–emitter, base and collector. Whenever a transistor is used one of these terminals is grounded and is a common terminal between the input and output circuit. A transistor can, therefore, be said to have four terminals-two of these are input terminals and two output terminals. A transistor is thus said to be a *two-port network*.

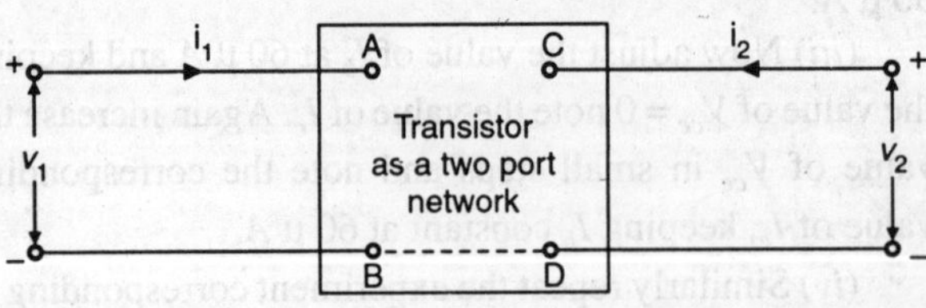

Fig. 51.11

A two port network is shown in Fig. 51.11. The input voltage and current are labelled as v_1 and i_1 and output voltage and current as v_2 and i_2. The currents are assumed *positive* when they *enter* the transistor and *negative* when they *leave* the transistor. In the above case i_1 and i_2 are both shown as flowing into the transistor and thus both are *positive.* Input and output voltage are assumed *positive* from *upper to lower* terminals. These are *standard* conventions and do not necessarily correspond to the actual directions and polarities. A voltage is taken as *negative* from the *lower to upper* terminal.

This circuit deals with two voltages v_1 and v_2 and two currents i_1 and i_2. Choosing any two out of these four as independent variables two equations for the circuit can be written from which the other two variables can be determined. For our study we shall take i_1 and v_2 as independent variables and i_2 and v_1 as dependent variables. Writing i_2 and v_1 as functions of i_1 and v_2, we have

$$v_1 = f_1(i_1, v_2) \quad \text{and} \quad i_2 = f_2(i_1, v_2)$$

Small changes in v_1 and i_2 due to small changes in i_1 and v_2 are given by

$$dv_1 = \frac{\partial v_1}{\partial i_1} di_1 + \frac{\partial v_1}{\partial v_2} dv_2 \qquad ...(i)$$

$$di_2 = \frac{\partial i_2}{\partial i_1} di_1 + \frac{\partial i_2}{\partial v_2} dv_2 \qquad ...(ii)$$

Generally, we are interested in developing only the *ac* equivalent circuit with operation over the quasi-linear portion of the volt-ampere (*VI*) characteristics. The signals considered are small-not driving the transistor into saturation or cut off. We can, therefore, take the partial derivatives as constants and write equations (*i*) and (*ii*) as

$$V_1 = h_{11} I_1 + h_{12} V_2 \qquad ...(iii)$$

and

$$I_2 = h_{21} I_1 + h_{22} V_2 \qquad ...(iv)$$

where V_1 and I_1 are *r.m.s.* values of applied *ac signal and* V_2 and I_2 the *r.m.s.* values of corresponding output *ac* signal voltage and current. h_{11}, h_{12}, h_{21} and h_{22} are known as *hybrid parameters* or *h-parameters*. The various *h*-parameters are defined as.

$$h_{11} = h_i = \left.\frac{\partial v_1}{\partial i_1}\right|_{v_2} = \left(\frac{V_1}{I_1}\right)_{V_2=0}$$ = short circuit input impedance for $V_2 = 0$ *i.e.* output shorted to *a.c.* signal

$$h_{21} = h_f = \left.\frac{\partial i_2}{\partial i_1}\right|_{v_2} = \left(\frac{I_2}{I_1}\right)_{V_2=0}$$ = short circuit *a.c.* forward current transfer ratio for $V_2 = 0$ *i.e.* output shorted to *a.c.* signal

$$h_{12} = h_r = \left.\frac{\partial v_1}{\partial v_2}\right|_{i_1} = \left(\frac{V_1}{V_2}\right)_{I_1=0}$$ = open circuit reverse voltage transfer ratio for $I_1 = 0$ *i.e.* input terminals open

$$h_{22} = h_o = \left.\frac{\partial i_2}{\partial v_2}\right|_{i_1} = \left(\frac{I_2}{V_2}\right)_{I_1=0}$$ = open circuit output admittance for $I_1 = 0$ *i.e.* input terminals open.

For common base, (*CB*) common emitter (*CE*) and common collector (*CC*) connection respectively

$$h_{11} = h_i = h_{ib}, h_{ie}, h_{ic}$$
$$h_{21} = h_f = h_{fb}, h_{fe}, h_{fc}$$
$$h_{12} = h_r = h_{rb}, h_{re}, h_{rc}$$

or

$$h_{22} = h_o = h_{ob}, h_{oe}, h_{oc}$$

51.7. Determination of h-parameters. For a transistor in **common base configuration** as shown in Fig. 51.12.

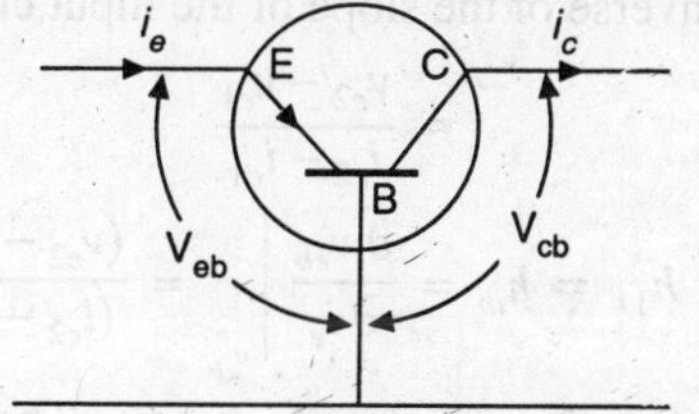

Fig. 51.12

$$i_1 = i_e \qquad i_2 = i_c$$
$$v_1 = v_{eb} \qquad v_2 = v_{cb}$$

$$\therefore \qquad h_{11} = h_{ib} = \left.\frac{\partial v_1}{\partial i_1}\right|_{v_2} = \left.\frac{\partial v_{eb}}{\partial i_e}\right|_{v_{cb}}$$

$$h_{21} = h_{fb} = \left.\frac{\partial i_2}{\partial i_1}\right|_{v_2} = \left.\frac{\partial i_c}{\partial i_e}\right|_{v_{cb}}$$

$$h_{12} = h_{rb} = \left.\frac{\partial v_1}{\partial v_2}\right|_{i_1} = \left.\frac{\partial v_{eb}}{\partial v_{cb}}\right|_{i_e}$$

$$h_{22} = h_{ob} = \left.\frac{\partial i_2}{\partial v_2}\right|_{i_1} = \left.\frac{\partial i_c}{\partial v_{cb}}\right|_{i_e}$$

Experiment 51.3. To calculate the h-parameters from the static characteristics of a transistor in common base configuration.

Apparatus. Same as in **Expt. 51.1.**

Procedure 1. Draw *three* **input** and *three* **output characteristics** for a transistor in common base configuration as in **Expt. 51.1.**

2. (*i*) *To calculate* h_{11} *or* h_{ib}. A set of three *input characteristics* ie. graph between *emitter current* i_e and *emitter base voltage* v_{eb} at constant value of *collector base voltage* $v_{cb} = v_c$, v_{c1} and v_{c_2} is shown in Fig. 51.13.

Find the *Q* point of the characteristic for $v_{cb} = v_c$. Draw a tangent to the characteristic at *Q*. Take two points *A* and *B* on either side of *Q* and very close to it. Draw perpendiculars Ai_{e1} and Bi_{e2} on i_e (*Y*) axis and Av_{e1} and Bv_{e2} on $v_{eb} = v_e$ (*X*) axis. The inverse of the slope of this input characteristic gives the value of $h_{11} = h_{ib}$

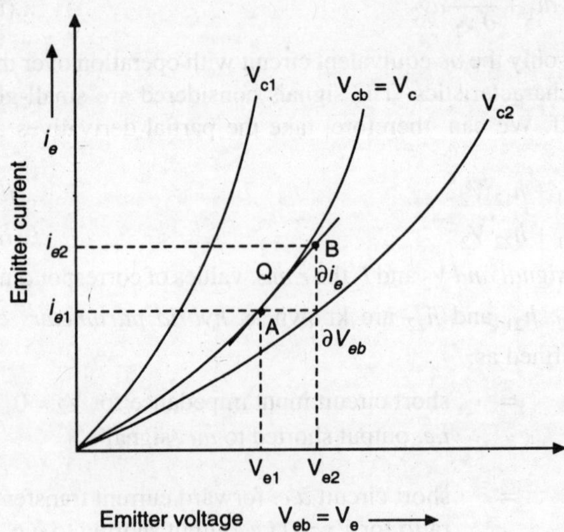

Fig. 51.13

Now $\partial i_e = i_{e2} - i_{e1}$

and $\partial v_{eb} = v_{e2} - v_{e1}$

∴ Inverse of the slope of the input characteristic

$$= \frac{v_{e2} - v_{e1}}{i_{e2} - i_{e1}}$$

and $$h_{11} = h_{ib} = \left.\frac{\partial v_{eb}}{\partial i_e}\right|_{v_{cb}} = \frac{(v_{e2} - v_{e1})}{(i_{e2} - i_{e1})}.$$

(*ii*) *To calculate* h_{12} *or* h_{rb}. The same set of input characteristics is again shown in Fig. 51.14. Draw a horizontal line parallel to *X* (v_{eb}) axis through the Quiescent point *Q* cutting the three input characteristics at points A_1, *Q* and A_2 respectively. If v_{e1} is the value of emitter voltage corresponding to A_1 and v_{e2} that corresponding to A_2 then

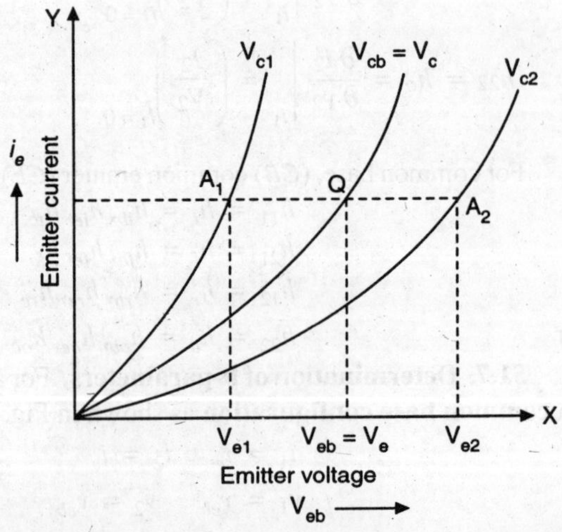

Fig. 51.14

the ratio of the difference $(v_{e2} - v_{e1})$ to the difference in the corresponding labelled values of collector voltage v_{c2} and v_{c1} *i.e.* $(v_{c2} - v_{c1})$ gives h_{12} or h_{rb}

$$\therefore \qquad h_{12} = h_{rb} = \left.\frac{\partial v_{eb}}{\partial v_{cb}}\right|_{i_e} = \frac{(v_{e2} - v_{e1})}{(v_{c2} - v_{c1})}.$$

(*iii*) *To calculate* h_{21} *or* h_{fb}. The value of h_{21} or h_{fb} can be calculated from the *output characteristics.* A set of three output characteristics *i.e.* a graph between collector current i_c and collector base voltage v_{cb} at constant values of emitter current i.e. i_e, i_{e1} and i_{e2} is shown in Fig. 51.15. Draw a vertical line parallel to Y-axis (i_c axis) through the quiescent point Q cutting the three output characteristics at the points A_1, Q and A_2 as shown in Fig. 51.15. If i_{c1} is the value of collector current corresponding to A_1 and i_{c2} that corresponding to A_2, then the ratio of the difference $(i_{c2} - i_{c1})$ to the difference of the corresponding labelled values of emitter current i_{e2} and i_{e1} *i.e.* $(i_{e2} - i_{e1})$ gives h_{21} or h_{fb}

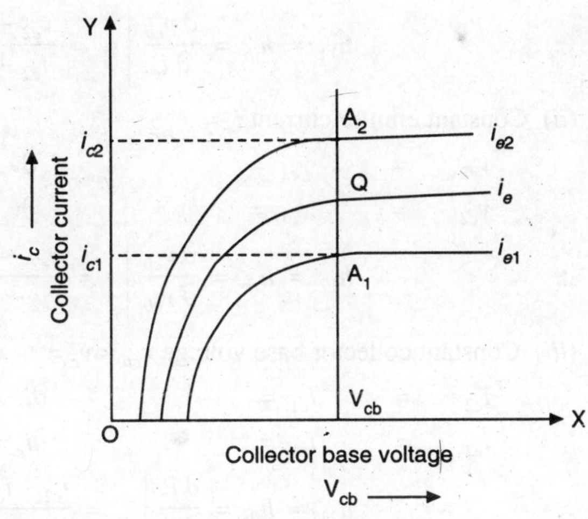

Fig. 51.15

$$\therefore \qquad h_{21} = h_{fb} = \left.\frac{\partial i_c}{\partial i_e}\right|_{v_{cb}} = \frac{(i_{c2} - i_{c1})}{(i_{e2} - i_{e1})}.$$

(*iv*) *To calculate* h_{22} *or* h_{ob}. The value of h_{22} or h_{ob} can also be calculated from the output characteristics.

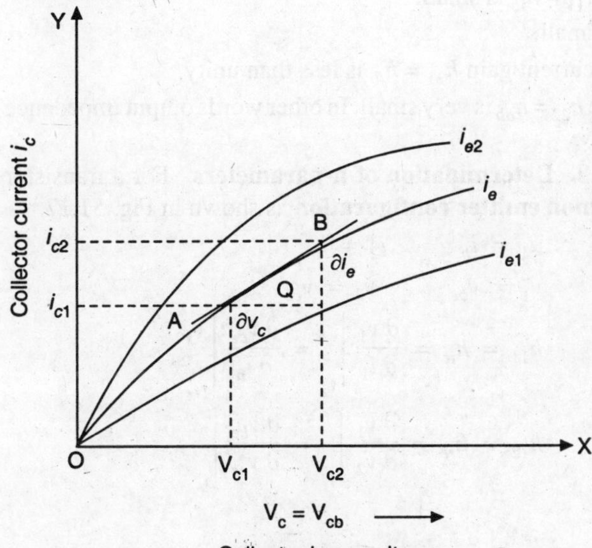

Fig. 51.16.

Draw a tangent to one of the characteristics for i_e = constant at the Quiescent point Q. Take two points A and B on either side of Q and very close to it. Draw perpendicular Ai_{c1} and Bi_{c2} on i_c (Y) axis and Av_{c1} and Bv_{c2} on $v_{cb} = v_c$ (X) axis. The slope of the output characteristic gives the value of h_{22} or h_{ob}

Now $\qquad \partial i_c = i_{c2} - i_{c1}$

and $\qquad \partial v_{cb} = v_{c2} - v_{c1}$

$\therefore$ Slope of the output characteristics

$$= \left.\frac{\partial i_c}{\partial v_{cb}}\right|_{i_e} = \frac{(i_{c2} - i_{c1})}{(v_{c2} - v_{c1})}$$

$$\therefore \qquad h_{22} = h_{ob} = \left.\frac{\partial i_c}{\partial v_{cb}}\right|_{i_e} = \frac{(i_{c2} - i_{c1})}{(v_{c2} - v_{c1})}.$$

Observations (*i*) Constant collector base voltage $v_{cb} = v_c =$

$v_{e2} = \quad v_{e1} = \quad \therefore \quad \partial v_{eb} = v_{e2} - v_{e1} =$

$i_{e2} = \quad i_{e1} = \quad \therefore \quad \partial i_e = i_{e2} - i_{e1} =$

$$\therefore \qquad h_{11} = h_{ib} = \left.\frac{\partial v_{eb}}{\partial i_e}\right|_{v_{cb}} = \frac{v_{e2} - v_{e1}}{i_{e2} - i_{e1}} =$$

(*ii*) Constant emitter current $i_e =$

$v_{e2} = \quad v_{e1} = \quad \partial v_{eb} = v_{e2} - v_{e1} =$

$v_{c2} = \quad v_{c1} = \quad \partial v_{cb} = v_{c2} - v_{c1} =$

$$\therefore \qquad h_{12} = h_{rb} = \left.\frac{\partial v_{eb}}{\partial v_{cb}}\right|_{i_e} = \frac{v_{e2} - v_{e1}}{v_{c2} - v_{c1}}$$

(*iii*) Constant collector base voltage $v_{cb} = v_c =$

$i_{c2} = \quad i_{c1} = \quad \partial i_c = i_{c2} - i_{c1} =$

$i_{e2} = \quad i_{e1} = \quad \partial i_e = i_{e2} - i_{e1} =$

$$\therefore \qquad h_{21} = h_{fb} = \left.\frac{\partial i_c}{\partial i_e}\right|_{v_{cb}} = \frac{i_{c2} - i_{c1}}{i_{e2} - i_{e1}} =$$

(*iv*) Constant emitter current $i_e =$

$i_{c2} = \quad i_{c1} = \quad \partial i_c = i_{c2} - i_{c1} =$

$v_{c2} = \quad v_{c1} = \quad \partial v_{cb} = v_{c2} - v_{c1} =$

$$\therefore \qquad h_{22} = h_{ob} = \left.\frac{\partial i_c}{\partial v_{cb}}\right|_{i_e} = \frac{i_{c2} - i_{c1}}{v_{c2} - v_{c1}} =$$

Conclusions. 1. The input impedance $h_{11} = h_{ib}$ is small.

2. The reverse voltage gain $h_{12} = h_{rb}$ is small.

3. The forward current transfer ratio or current gain $h_{21} = h_{fb}$ is less than unity.

4. The out put admittance or conductance $h_{22} = h_{ob}$ is very small. In other words output impedance is very high.

51.8. Determination of h-parameters. For a transistor in **common emitter configuration** as shown in Fig. 51.17.

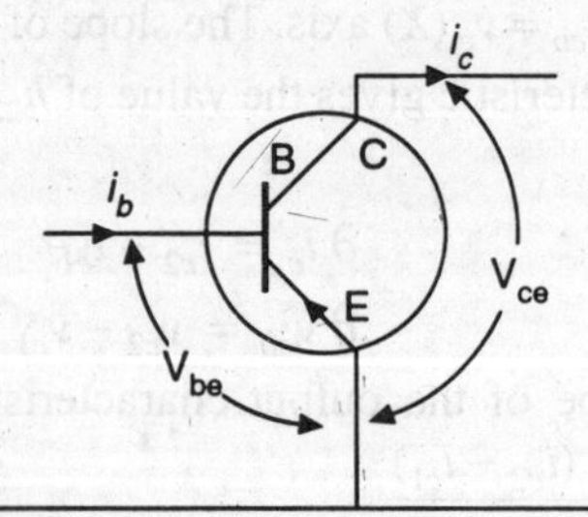

Fig. 51.17

$i_1 = i_b \qquad i_2 = i_c$

$v_1 = v_{be} \qquad v_2 = v_{ce}$

$$h_{11} = h_{ie} = \left.\frac{\partial v_1}{\partial i_1}\right|_{v_2} = \left.\frac{\partial v_{be}}{\partial i_b}\right|_{v_{ce}}$$

$$h_{12} = h_{re} = \left.\frac{\partial v_1}{\partial v_2}\right|_{i_1} = \left.\frac{\partial v_{be}}{\partial v_{ce}}\right|_{i_b}$$

$$h_{21} = h_{fe} = \left.\frac{\partial i_2}{\partial i_1}\right|_{v_2} = \left.\frac{\partial i_c}{\partial i_b}\right|_{v_{ce}}$$

$$h_{22} = h_{oe} = \left.\frac{\partial i_2}{\partial v_2}\right|_{i_1} = \left.\frac{\partial i_c}{\partial v_{ce}}\right|_{i_b}$$

Experiment 51.4. To calculate the h-parameters from the static cnaracteristics of a transistor in common emitter configuration.

Apparatus same as in **Expt. 51.2.**

Procedure 1. Draw *three* **input** and *three* **output characteristics** for a transistor in *common emitter* configuration as in **Expt. 51.2.**

2. *To calculate h_{11} or h_{ie}.* A set of three *input characteristics* i.e. graph between base current i_b and base emitter voltage v_{be} at constant values of collector emitter voltage $v_{ce} = v_c$, v_{c1} and v_{c2} is shown in Fig. 51.18. Find the Q point of the characteristic of $v_{ce} = v_c$. Draw a tangent to the characteristic at Q. Take two points A and B on either side of Q and very close to it. Draw perpendiculars Ai_{b1} and Bi_{b2} on i_b (Y) axis and Av_{b1} and Bv_{b2} on $v_{be} = v_b$ (X) axis. The inverse of the slope of this input characteristic gives the value of $h_{11} = h_{ie}$

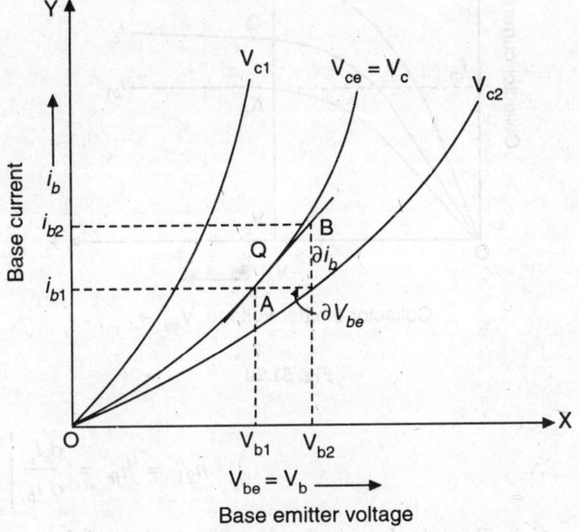

Fig. 51.18

Now $\quad \partial i_b = i_{b2} - i_{b1}$

and $\quad \partial v_{be} = v_{b2} - v_{b1}$

$\therefore$ Inverse of the slope of the input characteristic $= \dfrac{v_{b2} - v_{b1}}{i_{b2} - i_{b1}}$

and $$h_{11} = h_{ie} = \left.\frac{\partial v_{be}}{\partial i_b}\right|_{v_{ce}}$$

$$= \frac{v_{b2} - v_{b1}}{i_{b2} - i_{b1}}$$

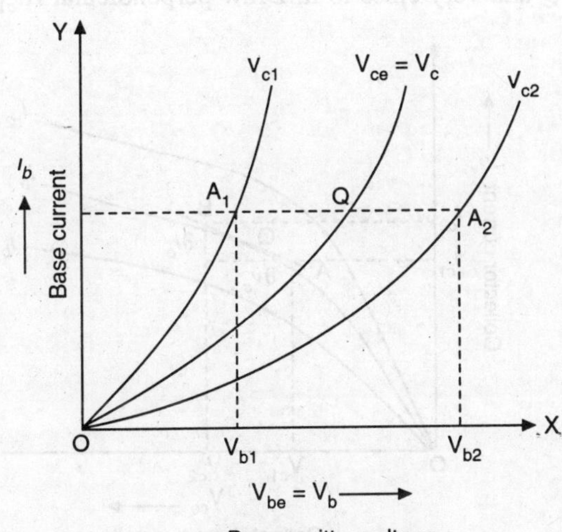

Fig. 51.19

(*ii*) *To calculate h_{12} or h_{re}.* The same sei of input characteristics is again shown in Fig. 51.19. Draw a horizontal line parallel to the X-axis (v_{be} axis) through the quiescent point Q cutting the three input characteristics at points A_1, Q and A_2 respectively. If v_{b1} is the value of base voltage corresponding to A_1 and v_{b2} that corresponding to A_2, then the ratio of the difference $(v_{b2} - v_{b1})$ to the difference in the corresponding labelled values of collector voltage v_{c2} and v_{c1} *i.e.* $v_{c2} - v_{c1}$ gives h_{12} or h_{re}.

$$\therefore \qquad h_{12} = h_{re} = \left.\frac{\partial\, v_{be}}{\partial\, v_{ce}}\right|_{i_b} = \frac{v_{b2} - v_{b1}}{v_{c2} - v_{c1}}.$$

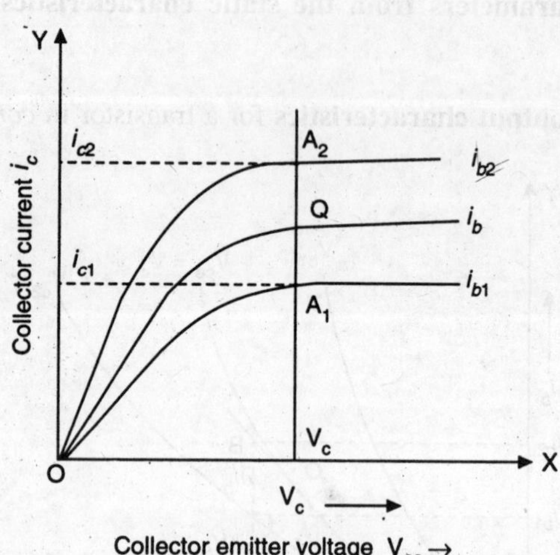

Fig. 51.20

(*iii*) *To calculate h_{21} or h_{fe}.* The value of h_{21} or h_{fe} can he calculated from the output characteristics. A set of three output characteristics *i.e.* a graph between collector current i_c and collector emitter voltage v_{ce} at constant values of base current i_b, i_{b1} and i_{b2} is shown in Fig. 51.20.

Draw a vertical line parallel to *Y*-axis (i_c-axis) through the quiescent point *Q* cutting the output characteristics at points A_1, *Q* and A_2 as shown in Fig. 51.20. If i_{c1} is the value of collector current corresponding to A_1 and i_{c2} that corresponding to A_2 then the ratio of the difference $(i_{c2} - i_{c1})$ to the difference of the corresponding labelled values of base current i_{b2} and i_{b1} *i.e.* $(i_{b2} - i_{b1})$ gives h_{21} or h_{fe}.

$$\therefore \qquad h_{21} = h_{fe} = \left.\frac{\partial\, i_c}{\partial\, i_b}\right|_{v_{ce}} = \frac{i_{c2} - i_{c1}}{i_{b2} - i_{b1}}.$$

(*iv*) *To calculate h_{22} or h_{oe}.* The value of h_{22} or h_{oe} can also be calculated from the output characteristics. Draw a tangent to one of the characteristics for i_b = constant at the Quiescent point *Q*. Take two points *A* and *B* on either side of *Q* and very close to it. Draw perpendicular Ai_{c1} and Bi_{c2} on i_c (*Y*-axis) and Av_{c1} and Bv_{c2} on v_{ce} (*X*) axis.

The slope of the output characteristic gives the value of $h_{22} = h_{oe}$.

Now $\qquad \partial\, i_c = i_{c2} - i_{c1}$

and $\qquad \partial\, v_{ce} = v_{c2} - v_{c1}$

$\therefore$ Slope of the output characteristics

$$= \left.\frac{\partial\, i_c}{\partial\, v_{ce}}\right|_{i_b}$$

$$\therefore \quad h_{22} = h_{oe} = \left.\frac{\partial\, i_c}{\partial\, v_{ce}}\right|_{i_b} = \frac{i_{c2} - i_{c1}}{v_{c2} - v_{c1}} =$$

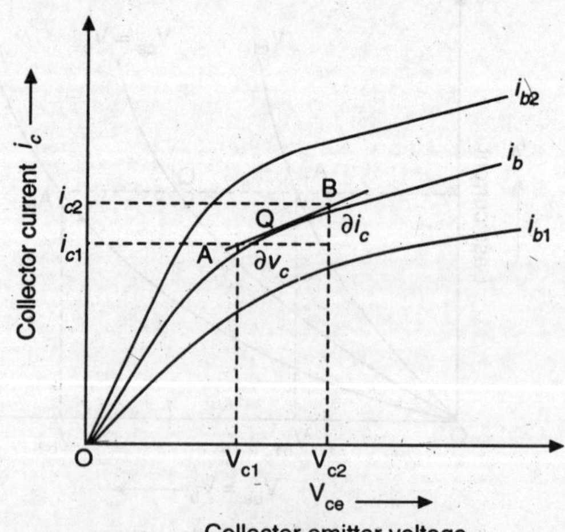

Fig. 51.21

Observations (*i*) Constant collector emitter voltage $v_{ce} = v_c =$

$$v_{b2} = \quad v_{b1} = \quad \therefore \; \partial v_{be} = v_{b2} - v_{b1} =$$

$$i_{b2} = \quad i_{b1} = \quad \therefore \; \partial i_b = i_{b2} - i_{b1} =$$

$$\therefore \quad h_{11} = h_{ie} = \left.\frac{\partial v_{be}}{\partial i_b}\right|_{v_{ce}} = \frac{v_{b2} - v_{b1}}{i_{b2} - i_{b1}} =$$

(*ii*) Constant base current i_b =

$$v_{b2} = \quad v_{b1} = \quad \partial v_{be} = v_{b2} - v_{b1} =$$

$$v_{c2} = \quad v_{c1} = \quad \partial v_{ce} = v_{c2} - v_{c1} =$$

$$\therefore \quad h_{12} = h_{re} = \left.\frac{\partial v_{be}}{\partial v_{ce}}\right|_{i_b} = \frac{v_{b2} - v_{b1}}{v_{c2} - v_{c1}} =$$

(*iii*) Constant collector emitter voltage $v_{ce} = v_c$ =

$$i_{c2} = \quad i_{c1} = \quad \partial i_c = i_{c2} - i_{c1} =$$

$$i_{b2} = \quad i_{b1} = \quad \partial i_b = i_{b2} - i_{b1} =$$

$$\therefore \quad h_{21} = h_{fe} = \left.\frac{\partial i_c}{\partial i_b}\right|_{v_{ce}} = \frac{i_{c2} - i_{c1}}{i_{b2} - i_{b1}} =$$

(*iv*) Constant base current i_b =

$$i_{c2} = \quad i_{c1} = \quad \partial i_c = i_{c2} - i_{c1} =$$

$$v_{c2} = \quad v_{c1} = \quad \partial v_{ce} = v_{c2} - v_{c1} =$$

$$\therefore \quad h_{22} = h_{oe} = \left.\frac{\partial i_c}{\partial v_{ce}}\right|_{i_b} = \frac{i_{c2} - i_{c1}}{v_{c2} - v_{c1}} =$$

Conclusions. Same as in **Expt. 51.3** except that $h_{21} = h_{fe}$ the forward current transfer ratio or current gain is very high >> 1.

Experiment 51.5. To determine the hybrid parameters of a transistor in common emitter configuration at low frequency.

Apparatus. A *n-p-n* transistor, resistors 1 K Ω, 50 K Ω, 100 Ω and 1 Meg Ω variable, capacitors

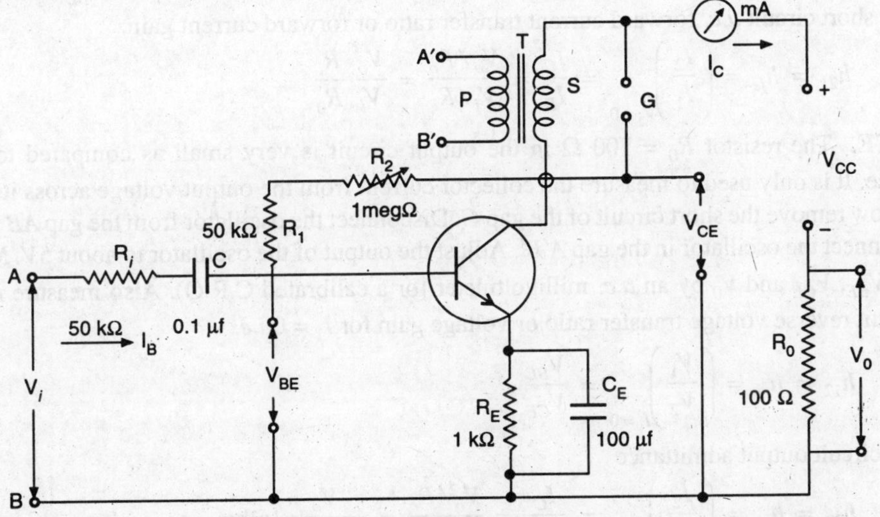

Fig. 51.22

100 μf, 0.1 μf, a battery, a step down transformer, An audio frequency signal generator (upto a few KHz), *a.c.* millivoltmeter or a calibrated C.R.O. *a.c.* milliammeter, *d.c.* milliammeter etc.

Theory. For the determination of, *h*-parameters the following two conditions must be satisfied
(*i*) output shorted to *a.c.* signal
(*ii*) input terminals open to *a.c.* as explained in **Art. 51.6.**

These conditions can be met with in the circuit given in Fig. 51.22 and *h*-parameters can be determined. The circuit consists of an *n-p-n* transistor in *common emitter configuration.* A *d.c.* source consisting of a battery of a few cells is connected between the terminals marked V_{cc} with the *positive* and *negative* terminals as shown. A resistor $R_E = 1$ K Ω and capacitor $C_E = 100\ \mu f$ in parallel acts as the emitter bias and a resistor R_1 (50 K Ω) in series with a variable resistor R_2 of 1 Megohm acts as base bias with the help of single *d.c.* source V_{cc}.

When the gap *G* in the collector circuit is short circuited the circuit fulfils the first condition "*Output shorted to a.c. signal i.e.*" $V_2 = 0$ by reducing the load of the secondary *S* of the transformer to zero. A 1 KHz oscillator (capable of giving signals from mV to a few volts) connected in the gap *AB* serves as an *a.c.* source.

If the gap *G* is not short circuited and the oscillator is connected in the gap *A′B′* (instead of the gap *AB*) the second condition "*input terminals open to a.c. i.e.* $I_1 = 0$" is satisfied.

Procedure 1. Draw a diagram showing the scheme of connections as in Fig. 51.22 and make necessary connections accordingly.

2. Adjust the variable resistance R_2 to obtain a *d.c.* quiescent condition $I_c = 0.5$ mA approximately. Note the corresponding value of V_{CE}.

3. Short circuit the gap *G* to make $V_2 = 0$. Connect a 1 KHz oscillator in the gap *AB*. Adjust the value of V_i between 10-20 mV so that the value of I_c does not change markedly. Note the voltages V_i, V_{BE} and V_0 by an *a.c.* millivoltmeter [or a calibrated C.R.O] and the values of resistors R_i and R_0, then

Short circuit input impedance for $V_2 = 0$ *i.e.*,

$$h_{11} = h_{ie} = \left(\frac{V_1}{I_1}\right)_{V_2=0} = \frac{V_{BE}}{I_B} = \frac{V_{BE}}{V_i/R_i} = \frac{V_{BE}}{V_i} R_i \text{ ohm.}$$

NOTE. In the above calculation it is assumed that the reactance of the capacitor *C* is negligibly small as compared to the resistance R_i which is of the order of 50 K Ω and $V_i >> V_{BE}$.

Also short circuit *a.c.* forward current transfer ratio or forward current gain.

$$h_{21} = h_{fe} = \left(\frac{I_2}{I_1}\right)_{V_2=0} = \frac{I_C}{I_B} = \frac{V_o/R_o}{V_i/R_i} = \frac{V_o}{V_i}\frac{R_i}{R_0}$$

NOTE. The resistor $R_0 = 100$ Ω in the output circuit is very small as compared to output impedance. It is only used to measure the collector current from the output voltage across it.

4. Now remove the short circuit of the gap *G*. Disconnect the oscillator from the gap *AB* to make $I_1 = 0$. Connect the oscillator in the gap *A′B′*. Adjust the output of the oscillator to about 5*V*. Measure voltages V_{BE}, V_{CE} and V_0 by an *a.c.* millivoltmeter [or a calibrated C.R.O). Also measure R_o, then open circuit reverse voltage transfer ratio or voltage gain for $I_1 = 0$ *i.e.*

$$h_{12} = h_{re} = \left(\frac{V_1}{V_2}\right)_{I_1=0} = \frac{V_{BE}}{V_{CE}}$$

and open circuit output admittance

$$h_{22} = h_{oe} = \left(\frac{I_2}{V_2}\right)_{I_1=0} = \frac{I_C}{V_{CE}} = \frac{V_o/R_o}{V_{CE}} = \frac{V_o}{V_{CE}\ R_o} \text{ mho.}$$

5. Repeat the experiment three times with different frequencies from the oscillator in the low frequency range.

Observations (*a*) Output shorted to *a.c.* signal *i.e.* $V_2 = 0$

No.	Frequency	Input voltage V_i	Base emitter voltage V_{BE}	Output voltage V_o	Input Resistance $R_i\ \Omega$	Output resistance $R_o\ \Omega$	$h_{11} = h_{ie}$ $\frac{V_{BE}}{V_i} R_i$ ohm	$h_{21} = h_{fe}$ $= \frac{V_o}{V_i} \frac{R_i}{R_o}$
1								
2								
3								

(*b*) Input terminals open *i.e.* $I_1 = 0$

No.	Frequency	Base emitter voltage V_{BE}	Collector emitter voltage V_{CE}	Output voltage V_o	Output Resistance R_o	$h_{12} = h_{re} = \frac{V_{BE}}{V_{CE}}$	$h_{22} = h_{oe} = \frac{V_o}{V_{CE} R_o}$ mho
1							
2							
3							

Precautions. 1. Use low frequency oscillator only.

2. h_{fe} and h_{re} are dimensionless quantities.

3. h_{ie} is in ohms and h_{oe} in mho.

Oral Questions

1. Transistor (*i*) What is a transistor? (*ii*) What are two types of transistors? (*iii*) Distinguish between *P-N-P* and *N-P-N* transistors. (*iv*) What are majority charge carriers in each of the two transistors? (*iv*) What are the three transistor configurations? (*v*) What is the relation between emitter current, collector current and base current? (*vi*) Compare the function of emitter, base and collector of a transistor with cathode, grid and plate of a triode valve. (*vii*) Can you interchange the emitter and collector terminals of a transistor? If not why? (*viii*) Why is the emitter base circuit forward biased and collector base circuit reverse biased? (*ix*) What is the origin of the word transistor? (*x*) What are the advantages of a transistor over triode valve?

2. Hybrid parameters. (*i*) What are hybrid or *h*-parameters? (*ii*) Why are they so called? (*iii*) How many *h*-parameters do we have? (*iv*) For a common base circuit what does h_{ib}, h_{rb}, h_{fb} and h_{ob} represent? (*v*) For a common emitter circuit what does h_{ie}, h_{re}, h_{fe} and h_{oe} represent? (*vi*) How can you determine *h*-parameters from various transistor characteristics? (*vii*) What are the dimensions of *h*-parameters, h_{fe} and h_{re}? (*viii*) What are the units of *h*-parameters h_{ie} and h_{oe}? (*ix*) State the two conditions necessary for introduction of *h*-parameters. (*x*) Which *h*-parameter represents short circuit input impedance? (*xi*) Which *h*-parameters represents open circuit output admittance. (*xii*) Which h-parameter represents forward current transfer ratio or current gain? (*xiii*) Which *h*-parameter represents reverse voltage transfer ratio?

52

Field Effect Transistor

52.1. Field effect transistor. A field effect transistor (F.E.T.) is a three terminal semi-conductor device, the three terminals being known as (*i*) Gate *G*, (*ii*) *Source S* and (*iii*) *Drain D*, in which the current through the *channel* connecting the two terminals *source* and *Drain* is controlled by the voltage applied to the third terminal *Gate*.

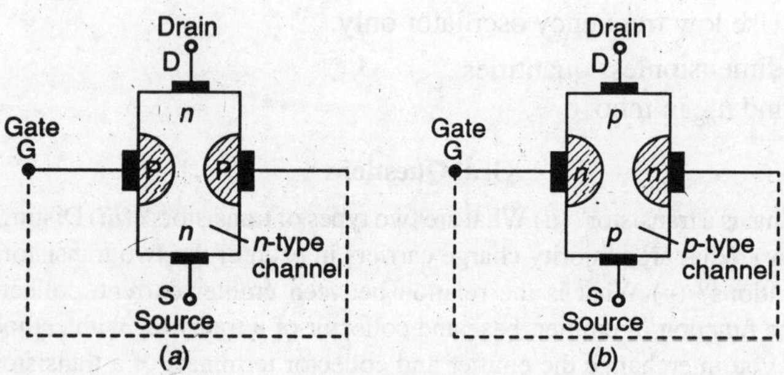

Fig. 52.1

The channel is a *p*-type or *n*-type silicon bar having two *p-n* junctions on the sides. A F.E.T having *n*-type channel is shown in Fig. 52.1 (*a*) and the one having a *p*-type channel is shown in Fig. 52.1 (*b*). The two *p-n* junctions forming *diodes* are connected internally (shown by dotted lines) and the common terminal forms the *Gate*. The *Drain* and *Source* terminals are taken out from the two ends of the channel bar. The *Drain* and *Source* terminals are inter changeable.

Working. Let us consider an *n*-channel *FET*. A voltage V_{GS} can be applied between the *Gate G* and the *Source S* and a voltage V_{DS} between the *Source S* and the *Drain D*. Suppose to begin with the Gate voltage is *zero*. When a voltage V_{DS} is applied between the *Drain D* and the *Source S* as shown by connecting the *negative* terminal of the battery to the *Source S* and *positive* to the *Drain D* electrons flow from the *Source* to the *Drain*. The two *p-n* junctions at the side of the bar establish *depletion layers* and these electrons flow through a channel between the depletion layers. The size of the depletion layers determines the *width* of the channel and hence the *Drain* current I_D through the bar.

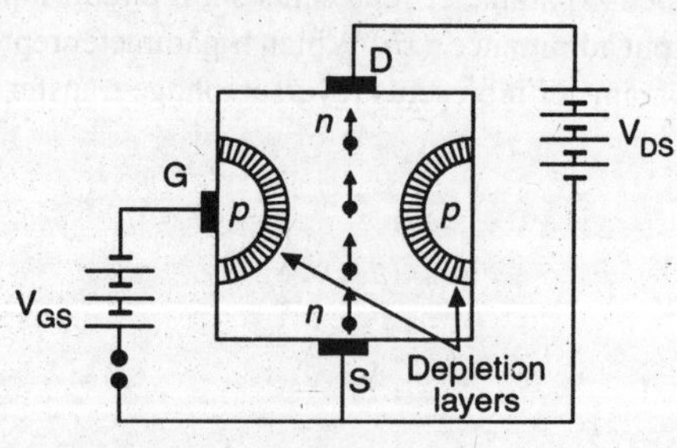

Fig. 52.2

At first the *Drain* current I_D increases rapidly with *Drain-Source voltage* V_D but finally attains a constant value. The *Drain-Source* voltage V_{DS} at which I_D becomes a constant is known as *pinch off voltage* because any further increase in *Drain-Source* voltage does not give rise to any increase in *Drain* current.

When a reverse voltage V_{GS} is applied between the *Gate* and the *Source* as shown the width of the depletion layers increases, thereby decreasing the area of cross-section of the conducting channel and thus increasing its resistance. Due to the increase in resistance the *Drain* current (*i.e.* the flow of electrons from the *Source* to the *Drain* decrease. The value of *Drain* current I_D for a given *Drain Source* voltage (V_{DS}) depends upon the *Gate-Source* voltage (V_{GS}). The more negative is V_{GS} the smaller is the value of I_D. Further I_D attains a constant maximum value for a smaller value of V_{DS} or pinch off voltage.

Thus we find that the value of *Drain* current I_D from *Source* to *Drain* can be controlled by the *potential* applied to the *Gate i.e.* the *Electric Field* between the *Gate* and the *Source*. This is why such a transistor is known as a *Field effect transistor*.

NOTE. If we have a *p*-channel FET as shown the positive of the battery V_{DS} is applied to the *Source* and positive of the battery V_{GS} is applied to the *Gate* to make the Gate *G* reverse biased with respect to the *Source S*. The electric current flows from *S* to *D* through the channel and the charge carriers are the *holes* which also flow from *S* to *D*. Thus the *charge carriers* in a *n*-channel EFT are *electrons* whereas in a *p*-channel FET these are *holes i.e.* in a Field Effect Transistor current conduction is by *one type of carrier* only *i.e. electrons* or *holes*.

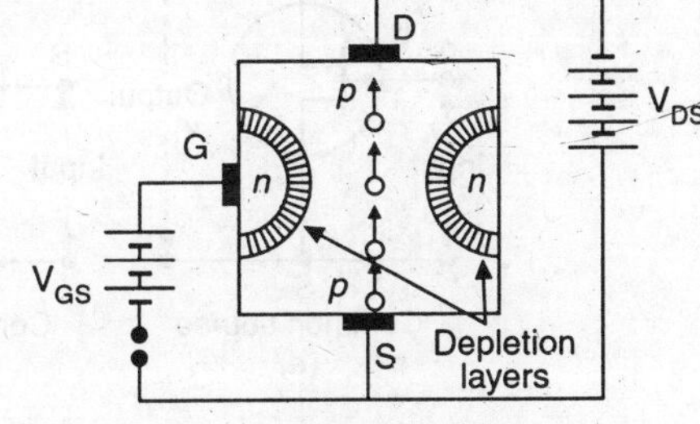

Fig. 52.3

A Field Effect Transistor is schematically shown as under. The vertical line represents the channel. The *Source* and *Drain* are connected to this line. If the channel is *n*-type Fig. 52.4 (*a*) the arrow on the Gate is drawn from the *Gate* to the channel whereas for a *p*-type channel Fig. 52.4 (*b*) the arrow is drawn from the channel to the *Gate*.

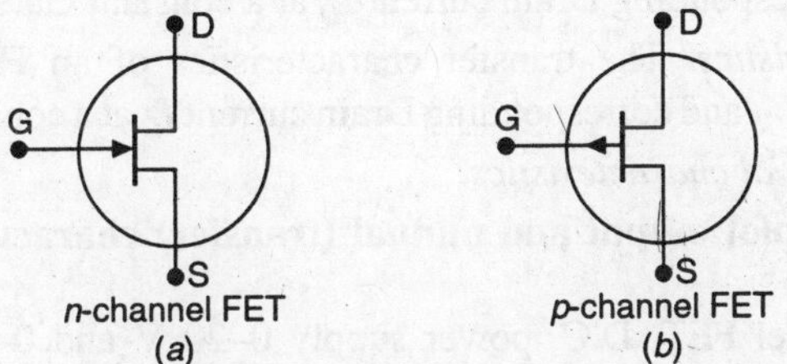

Fig. 52.4

Comparing a Field effect transistor FET with a Bipolar Junction Transistor BJT we find that the Source, Gate and Drain of a FET correspond to Emitter, Base and Collector of a BJT Fig. 52.5 (*a*) and (*b*). Further as the *input circuit i.e. Gate* to Source is reverse biased it has a *high input impedance* whereas the input circuit of a BJT *i.e.* Base to Emitter is *forward biased* and, therefore, has a *low input impedance*. FET is a *voltage operated device i.e.* input voltage controls the output current and a BJT is a *current operated device i.e.* input current controls the output current.

Like the BJT, the FET can also be used in *three* basic configurations *i.e. Common Source, Common Gate* and *Common Drain* as shown in Fig. 52.6 (*a*), (*b*) and (*c*).

These respectively correspond to the *Common emitter, Common base* and *Common collector* configuration of a BJT amplifier.

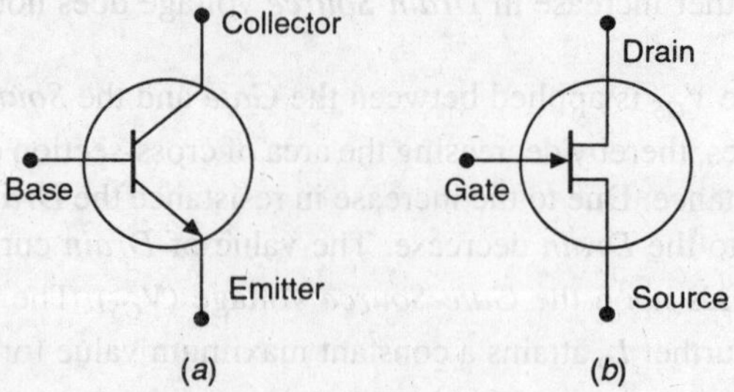

Fig. 52.5

For a FET there is a phase-shift of π in case of Common Source configuration but no phase shift takes place in the other two cases. A FET has a high input impedance and a high output impedance.

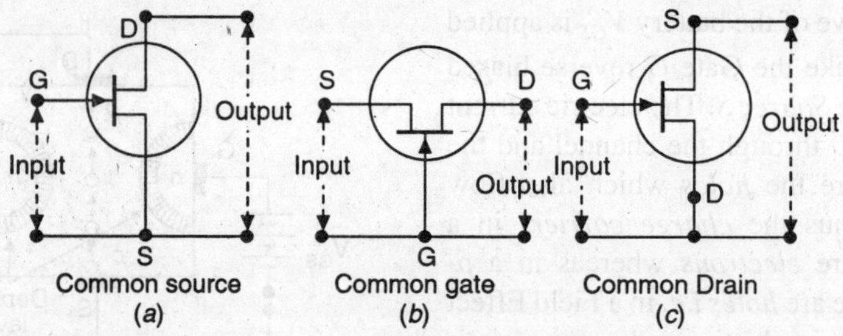

Fig. 52.6

It is a voltage controlled device and there are only one kind of charge carriers i.e. electrons in *n*-type and holes in a *p*-type channel.

52.2. Characteristics of a Field Effect Transistor.

(*i*) *Output characteristics*. The output characteristics of an FET is the curve between Drain-Source *voltage* V_{DS} and corresponding Drain current I_D at a constant Gate-Source voltage V_{GS}.

(*ii*) *Transfer characteristics*. The transfer characteristics of an FET is the curve between Gate-Source bias (voltage) V_{GS} and corresponding Drain current I_D at a constant Drain-Source voltage V_{DS}. It is also known as *mutual characteristics*.

Experiment 52.1. To plot output and mutual (transfer) characteristics of a Field Effect Transistor.

Apparatus. A *n*-channel FET, D.C. power supply 0–20 V and 0–12 V, Two voltmeters, a milli-ammeter, two capacitors each 50 μF (50 V).

Procedure. 1. Draw a diagram showing the scheme of connections as in Fig. 52.7 and connect accordingly.

Note 1. The choice of power supply, volt meter and milli-ammeter depends upon the characteristics of the FET selected for the experiment.

Note. As electronic power supplies have been used for applying the Drain source voltage and Gate Source bias a capacitor of 50 μF capacitance has been connected in each circuit to stop the ripple.

2. (*a*) **Output characteristics** Keeping the Gate-Source bias zero by short circuiting the Gate to the Source slowly increase the Drain-Source voltage from zero upward in steps of 0.5 to 1 volt and note the values of Drain source voltage V_{DS} and corresponding values of Drain current I_D till the Drain

current becomes constant and does not increase further for at least three steps of increasing Drain-Source voltage.

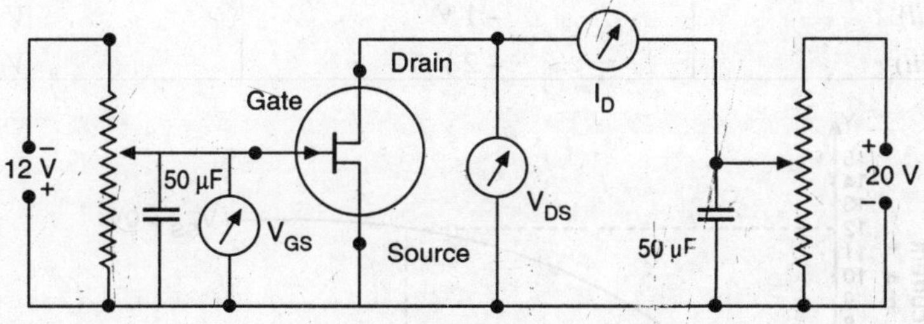

Fig. 52.7

3. Now keep the Gate-Source bias at $-1V$ and again note the values of Drain-Source voltage V_{DS} from zero upwards and corresponding values of Drain current I_D till three constant readings of I_D are obtained.

4. Similarly repeat with a Gate-Source bias of $-2V$ and $-3V$ (or in smaller steps of 0.5 V in each case increasing the Drain-Source voltage V_{DS} from zero upward till three constant values of I_D are obtained.

5. (*b*) **Transfer characteristics.** Keep the Drain-Source voltage at +1 volt or at any suitable value in accordance with the data of the FET used. Slowly increase the Gate-Source bias V_{GS} from zero towards increasing *negative* values in steps of 0.5 to 1 volt. Note the values of V_{GS} and corresponding values of Drain current I_D. It is seen that as the negative Gate Source bias increases the Drain current I_D decreases. Hence go on increasing the negative Gate Source bias till the current I_D becomes zero.

6. Similarly repeat with Drain Source voltage of +2 V and +3 V or other suitable values according to FET data.

Observations. (*a*) *Output characteristics*

Sl. No.	$V_{GS} = 0V$		$V_{GS} = -1V$		$V_{GS} = -2V$		$V_{GS} = -3V$	
	V_{DS}	I_D	V_{DS}	I_D	V_{DS}	I_D	V_{DS}	I_D
1								
2								
3								
4								
5								
.								
.								
.								
15								

7. For output characteristics draw graphs between Drain current I_D taken along the Y-axis and Drain-Source voltage V_{DS} taken along the X-axis for various values of Gate-Source voltage V_{GS} on the same graph paper as shown in Fig. 52.8.

From the graph find the value of *pinch off voltage i.e.* the value of V_{DS} for which I_D attains a constant maximum value for each value of Gate Source bias and record.

Sl. No.	*Gate Source bias* V_{GS}	*Pinch off voltage*
(*i*)	0 *V*	*V*
(*ii*)	– 1 *V*	*V*
(*iii*)	– 2 *V*	*V*

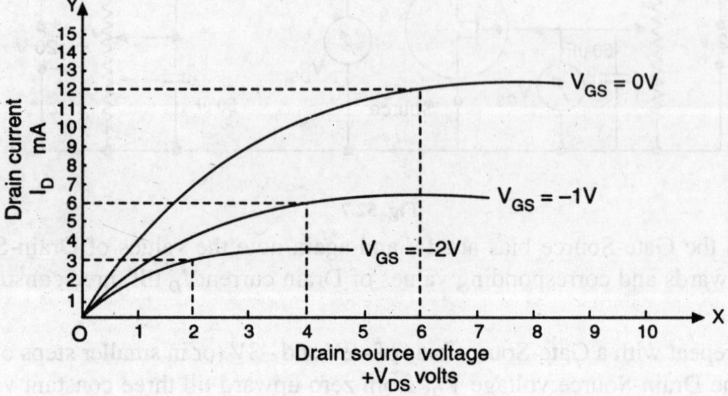

Fig. 52.8

As the value of *reverse* Gate Source bias increases pinch off takes place at a smaller Drain Source voltage.

(*b*) ***Transfer characteristics*** or ***mutual characteristics***. [Gate source bias V_{GS}, Drain current I_D graph].

Sl. No.	$V_{DS} = 3\ V$		$V_{DS} = 2\ V$		$V_{DS} = 1\ V$	
	V_{GS}	I_D	V_{GS}	I_D	V_{GS}	I_D

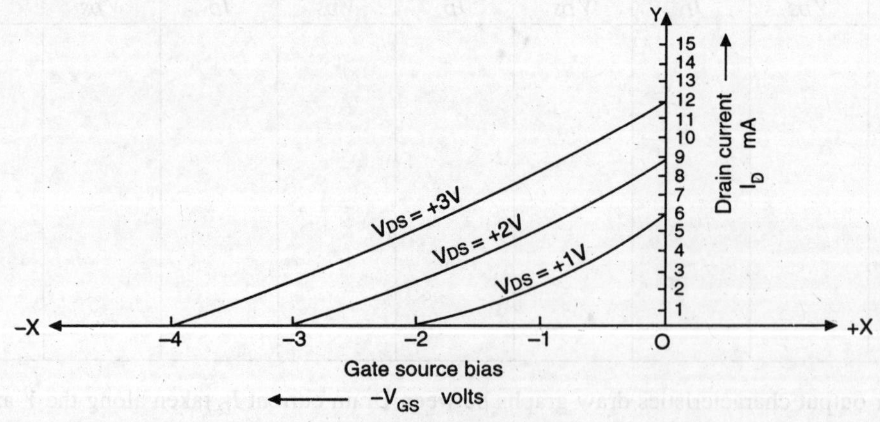

Fig. 52.9

8. For transfer characteristics, draw graphs between Drain current I_D taken along the *Y*-axis and Gate Source bias V_{GS} taken along –*X*-axis for various values of Drain Source voltage V_{DS} on the same graph paper as shown in Fig. 52.9.

The end points of the transfer characteristic curves are

(1) $V_{GS} = 0$ and $I_D = I_{DSS}$ *i.e.* the maximum saturation value of Drain current.

(2) $V_{GS} = V_{GS(off)}$ i.e. the value of negative Gate Source bias for which drain current I_D is cut off and $I_D = 0$.

Note. When the drain current in cut off $V_{GS} = -V_P$, the pinch off voltage.

Note the value of I_{DSS} and $V_{GS(off)}$ for different values of V_{DS}.

Sl. No.	*Drain Source Voltage* (V_{DS})	*Saturation Drain Current* (I_{DSS})	*Cut off value of Gate Source bias* V_{GS} *(off)*

Precautions. 1. Note the maximum rating of FET and the temperature upto which it can be used.

2. Note whether the FET is *n*-channel or *p*-channel device and apply the Gate Source bias and Drain Source voltage accordingly.

3. For an electronic D.C. power supply use 50 μF capacitor to stop the ripple.

Oral Questions

Field Effect Transistor (*i*) What is the construction of a Field Effect Transistor? (*ii*) Why is it so called? (*iii*) How does it differ from a bi-junction transistor? (*iv*) What is the function of the Gate, Source and Drain? (*v*) How do their functions compare with the emitter, base and collector of a BJT. (*vi*) What is the difference between *p*-channel and *n*- channel FET? (*vii*) What is the material of which the channel is made? (*viii*) What do you mean by depletion layer? (*ix*) What determines the width of the channel? (*x*) What do you mean by pinch off voltage? (*xi*) On what factors does it depend? (*xii*) What are the charge carriers in a *p*-channel FET and in a *n*-channel FET. (*xiii*) What are the three configurations in which a FET can be used? (*xiv*) Explain what you mean by output and mutual characteristics of an FET?

53

Oscillators

53.1. Wave meter or frequency meter. A *heterodyne type* of wave meter (also known as *beat frequency type*) consists of an R.F. oscillator having a calibrated frequency scale also called *calibrated signal generator*. The oscillator frequency depends upon the value of inductance L and capacitance C of the tank circuit given by $\frac{1}{2\pi\sqrt{LC}}$. The frequency can be varied in *steps* by using a number of coils. A given frequency range can be selected by using a *selector switch* which when operated introduces a coil (or coils) of suitable value of inductance L, within this frequency band (range). Thus selected frequency can be continuously varied from a certain minimum to a maximum value by changing the capacitance of a variable (gang) capacitor. The frequency of the oscillations generated by the oscillator is given by the calibrated scale. The output of the oscillator is suitably amplified.

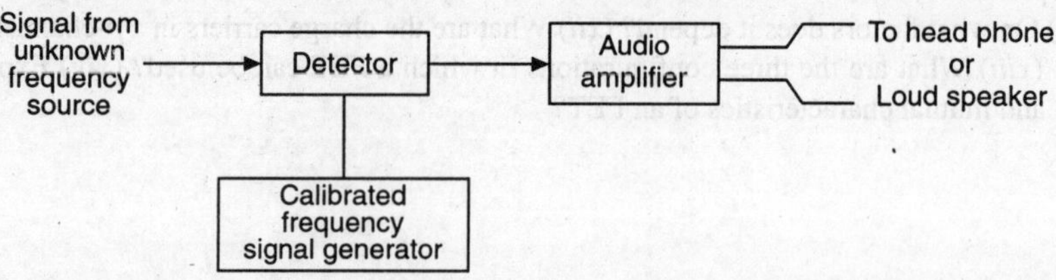

Fig. 53.1

To find the unknown frequency f_u, the signal from the unknown frequency source as well as that of known frequency f_w from the calibrated signal generator are mixed in a non-linear device such as a rectifier or diode known as *detector* as shown in Fig. 53.1. The two frequencies produce a *beat note*

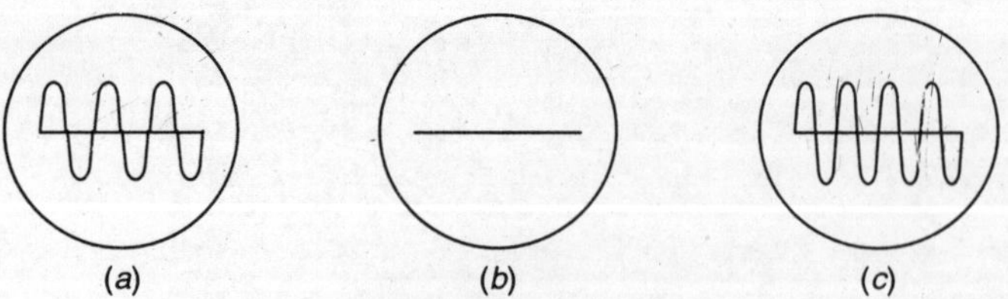

Fig. 53.2

which consists of the difference $(f_w - f_u)$ and sum $(f_w + f_u)$ of the two frequencies. When the difference between the two frequencies lies in the *audio range* (30 to 15000 Hz) the beat note can be heard by using a head phone or a loud speaker after suitable amplification by an audio amplifier as shown.

The frequency of the signal from the calibrated signal generator is slowly varied so that the frequency difference goes on dereasing and finally approaches *zero.* The two signals are equal in frequency at zero beat. Zero beat cannot be heard but can be located between two points on the calibrated frequency scale which produces the lowest audible frequency. The mean of these two values gives the frequency for zero beat and is equal to the unknown frequency.

For very accurate work we can use an oscilloscope in place of the headphone or speaker as a **null** indicator. The heterodyne beat pattern on the oscilloscope are shown in Fig 53.2. Zero beat is shown at (*b*). At (*a*) and (*c*) are shown audio-sine wave forms which are observed on either side of zero beat.

53.2. Oscillatory circuit. *A circuit which produces electrical oscillations of any desired frequency is known as an oscillatory circuit.*

A simple oscillatory circuit consists of a capacitor C and an inductor L placed in parallel (known as *tank circuit*) connected to a battery as shown. When the key K_1 is switched on the capacitor C is charged with the upper plate at a positive potential and the lower plate at a negative potential. If now the key K_1 is switched off, the position is as shown in Fig. 53.4 (*a*) (where battery has not been shown).

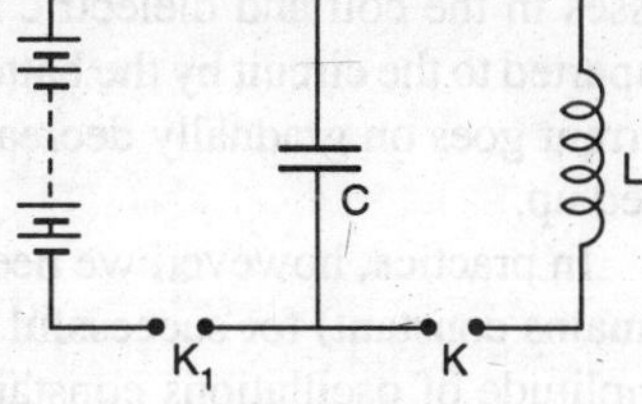

Fig. 53.3

In this position, the lower plate has an *excess*, and the upper plate a *deficit* of electrons. This gives rise to an electric field across the capacitor plates in the direction shown in Fig. 53.4. (*a*) and the capacitor has electrostatic energy given by $\frac{1}{2}CV^2$ where V is the potential difference between the plates and C the capacitance of the capacitor.

When the key K is closed the capacitor discharges itself through the inductor and there is a flow of electrons in a direction indicated by the arrow head. The current flow sets up a magnetic field around the inductor coil. Due to the inductive effect the current builds up slowly upto a maximum value which is attained when the capacitor is fully discharged. At this instant the electrostatic energy is zero but due to the maximum current flow the magnetic field energy around the inductor coil is maximum given by $\frac{1}{2}LI^2$ where I is the current and L the self-inductance of the coil. In other words, the electrostatic energy across the capacitor has been completely converted into magnetic energy around the coil. [Fig. 53.4 (*b*)]

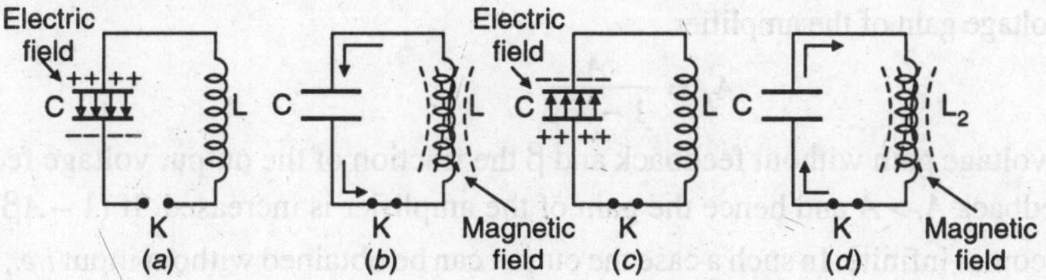

Fig. 53.4

As the capacitor is fully discharged the magnetic field will start decreasing. As the magnetic flux linked with the coil decreases, an induced e.m.f. is set up in it which in accordance with Lenz's law makes the current flow in the *same* direction. This current recharges the capacitor in the opposite direction with its upper plate negative and the lower plate positive as shown in Fig. 53.4. (*c*). Finally, the magnetic field fully collapses. At this stage the magnetic field energy is zero and the electrostatic energy is again maximum.

As soon as the magnetic field is zero, the capacitor which is fully recharged begins to discharge, due to which a current flows in the opposite direction and a magnetic field is again set up around the inductor coil. The magnetic field energy becomes a maximum when the electric field energy is zero [Fig. 53.4(*d*)].

This sequence of charge and discharge results in an alternating motion of electrons in the LC circuit thus setting up an oscillating current. The energy alternately changes from electric to magnetic and is also alternately stored in the electric field across the capacitor and the magnetic field around the inductor. This interchange of energy between L and C is repeated over and over again giving rise to electrical oscillations.

The frequency of oscillations produced in the tank circuit is given by

$$f = \frac{1}{2\pi \sqrt{LC}}$$

which is also the natural frequency of an LC circuit.

The oscillations produced in such a circuit are **damped oscillations** *i.e.* the amplitude of successive oscillations goes on decreasing and finally becomes zero as shown in Fig. 53.5. This is because in a practical LC tank circuit there is a radiation loss in addition to the resistive and other losses in the coil and dielectric losses in the capacitor. In each cycle a part of the original energy imparted to the circuit by the battery is used up to over come these losses. The amplitude of oscillating current goes on gradually decreasing and finally becomes zero when whole of the energy has been used up.

In practice, however, we need *continuous undamped oscillations* (oscillations whose amplitude remains constant) for successful working of most of electronic equipments. In order to maintain the amplitude of oscillations constant, the *appropriate* amount of energy must be supplied to the tank circuit at *proper* time intervals in *correct* phase to make up for the loss. This is generally done by supplying D.C. power to an electronic device say a *triode* or a *transistor circuit* which converts it to necessary A.C. power and supplies to the tank circuit.

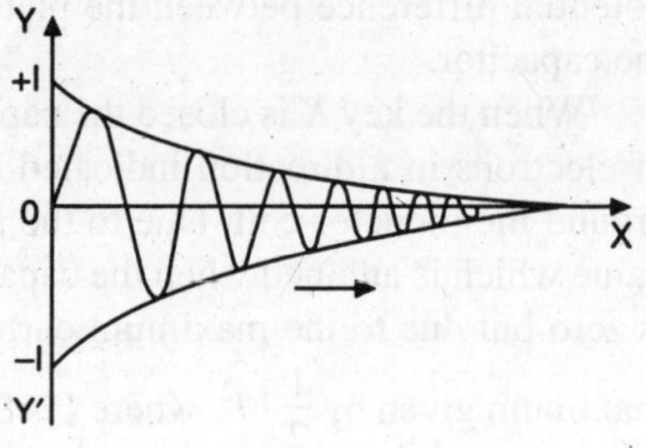

Fig. 53.5

53.3. Feedback for oscillation. *An oscillator is basically an amplifier provided with a feedback circuit.* It amplifies the oscillations of an LC circuit known as the *tank circuit* and renders them *undamped* by feeding the *appropriate amount* of energy in *correct phase,* back to the tank circuit.

In oscillating circuits *positive* feedback is necessary. This positive feedback is in *phase* with the input signal and increases the gain of the amplifier. For positive feedback, the voltage gain of the amplifier.

$$A_f = \frac{A}{1 - A\beta}$$

where A is the voltage gain without feedback and β the fraction of the output voltage feedback. Thus for positive feedback $A_f > A$ and hence the gain of the amplifier is increased. If $(1 - A\beta) = 0$ or $A\beta = +1$, the gain becomes infinite. In such a case the output can be obtained without input *i.e.*, the amplifier provides its own input and, therefore, becomes an *oscillator*.

53.4. Barkhausen criterion. Suppose an input voltage V_i is applied to an amplifier without feedback and V_0 is the output voltage, then

$$\text{Voltage gain without feedback} \quad A = \frac{V_0}{V_i}$$

If now a feedback voltage V_f is applied, then

$$A = \frac{V_0}{V_i + V_f}$$

or
$$A\,(V_i + V_f) = V_0 \quad ...(i)$$

As β is the fraction of the output voltage feedback *i.e.*,
$$V_f = \beta V_0$$

Substituting in Eq. (*i*), we get
$$A\,(V_i + \beta V_0) = V_0$$

or
$$V_i = \frac{V_0\,(1 - A\beta)}{A}$$

When
$$(1 - A\beta) = 0 \text{ or } A\beta = 1,\ V_i = 0$$

i.e., an output voltage is obtained without the input signal. In such a situation the amplifier works as an oscillator.

Hence the condition that the circuit may oscillate is
$$(1 - A\beta) = 0$$

or
$$A\,\beta = +1$$

This is known as **Barkhausen criterion** for sustained oscillations.

53.5. Condition for feedback amplifier to work as oscillator. The conditions that a feedback amplifier may work as an oscillator are

(*i*) The feedback is *positive i.e.*, the phase shift of $A = -$ phase shift of β and the total phase difference around the loop is an integral multiple of 2π or zero.

(*ii*) The fraction of the output voltage fed back β = the reciprocal of the voltage gain without fed back $\frac{1}{A}$ *i.e.*, the overall loop gain Aβ is unity.

VALVE OSCILLATORS

53.6. Tuned grid oscillator. A basic circuit using a triode to produce undamped oscillations, known as *tuned grid* or *tickler feed-back oscillator* is shown in Fig. 53.6.

The tank circuit consisting of an inductor L and a variable capacitor C in parallel, is connected between the grid G and the cathode of a triode through a parallel combination of a grid leak capacitor C_g and resistor R_g. An *H.T.* battery is connected between the cathode and the plate P through another coil L_1 known as tickler coil. The *feedback* from the (output) plate circuit to the tank circuit connected to the grid takes place through the tickler coil L_1 which is inductively coupled to the coil L. Another coil L_2 is also inductively coupled with L, one end of which is connected to the aeriel and the other to the earth through a variable capacitor. A capacitor C_b is connected in parallel with the *H.T.* battery E_{bb}.

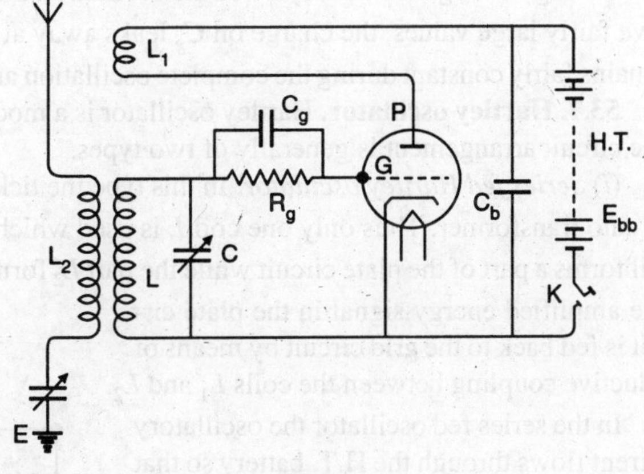

Fig. 53.6

When the cathode is heated, the emitted electrons collect round it forming a negative space charge. As soon as the key K in the plate circuit is pressed the plate attains a positive potential, attracts the electrons and a current starts in the coil L_1. The growth of the plate current through L_1 sets up an induced *e.m.f.* in the coil L equal to $-M\frac{di}{dt}$ where M is the co-efficient of mutual induction between the two coils and $\frac{di}{dt}$ is the rate of growth of current in L_1. By connecting the coil L properly, this *e.m.f.* is made to increase the grid potential and thereby the plate current. Increase in plate current induces more e.m.f. in the coil L thus further increasing the grid potential. This causes the plate current to increase further. The process continues till the plate current becomes saturated and cannot increase further. As the plate current approaches the saturation value, $\frac{di}{dt}$ goes on decreasing and finally becomes zero. The increase in the plate current upto this point has been caused by the induced voltage set up in the grid circuit. When $\frac{di}{dt} = 0$ the induction stops and the plate current begins to fall. This decreasing current causes and induced voltage in the opposite direction in the grid circuit. This further decreases the plate current. The fall in plate current makes the grid more and more negative until the current is cut off *i.e.* becomes zero. The condition now is the same as it was in the beginning. The current therefore, again begins to increase and the whole cycle is repeated. *Thus continuous oscillations of constant amplitude are generated. The energy for maintenance of oscillations is drawn from the H.T. battery*, The capacitor C_b provides a low reactance path to the *a.c.* component of the plate current and does not allow it to pass through the battery.

Function of C_g and R_g. The resistance R_g known as the *grid leak resistance* and the *grid leak capacitor* C_g provide the necessary negative bias required for the operation of the tube and makes the oscillator self-starting.

During the working of the tube, the grid becomes positive for the time of the positive half cycle of the oscillation and draws electrons from the cathode. An electronic current, known as *grid current*, thus begins to flow from the cathode to the grid and then through the external circuit consisting of the resistor R_g and the inductance L back to the cathode. This provides a potential drop across R_g and the end of R_g connected to the grid is thus *negatively biased* by an amount equal to the potential drop across R_g. As R_g is in parallel with C_g the voltage developed across R_g during the flow of grid current also charges the grid capacitor to the same potential.

During the negative half cycle of the oscillation the grid current stops flowing and if R_g and C_g have fairly large values, the charge on C_g leaks away at a very slow rate. The voltage across C_g thus remains fairly constant during the complete oscillation and provides a steady negative bias to the grid.

53.7. Hartley oscillator. Hartley oscillator is a modified form of the *tickler feedback oscillator*. The circuit arrangement is generally of two types.

(*i*) *Series fed Hartley oscillator.* In this type the tickler coil L_1 is a part of coil L which becomes an auto transformer. Thus only one coil L is used which is suitably tapped such that a part L_1 of the coil forms a part of the plate circuit while the part L_2 forms a part of grid circuit as shown in Fig. 53.7. The amplified energy signal in the plate circuit is fed back to the grid circuit by means of inductive coupling between the coils L_1 and L_2.

In the series fed oscillator the oscillatory current flows through the H.T. battery so that the *R.F/AF* and the *D.C.* paths are the same. To remove this a capacitor C_b is placed in parallel with H.T. supply (shown by dotted line)

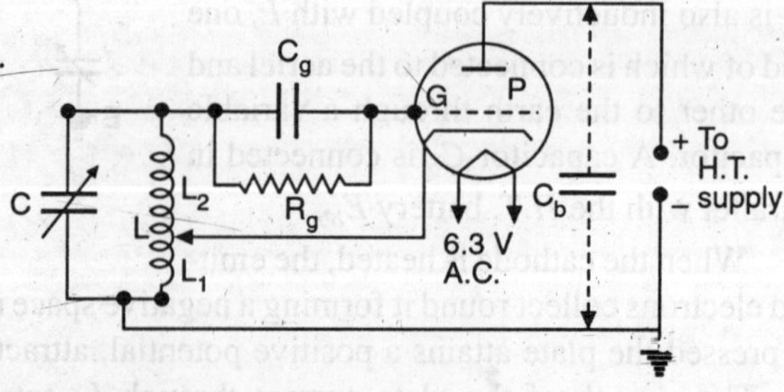

Fig. 53.7

(*ii*) *Parallel or shunt fed Hartley oscillator.* In this type, the plate circuit is divided into two parallel paths. One branch connects the plate to the positive of the battery through an *R.F.* choke and provides the path for direct current keeping the alternating current out. The other branch connects the plate to the LC tank circuit through a capacitor and provides the path for alternating current keeping the direct current out.

TRANSISTOR OSCILLATORS

53.8. Transistor oscillators. We shall describe (*i*) a tuned collector oscillator and (*ii*) Hartley oscillator

(*i*) **Tuned collector oscillator.** A simple circuit for a *tuned collector oscillator* using a *p-n-p* transistor is shown in Fig. 53.8. It has a tuned LC network (tank circuit) in the collector circuit connected between the collector and $-V_{CC}$. This is why it is known as tuned collector oscillator. The frequency of oscillations is given by

$$f = \frac{1}{2\pi\sqrt{LC}}$$

Any desired frequency can be obtained by adjusting the value of L and C.

Principle of operation. When the current is switched on, a small signal is momentarily induced in the LC circuit and gives rise to oscillations in it. These oscillations induce an *e.m.f.* in the coil L_S (in the base circuit) inductively coupled to the coil L with a phase shift of 180° due to transformer action. This induced *e.m.f.* feeds the voltage back to the LC circuit to supply energy to maintain the oscillations.

The circuit is essentially a common emitter (CE) amplifier. A further phase shift of 180° is, therefore, introduced between its base-emitter and collector circuit due to transistor operation in CE mode. As a result the energy feedback to the tank circuit is in phase with the generated oscillations.

The resistors R_1, R_2 and R_E and the supply voltage V_{CC} establish the *d.c* operating point of the transistor. C_E known as emitter bypass capacitor provides a low impedance path to the oscillations. The resistance R_B reduces the input non-linear distortion, prevents loading of the collector circuit by the low input resistance and controls the amount of feedback to the value just required to maintain the oscillations.

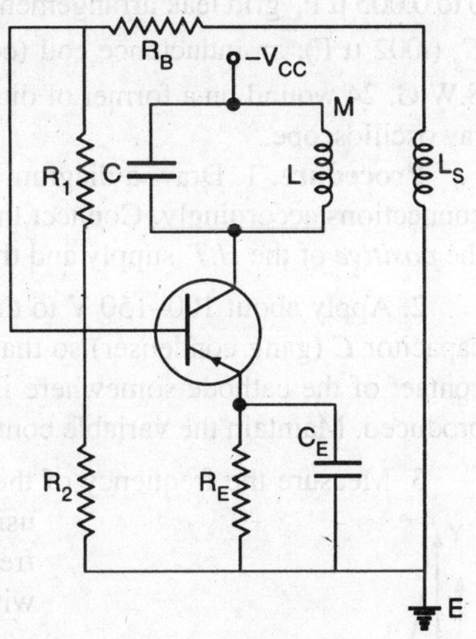

Fig. 53.8

53.9. Hartley oscillator. A simple Hartely oscillator circuit using a *n-p-n* transistor in CE mode is shown in Fig. 53.9. The tank circuit consists of a capacitor C and a tapped inductor coil AB of inductance $L = L_1 + L_2$. The coil is tapped at a point O dividing it into two parts OA of inductance L_1 and OB of inductance L_2. The tapped coil acts as a sort of auto transformer in which L_1 and L_2 are inductively coupled so that the extreme points A and B are 180° out of phase. In addition a further phase shift of 180° between input and output voltages is produced by the transistor used in CE mode. Thus the total phase difference around the loop is zero or 360° and the voltage V_1 appearing across the inductance

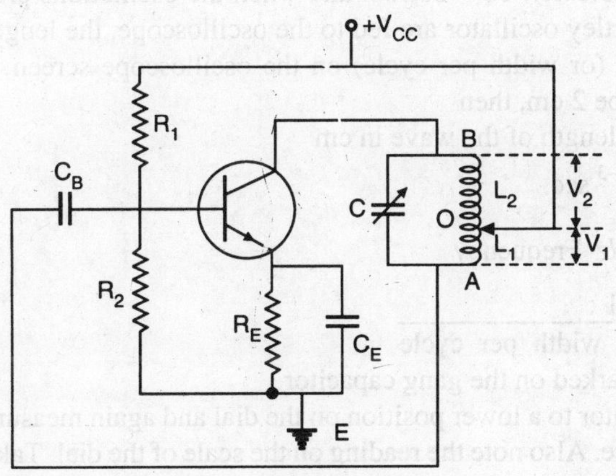

Fig. 53.9

L_1 fed back to the LC circuit is in phase with the output voltage. This makes up for the loss of energy and maintains the oscillations. The frequency of oscillations is given by

$$f = \frac{1}{2\pi}\frac{1}{\sqrt{C(L_1+L_2)}} = \frac{1}{2\pi\sqrt{LC}}$$

The resistors R_1, R_2, R_E and the supply voltage V_{CC} establish the *d.c* operating point. Capacitor C_B is the blocking capacitor and C_E is the bypass capacitor.

Experiment 53.1. To set up a valve type Hartley oscillator and calibrate the gang condenser (or capacitor).

Apparatus. A triode valve (6 J5 or 6C5) with base sockets provided with terminals, a power supply with H.T. and 6.3 volt *AC* supply terminals, a variable capacitor (gang capacitor) *C* of capacity 0 to 0.005 μ F, grid leak arrangement consisting of high resistance R_g (33 K Ω) and a fixed capacitor C_g (.002 μ F), an inductance coil (consisting of 50-70 turns of double cotton covered copper wire S.W.G. 24 wound on a former of diameter about 2.5 cm), a frequency (or wave) meter or a cathode ray oscilloscope.

Procedure. 1. Draw a diagram showing the scheme of connections as in Fig. 53.7. and make connections accordingly. Connect the filament of the triode to 6.3 *V A.C.* terminals, the plate *P* to the *positive* of the *H.T.* supply and the negative of the *H.T.* to the earth.

2. Apply about 100–150 V to the plate of the triode, and adjust the capacitance of the variable capacitor *C* (gang condenser) so that it has its maximum value. Adjust the position of the variable contact of the cathode somewhere in the middle of the air core inductor *L* so that oscillations are produced. Maintain the variable contact in the same position throughout the experiment.

3. Measure the frequency of the oscillations produced in the tank circuit of Hartley oscillator using a frequency (or wave) meter or a cathode ray oscillope. The frequency meter gives the frequency directly and the frequency with a cathode ray oscilloscope is determined as follows:-

Y
Frequency ⟶
O
Dial scale reading ⟶
X

Fig. 53.10

The oscilloscope with a calibrated time base is used to measure the period of the wave form and hence the frequency of the oscillations produced. Suppose the time base of the oscilloscope is set at the *calibration* 10^{-3} *sec/cm* and when the oscillations produced by Hartley oscillator are fed to the oscilloscope, the length of the waves (or width per cycle) on the oscilloscope screen is measured to be 2 cm, then

Period of the wave T = Time in sec/cm × length of the wave in cm

$$= 10^{-3} \times 2 = 2 \times 10^{-3} \text{ sec.}$$

$$\therefore \quad f = \frac{1}{T} = \frac{10^3}{2} = 500 \; Hz \text{ Frequency}$$

$$\text{Hence frequency} \quad f = \frac{1}{\text{Time per cm} \times \text{width per cycle}}$$

Note the reading on the scale of the dial marked on the gang capacitor.

4. Now turn the knob of the variable capacitor to a lower position on the dial and again measure the frequency by one of the methods given above. Also note the reading on the scale of the dial. Take a number of readings corresponding to various positions of the knob of the capacitor.

Observations.

Sl. No.	Dial scale reading of gang capacitor C	Frequency by frequency meter	Frequency by oscilloscope		
			Time base Time/cm	Width per cycle	Frequency

5. Plot a graph between dial scale reading of the gang capacitor taken along the X-axis and corresponding frequency along the Y-axis. The graph gives the calibration curve for the Hartley oscillator. (Fig. 53.10)

Precautions 1. The inductor coil should be wound on a former of constant and uniform radius.

2. The winding on the inductor coil should be uniform and equally spaced.

3. Capacitors and resistors of proper values according to the data of the valve or transistor should be used.

Experiment 53.2. To study a transistorised Hartley oscillator and calibrate the variable capacitor.

Apparatus. *N-P-N* transistor (S.G 221 or equivalent), 15 volt *d.c.* battery, A variable capacitor C 0–0.02 μf, variable inductor 0–20 mH, Resistors $R_E = 2.2$ K Ω, $R_1 = 33$ K Ω, $R_2 = 10$ K Ω, capacitors $C_E = 0.1$ μf, $C_B = 10$ μf, a frequency meter (or wave meter), V.T.V.M or C.R.O etc.

Procedure 1. Draw a diagram showing the scheme of connections as in Fig. 53.11 and make connections accordingly. [The scheme of connections shown in Fig. 53.11 is a slight modification of Fig. 53.9 for convenience in performing the experiment].

2. Put in the key K. Adjust the position of the variable contact on the inductor L in the tank circuit so that oscillations are produced. Measure the output of the oscillator using a sensitive *a.c.* millivoltmeter, V.T.V.M or a C.R.O. (*calibrated*) Maintain the variable contact in the same position throughout the experiment.

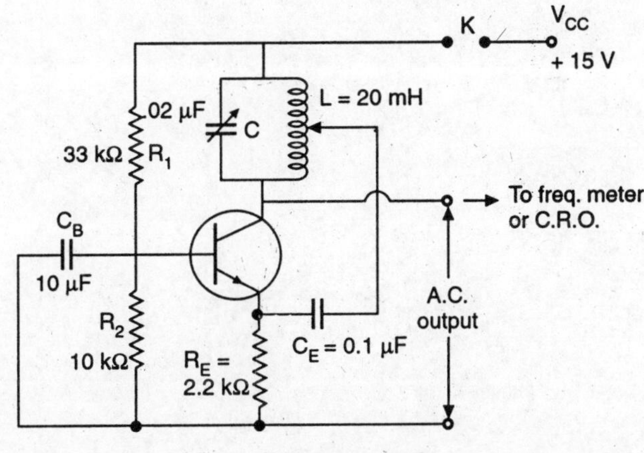

Fig. 53.11

3. Now connect the A.C. output terminals of oscillator to a frequency meter (or calibrated C.R.O) and proceed as in step 3 and 4 of **Expt. 53.1.**

Observations. A.C. output of the oscillator =

Table Same as in **Expt. 53.1**

Graph Same as in **Expt. 53.1** step 5

Precautions. Same as in **Expt. 53.1** and

4. Measure A.C. output voltage with a sensitive V.T.V.M.

Oral Questions

1. Oscillatory circuit (*i*) What is an oscillatory circuit? (*ii*) How does an *LC* circuit produce electrical oscillations ?(*iii*) Are these oscillations of constant amplitude or damped? (*iv*) What is the natural frequency of oscillations of a circuit having an inductance *L* and capacitance *C* ? (*v*) What methods do you employ to keep the amplitude of oscillations constant? (*vi*) What is the function of the triode or transistor in an oscillatory circuit? (*vii*) What are radio frequency waves? (*viii*) Does a radio station broadcast at R.F. or A.F? (*ix*) What is Barkhausen criterion?

2. Valve and transistor oscillator (*i*) What is a Hartley oscillator? (*ii*) Can you design a Hartley oscillator with a triode as well as with a transistor ? (*iii*) What is the function of a gang capacitor? (*iv*) What type of feedback is being used in a Hartley oscillator ? (*v*) Is there any difference in the working of a series fed and a parallel fed Hartley oscillator ? (*vi*) What is a tuned grid valve oscillator? (*vii*) What is a tuned collector (transitor) oscillator?

3. Wavemeter (*i*) What is a Wave meter or frequency meter ? (*ii*) What is its underlying principle ? (*iii*) How do you locate the zero beat position ? (*iv*) Can you use an oscilloscope to locate the zero beat position? (*v*) What type of trace you get on the oscilloscope at zero beat ?

54

Transistor Amplifiers

54.1. Transistor gain. A transistor can be used as an amplifier. There may be a current gain, voltage gain, power gain and resistance gain.

Current gain. Current gain is also known as *current amplification factor*. For each of the three transistor configurations already described two current gains can be defined.

(*i*) *d.c. current gain*. It is defined as the ratio of *d.c.* output current I_0 and the corresponding *d.c.* input current I_i.

$$\therefore \qquad d.c. \text{ current gain} = \frac{I_0}{I_i}$$

(*ii*) *a.c. current gain*. It is defined as the ratio of *change* in output current ΔI_0 and the corresponding *change* in input current ΔI_i

$$\therefore \qquad a.c. \text{ current gain} = \frac{\Delta I_0}{\Delta I_i}$$

For a *common base* mode the input current is the emitter current and output current is the collector current.

$$\therefore \qquad d.c. \text{ current gain } \alpha_{d.c.} = -\left[\frac{I_c}{I_e}\right]_{V_{cb}}$$

where I_e is the *d.c.* input emitter current and I_c the *d.c.* output collector current corresponding to a specific collector voltage V_{cb} relative to the base.

$$a.c. \text{ current gain } \alpha_{a.c.} = -\left[\frac{\Delta I_c}{\Delta I_e}\right]_{V_{cb}} = -\left[\frac{dI_c}{dI_e}\right]_{V_{cb}}$$

where ΔI_c (or dI_c) is the change in output collector current corresponding to a change ΔI_e (or $d I_e$) in input emitter current keeping collector to base voltage V_{cb} constant.

$\alpha_{d.c.}$ is very nearly equal to $\alpha_{a.c.}$. Therefore, we do not generally distinguish between them and denote the current gain as α. Unless otherwise specified α means *a.c.* current gain. This is known as α-*characteristic* of a transistor.

For a *common emitter* mode the input current is the base current and the output current is the collector current.

$$\therefore \qquad d.c. \text{ current gain } \beta_{d.c} = \left(\frac{I_c}{I_b}\right)_{V_{ce}}$$

where I_b is the *d.c.* input base current and I_c the *d.c.* output collector current corresponding to a specific collector voltage V_{ce} relative to the emitter.

$$a.c.\text{ current gain } \beta_{a.c.} = \left(\frac{\Delta I_c}{\Delta I_b}\right)_{V_{ce}} = \left(\frac{d I_c}{d I_b}\right)_{V_{ce}}$$

where ΔI_c (or $d I_c$) is the change in output collector current corresponding to a change ΔI_b (or dI_b) in input base current keeping the collector emitter voltage constant.

$\beta_{d.c.}$ is very nearly equal to $\beta_{a.c.}$. Therefore, we denote the current gain of a common emitter amplifier by β. Unless otherwise specified β means *a.c.* current gain. This is known as β-*characteristic* of a transistor.

For a *common collector* mode the input current is the base current and the output current is the emitter current.

$$\therefore \quad d.c.\text{ current gain } \gamma_{d.c.} = \left[\frac{I_e}{I_b}\right]_{V_{ec}}$$

where I_b is the *d.c.* input base current and I_e the *d.c* output emitter current corresponding to a specific emitter voltage V_{ec} relative to the collector.

$$a.c.\text{ current gain } \gamma_{a.c.} = -\left[\frac{\Delta I_e}{\Delta I_b}\right]_{V_{ec}} = -\left[\frac{d I_e}{d I_b}\right]_{V_{ec}}$$

where ΔI_e (or $d I_e$) is the change in output emitter current corresponding to a change ΔI_b (or $d I_b$) in input base current keeping emitter collector voltage V_{ec} constant.

As in the case of α and β, we denote the gain of a common collector amplifier by γ and unless specified otherwise we mean *a.c.* current gain. This is known as the γ-*characterstic* of the transistor.

54.2. Relation between β and α. We have proved that

$$I_e = -(I_b + I_c)$$

$$\therefore \quad \Delta I_e = -\Delta I_b - \Delta I_c$$

or

$$\Delta I_b = -\Delta I_e - \Delta I_c$$

$$\therefore \quad \beta = \frac{\Delta I_c}{\Delta I_b} = \frac{\Delta I_c}{-\Delta I_e - \Delta I_c} = \frac{\dfrac{-\Delta I_c}{\Delta I_e}}{1 + \dfrac{\Delta I_c}{\Delta I_e}} = \frac{\alpha}{1-\alpha}$$

The value of α is of the order of 0.95 to 0.99 (*i.e.* it is approximately unity) but the value of β is quite large. For example, for $\alpha = 0.98$

$$\beta = \frac{\alpha}{1-\alpha} = \frac{0.98}{1-0.98} = 49$$

The current gain in a common emitter is, therefore, very high.

Relation between γ and α. We have

$$\gamma = -\frac{\Delta I_e}{\Delta I_b}$$

and

$$\Delta I_e = -\Delta I_b - \Delta I_c$$

or

$$1 = -\frac{\Delta I_b}{\Delta I_e} - \frac{\Delta I_c}{\Delta I_e}$$

$$= \frac{1}{\gamma} + \alpha$$

or $$\gamma = \frac{1}{1-\alpha}$$

54.3. Voltage gain. Voltage gain is defined as the ratio of output voltage V_0 to input voltage V_i

or $$\text{Voltage gain} = \frac{V_0}{V_i}$$

Both the voltages are considered either at their peak value or peak to peak value or *r.m.s.* values and we shall restrict ourselves to sinusoidal signals only.

The collector circuit being reverse biased has a very high internal resistance. It, therefore, permits a very high load resistance R_L in the collector circuit without affecting the circuit. This is in fact, the basis of *voltage amplification* in a transistor.

For *common base* amplifier the input voltage is emitter to base voltage V_{eb} and output voltage is collector to base voltage V_{cb}

$$\therefore \quad \text{Common base voltage gain} = \frac{V_0}{V_i} = \frac{V_{cb}}{V_{eb}} = \frac{-I_c R_L}{I_e R_i} = \alpha \frac{R_L}{R_i}$$

where R_L is the output resistance in the collector base circuit and R_i is the input resistance in the emitter base circuit.

$$\therefore \quad \text{Voltage gain} = \text{current gain} \times \frac{\text{Output load resistance}}{\text{Input resistance}}$$

$$= \text{current gain} \times \text{resistance gain.}$$

For *common emitter* amplifier the input voltage is base to emitter voltage V_{be} and output voltage is collector to emitter voltage V_{ce}.

$$\therefore \quad \text{Common emitter voltage gain} = \frac{V_0}{V_i} = \frac{V_{ce}}{V_{be}} = \frac{I_c R_L}{I_b R_i} = \beta \frac{R_L}{R_i}$$

where R_L is the output load resistance in the collector emitter circuit and R_i is the input resistance in the base emitter circuit. In this case again

$$\text{Voltage gain} = \text{current gain} \times \frac{\text{Output load resistance}}{\text{Input resistance}}$$

$$= \text{current gain} \times \text{resistance gain.}$$

In common base configuration the voltage gain is quite high and in common emitter configuration it is very high.

54.4. Resistance gain. *Resistance gain of a transistor is defined as the ratio of the out put resistance R_0 to the input resistance R_i.*

For a *common base amplifier*

$$\text{Resistance gain} = \frac{R_0}{R_i} = \frac{\text{Collector base internal resistance}}{\text{Emitter base internal resistance}}$$

For a *common emitter amplifier*

$$\text{Resistance gain} \quad \frac{R_0}{R_i} = \frac{\text{Collector emitter internal resistance}}{\text{Base emitter internal resistance}}.$$

54.5. Power gain. It is defined as the ratio of output signal power to input signal power.

or $$\text{Power gain} = \frac{\text{Output power } P_0}{\text{Input power } P_i}$$

For *common base* amplifier the input power is the power of the signal applied to emitter base circuit and output power is the power of the signal obtained across the collector base circuit.

For *common emitter* amplifier the input power is the power of the signal applied to the base emitter circuit and output power is the power of the signal obtained across the collector emitter circuit.

As Power = voltage × current

∴ Power gain = voltage gain × current gain

In common base configuration the power gain is slightly less than the voltage gain as current gain α is less than one.

In common emitter configuration the power gain is very very high.

54.6. D.C. Load line. Out of the common base and common emitter amplifiers, it is the common emitter amplifier which is mostly used. We shall, therefore, explain the meaning of load line with reference to a common emitter amplifier.

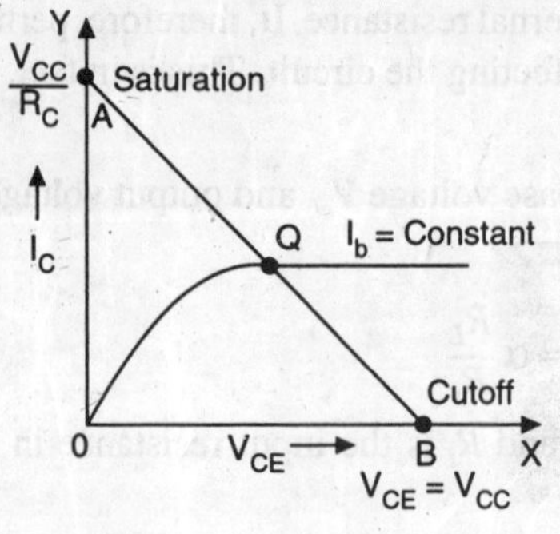

Fig. 54.1

The I_c, V_{ce} collector current and collector to emitter (ground) voltage characteristic is shown in Fig. 54.1.

When no *a.c.* signal is fed to the input circuit of a common emitter amplifier, net collector voltage

$$V_{ce} = V_{cc} - I_c R_c$$

where R_c is the load resistance in the output collector-emitter circuit.

Note. It is sometimes also denoted as R_L.

When $I_c = 0$ $\qquad V_{ce} = V_{cc}$

Under this condition the transistor is said to be in the *cut off stage*, because the transistor does not conduct any current. This gives the point *B* on the $I_c - V_{ce}$ graph, and the net collector voltage is denoted as V_{ce}.

When $V_{ce} = 0$ $\qquad I_c = \frac{V_{cc}}{R_c}$

Under this condition the transistor is said to be in the *saturation stage* because it carries the maximum collector current I_c (sat). This gives the point *A* on the $I_c - V_{CE}$ graph.

The line *AB* joining the two points is known as *D.C. load line*.

It is a straight line, because if we consider the relation

$$V_{ce} = V_{cc} - I_c R_c$$

We have
$$I_c = \frac{-V_{ce}}{R_c} + \frac{V_{cc}}{R_c}$$

As $\frac{V_{cc}}{R_c}$ = a constant, this equation is of the form $y = mx + c$ and hence is a straight line.

54.7. Operating point. It is a point on the *D.C.* load line, which represents values of collector current I_c and collector to emitter voltage V_{ce} that exist in a transistor circuit, when no input signal is applied.

It is called the operating point, because variation in I_c and V_{ce} takes place about this point, when a signal is applied. It is also called *quiescent* (*silent*) point or *Q*-point, because this is the point on I_c–V_{ce} characteristic, when the transistor is *silent i.e.* zero signal.

The I_c–V_{ce} characteristic of a transistor for a given constant value of I_b, is shown in Fig. 54.1. The I_c–V_{ce} conditions for zero signal must be represented by some point on this characteristic, as well as on the load line *i.e.* the point Q where the D.C. load line AB intersects the I_c–V_{ce} characteristic gives the Q-point or operating point.

54.8. Classification of amplifiers. One method of classification of amplifiers is according to their *mode of operation.* According to the mode of operation, the amplifiers are classified into four classes known as *class A, class B, class AB* and *class C.*

Class A amplifier. In class *A* operation the bias voltage and amplitude of the externally applied *a.c.* signal is so adjusted that the output current is never zero during the complete cycle of input *a.c.* signal. In class *A* transistor amplifier operating in *CE* mode *the collector current flows at all time during the full cycle of input signal* i.e. 360°. For class *A*-amplificator, the transistor is so biased that

(*i*) Its Q–point lies at the centre of the load line.

(*ii*) Amplitude of the input signal is such that the transistor operates over the linear portion of the load line.

(*iii*) Output current flows for the entire cycle of the input signal.

Class B amplifier. In class *B* operation the bias voltage and amplitude of the externally applied *a.c.* signal are so adjusted that the output current flows for just one half cycle of the input *a.c.* signal. In class *B* transistor amplifier operating in *CE* mode the *collector current* flows only for one half (say positive half) of the input signal *i.e.* 180°.

Class AB amplifier. In class *AB* operation the bias voltage and amplitude of the *a.c.* input signal are such that the output (collector) current flows for more than half but less than full cycle of the input *a.c.* signal *i.e.* more than 180° but less than 360°.

Class C amplifier. In class *C* operation the bias voltage and amplitude of the *a.c.* input signal are such that the output (collector) current flows for less than half the cycle of the *a.c.* input signal *i.e.* less than 180°.

54.9. Common base amplifier. The basic circuit for a common base amplifier using an *N-P-N* transistor is shown in Fig. 54.2.

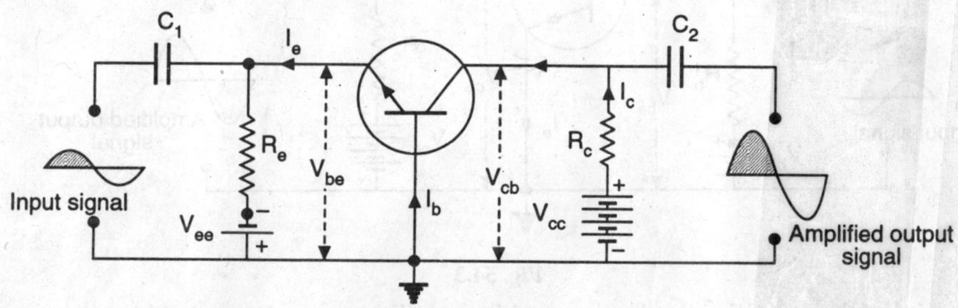

Fig. 54.2

Note. An *N-P-N* transistor is being used to explain the basic principle because of its simplicity. In this circuit the base terminal is common to both the *input i.e.* emitter base circuit to which the *a.c.* signal to be amplified is applied and the *output i.e.* collector base circuit from which the amplified signal is taken out. The input circuit *i.e.* the emitter base junction is *forward biased* by V_{ee} *and the output circuit i.e.* the collector base junction is *reverse biased by* V_{cc}.

The Q-point of the amplifier *i.e.* the *d.c.* working conditions are determined by the batteries alongwith the resistors R_e and R_c. In other words the values of I_c, I_e, V_{be} and V_{cb} depend upon the values of V_{ee}, V_{cc}, R_e and R_c chosen.

The voltage V_{cb} is given by the relation

$$V_{cb} = V_{cc} - I_c R_c \quad \text{...(i)}$$

When no signal is applied to the input circuit the output voltage and current just correspond to the *Q*-point and there is no output signal. The *a.c.* signal is applied to the input *emitter-base* circuit via the coupling capacitor C_1. It is supposed that C_1 offers almost zero reactance to the input signal.

The positive half cycle of the applied *a.c.* signal *decreases the forward bias* V_{be} (as V_{be} is negative w.r.t. the ground). This in turn decreases the base current I_b. As $I_c = \beta I_b$, the collector current I_c also decreases considerably, thereby reducing the potential drop $I_c R_c$ across the collector resistance R_c. As

$$V_{cb} = V_{cc} - I_c R_c$$

the collector to base voltage V_{cb} increases. Thus we get an *amplified positive* half cycle in the output circuit. As the positive half cycle in the input circuit produces a positive half cycle in the output circuit, the output voltage is *in phase* with the input voltage. Similarly when the negative half cycle is applied in the input circuit an *amplified negative* half cycle is obtained in the output circuit.

In a common base amplifier current gain α is nearly *one*. The input resistance (impedance) is *very low* (30–150 Ω). Output resistance (impedance) is *very high* (of the order of 500 KΩ). The voltage gain is very large (about 1500) and high power gain (about 30 *dB*). A common base amplifier is used in matching a low impedance source with a high impedance load.

54.10. Common emitter amplifier. The basic circuit for a common emitter *N-P-N* transistor amplifier is shown in Fig. 54.3. In this circuit the emitter terminal is common to both the *input i.e.* the base-emitter circuit to which the *a.c.* signal to be amplified is applied and the *output i.e.* the collector emitter circuit from which the amplified signal is taken out. The input circuit *i.e.* the base-emitter junction is *forward biased* by V_{bb} and the output circuit *i.e.* the collector emitter junction is *reverse biased* by V_{cc}.

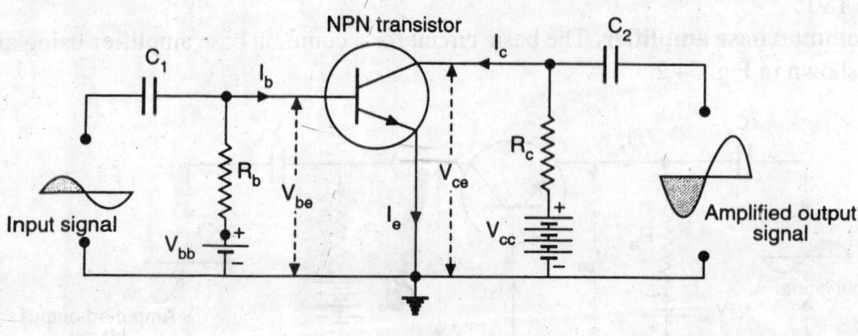

Fig. 54.3

The *Q* point of the amplifier *i.e.* the *d.c.* working conditions are determined by V_{cc} together with the resistances R_b and R_c. When no signal is applied to the input circuit, the base current I_b is given by

$$I_b = \frac{V_{be}}{R_b} \cong \frac{V_{bb}}{R_b}$$

The voltage V_{ce} is given by the relation

$$V_{ce} = V_{cc} - I_c R_c \quad \text{...(i)}$$

The *a.c.* signal is applied to the input *base-emitter* circuit via the coupling capacitor C_1. The positive half cycle of the applied *a.c.* signal *increases* the forward bias V_{be} (as V_{be} is already *positive* w.r.t. the ground). This increase in forward bias of the base-emitter junction increases the base current I_b. As $I_c = \beta I_b$, the collector current also increases considerably, thereby increasing the potential drop $I_c R_c$ across the collector resistance R_c. As

$$V_{ce} = V_{cc} - I_c R_c$$

the collector to emitter voltage V_{ce} *decreases*. The decrease in V_{ce} corresponding to the *positive half* of the input cycle gives the *negative half* of the output cycle and vice-versa. In other words in a common emitter amplifier there is a *phase reversal* of the signal *i.e.* the output signal has a phase difference of π w.r.t. the input signal.

A common emitter transistor amplifier has a very high current gain $\beta = \frac{1}{1-\alpha}$. It has a very high voltage gain (about 1500) highest power gain (about 40 *dB*). The input resistance is moderately low (1–2 KΩ) and the output resistance is moderately high (about 50 K Ω).

54.11. Frequency response. The voltage gain of an amplifier varies with signal frequency. *The curve between voltage gain and input frequency of an amplifier is known as frequency response.* Fig. 54.4 shows the frequency response of a typical amplifier. The curve is usually plotted on a *semi-log graph paper* with frequency on logarithmic scale so as to accommodate large frequency range. It is seen from the graph that the voltage gain is constant only for a limited band of frequency. This range of frequency is called the *mid-frequency range* and the gain is called *mid-band gain.* On both sides of the mid frequency range the gain decreases. For very low and for very high frequencies the gain of the amplifier reduces to almost zero. The frequency response of an amplifier is measured in terms of *bandwidth* β.

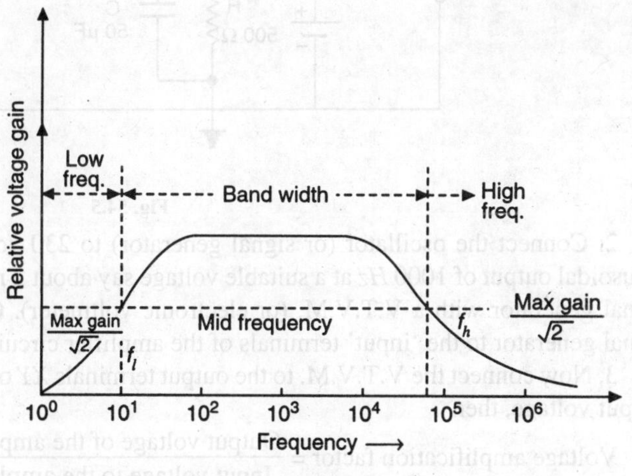

Fig. 54.4

54.12. Bandwidth. *It is defined as the frequency band in which the amplifier gain remains above 0.707 of the mid frequency gain.* The low frequency f_l at which the gain falls to $1/\sqrt{2}$ *i.e.* (0.707) of its mid-frequency value is knows as *lower cut off frequency.* The high frequency f_h at which the gain falls to $1/\sqrt{2}$ *i.e.* (0.707) of its mid frequency value is known as *higher cut off frequency.*

$$\therefore \qquad \text{Bandwidth } \beta = f_h - f_l$$

The bandwidth measurement is done by first measuring the amplifier gain in mid-frequency region (1 *KHz*). Keeping the input signal constant, the frequency is varied on the low as well as high frequency side till the gain falls to 0.707 of the gain at 1 *KHz*. To obtain the frequency response curve the amplifier gain is plotted as a function of input signal frequency.

To increase the mid frequency band f_l is lowered and f_h is increased by selecting suitable circuit components/parameters.

Note. In the experiment 'To study transistor as voltage amplifier in common base configuration' we shall use a *P-N- P* transistor and in the experiment 'To study transistor as a voltage amplifier in common emitter configuration' we shall use an *N-P-N* transistor to familiarise the students with the use of both the types.

Experiment 54.1. To find the voltage gain and to study the frequency response of a common base transistor voltage amplifier.

Apparatus. Audio-frequency signal generator, stabilised power supply with 1.5 volt and 9 volt tappings, a V.T.V.M., capacitors and resistors of suitable values, a transistor (*P- N-P* type).

Procedure. 1. Draw a diagram showing the scheme of connections and connect accordingly as in Fig. 54.5.

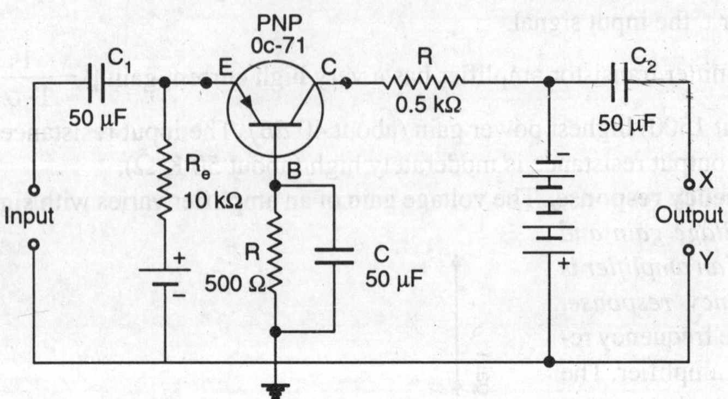

Fig. 54.5

2. Connect the oscillator (or signal generator) to 230 volts *A.C.* mains and set it to supply a sinusoidal output of 1000 *Hz* at a suitable voltage say about 2 mV. Measure the output voltage of the signal generator with a V.T.V.M. (or electronic voltmeter). Connect the 'output' terminals of the signal generator to the 'input' terminals of the amplifier circuit.

3. Now connect the V.T.V.M. to the output terminals *XY* of the amplifier circuit and measure the output voltage, then

$$\text{Voltage amplification factor} = \frac{\text{Output voltage of the amplifier}}{\text{Input voltage to the amplifier}}$$

4. **To study the voltage gain.** Change the output of the signal generator (*i.e.* input to the amplifier circuit) in small steps and in each case measure the input voltage to the amplifier circuit and its output voltage.

5. **To study the frequency response.** Keeping the output voltage from the signal generator constant at a suitable level, first change the frequency to the lower side to say about 50 *Hz* and note the output voltage of the amplifier at intervals of 100 *Hz* first and then at intervals of 50 *Hz*. Again keeping the voltage from the signal generator constant at the *same* level change the frequency to the higher side to about 5 *KHz* at intervals of 200–500 *Hz*.

Again measure the output voltage of the signal generator with a V.T.V.M.

6. Calculate the voltage gain corresponding to various values of frequency and plot a graph between voltage gain and frequency on a *semi log* graph paper as shown in Fig. 54.4. From the graph locate two points where the voltage gain is 0.707 of the maximum voltage gain on the lower as weli as higher frequency side and find the lower cut off frequency f_l and higher cut off frequency f_h and hence the bandwidth.

7. Also plot a graph by taking log of frequency ($\log_{10} f$) along the *X*-axis and corresponding output voltage along *Y*-axis on a regular graph paper. The graph shows the variation of output voltage with frequency.

Observations (*a*) Voltage gain.

Transistor used =

Capacitor $C_1 =$ μ*F* Capacitor $C_2 =$ μ*F*

Capacitor $C =$ μ*F*

Resistor $R_e =$ Ω Resistor $R_c =$ Ω

Resistor $R =$ Ω

Constant frequency of the input signal = 1000 Hz

Sl. No.	*Input signal voltage* V_i	*Amplified output voltage* V_0	*Voltage gain* $A_v = \frac{V_0}{V_i}$
1			
2			
...			
...			
...			

Mean voltage gain at 1000 *Hz* =

(*b*) **Variation of output voltage with frequency**

Constant value of input signal voltage $V_i =$ volts

No.	*Input signal; Frequency f (Hz)*	$\log_{10} f$	*Output voltage* V_0	*Voltage Gain* $A_v = \frac{V_0}{V_i}$
	50			
	100			
	...			
	250			
	300			
	400			
	...			
	...			
	...			
	800			
	1000			
	...			
	...			
	2000			
	2500			
	...			
	5000			

Maximum voltage gain =

(From the graph between voltage gain and frequency.)

0.707 of maximum voltage gain =

Higher cut off frequency $f_h =$ *Hz*

Lower cut off frequency $f_l =$ *Hz*

Bandwidth $\beta = f_h - f_l =$ *Hz*

Also find the above values from the graph between output voltage and $\log_{10} f$.

Precautions. 1. The emitter base and collector base circuits should be correctly biased.

2. The output of the *A.F.* signal generator should be set at 1000 *Hz* and at a low power of 2 mV.

3. To study the voltage gain the output of the *A.F.* signal generator should be increased in small steps.

4. To study the voltage gain with frequency, the frequency of the signal generator must be taken above as well as below 1000 *Hz*.

5. While taking the reading frequency must be changed in appropriate steps first of 50 *Hz*, then 100 *Hz*, 200 *Hz* and finally 500 *Hz*.

Experiment. 54.2. To find the voltage gain and to study the frequency response of a common emitter transistor voltage amplifier.

Apparatus. Transistor (Type *N-P-N*; SG 104 or BC 109), two resistances of 33 *K* ohm, one resistance of 100 *K* ohm, one resistance 10 *K* ohm (each 1/2 watt), three capacitors 10 μ*fd* (6.3 V) each, a signal generator, a *D.C.* source (4.5 V) and a vacuum tube voltmeter (V.T.V.M.).

Procedure. 1. Draw a diagram showing the scheme of connections as in Fig. 54.6 and make connections accordingly.

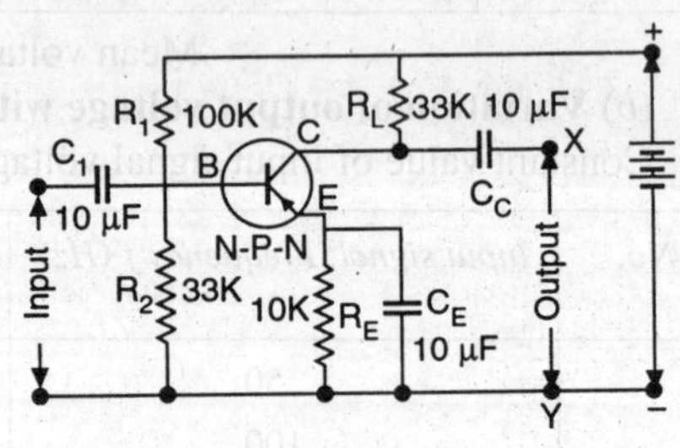

Fig. 54.6

Explanation of the circuit arrangement. The input from the audiofrequency oscillator is fed to the circuit through the capacitor C_1 between base and emitter terminals *B* and *E* shown as 'INPUT' in the diagram. The bias resistors R_1 and R_2 act as voltage divider net work and supply the necessary base bias to the transistor. The capacitor C_1 with exceedingly large impedance to *D.C.*, ensures that the oscillator connection does not alter the *D.C.* bias of the base emitter junction but allows varying *A.C.* voltage across it. The emitter base junction is *forward* biased by connecting the *negative* of the *D.C.* source to the *emitter* and positive to the base of the *N-P-N* transistor used in the circuit *i.e.*, the potential of *B* is *positive* relative to *E*. [If a *P-N-P* transistor is used the emitter base junction is forward biased by connecting the positive of the *D.C.* source to the emitter and negative to the base.] The collector base circuit is reverse biased as the potential of *B* is *negative* relative to *C*. The current in the collector circuit causes a voltage drop to appear across the load resistance R_L which produces the output across *XY*. The resistance R_E and the capacitor C_E do not permit minor variations in the characteristics of the transistor or rapid changes in voltage to effect the *D.C.* voltage conditions and performance of the transistor. The capacitor C_E prevents undesirable feedback of the amplified signal to base emitter circuit. The resistance R_E stabilises the circuit for excessive temperature rises because if the collector current rises the current through R_E increases. This lowers the potential difference between *E* and *B* so that the collector current is automatically lowered. The output is measured between *XY* marked 'OUTPUT' *i.e.*, between the collector and emitter terminals *C* and *E* through the coupling capacitor C_c.

2. Steps 2 to 7, same as in **Expt. 54.1. Observations, calculations and precautions, same as in Expt. 54.1.**

54.13. Two stage R.C. coupled transistor amplifier. When a large amplification of a signal is required and the same cannot be achieved by a single stage amplifier, two or more stages are joined to form a composite or cascade or multistage amplifier.

An interstage transformer may be used to connect the two stages. In transformer coupling, the *D.C.* isolation between the collector of the first stage and the base of the second stage is brought about automatically. But as the size of the transistor amplifier has to be kept very small, the large size of the coupling transformers is a serious disadvantage.

A direct coupling of the two stages has the disadvantage of disturbing *D.C.* conditions because the *D.C.* collector potential of the first stage is directly transmitted to the base of the following stage.

Hence it is a universal practice in all small scale transistor amplifiers to use resistance capacitance coupling. As already explained the capacitance isolates the *D.C.* between the collector of the first stage and base of the second stage.

Frequency response. The frequency response of an R.C. coupled transistor amplifier is similar to the R.C. coupled valve amplifier. Due to the presence of the coupling capacitor in a common emitter amplifier; at low frequencies the gain is reduced as the capacitor offers a high impedance at low frequencies. At higher frequencies the gain is again reduced due to internal transistor capacitances and due to low impedance of parasitic capacitances.

The cascading of two stages further reduces the overall bandwidth of the amplifier. The frequency response of the composite circuit may be considerably poorer than that of the individual stage.

Experiment 54.3. To study the voltage gain and frequency response of a two stage R.C. coupled transistor amplifier.

Apparatus. Two *P-N-P* transistors (*AC* 125), two resistances 120 K ohm, two resistance 12 K ohm, two resistances 560 ohm, one resistance 1 K ohm, one resistance 2.7 K ohm, three capacitors 0.47 *µfd*, two capacitors 1 *µfd*, 12 *V D.C.* supply, signal generator, a vacuum tube voltmeter (V.T.V.M.)

Procedure. 1. Draw a diagram showing the scheme of connections as in Fig. 54.7 and make connections accordingly. Join the points *A* and *B* to connect the two stages. Connect the output from the signal generator (at 1000 *Hz* and 2 *mV*) between the 'INPUT' terminals *I* and *G*. Measure the output voltage between the output terminals with a vacuum tube voltmeter. Change the frequencies on the lower as well as higher side (from 50 *Hz*, to 50 *KHz*) and note the output voltage corresponding to each frequency.

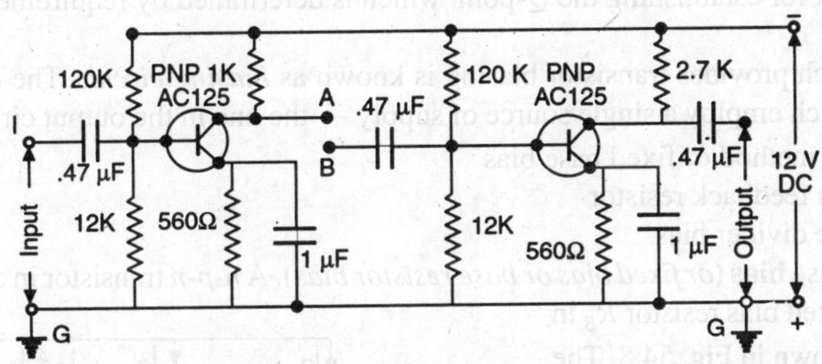

Fig. 54.7

Measure the input voltage supplied to the amplifier between the terminals *I* and *G* with *V.T.V.M.* Calculate the two stage voltage gain and draw a graph between voltage gain and frequency. From the graph find lower and higher cut-off frequencies f_l and f_h.

2. Again connect the output from the signal generator at 1000 *Hz* (middle of the mid-frequency region) and note the input voltage between the terminals *I* and *G*, output of the first stage between the terminals *A* and *G*, output of both the stages between the output terminals with a *V.T.V.M.*

Observations (*i*) Input voltage =

Sl. No.	*Frequency cycles/sec*	*Output of two stages*	*Voltage gain*

Maximum voltage gain (from the graph) =

Higher cut off frequency $f_h =$ Hz
Lower cut off frequency $f_l =$ Hz
Bandwidth $\beta = f_h - f_l =$ Hz
(*ii*) Input voltage =
Output voltage of the first stage =
Output voltage of two stages =
Voltage gain of the first stage

$$k_1 = \frac{\text{Output voltage of the first stage}}{\text{Input voltage}}$$

Voltage gain of the second stage =

$$k_2 = \frac{\text{Output voltage of two stages}}{\text{Output voltage of the first stage}}$$

Voltage gain of the two stage amplifier =

$$k = \frac{\text{Output voltage of two stages}}{\text{Input voltage}}$$

Precautions. 1. Same as in **Expt. 54.1.**
2. To connect the two stages points *A* and *B* should be properly connected or soldered.

54.14. Transistor biasing. The proper flow of zero signal collector current and the maintenance of proper collector-emitter voltage during the passage of the signal is known as *transistor biasing*.

For proper transistor biasing the base-emitter junction is suitably forward biased and collector emitter junction reverse biased during the application of the signal. In addition, amount of bias required is important for establishing the *Q*-point which is determined by requirements of transistor operation.

The circuit which provides transistor biasing is known as *biasing circuit*. The commonly used biasing methods which employ a single source of supply — the one in the output circuit *i.e.* V_{cc}, are

(*i*) Base resistor method or fixed base bias
(*ii*) Biasing with feedback resistor
and (*iii*) Voltage divider bias.

54.15. Fixed base bias (*or fixed bias or base resistor bias*). A *n-p-n* transistor in common emitter (*CE*) mode with a fixed bias resistor R_B in the base circuit is shown in Fig. 54.8. The forward bias for the emitter-base junction and reverse bias for collector-base junction is taken from the same source of *e.m.f* V_{CC}. The requried value of zero signal or quiescent base current I_B is obtained by selecting the proper values of base resistor R_B.

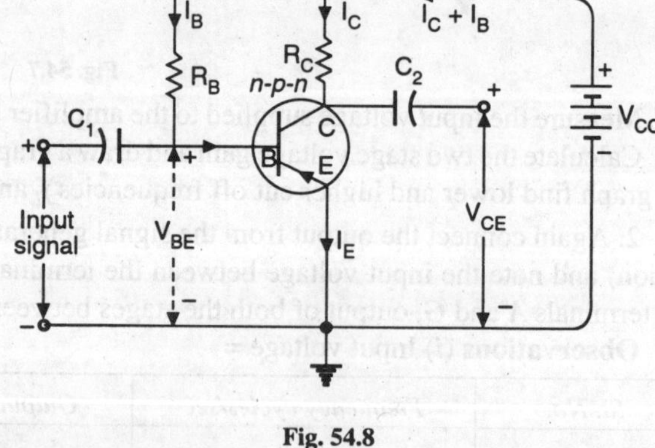

Fig. 54.8

Base resistance. In the base bias circuit the value of quiscent current is given by

$$I_B = \frac{V_{CC} - V_{BE}}{R_B}$$

∴ Value of base resistance required $R_B = \dfrac{V_{CC} - V_{BE}}{I_B}$

As V_{CC} and I_B are known and V_{BE} can be seen from the transistor manual, R_B can be calculated.

The voltage V_{BE} between base and emitter is of the order of milli-volts and can be neglected as compared to V_{CC}.

$$\therefore \qquad R_B = \frac{V_{CC}}{I_B}$$

V_{CC} is a fixed quantity and the value of I_B can be suitably chosen. R_B can, therefore, be calculated directly. For this reason, this method is known as *fixed base bias* method. I_B is independent of quiescent collector current I_C and is a constant.

54.16. Biasing with feedback resistor or collector to base bias. A *n-p-n* transistor in *CE* mode with feedback resistor R_B is shown in Fig. 54.9. In this case the resistance R_B is connected between the base and the collector (collector to base bias) rather than between the base and the positive terminal of the battery as in fixed base bias circuit. A single source V_{CC} is used to make emitter base junction forward biased and collector base junction reverse biased.

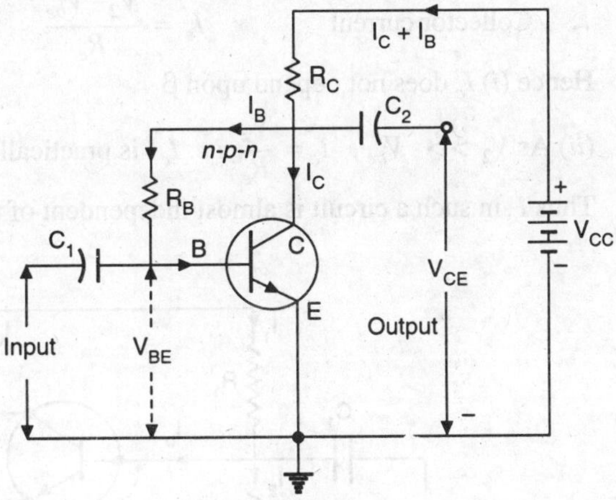

Now $\quad V_{CC} = (I_C + I_B)\, R_C + V_{CE}$

Also $\quad V_{CE} = I_B R_B + V_{BE}$

Fig. 54.9

$$\therefore \qquad R_B = \frac{V_{CC} - V_{BE} - (I_C + I_B)\, R_C}{I_B} \qquad ...(i)$$

As $I_B << I_C$, I_B may be neglected as compared to I_C and putting $I_C = \beta\, I_B$, we have

$$R_B = \frac{V_{CC} - V_{BE} - I_C\, R_C}{I_B} = \frac{V_{CC} - V_{BE} - \beta\, I_B\, R_C}{I_B} \qquad ...(ii)$$

Again $V_{BE} << V_{CC}$. Therefore, neglecting V_{BE} as compared to V_{CC}

$$R_B = \frac{V_{CC} - \beta\, I_B\, R_C}{I_B} = \frac{V_{CC}}{I_B} - \beta\, R_C \qquad ...(iii)$$

from which R_B can be calculated knowing V_{CC}, I_B, R_C and β.

These two methods are generally not used due to various disadvantages.

54.17. Voltage divider bias. This is the most widely used method of providing biasing to a transistor.

Two resistances R_1 and R_2 are connected across the supply voltage V_{CC} and provide biasing. The resistance R_e in the emitter circuit provides stabilisation (alongwith capacitor C_3). The name *voltage divider* is derived from the fact that resistances R_1 and R_2 form a potential divider across V_{cc}. The voltage drop across R_2 provides a forward bias to the base emitter junction due to which there is a base current and hence collector current even in the zero signal condition.

Let I_1 be the current through R_1. As base current I_b is very small it is assumed that current through R_2 is also I_1 *i.e.*

$$I_2 \cong I_1$$

$$\therefore \qquad I_1 = \frac{V_{cc}}{R_1 + R_2} \text{ and}$$

voltage across $R_2 = V_2 = I_2 R_2 = I_1 R_2 = \left(\dfrac{V_{cc}}{(R_1 + R_2)}\right) R_2$

Also $V_2 = V_{be} + V_e = V_{be} + I_e R_e$

$\therefore$ $I_e = \dfrac{V_2 - V_{be}}{R_e}$

As I_b is negligible, I_e is very nearly equal to I_c

$\therefore$ Collector current $I_c = \dfrac{V_2 - V_{be}}{R_e}$

Hence (*i*) I_c does not depend upon β

(*ii*) As $V_2 >> V_{be}$, $I_c = \dfrac{V_2}{R_e}$ *i.e.* I_c is practically independent of V_{be}.

Thus I_c in such a circuit is almost independent of transistor parameters.

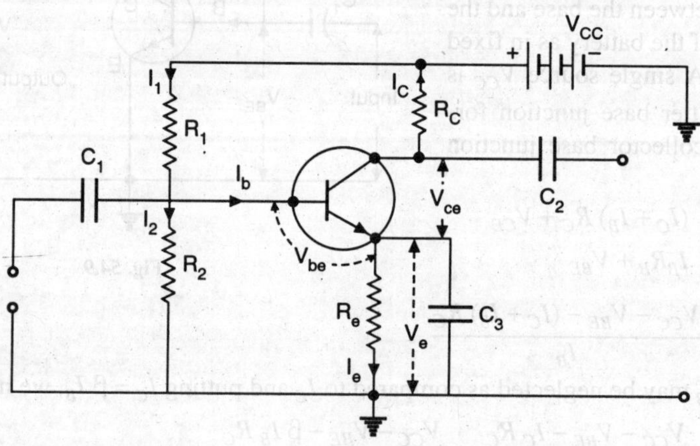

Fig. 54.10

Value of collector-emitter voltage V_{ce}.

$$V_{cc} = I_c R_c + V_{ce} + I_e R_e$$

$$= I_c R_c + V_{ce} + I_c R_e \qquad [\because I_e \cong I_c]$$

$$= I_c (R_c + R_e) + V_{ce}$$

or $$V_{ce} = V_{cc} - I_c (R_c + R_e)$$

Stabilisation. The emitter resistance R_e provides excellent stabilisation as explained below :

Suppose the collector current I_c increases due to rise in temperature, then the voltage drop across $R_e = (I_e R_e) = I_c R_e$ will increase.

As the voltage drop across R_2 is independent of I_c (or I_e), V_{be} will decrease according to the relation

$$V_2 = V_{be} + I_e R_e$$

The decrease in V_{be} causes I_b to decrease. The reduced value of I_b causes I_c also to decrease and restore to its original value, because $I_c = \beta I_b + I_{ceo}$ where I_{ceo} is the current from collector to emitter when $I_b = 0$ *i.e.* the base circuit is open.

Experiment 54.4. To study a common emitter amplifier for normal class – A operation (*a*) without bias stabilisation

(b) with bias stabilisation

(c) To study the effect of emitter by-pass capacitor on amplifier gain.

Apparatus. A variable low voltage *D.C.* power supply, an electronic voltmeter (*EVM*), Audio-frequency sine wave signal generator, an oscilloscope, resistors 220 K Ω, 1 K Ω, 8.2 K Ω, 18 K Ω, 560 ohm, two capacitors – 25 μ *F*, and one 100 μ *F*, a suitable transistor with appropriate socket (2 N 6004 or equivalent).

Procedure. (*a*) *Without bias stabilisation* 1. Draw a diagram as in Fig. 54.11 to set up a common emitter amplifier for normal class *A* operation and connect accordingly.

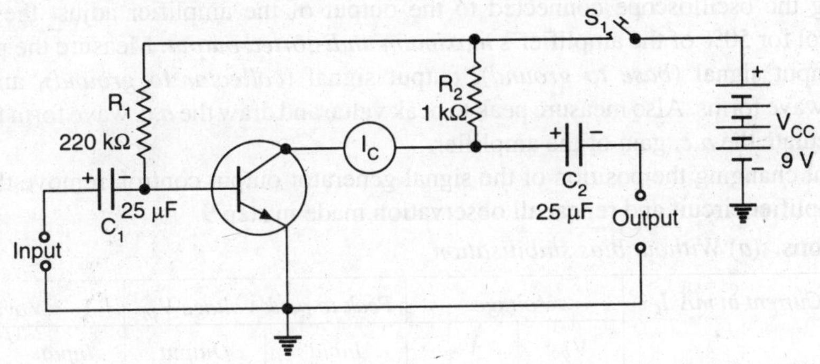

Fig. 54.11

2. Switch on the power supply by closing S_1. Measure the collector current I_c. Also measure base to emitter voltage V_{be} and collector to emitter voltage V_{ce} with the help of a *EVM*.

3. Connect an audio signal generator set at a frequency 1000 *Hz* at minimum output level to the input terminals of the amplifier circuit. Connect the output terminals of the amplifier circuit to the VERTICAL INPUT terminals of an oscilloscope. Adjust the oscilloscope for obtaining proper pattern. Increase the output of the audio signal generator just below the point of distortion so that maximum undistorted sine wave appears on the oscilloscope screen. Measure the peak to peak amplitude of the output wave form and the peak to peak amplitude of the input wave form in this very position. Also draw the input and output wave forms.

4. Again measure the collector current I_c, base to emitter voltage V_{be} and collector to emitter voltage V_{ce}. Switch off the power supply.

(*b*) *With bias stabilisation.*

5. Draw a diagram showing the scheme of connections as shown in Fig. 54.12 and connect accordingly.

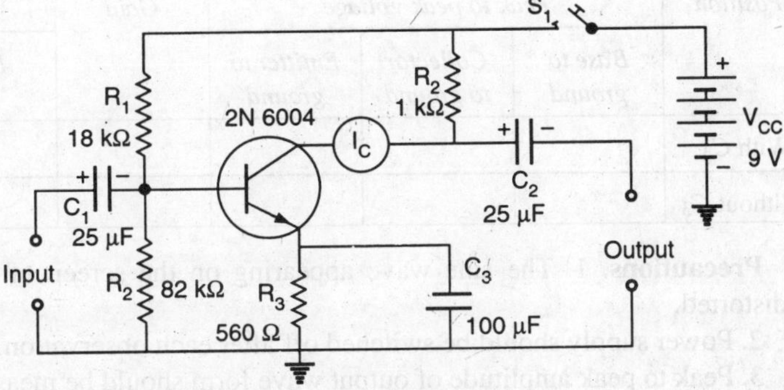

Fig. 54.12

6. Switch on the power supply by closing S_1. Measure the collector current I_c. Also measure base to emitter voltage V_{be} and collector to emitter voltage V_{ce} with the help of an *EVM*.

7. Connect an audio signal generator set at a frequency of 1000 *Hz* at minimum output level to the input terminals of the amplifier circuit. Connect the output terminals of the amplifier circuit to the VERTICAL INPUT terminals of the oscilloscope. Adjust the oscilloscope for obtaining proper

pattern. Adjust the audio signal generator control so as to obtain maximum undistorted output from the amplifier. Draw the input and output wave forms. Measure the peak to peak amplitude of the output and input wave forms. Also measure collector current I_c, base to emitter voltage V_{be} and collector to emitter voltage V_{ce}.

8. Reduce the input signal to the *minimum* value for undistorted output as observed on the oscilloscope. In this position draw the output and input wave forms. Measure peak to peak amplitude of the output and input signal voltages and values of I_c, V_{be} and V_{ce}.

(c) *Effect of emitter by-pass capacitor on amplifier gain.*

9. Keeping the oscilloscope connected to the output of the amplifier adjust the *A.F.* signal generator control for 50% of the amplifier's *maximum undistorted output*. Measure the peak to peak value of the input signal (*base to ground*), output signal (*collector to ground*), and draw the corresponding wave forms. Also measure peak to peak value and draw the *a.c.* wave form from emitter to ground. Calculate the *a.c.* gain of the amplifier.

10. Without changing the position of the signal generator output control, remove the capacitor C_3 from the amplifier circuit and repeat all observation made in step 9.

Observations. (a) *Without bias stabilisation*

Position	*Current in mA I_c*	*Voltage*		*Peak to peak voltage V_{pp}*		*Waveform*	
		V_{be}	V_{ce}	*Input*	*Output*	*Input*	*Output*
No Signal							
at 1000 *Hz*							

(b) *With bias stabilisation*

Position	*Current in mA I_c*	*Voltage*		*Peak to peak voltage V_{pp}*		*Waveform*	
		V_{be}	V_{ce}	*Input*	*Output*	*Input*	*Output*
No Signal							
1000 *Hz* at Max.							
1000 *Hz* at Min.							

(c) *Emitter bye-pass capacitor and gain.*

Position	*Peak to peak voltage*			*Gain*	*Wave form to ground*		
	Base to ground	*Collector to ground*	*Emitter to ground*		*Base*	*Collector*	*Emitter*
With C_3							
Without C_3							

Precautions. 1. The sine wave appearing on the screen of the oscilloscope should remain undistorted.

2. Power supply should be switched off after each observation.
3. Peak to peak amplitude of output wave form should be measured.
4. While removing the capacitor the position of the signal generator should not be disturbed.

Oral Questions

1. Transistor gain. (*i*) What do you mean by current gain of a transistor amplifier ? (*ii*) Distinguish between d.c. current gain and a.c. current gain. (*iii*) Are the two equal? If so, explain, why? (*iv*) What is current gain in common base configuration and common emitter configuration? (*v*) What is the relation between the two? (*vi*) Which of these is greater and why? (*vii*) Define voltage gain and power gain of a common emitter transistor amplifier. (*viii*) What is load line? (*ix*) Define *Q*-point of a common emitter amplifier. (*x*) What are different types of amplifications? (*xi*) What do you mean by class A amplification.

2. Common base and common emitter amplifiers (*i*) what is the difference between a common base and a common emitter amplifier? (*ii*) Which of these is generally used and why? (*iii*) What do you mean by higher cut off frequency and lower cut off frequency of an amplifier ? (*v*) What is bandwidth?

3. Two stage R.C. coupled amplifier. (*i*) What are the various methods of coupling used for amplifiers ? (*ii*) What is R.C. coupling? (*iii*) Why are two amplification stages coupled? (*iv*) What is the total voltage gain with a two stage R.C. coupled amplifier? (*v*) Is there any difference in coupling for valve or transistor amplifier stages ?

4. Transistor biasing (*i*) Give three methods of transistor biasing? (*ii*) Why transistor biasing is required? (*iii*) Which method of transistor biasing is commonly used and why?

55

Voltage Multiplier

55.1. Voltage multiplier circuit. It is the circuit which is capable of delivering twice, three times, four time etc of *d.c.* voltage produced by a single half wave rectifier. To understand the basic working principle of a voltage multiplier consider the circuit shown in Fig. 55.1.

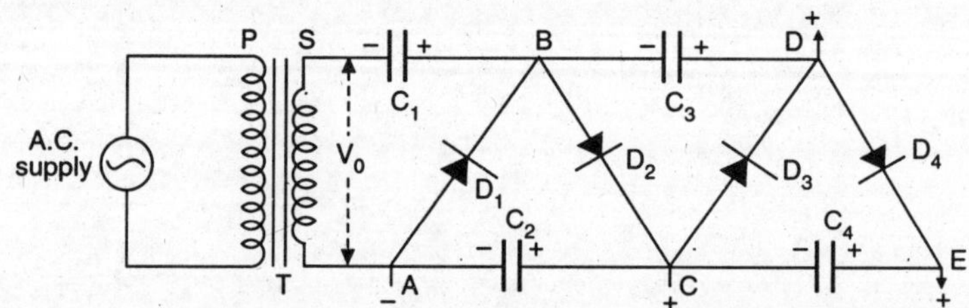

Fig. 55.1

Let V_0 be the peak value of voltage across the secondary *S* of the transformer when the *A.C.* mains supply is connected to the primary *P*. Diodes D_1, D_2, D_3 and D_4 are half wave rectifiers and C_1, C_2, C_3 and C_4 are the capacitors connected as shown.

Let us first consider that only the rectifier D_1 and capacitor C_1 are present. Suppose during the first half cycle of the alternating current the end *A* of the secondary of the transformer is *positive* and the end *S* is *negative*. In such a case an electric current flows through the rectifier D_1 and the capacitor C_1 is charged to the peak voltage V_0 as shown. During the second half cycle of the alternating current the end *A* of the secondary of the transformer is *negative* and the end *S* is *positive*. In such a case no electric current flows through the rectifier D_1 and so the voltage at *B* is still $+ V_0$ whereas the voltage at *A* is $-V_0$. In other words the potential difference between the points *B* and *A* is $2\ V_0$ or the potential of *B* is $2\ V_0$ with respect to *A*.

Let us now consider that the rectifier D_2 and capacitor C_2 are also present. During the second half cycle the current flows through the rectifier D_2 with the result that the capacitors C_1 and C_2 share their charges equally supposing that their capacitances are equal.

During the third half cycle of the alternating current the end *A* is again *positive* and, therefore, the capacitor C_1 is again charged to the peak voltage V_0. As no current flows through D_2 there is no change in the voltage of capacitor C_2.

During the fourth half cycle of the alternating current the capacitor C_2 again shares its charge with capacitor C_1 and its voltage increases. If the process is repeated for a few cycles the capacitor

C_2 also gets fully charged to the peak voltage V_0, supposing there is no leakage of charge either in the capacitors or in other parts of the circuit.

Finally an equilibrium condition is reached where no current flows through either the rectifier D_1 or through D_2. At this stage

Potential of the point $C = 2\ V_0$ w.r.t the point A

$\therefore$ Potential difference between C and $A = +2\ V_0$

This arrangement is, therefore, known as a **voltage doubler.**

Let us now consider that the rectifier D_3 and the capacitor C_3 are also included in the circuit. The above process is again repeated so that the capacitor C_3 is also fully charged. In such a case potential of the point D is $+ V_0$ with respect to the point C and $+ 3\ V_0$ with respect to the point A. The arrangement is, therefore, known as **voltage tripler.**

Similarly if the rectifier D_4 and capacitor C_4 are also included in the circuit, the final potential of the point E will be $4\ V_0$ w.r.t the point A. In other words, by connecting more rectifiers and capacitors in cascade, the potential can be raised to any multiple of V_0 at least theoretically though there may be many practical limitations and difficulties.

Experiment 55.1. To study voltage doubler and tripler circuits

Apparatus. A step down transformer (secondary tappings 6, 9, 12 *V*) (current rating 0.2 A), three junction diodes, 3 capacitors (32 μ F each electrolytic type) *A.C.* voltmeter, *D.C.* voltmeter, *D.C.* milli-ammeter, a resistance box (or 500 Ω resistors connected in series, connecting wires etc.

Procedure. 1. Draw a diagram showing the scheme of connections as in Fig. 55.2. and connect accordingly.

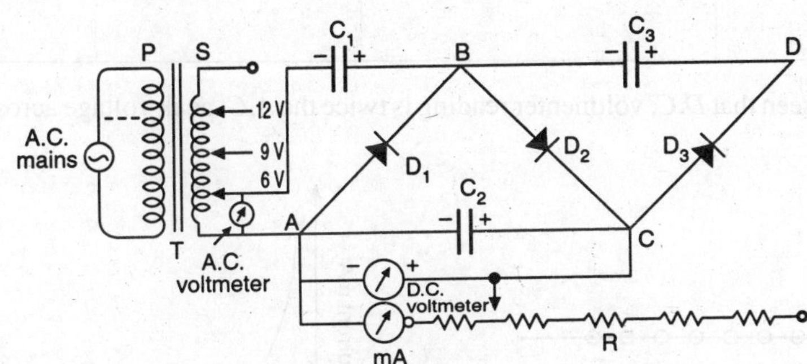

Fig. 55.2

Connect the variable contact to 6 *V* tapping of the secondary of the transformer *T*. Switch on the *A.C.* mains and measure the voltage across this terminals with an *A.C.* voltmeter. Remove the voltmeter.

2. **To study voltage doubler arrangement.** Adjust the value of *R* to about 500 ohm and note the reading of the *D.C.* voltmeter connected between the points *A* and *C* and the reading of the milliammeter. Increase the value of resistance *R* in small steps and note the corresponding reading of the *D.C.* voltmeter and milli-ammeter. It is seen that the voltmeter reading remains almost constant whereas the milliammeter reading decreases with increasing values of *R*.

3. Now connect the variable contact to 9 *V* tapping and repeat the experiment. Similarly repeat with 12 *V* tapping.

(*b*) **To study voltage tripler arrangement.** To study the voltage tripler arrangement connect *D.C.* voltmeter the resistance *R* and milliammeter between the points *A* and *D*. Repeat steps 2 and 3.

Observations (*a*) voltage doubler arrangement

R.M.S. voltage across the secondary coil

$E_{r.m.s}$ = (i) (ii) (iii) volts

Peak voltage across secondary coil

$E_{peak} = \sqrt{2}\, E_{r.m.s}$ (i) (ii) (ii) volts

Sl. No.	E_{peak}	D.C. Voltmeter reading volts	Resistance R ohm	Ammeter reading mA
1	(i)			
2				
.				
.				
6				
1	(ii)			
2				
.				
.				
6				
1	(iii)			
2				
.				
.				
.				
6				

Result. It is seen that *D.C.* voltmenter reading is twice the *A.C.* peak voltage across the secondary coil.

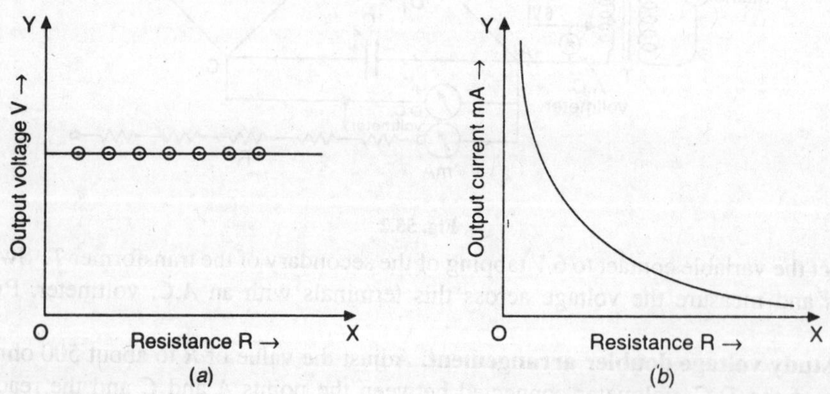

Fig. 55.3

4. Plot a graph between (*a*) Output voltage (*D.C.* voltmeter reading) and resistance *R* and (*b*) Output current (milli-ammeter reading) and *R*. The graph is shown in Fig. 55.3 (*a*) and (*b*)

(*b*) **Voltage tripler arrangement.** Record the observations as in (*a*).

Result. It is seen that the *D.C.* voltmeter reading is thrice the *A.C.* peak voltage across the secondary coil.

Precautions. 1. The output voltage should be measured with a *V.T.V.M.* instead of a an ordinary voltmeter.

2. Electrolytic type capacitors of 50 V rating at least should be used.
3. A resistor must always be connected in the output circuit so that there is no excessive current.
4. The diodes must be connected in the proper direction.

Oral Questions

1. Voltage multipliers (*i*) What do you mean by a voltage doubler and a voltage tripler? (*ii*) Can you multiply the voltage more than three times? (*iii*) What kind of capacitors are used in such circuits? (*iv*) Explain the principle of a voltage doubler arrangement. (*v*) As the voltage is doubled do you think the energy is also doubled? (*vi*) What is the difference between a voltage multiplier and a step-up transformer?

56

Silicon Controlled Rectifier

56.1. Silicon controlled rectifier. A silicon controlled rectifier is a semi-conductor device made of silicon which combines the features of a *rectifier* and a *transistor*. It consists of a *P-N* junction diode added to an *N-P-N* junction transistor forming a *three* junction *P-N-P-N* device. It has three *terminals*. One terminal attached to the outer *P*-type layer is called the *Anode A*, the second terminal attached to the outer *N*-type layer is called the *Cathode K* and the third terminal attached to the *Base* of the *N-P-N* transistor section is called the *Gate G*.

A silicon controlled rectifier is also known as *Thyristor* which is the solid state equivalent of Thyratron. The gate, anode and cathode of *S.C.R.* respectively correspond to grid, plate and cathode of Thyratron.

Working of S.C.R. The working of *S.C.R.* can be studied under the following heads

(*i*) *When Gate is open*. The *S.C.R.* circuit with gate open *i.e.* no voltage applied to the Gate is shown in Fig. 56.1 (*a*).

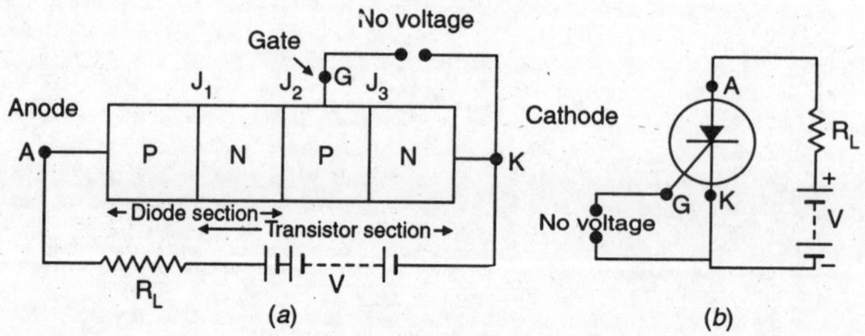

Fig. 56.1

It is seen that the junction J_1 and J_3 are *forward biased* whereas the junction J_2 is reverse biased. Hence under the above conditions no current flows through the *S.C.R.* and the load resistance R_L for small values of the applied voltage. In this position the *S.C.R.* is in the *cut off* stage. If the applied voltage is slowly increased a stage is reached when the reverse biased junction J_2 breaks down and the whole applied voltage *V* appears as reverse bias across the junction J_2 since the junctions J_1 and J_3 are forward biased. The *S.C.R.* now conducts heavily and is said to be in the *ON* stage.

The minimum applied voltage at which S.C.R. conducts heavily without Gate voltage is called **break over voltage**

(*ii*) *Gate positive with respect to cathode*. The *S.C.R.* can be made to conduct heavily at smaller applied voltages by applying a small positive potential to the Gate as shown in Fig. 56.2. The junction

J_3 is still forward biased whereas the junction J_2 is reverse biased. The electrons from the N- type layer start moving towards the left as shown where as the holes move from P-type layer towards the right. As a result the electrons from junction J_3 are attracted across the junction J_2 and a Gate current starts flowing. As soon as the Gate current flows the anode current increases. The increased anode current makes more electrons available at the junction J_2. This process continues and in a small time the junction J_2 breaks down and the *S.C.R.* starts conducting heavily.

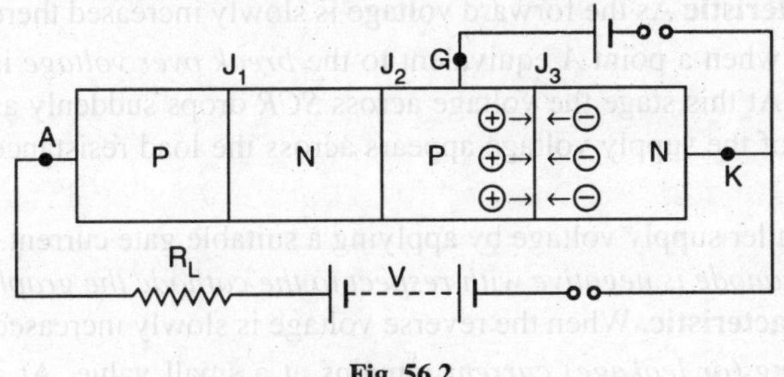

Fig. 56.2

Once the *S.C.R.* starts conducting the Gate loses all control. Even if the Gate voltage is removed the anode current does not decrease at all. To stop conduction *i.e.* to bring *S.C.R.* in the OFF condition the only method is to reduce the applied voltage to almost zero. At this point the internal transistor comes out of saturation and *opens* the *S.C.R.* The anode current under this condition is very small and is called the *holding current. The maximum anode current, gate being open, at which the S.C.R. is turned OFF from ON condition is called* **holding current.**

Remember. *(i)* A *S.C.R* has only two states. Either it *does not conduct* or it *conducts heavily.* There is no intermediate state. Thus *S.C.R* acts like a *switch.*

(ii) There are two methods to turn *ON* the *S.C.R.* In the *first* method the Gate is kept OPEN and the supply voltage is made equal to the *break over voltage*. In the *second* method the supply voltage is kept below the break over voltage and a small voltage is applied to the Gate to turn *ON* the *SCR.*

The second method is most commonly used to switch *ON* the *SCR.*

(iii) To OPEN the *SCR i.e.* to make it non-conducting the supply voltage is reduced to ZERO.

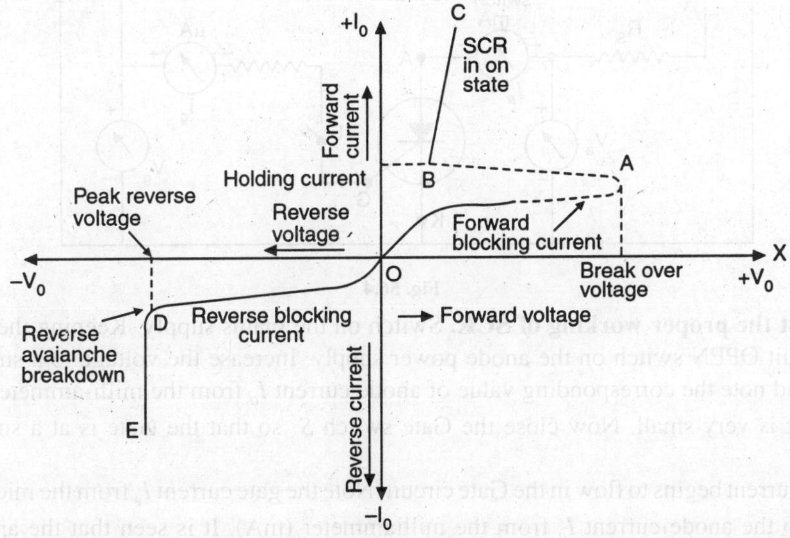

Fig. 56.3

56.2. V-I characteristics of the SCR. It is a graph between anode-cathode voltage V_0 and anode current I_0 of the *SCR* at constant gate current. The forward as well as reverse characteristics of the *SCR* when no gate voltage is applied are shown in Fig. 56.3.

(*i*) **Forward characteristic.** *When the anode is positive with* **respect** *to the cathode the graph between* V_0 and I_0 is called **forward characteristic** As the forward voltage is slowly increased there is a small **blocking (leakage) current** and when a point *A* equivalent to the *break over voltage* is reached the *SCR* starts conducting heavily. At this stage the voltage across *SCR* drops suddenly as shown by the dotted line *AB* because most of the supply voltage appears across the load resistance R_L.

NOTE. The *SCR* can close at much smaller supply voltage by applying a suitable gate current.

(*ii*) **Reverse characteristics.** *When the anode is negative with respect to the cathode the graph between* V_0 *and* I_0 *is known as* **reverse characteristic.** When the reverse voltage is slowly increased the anode current known as *reverse blocking (or leakage) current* remains at a small value. At a certain value (corresponding to *D*) *reverse avalanche breakdown* occurs and the *SCR* starts conducting heavily in the reverse direction as shown by the part *DE* of the graph. This voltage is known as *reverse avalanche (or breakdown) voltage or peak reverse voltage*.

Experiment 56.1. To draw the voltage current characteristics of a silicon controlled rectifier.

Apparatus. A silicon controlled rectifier (*SCR*) a stabilised *D.C.* power supply having continuously variable tappings 0–50 *V* for Anode and 0-1.5 *V* for Gate, voltmeters 0-50 *V*, 0-1.5 *V*, a milliammeter 0-50 m A, a micro-ammeter 0-100 μ A, resistors of suitable values, switches etc.

Procedure 1. Draw a diagram showing the scheme of connections as in Fig. 56.4. and connect accordingly.

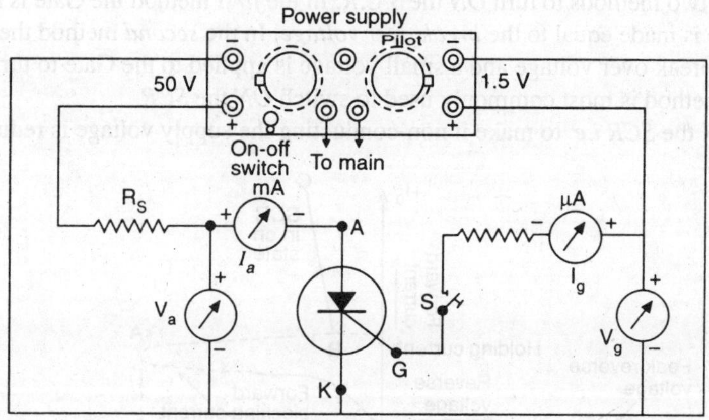

Fig. 56.4

2. **To test the proper working of SCR.** Switch on the mains supply. Keeping the switch *S* in the Gate circuit OPEN switch on the anode power supply. Increase the voltage to a small value of about 10 *V* and note the corresponding value of anode current I_a from the milli-ammeter (mA). The anode current is very small. Now close the Gate switch *S* so that the Gate is at a small positive potential.

A small current begins to flow in the Gate circuit. Note the gate current I_g from the micro-ammeter (μA) and also the anode current I_a from the milliammeter (mA). It is seen that the anode current suddenly increases *i.e.* the *SCR* is now ON.

Open the Gate switch *S* again. The anode current still has a large value. The above sequence of operations shows that the *SCR* is working properly. Switch off the anode power supply and the Gate voltage.

3. **Forward characteristics.** Keeping the anode power supply and Gate voltage switched off note the leakage current, if any in the anode circuit. Keeping the switch *S* in the Gate circuit OPEN,

switch on the anode power supply. Now slowly increase the anode voltage in very small steps (say 0.1 *V*). Note the anode voltage V_a and the corresponding values of I_a *i.e.* the forward leakage (or blocking) current.

4. As the anode voltage approaches the *break over* point the *SCR* begins to conduct. Note the anode voltage V_a at which break over takes place and also note the corresponding value of anode current I_a.

At this stage the reading of the voltmeter in the anode circuit suddenly drops as most of the supply voltage appears across the load resistance R_s. The value of the current in the *SCR* at this *ON* state depends upon the value of R_s.

5. Close the switch *S* in the Gate circuit and apply a small *D.C.* voltage (say 0.2 *V*) to the Gate. Repeat steps 3 and 4.

Similarly repeat steps 3 and 4 with different values of Gate voltage.

It is seen that with increasing values of Gate voltage, the break over voltage decreases *i.e.* the break over takes place at smaller values.

Reverse characteristics. To draw the reverse characteristics, reverse the polarity of the anode power supply and again repeat steps 3, 4, & 5.

Observations.

FORWARD CHARACTERISTICS								
No.	*Gate Voltage* $V_g = 0$		*Gate Voltage* $V_g = V_1$		*Gate Voltage* $V_g = V_2$		*Gate Voltage* $V_g = V_3$	
	Anode voltage V_a*, volts*	*Anode current* I_a*, mA*	*Anode voltage* V_a*, volts*	*Anode current* I_a*, mA*	*Anode voltage* V_a*, volts*	*Anode current* I_a*, mA*	*Anode voltage* V_a*, volts*	*Anode current* I_a*, mA*
1								
2								
...								
...								
REVERSE CHARACTERISTICS								
1								
2								
...								
...								

7. Plot graphs by taking the anode voltage along the *X*-axis and anode current along *Y*–axis separately for different values of Gate voltage. The graph for Gate voltage $V_g = 0$ is shown in Fig. 56.3. From the graphs note the following observations.

No.	*Gate Voltage*	*Break over voltage*	*Holding current*	*Reverse peak voltage*
	$V_g = 0$			
	$V_g = V_1$			
	$V_g = V_2$			
	$V_g = V_3$			

Precautions. 1. The working of *SCR* must be checked before performing the experiment.

2. Anode potential must be increased in small steps of 0.1 volt.

3. The data given on the *SCR* must be carefully noted and values of resistances selected and voltages must be applied accordingly.

4. The ammeters and voltmeters must be connected with proper polarities.

Oral Questions

1. Silicon controlled rectifier *(i)* What is a silicon controlled rectifier? *(ii)* What is the material of which it is made? *(iii)* How many junctions does an SCR have? *(iv)* Explain the terms anode, cathode and Gate as applied to S.C.R. *(v)* What is a thyristor? *(vi)* What is the thermionic equivalent of Thyristor? *(vii)* Define break over voltage. *(viii)* How does break over voltage change with Gate voltage or gate current? *(ix)* Why do we say that an SCR is a combination of a diode and a transistor? *(x)* What do you mean when you say the SCR is ON and SCR is OFF? *(xi)* What is holding current? *(xii)* What are the different ways in which S.C.R. can be turned OFF? *(xiv)* Why is SCR called a switch? *(xv)* Does the SCR have any state other than ON or OFF?

57

Radio Transmission and Reception

57.1. Transmission and reception of radio waves. Radio transmission involves sending of audio signals like music, speech etc from a broadcasting station over large distances and reception by a radio receiver. This is achieved by superimposing electrical audio-signal on high frequency carrier waves. The process is known as *modulation* and the resultant wave is called *modulated carrier wave*. At the radio receiver the audio signal is separated from the carrier wave. The process is known as *de-modulation* or *detection*. The signal is then *amplified* and reproduced by a loud speaker.

There are two types of modulation processes commonly in use.

(*i*) Amplitude modulation (*ii*) Frequency modulation.

As most of the radio broadcasts have amplitude modulation, radio recievers in common use are also amplitude modulated recievers.

57.2. Super hetrodyne reciever. All modern radio recievers make use of super hetrodyne circuit. The block diagram of a super hetrodyne receiver is shown in Fig. 57.1.

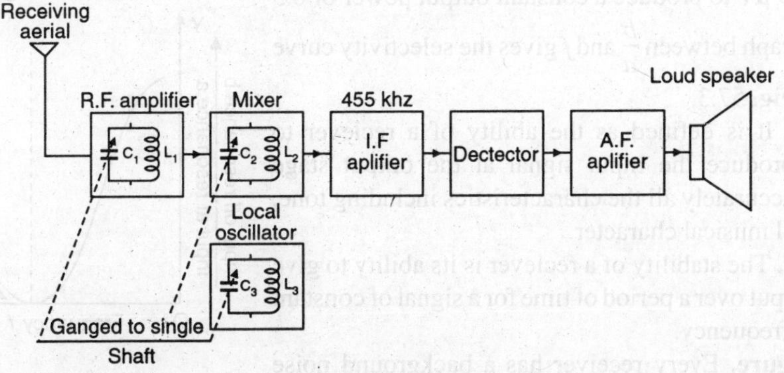

Fig. 57.1

The *R.F.* amplifier stage uses a tuned parallel circuit with inductance L_1 and a variable capacitor C_1. The radio waves from various broadcasting stations are intercepted by the recieving aerial out of which the desired frequency (or station) is selected by this stage and amplified.

The mixer stage combines the amplified *R.F.* signal with the output of a local oscillator thereby producing an intermediate frequency (*I.F.*) given by

I.F. = Oscillator frequency – Radio frequency (*R.F.*)

The *I.F.* is always kept at 455 *KHz*. For this purpose the capacitor C_1 of *R.F.* amplifier, C_2 of the mixer and C_3 of the local oscillator are all ganged to the same shaft.

The production of fixed intermediate frequency *I.F.* = 455 *KHZ* is the most important feature of the super hetrodyne circuit. At this fixed frequency the receiver gives maximum stability, selectivity and sensitivity with proper fidelity, and least noise figure – the properties which are known as characteristics of a radio receiver.

The *I.F.* amplifier stage amplifies the *I.F.* signal which still contains the audio signal. The detector stage separates the audio signal from the *I.F.* The *A.F.* amplifier stage amplifies the audio signal so that it is sufficiently strong to operate the loud speaker.

57.3. Characteristics of a radio receiver. The important characteristics of a radio receiver are: (*i*) Sensitivity (*ii*) Selectivity (*iii*) Fidelity (*iv*) Stability and (*v*) Noise figure.

(*i*) **Sensitivity.** Sensitivity of an amplitude modulated broadcast receiver is defined as *'The amplitude of the carrier wave (in micro volts μV) modulated 30% at 400 cycles per second which when applied to the input terminals of the reciever through a standard dummy antenna develops an output of 0.5 watt in a load resistance of suitable value substituted, in place of the loud speaker.*

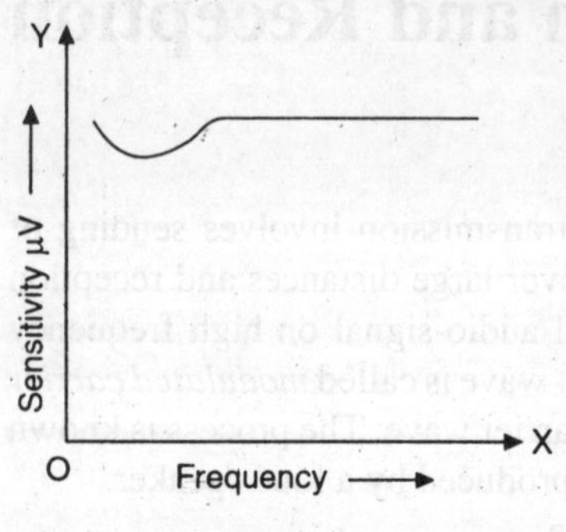

Fig. 57.2

A graph between sensitivity and corresponding frequency at resonance is shown in Fig. 57.2.

(*ii*) **Selectivity.** The selectivity of a radio receiver is its ability to separate the frequency of the desired broadcasting radio-station from the frequencies of the undesired stations.

It is expressed in the form of a curve plotted between the frequencies off the resonance versus the ratio of the amplitude of the input signal (in micro volts, μ*V*) which produces a constant output of (0.5 watt) at a frequency off the resonance position to the amplitude of the input signal to produce the same output (0.5 W) at the resonant frequency.

If the amplitude of the input signal at resonant frequency is *a* μ*V* and for another frequency *f* off resonance is *b* μ*V* to produce a constant output power of 0.5 *W*, then the graph between $\frac{b}{a}$ and *f* gives the selectivity curve as shown in Fig. 57.3.

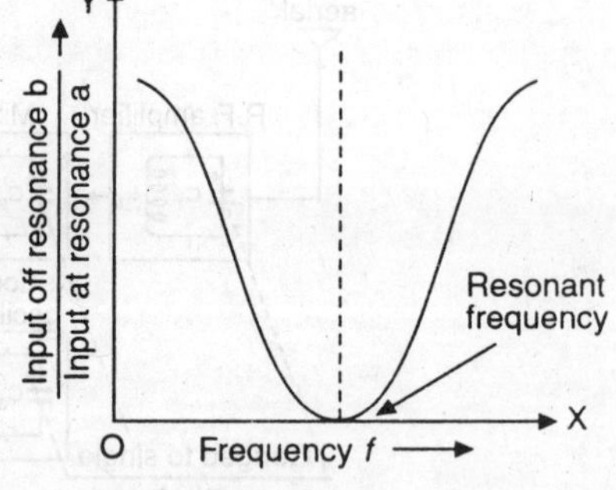

Fig. 57.3

Fidelity. It is defined as the ability of a reciever to faithfully reproduce the input signal at the output stage maintaining accurately all the characteristics including tone, sweetness and musical character.

Stability. The stability of a reciever is its ability to give a constant output over a period of time for a signal of constant strength and frequency.

Noise figure. Every receiver has a background noise associated with it. As a result signals weaker than this noise level cannot be detected by the receiver even after amplification.

The ratio of the output power or voltage for the specified modulated carrier at the input, to the power or voltage output of the noise is known as signal to noise ratio.

The noise figure determines the smallest power that may be received without being downed into the receiver noise. The noise figure is generally expressed as the power ratio as defined above or in decibles.

Experiment 57.1. To study the sensitivity and selectivity of a given radio receiver using a dummy antenna.

Apparatus. A radio receiver, standard signal generator, *A.F.* power output meter or *V.T.V.M.*, a resistor having a resistance equal to the impedance of the loud speaker of the given receiver, a dummy antenna etc.

Procedure. 1. Draw a diagram showing the scheme of connections as in fig. 57.4. and connect accordingly.

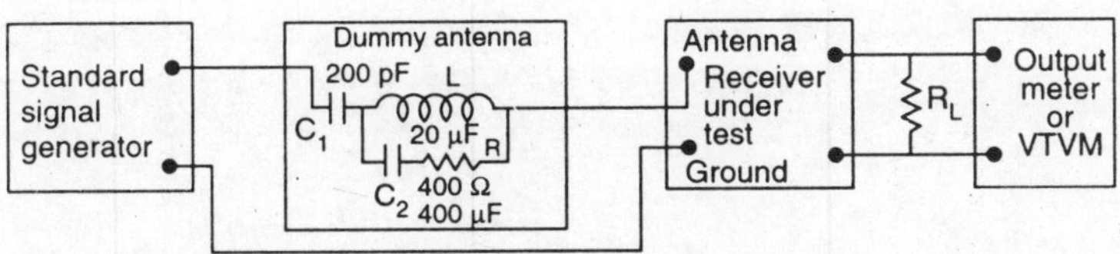

Fig. 57.4

Note. The dummy antenna corresponds to an aerial antenna having an effective height of 4 metre. A typical circuit of a dummy antenna is shown within the block diagram with values of various *L*, *C* and *R* components.

2. (*a*) **Sensitivity** Disconnect the loud speaker of the radio receiver and connect to the same points a resistor R_L having a resistance equal to the impedance of the speaker at 400 *Hz*. Also disconnect the automatic volume control (A.V.C.) if it is there.

3. Apply a signal from the *R.F.* standard signal generator at a suitable radio frequency in the medium wave band (say 1000 *K Hz*) modulated by 400 *Hz* audio frequency having 30% depth of modulation. Tune the radio receiver to the frequency of the signal generator and adjust the input of the signal generator to get an output of 0.5 W across the resistor R_L as measured by the A.F. output power meter with volume control set at maximum.

Note the input of the signal generator using a microvoltmeter. Repeat for a number of frequencies in the medium wave band.

Change over to short wave band and repeat the experiment.

4. Plot a graph between sensitivity *i.e.* input from the *R.F.* standard signal generator in μ*V* taken along the *Y*-axis and the corresponding frequency in *KHz* taken along the *X*-axis. A typical graph is shown in Fig. 57.2.

5. (*b*) **Selectivity** Tune the radio receiver to a frequency 1000 *KHz* (or any other suitable frequency) from the *R.F.* standard signal generator. This is referred to as *resonant frequency*. Adjust the input from the signal generator to get an output of 0.5W across the resistor R_L. Note the input from the standard signal generator in μV. *Do not disturb the tuning of the receiver during the experiment*.

6. Now change the frequency of the modulated signal from the standard *RF* signal generator by a small step of about 10 *KHz* above the resonant frequency and adjust the strength of the signal so that the output across the resistor R_L remains constant at 0.5 W as measured by the output power meter or *V.T.V.M*. Note the input from the standard signal generator in μ*V*. Repeat by taking a number of observations first in increasing steps of 10 *KHz* above the resonant frequency and then in decreasing steps of 10 *KHz* below the resonant frequency.

7. Plot a graph between frequency of the input signal from the standard *R.F.* signal generator taken along the *X*-axis and ratio of input power (in μ*V*) at that frequency to the input power (in μ*V*) at resonant frequency to give a constant output of 0.5 W. The graph is as shown in Fig. 57.3.

Observations. (*a*) Sensitivity.

Constant output power of the receiver = 0.5 W

Sl. No.	*Medium wave band*		*Short wave band*	
	Frequency of input signal KHz	*Amplitude of input signal* μ *V*	*Frequency of input signal KHz*	*Amplitude of input signal* μ *V*
1				
.				
.				
.				
.				
.				
.				
8				

(*b*) **Selectivity.**

Constant output power of the receiver = 0.5 W

	Medium wave band			*Short wave band*		
	Resonant Frequency = KHz			*Resonant Frequency = KHz*		
	Input signal at resonant frequency (a) = μV			*Input signal at resonant frequency (a′) = μV*		
	Frequency off resonance KHz	*Input signal (b)* μ *V*	*Ratio (b/a)*	*Frequency off resonance KHz*	*Input signal (b′)* μ *V*	*Ratio (b′/a′)*
1						
2						
.						
.						
.						
.						
.						
8						

Precautions. 1. A standard dummy antenna should be used for this experiment.

2. The output of the radio receiver should always be kept fixed at 0.5 watt as indicated by a *V.T.V.M.* or a power meter.

3. In the sensitivity experiment the radio receiver must always be tuned to the changing frequency from the *R.F.* oscillator for resonance position.

4. In selectivity experiment the resonant frequency should be kept fixed throughout the experiment.

5. The loud speaker of the radio receiver must be replaced by a resistor having the same resistance as the impedance of the speaker.

Oral Questions

1. Radio transmission and reception (*i*) What is a radio broad-casting station? (*ii*) The frequency at which a radio station broadcasts is known as what? (*iii*) What is modulation and depth of modulation? (*iv*) What do you mean by demodulation? (*v*) What is detection? (*vi*) What are the various stages in a super heterodyne receiver? (*vii*) What is intermediate frequency (*I.F.*)? (*viii*) Does it have a standard fixed value, if so what is it? (*ix*) What is the advantage of *I.F.* stage?

2. Characteristics of a radio-receiver. (*i*) What are the important characteristics of a radio-receiever ? (*ii*) Define sensivity. (*iii*) Define selectivity (*iv*) How does selectivity vary with frequency? (*v*) Define fidelity and noise figure. (*vi*) What is a dummy antenna?

58

Operational Amplifiers

58.1. Operational amplifier. The operational amplifier, in short, known as *OP AMP*, is a direct-coupled, high-gain, differential amplifier. The term operational amplifier was originally used for *dc* amplifiers which perform mathematical operations like addition, subtraction, integration and differentiation in analog computers. *OP AMPs* also find application in voltage regulators, active filters, instrumentation, analog to digital and digital to analog converters etc.

Operational amplifiers use negative feedback. The performance of the amplifier with negative feedback is solely controlled by feedback elements and does not depend upon the characteristics of the transistors, resistors and capacitors which combine to form the amplifier. The feedback elements being generally passive, the circuit is highly stable. We shall not deal with the internal circuit configuration of the *OP AMP* but only with its terminal properties.

An ideal operational amplifier has an infinite voltage gain, infinite band width, infinite input impedance, zero-out put impedance, and perfect balance *i.e.*, the output is zero when equal voltages are present at the two input terminals and its characteristics do not change with temperature.

Operational amplifiers, in actual practice, have a high (not infinite) low frequency voltage gain of the order of 10^3 to 10^5 and finite band width. The gain is constant upto a few hundred kilo-Hertz and then decreases with increase in frequency. The input impedance is about 150 K Ω to a few hundred M Ω and output impedance about 0.75 to 100 Ω. With practical *OP AMPs* perfect balance cannot be achieved.

58.2. Circuit symbol. The circuit symbol of *OP AMP* is shown in Fig. 58.1. Terminals *a* and *b* are the input terminals and the terminal *c* is the output terminal. Terminal *a* marked negative (–) is called the *inverting input terminal.* The negative sign indicates that a signal applied at the terminal *a* will appear at the terminal *c* with polarity opposite to that at the terminal *a*. The terminal *b* marked positive (+) is the *non-inverting input terminal.* This means that the signal applied to the terminal *b* appears at the output terminal *c* with the same polarity as at *b*. The output signal voltage is proportional to the *difference* between the two signal voltages applied at the two input terminals. The constant of proportionality is the *gain* of the amplifier denoted by *A*, which is a real constant approaching infinity in an ideal case.

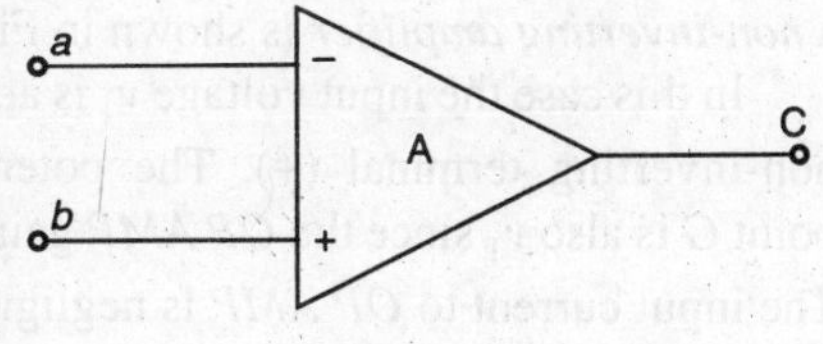

Fig. 58.1

Some other applications of *OP AMPs* are (*i*) Inverting amplifier (*ii*) Scale changer (*iii*) Phase shifter, (*iv*) Non-inverting amplifier, (*v*) Summing amplifier or adder, (*vi*) Voltage to current convertor (*vii*) Integrator and (*viii*) Differentiator.

58.3. Inverting amplifier. The circuit of a basic inverting amplifier is shown in Fig. 58.2, R_1 is the *input resistor* and R_f the *feedback resistor.*

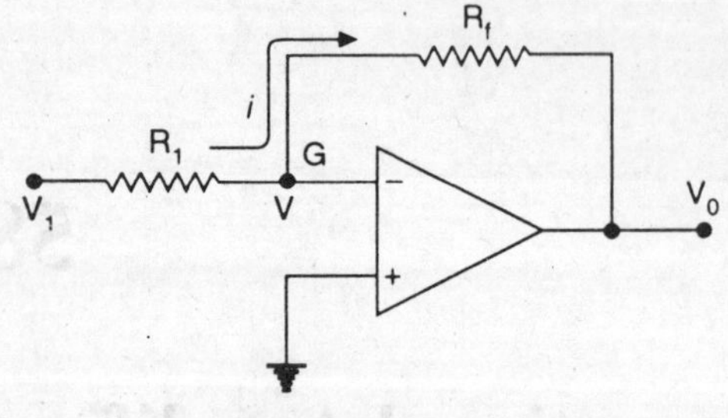

Fig. 58.2

The non inverting input terminal (+) is grounded. v_1 is the input voltage applied through R_1 to the inverting terminal (–) and v_0 is the output voltage. As the gain A of the amplifier (or the open loop gain) is very large the voltage v at the inverting input terminal (–) is very small *i.e.*, very close to the ground potential. In other words, the point G though not connected to the ground is *virtually* at ground potential irrespective of the magnitude of v_1 and v_0. The current i flowing through R_1 is given by

$$i = \frac{v_1 - v}{R_1}$$

Assuming that the *OP AMP* is ideal having an infinite input resistance the current will flow through R_f and not through the *OP AMP*. Applying Kirchhoff's law at the point G

$$\frac{v_1 - v}{R_1} = \frac{v - v_0}{R_f}$$

As the point G is virtually at grounded potential $v = 0$

$$\therefore \qquad \frac{v_1}{R_1} = \frac{-v_0}{R_f}$$

The ratio of the output voltage v_0 and the input voltage v_1 is the gain of the amplifier

$$\therefore \quad \text{Gain of the inverting amplifier} \quad \frac{v_0}{v_1} = -\frac{R_f}{R_1} \quad \text{or} \quad v_0 = -v_1 \frac{R_f}{R_1}$$

Denoting v_1 as v_i and R_1 as R_i we have $v_0 = -v_i \dfrac{R_f}{R_i}$

Thus the voltage gain is given by the ratio of feedback resistance R_f to the input resistance R_L. *The negative sign indicates that output voltage is inverted with respect to the input voltage.* Hence the name *'inverting amplifier'*

58.4. Non-inverting amplifier. The circuit of a *non-inverting amplifier* is shown in Fig. 58.3.

In this case the input voltage v_1 is applied to the non-inverting terminal (+). The potential of the point G is also v_1 since the *OP AMP* gain is infinite. The input current to *OP AMP* is negligible. Applying Kirchhoff's law at the point G, we have

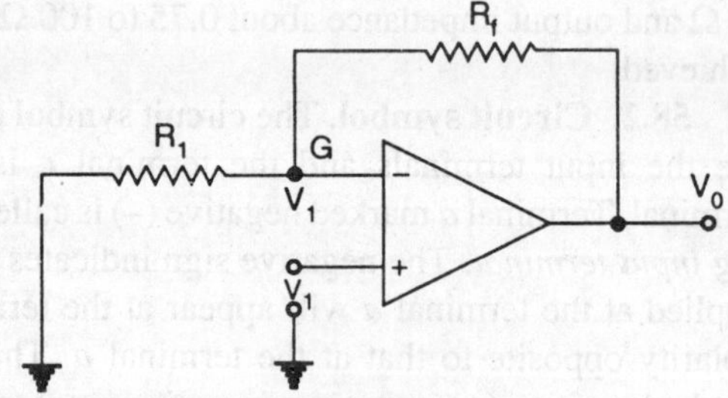

Fig. 58.3

$$\frac{v_1}{R_1} = \frac{v_0 - v_1}{R_f}$$

$$\therefore \quad \text{Gain of the amplifier} \quad \frac{v_0}{v_1} = \frac{R_f + R_1}{R_1} = 1 + \frac{R_f}{R_1}$$

Thus in the case of non-inverting amplifier the gain is 1 + the ratio of the resistance R_f and R_1. Further the *positive* sign shows that the output voltage is in phase with the input voltage *i.e.*, it is a *non-inverting amplifier.*

Denoting v_1 as v_i and R_1 and R_i we have $\frac{v_0}{v_i} = 1 + \frac{R_f}{R_i}$

Experiment 58.1. To study the inverting and non-inverting amplifier.

Apparatus. An operational amplifier having two input terminals (*a*) maked (–), (*b*) marked (+) and one output terminal (*c*), resistances 1 K Ω, 5 K Ω and 10 K Ω, an oscillator, a multimeter etc.

(*a*) **Inverting operational amplifier**

Procedure 1. *If an oscillator is available,* draw a diagram showing the scheme of connections as in Fig. 58.4 and make connections accordingly.

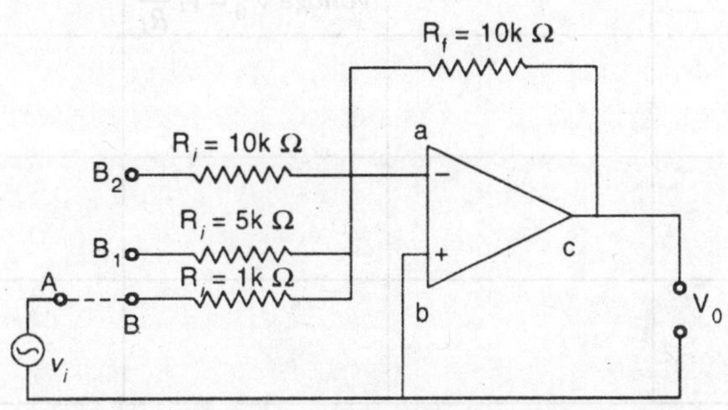

Fig. 58.4

Alternatively, *if an oscillator is not available*, draw a diagram showing the scheme of connections as in Fig. 58.5 and make connections accordingly.

2. If the oscillator is available apply a small suitable signal (.5 V at 500 Hz) from the oscillator.

If the oscillator is not available adjust the value of the supply to 0.5 V with the help of the adjusting knob provided with the supply which may be fitted on the experimental panel board. In this case the *positive* (+) of the supply is connected to the *inverting* terminal (–) of the *OP AMP* through the key *AB* and the negative (–) of the supply to the non-inverting terminal of the *OP AMP*.

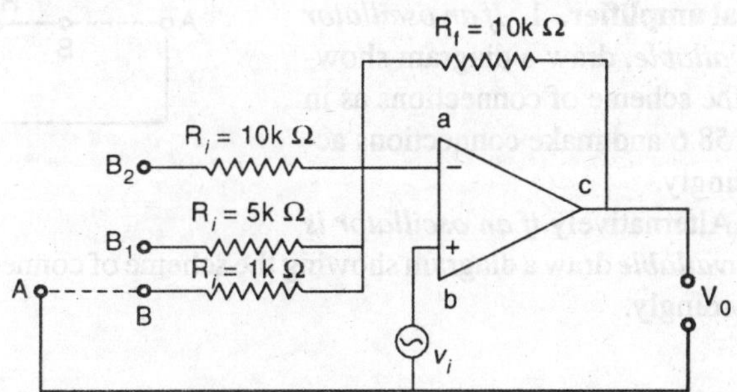

Fig. 58.5

3. Insert the key between *A* and *B* so that R_i = 1 K Ω. Measure the input voltage v_i and output voltage v_0 with the help of (suitable) milli-voltmeter usually fitted on the experimental panel board or by using a suitable electronic multimeter.

4. Take out the key and insert the key between *A* and B_1 such that R_i = 5 K Ω. Again measure the input and the output voltages v_i and v_0.

5. Take out the key and insert between *A* and B_2 such that R_i = 10 K Ω. Again measure v_i and v_0.

6. Take at least *three* observations by changing the input voltage to 0.4 V and 0.6 V and measure corresponding output voltages for R_i = 1 K Ω, 5 K Ω and 10 K Ω as explained above.

Observations Feedback resistance R_f =

S. No.	*Input Resistance* R_i	*Input voltage* v_i	*Output voltage* v_0	*Theoretical output voltage* $v'_0 = v_i \frac{R_f}{R_i}$	*Difference* $v_0 - v'_0$
1	1 K Ω 5 K Ω 10 K Ω	v_i =			
2	1 K Ω 5 K Ω 10 K Ω	v_i =			
3	1 K Ω 5 K Ω 10 K Ω	v_i =			

Result Difference between $v_0 - v'_0$ is small *i.e.* the value of v_0 (observed) is very nearly equal to the value of calculated output voltage v'_0.

(*b*) **Non-inverting operational amplifier.** 1. *If an oscillator is available*, draw a diagram showing the scheme of connections as in Fig. 58.6 and make connections accordingly.

Fig. 58.6

Alternatively *if an oscillator is not available* draw a diagram showing the scheme of connections as in Fig. 58.7 and make connections accordingly.

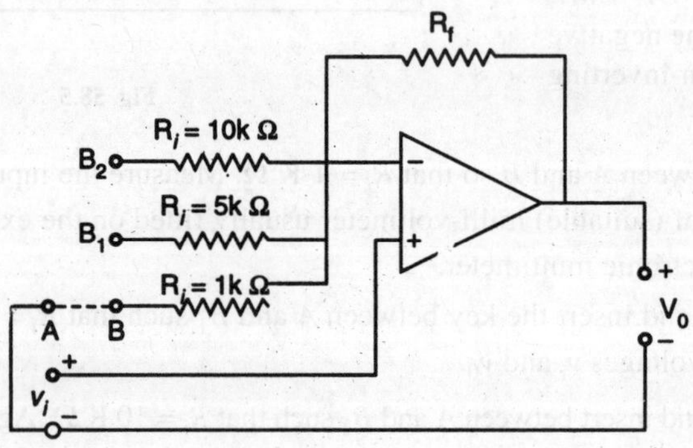

Fig. 58.7

2. If the oscillator is available apply a small suitable signal (0.5 V at 500 Hz) from the oscillator. If the oscillator is not available adjust the value of the supply to 0.5 V with the help of adjusting knob provided with the supply which may be fitted on the experimental panel board. In this case the positive (+) of the supply is connected to the non inverting terminal (+) of the *OP AMP* and the negative to the inverting terminal of the *OP AMP* through the key *AB*.

3. Insert the key between A and B so that $R_i = 1\ K\ \Omega$. Measure the input voltage v_i and the output voltage v_0 with the help of milli-voltmeter (usually fitted on the experimental panel board) or by using suitable electronic multimeter.

4. Take out the key and insert the key between A and B_1 such that $R_i = 5\ K\ \Omega$. Again measure the input and output voltages v_i and v_0.

5. Take out the key and insert between A and B_2 such that $R_i = 10\ K\ \Omega$. Again measure v_i and v_0.

6. Take at least *three* observations by changing the input voltage to 0.4 and 0.6 V and measure the corresponding output voltage for $R_i = 1\ K\ \Omega$, $5\ K\ \Omega$ and $10\ K\ \Omega$ as explained above.

Observations

Feedback Resistance $R_f =$

S. No.	*Input Resistance* R_i	*Input voltage* v_i	*Output voltage* v_0	*Theoretical output voltage* $v'_0 = v_i\left(1 + \frac{R_f}{R_i}\right)$	*Difference* $v_0 - v'_0$
1	1 K Ω 5 K Ω 10 K Ω	$v_i =$			
2	1 K Ω 5 K Ω 10 K Ω	$v_i =$			
3	1 K Ω 5 K Ω 10 K Ω	$v_i =$			

Result Difference $v_0 - v'_0$ is small *i.e.* v_0 is very nearly equal to the calculated value of output voltage v'_0.

Precautions 1. The input voltage v_i must remain constant during the period of observation.

2. Suitable electronic voltmeter [sensitive and of proper range) should be used in both cases *i.e.* when the experiment is performed using an oscillator or when the experimental is performed using a suitable D.C. voltage supply.

3. All the resistances used such as R_i and R_f must have accurate value *i.e.* their tolerance should be small.

Oral Questions

1. **Operational amplifiers.** (*i*) What is an operational amplifier? (*ii*) What are the main uses of an operational amplifier? (*iii*) How many input terminals does an *OP AMP* have? (*iv*) What is the function of non inverting terminal? (*v*) What is the function of inverting terminal? (*vi*) How many output terminals are there? (*vii*) What is the gain of an inverting *OP AMP*? (*viii*) What is the gain of a non-inverting *OP AMP*?

B. Sc. Practical Physics
Additional Experiments

HEAT AND STATISTICAL PHYSICS

59

Thermal Expansion of Crystals

(Interference method)

59.1. Thermal Expansion of a Crystal. Crystals are generally available in small dimensions. Their total thermal expansion is, therefore, quite small. To study the thermal expansion of crystals, therefore, optical interference methods like that of Newton's rings are generally employed. The method is capable of great accuracy as it can measure the expansion of the order of 10^{-6} cm.

59.2. Experimental arrangement. To study the thermal expansion of a crystal available in the form of a thin plate, the experimental arrangement making use of Newton's fringes is shown in Fig. 59.1. *FF* is a metallic frame carrying three levelling screws (only two shown in the Figure) and an optically plane plate *PP* with its *optically plane upper* surface. A plano convex lens *L* of large focal length (about 200 cm) is placed as shown with its curved surface downward and the levelling screws are adjusted in such a way that a thin air film is formed between the lower curved surface of *L* and upper face of *C*. *S* is a broad source of monochromatic light (say a sodium lamp). Light from this source rendered parallel by the lens *L'* is reflected vertically downward by the glass plate *G*. The light beams reflected from the lower curved surface of *L* and upper optically plane surface of *C* produce interference thereby giving rise to Newton's fringes. A long focus microscope *M* is focussed on the upper surface of crystal *C* to view the Newton's rings pattern. To obtain a good system of Newton's rings, suitably adjust the position of *L*, *G* and *M*.

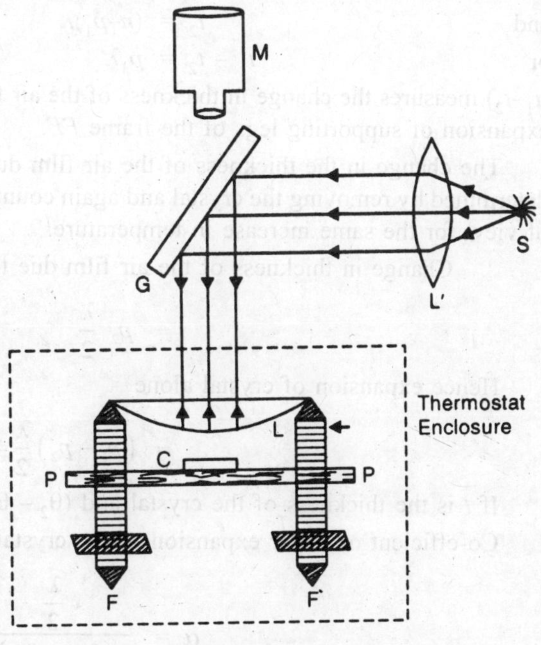

Fig. 59.1

The system carrying the crystal C and lens L can be placed in a thermostat which can be maintained at constant temperature at different suitable values.

59.3. Theory. According to the theory of formation of Newton's fringes viewed normally by reflected light any point in the field of view appears dark or bright according as the path difference

$$2\mu t = n\lambda \qquad \text{(For dark ring)}$$

or

$$2\mu t = (2n+1)\frac{\lambda}{2} \qquad \text{(For bright ring)}$$

where t is the thickness of air film between the crystal and the lens L. Taking $\mu = 1$ (for air), we have

$$2t = n\lambda \qquad \text{(For dark ring)} \qquad ...(i)$$

and

$$2t = (2n+1)\frac{\lambda}{2} \qquad \text{(For bright ring)} \qquad ...(ii)$$

If the thickness of the air film between the crystal and lens L is made to change due to thermal expansion of the crystal, the interference bands will shift at the rate of one band width (*i.e.*, distance between two successive dark or bright bands) for every change of thickness of air film by half a wavelength $\left(\frac{\lambda}{2}\right)$ of light used.

When the temperature of the system is slowly increased from θ_1 to θ_2 with the help of the thermostat arrangement, let p_1 be the number of bands that cross the field view with reference to the cross wire till the temperature becomes constant and no more fringes move across the field of view. If t_1 and t_2 are the thicknesses of air film at the initial and final temperatures, then according to *Eq* (*i*) for dark fringes

$$t_1 = n\lambda$$

and

$$t_2 = (n-p_1)\lambda$$

or

$$t_1 - t_2 = p_1\lambda$$

(t_1-t_2) measures the change in thickness of the air film due to the expansion of the crystal and the expansion of supporting legs of the frame FF.

The change in the thickness of the air film due to the expansion of supporting legs alone is determined by removing the crystal and again counting the number of fringes p_2 that cross the field of view for the same increase in temperature.

$\therefore$ Change in thickness of the air film due to expansion of supporting legs alone

$$= p_2\frac{\lambda}{2}$$

Hence expansion of crystal alone

$$= (p_1 - p_2)\frac{\lambda}{2} = x\frac{\lambda}{2}$$

If l is the thickness of the crystal and $(\theta_2 - \theta_1)$°C is the rise in temperature, then

Co-efficient of linear expansion of the crystal

$$\alpha = \frac{x\frac{\lambda}{2}}{l(\theta_2 - \theta_1)} = \frac{x\lambda}{2l(\theta_2 - \theta_1)} \text{ per}°C$$

Experiment 59.1. To study the thermal expansion of a crystal by optical interference method and to find the value of co-efficient of linear expansion of the material of a crystal in the form of a thin slice of uniform thickness.

Apparatus. A crystal (in the form of a thin slice), a high powered long focus travelling microscope, two optically plane transparent glass plates, a double convex lens (short focus), a plano convex lens-long focus (*F. L* 200 cms), a metallic frame having three levelling screws attached to a polished plate over which can be placed a plane glass plate and the crystal. A thermostatic enclosure with temperature recording device and a screw gauge.

Procedure. 1. Set up the apparatus for obtaining Newton's rings as explained in Art. 59.2 under the head '**Experimental arrangement**' so that a clear and sharp pattern of Newton's rings is oblained. Place the crystal slice *C* on the plate *PP* and again obtain Newton's fringes. Enclose the metallic frame with plano convex lens *L*, crystal slice *C* within the thermostatic enclosure, the temperature of which can be kept constant and accurately recorded.

2. Note the initial temperature θ_1°*C* which may be taken to be the room temperature.

3. Now, slowly raise the temperature and go on counting the number of rings that shift. As the temperature rises, interference fringes will move past the cross wire. When the final constant temperature θ_2°*C* is reached the interference rings will stop shifting. Count the total number of rings that shift till all is steady. Let it be p_1.

4. Record the final constant temperature θ_2°*C*.

5. Now remove the crystal *C*. Let the system cool down to the initial temperature θ_1°*C*. Again obtain Newton's rings with the lens and plate *PP* only. Repeat the above experiment and count the number of fringes that shift for the same rise of temperature $(\theta_2 - \theta_1)$°*C*. Let it be p_2.

6. Record the final constant temperature θ_2°*C* again.

7. Measure the thickness *l* of the crystal using a micrometer screw gauge.

Observations and Record.

Initial temperature θ_1°*C* =

Final temperature (constant temperature at which the thermostat is maintained)

(*i*) when the crystal *C* is in position =

(*ii*) when the crystal *C* is removed =

Mean final temperature θ_2°*C* =

Wavelength of sodium light = 5893 *A*° = 5893×10^{-8} cm

Number of fringes shifting with the crystal,

p_1 =

Number of fringes shifting without the crystal,

p_2 =

∴ Expansion of the crystal $= (p_1 - p_2)\frac{\lambda}{2} = x\frac{\lambda}{2}$

Least count of the screw gauge = mm = cm

Thickness of the crystal *l* = mm = cm.

∴ Co-efficient of linear expansion of the crystal

$$\alpha = \frac{x\lambda}{2l(\theta_2 - \theta_1)} = \quad \text{per°}C.$$

Precautions. 1. Counting of fringes should be stopped only when the fringes do not shift further and a steady state has been reached.

2. Correction for expansion of levelling screws of the metallic frame must be applied.

3. The crystal should be in the form of a thin slice of uniform thickness.

4. The final temperature must be kept constant with the help of the thermostat arrangement for a sufficiently long time.

5. Observe all precautions given in **Experiment 17.1** *'To find the wavelength of sodium light by measuring the diameter of Newton's rings'*.

Oral Questions

(*i*) What is co-efficient of linear expansion? (*ii*) Does every crystal have the same co-efficient of linear expansion in all the three mutually perpendicular directions? (*iii*) What is an isotropic crystal? (*iv*) What is that crystal called which has different values of co-efficient of linear expansion in different directions of co-ordinate axes? (*v*) What is the value of minimum expansion that can be measured with Newton's rings arrangement?

NOTE. Also see **'Oral Questions'** at the end of **Chapter 17** under the head **'Newton's Rings'.**

60

Adiabatic Expansion of a Gas

60.1. Two specific heats of a gas. When a gas is heated, ordinarily there is an increase in volume as well as pressure in addition to the rise of temperautre. For the sake of simplicity, either the volume or the pressure may be kept constant. Therefore a gas has two specific heats.

(*i*) *Specific heat at constant volume,* and

(*ii*) *Specific heat at constant pressure.*

The specific heat of a gas at constant volume is the amount of heat required to raise the temperature of a unit mass of the gas through 1°C keeping the volume constant.

If dQ is the amount of heat required to raise the temperature of a unit mass of the gas through dT°C, keeping the volume constant, then

$$\text{Specific heat at constant volume} \quad c_v = \left(\frac{dQ}{dT}\right)_v$$

The specific heat of a gas at constant pressure is the amount of heat required to raise the temperature of a unit mass of the gas through 1°C, keeping the pressure constant.

If dQ is the amount of heat required to raise the temperature of a unit mass of the gas through dT°C keeping the pressure constant, then

$$\text{Specific heat at constant pressure} \; c_p = \left(\frac{dQ}{dT}\right)_p$$

60.2. c_p is > c_v. In the first case, the heat supplied is used to increase the speed and the kinetic energy of the molecules. This results in a rise of temperature and an increase in the pressure of the gas. But as the volume is kept constant the gas does no external work.

In the second case, since the pressure is kept constant, therefore, the volume increases. The heat supplied now does two things :

(*i*) it raises the temperature of the gas, and

(*ii*) it does work in expanding the gas against the external pressure.

The amount of heat required to raise the temperature of the gas in this case is the same as in the first case when the volume is kept constant but in addition heat energy has to be supplied to do external work. *Hence the specific heat at constant pressure is greater than the specific heat at constant volume, by an amount which is thermal equivalent of the work done in expanding the gas against external pressure.*

60.3. Ratio of two specific heats. The ratio of specific heat of a gas at constant pressure to its specific heat at constant volume is denoted by γ, *i.e.*,

$$\gamma = \frac{c_p}{c_v}$$

The value of γ depends upon the atomicity of a gas. For a monoatomic gas $\gamma = 1.66$. For a diatomic gas (including air) $\gamma = 1.4$. For a triatomic gas with six degrees of freedom $\gamma = 1.33$ and with seven degrees of freedom $\gamma = 1.29$.

60.4. Isothermal change. *When a change in pressure and volume of a gas takes place but the temperature remains constant, the change is said to be isothermal.*

For an isothermal change $\quad PV = \text{constant}$

or $\quad P_1V_1 = P_2V_2$

Adiabatic change. *When a change in pressure and volume of a gas takes place but no heat is allowed to enter or leave it, the change is said to be adiabatic.*

For an adiabatic change $\quad PV^\gamma = \text{constant}$

or $\quad P_1V_1^\gamma = P_2V_2^\gamma$

This relation is in terms of P and V. In terms T and V, the same relation becomes

$$TV^{\gamma-1} = \text{constant}$$

or $\quad T_1V_1^{\gamma-1} = T_2V_2^{\gamma-1}$

Again in terms of T and P the relation is

$$T^\gamma P^{1-\gamma} = \text{constant}$$

or $\quad T_1^\gamma P_1^{1-\gamma} = T_2^\gamma P_2^{1-\gamma}$

60.5. Adiabatic expansion of a gas. To study the adiabatic expansion of a gas, we shall describe Clement and Desorme's experiment and find the value of γ– the ratio of c_p, the specific heat at constant pressure to c_v, the specific heat at constant volume when a gas (say air) undergoes an adiabatic expansion.

Clement and Desorme's experiment. Clement and Desorme's apparatus consists of a large, thick-walled glass vessel A having a capacity of 20-30 litres fitted in a thick-walled wooden box packed with cotton wool or some other similar insulating material to completely insulate it. Three tubes are connected at the top end of the vessel through a brass cap fitted around the neck of the bottle as shown in Fig. 60.1. One tube is connected to a manometer M filled with low density, non hygroscopic oil, having low vapour pressure at room temperature or sulphuric acid, the second to a stop cock S which can be suddenly opened and closed and the third to a stop cock S_1. Dry air can be compressed into the vessel A though S_1, or S_1 can be connected to a reservior containing any gas like Oxygen, Nitrogen etc. To keep the gas within the flask completely dry, a small crucible containing some strong sulphuric acid is placed at the bottom of the flask (not shown in the Fig.).

Working. The stop cock S is closed and S_1 is opened. Dry air is compressed into the vessel A through S_1. The stop cock S_1 is closed when the pressure of air inside the vessel is slightly greater than the atmospheric pressure. The apparatus is given some rest so that the air inside the flask acquires the temperature of the surrounding.

If H is the difference in level of the two limbs of the manometer when steady state is reached and P_0 the atmospheric pressure (in terms of height of mercury column), then

Pressure of air inside the vessel $P_1 = P_0 + H\frac{\rho}{d}$

where $\quad \rho$ = density of oil or liquid in the manometer at room temperature.

d = density of mercury

The stop cock S is now suddenly opened and closed just at the moment when the level of the liquid on the two sides of the manometer is the same. In this process, the air inside the vessel expands adiabatically as the flask is well insulated so that no heat is either gained or lost. Some quantity of air escapes to the atmosphere and the temperature of air inside the vessel falls due to adiabatic expansion. The air inside the vessel is allowed to gain heat from the surroundings and it finally attains the temperature of the surroundings. The pressure of the air inside the vessel increases to P_2 and the difference in levels of the two sides of the manometer is h, so that

$$P_2 = P_0 + \frac{h\rho}{d}$$

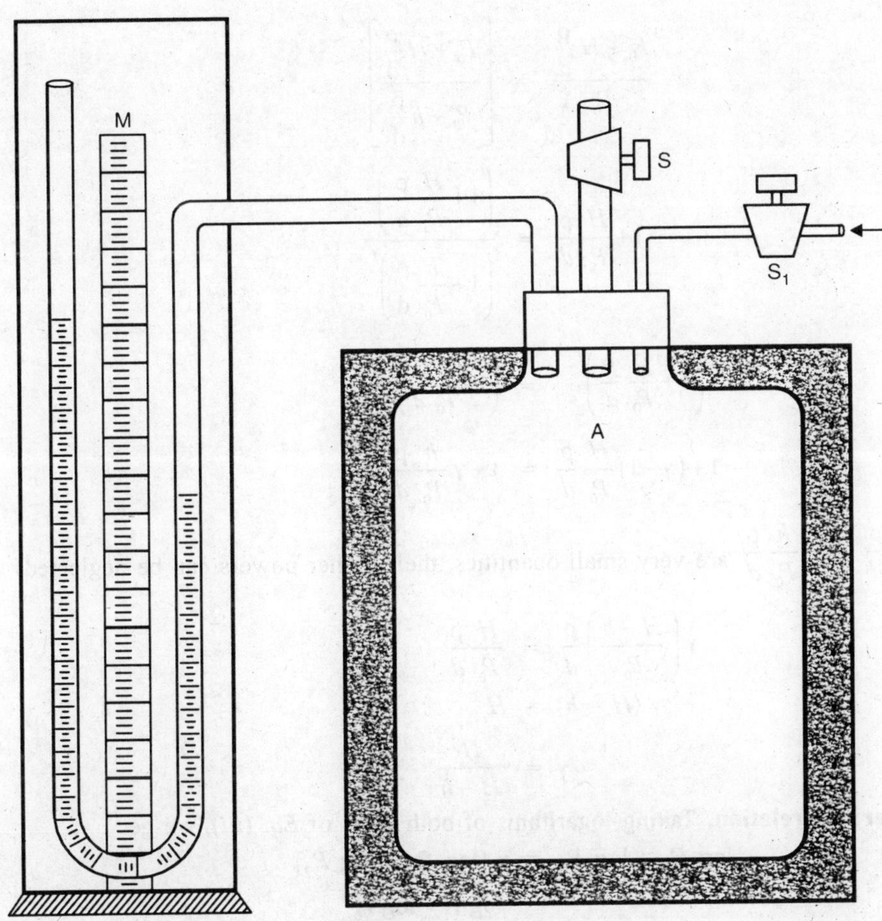

Fig. 60.1

60.6. Theory. When the stop cock S is suddenly opened and closed, the mass of air left in the vessel at the end has expanded adiabatically from a volume V_1 (less than the volume of the vessel) at a pressure P_1 to a volume V_2 (equal to the volume of the vessel) at the atmospheric pressure P_0

$$\therefore \quad P_1 V_1^{\gamma} = P_0 V_2^{\gamma}$$

or
$$\frac{P_1}{P_0} = \left(\frac{V_2}{V_1}\right)^{\gamma} \quad ...(i)$$

When the air finally attains the room temperature, its volume remains the same V_2 but pressure changes to P_2. Thus the change from a volume V_1 at a pressure P_1 to a volume V_2 at a pressure P_2 takes place at the *same* temperature (room temperature) and is, therefore, an isothermal change

$$\therefore \quad P_1 V_1 = P_2 V_2$$

or
$$\frac{V_2}{V_1} = \frac{P_1}{P_2} \qquad ...(ii)$$

From (*i*) and (*ii*), we get

$$\frac{P_1}{P_0} = \left(\frac{P_1}{P_2}\right)^{\gamma} \qquad ...(iii)$$

or
$$\frac{P_0 + H\dfrac{\rho}{d}}{P_0} = \left[\frac{P_0 + H\dfrac{\rho}{d}}{P_0 + h\dfrac{\rho}{d}}\right]^{\gamma}$$

or
$$1+\frac{H}{P_0}\frac{\rho}{d} = \frac{\left(1+\dfrac{H}{P_0}\dfrac{\rho}{d}\right)^{\gamma}}{\left(1+\dfrac{h}{P_0}\dfrac{\rho}{d}\right)^{\gamma}}$$

or
$$\left(1+\frac{H}{P_0}\frac{P}{d}\right)^{\gamma-1} = \left(1+\frac{h}{P_0}\frac{\rho}{d}\right)^{\gamma}$$

or
$$1+(\gamma-1)\frac{H}{P_0}\frac{\rho}{d} = 1+\gamma\frac{h}{P_0}\frac{\rho}{d}$$

(as $\dfrac{H}{P_0}\dfrac{\rho}{d}$ and $\dfrac{h}{P_0}\dfrac{\rho}{d}$ are very small quantities, their higher powers can be neglected)

$$\therefore \quad \gamma\left(\frac{H-h}{P_0}\right)\frac{\rho}{d} = \frac{H}{P_0}\frac{\rho}{d}$$

or
$$\gamma\,(H-h) = H$$

or
$$\gamma = \frac{H}{H-h} \qquad ...(iv)$$

Alternate relation. Taking logarithms of both sides of *Eq.* (*iii*), we get

$$\log P_1 - \log P_0 = \gamma\,[\log P_1 - \log P_2]$$

or
$$\gamma = \frac{\log P_1 - \log P_0}{\log P_1 - \log P_2} \qquad ...(v)$$

Experiment. 60.1. To study the adiabatic expansion of a gas and hence to find the value of γ – the ratio of specific heat at constant pressure to specific heat at constant volume for air using Clement and Desorme's apparatus.

Apparatus. Clement and Desorme's apparatus, a compression pump, a thermometer, plumb line and telescope.

Procedure. 1. Set the Clement and Desorme's apparatus on the table at a convenient place. With the help of a plumb line set the manometer to be vertical. Focus the telescope on the manometer scale so that the liquid level in the two limbs is clearly visible and the horizontal cross wire is parallel to the graduations marked on it.

2. Close the stop cock S. Open the stop cock S_1. Connect this end of the apparatus to an air compression pump and with its help fill the vessel A with *dry* air at a pressure higher than the atmospheric pressure so that the liquid levels in the two limbs of the manometer stand at a difference of about 10-15 cms. Close the stop cock S_1.

3. Wait for some time so that the air inside the vessel comes to the steady temperature of the surroundings. This will happen when the liquid level in the two limbs of the manometer remains constant. When the steady state is reached, note the level of the liquid in each of the two limbs of the monometer M and find the difference. Let it be H.

4. Suddenly open the stop cock S and close it just when the height of the liquid in the two limbs of the manometer is at the same level.

5. Wait for some time again (say about 30 minutes) to allow the air inside the vessel (which had cooled due to adiabatic expansion) to come to the temperature of the surroundings. At this stage the liquid in the manometer will stop rising further and a steady state will be reached. Again note the level of the liquid in each of the two limbs of the manomeeter and find the difference. Let it be h.

6. Note the room temperature.

7. Repeat the experiment a number of times.

8. Note the atmospheric pressure from Fortin's barometer (For alternative method)

Observations and Record. Room temperature $t =$ °C

S.No.	Reading of manometer						
	When air is compressed			When air is made to expand adiabatically			$\gamma = \frac{H}{H-h}$
	Left limb	Right limb	Diff. H	Left limb	Right limb	Diff. h	
1.							
2.							
3.							

Mean value of $\gamma =$

Standard value of γ for air at t °C = [From constant tables]

Alternate method. Atmospheric pressure (from Fortin's barometer) = P_0 = cm of Hg.

Density of liquid ρ = gm/cc (From constant tables)

Density of mercury d = gm/cc (From constant tables)

S.No.	Diff. H (From last table)	$P_1 = P_0 + H\frac{\rho}{d}$	Diff. h (From last table)	$P_2 = P_0 + h\frac{\rho}{d}$	$\gamma = \frac{\log P_1 - \log P_0}{\log P_1 - \log P_2}$
1.					
2.					
3.					

Mean $\gamma =$

Precautions. 1. All the stop cocks should be perfectly air tight.

2. The oil used in the manometer should be non-hygroscopic and should have low density and low vapour pressure.

3. To keep the air in the vessel dry some strong sulphuric acid should be placed in it.

4. The difference in levels of the two limbs of the manometer when the air is compressed into the vessel should be about 10-15 cm.

5. The levels of the liquid in the two limbs of the manometer should be noted when a steady state has been reached.

Sources of error. When the stop cock is opened, a series of oscillations are set up. This is shown by the up and down movement of the liquid in the manometer. Therefore, the exact moment when the stop cock should be closed is not known. Thus the pressure in the vessel may not be equal to the almospheric pressure when the stop cock is closed. For this reason the tube carrying the stop cock *S′* is made wide enough to ensure sudden expansion and produce proper adiabatic cooling. This further helps to avoid oscillations which generally occur with a narrow tube.

Oral questions

1. What are the two specific heats of a gas? 2. Define specific heat at constant pressure and specific heat at constant volume. 3. Which is greater C_p or C_v and why? 4. What is an isothermal change? 5. What is an adiabatic change? 6. State the relation between (*a*) pressure and volume (*b*) temperature and volume and (*c*) temperature and pressure for a gas when it undergoes an adiabatic change. 7. What is γ? 8. What is its value for a diatomic gas? 9. Can the value of γ be less than one? 10. What is the ratio of adiabatic elasticity to isothermal elasticity of a gas.

61

Total Radiation Law

(Temperature dependence of radiation)

61.1. Stefan's law. The rate of emission of heat energy by a hot body depends upon its temperature and is given by Stefan's law. According to this law *'The rate of emission of radiant energy by a unit area of a perfectly black body is directly proportional to the fourth power of its absolute temperature' i.e.*

$$R \propto T^4 \quad \text{or} \quad R = \sigma T^4$$

where σ is a constant called *Stefan's constant.*

If a perfectly black body at an absolute temperature T_1 is surrounded by another black body at an absolute temperature T_2, then the net rate of loss (or gain) of heat energy per unit area of the surface is given by

$$R \propto \left(T_1^4 - T_2^4\right)$$

or

$$R = \sigma \left(T_1^4 - T_2^4\right)$$

The value of Stefan's constant $\sigma = 5.76 \times 10^{-8}\ J\ m^{-2}\ \text{sec}^{-1}\ °K^{-4}$

61.2. Determination of Stefan's constant. The apparatus consists of a hollow metallic sphere *H* blackened from inside and placed inside a rectangular box *S′* which serves as a steam chamber. The whole arrangement rests on a wooden platform *AB* which has a small hole 2-3 cm in diameter exactly at the centre of the hemisphere.

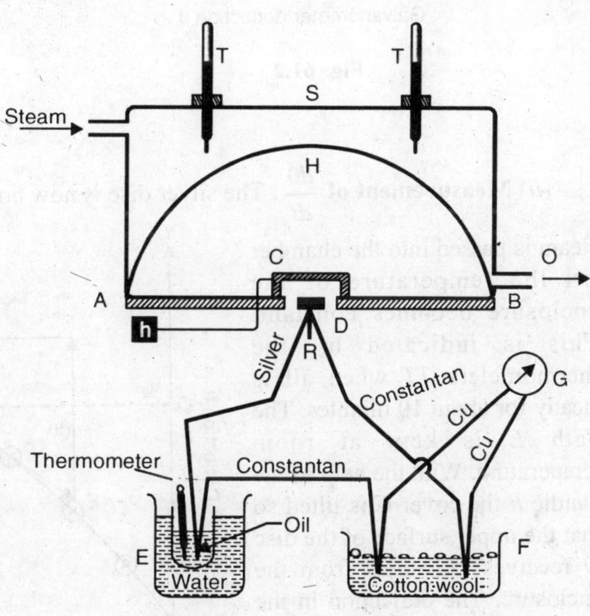

Fig. 61.1

A small silver disc *D* whose upper surface is coated with lamp black can be put in position into the hole with the help of an ebonite rod *R* fixed to it. The silver disc *D* can be covered and uncovered from the radiations of the enclosure by an ebonite cover *C*. This can be arranged from outside with the help of a handle *h*.

One junction of a silver constantan thermocouple is soldered to the lower surface of the disc while the other junction is kept in a test tube containing oil kept in a water bath *E*. A thermometer is also placed in the test tube to note the temperature. The two constantan wires are further soldered to two copper wires which lead to sensitive galvanometer *G*. To prevent any disturbing effect due to the difference of temperature in the leads, the wires leading to the galvanometer are packed in cotton wool in a box *F*. A thermometer is also placed in the box (not shown).

The experiment is performed in two steps:

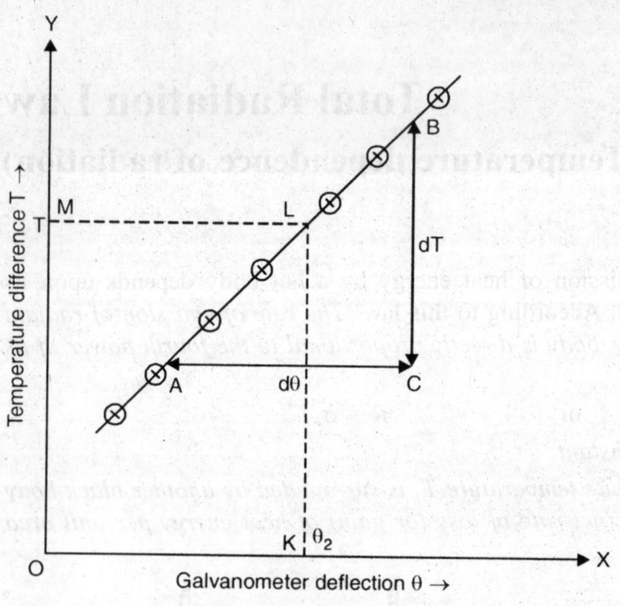

Fig. 61.2

(*i*) **Calibration of the thermocouple.** Before passing steam in the chamber, the silver disc is placed in central hole and covered by the ebonite cover *C*. Room temperature of the water bath is noted. The water bath is now heated and the deflection in the galvanometer due to thermoelectic current as well as the temperature of the bath are simultaneously noted at regular intervals.

A graph is plotted between galvanometer deflection θ taken along the *X*-axis and temperature difference *T* (Difference between the temperature of the water bath and room temperature) taken along the *Y*-axis. The graph is a straight line as shown in Fig. 61.2. From the graph

$$\frac{dT}{d\theta} = \frac{BC}{AC} = \text{slope of the curve.}$$

(*ii*) **Measurement of $\frac{d\theta}{dt}$.** The silver disc is now completely covered with ebonite cover *C* and steam is passed into the chamber till the temperature of the enclosure becomes constant. This is indicated by the thermometers *TT* when all is steady for about 10 minutes. The bath *E* is kept at room temperature. With the help of the handle *h* the cover *C* is tilted so that the upper surface of the disc *D* receives radiations from the enclosure. The deflection in the galvanometer is observed at regular intervals of time (say 10 sec.). A graph is plotted between

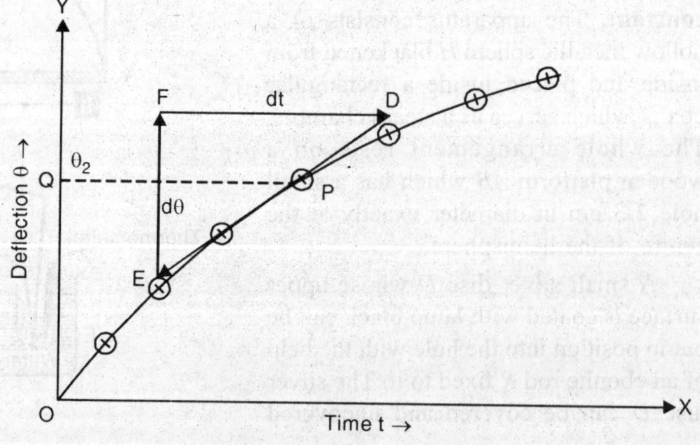

Fig. 61.3

time t taken along the X-axis and galvanometer deflection θ taken along the Y-axis as shown in Fig. 61.3.

If we draw a tangent DE to the curve at the point P, then

$$\frac{d\theta}{dt} = \frac{EF}{DF}$$

$$\therefore \quad \frac{dT}{dt} = \frac{dT}{d\theta} \cdot \frac{d\theta}{dt} = \frac{BC}{AC} \times \frac{EF}{DF}$$

To find the temperature of the disc corresponding to the point P, the deflection θ_2 is found out by drawing PQ parallel to X-axis as shown in Fg. 61.3. Again corresponding to this deflection θ_2, the temperature T_2 is found as shown in Fig. 61.2.

61.3. Theory. When steam is passed into the steam chamber S and the thermometers show a constant temperature, the inner surface of the hemisphere H acts as a black body radiator at the constant temperature of steam say T_1 K.

When the disc D is placed in position and its ebonite cover C is removed, it absorbs heat radiation emitted by the black body so that its temperature increases continuously. Let $T_2 K$ be the temperature of the disc at any instant. With the rise of temperature of the disc, it radiates heat in turn but its rate of absorption will be greater than its rate of emission. If R_1 and R_2 are the amounts of heat radiation absorbed and emitted by the disc respectively per unit area per second (in units of work), then according to Stefan's law

$$R_1 = \sigma T_1^4$$

and

$$R_2 = \sigma T_2^4$$

where σ is Stefan's constant.

If A is the area of the disc, then

Heat gained by the disc per second

$$= \frac{(R_1 - R_2)A}{J} \text{ cals}$$

If m is the mass of the disc, s the specific heat of silver and $\frac{dT}{dt}$ the rate of change of temperature, then

$$ms\frac{dT}{dt} = \frac{(R_1 - R_2)A}{J} = \frac{\sigma\left(T_1^4 - T_2^4\right)A}{J}$$

or

$$\sigma = \frac{Jms}{\left(T_1^4 - T_2^4\right)A} \frac{dT}{dt}$$

Experiment 61.1. To find the value of Stefan's constant.

Apparatus. Stefan's constant apparatus, a silver disc about 2.5 cms in diameter, steam generator, water bath with a test tube containing oil, two half degree thermometers, a $\frac{1}{10}$ degree thermometer, a stop clock, vernier callipers, balance and weight box.

Procedure. 1. Set up the apparatus for finding the value of Stefan's constant as explained under **'Determination of Stefan's constant'.**

2. Calibrate the thermocouple as explained under the head **'Calibration of thermocouple'**

3. Taking the galvanometer deflection θ along the X-axis and temperature difference T between the two junctions of silver constantan thermocouple along the Y-axis, draw a graph between θ and T (Fig. 61.2). The graph is a straight line. Take two points A and B on the graph. Draw AC parallel to X-axis and BC perpendicular to it, then $BC = dT$ and $AC = d\theta$. Find the value of $\frac{dT}{d\theta}$ from the graph

$$\frac{dT}{d\theta} = \frac{BC}{AC}.$$

4. Now cover the silver disc D completely with the ebonite cover C with the help of the handle h. Pass steam through the steam chamber till the temperature of the enclosure becomes constant. This will take about ½ hour and will be indicated by the temperature recorded by the two thermometers TT remaining constant for about 10 minutes. Replace the water bath E with fresh water at room temperature. Keep the stop clock and galvanometer close to each other so that readings can be taken on both simultaneously. Remove the ebonite cover C from above the silver disc. Immediately start the stop clock and go on recording the deflection of the galvanometer θ and corresponding time t at intervals of 10-15 seconds, till the deflection becomes constant.

5. Taking time t along the X-axis and galvanometer deflection θ along the Y-axis, plot a graph between θ and t. The graph is shown in Fig. 61.3. Take a point P on the graph and draw a tangent DPE to the graph at the point P. Draw DF parallel to X-axis and EF perpendicular to it, then

$$\frac{d\theta}{dt} = \frac{EF}{FD}$$

6. From P draw PQ parallel to X-axis meeting the Y-axis at Q. From the graph note the galvanometer deflection θ_2 corresponding to the point Q.

7. Now locate the point K corrosponding to the value of deflection θ_2 on the graph of Fig. 61.2 and draw KL parallel to Y-axis meeting the graph at L. From L draw LM parallel to X-axis, then the temperature T corrosponding to the point M gives the temperature difference. Thus the temperature of the disc corresponding to deflection θ_2 *i.e.* $T_2 = T +$ room temperature.

8. Weigh the silver disc to find its mass m.

9. Measure the diameter of the disc with a vernier callipers in two directions at right angles.

Observations and Record. Mass of silver disc $m =$ gm. Specific heat of silver $s = 0.056$

Area of silver disc

Sl. No.	*Diameter*						*Average diameter*	*Radius r*	*Area* $= \pi r^2$
	1st direction			*2nd direction*					
	Main scale	*Vernier scale*	*Total*	*Main scale*	*Vernier scale*	*Total*			
1.									
2.									
3.									

Mean area of the disc $A =$ cm^2

Room temperature = °C

Variation of thermo *e.m.f* with temperature

Sl. No.	*Temp. of hot junction °C*	*Temp. of cold junction °C*	*Diff. of temp.°C*	*Diff. of temp. °K*	*Galv. Deflection θ*
1.					
2.					
3.					
4.					
5.					
6.					
7.					
8.					
9.					

NOTE: The difference of temperature in °*K* = difference in temperature in °*C*

$$\frac{dT}{d\theta} \text{ from the graph } = \frac{BC}{AC} =$$

Variation of galvanometer deflection with time

Sl. No.	*Time t*	*galvanometer Deflection θ*	*Sl. No.*	*Time t*	*galvanometer Deflection θ*
1.			6.		
2.			7.		
3.			8.		
4.			9.		
5.			10.		

$$\frac{d\theta}{dt} \text{ from the graph } = \frac{EF}{DF} =$$

$$\therefore \quad \frac{dT}{dt} = \frac{BC}{AC} \times \frac{EF}{DF} =$$

Temperature of steam T_1 = °*C* = °*K*

Deflection θ_2 corresponding to point *P* (from *t*, θ graph) =

Temperature *T* corresponding to deflection θ_2 (from *T*, θ graph)

Temperature T_2 = *T* + Room temperature = °*C* = °*K*

Stefan's constant $\sigma = \frac{Jms}{\left(T_1^4 - T_2^4\right)A} \frac{dT}{dt}$ $Jm^{-2}\ \sec^{-1}\ K^{-4}$

Precautions. 1. The thermocouple wires leading to the galvanometer must be packed in cotton wool to prevent any disturbing effect.

2. The galvanometer must be sensitive enough to give deflection corrosponding to small thermo *e.m.f's* and must have a low resistance.

3. While calibrating the thermocouple, the silver disc must remain shielded with ebonite cover.

4. One junction of the silver constantan thermocouple must be soldered to the silver disc and the other immersed in oil kept in a test tube in the water bath. The junction should not be directly placed in water.

5. The silver disc should be uncovered only when the temperature of steam chamber remains constant for some time.

6. The reading of stop clock and galvanometer must be started as soon as the silver disc is uncovered and time and deflection must be simultaneously recorded at regular intervals.

7. All temperatures involved in the calculation (T_1 and T_2) must be taken in degrees Kelvin.

Oral Questions

(*i*) State Stefan's law. (*ii*) What is the value of Stefan's constant in *S.I* units? (*iii*) What is a black body? (*iv*) Define emissive power and absorptive power of a body. (*v*) On what factors does the emissive power of a black body depend? (*vi*) State Wien's displacement law. (*vii*) As the temperature of a body is raised, does the wave length corresponding to maximum intensity of radiation shift to longer or shorter wave length side? (*viii*) What is solar constant? (*ix*) What is the value of solar constant? (*x*) Which statistics applies to a photon gas?

OPTICS AND LASER

62

Principal Points of a Lens System

62.1. Cardinal points. There are six cardinal points of a lens system. These are (*i*) *Two focal points* (*ii*) *Two principal (or unit points)* and (*iii*) *Two nodal points.*

Two focal points. An optical lens system is shown in Fig. 62.1. $X'X$ is the optical or the principal axis of the system. An incident ray $O_1 A_1$ parallel to the principal axis from the top of an object OO_1 after refraction through the lens system travels along $A_1 B_1$ and then emerges along $B_1 I_1$, passing through the principal axis at the point F_2. Similarly a ray of light $O_1 A_2$ passing through the principal axis at the point F_1 after refraction through the lens system travels along $A_2 B_2$ and finally emerges along $B_2 I_1$, parallel to the principal axis. II_1, is the real image of OO_1.

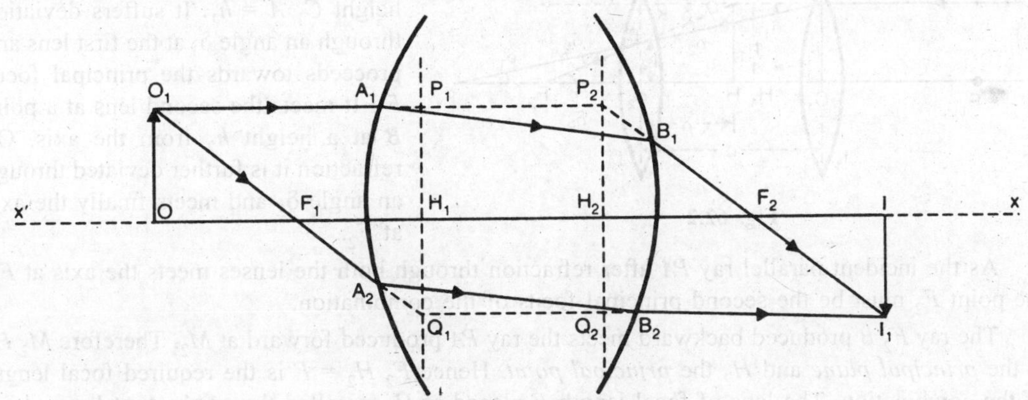

Fig. 62.1

Hence the **first focal point** *is a point on the principal axis of a lens system such that a ray of light passing through it is rendered parallel to the principal axis after refraction through the lens system.* Thus F_1 is the first focal point.

The **second focal point** *is a point on the principal axis of a lens system such that a ray of light parallel to the principal axis after refraction through the lens system passes through it.* Thus F_2 is the second focal point.

Focal planes. *The planes perpendicular to the principal axis and passing through F_1 and F_2 are called the first and second focal planes respectively.*

Two principal points. The incident ray O_1A_1 and the emergent ray B_1I_1 when produced meet at P_2. *A plane P_2Q_2 passing through P_2 and perpendicular to the optical or principal axis is called the second principal plane.*

The point H_2 where the second principal plane intersects the optical or principal axis is called the **second principal point.**

Similarly the incident ray O_1A_2 and the emergent ray B_2I_1 when produced meet at Q_1. *A plane P_1Q_1 passing through Q_1 and perpendicular to the optical or principal axis is called the first principal plane.*

The point H_1 where the first principal plane intersects the optical or principal axis is called the **first principal point.**

The principal points give the two locations on the principal axis at which a single lens optically equivalent to the lens system may be placed.

Focal length. *The distance H_2F_2 between the second principal point and the second focal point is called the* **second focal length** *of the lens system.*

The distance H_1F_1 between the first principal point and the first focal point is called the **first focal length** *of the lens system.*

Two nodal points. *The nodal points are a pair of points in an optical system such that a ray of light directed towards one of these points after refraction through the optical system emerges out from the second nodal point in a direction parallel to the incident ray.*

62.2. Equivalent focal length of two thin lenses. Consider two thin lenses L_1 and L_2 of focal lengths f_1 and f_2 placed on a common axis separated by a distance d.

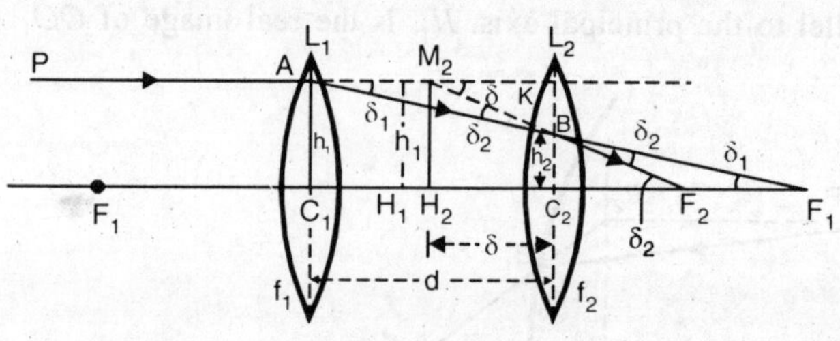

Fig. 62.2

Consider a ray PA pararallel to the axis incident at A on the lens L_1 at a height $C_1\ A = h_1$. It suffers deviation through an angle δ_1 at the first lens and proceeds towards the principal focus F_1. It meets the second lens at a point B at a height h_2 from the axis. On refraction it is further deviated through an angle δ_2 and meets finally the axis at F_2.

As the incident parallel ray PA after refraction through both the lenses meets the axis at F_2, the point F_2 must be the second principal focus of the combination.

The ray $F_2\ B$ produced backward meets the ray PA produced forward at M_2. Therefore $M_2\ H_2$ is the *principal plane* and H_2 the *principal point.* Hence $F_2\ H_2 = F$ is the required focal length of the combination. The lens of focal length F placed at H_2 is called the **equivalent lens** which can replace the combination.

Deviation produced by the equivalent lens

$$\delta = \tan\delta = \frac{M_2H_2}{H_2F_2} = \frac{h_1}{F}$$

Deviation produced by the first lens

$$\delta_1 = \tan\delta_1 = \frac{AC_1}{C_1F_1} = \frac{h_1}{f_1}$$

As the lenses are very thin the ray AB is almost parallel to the principal axis

$\therefore$ Deviation produced by the second lens

$$\delta_2 = \tan \delta_2 = \frac{BC_2}{C_2F_2} = \frac{h_2}{f_2}$$

Now
$$\delta = \delta_1 + \delta_2$$

$\therefore$
$$\frac{h_1}{F} = \frac{h_1}{f_1} + \frac{h_2}{f_2} \qquad ...(i)$$

Now, in Fig. 62.2
$$h_2 = C_2 K - BK$$

But
$$\frac{BK}{AK} = \delta_1 = \frac{h_1}{f_1}$$

$\therefore$
$$BK = AK\frac{h_1}{f_1} = \frac{dh_1}{f_1}$$

and
$$C_2K = h_1$$

$\therefore$
$$h_2 = h_1 - d\frac{h_1}{f_1} = h_1\left(1 - \frac{d}{f_1}\right) \qquad ...(ii)$$

Substituting the value of h_2 in (i), we get

$$\frac{h_1}{F} = \frac{h_1}{f_1} + \frac{h_1}{f_2}\left(1 - \frac{d}{f_1}\right)$$

or
$$\frac{1}{F} = \frac{1}{f_1} + \frac{1}{f_2} - \frac{d}{f_1 f_2}$$

or
$$F = \frac{f_1 f_2}{f_1 + f_2 - d} = -\frac{f_1 f_2}{\Delta}$$

where $\Delta = d - (f_1 + f_2)$ is known as **optical separation** or **optical interval** between the two lenses.

62.3. Principal points. To find the position of first and second principal point we have to first find the position of first and second focal point.

Position of focal points. In Fig. 62.2, triangles $M_2H_2F_2$ and $BC_2\,F_2$ are similar

$\therefore$
$$\frac{C_2F_2}{H_2F_2} = \frac{BC_2}{M_2H_2} = \frac{h_2}{h_1}$$

or
$$C_2F_2 = H_2F_2\,\frac{h_2}{h_1}$$

But H_2F_2 = focal length of the combination $= F$

and according to Eq. (ii)
$$\frac{h_2}{h_1} = 1 - \frac{d}{f_1}$$

$\therefore$
$$C_2F_2 = F\left(1 - \frac{d}{f_1}\right)$$

Thus C_2F_2 gives the position of **second focal point.**

Similarly the position of the **first focal point** will be given by

$$C_1F_1 = -F\left(1-\frac{d}{f_2}\right).$$

Position of principal points In Fig. 62.2

$$C_2H_2 = H_2F_2 - C_2F_2 = F - F\left(1-\frac{d}{f_1}\right)$$

or

$$C_2H_2 = F\frac{d}{f_1}$$

If we denote C_2H_2 by β, then conventionally β is negative

$$\therefore \qquad \beta = F\frac{d}{f_1} \qquad \text{or} \qquad \beta = -F\frac{d}{f_1}$$

This gives the position of **second principal point.**

Similarly the position of the first principal point is given by

$$\alpha = C_1H_1 = F\frac{d}{f_2}$$

Experiment. 62.1. To determine the principal points of a lens system.

Aparatus, Procedure, observations and **calculations** same as in **Expt. 12.2.**

Calculation of position of principal points

Focal length of the first lens $f_1 =$ cms

Focal length of the second lens $f_2 =$ cms

Distance between the two lenses $d =$ cms

Focal length of the combination $F =$ cms

Distance of the first focal point from the optical centre of the first lens

$$L_1 = \alpha = C_1H_1 = F\frac{d}{f_2} = \qquad \text{cms}$$

Distance of the second focal point from the optical centre of the second lens

$$L_2 = \beta = -F\frac{d}{f_1} = \qquad \text{cms.}$$

Precautions. Same as in **Expt. 12.2**

Oral Questions

Same as in **Chapter 12** under the head **'Cardinal points and nodal slide'.**

63

Absorption Spectra of Iodine

63.1. Origin of spectra. According to Bohr's atomic model, the atom consists of a central nucleus having a net positive charge and a suitable number of electrons revolving around it, in more or less circular orbits so that the atom, on the whole, is neutral.

The number of protons in the nucleus is equal to the atomic number Z and the number of neutrons in it is equal to $(A–Z)$ where A is the mass number. The number of electrons revolving around the nucleus is also equal to Z. These electrons revolve in various orbits. The maximum number of electrons that an orbit can accommodate is $2n^2$ where n is the number of the orbit. For example, the innermost i.e., first orbit ($n = 1$) can have a maximum of $2 \times 1^2 = 2$ electrons, the second ($n = 2$), $2 \times 2^2 = 8$ electrons and so on. A definite amount of energy is associated with each orbit. An electron possesses a smaller amount of energy in the inner orbit and a greater amount of energy in an outer orbit. This is why an electron passes from an inner orbit to an outer orbit when energy is supplied to the atom. On the other hand, when an electron jumps back from an outer orbit to an inner orbit, it emits radiation of energy equal to the difference of energy between the two orbits.

For example, when an electron falls from an outer orbit n_2 into an inner orbit n_1, the difference in energies of the two orbits is emitted as light of frequency ν, given by

$$h\nu = W_{n_2} - W_{n_1} \qquad ...(i)$$

where W_{n_2} is the energy of the electron in n_2th (outer) orbit and W_{n_1} the energy of the electron in n_1th (inner) orbit, h being Planck's constant.

63.2. Emission spectra. When a blackened platinum wire is electrically heated, it becomes *dull red* at about 500°C, *cherry red* at about 900°C, *orange red* at 1100°C, *yellow* at 1250°C and finally *white* at 1600°C. This is because when a substance is heated to a high temperature its atoms collide violently with each other. Due to this, electrons in the atom are raised to higher energy orbits. When one such electron falls back, light of definite frequency (or wavelength) is emitted according to relation (*i*). However, the light emitted by a very hot body has a number of wavelengths in it, each wavelength corresponding to energy difference between two orbits. This is known as **emission spectrum**. If the wavelength extends over almost the whole of visible range it is called **continuous emission spectrum**.

63.3. Absorption spectrum. Certain substances when interposed in the path of light coming from a source giving out a continuous emission spectrum, absorb some portions of the spectrum of particular wavelengths from it. In such a case the spectrum obtained contains *dark bands* or *lines* in places of the positions of the wavelength absorbed. These are called **absorption spectra** of these substances.

An absorption spectrum in which absorption takes place in the form of broad dark bands is called **band absorption spectrum** *or* **continuous absorption spectrum**.

An absorption spectrum in which absorption takes place in the form of well-defined dark lines or selected wavelengths is called **line absorption spectrum**.

An arrangement to obtain line absorption spectrum of sodium vapour is shown in Fig. 63.1. Light from an arc lamp is made to pass through vapours of sodium produced by burning an

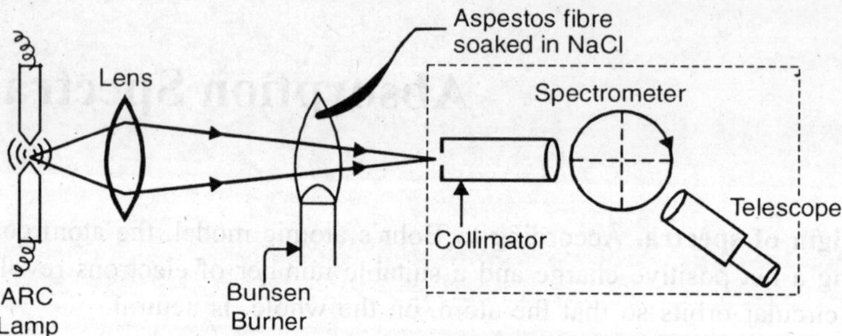

Fig. 63.1

asbestos fibre soaked in NaCl solution. This light is examined by a spectrometer. It is observed that the continuous emission spectrum of the arc lamp is crossed by two dark lines in the yellow region. These dark lines are located exactly in the same place as occupied by them in the emission spectrum of sodium. Thus we find that sodium vapours at lower temperature absorb from white light the same particular wavelength which it would emit in its emission spectrum at a higher temperature. This is known as '*absorption spectrum of sodium*' or '*reversal of sodium lines*'. This phenomenon can be explained on the basis of Kirchoff's law. This law states '*A substance which emits light of certain wavelength when heated to incandescence absorbs the light of the same wavelength* (*or colour*) *when cold if light from a source at a higher temperature is made to pass through it.*' Thus it is the light corresponding to yellow lines of sodium in the arc light spectrum at the higher temperature which is absorbed by the sodium vapour in the flame at the lower temperature that we get dark lines in their place. In fact, these lines are not dark but comparatively less bright than the rest of the spectrum.

Narrow, *sharp* lines are observed in the case of absorption spectra of gases or vapours. On the other hand, there are many substances, solids or liquids, such as potassium salts and aniline dyes which produce absorption spectra consisting of *dark bands*. In such cases, the *number*, *width* and *position* of the band are characteristic of the absorbing substance.

Absorption spectroscopy can, therefore, be made use of for the identification of dyes, drugs, oils, glasses etc.

63.4. Absorption spectra of iodine. The total energy of a molecule in the excited state is given by

$$E = E_e + E_v + E_r$$

where E_e is the *excitation energy of the electron* in the molecule, E_v the *energy due to vibration* of the molecule and E_r the *energy due to rotation* of the molecule.

The electronic energy levels in a molecule are separated by energy intervals of the order of 1 eV – 10 eV. Each electron energy level may have a number of vibrational energy levels which are separated by energy intervals of about 0.1 eV. Each vibrational energy level may have a number of rotational energy levels separated by energy intervals of the order 10^{-3} eV.

The iodine molecule also has an electronic state. The transition between its excited state and ground state falls within the visible region of the spectrum. In Iodine also, besides the ground and excited electronic levels, the vibrational and rotational energy levels are present. These vibrational and rotational energy levels are present very close to the ground as well as excited levels. Let us consider the vibrational energy levels. These energy levels are *quantised* and are *equally spaced.* Suppose the quantum numbers for various vibrational levels close to the ground state are represented as $n' = 0, 1, 2, 3, 4, ...$ and for levels close to the excited state are represented as $n'' = 0, 1, 2, 3, 4, ...$ The level for $n' = 0$ corresponds to electronic ground state. Similarly $n'' = 0$ corresponds to the electronic excited state. The transition between $n'' = 0$ to $n' = 0$ gives the energy gap between the two electronic levels.

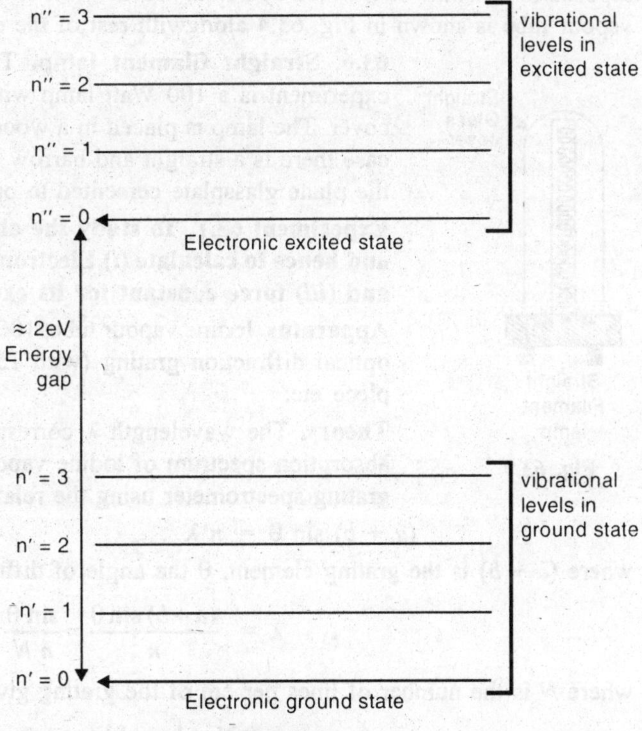

Fig. 63.2

Now, at ordinary room temperature *all the molecules of iodine reside in the ground state* $n' = 0$. If a beam of light having continuous spectrum is made to pass through iodine vapour, energy $h\nu$ which corresponds to $n' = 0$ and any of the n'' levels may be absorbed. *Hence the continuous spectrum of white light is superposed by a series of dark lines.* But since rotational energy levels are also superposed on the vibrational and electronic energy levels, the lines spread in the form of *narrow bands.* The wavelength corresponding to these bands are determined and the energy level diagram is plotted.

The energy gap is found with reference to the **highest wavelength.** The vibrational excited levels are also plotted which correspond from higher to lower wavelength side. The dissociation limit refers to the lowest wavelength i.e., corresponding to $n'' = \infty$. The difference in wavelength corresponding to successive vibrational levels may now be found out by measurement. Fig. 63.2 shows the electronic and vibrational energy levels of iodine molecule. The gap between ground state and excited state levels in electronic configuration is about 2 eV. This corresponds to a wavelength $\lambda = \frac{hc}{E}$. On calculation it comes out to be 6.21×10^{-7} m = 6210 Å which lies in the visible region of the spectrum. From the difference in wavelength corresponding to successive vibrational levels, the value of change of wave number $\Delta\bar{\nu}$ for two successive vibrational levels can be found out and the *force constant* for iodine molecule in excited state can be calculated.

To perform the experiment, we make use of an *iodine vapour tube* and a *straight filament lamp* which are described in detail.

63.5. Iodine vapour tube. To obtain the absorption spectrum of iodine vapour, a glass tube of uniform bore of 3 cm diameter and length about 1 meter is used. The tube is provided with a small inlet at its middle to introduce *iodine specks* in the tube. The inlet is closed by glass stopper or a tight rubber cork after introducing the iodine specks, which form thick iodine vapours for absorption of proper wavelength of white light. A plane glass plate of 4 cm diameter is cemented by some adhesive material like *araldite* at one end of the tube. To the other end of the tube a double convex lens of short focal length is cemented by some *transparent adhesive material of almost the same refractive index as that of glass.* It is ensured that the system is air tight and no iodine vapours escape out after introducing the iodine specks and putting the stopper in position in the tube.

The iodine specks vaporise and the vapours develop sufficient vapour pressure at 20°C (or room temperature) to give good absorption bands. However, *it is better that the tube is slightly exposed to sunshine or warmed by a heat source if the temperature is low as in winter*. The iodine vapour tube is shown in Fig. 63.4 alongwith rest of the experimental arrangement.

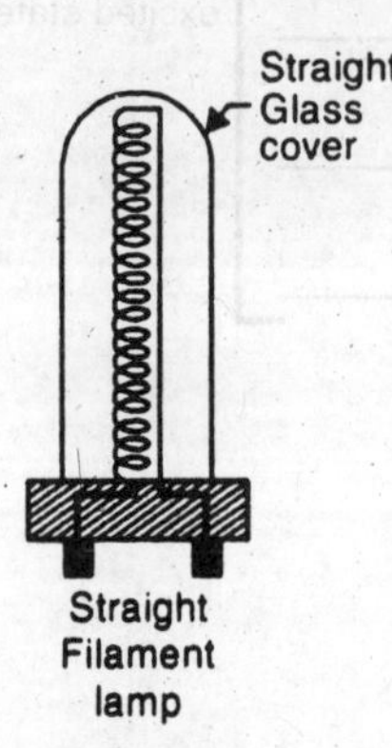

Fig. 63.3

63.6. Straight filament lamp. The source of white light used in the experiment is a 100 Watt lamp with straight filament and a straight glass cover. The lamp is placed in a wooden or metallic case. On one side of the case there is a straight and narrow vertical slot to allow the light to fall on the plane glassplate cemented to one end of the iodine vapour tube.

Experiment 63.1. To study the absorption spectrum of iodine vapours and hence to calculate (*i*) Electronic energy gap, (*ii*) vibrational energies and (*iii*) force constant for its excited state.

Apparatus. Iodine vapour tube, 100 W straight filament lamp, spectrometer, optical diffraction grating (with 15000 lines per inch) reading lamp, eye piece etc.

Theory. The wavelength λ corresponding to the absorption bands in the absorption spectrum of iodine vapour can be measured with the help of a grating spectrometer using the relation.

$$(a + b) \sin \theta = n \lambda$$

where $(a + b)$ is the grating element, θ the angle of diffraction and n the order of spectrum

$$\therefore \quad \lambda = \frac{(a+b)\sin\theta}{n} = \frac{\sin\theta}{nN} \quad ...(i)$$

where N is the number of lines per cm of the grating given by

$$\frac{\text{Number of lines per inch}}{2.54}$$

Relation (*i*) gives the value of λ in cms. The number of lines per inch of the grating is given by the makers.

Electronic energy gap. The energy gap between the first excited electronic level and the ground level corresponding to it is given by

$$E_1 = h \times \text{lowest frequency of the absorption band}$$

$$= h \times \frac{c}{\text{highest wave length of the absorption band}}$$

$$= hc\,\overline{\nu_1}$$

where $\overline{\nu_1}$ is the wave number corresponding to the **highest wavelength**, in cm, h is Planck's constant in erg sec. and c the velocity of light in cm s^{-1}.

Vibrational energy. If $\overline{\nu_2}$ is the wave number corresponding to the next absorption band, then energy gap E_2 between the first vibrational level of the first excited electronic level and the ground level is given by

$$E_2 = h\,c\,\overline{\nu_2}$$

$\therefore$ Energy difference between two consecutive vibrational levels is given by

$$E_2 - E_1 = h\,c\left(\overline{\nu_2} - \overline{\nu_1}\right)$$

$$= h\,c\,\overline{\Delta\nu} \qquad ...(ii)$$

Force constant. Now $E_2 - E_1$ in vibrational spectra is given by

$$E_2 - E_1 = \Delta E = \frac{h}{2\pi}\sqrt{\frac{k}{m'}} \qquad ...(iii)$$

where k is force constant and m' is the reduced mass of the iodine molecule.

Hence from (*ii*) and (*iii*), we get

$$h\,c\,\overline{\Delta\nu} = \frac{h}{2\pi}\sqrt{\frac{k}{m'}}$$

or Force constant $\quad k = 4\pi^2 m' (c\,\overline{\Delta\nu})^2 \qquad ...(iv)$

The reduced mass of a molecule $m' = \dfrac{m_1 \times m_2}{m_1 + m_2} \times 1.7 \times 10^{-24}$ gm

For iodine $\quad m_1 = m_2 = 127$

$$\therefore \quad m' = \frac{127 \times 127}{127 + 127} \times 1.7 \times 10^{-24} = 1.08 \times 10^{-22} \text{ gm.}$$

If $c = 3 \times 10^{10}$ cm sec^{-1}, then k is in dyne/cm

Procedure. 1. Find the vernier constant and least count of each of the two verniers fixed to the spectrometer.

2. Set up the spectrometer and grating on the prism table. Adjust the grating normal to the incident light and arrange the telescope to receive the first order spectrum on either sides of the normal position.

All these steps have been explained in Expt. 18.2. **'To determine the wavelength of sodium light using a plane diffraction grating'**.

3. Take the 100 W straight filament lamp, enclose it in a case having a straight vertical slot to allow light to come out of it to serve as a source of white light.

4. *Keep the tube containing iodine vapour between the lamp and the collimator slit with its lens side towards the collimator and plane glass plate side towards the lamp.* Hold the iodine vapour tube suitably in a clamp and adjust its position so that light coming out of it is condensed on the collimator slit as shown in Fig. 63.4. Observe the absorption spectrum through the telescope.

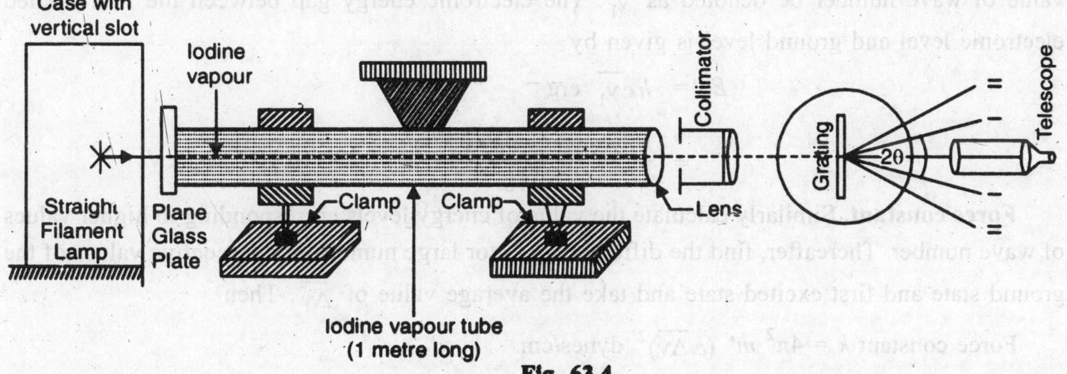

Fig. 63.4

You will see a number of dark lines (or bands) in the background of a continuous emission spectrum. Record the direct slit reading. Rotate the telescope in anticlockwise direction and set the cross wire on successive dark lines (or bands). In each case record the reading on both the verniers of the spectrometer. Repeat for the absorption lines (or bands) in the clockwise direction.

Find the difference between readings for a line to line, vernier to vernier for first order on two sides of the direct reading. This gives 2θ for each line separately. Half of it *i.e.*, θ gives the angle of diffraction. Calculate the wavelength λ corresponding to each value of θ for $n = 1$ using the relation $\lambda = \dfrac{\sin\theta}{N}$.

Observations and Record

Main scale division of the spectrometer =

Number of divisions on the vernier =

Least count =

Number of lines in the grating (per inch) =

Number of lines per cm, N =

Wavelength

Order of spectrum	*Colour of spectral band*	*Position of Telescope*				*Difference between the readings 2θ*		*Mean 2θ*	*Mean θ*
		Left side		*Right side*					
		Vernier V_1	*Vernier* V_2	*Vernier* V_1	*Vernier* V_2	V_1	V_2		

First order

$$\text{Wavelength } \lambda = \frac{\sin\theta}{N} = \text{cm} = \text{Å}$$

Electronic energy gap. In the same manner find the wavelengths corresponding to other absorption bands. Calculate the wave number corresponding to each wavelength. Let the **lowest** value of wave number be denoted as $\overline{\nu}_1$. The electronic energy gap between the first excited electronic level and ground level is given by

$$E_1 = hc\overline{\nu}_1 \text{ erg}$$

$$= \frac{6.62\times10^{-27}\times3\times10^{10}\,\overline{\nu}_1}{1.6\times10^{-12}} = 12.416\times10^{-5}\,\overline{\nu}_1 \text{ eV.}$$

Force constant. Similarly calculate the value of energy levels corresponding to higher values of wave number. Thereafter, find the difference $\overline{\Delta\nu}$ for large number of consecutive values of the ground state and first excited state and take the average value of $\overline{\Delta\nu}$. Then

Force constant $k = 4\pi^2\, m'\, (c\,\overline{\Delta\nu})^2$ dynes/cm

Take m' = reduced mass of iodine molecule = 1.08×10^{-22} gm as calculated above.

Precautions. 1. Same as in **Expt. 18.2.**

2. The stopper for closing the iodine vapour tube should be air tight.

3. The vapour tube should be exposed to sunlight or slightly warmed, if the temperature is low.

4. The plate glass side of the iodine vapour tube should be towards the source of light and the convex lens side towards the collimator of the spectrometer.

5. To calculate the energy gap the lowest value of wave number $\overline{v_1}$ should be used.

Oral Questions

(*i*) What is the origin of spectra? (*ii*) What is emission spectrum? (*iii*) Name the various classes of emission spectra. (*iv*) What is absorption spectrum? (*v*) Name the various classes of absorption spectrum. (*iv*) Is band spectrum an atomic or molecular spectrum? (*vii*) Name some substances which give line spectra. (*viii*) Is line spectrum an atomic or molecular spectrum? (*ix*) Name some substances which give molecular spectra (*x*) What are the various types of molecular spectra? (*xi*) Distinguish between rotational, vibrational and electronic spectra of molecules. (*xii*) What is order of energy of electronic, vibrational and rotational levels? (*xiii*) What is reduced mass? (*xiv*) What is force constant? (*xv*) State Kirchoff's law which explains absorption spectrum.

64

Divergence and Wavelength of Laser

Note. For **Principle of Laser** and description and working of a **Ruby Laser** and **Helium-Neon Laser** See **Chapter 41**.

Experiment. 64.1 **To measure the divergence of a given Laser source.**

Apparatus. He-Ne Laser Source, a stand, a screen, a m.m graph paper or a pair of dividers and m.m. scale and a measuring tape (or metre rod).

Theory. Divergence is defined as the spread of the laser beam with distance. *It is measured by the angle in radians subtended by the laser light spot at the point of origin.*

The divergence of a laser beam is extremely small as compared to the conventional light sources. The typical divergence of a He-Ne laser is of the order of 1 milli-radian.

If we consider the laser source as a *point source*, then the angle θ_0 subtended by the laser spot of diameter ω_1 formed at a distance Z from the source is given by

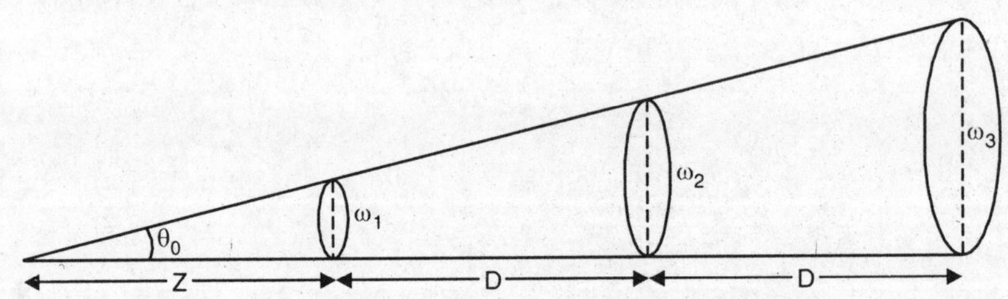

Fig. 64.1

$$\theta_0 = \frac{\omega_1}{Z} \quad \text{or} \quad \omega_1 = \theta_0 Z$$

or $$\omega_1^2 = \theta_0^2 Z^2$$

Similarly, if ω_2 and ω_3 are the diameters of the laser spot formed at distances $(Z + D)$ and $(Z + 2D)$ respectively from the origin, Then

$$\theta_0 = \frac{\omega_2}{Z + D} \quad \text{or} \quad \omega_2 = \theta_0(Z + D)$$

or
$$\omega_2^2 = \theta_0^2 (Z + D)^2$$

and
$$\theta_0 = \frac{\omega_3}{Z+2D} \text{ or } \omega_3 = \theta_0(Z + 2D)$$

or
$$\omega_3^2 = \theta_0^2 (Z + 2D)^2$$

But the laser source cannot be a point source. It has a finite size though small. If ω_0^2 is the size of the laser source even at the origin, then

$$\omega_1^2 = \omega_0^2 + \theta_0^2 Z^2$$

$$\omega_2^2 = \omega_0^2 + \theta_0^2 (Z+D)^2$$

and
$$\omega_3^2 = \omega_0^2 + \theta_0^2 (Z+2D)^2$$

Adding $\omega_1^2 + \omega_3^2$ and subtracting $2\omega_2^2$ from it, we have

$$\omega_1^2 + \omega_3^2 - 2\omega_2^2 = \omega_0^2 + \theta_0^2 Z^2 + \omega_0^2 + \theta_0^2 Z^2 + 4\theta_0^2 D^2 + 4\theta_0^2 ZD$$

$$-2\omega_0^2 - 2\theta_0^2 Z^2 - 2\theta_0^2 D^2 - 4\theta_0^2 ZD$$

$$= 2\theta_0^2 D^2$$

or
$$\theta_0^2 = \frac{\omega_1^2 + \omega_3^2 - 2\omega_2^2}{2D^2}$$

$$\therefore \quad \theta_0 = \frac{\left[\omega_1^2 + \omega_3^2 - 2\omega_2^2\right]^{1/2}}{\sqrt{2}\,D}$$

Procedure. 1. Arrange the He-Ne source on a suitable stand at one end of the laboratory room which may be partially dark. With a piece of chalk draw a straight line from one end to the other end of the room.

2. Paste a m.m. graph paper on the screen and place it in front of the source so that light falls normally on it and a bright spot of laser light of about 5 m.m. diameter is formed on the screen. See that the plane of the screen is perpendicular to the line drawn in the room. Count the number of m.m. squares along the vertical and horizontal diameters of the spot and record it. Alternatively measure the diameter with a pair of dividers and a m.m. scale.

3. With the help of a measuring tape (or a metre scale) measure the distance between the laser source and the screen. This distance is Z.

4. Now move the screen away from the source so that the spot size increases and place the screen at a distance where the diameter of the spot of light is at least doubled. Again measure the diameter of the spot in two directions at right angles either on the graph or with a pair of dividers and m.m. scale. Also measure the distance between the laser source and the screen. This distance is $(Z + D)$. Calculate the value of D.

5. Now move the screen away from the source through a further distance D and measure the diameter of the spot of light as already explained and also the distance D.

6. Take 3 observations by changing Z and D.

Observations and Record

No.	Distance cms		Spot size			$\theta_0 = \frac{\left[\omega_1^2 + \omega_3^2 - 2\omega_2^2\right]^{1/2}}{\sqrt{2}\,D}$
		D	Vertical	Horizontal	Mean ω (cms)	
1	$Z =$ $Z + D =$ $Z + 2D =$	$D =$			$\omega_1 =$ $\omega_2 =$ $\omega_3 =$	
2	$Z =$ $Z + D =$ $Z + 2D =$	$D =$			$\omega_1 =$ $\omega_2 =$ $\omega_3 =$	
3	$Z =$ $Z + D =$ $Z + 2D =$	$D =$			$\omega_1 =$ $\omega_2 =$ $\omega_3 =$	

Precautions. 1. The laser tube axis should be horizontal.

2. The laser source should be switched on only while taking the observation and immediately switched off thereafter.

3. The spot size to begin with should be about 5 mm.

4. The diameter of the spot should be measured in two directions at right angles.

64.1. Wavelength of a Laser Source. The wavelength of a laser source can be determined by one of the following methods:

1. Double slit interference **Experiment 41.1**

2. Michelson's interferometer **Experiment. 71.1**

3. Transmission diffraction grating.

Transmission diffracting grating See Art. 18.3. (*Chapter 18*).

Experiment 64.2. (*a*) To determine the wavelength of laser light by using transmission diffraction grating. (*b*) Given the wavelength of laser light to determine the number of lines in a transmission diffraction grating.

Apparatus. A He-Ne laser source, a transmission diffraction grating, a grating stand, screen, measuring tape and a m.m. scale.

Theory. When a parallel beam of monochromatic light is incident normally on a grating, the transmitted light gives rise to primary maxima in certain directions given by the relation

$$(a + b) \sin \theta_n = n\lambda$$

where a is the width of a transparency, b that of opacity, θ_n the *angle of diffraction for the nth order maximum* and λ the wavelength of light.

For the first order spectrum, $(a + b) \sin \theta_1 = \lambda$

For the second order spectrum, $(a + b) \sin \theta_2 = 2\lambda$

and so on.

Procedure. 1. Place the He–Ne Laser source at one end of the table. *Take out the grating carefully from the box, holding it from the edge and without touching its surface*, mount the grating on its stand. Place the grating with stand at a suitable distance from the source so that laser light falls normally on it.

2. Place a screen behind the grating at a suitable distance so that the light beam after passing through the grating produces several bright spots on the screen as shown in Fig. 64.2. Adjust the position of the screen so that the diffraction spots are sharp and bright. If necessary, adjust the grating also.

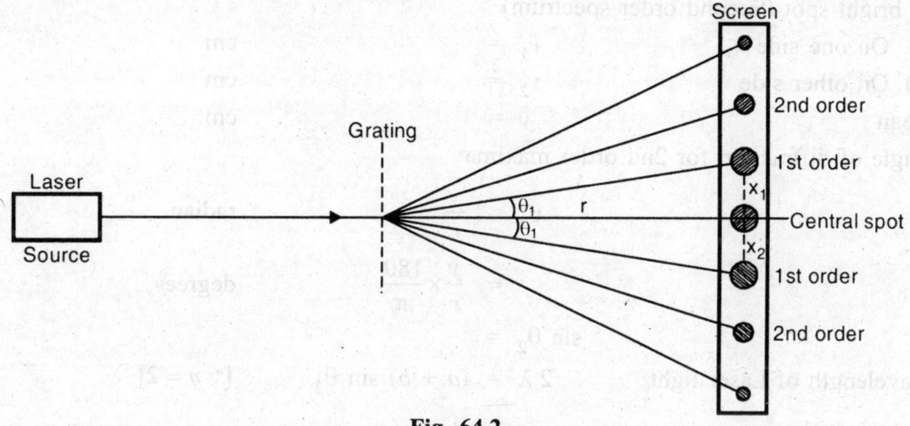

Fig. 64.2

3. The brightest spot is the central maxima and symmetrically situated on both sides of the central maxima there are several bright spots of diminishing intensity. The first bright spot on either side is the *first order spectrum.* Measure the distance between the centre of central maxima and the centres of first bright spot (*First order spectrum*) on either side of the central maxima.

4. The second brightest spot on either side of the central maxima is the *second order spectrum.* Measure the distance between the centre of the central maxima and centres of second bright spot (*Second order spectrum*) on either side of the central maxima.

5. Measure the distance of the centre of the central bright spot on the screen from the centre of the grating.

Observations and Record

(*a*) **Number of lines on the grating** $= N \quad =$ per inch

$$= \frac{N}{2.54} \quad = \text{ per cm}$$

$\therefore$ Grating element $(a + b)$ $= \dfrac{2.54}{N} \quad =$ cm

Distance between centre of central bright maxima and centre of diffraction grating

$r =$ cm.

First order spectrum. Distance between the centre of central bright spot and centre of first bright spot (1st order spectrum)

(*i*) On one side $x_1 =$ cm

(*ii*) On other side $x_2 =$ cm

Mean $x =$ cm

Angle of diffraction for 1st order maxima

$$\theta_1 = \frac{x}{r} \quad \text{radian}$$

$$= \frac{x}{r} \times \frac{180}{\pi} \quad \text{degree}$$

$$\sin \theta_1 =$$

Wavelength of Laser light $\lambda = (a + b) \sin \theta_1$ $[\because n = 1]$

$= \quad cm = \quad Å$

Second order spectrum. Distance between the centre of central bright spot and centre of second bright spot (Second order spectrum)

(*i*) On one side $y_1 =$ cm

(*ii*) On other side $y_2 =$ cm

Mean $y =$ cm

Angle of diffraction for 2nd order maxima

$$\theta_2 = \frac{y}{r} \quad \text{radian}$$

$$= \frac{y}{r} \times \frac{180}{\pi} \quad \text{degrees,}$$

$\therefore$ $\sin \theta_2 =$

Wavelength of Laser light $2\lambda = (a + b) \sin \theta_2$ $[\because n = 2]$

or $\lambda = \frac{1}{2}(a+b)\sin\theta_2$

$= \quad cm = \quad Å$

(*b*) **Number of lines on the diffraction grating.** Given for He–Ne Laser $\lambda = 6328$ Å

From First Order Spectrum

Grating element $(a + b) = \frac{\lambda}{\sin \theta_1} = $ cm

$\therefore$ Number of lines per cm $= \frac{1}{(a+b)} = $ cm^{-1}

From Second Order Spectrum

Grating element $(a + b) = \frac{2\lambda}{\sin \theta_2} = $ cm

$\therefore$ Number of lines per cm $= \frac{1}{(a+b)} = $ cm^{-1}

Average number of lines per cm. of the grating $= m =$ cm^{-1}

Precautions. 1. The Laser tube axis should be horizontal.

2. The Laser source should be switched on only while taking the observation and switched off immediately thereafter.

3. The grating should be held from the edges and the ruled surface should not be touched.

4. The light should fall on whole of the grating surface.

Oral Questions

Divergence of Laser. (*i*) What do you mean by divergence of a Laser beam? (*ii*) In what units is it measured? (*iii*) What is the order of divergence of a typical Laser beam? (*iv*) How do you remove the error due to finite size of laser source? (*v*) For more questions see **Chapter 41.**

Wave length of Laser Source. (*i*) What is diffraction? (*ii*) What is diffraction grating? (*iii*) How many lines per inch are there in the grating you used? (*iv*) What is the relation between wavelength and angle of diffraction for *n*th order spectrum? (*v*) How many orders of spectra were visible with the He–Ne source? (*vi*) How do you verify that θ_2 is the angle of diffraction for the second order when θ_1 is the angle of diffraction for the first order?

65

Interference

Experiment 65.1. To determine angle of wedge using interference method. This experiment is an exercise to **Expt. 17.3** '*To determine the thickness of a thin plate or a thin paper by measuring the width of interference fringes in an air wedge*'. The **procedure** and **theory** is exactly the same. It has been proved in the theory of that experiment that

$$\tan\theta = \frac{\lambda}{2}\cdot\frac{1}{\beta}$$

where θ is the angle of the air wedge, λ the wavelength of light used and β the fringe width

As θ is small, $\tan\theta = \theta$

$$\therefore \quad \text{Angle of the wedge } \theta = \frac{\lambda}{2\beta}$$

Precautions etc. Same as in **Expt. 17.3.**

WAVES AND VIBRATIONS

66

Damped Oscillator

66.1. Oscillatory motion. A motion which repeats itself after regular intervals of time is called a *periodic motion.*

If a body in periodic motion executes to and fro motion about a fixed reference point, it is said to have an oscillatory motion.

66.2. Simple harmonic motion. *When a body moves such that its acceleration is always directed towards a certain fixed point and varies directly as its distance from that point, the body is said to execute simple harmonic motion.*

For such a motion to take place the force acting on the body should be directed towards the fixed point and should also be proportional to the *displacement i.e.,* the distance from the fixed point. The function of the force is to bring the body back to its equilibrium position and hence this force is often known as *restoring force.*

66.3. Differential equation of simple harmonic motion. Consider a particle of mass m executing simple harmonic motion. If y be the displacement of the particle from equilibrium position at any instant t, the restoring force F acting on the particle is given by

$$F \propto y \quad \text{or} \quad F = -sy$$

where s is the *force constant* of proportionality or *stiffness* or *spring constant.* The negative sign is used to indicate that the direction of the force is opposite to the direction of increasing displacement.

Force constant s is defined as the restoring force per unit displacement or

$$s = \frac{F}{y}$$

Its units are *Newton per metre.*

If $\frac{d^2y}{dt^2}$ is the acceleration of the particle at time t, then

$$m\frac{d^2y}{dt^2} = -sy \quad \text{or} \quad \frac{d^2y}{dt^2} + \frac{s}{m}y = 0$$

Substituting $\frac{s}{m} = \omega^2$ we get $\frac{d^2y}{dt^2} + \omega^2 y = 0.$

This is the general *differential equation* of motion of a *simple harmonic oscillator.*

The solution of this equation is

$$y = a \sin(\omega t + \phi)$$

where ϕ is a constant of integration and a the *amplitude* of oscillation. The term $(\omega t + \phi)$ represents the total phase of the particle at time t and ϕ is known as the *initial phase.*

66.4. Damping. The amplitude of a vibrating string, a sounding tuning fork or an oscillating pendulum goes on gradually decreasing. It is because some energy is inevitably lost due to resistive forces. For example, in the case of a simple pendulum energy is lost due to friction at the supports and resistance of air. The amplitude of vibration decreases gradually and ultimately the pendulum comes to rest.

The decay of amplitude with time is called damping.

66.5. Differential equation of damped simple harmonic motion. If the velocity of the vibrating body is small, then the retarding force due to the medium is proportional to the first power of velocity. In such a case there are two forces acting on the vibrating particle.

(*i*) The restoring force proportional to displacement given by $-sy$ where s is a constant known as *stiffness* or *spring constant.* The negative sign shows that the direction of restoring force is opposite to that of displacement, and

(*ii*) A retarding force proportional to velocity given by $-r\frac{dy}{dt}$ where r is another constant known as *damping constant.* The negative sign again shows that the retarding force also acts opposite to the direction of motion of the body.

If m is the mass of the vibrating particle and its acceleration $\frac{d^2y}{dt^2}$ when it has a displacement y and is moving with a velocity $\frac{dy}{dt}$, then

$$m\frac{d^2y}{dt^2} = -r\frac{dy}{dt} - sy$$

or

$$m\frac{d^2y}{dt^2} + r\frac{dy}{dt} + sy = 0 \qquad ...(i)$$

The equation represents the motion of the particle under various forces mentioned above and the vibrations are *damped.*

Dividing (*i*) by m and putting $\frac{r}{m} = 2k$ and $\frac{s}{m} = \omega_0^2$ we get

$$\frac{d^2y}{dt^2} + 2k\frac{dy}{dt} + \omega_0^2 y = 0 \qquad ...(ii)$$

In this case k is called the *damping coefficient* and $2k$ gives the *force due to the resistance of the medium per unit mass per unit velocity.*

Eq. (*ii*) is known as *differential equation* of damped simple harmonic motion.

When the damping is *light*; *i.e.*, $k < \omega_0$ the solution of the equation is

$$y = A_o e^{-kt} \sin(\omega' t + \phi) \qquad ...(iii)$$

where the amplitude is given by $A = A_o e^{-kt}$ and ω' is the new angular frequency given by

$$\omega' = \sqrt{\omega_0^2 - k^2}$$

which is less than $\omega_0 = \sqrt{\dfrac{s}{m}}$

66.6. Coefficient of damping *k*. From Eq. (*iii*) we find that the amplitude of damped simple harmonic motion decays exponentially with time *i.e.,* the amplitude of successive oscillations goes on decreasing. If A_o is the amplitude to begin with, then the amplitude of the *n*th vibration is given by

$$A_n = A_o\, e^{-kt}$$

where t is the time taken by n vibrations. Taking natural logarithms on both sides, we have

$$\log_e A_n = -kt + \log_e A_o$$

From this equation we find that graph between $\log_e A_n$ and t is a straight line. The slope of the graph is given by

$$-k = \frac{\log_e A_n - \log_e A_m}{t_n - t_m} = \frac{\Delta \log_e A_n}{\Delta t}$$

where A_n is the amplitude of the nth vibration after a time t_n and A_m that of the mth vibration after a time t_m

or
$$k = \frac{-2.3026\, \Delta \log_{10} A_n}{\Delta t}$$

Therefore, for convenience we draw a graph between $\log_{10} A_n$ and t.

66.7. Relaxation time. *Relaxation time is defined as the time interval during which the amplitude of the oscillator decays to* $\frac{1}{e}$ *times its initial value.*

The amplitude of the nth damped oscillation is given by

$$A_n = A_o e^{-kt}$$

If τ is the relaxation time, then after further time τ,

$$A'_n = A_o\, e^{-k(t+\tau)} = \frac{A_n}{e} \quad \text{(by definition)}$$

$$\therefore \quad \frac{A_n}{A_n/e} = \frac{A_o\, e^{-kt}}{A_o\, e^{-k(t+\tau)}}$$

or
$$e = e^{k\tau}$$

$$\therefore \quad k\tau = 1$$

or
$$\tau = \frac{1}{k}$$

Hence relaxation time τ is the reciprocal of the co-efficient of damping k.

66.8. Quality factor. *The quality factor Q is defined as the total angle in radians through which the damped oscillator moves during the relaxation time τ.*

The damped oscillator completes one vibration in time period T and in this time traces an angle 2π. Therefore, the angle traced out by the oscillator in time τ, the relaxation time is $\frac{2\pi}{T}\tau$.

Hence Quality factor $Q = \dfrac{2\pi}{T}\tau$

66.9. Logarithmic decrement λ. *It is defined as the natural logarithm of the ratio between two successive amplitudes on the same side of the mean position of a damped oscillator separated by one full period.*

$$\therefore \qquad \lambda = \log_e \frac{A}{A_T}$$

where A is the amplitude at a time t and A_T the amplitude a time $t + T$.

Now $$A = A_o\, e^{-kt} \text{ and } A_T = A_o\, e^{-k(t+T)}$$

$$\therefore \qquad \lambda = \log_e \frac{A}{A_T} = \log_e \frac{A_o\, e^{-kt}}{A_o\, e^{-k(t+T)}}$$

$$= \log_e e^{kT} = kT$$

Also $$k = \frac{1}{\tau}$$

$$\therefore \qquad \lambda = \frac{T}{\tau}$$

66.10. Simple pendulum. (*used in the experiment*) *An ideal simple pendulum consists of a heavy point mass suspended from a rigid support by a weightless, inextensible and perfectly elastic string.* In actual practice it is not possible to have a simple pendulum which satisfies all the above conditions. We can neither have a heavy point mass nor a weightless inextensible string. An actual pendulum consists of a heavy metallic bob suspended by a long silk thread from a rigid support S as shown in Fig. 66.1. A graduated scale is fixed slightly above the bob and the bob oscillates parallel to this scale. When the bob oscillates, its amplitude goes on gradually decreasing due to viscous drag of the air and finally the pendulum stops after oscillating through a number of vibrations.

Experiment 66.1. To measure the logarithmic decrement, co-efficient of damping, relaxation time and quality factor of a damped simple pendulum.

Apparatus. A simple pendulum with scale (as described above) with a brass bob, an aluminium bob and an iron bob (either of the same mass or of the same diameter as the brass bob), a silken thread (about 2 metres long) and a stop watch.

Procedure. 1. Take the silken thread, attach its one end to the brass bob and set up the simple pendulum with scale (as already described) so that the length of the pendulum is about 150 cm as shown in Fig. 66.1.

2. Give a displacement of about 75 cms to the bob on one side and let it oscillate freely. The thread and the bob should not touch the scale behind it.

3. When the amplitude of the bob slowly decreases to about 50 cm, note it. Let it be A_o.

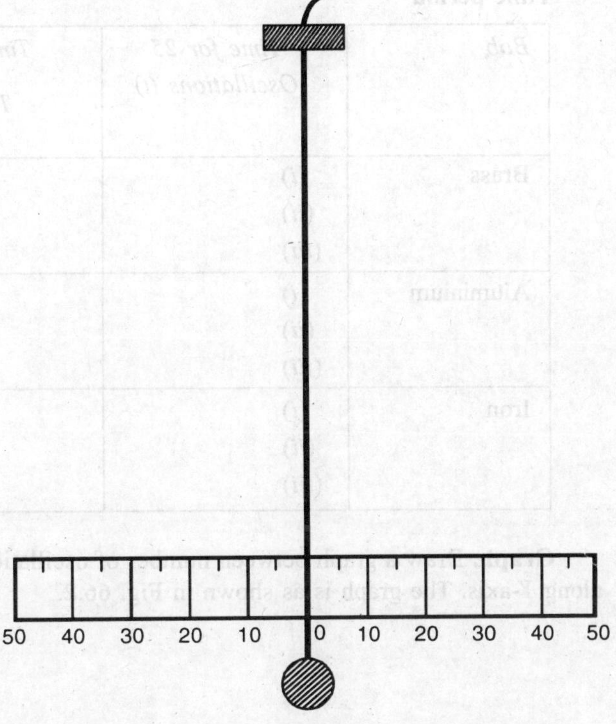

Fig. 66.1

4. Now go on nothing the amplitude after every five oscillations *i.e.,* after 5, 10, 15 oscillations and so on till the amplitude becomes about 10 cms.

5. As the amplitude of the bob is now small, start the stop watch and note the time taken for 25 oscillations. Take at least *three* observations. While noting the time for 25 oscillations also go on noting the amplitude after every five oscillations.

6. Repeat the experiment with aluminium bob and then with iron bob.

Observations and Record

Amplitude

S. No.	*No. of Oscillations n*	*Amplitude A_n for*			*$\log_{10} A_n$ for*		
		Brass bob	*Aluminium bob*	*Iron bob*	*Brass bob*	*Aluminium bob*	*Iron bob*
	0						
	5						
	10						
	15						
	...						
	...						
	...						
	65						
	70						
	75						

Time period

Bob	*Time for 25 Oscillations (t)*		*Time period $T = \frac{t}{25}$*	*Mean time period*
Brass	(*i*)			
	(*ii*)			T_1 =
	(*iii*)			
Aluminium	(*i*)			
	(*ii*)			T_2 =
	(*iii*)			
Iron	(*i*)			
	(*ii*)			T_3 =
	(*iii*)			

Graph. Draw a graph between number of oscillations *n* taking along the *X*-axis and $\log_{10} A_n$ along *Y*-axis. The graph is as shown in Fig. 66.2.

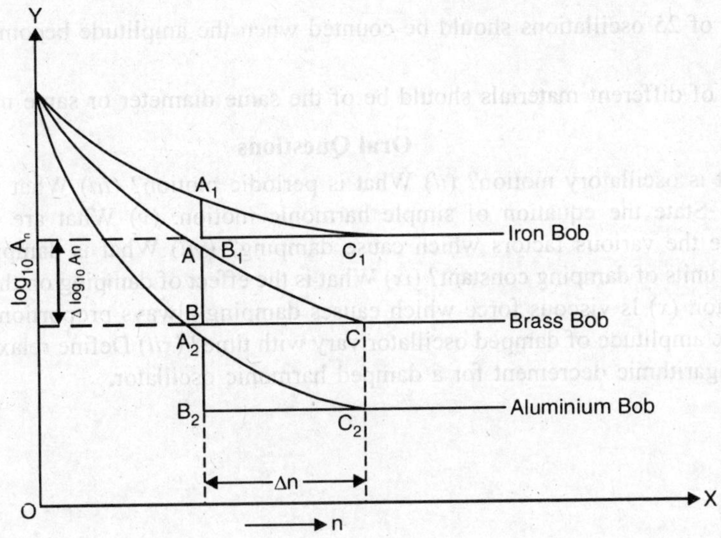

Fig. 66.2

Calculations.

(*i*) **Brass bob**

$$\text{Slope of the curve} = \frac{\Delta \log_{10} A_n}{\Delta n} = \frac{AB}{BC}$$

where Δn is the number of oscillations between B and C.

$$\therefore \quad \frac{\Delta \log_{10} A_n}{\Delta t} = \frac{\Delta \log_{10} A_n}{\Delta n\, T_1} = \frac{AB}{BC} \cdot \frac{1}{T_1}$$

where T_1 is the time period of the pendulum with brass bob.

and Co-efficient of damping for Brass

$$= k_{\text{Brass}} = -\frac{2.3026}{T_1} \frac{AB}{BC}$$

$$= + \qquad s^{-1}$$

$$\text{Relaxation time } \tau_{\text{Brass}} = \frac{1}{k_{\text{Brass}}} = s$$

$$\text{Quality factor } Q_{\text{Brass}} = \frac{2\pi}{T_1} \tau_{\text{Brass}} =$$

$$\text{Logarithmic decrement } \lambda_{\text{Brass}} = \frac{T_1}{\tau_{\text{Brass}}} =$$

Similarly calculate the values for the pendulum with iron and aluminium bob.

k_{iron} = ; τ_{iron} = ; Q_{iron} = ; λ_{iron} =

k_{Al} = ; τ_{Al} = ; Q_{Al} = ; λ_{Al} =

Precautions. 1. The bob or thread of the pendulum should not touch the scale behind it.

2. Initial amplitude of the bob should not be less than 60–70 cm.

3. Length of the silk thread of the pendulum should be 150–200 cms.

4. Time of 25 oscillations should be counted when the amplitude becomes small *i.e.*, about 10 cms.

5. Bobs of different materials should be of the same diameter or same mass.

Oral Questions

(*i*) What is oscillatory motion? (*ii*) What is periodic motion? (*iii*) What is simple harmonic motion? (*iv*) State the equation of simple harmonic motion. (*v*) What are damped vibration? (*vi*) What are the various factors which cause damping? (*vii*) What is damping constant? (*viii*) What are the units of damping constant? (*ix*) What is the effect of damping on the natural frequency of an oscillator. (*x*) Is viscous force which causes damping always proportional to velocity? (*xi*) How does the amplitude of damped oscillator vary with time? (*xii*) Define relaxation time, Quality factor and logarithmic decrement for a damped harmonic oscillator.

ELECTRICITY

67

Efficiency of an Electric Device

67.1. Efficiency of an electric device. *Efficiency η of an electric device is defined as the ratio of output power to input power.*

or
$$\text{Efficiency } \eta = \frac{\text{Output power}}{\text{Input power}}$$

Experiment 67.1. To find the efficiency of an electric kettle (or heating element) under varying input voltages.

Apparatus. Electric kettle (or heater element), A.C. voltmeter (0–250 volt), A.C. ammeter (0–10 Amp), rheostat capable of working on 220 V supply, (or a variable transformer), plastic covered (and well insulated), connecting wires, physical balance, spring balance. weight box, $\frac{1}{2}$°C thermometer, stop watch etc.

Theory. Suppose the mass of the electric kettle is M and S is the specific heat of the material of which the kettle is made. It is filled with m gms of water at a temperature θ_1°C. If a current I Amp. is passed through the kettle at a voltage V, then

$$\text{Input power} = VI \text{ Watt}$$

When the current flows through the kettle for a time t seconds and the temperature of the kettle and its contents rises from θ_1°C to θ_2°C, then

Total heat gained by the kettle and water

$$H = MS(\theta_2 - \theta_1) + m(\theta_2 - \theta_1)$$
$$= (MS + m)(\theta_2 - \theta_1) \text{ cals}$$

$\therefore$ Work done by the kettle in time t

$$= JH = 4.2\,(MS + m)(\theta_2 - \theta_1) \text{ Joule}$$

$$\text{Output power} = \frac{4.2(MS+m)(\theta_2 - \theta_1)}{t} \text{ Watt}$$

and percentage efficiency of the kettle

$$= \frac{\text{Output power}}{\text{Input power}} \times 100$$

$$= \frac{4.2(MS+m)(\theta_2-\theta_1)}{t \times V\,I} \times 100$$

Procedure. 1. Find the weight of the kettle with a spring balance.

2. Weigh a known quantity of water with a physical balance and put this water in the kettle.

3. Suspend a half degree thermometer in the kettle and note the initial temperature θ_1°C.

4. Make the connections as shown in Fig. 67.1(*a*) if using a rheostat and as shown in Fig. 67.1(*b*) if using a variable transformer.

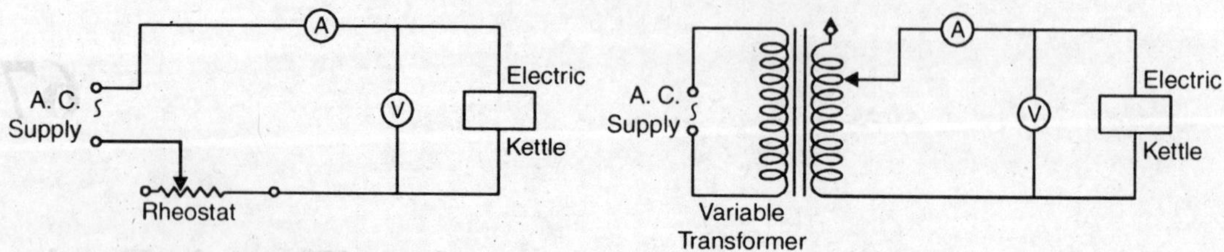

Fig. 67.1 (*a*) **Fig. 67.1 (*b*)**

5. Switch on the mains supply and simultaneously start the stop watch. Note the reading of the ammeter and voltmeter.

6. Pass electric current till there is an appreciable rise in temperature of water and kettle.

7. Switch off the mains supply and simultaneously stop the stop watch.

8. Immediately note the temperature θ_2°C of water (and kettle).

9. Repeat the experiment by changing the input voltages.

Observations and record.

Mass of electric kettle M = gms

Mass of water taken in the kettle m = gms

Specific heat of material of the kettle S =

Initial temperature of water θ_1 = °C

Ammeter reading I = Amp.

Voltmeter reading V = Volt.

Time for which current is passed t = sec.

Final temperature of water θ_2 = °C.

Calculations.

Total heat gained by water and kettle = $4.2\,(MS + m)\,(\theta_2 - \theta_1)$ Joule

$$\therefore \quad \text{Output power} = \frac{4.2(MS+m)\,(\theta_2-\theta_1)}{t} = \text{Watt}$$

$$\text{Input power} = VI$$

$$\therefore \quad \text{Efficiency of the kettle} = \eta = \frac{4.2\,(MS+m)\,(\theta_2-\theta_1)}{t \times VI} \times 100$$

Similarly record observations for two more values of input voltages.

Precautions. 1. Well insulated plastic covered wires should be used for making electrical connections.

2. Time should be recorded from the instant current is switched on and stopped as soon as the current is switched off.

Oral Questions

(*i*) Define efficiency of an electric kettle? (*ii*) What is the unit in which electric power is measured? (*iii*) Is it possible to have an efficiency of 100% for an electric device. (*iv*) Which method is preferred, using a rheostat or using a variable transformer? (*v*) How do you make allowance for the insulating handle of the kettle?

MAGNETIC PROPERTIES OF MATERIALS

68

Magnetic Susceptibility

Note. For definition of '*Magnetic materials, paramagnetic, ferromagnetic and diamagnetic substances; Magnetising field H, Intensity of magnetisation I, Magnetic susceptibility (volume susceptibility and mass susceptibility)*' See **Chapter 44 Art. 44.1, 44.2, 44.3** and **44.4**.

68.1. Molar Susceptibility. *If M is the molecular weight of the material, then*

$$\text{Molecular (or molar) susceptibility } \chi_{molar} = \frac{\chi}{\rho} M .$$

68.2. Determination of Magnetic Susceptibility—Quincke's Method. When a paramagnetic or a ferromagnetic liquid is taken in a U-tube it keeps its level in both the limbs. If one limb of the U-tube is placed in a strong magnetic field, the liquid in this limb shows a rise in level. This property is used to measure the susceptibility of a liquid by Quincke's method.

A Quincke's tube is a U-shaped glass tube with one limb very narrow and the other one ending into a wide funnel as shown in Fig. 68.1. The narrow limb is placed between the pole pieces *NS* of an electromagnet such that the meniscus of the liquid lies symmetrically between the two pole pieces where the field is uniform while the wider limb is placed slightly away in the region where the field is negligible.

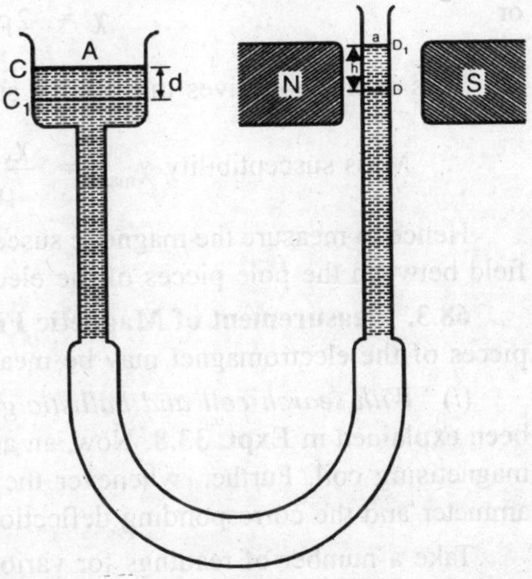

Fig. 68.1

When no magnetic field is applied, suppose the liquid in the funnel stands at *C* and in the narrow tube at *D*. When the magnetic field is applied the (paramagnetic or ferromagnetic) liquid in the narrow tube rises to D_1 through a height $h = DD_1$ and the liquid in the funnel falls from *C* to C_1 through a depth *d* so that

Total difference in height $= h + d$

Suppose the area of cross-section of the narrow tube $= a$

and area of cross-section of the funnel $= A$

then $$Ad = a\,h \quad \text{or} \quad d = \frac{a}{A}h$$

Hence total difference in height $= h + d = h + \frac{a}{A}h = h\left[1 + \frac{a}{A}\right]$

As $a << A$, $\frac{a}{A}$ can be neglected as compared to 1 and the net rise of the liquid column in the narrow tube can be taken to be $= h$.

$\therefore$ Volume of the liquid corresponding to a height h ; $V = a\,h$

and weight of the liquid column $= a\,h\,\rho\,g$ ...(*i*)

where ρ is the density of the liquid and g acceleration due to gravity. The downward weight of this liquid column is balanced by the upward force on the liquid due to the applied magnetic field of strength H between the pole pieces.

The force acting on a liquid column of height dx, is given by

$$dF = \frac{1}{2}\chi\frac{d}{dx}(H^2 a\,dx)$$

where χ is the magnetic susceptibility, H the strength of the magnetic field and $a\,dx$ = the volume of the liquid.

$\therefore$ Force acting on the entire liquid column $= \frac{1}{2}\chi\int_0^H \frac{d}{dx}(H^2 a dx)$

$$= \frac{1}{2}\chi\,a\,H^2 \qquad ...(ii)$$

From (*i*) and (*ii*), we have

$$\frac{1}{2}\chi\,a\,H^2 = a\,h\,\rho\,g$$

or $$\chi = 2\rho g\left(\frac{h}{H^2}\right)$$

This expression gives the volume susceptibility (χ_{vol})

$\therefore$ Mass susceptibility $\chi_{mass} = \frac{\chi_{vol}}{\rho} = \frac{2hg}{H^2}$

Hence to measure the magnetic susceptibility we have to measure the strength of the magnetic field between the pole pieces of the electromagnet.

68.3. Measurement of Magnetic Field. The strength of the magnetic field between the pole pieces of the electromagnet may be measured by one of the following methods:

(*i*) *With search coil and ballistic galvanometer using mutual inductance.* This method has been explained in **Expt. 33.8**. Now, an additional ammeter is also introduced in the circuit of the magnetising coil. Further, whenever the key K_1 is switched on, the current I_2 indicated by the ammeter and the corresponding deflection θ_2 of the ballistic galvanometer is immediately noted.

Take a number of readings for various values of current I_2 and corresponding deflection θ_2. Calculate the value of magnetic field H (In previous experiment it has been indicated by B) corresponding to various values of I_2 (or deflection θ_2) using the formula given in that experiment.

(*ii*) *With search coil and calibrated fluxmeter.* The connections for this method are shown in Fig. 68.2. To find the value of the magnetic field H between the pole pieces, place the search coil in the centre of the space between the pole pieces of the electromagnet with the face of the coil perpendicular to the magnetic lines of force. Level the screws of the fluxmeter so that for a given current the deflection in the fluxmeter is a maximum. Allow a suitable current I to flow in the electromagnet circuit and note the deflection θ in the fluxmeter. Take a number of readings for different values of current in the electromagnet circuit.

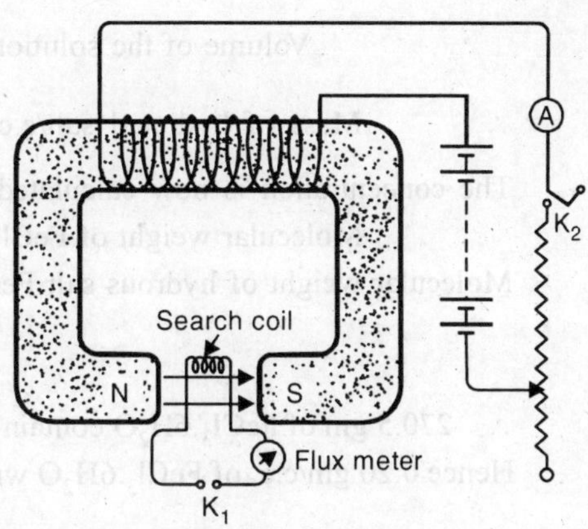

Fig. 68.2

If θ = Deflection produced in the fluxmeter pointer for the current I

b = Magnetic Flux in Maxwell turns per division deflection of fluxmeter

n = number of turns of search coil

and a = mean area of search coil, then

∴ Total flux passing through the search coil = $\theta \times b$ Maxwell turns and strength of the magnetic field in which the search coil is placed

$$H = \frac{\text{Total flux}}{\text{Area of search coil} \times \text{number of turns}} = \frac{\theta b}{a\,n}$$

Experiment 68.1. To find mass susceptibility of

(*i*) $FeCl_3.6H_2O$ (Hydrous Ferric Chloride)

(*ii*) Anhydrous $FeCl_3$ and

(*iii*) Molecular (or molar) susceptibility by Quincke's methods.

Apparatus. Quincke's apparatus consisting of an electromagnet which can produce very high magnetic field, a U-tube, rubber tubing, a travelling microscope, $FeCl_3.6H_2O$ crystals, beaker, graduated cylinder, physical balance, weight box and an arrangement for measuring the strength of magnetic field using a search coil and calibrated fluxmeter.

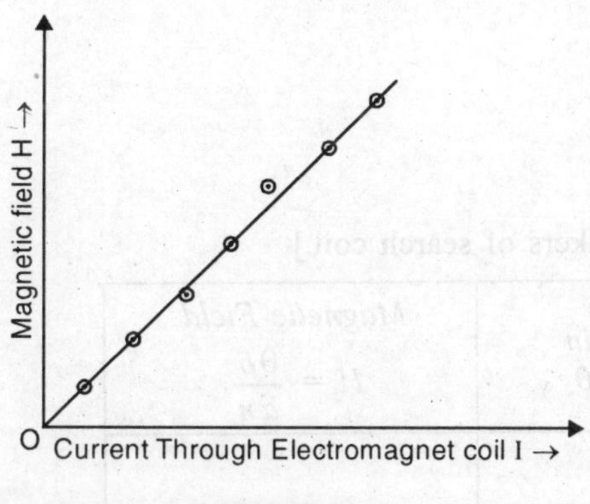

Fig. 68.3

Procedure. 1. Draw the circuit diagram according to the Fig. 68.2 and make connections accordingly for the measurement of the strength of the magnetising field between the pole pieces of the electromagnet by using *search coil and calibrated flux meter.* Take a number of readings for various values of current I passing through the magnetising coil of the electromagnet (and corresponding values of deflection θ in the fluxmeter).

2. Calculate the value of magnetising field H for each value of current I using corresponding values of deflection θ in the fluxmeter and applying the relation $H = \frac{\theta b}{an}$ as explained in **Art 68.3 (*ii*)**. Taking current I along the X-axis and H along the Y-axis, plot a graph between I and H. The graph is a straight line as shown in Fig. 68.3. This graph is known as *Calibration curve* of the *fluxmeter.* The value of magnetising field H can be found from this curve for any value of current I.

3. Dissolve a weighed quantity of hydrous salt ($FeCl_3.6H_2O$), say 20 gms in water to make up 100 c.c. of solution, then

Volume of the solution $= 100$ c.c.

$\therefore$ Mass of hydrated salt/c.c. $= \frac{20}{100} = 0.20$ gm/c.c.

The concentration is now calculated as under:–

Molecular weight of $FeCl_3 = (56 + 3 \times 35.3) = 162.5$ gm

Molecular weight of hydrous salt $FeCl_3.6H_2O$

$= 162.5 + 6(2 + 16) = 162.5 + 108$

$= 270.5$ gm.

$\therefore$ 270.5 gm of $FeCl_3.6H_2O$ contain 162.5 gm of $FeCl_3$.

Hence 0.20 gm/c.c. of $FeCl_3.6H_2O$ will contain

$$\frac{162.5 \times 0.20}{270.5} \text{ gm/cc}$$

or concentration $$m = \frac{162.5 \times 0.20}{270.5} \text{ gm/c.c.} = 0.12 \text{ gm/c.c.}$$

4. Thoroughly clean the Quincke's tube and fill it with the 20% solution of ($FeCl_3.6H_2O$) prepared as given above and insert the narrow limb of the U-tube vertically between the pole pieces of the electromagnet. Adjust the funnel limb so that when the electromagnet is energised the liquid meniscus lies in the central region of the uniform magnetic field between the pole pieces. Illuminate the meniscus with an electric lamp (if necessary) and view it through a travelling microscope. Bring the horizontal cross-wire of the microscope on the meniscus and note the reading when no current is passing through the magnetising coil.

5. Now switch on the current and note the value of current. Again bring the horizontal cross-wire of the microscope on the liquid meniscus and note the reading.

6. Repeat the experiment for different values of magnetising current. It is preferable to use the same values of current for which magnetic fields have been calculated while plotting the graph between I and H.

7. Take observations for a number of different values of concentration.

Observations and Record

Calibration of Fluxmeter

Number of turns of search coil $= n =$

Mean area of the search coil $= a =$

Number of Maxwell turns per division $= b =$

[**Note.** These data are generally provided by the makers of search coil.]

S. No.	*Magnetising current I Amp*	*Deflection in fluxmeter θ*	*Magnetic Field* $H = \frac{\theta b}{a n}$
1			
2			
3			
4			
..			
..			
..			

Height of liquid column

S. No.	*Magnetising current I amp*	*Value of H (From graph)*	H^2	*Position of liquid meniscus*		*Height of Liquid column h*
				Without field	*With field*	
1						
2						
3						
4						
..						
..						
..						
..						

8. Plot a graph between H^2 taken along the X-axis and height of liquid column h taken along the Y-axis. The graph is a straight line passing through the origin as shown in Fig. 68.4. The slope of the graph gives the value of $\frac{h}{H^2}$.

$$\therefore \qquad \frac{h}{H^2} = \tan\alpha = \frac{PQ}{OQ}$$

Calculations. Mass susceptibility

$$\chi_{mass} = \frac{2hg}{H^2} = 2g\frac{h}{H^2} = 2g\frac{PQ}{OQ}$$

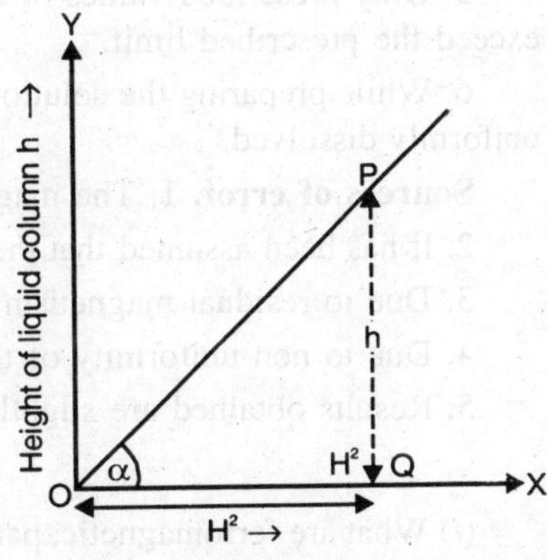

Fig. 68.4

Mass susceptibility of anhydrous salt. To find the mass susceptibility of anhydrous $FeCl_3$ we have to first find the susceptibility of water.

Mass susceptibility of water χ_{water}. 9. To find the mass susceptibility of water take at least five readings for different concentrations of $FeCl_3.6H_2O$ in solution. Calculate the mass susceptibility for each concentration and record in a tabular form.

Observations.

S. No.	*Concentration m gm/cc*	*Susceptibility χ_{mass}*
1		
2		
3		
4		
5		

10. Plot a graph between concentration m and susceptibility χ_{mass}. The graph is as shown in Fig. 68.5. Produce the graph to cut the χ_{mass} (YY') axis at R, then the intercept OR gives the value of χ_{mass} for zero concentration *i.e.*, for water only.

$\therefore \qquad \chi_{water} = \text{Intercept } OR$

Mass susceptibility of anhydrous $FeCl_3$. The susceptibility of anhydrous $FeCl_3$ is given by the relation

$$\chi_{anhyd} = \frac{\chi_{mass} - (1-m)\,\chi_{water}}{m}$$

where χ_{mass} and m are the values for a particular observation.

Molecular susceptibility. Molecular susceptibility

$$\chi_{molar} = M\,\chi_{anhyd}$$

where M = Molecular weight of $FeCl_3$ = 162.5.

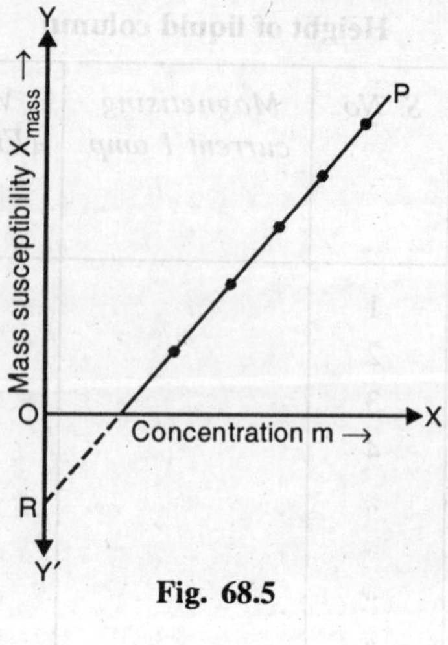

Fig. 68.5

Precautions. 1. The fluxmeter should be properly levelled.

2. While calibrating the fluxmeter one reading must be taken when the current is on and another when the current is off.

3. The joints between the rubber tubing and the glass capillary must be tight so that there is no leakage of solution from the joints.

4. The initial level of the liquid meniscus in the tube should be noted for each observation as it is liable to change due to residual magnetism.

5. Only prescribed values of magnetising current should be used and the current should not exceed the prescribed limit.

6. While preparing the solution it should be seen that the whole of the salt is completely and uniformly dissolved.

Sources of error. 1. The magnetic field of the earth has not been taken into consideration.

2. It has been assumed that the magnetic field at the wider end is zero but in fact it is not so.

3. Due to residual magnetism the rise in the capillary is not correct.

4. Due to non-uniformity of the narrow limb bore error due to surface tension may occur.

5. Results obtained are slightly less than the actual value due to evaporation of water.

Oral Questions

(*i*) What are ferromagnetic, paramagnetic and diamagnetic substances? (*ii*) What do you mean by intensity of magnetisation? (*iii*) Define magnetic susceptibility of a substance. (*iv*) Distinguish between volume susceptibility and mass susceptibility. (*v*) What is molar susceptibility? (*vi*) Can susceptibility have a zero, positive or negative value? (*vii*) If yes, what are those substances which have positive susceptibility and what are those substances which have negative susceptibility called? (*viii*) What is the effect of magnetic field on the level of paramagnetic and diamagnetic liquids filled in a tube when placed in a magnetic field? (*ix*) What is a search coil? (*x*) What is a fluxmeter? (*xi*) A search coil of n turns and area of cross-section a connected to a fluxmeter of b maxwell turns per unit deflection when placed in a magnetic field of strength of H produces a deflection θ. What is the value of H? (*xii*) Can you determine the magnetic susceptibility of solids like wood, ebonite etc. (*xiii*) If yes, by which method?

ELECTRONICS

69

Logic Circuits

69.1. Binary number system. The binary number system is a code which uses only two digits 0 and 1. Each binary digit (0 or 1) is called a *bit.* A string of four bits is called a *nibble* and eight bits make a *byte.* The base of binary number systems is 2 instead of 10, which is the base of decimal number system.

Decimal to binary conversion. To convert a number in the decimal system to binary system, '*Divide progressively the decimal number by 2 and write down the remainder after each division. Continue the process till you get a quotient of 0 and remainder of 1. The remainders taken in the reverse order form, the binary number.*'

The following example will make the process clear.

Example. *Convert decimal number* (*i*) *25 and* (*ii*) *37 to equivalent binary number.*

(*i*) Binary representation of 25

2	25
2	12 – 1
2	6 – 0
2	3 – 0
2	1 – 1
	0 – 1

The remainders in the reverse order are 1, 1, 0, 0, 1

∴ Binary representation of 25 is $(11001)_2$

(*ii*) Binary representation of 37

2	37
2	18 – 1
2	9 – 0
2	4 – 1
2	2 – 0
2	1 – 0
	0 – 1

The remainders in the reverse order are 1, 0, 0, 1, 0, 1

∴ Binary representation 37 is $(100101)_2$

The number 25 in decimal system is represented as $(25)_{10}$ and in binary system as $(11001)_2$. Similarly the number 37 is represented in decimal system as $(37)_{10}$ and in binary system as $(100101)_2$.

Binary to decimal conversion. In *binary* representation of a number the *first bit* is the MOST SIGNIFICANT BIT (MSB) and the *last bit* is the LEAST SIGNIFICANT BIT (LSB). For example, in the binary representation of 37, we have

$$\text{MSB} \leftarrow 100101 \rightarrow \text{LSB}$$

To convert a number in the binary system to decimal system '*Multiply the Least Significant Bit (LSB) by 2^0, the next by 2^1, next by 2^2 and so on till the Most Significant Bit (MSB) is reached, and then add all the products of multiplication.*'

The following example will make it clear.

Example. *Convert the binary number $(100101)_2$ into decimal system.*

$$\begin{aligned}(100101)_2 &= \text{MSB} \rightarrow 100101 \leftarrow \text{LSB}\\ &= 1 \times 2^0 + 0 \times 2^1 + 1 \times 2^2 + 0 \times 2^3 + 0 \times 2^4 + 1 \times 2^5\\ &= 1 + 0 + 4 + 0 + 0 + 32 = 37 \text{ or } (37)_{10}\end{aligned}$$

In binary to decimal conversion all positions containing 0 (zero) can be ignored and we need to add the decimal values of positions where 1 appears. The following example will make it clear.

Example. *Convert the binary number $(110011)_2$ into decimal system*

$$\begin{aligned}(110011)_2 &= \text{MSB} \rightarrow 110011 \leftarrow \text{LSB}\\ &= 1 \times 2^0 + 1 \times 2^1 + 0 \times 2^2 + 0 \times 2^3 + 1 \times 2^4 + 1 \times 2^5\\ &= 1 + 2 + 16 + 32 = 51 \text{ or } (51)_{10}\end{aligned}$$

Here, we have ignored all positions where 0 (zero) appears.

69.2. Logic gate. *A digital circuit which either allows a signal to pass through or stop it is called a gate.* It can have one or more input signals but only one output signal. Actual logic gates are made with diodes and transistors.

There are three types of basic logic gates:

(*i*) OR gate (*ii*) AND gate and (*iii*) NOT gate.

69.3. Truth table. *Truth table of a logic gate is a table which shows all input combinations and the corresponding output possibilities.*

69.4. Boolean algebra. Boolean algebra is based on binary nature of logic, such as HIGH or LOW; ON or OFF; YES or NO; TRUE or FALSE etc. The binary number 1 is used to denote HIGH and binary number 0 to denote LOW values.

Boolean expression. It is the logical statement that logic gates follow.

(*i*) The function of an OR gate is represented by Boolean expression

$$y = A + B$$

and is read as 'y equals A OR B'

(*ii*) The function of an AND gate is represented by Boolean expression

$$y = A \, . \, B$$

and is read as 'y equals A AND B'

(*iii*) The function of a NOT gate is represented by Boolean expression

$$y = \overline{A}$$

and is read as 'y equals NOT A'

The NOT operation is also called *negation* or *inversion*.

69.5. The OR gate. The OR gate has two inputs A and B and one output y.

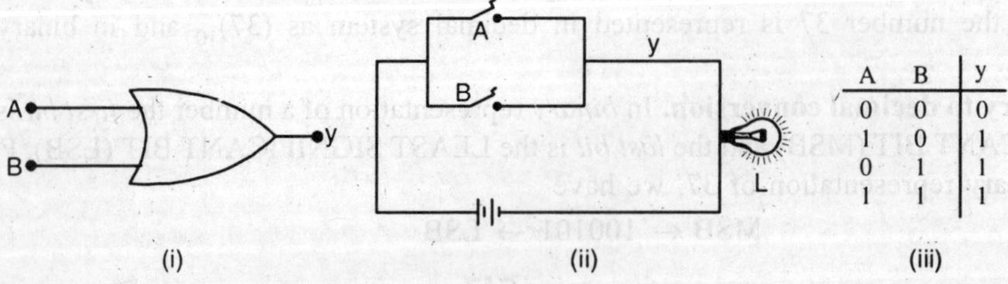

A	B	y
0	0	0
1	0	1
0	1	1
1	1	1

Fig. 69.1

Logic symbol. The OR gate is represented by the logic symbol shown in Fig. 69.1(*i*).

Boolean expression. The OR gate combines the inputs A and B with the output y according to Boolean expression

$$y = A + B \quad i.e., \quad y \text{ equals } A \text{ OR } B.$$

Working. The working of the OR gate is represented by the electrical circuit shown in Fig. 69.1(*ii*).

(*i*) When the switch A is OPEN and B is also OPEN ($A = 0, B = 0$) the lamp will not glow ($y = 0$)

(*ii*) When A is CLOSED and B is OPEN ($A = 1, B = 0$) the lamp will glow ($y = 1$)

(*iii*) When A is OPEN and B is CLOSED ($A = 0, B = 1$) the lamp will glow ($y = 1$)

(*iv*) When A is CLOSED and B is also CLOSED ($A = 1, B = 1$) the lamp will glow ($y = 1$).

Truth table. All the four possible inputs and corresponding output of the OR gate are shown in the TRUTH TABLE [Fig. 69.1(*iii*)]. Thus we find that the lamp glows when switch A OR B is CLOSED. This is why the electrical circuit shown in Fig. 69.1(*ii*) is known as OR gate.

69.6. The AND Gate. The AND gate has two inputs A and B and one output y.

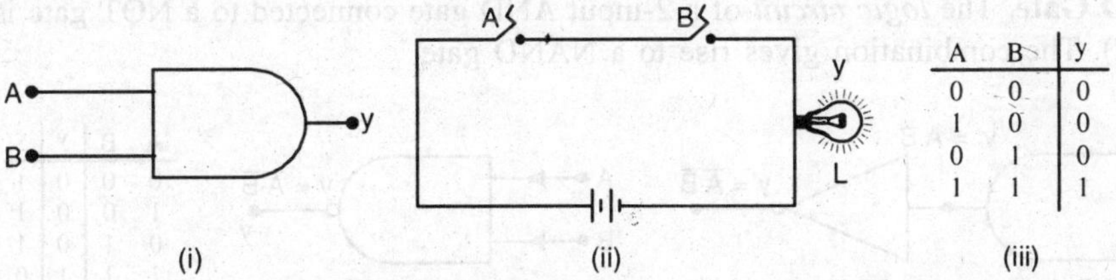

A	B	y
0	0	0
1	0	0
0	1	0
1	1	1

Fig. 69.2

Logic symbol. The AND gate is represented by the logic symbol shown in Fig. 69.2(*i*).

Boolean expression. The AND gate combines the inputs A and B with the output y according to Boolean expression

$$y = A \,.\, B \quad \text{or} \quad y \text{ equals } A \text{ AND } B.$$

Working. The working of the AND gate is represented by the electrical circuit shown in Fig. 69.2(*ii*).

(*i*) When the switch A is OPEN and B is also OPEN ($A = 0; B = 0$) the lamp will not glow ($y = 0$)

(*ii*) When A is CLOSED and B is OPEN ($A = 1; B = 0$) the lamp will **not** glow ($y = 0$)

(*iii*) When A is OPEN and B is CLOSED ($A = 0, B = 1$) the lamp will **not** glow ($y = 0$)

(*iv*) When A is CLOSED and B is also CLOSED ($A = 1; B = 1$) the lamp will glow ($y = 1$).

Truth table. All the FOUR possible inputs and corresponding output of the AND gate are shown in the TRUTH TABLE Fig. 69.2(*iii*). Thus we find that the lamp glows when the switch A AND B are CLOSED. This is why the electrical circuit shown in Fig. 69.2(*ii*) is known as AND gate.

69.7. The NOT gate. The NOT gate has one input A and one output y.

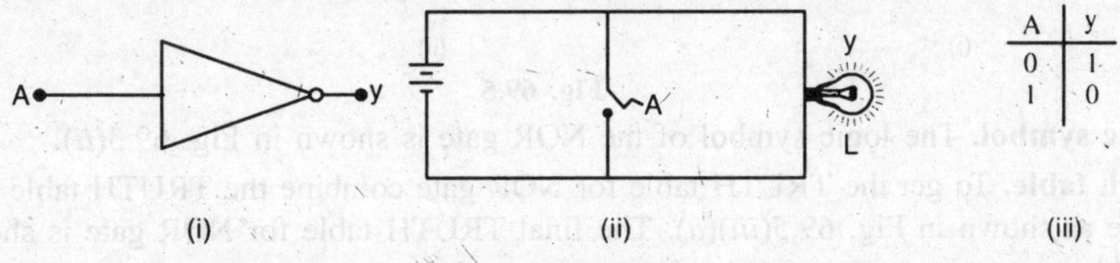

A	y
0	1
1	0

Fig. 69.3

Logic symbol. The NOT gate is represented by the logic symbol shown in Fig. 69.3(*i*).

Boolean expression. The NOT gate combines the inputs A with the output y according to Boolean expression

$$y = \overline{A} \text{ i.e., } y \text{ equals NOT } A.$$

Working. The working of the NOT gate is represented by the electrical circuit shown in Fig. 69.3(*ii*).

(*i*) When switch A is OPEN ($A = 0$) the lamp will glow ($y = 1$)

(*ii*) When switch A is CLOSED ($A = 1$) the lamp will not glow ($y = 0$)

Truth table. The two possible input output combinations are shown in the TRUTH table Fig. 69.3(*iii*).

This shows that the lamp glows only when the switch A is NOT CLOSED. This is why the electrical circuit shown in Fig. 69.3(*ii*) is known as a NOT gate.

The NOT gate is also known as *Invertor.*

69.8. Combination of Gates. By combining OR, AND and NOT gates we can make a large number of complicated digital circuits. We shall, however, discuss only three such combinations *i.e.*, NAND, NOR and XOR gates.

NAND Gate. The *logic circuit* of a 2-input AND gate connected to a NOT gate is shown in Fig. 69.4(*i*). The combination gives rise to a NAND gate.

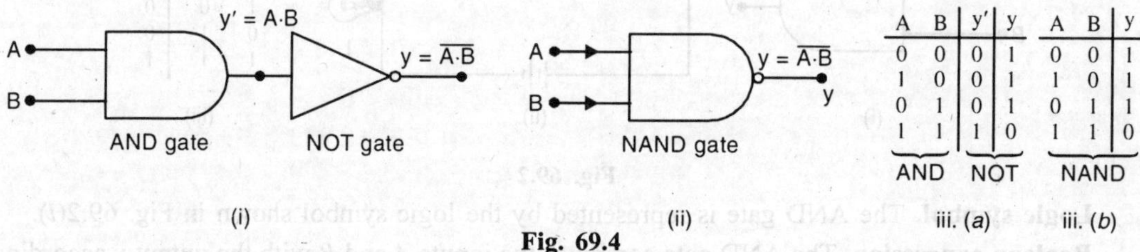

A	B	y′	y
0	0	0	1
1	0	0	1
0	1	0	1
1	1	1	0
AND		NOT	

A	B	y
0	0	1
1	0	1
0	1	1
1	1	0
NAND		

Fig. 69.4

Logic symbol. The logic symbol of the NAND gate is shown in Fig. 69.4(*ii*).

Truth table. To get the TRUTH table for NAND gate combine the TRUTH table for AND and NOT gate as shown in Fig. 69.4(*iii*) (*a*). The final TRUTH table for NAND gate is shown in Fig. 69.4(*iii*) (*b*).

Boolean expression. The Boolean expression for NAND gate is

$$y = \overline{A \cdot B}$$

NOR Gate. The *logic circuit* of a 2-input OR gate connected to a NOT gate is shown in Fig. 69.5(*i*). The combination gives rise to a NOR gate.

y′ = A+B

A

B

y = $\overline{A+B}$

OR gate

NOT gate

(i)

A

B

y = $\overline{A+B}$

y

NOR gate

(ii)

A	B	y′	y
0	0	0	1
1	0	1	0
0	1	1	0
1	1	1	0
OR gate		NOT	

(a)

A	B	y
0	0	1
1	0	0
0	1	0
1	1	0
NOR		

(b)

(iii)

Fig. 69.5

Logic symbol. The logic symbol of the NOR gate is shown in Fig. 69.5(*ii*).

Truth table. To get the TRUTH table for NOR gate combine the TRUTH table for OR and NOT gate as shown in Fig. 69.5(*iii*)(*a*). The final TRUTH table for NOR gate is shown in Fig. 69.5(*iii*)(*b*).

Thus we find that in a NOR gate, if all the inputs are at logic 0, the output y will be at logic 1. If any input is at logic 1, the output y will be at logic 0.

Boolean expression. The Boolean expression for NOR gate is

$$y = \overline{A+B}$$

Exclusive OR (XOR) Gate. The logic circuit of an XOR gate is shown in Fig. 69.6(*i*). It is obtained by using two NOT gates, two AND gates and one OR gate.

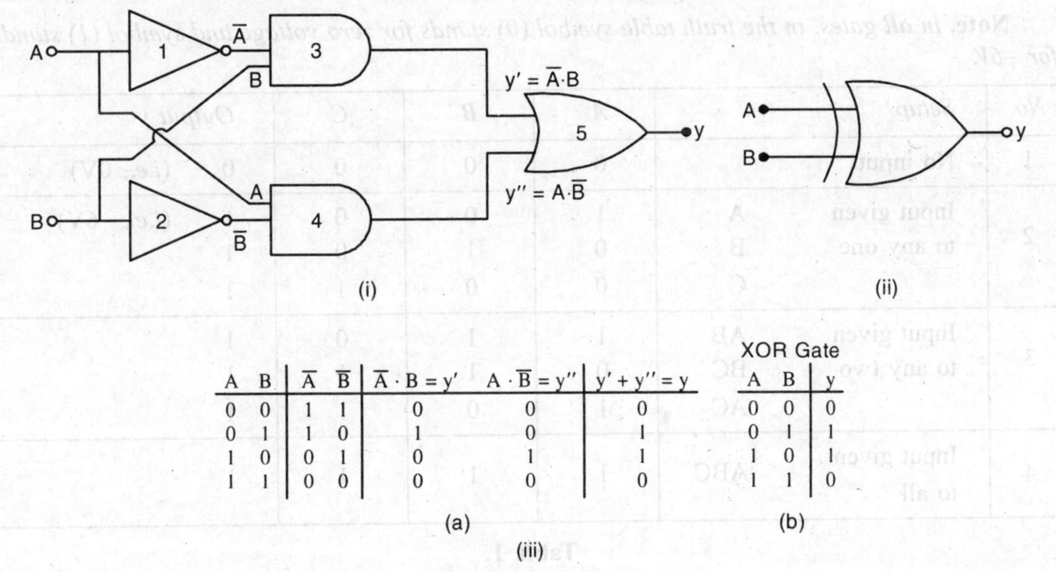

A	B	$\overline{A}$	$\overline{B}$	$\overline{A}\cdot B = y'$	$A\cdot\overline{B} = y''$	$y' + y'' = y$
0	0	1	1	0	0	0
0	1	1	0	1	0	1
1	0	0	1	0	1	1
1	1	0	0	0	0	0

XOR Gate

A	B	y
0	0	0
0	1	1
1	0	1
1	1	0

Fig. 69.6

The logic symbol of XOR gate is shown in Fig. 69.6(*ii*).

Truth table. The NOT gate 1 inverts A to $\overline{A}$ and NOT gate 2 inverts B to $\overline{B}$. The AND gate 3 has two inputs $\overline{A}$ and B which give the output $y' = \overline{A} \cdot B$. AND gate 4 has two inputs A and $\overline{B}$ which give the output $y'' = A \cdot \overline{B}$. The inputs to the OR gate (5) are y' and y'' which give the output $y = y' + y''$. The final TRUTH table for an XOR gate is shown in Fig. 69.6(*iii*) (*b*).

Boolean expression. $y = y' + y'' = \overline{A} \cdot B + A \cdot \overline{B}$.

Thus we find that a XOR gate justifies its definition because a two input circuit provides an output when one input or the other is present but not when both the inputs are present.

Experiment 69.1. To study (*a*) OR, (*b*) AND, (*c*) NOR and (*d*) NAND gates using transistors.

Apparatus. Transistors 2(AC 125), diodes, Resistances 15 kΩ, 2.2 kΩ, 220 kΩ, 150 kΩ, 47 kΩ, 1.5 kΩ etc., 6V d.c. supplies and measuring instrument like voltmeter etc.

(*a*) OR gate.

Procedure. 1. Draw the circuit diagram as shown in Fig. 69.7 and make connections accordingly.

2. Without applying any input *i.e.,* zero input note the output. It will be 0 (zero).

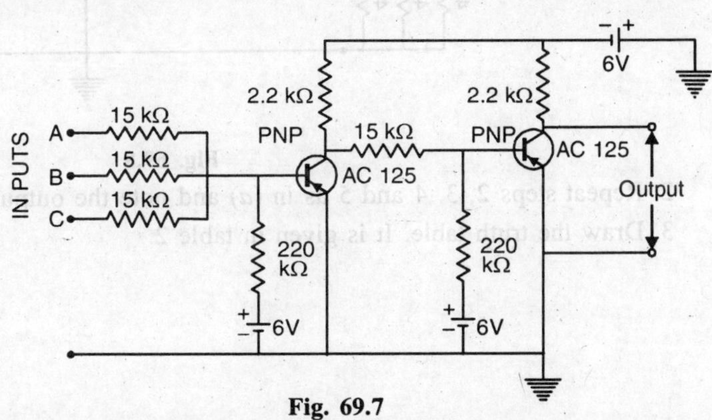

Fig. 69.7

3. Now apply input (–6V) to either *A*, or *B*, or *C* and note the output at the output terminals. It will be 1 (–6V).

4. Apply input (–6V) to any two at a time *i.e.*, (*AB* or *BC* or *AC*) and note the output. The output will be 1(–6V).

5. Now apply input (–6V) to all the three (*A* and *B* and *C*) and note the output. The output will be 1(–6V).

6. Draw the truth table. It is given in table 1.

Note. *In all gates, in the truth table symbol (0) stands for zero voltage and symbol (1) stands for –6V.*

No.	*Setup*		*A*	*B*	*C*	*Output y*	
1	No input		0	0	0	0	(*i.e.*, 0V)
2	Input given to any one	A	1	0	0	1	(*i.e.*, –6V)
		B	0	1	0	1	
		C	0	0	1	1	
3	Input given to any two	AB	1	1	0	1	
		BC	0	1	1	1	
		AC	1	0	1	1	
4	Input given to all	ABC	1	1	1	1	

Table 1.

(*b*) AND gate

Procedure. 1. Draw the circuit diagram as shown in Fig. 69.8 and make connections accordingly.

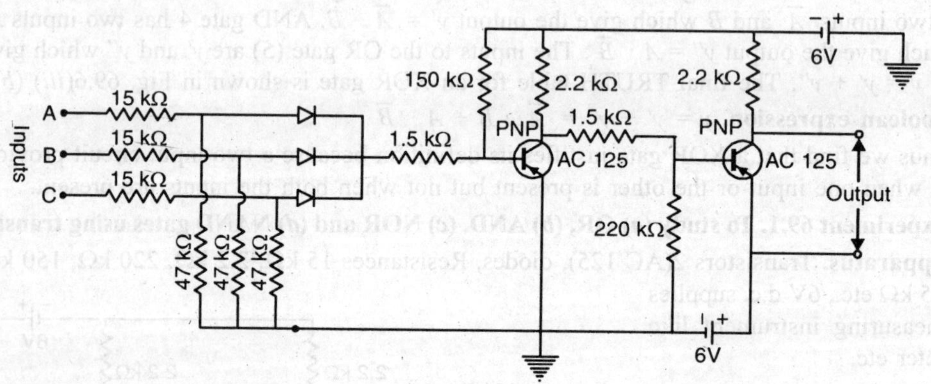

Fig. 69.8

2. Repeat steps 2, 3, 4 and 5 as in (*a*) and note the output in all the four cases.

3. Draw the truth table. It is given in table 2.

No.	Setup		A	B	C	Output y
1	No input		0	0	0	0 (i.e., 0V)
2	Input given to any one	A	1	0	0	0
		B	0	1	0	0
		C	0	0	1	0
3	Input given to any two	AB	1	1	0	0
		BC	0	1	1	0
		AC	1	0	1	0
4	Input given to all	ABC	1	1	1	1 (i.e., – 6V)

Table 2.

(c) NAND gate

Procedure. 1. Draw the circuit diagram as shown in Fig. 69.9 and make connections accordingly.

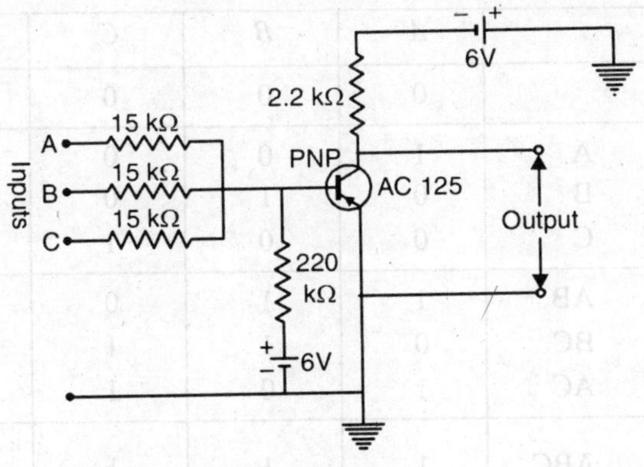

Fig. 69.9

2. Repeat steps 2, 3, 4 and 5 as in (a) and note the output in all the four cases.

3. Draw the truth table. It is given in table 3.

No.	Setup		A	B	C	Output y
1	No input		0	0	0	1 (i.e., – 6V)
2	Input given to any one	A	1	0	0	1
		B	0	1	0	1
		C	0	0	1	1
3	Input given to any two	AB	1	1	0	1
		BC	0	1	1	1
		AC	1	0	1	1
4	Input given to all	ABC	1	1	1	0 (i.e., 0V)

Table 3.

(*d*) NOR gate

Procedure. 1. Draw the circuit diagram as shown in Fig. 69.10 and make connections accordingly.

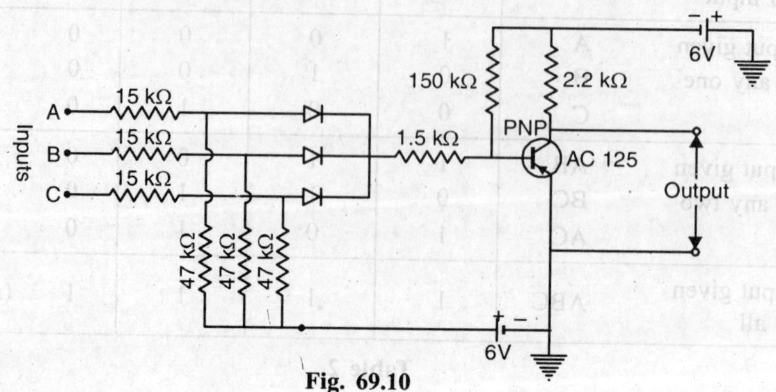

Fig. 69.10

2. Repeat steps 2, 3, 4 and 5 as in (*a*) and note the output in all the four cases.
3. Draw the truth table. It is given in table 4.

No.	*Setup*		*A*	*B*	*C*	*Output y*
1	No input		0	0	0	1 (*i.e.*, – 6V)
2	Input given to any one	A	1	0	0	0
		B	0	1	0	0
		C	0	0	1	0
3	Input given to any two	AB	1	1	0	0
		BC	0	1	1	0
		AC	1	0	1	0
4	Input given to all	ABC	1	1	1	0 (*i.e.*, 0V)

Table 4.

Oral Questions

(*i*) What is binary number system? (*ii*) How do you convert a decimal number into binary number? (*iii*) What is logic gate? (*iv*) What is truth table of a logic gate? (*v*) Draw the logic symbol of an OR gate. (*vi*) Give the truth table of an OR gate with two inputs? (*vii*) Draw the logic symbol of an AND gate. (*viii*) Give the truth table of an AND gate with two inputs. (*ix*) Draw the logic symbol of a NOT gate. (*x*) Give the truth table of a NOT gate with two inputs. (*xi*) Draw the logic symbol of a NAND gate. (*xii*) Give the truth table of a two input NAND gate. (*xiii*) Draw the logic symbol of a NOR gate. (*xiv*) Give the truth table of a two input NOR gate. (*xv*) Draw the logic symbol of an XOR gate. (*xvi*) Give the logic symbol of two input XOR gate. (*xvii*) Why is XOR gate so called?

PUNJAB TECHNICAL UNIVERSITY

Experiments in Physics for B. Tech Course

1. To study the magnetic field of a circular coil carrying current.
   See **Expt. 25.1**.
2. To find out the frequency of A.C. mains using electric vibrator.
   See **Expt. 34.5**.
3. To study laser beam characteristics like wavelength and divergence.
   See **Expt. 64.1** For divergence of a laser beam and
   Expt. 64.2 For wavelength using a diffraction grating.
4. Study of diffraction using laser beam and thus to determine grating element.
   See **Expt. 64.2(*a*)** and **Expt. 64.2(*b*)**.
5. To find out the susceptibility of $FeCl_3$ by Quincke's method.
   See **Expt. 68.1**.
6. To find polarisability of a dielectric substance.
   See **Expt. 70.1**.
7. To study laser interference using Michelson's interferometer.
   See **Expt. 71.1**.
8. To study characteristics of a rectangular wave guide.
   See **Expt. 72.1**.
9. Determination of numerical aperture of an optical fibre.
   See **Expt 73.1**
   and **Expt. 73.2**.
10. To determine attenuation and propagation loss in an optical fibre.
    See **Expt. 73.3**
    and **Expt. 73.4**.

70

Polarisability of a Dielectric

70.1. Polarisability of a dielectric substance. When a non-polar molecule is placed in an electric field, the centre of the positive charge moves in the direction of the field and the centre of negative charge in an opposite direction. This separation of positive and negative charges continues till the force on either of them due to the external field is completely balanced by the internal forces arising due to their relative displacements. The molecule develops a dipole moment known as *induced dipole moment.* Such a molecule is said to the *polarised*. The induced dipole moment disappears as soon as the electric field is removed.

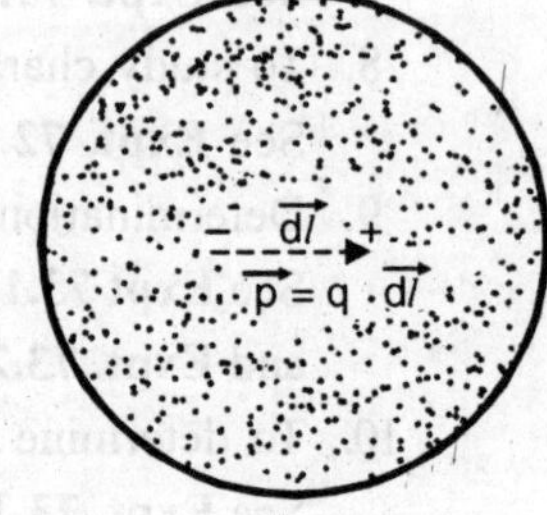

Fig. 70.1

If $\vec{dl}$ is the separation between the centres of positive and negative charges and q is the positive or negative charge on the atom, then

$$\text{Atomic dipole moment } \vec{p} = q\,\vec{dl}$$

Thus atomic dipole moment is the product of the positive or negative charge on the atom and the distance between the positive and negative charges in the direction of the applied electric field.

The net induced electric dipole moment $\vec{p}$ for an atom is proportional to the strength of the applied electric field $\vec{E}$ and in S.I. units, the ratio of $\vec{p}$ to $\varepsilon_o \vec{E}$ is known as atomic polarisability α.

$$\therefore \quad \text{Atomic polarisability } \alpha = \frac{\vec{p}}{\varepsilon_o \vec{E}}$$

According to Claussius Mossoti equation

$$\frac{N\alpha}{3} = \frac{\varepsilon_r - 1}{\varepsilon_r + 2} = \frac{k-1}{k+2}$$

where N is the number of molecules per unit volume of the dielectric and ε_r = relative permittivity = k = dielectric constant.

$N\,\alpha$ is known as *polarisability*

$$\text{Polarisability} = \frac{3(k-1)}{k+2}$$

To find the value of polarisability of a dielectric substance we experimentally determine the value of k the dielectric constant.

70.2. Parallel resonant circuit. Consider a circuit having a capacitance C (Farad) and inductance L (Henry) in parallel. If the resistance of the circuit is negligible, then the natural frequency of this circuit.

$$f = \frac{1}{2\pi\sqrt{LC}} \text{ Hertz}$$

If a high frequency alternating potential difference given by $E = E_0 \sin \omega t$ is applied to such a circuit, then

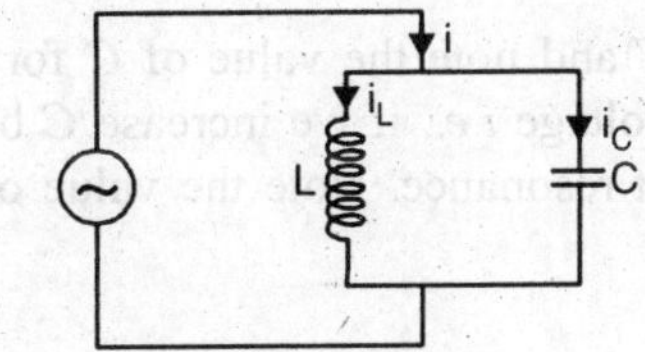

Fig. 70.2

current through L, $\quad i_L = \dfrac{E_0}{L\omega} \sin(\omega t - \pi/2)$

and current through C, $\quad i_C = \dfrac{E_0}{1/C\omega} \sin(\omega t + \pi/2)$

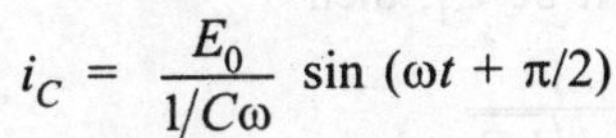

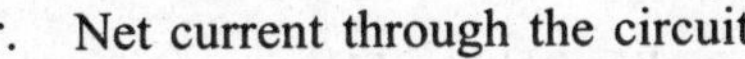

$\therefore$ Net current through the circuit

$$i = \frac{E_o}{L\omega} \sin(\omega t - \pi/2) + E_o\, C\omega \sin(\omega t + \pi/2)$$

$$= E_o \left[C\omega - \frac{1}{L\omega} \right] \cos \omega t$$

If $C\omega = \dfrac{1}{L\omega}$, the net current in the circuit will be zero or minimum. This will happen, when

$$\omega = \frac{1}{\sqrt{LC}} \quad \text{or} \quad f = \frac{1}{2\pi\sqrt{LC}}$$

i.e., the frequency of the A.C. supply is equal to the natural frequency of the LC circuit.

At this stage resonance is said to take place. Although the net current in the circuit at resonance is minimum (or zero) the current through L or through C is maximum and the two currents have a phase difference of 180° between them.

Determination of dielectric constant. To determine the value of k by resonance method, the circuit is shown in Fig. 70.3.

Experiment 70.1. (*a*) To determine the dielectric constant of a solid by resonance method, (*b*) To find out the polarisability of a dielectric substance.

Apparatus. A variable radio-frequency calibrated oscillator with variable output power, a calibrated variable capacitor with capacitance from 50 to 1000 pF, a circular parallel plate capacitor whose dielectric can be removed (or inserted) without changing the distance between the plates, a sensitive V.T. V.M or an A.C. millivoltmeter, an inductive coil L of a few turns of 18 S.W.G. enamelled copper wire having a diameter of about 1.5 cm.

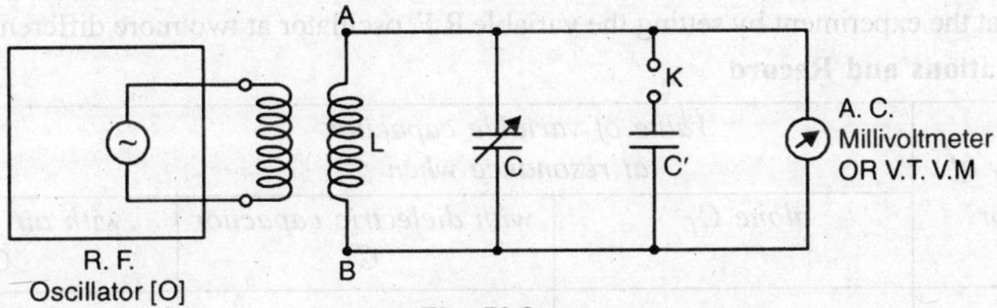

Fig. 70.3

Experimental arrangement. The experimental arrangement is shown in Fig. 70.3. O is an R.F. oscillator fitted with a calibrated frequency dial. The inductance coil L is connected to the output terminals of the oscillator. The two ends A and B of the coil L are connected to the calibrated variable capacitor C, a capacitor C' of unknown capacitance and an A.C. millivoltmeter (or V.T.V.M.) all placed in parallel.

Procedure and Theory. 1. Switch on the R.F. oscillator. Set it at the lowest frequency. Vary the value of the graduated variable capacitor C and note the resonance. Now, set the oscillator on the highest frequency, again vary C and note the resonance. This shows that resonance will take place for the whole range of the oscillator from the lowest to the highest value.

2. Disconnect the capacitor C'. Set the oscillator at a convenient frequency say $f = \dfrac{\omega}{2\pi}$. Vary C and note the value of C for which the V.T.V.M. or (A.C. millivoltmeter) gives the maximum voltage *i.e.*, if we increase C beyond this value, the voltage begins to decrease. This is the value at resonance. Note the value of C. Let it be C_1, then

$$\omega = \frac{1}{2\pi\sqrt{LC_1}} \quad ...(i)$$

3. Now include the unknown capacitor C' (with dielectric disc in it) and having a capacitance C_D (say). Repeat the experiment by varying C to again obtain maximum voltage or resonance. Let the reading of the variable capacitor now be C_2, then

$$\omega = \frac{1}{2\pi\sqrt{L(C_D + C_2)}} \quad ...(ii)$$

4. Now carefully remove the dielectric disc from the capacitor C' without changing the distance between the plates. Repeat the experiment again by varying C to obtain maximum voltage or resonance. Let the reading of the variable capacitor now be C_3, then

$$\omega = \frac{1}{2\pi\sqrt{L(C_A + C_3)}} \quad ...(iii)$$

where C_A is the capacitance of the capacitor C' with air as dielectric. From (*i*) and (*ii*), we have

$$\frac{1}{2\pi\sqrt{LC_1}} = \frac{1}{2\pi\sqrt{L(C_D + C_2)}}$$

or $$C_D + C_2 = C_1 \quad \text{or} \quad C_D = C_1 - C_2$$

From (*i*) and (*iii*), we have

$$\frac{1}{2\pi\sqrt{LC_1}} = \frac{1}{2\pi\sqrt{L(C_A + C_3)}}$$

or $$C_A + C_3 = C_1 \quad \text{or} \quad C_A = C_1 - C_3$$

$\therefore$ $$\text{Dielectric constant } k = \frac{C_D}{C_A} = \frac{C_1 - C_2}{C_1 - C_3}$$

5. Repeat the experiment by setting the variable R.F. oscillator at two more different frequencies.

Observations and Record

Frequency of oscillator f	*Value of variable capacitor at resonance when*		
	alone C_1	*with dielectric capacitor* C_2	*with air capacitor* C_3
1 =			
2 =			
3 =			

$$\text{Dielectric constant } k = \frac{C_1 - C_2}{C_1 - C_3}$$

(1)

(2)

(3)

Mean $k =$

(*b*) Polarisability $N\alpha = \dfrac{3(k-1)}{(k+2)} =$

Precautions. 1. Resonance should be checked for whole frequency range by noting the resonance for the lowest as well as the highest frequency of the R.F. oscillator.

2. Resonance must be sharp.

3. The dielectric disc must fit tightly into the capacitor and should be removed carefully without disturbing the capacitor.

Oral Questions

(*i*) Define atomic polarisability. (*ii*) Define relative permittivity or dielectric constant. (*iii*) What are units of dielectric constant? (*iv*) State Claussius Mossoti relation. (*v*) What is the natural frequency of an LC circuit? (*vi*) What is resonance? (*vii*) What is the condition for resonance? (*viii*) What is the phase difference between voltage across C and voltage across L when resonance takes place in an LC circuit having L and C in parallel? (*ix*) What is net current in the parallel LC circuit at resonance? (*x*) What is a Farad? (*xi*) What is a microfarad? (*xii*) In what units L is measured?

71

Michelson's Interferometer

71.1. Michelson's interferometer. The underlying principle of Michelson's interferometer is as under: '*The amplitude of the light beam from an extended source is divided into two parts of equal intensities by partial reflection and refraction. These beams are sent in two directions at right angles and are brought together after reflection from plane mirrors to produce interference fringes.*'

Construction. The main optical parts consist of two highly polished *front silvered plane mirrors. M_1 and M_2* and two *plane parallel* glass plates G_1 and G_2 of the same thickness. The plate G_1 is *half silvered* at the back so that the incident beam is divided into a reflected and a transmitted beam of equal intensity. The plates G_1 and G_2 are held parallel to each other and are inclined at an angle of 45° to the mirrors M_1 and M_2 which are mutually perpendicular to each other. The mirror M_1 is mounted on a carriage and can be moved exactly parallel to itself with the help of a micrometer screw fitted with a graduated drum capable of reading displacement of the order of 10^{-5} cm. The mirrors M_1 and M_2 are provided with levelling screws at their backs with the help of which these can be made perfectly perpendicular to the direction of the two beams. The interference bands are observed in the field of view of the telescope *T*. (See Fig. 71.1).

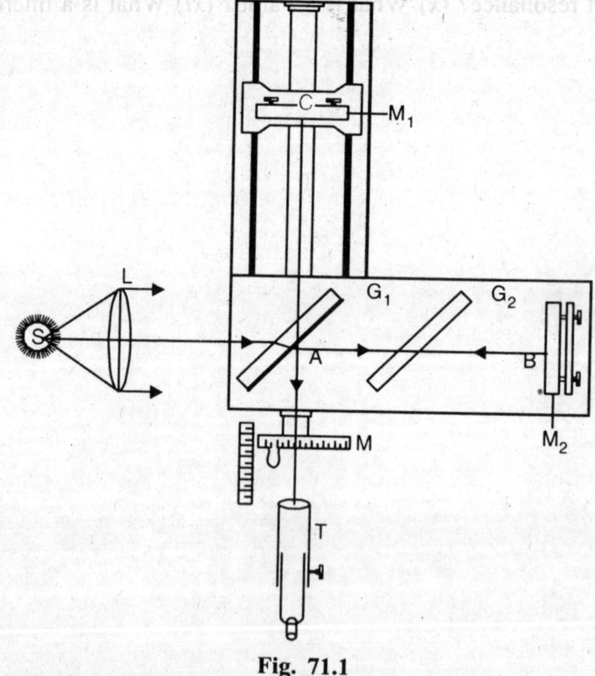

Fig. 71.1

Least count. The chief micrometer screw *M* consists of a *large wheel* and a *small fine screw wheel.* The large wheel has 100 graduations on it. As this large wheel is given one full rotation, the screw moves through 1 mm on the linear scale. Therefore, each graduation of the large wheel

= 0.01 mm. The fine screw wheel also has 100 graduations on it and for one full rotation of the fine screw wheel only one division of the large wheel is moved.

$$\therefore \qquad \text{Least count} = \frac{1}{100} \times 0.01\,\text{mm}$$

$$= 10^{-4}\ \text{mm} = 10^{-5}\ \text{cm.}$$

Working. Light from the source S is rendered parallel by means of a collimating lens L and is made to fall on the glass plate G_1. It is partly reflected at the back surface of G_1 along AC and partly transmitted along AB. The ray AC is received by the plane mirror M_1 normally, so that it is reflected along the same path and emerges out along AT. The transmitted ray AB is received by the mirror M_2 normally, reflected along the same path and then moves along AT after reflection at the back surface of G_1. Thus two beams along AT are produced from a single source by division of the amplitude. These two beams produce interference under suitable conditions.

Compensating plate. It is clear from Fig. 71.1 that a ray starting from the source and suffering reflection at the mirror M_1 traverses the glass plate G_1 thrice, whereas the ray reflected from the mirror M_2 traverses the glass plate G_1 only once. To compensate for this an exactly similar glass plate G_2 is introduced in the path AB parallel to the glass plate G_1.

Adjustment. Make the distance of the mirrors M_1 and M_2 from the glass plate G_1 to be very nearly equal.

To make the incident light from the source parallel, place a tin sheet with a fine hole in front of the source S just opposite to the bright part of it. Adjust the hole and source of light in line with the centres of the glass plates G_1 and G_2 and the mirror M_2. Now view the images of the pin hole with the naked eye in the direction AT. Ordinarily four pin hole images as shown in Fig. 71.2 will be seen. Two images are bright (strong) and two are less bright (weak). Now adjust the screws at the backs of the mirrors M_1 and M_2 such that the two weaker images merge with the stronger ones and only two images remain in the field of view. Now remove the tin sheet with the pin hole and if the two paths are exactly equal the field of view will be totally dark.

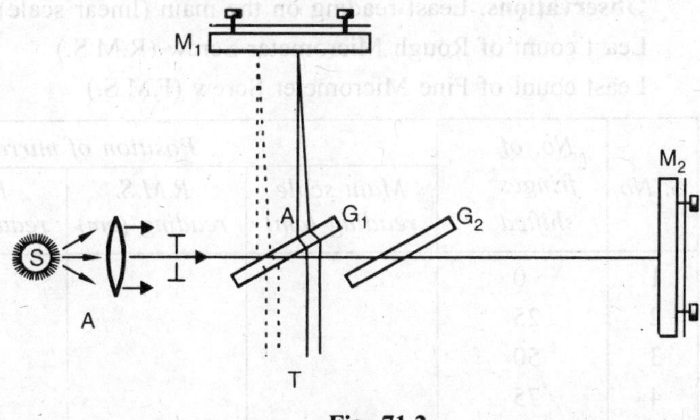

Fig. 71.2

Now slightly move the mirror M_1 parallel to itself. Circular interference fringes will be visible in the field of view. Now focus the telescope at the centre and view the fringes through it for measurement.

Experiment 71.1. To measure the wavelength of light from the given source using Michelson's interferometer.

Apparatus. Michelson's interferometer, source of light (sodium vapour lamp or a Laser source), a tin sheet with a fine hole, a convex lens etc.

Theory. The interferometer is adjusted to obtain circular fringes in the field of view of the telescope. If the mirrors M_1 and M_2 are equidistant from the glass plate G_1, the field of view will be perfectly dark. The position of the mirror M_1 is adjusted till a particular bright fringe appears in the field of view of the telescope with its centre coinciding with the cross-wire. When the mirror M_1 is moved backward or forward, each fringe in the focal plane of the telescope is displaced parallel to itself. When the mirror M_1 is moved through a distance $\lambda/2$, the path-difference changes by λ and the position of a particular bright fringe is taken by the next bright fringe. If n is the number of fringes that move across the field of view when the mirror is displaced through a distance l, then

$$n\frac{\lambda}{2} = l$$

or $$\text{Wavelength } \lambda = \frac{2l}{n}$$

Procedure. 1. Find the least count of Michelson's interferometer as explained under the head **Least count.**

2. Obtain clear and bright circular fringes in the field of view of the telescope as explained under the head **Adjustment.**

3. When the centre of the cross-wires (as seen through the telescope) lies on a bright fringe, note and record the main (or linear) scale reading, R.M.S. reading and F.M.S. reading.

4. Now move the *Fine Micrometer Screw* (F.M.S.) slowly and go on counting the number of bright fringes that shift across the field of view. (The fringes will go either rising or sinking). When a large number of fringes say 25 have shifted again, make the centre of the cross wire (as seen through the telescope) lie on a bright fringe. Note and record the reading on the main (linear) scale, R.M.S. and F.M.S.

5. Similarly repeat the process after shifting another 25 fringes and so on. Take about 20 readings.

Observations. Least reading on the main (linear scale) = mm = cm

Least count of Rough Micrometer Screw (R.M.S.) = mm = cm

Least count of Fine Micrometer Screw (F.M.S.) = mm = cm

S. No	No. of fringes shifted	Position of mirror M_1				Difference for 100 fringes
		Main scale reading (cm)	R.M.S. reading (cm)	F.M.S. reading (cm)	Total cm	
1	0					
2	25					
3	50					
4	75					
...						
...						
...						
...						
19	475					
20	500					

Mean difference for 100 fringes = l = cm.

$\therefore$ Wavelength of light $\lambda = \frac{2l}{n} =$ cm = Å

NOTE. Here $n = 100$

Precautions. 1. Glass plates G_1, G_2 and mirrors M_1, M_2 are front silvered. These should not be touched or cleaned.

2. The screws behind the mirror M_2 should be rotated through a very small angle.

3. The micrometer screw should always be moved in the same direction to avoid error due to backlash.

4. There should be no movement near the experiment as the fringes are likely to be disturbed.

5. There should be no linear or lateral displacement of fringes.

Oral Questions

(*i*) What is interference? (*ii*) What do you mean by interferometer? (*iii*) What are the two types of interference phenomenon? (*iv*) In Michelson's interferometer does interference take place by 'division of wave front' or 'division of amplitude'? (*v*) Why is the plate G_1 half silvered? (*vi*) What is the function of plate G_2? (*vii*) Why is it called the compensating plate? (*viii*) What is the angle between the two mirrors M_1 and M_2? (*ix*) What is the shape of fringes that you get? (*x*) Where are the circular fringes formed? (*xi*) What are localised fringes? (*xii*) When the mirror is moved through a distance $\frac{\lambda}{2}$, how many interference rings appear or disappear? (*xiii*) Can you measure the difference between two wavelengths by Michelson's interferometer? (*xiv*) What is the least count of the Michelson's interferometer you are using?

72

Characteristics of Rectangular Wave Guide

72.1. Wave guide. At frequencies higher than 3 GHz (1 GHz = 10^9 Hz), transmission of electromagnetic waves along transmission lines and cables becomes difficult mainly due to losses that occur both in the solid dielectric needed to support the conductor and in the conductors themselves. A metallic tube can be used to transmit electromagnetic waves at these frequencies.

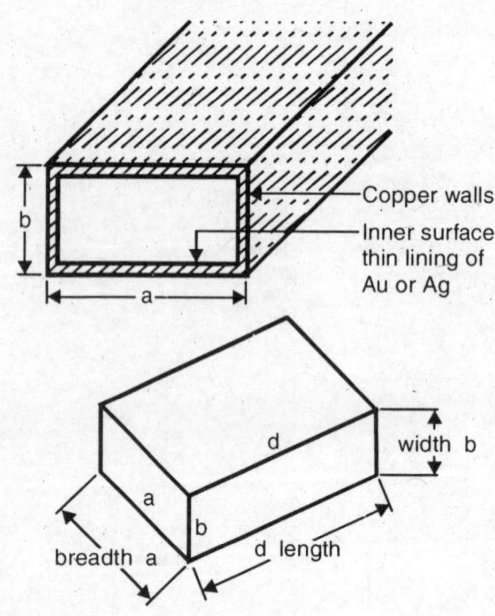

Fig. 72.1

A hollow metallic tube of uniform cross-section for transmitting electromagnetic waves by successive reflections from the inner walls of the tube is called a wave guide.

The wave guides (WG) are generally air filled. Any shape of cross-section of a wave guide can support electromagnetic waves. But since irregular shapes are difficult to analyse, these are rarely used. *Rectangular* and *circular* waveguides are very common. A rectangular wave guide is shown in Fig. 72.1. It has copper walls and its inner surface has a thin lining of gold or silver. The *breadth* of the wave guide is *a*, width *b* and length *d*.

72.2. Propagation of electromagnetic waves through a wave guide. The electromagnetic waves inside a wave guide can have an infinite number of patterns which are called **modes**. An electromagnetic wave consists of electric and magnetic fields which are always perpendicular to each other. The fields in the wave guide which make up these mode patterns must obey certain physical laws. At the surface of a conductor, the electric field cannot have a component parallel to the surface. This indicates that the electric field must always be perpendicular to the surface of the conductor. The magnetic field, on the other hand, is always parallel to the surface of the conductor and cannot have a component perpendicular to it at the surface.

72.3. Transverse Electric (TE) and Transverse Magnetic (TM) waves. In general, there are two kinds of modes in a wave guide. In the first type, the electric field is always transverse to the direction of propagation and is called the **Transverse Electric** or **TE** wave.

In the second type, the magnetic field is always transverse to the direction of propagation and is called **Transverse Magnetic** or **TM** wave. Thus in a *TE* mode, no electric line is in the direction of propagation *i.e.,* $\mathbf{E_Z = 0}$ but $\mathbf{H_Z \neq 0}$ if *Z* is the direction of propagation. In a TM mode no magnetic line is in the direction of propagation *i.e..,* $\mathbf{H_Z = 0}$ but $\mathbf{E_Z \neq 0}$.

72.4. TE waves in rectangular wave guides. The TE_{mn} is a general mode in a wave guide characterized by $E_Z = 0$. In other words, the *Z*-component of the magnetic field, H_Z must exist in order to have energy transmission in the guide.

For a rectangular wave guide for which breadth = *a* along *X*-axis, width *b* along *Y*-axis and wave propagates along *Z*-axis, on solving the differential equations for a *TE* wave we get

$$H_z = C \cos\left(\frac{m\pi}{a}\right) x \cos\left(\frac{n\pi}{b}\right) y\, e^{(j\omega t - \gamma z)}$$

where symbols have their usual meaning.

The other field components of *E i.e.,*

E_x contains the terms $\cos\left(\frac{m\pi}{a}\right) x \sin\left(\frac{n\pi}{b}\right) y$ and

E_y contains the terms $\sin\left(\frac{m\pi}{a}\right) x \cos\left(\frac{n\pi}{b}\right) y$

Similarly the other field components of *H i.e.,*

H_x contains the terms $\sin\left(\frac{m\pi}{a}\right) x \cos\left(\frac{n\pi}{b}\right) y$ and

H_y contains the terms $\cos\left(\frac{m\pi}{a}\right) x \sin\left(\frac{n\pi}{b}\right) y$

TE modes in Rectangular wave guides. As stated above TE_{mn} is the general mode and specific modes are given by various values of *m* and *n*

(*a*) $\mathbf{TE_{00}}$ **mode.** $m = 0$, $n = 0$.
All field components vanish. Therefore it cannot exist.

(*b*) $\mathbf{TE_{01}}$ **mode.** $m = 0$, $n = 1$
$\therefore E_y = 0$, $H_x = 0$, E_x and H_y exist.

(*c*) $\mathbf{TE_{10}}$ **mode.** $m = 1$, $n = 0$
$\therefore E_x = 0$, $H_y = 0$, E_y and H_x exist

(*d*) $\mathbf{TE_{11}}$ **mode.** $m = 1$, $n = 1$
All field components exist.

72.5. Cut off wavelength. The cut off wavelength $\lambda_{c,\,mn}$ is given by

$$\lambda_{c,mn} = \frac{2ab}{\sqrt{m^2 b^2 + n^2 a^2}}$$

All wavelengths greater than λ_c are attenuated and those less than λ_c are allowed to propagate inside the wave guide.

For TE_{10} mode $\lambda_c = 2a$

where *a* is the breadth or broad dimension of the wave guide. TE_{10} mode is the *dominant mode* in rectangular wave guides since for this mode λ_c has the maximum value equal to $2a$.

72.6. Wavelength in the wave guide. The wavelength in the wave guide λ_g is different from the wavelength in free space. In fact, it is related to the free space wavelength λ_o and cut off wavelength λ_c by

$$\frac{1}{\lambda_g^2} = \frac{1}{\lambda_o^2} - \frac{1}{\lambda_c^2}$$

Experiment 72.1. To find the frequency and wavelength in a rectangular wave guide working on TE_{10} mode.

Apparatus. Klystron tube with Klystron mount, Klystron power supply, Isolator, frequency meter, variable attenuator, slotted line, tunable probe, VSWR meter, movable short, matched termination.

Before giving the procedure, we give a brief description of various pieces of apparatus to be used.

1. Klystron. A Klystron is a vacuum tube that can be used either as a generator or as an amplifier of power at micro-wave frequencies. There is a two–cavity Klystron, multi–cavity Klystron and a reflex Klystron.

A reflex Klystron is the one which is most widely used. It is a single cavity, variable frequency micro-wave generator of low power and low efficiency.

It consists of an electron gun, a filament surrounded by cathode and a focussing electrode at cathode potential as shown in Fig. 72.2. The electron beam is accelerated towards the anode cavity at positive potential. After passing the gap in the cavity, the electrons travel towards a repeller electrode which is at a high negative potential V_R. The electrons never reach the repeller because of the negative field and are returned back towards the gap. Under suitable conditions the electrons give more energy to the gap than they took from the gap on their forward journey and oscillations are sustained.

It is assumed that oscillations are set up in the tube initially due to noise or switching transients and these oscillations are sustained by device operation.

The R.F. voltage that is produced across the gap by cavity oscillations act on the electron beam to cause velocity modulation. Suppose an electron passes through the gap when the gap voltage is zero and going negative. Let us call it, the *reference electron* e_R. The velocity of this electron is not affected by the gap voltage. It moves towards the repeller and gets reflected by the negative voltage of the repeller. It returns and passes through the gap a second time.

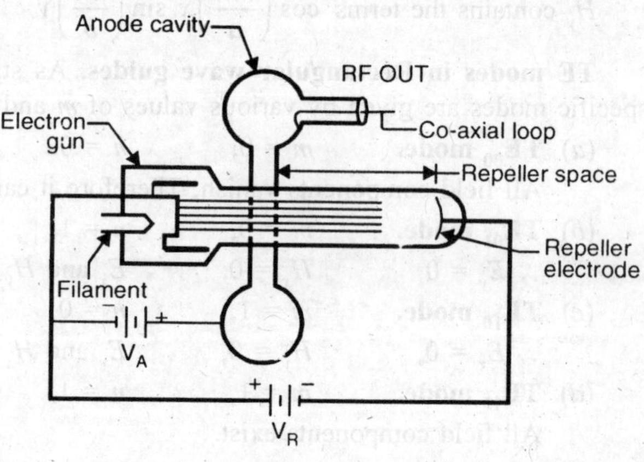

Fig. 72.2

An electron which passes through the gap before the reference electron e_R is called the *early electron* e_E. It experiences maximum positive voltage across the gap and is accelerated. It, therefore, moves with greater velocity and penetrates deep into the repeller space. The return time for electron e_E is greater as its depth of penetration into repeller space is more. Hence the electron e_E and e_R appear at the gap for the second time at the same instant.

An electron which passes through the gap later than the reference electron e_R is called *late electron* e_L. It experiences maximum negative voltage and moves with a retarding velocity. Its return time is shorter as its penetration into the repeller space is less. It also catches up with the electrons e_R and e_E, thus forming a bunch.

Bunches occur once per cycle, centred around the reference electron e_R and these bunches transfer maximum energy to the gap to produce sustained oscillations. For oscillations to be sustained the time taken by the electron to travel into repeller space and back to the gap (called transit time) must have an optimum value given by

$$T = n + \frac{3}{4} \text{ cycles}$$

where n is any integer.

The Klystron tube is mounted on the Klystron mount.

2. Klystron power supply. It is used to work the Klystron tube. It gives regulated power output. Various switches and controls of most of the power supplies are :

(*i*) ON, OFF switch

(*ii*) (*a*) BEAM VOLTAGE knob. It adjusts the variable accelerating voltage applied to the electron beam.

(*b*) REPELLER VOLTAGE knob. It adjusts the variable retarding voltage applied to the repeller electrode.

(*iii*) METER. The meter can measure current in mA and voltage (in volts).

(*iv*) SELECT DISPLAY switch. It is put on V for voltage measurement, on *C* for current measurement and on REP for repeller voltage measurement by the meter.

(*v*) MOD switch. It has three modes

(*a*) CW for continuous wave

(*b*) AM for amplitude modulation

(*c*) FM for frequency modulation.

(*vi*) (*a*) AM *amplitude and Frequency knob.* It controls the amplitude and frequency for amplitude modulation.

(*b*) FM *amplitude and Frequency knob.* It controls the amplitude and frequency for frequency modulation.

3. Isolator. An isolator is a two port device which provides very *small* amount of attenuation for transmission from *port* (1) to *port* (2) but provides *maximum* attenuation from *port* (2) to *port* (1). This requirement is very much desirable when we want to match a source with a variable load.

In most micro-wave generators, the output amplitude and frequency tend to fluctuate very significantly with changes in load impedance. This is due to mismatch of generator output to the load resulting in reflected wave from load. But these reflected waves should not be allowed to reach the microwave generator which will cause amplitude and frequency instability of the microwave generator.

When *isolator* is inserted between the generator and the load, the generator is coupled to the load with zero attenuation and reflections, if any, from the load side are completely absorbed by the isolator without affecting the generator output. Hence the generator appears to be matched to all loads in the presence of the isolator so that there is no change in frequency and output power due to variation in load.

4. Frequency meter. Microwave frequency can be measured by either electronic or mechanical techniques. The electronic techniques are more accurate. In such techniques, high frequency heterodyne systems are used. Here, the unknown frequency f_{un} is compared with harmonics of a known lower frequency f_c by use of a low frequency generator, a harmonic generator and a mixer as shown in Fig 72.3.

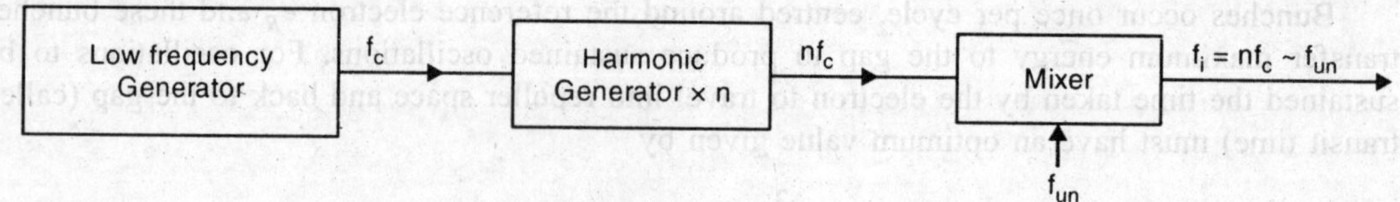

Fig. 72.3

The intermediate frequency f_i is the difference of $n f_c$, the nth harmonic of low frequency f_c and unknown frequency f_{un}. As f_i and $n f_c$ are known, f_{un} can be measured.

5. Variable Attenuators. For perfect matching we sometimes require that microwave power in a wave guide be completely absorbed without any reflection. For this we make use of attenuators. Attenuators are also used for measuring power gain or loss in dB's, for providing isolation between instruments, for reducing the power input to a particular stage to prevent over loading. Attenuators are further used for providing the signal generators with a means of calibrating their outputs accurately so that precise measurements could be made.

Attenuators are of two types:

Fixed Attenuators are used where fixed amount of attenuation is to be provided.

Variable Attenuators provide continuous or step-wise variable attenuation. For rectangular wave guides these attenuators can be *flap-type* or *vane-type.*

6. Slotted line. A coupling probe moving along the wave guide can be used to detect the standing wave pattern present inside the wave guide. It consists of a slotted section of a transmission line *i.e.,* wave guide, a travelling probe carriage and facility for attaching detecting instruments. A slot is made in the centre of the broad face of the guide parallel to the axis of the wave guide. For the dominant mode travelling inside the wave guide the slot does not radiate any power. A small *probe* inserted through the slot senses the relative field strength of the standing wave pattern inside the wave guide. The probe is mounted on a carriage plate which moves on the top surface of the wave guide and is connected to a crystal detector. As the probe is moved along the wave guide the crystal detector gives an output voltage proportional to the standing wave pattern inside the wave guide at that point. As the crystal is a square law device, the square root of the ratio of the maximum output to minimum output gives the *voltage standing wave ratio* (VSWR). For noting the position, the precision built probe carriage has a centimeter scale with a vernier reading of 0.1 mm least count.

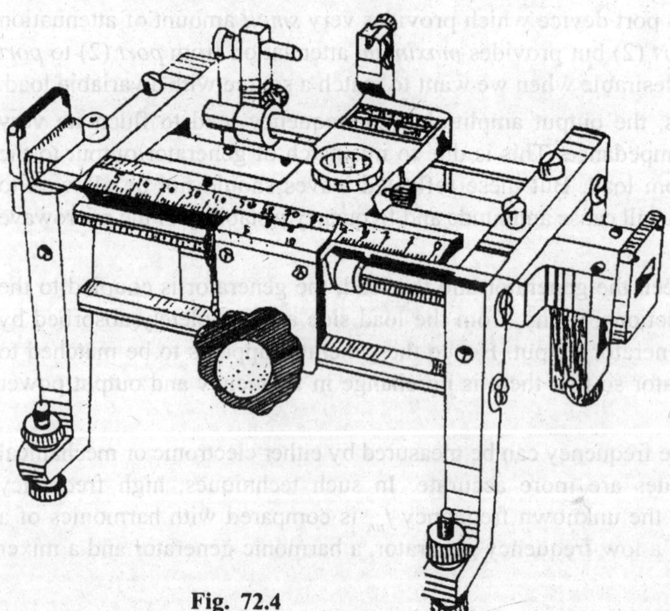

Fig. 72.4

The probe extends into the slot coming quite close to the inner conductor of the line but not touching it. This loose coupling between line and probe is adequate for measurement purposes. The probe must be quite thin as compared to the wavelengths and the depth also must be small enough to avoid any field distortion.

The slotted line carriage with tunable detector (Fig. 72.4) is used to measure the standing wave ratio using a VSWR meter.

7. Voltage Standing Wave Ratio (VSWR). The magnitude of the standing waves can be measured in terms of standing wave ratio. A moving detector along the line can indicate the voltage maximas (V_{max}) and minimas (V_{min}). The ratio of these two is called *Voltage Standing Wave Ratio* or VSWR denoted by S

$$\therefore \qquad S = \frac{|V_{max}|}{|V_{min}|}$$

The value of S lies between 1 and ∞

For a short circuit load $\quad S = \infty$

For an open circuit load $\quad S = \infty$

For a matched load $\quad S = 1.$

VSWR meter. A VSWR meter, basically consists of a high gain, high Q, low noise voltage amplifier normally tuned at a fixed frequency (1 kHz) at which the micro-wave signal is modulated. The VSWR meter uses the detected signal out of the micro-wave detector as its input, amplifies the same and provides the output on a calibrated voltmeter. The meter itself can be calibrated directly in terms of VSWR.

In this case, the probe carriage is moved to give maximum deflection on the VSWR meter by adjusting the pad. This full scale deflection (FSD) corresponds to VSWR of 1 as shown with Fig. 72.5. As an example, an F.S.D. of 10 mV corresponds to a VSWR of 1. The travelling probe is adjusted to get the minimum reading on the meter. If this corresponds to 5 mV, then

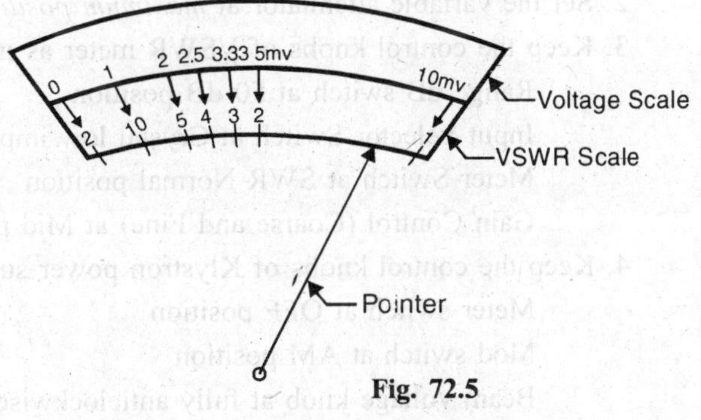

Fig. 72.5

$$\text{VSWR} = \frac{10 \text{ mV}}{5 \text{ mV}} = 2.$$

If it is 3.3 mV, VSWR = 3, if it is 2.5 mV, VSWR = 4. If it is 1 mV, VSWR = 10.

The actual VSWR meter has a gain control to adjust the reading to a desired value by fine or coarse adjusting knobs.

A VSWR meter has the following controls:–

(*i*) ON, OFF switch

(*ii*) VSWR meter with following scales

(*a*) SWR Normal. It is to be used when VSWR is between 1 and 4

(*b*) SWR Expanded. It is to be used when VSWR is less than 1.3.

(*c*) dB scale-normal

(*d*) dB scale expanded (between 0–2dB).

(*iii*) GAIN CONTROL switch. It has two knobs, one for *coarse* and the other for *fine* adjustment.

(*iv*) RANGE dB switch. It is used to adjust the overall gain in steps of 10, *i.e.,* 0, 10, 20, 30, 40, 50, 60, ..., dB.

(*v*) INPUT TERMINAL and INPUT SELECTOR switch. It is provided for different inputs, *i.e.,*

(*a*) Crystal (XTL) low currents 4.5 mA or

(*b*) High currents (8.75 mA) Bolometer bias.

8. Wave guide terminations. In a wave guide system, it is not possible to attach a fixed resistive load as a termination. Graphited sand at the end of the waveguide can dissipate energy to achieve termination. Alternatively a resistive rod placed at a point in the wave guide where electric field strength is maximum or a wedge of resistive material (powdered iron or carbon mixed with a binder deposited on a dielectric strip) in the form of a taper can act as termination. All these are *matched terminations* and almost no reflection occurs.

Procedure. 1. Set up the various components and equipment as shown in Fig. 72.6.

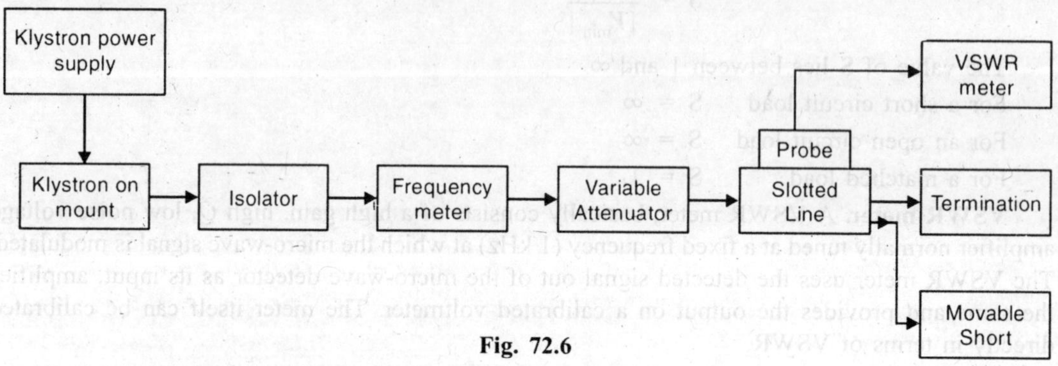

Fig. 72.6

2. Set the variable attenuator at *maximum position.*

3. Keep the control knobs of VSWR meter as indicated

Range dB switch at 50 dB position
Input Selector Switch at Crystal low impedance position
Meter Switch at SWR Normal position
Gain Control (Coarse and Fine) at Mid position

4. Keep the control knobs of Klystron power supply as indicated

Meter switch at OFF position
Mod switch at AM position
Beam voltage knob at fully anticlockwise position
Reflector (Repeller) voltage at fully clockwise position.
AM amplitude knob around fully clockwise position
AM frequency knob around mid-position.

5. Switch 'ON' the Klystron power supply, VSWR meter and Cooling Fan.

6. Turn the meter switch of the power supply to Beam voltage position. Set the beam voltage to 300 V with the help of beam voltage knob.

7. Now adjust the repeller (or reflector) voltage to get some deflection in VSWR meter. Adjust the deflection to maximum value using the AM amplitude and frequency control knobs of the power supply.

8. Tune the plunger of Klystron mount for maximum deflection. Tune the reflector (repeller) voltage knob for maximum deflection and also tune the probe for maximum deflection in VSWR meter.

9. Tune the frequency meter knob to get the 'dip' on the VSWR scale and note down the value of frequency directly from the frequency meter.

10. Replace the Termination with movable short and detune the frequency meter.

11. Move probe alongwith the slotted line, the deflection in the VSWR meter will vary. Move the probe to a minimum deflection position. To get more accurate reading, increase the VSWR meter range dB switch to higher position. Note and record the probe position

12. Move the probe to next minimum position and record the probe position again.

Twice the distance between two successive minimum positions obtained as above gives the guide wavelength λ_g.

13. Measure the wave guide inner broad dimension (breadth) 'a' with the help of a vernier callipers or a travelling microscope.

The cut off wave length $\lambda_c = 2a$ for the TE_{10} mode.

14. Repeat the experiment for at least three different values of frequency and repeat the observation for each frequency also three times.

Observations. Wave guide inner broad dimension (breadth) a = cm = m

$\therefore$ Cut off wavelength λ_c = $2a$ = m

S. No.	*Frequency (from frequency meter)*	*Position of minimum deflection*		*Guide wavelength*	*Mean* λ_g
		1st x_1	*2nd (next)* x_2	$\lambda_g = 2(x_2 - x_1)$	
1	f_1 =	(*i*) (*ii*) (*iii*)	(*i*) (*ii*) (*iii*)	(*i*) (*ii*) (*iii*)	
2	f_2 =	(*i*) (*ii*) (*iii*)	(*i*) (*ii*) (*iii*)	(*i*) (*ii*) (*iii*)	
3	f_3 =	(*i*) (*ii*) (*iii*)	(*i*) (*ii*) (*iii*)	(*i*) (*ii*) (*iii*)	

Frequency $f = \frac{c}{\lambda_a} = c\sqrt{\frac{1}{\lambda_g^2} - \frac{1}{\lambda_c^2}}$ = Hz

f_1 = f_2 = f_3 =

Compare the values with the values noted directly from the frequency meter.

Precautions. 1. Fan should be switched on immediately.

2. The dB switch should be moved to higher position to get an accurate reading on VSWR meter.

3. To find the value of a, inner broad dimension should be measured.

Oral Questions

(*i*) What is a wave guide? (*ii*) Why do we prefer a rectangular wave guide? (*iii*) What do you mean by TE_{10} mode? (*iv*) What is cut-off wavelength? (*v*) What is the value of cut-off wavelength for a rectangular waveguide of breadth a, width b and length d for TE_{10} mode? (*vi*) What is the relation between guide wavelength, cut off wavelength and wavelength of a micro-wave in free space? (*vii*) What is a Klystron? (*viii*) What do you mean by VSWR? (*ix*) What is the value of VSWR for a short circuit? (*x*) What is the value of VSWR for a matched load?

73

Fibre Optics

73.1. Optical fibre. *Optical fibre is a very thin glass or plastic conduit designed to guide light waves along the length of the fibre with the help of successive total internal reflections from the side walls of the fibre.*

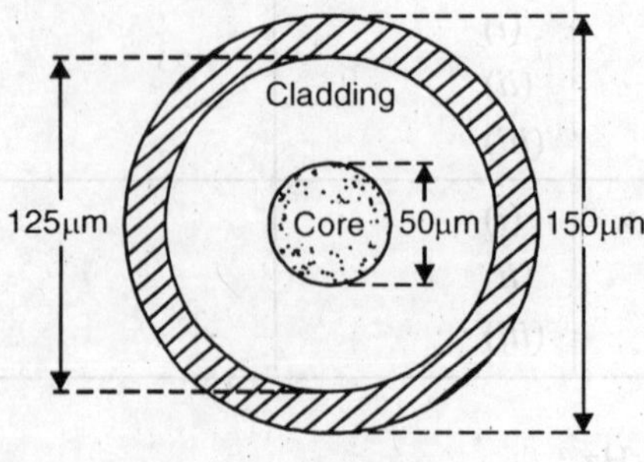

Fig. 73.1

It consists of a thin glass *core* of refractive index μ_1 (about 5–50 μm) in diameter surrounded by a glass coating called *cladding* of refractive index $\mu_2(\mu_1 > \mu_2)$ enclosed in a protective plastic jacket called the **sheath**.

The light gathering power of an optic fibre depends upon (*i*) core size and (*ii*) Numerical aperture the value of which is given by (*a*) acceptance angle and (*b*) relative refractive index difference.

73.2. Acceptance angle. Suppose *AB* is the optical fibre through which light is being sent. The end at which light enters the fibre is called launching end. Refractive index of core is μ_1 and that of cladding is $\mu_2(\mu_1 > \mu_2)$. Let the refractive index of the medium from which light enters the fibre be μ_a.

Consider a ray of light incident on the fibre at an angle of incidence *i*. This ray gets refracted at an angle *r* and strikes the core-cladding interface at an angle ϕ. If ϕ is equal to or more than the critical angle *c* between the core and the cladding then this ray of light will suffer total internal reflection. If *i* increases *r* increases and ϕ decreases (as $\phi = 90 - r$). Thus as *i* increases, ϕ may fall below the critical value.

Fig. 73.2

$\therefore$ Maximum value of *i* occurs when $\phi = c$

Now $$\frac{\mu_1}{\mu_a} = \frac{\sin i}{\sin r} \text{ and } \sin r = \sin(90 - \phi) = \cos\phi$$

$$\therefore \quad \sin i = \frac{\mu_1}{\mu_a}\sin r = \frac{\mu_1}{\mu_a}\cos\phi$$

When $\phi = c$, $\sin i(\max) = \dfrac{\mu_1}{\mu_a}\cos c$

But $\sin c = \dfrac{\mu_2}{\mu_1}$ $\therefore$ $\cos c = \sqrt{1-\left(\dfrac{\mu_2}{\mu_1}\right)^2} = \dfrac{\sqrt{\mu_1^2-\mu_2^2}}{\mu_1}$

$$\therefore \quad \sin i(\max) = \frac{\mu_1}{\mu_a}\frac{\sqrt{\mu_1^2-\mu_2^2}}{\mu_1} = \frac{\sqrt{\mu_1^2-\mu_2^2}}{\mu_a}$$

If the medium from which light enters the fibre is air, then $\mu_a = 1$

$$\therefore \quad \sin i(\max) = \sqrt{\mu_1^2-\mu_2^2}$$

If we denote i (max) by θ_a, we have

$$\sin\theta_a = \sqrt{\mu_1^2-\mu_2^2}$$

The angle θ_a *is called the acceptance angle* for the fibre. *It is defined as the maximum angle of incidence that a light ray can make with the axis of the fibre so that it may propagate through the fibre.*

73.3. Numerical aperture. *The numerical aperture N.A. is defined as the sine of the acceptance angle.*

Numerical aperture is a factor on which depends the light gathering ability of the fibre. It is a measure of the amount of light accepted by the fibre.

Experiment 73.1. To measure the numerical aperture of an optical fibre using a He–Ne Laser Source.

Apparatus. He–Ne Laser Source, 20X microscopic objective, Two fibre optic chucks, optical fibre, screen, graph paper, millimeter scale, measuring tape.

Procedure. 1. Mount both ends of the optical fibre on fibre optic chucks.

2. Couple the light from the He–Ne Laser source to one end of the optical fibre through a 20X microscope objective. This is the '*input end*'.

3. Place a white screen on which a m.m. graph paper has been pasted at some distance from the end other than the one which has been coupled to the light source. This end is known as '*output end*'.

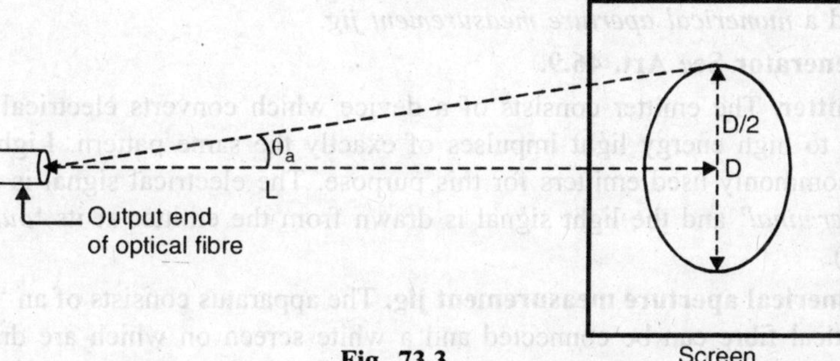

Fig. 73.3

4. Adjust the position of the screen such that it is perpendicular to the axis of the fibre. Move the screen towards or away from the output end of the optical fibre such that a circular spot is formed on the screen. Measure the distance between the output end of the optical fibre and the screen. Let it be L.

5. Measure the diameter of the circular spot in two directions at right angles either by counting the number of m.m. squares on the graph paper or by using a m.m. scale. Let the diameter be D. Then numerical aperture is given by

$$A = \sin\theta_a = \frac{D/2}{\sqrt{L^2 + (D/2)^2}} = \frac{D}{\sqrt{4L^2 + D^2}}$$

6. Repeat the procedure for different values of L and corresponding values of D.

Observations

S. No.	*Diameter of the spot*			*Distance between screen and output end of optical fibre L*	*Numerical aperture*
	Vertical	*Horizontal*	*Mean D*		$\sin\theta_a = \frac{D}{\sqrt{4L^2+D^2}}$

Mean $\sin\theta_a$ =

$\therefore$ Mean numerical aperture A =

Acceptance angle. $\theta_a = \sin^{-1} A$ =

Precautions. 1. Observe all precautions for use of a He–Ne Laser source.

2. The optical fibre must be connected through optic fibre chucks to avoid any leakage.

3. The light from the He–Ne source must be coupled to the optical fibre through a 20X microscopic objective.

73.4. Alternative method. The numerical aperture of an optical fibre can also be measured by using a *signal generator* which can produce a sine wave of frequency 1000 cycles per second, an *emitter* and a *numerical aperture measurement jig.*

Signal generator See **Art. 46.9.**

73.5. Emitter. The emitter consists of a device which converts electrical current impulses supplied to it to high energy light impulses of exactly the same pattern. Light emitting diodes (LED's) are commonly used emitters for this purpose. The electrical signal is fed to the emitter at its '*input terminal*' and the light signal is drawn from the emitter at its '*output terminal*' (or output socket).

73.6. Numerical aperture measurement jig. The apparatus consists of an '*input terminal*' to which the optical fibre can be connected and a white screen on which are drawn 4 concentric circles of diameter 10, 15, 20 and 25 mm respectively as shown in Fig. 73.4(b). The distance between the screen and the vertical board carrying the input terminal can be adjusted and measured on a horizontal scale as shown in Fig. 73.4(a).

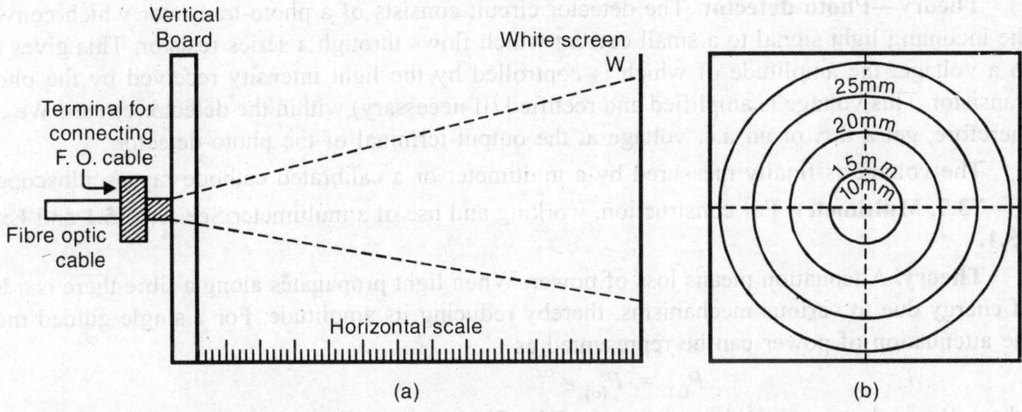

Fig. 73.4

Experiment 73.2. To measure the numerical aperture of an optical fibre using a signal generator, emitter (fitted on a board), Numerical aperture measurement jig etc.

Apparatus. As stated.

Procedure. 1. Set up the apparatus as shown in Fig. 73.5.

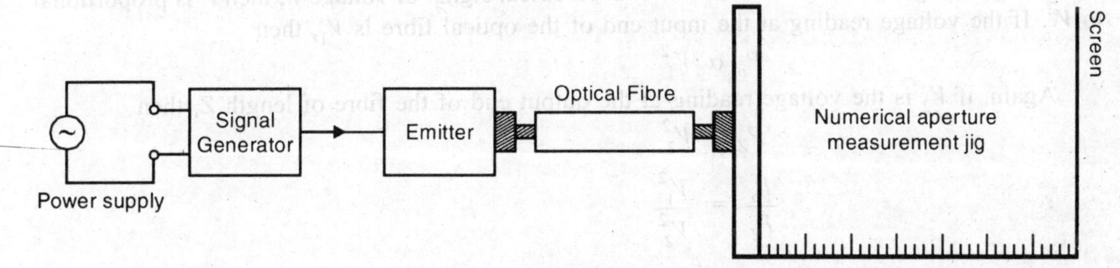

Fig. 73.5

2. Connect the mains power supply to the input of the frequency (signal) generator.

3. Connect the 1 kHz sine wave output of the frequency generator to the input of the emitter circuit. Adjust the amplitude of the signal to a convenient value say 5 V peak to peak.

4. Connect one end of the fibre optic cable to *output socket* of the emitter circuit and the other end to the *input socket* of the numerical aperture measurement jig. Hold the white screen facing the optical fibre such that its cut face is perpendicular to the axis of the fibre.

5. Now switch on the power supply and hold the screen with four concentric circles vertically at a suitable distance to make the bright spot from the fibre coincide with 10 mm circle. Record the distance of the screen from the fibre end. Let it be *L*. Also note the diameter of the spot. Let it be *D*.

6. Repeat the experiment by varying the distance between the screen and fibre optic cable by making the bright spot coincide with 15 mm, 20 mm, and 25 mm circle respectively and note the value of *L* and *D* in each case.

Observations. Same as in last experiment.

Experiment 73.3. To measure the attenuation or propagation loss in an optical fibre using He–Ne Laser Source.

Apparatus. He–Ne Laser, 20*X* microscopic objective, fibre chucks, optical fibre, photo-detector, digital multimeter etc.

Theory—Photo-detector. The detector circuit consists of a photo-transistor which converts the incoming light signal to a small current which flows through a series resistor. This gives rise to a voltage, the amplitude of which is controlled by the light intensity received by the photo-transistor. This voltage is amplified and rectified (if necessary) within the detector circuit. We can, therefore, get a **d.c.** or an **a.c.** voltage at the output terminal of the photo-detector.

The voltage is finally measured by a multimeter or a calibrated cathode ray oscilloscope.

73.7. Multimeter. For construction, working and use of a multimeter See **Art 35.1** and **Expt. 35.1.**

Theory. Attenuation means loss of power. When light propagates along a fibre there is a loss of energy due to various mechanisms, thereby reducing its amplitude. For a single guided mode the attenuation of power can be represented as

$$P_{(z)} = P_{(o)} e^{-\alpha z}$$

where $P_{(o)}$ is the power at the input end of the fibre, $P_{(z)}$ the power after traversing a distance Z in the fibre and α the absorption co-efficient. Thus loss in decibles per kilometer is given by

$$\text{Loss (dB/km)} = \frac{10}{Z} \log \frac{P_o}{P_z}$$

The losses in optical fibre are due to absorption, scattering, bending and micro-bending.

When the light pulse is converted into an electrical signal of voltage V, then P is proportional to V^2. If the voltage reading at the input end of the optical fibre is V_1, then

$$P_o \ \alpha \ V_1^2$$

Again, if V_2 is the voltage reading at the output end of the fibre of length Z, then

$$P_Z \ \alpha \ V_2^2$$

$$\therefore \qquad \frac{P_o}{P_Z} = \frac{V_1^2}{V_2^2}$$

Hence
$$\text{Loss } (dB/\text{Km}) = \frac{10}{Z} \log \frac{P_o}{P_Z} = \frac{10}{Z} \log \frac{V_1^2}{V_2^2}$$

$$= \frac{20}{Z} \log \frac{V_1}{V_2}$$

where Z is in Km.

Procedure. 1. Measure the length of the optical fibre. Let it be Z. Mount both ends of the optical fibre on fibre chucks.

2. Couple a $20X$ microscopic objective to the He–Ne Laser source. Place the photo-detector in such a manner that light from the microscopic objective is completely falling on it.

Set the multimeter in **d.c.** voltage mode and connect it to the output terminal of the photo-detector. Note the reading of voltage in the multimeter. Let it be V_1.

3. Couple one of the fibre ends to the $20X$ microscopic objective attached to the He–Ne Laser Source. This is the *input end.* Now place the photo–detector just at the *output end* of the fibre so that light coming out of this end falls completely on the detector. Let the reading of the multimeter connected to the output terminal of the photo-detector now be V_2.

4. Take 3 readings with one length of the fibre and repeat with one more length.

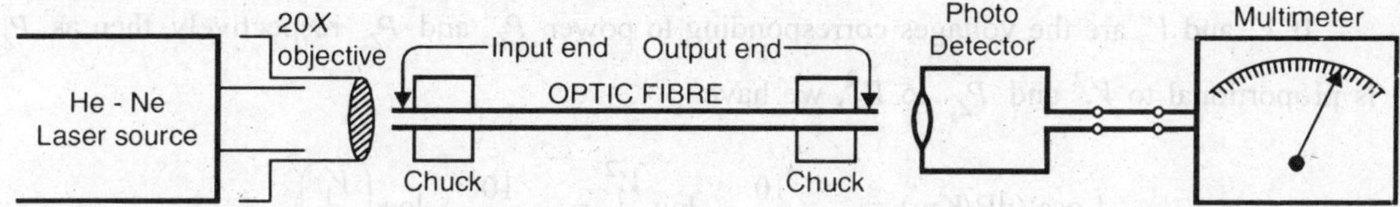

Fig. 73.6

Observations and Record

S. No.	*Length of optical fibre Z Km*	*Voltage at the input end V_1 volts*	*Voltage at the output end V_2 volts*	*Loss in dB/Km* = $\frac{20}{Z} \log \frac{V_1}{V_2}$
1. (*i*)				
(*ii*)				
(*iii*)				
2. (*i*)				
(*ii*)				
(*iii*)				

Precautions. Same as in **Expt. 73.1**.

73.8. Alternative method. The attenuation or propagation loss of an optical fibre can also be measured using a signal generator which can produce a sine wave of frequency 1000 cycles per second, an emitter, a photo-detector and a multimeter (or a cathode ray oscilloscope).

Experiment 73.4. To measure the propagation or attenuation loss in an optical fibre using a signal generator, emitter and photo-detector.

Apparatus. A signal generator, emitter (fitted on a board), photo-detector (fitted on a board) and two pieces of optical fibre 1 metre and 10 metre (or 5 metre) in length, fibre chucks, multimeter (or a cathode ray oscilloscope).

Theory. As already stated in the last experiment, for a single guided mode, the attenuation of power can be represented as

$$P_Z = P_o e^{-\alpha Z}$$

where P_o is the power at the input end of the fibre, P_z the power after traversing a distance Z in the fibre and α the absorption co-efficient. For a fibre of length Z_1, we have

$$P_{Z_1} = P_o e^{-\alpha Z_1}$$

and for a fibre of length Z_2, we have

$$P_{Z_2} = P_o e^{-\alpha Z_2}$$

$$\therefore \quad \frac{P_{Z_1}}{P_{Z_2}} = \frac{e^{-\alpha Z_1}}{e^{-\alpha Z_2}} = e^{-\alpha(Z_1 - Z_2)}$$

The loss in decibles per kilometer is given by

$$\text{Loss (dB/Km)} = \frac{10}{Z_1 - Z_2} \log \frac{P_{Z_1}}{P_{Z_2}}$$

If V_1 and V_2 are the voltages corresponding to power P_{Z_1} and P_{Z_2} respectively, then as P_{Z_1} is proportional to V_1^2 and P_{Z_2} to V_2^2, we have

$$\text{Loss (dB/Km)} = \frac{10}{(Z_1 - Z_2)} \log \frac{V_1^2}{V_2^2} = \frac{10}{(Z_1 - Z_2)} \log \left(\frac{V_1}{V_2}\right)^2$$

$$= \frac{20}{(Z_1 - Z_2)} \log \frac{V_1}{V_2}$$

Procedure. 1. Set up the apparatus as shown in Fig. 73.7.

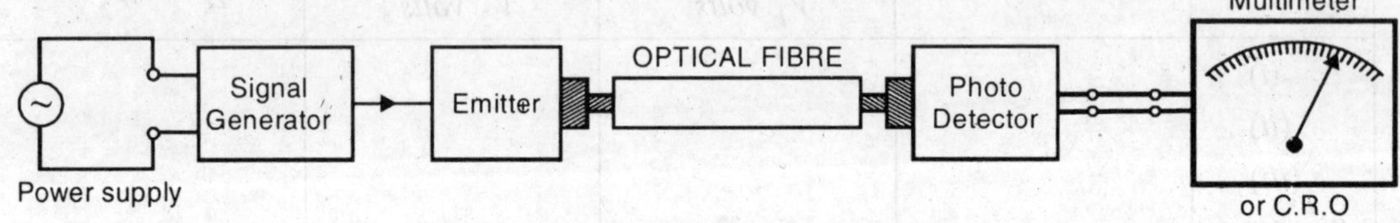

Fig. 73.7

2. Connect the mains power supply to the input of the frequency (signal) generator.

3. Connect the 1 KHz sine wave output of the frequency generator to the input of the emitter circuit. Adjust the amplitude of the signal to a convenient value.

4. Measure the length of the smaller piece of the fibre optic cable. Let it be Z_1.

5. Connect one end of this piece to the output socket of the emitter circuit and the other end to the input socket of the photo-detector. Connect the output socket of the photo-detector to the multimeter (or a cathode ray oscilloscope). Now switch on the power supply and note the voltage reading given by the multimeter (or cathode ray oscilloscope). Let it be V_1.

6. Now measure the length of the longer piece of the fibre optic cable. Let it be Z_2.

7. Replace the small piece of optical fibre by the longer piece. Proceeding as in step 5, again switch on the power supply. Note the reading of the multimeter (or C.R.O.) for the same amplitude of the signal from the frequency generator. Let it be V_2.

8. Take at least 3 readings with each cable.

Observations and Record

Length of the optical fibre (Smaller piece) Z_1 = m = Km

(Longer piece) Z_2 = m = Km.

S. No.	*Voltage at the output end*		*Loss in dB/Km* =
	Smaller piece V_1	*Longer piece* V_2	$\frac{20}{Z_2 - Z_1} \log \frac{V_1}{V_2}$
1.			
2.			
3.			

Mean Loss in *dB*/Km =

Oral Questions

(*i*) What is an optical fibre? (*ii*) What parts of optical fibre are known as core, cladding and sheath? (*iii*) What is the relation between refractive indices of core and cladding? (*iv*) What is acceptance angle of an optical fibre? (*v*) What is numerical aperture of an optical fibre? (*vi*) What is the relation between acceptance angle and numerical aperture? (*vii*) If the medium from which light enters the fibre is air, then what is the value of numerical aperture in terms of refractive indices of core and cladding? (*viii*) What is the function and construction of an emitter? (*ix*) What is the function and construction of a photo-detector? (*x*) What do you mean by attenuation or propagation loss in a fibre? (*xi*) In what units is it measured? (*xii*) What is the relation between P_o the power at the input end and P_z the power at the output end of a fibre of length Z? (*xiii*) What are the main uses of an optical fibre? (*xiv*) How does an optical fibre help in telecommunication?

TABLES OF CONSTANTS

1. Density of common substances.

(in gm per cc)

Substance	*Density*	*Substance*	*Density*	*Substance*	*Density*
Solid		Paraffin wax	0.87 – 0.93	Spirit	0.83
Aluminium	2.70	Sodium chloride	2.17	Milk	1.03
Brass	8.4 — 8.7	Sand (silver)	2.63	Olive oil	0.91 – 0.93
Copper	8.93	Silver	10.5	Turpentine oil	.87
Cork	.22—.26	Steel	7.7 – 7.9	**Gases**	at N.T.P.
Gold	19.32	Sugar	1.59	Air	0.001293
Glass (crown)	2.5—2.7	**Liquids**		Carbon dioxide	0.00191977
Glass (Flint)	2.9—4.5	Alcohol (ethyl)	.792	Helium	0.0001785
Glass (pyrex)	2.25	Glycerine	1.26	Hydrogen	0.0000899
Iron (cast)	7.1 — 7.7	Kerosene oil	.799	Nitrogen	0.001251
Iron (wrought)	7.8 — 7.9	Mercury	13.6	Oxygen	0.001429
Lead	11.37			Steam at (100°C)	0.000581

In S.I. units density is expressed in kg per m^3. To find the density in S.I. units multiply the values given in the above table by 10^3.

2. Elastic constants.

Metal	*Young's Modulus (Y)* Nm^{-2}	*Rigidity* η Nm^{-2}	*Poison's ratio* σ
Aluminium	7.50×10^{10}	2.67×10^{10}	0.34
Brass	$0.7 - 10.2 \times 10^{10}$	3.5×10^{10}	0.34 – 0.38
Copper	$12.4 - 12.9 \times 10^{10}$	4.55×10^{10}	0.34
Iron (wrought)	$19.9 - 20.0 \times 10^{10}$	$7.7 - 8.3 \times 10^{10}$	0.27
Steel	$19.5 - 20.6 \times 10^{10}$	7.9 to 8.9×10^{10}	0.28
Rubber	—	—	0.05

3. Surface tension and angle of contact :

Substance	*Surface tension in* Nm^{-1}	*Angle of constant with glass*
Water	73×10^{-3}	8° – 9°
Mercury	465×10^{-3}	137° – 20′
Turpentine	27.3×10^{-3}	17°

4. Viscosity of various substance in Deca-Poise or kg/m s.

Substance	15° C	20° C	50° C
Water	0.00114	0.00101	0.00055
Mercury	—	0.00156	0.00141
Pitch	51×10^8	1.3×10^9 (at 15°c)	—
Air	183×10^{-7} (at 23°C)	—	—
Hydrogen	89×10^{-7}	—	—
Oxygen	198×10^{-7}	—	—

5. Refractive index of substances.

For Sodium Light $\lambda = 5895 \times 10^{-10}$ m

Solid	*Refractive Index*	*Liquid*	*Refractive Index*
Glass (crown)	1.51 – 1.61	Water	1.333
Glass (flint)	1.58 – 1.96	Alcohol (ethyl)	1.362
Diamond	2.417	Canada Balsam	1.53
Mica	1.56 – 1.60	Cedar oil	1.516
Quartz (O-ray)	1.5443	Glycerine	1.47
Quatz (E-ray)	1.5334	Olive oil	1.46
Ice	1.31	Turpentine	1.47

6. Wavelength of spectral lines.

Sodium	Yellow	D_2	5890×10^{-10} m
	Yellow	D_1	5896×10^{-10} m
Potassium	Blue		4703×10^{-10} m
	Green		5452×10^{-10} m
	Yellow		5607×10^{-10} m
	Yellow		5912×10^{-10} m
	Orange		6031×10^{-10} m
	Red		7668×10^{-10} m
	Red		7702×10^{-10} m

7. Specific rotation.

Cane sugar 66.67 (Dextro-rotatory)
Turpentine 37 (Leave-rotatory)

8. E.M.F. of cells.

Cell	*E.M.F.* in volts	*Cell*	*E.M.F. in volts*
Daniel	1.08 – 1.09	Cadmium at 20°C	1.01854
Leclanche	1.45	Lead accumulator	1.9 – 2.2
Clark at 15°C	1.433	Edison Cell	1.45

9. Specific resistance and temperature co-efficient.

Metals

Element	*Sp. Resist. Ohm. m.* $\times 10^{-8}$	*Temp. Coefficient* $\alpha \times 10^{-4}$	*Element*	*Sp. Resist. Ohm. m.* $\times 10^{-8}$	*Temp. Coefficient* $\alpha \times 10^{-4}$
Aluminium	2.8	49	Mercury	95.8	8.9
Copper	1.78	42.8	Nickle	11.8	27
Iron	9.8	62	Gold	2.42	40
Lead	22.1	41	Silver	1.65	40
Zinc	5.9	35	Platinum	11	37

Alloys

Sl. No.	*Alloys*	*Specific Resistance Ohm. m* $\times 10^{-8}$	*Temp. Coefficient* $\alpha \times 10^{-4}$
1.	Constantan (58.80% Cu., 40% Ni 1.2 % Mn)	49.1	– 0.4 to – 0.1
2.	Manganin (85 % Cu, 12 % Mn,, 30 % Ni)	44.5	0.02 – 0.5
3.	Nichrome (67.5 % ni,, 15 % Cr,, 16 % Fe,, 1.5 % Mn)	110	1.7 – 3.5
4.	Brass	6.6	10
5.	Phosphor Bronze	5 – 10	35
6.	Platinoid	34.4	2.5
7.	German Silver	26.6	2.3 – 6

10. Electro-chemical equivalents

Element	*Chemical Equivalent*	*E.C.E in gm./coulomb*
Copper	63.57/2	0.0003294
Hydrogen	1.008/1	0.00001045
Silver	107.88/1	0.001118

11. E.M.F. of common thermocouples.

Values of a and b in the equation $E = at + bt^2$

Couple	*a*	*b*
Copper constantan	87.54×10^{-6}	0.0446×10^{-6}
Copper iron	13.40×10^{-6}	-0.137×10^{-6}

12. Value of dip B_H and *g*.

Place	*Angle of dip.*	*Horizontal component* $B_H \times 10^{-4}$ *tesla*	*Acceleration due to gravity (g)* ms^{-2}
Agra	39° – 57.8′	0.358	9.79
Ajmer	39° – 46′	0.358	9.78
allahabad	36° – 7.3′	0.370	9.79
Amritsar	47° – 21.3′	0.300	9.792

Place	*Angle of dip.*	*Horizontal component* $B_H \times 10^{-4}$ *tesla*	*Acceleration due to gravity (g)* ms^{-2}
Bangalore	9° – 21.2′	0.405	9.78
Bombay	35° – 56.5′	0.372	9.786
Calcutta	34° – 14′	0.389	9.788
Delhi	42° – 12.7′	0.350	9.792
Hyderabad (Deccan)	19° – 39.2′	0.397	9.789
Labore	47° – 23.9′	0.330	9.794
Lucknow	38° – 55.8′	0.362	9.79
Ludhiana	45° – 13.3′	0.335	9.792
Nagpur	42° – 09′	0.385	9.795
Patna	36° – 24′	0.373	9.784

13. Boiling point of water in C under various barometeric pressure.

Baromettric height mm	0	1	2	3	4	5	6	7	8	9
680	96.91	.95	.99	97.03	0.7	.11	.15	19	.23	.27
690	97.31	.35	.39	.43	.47	.51	.55	.59	.63	.67
700	97.71	.75	.79	.83	.83	.87	.91	.95	.98	98.02
710	98.10	.14	.18	.22	.26	.30	.34	.37	.41	.45
720	98.49	.53	.57	.61	.64	.68	.72	.76	.80	.84
730	98.87	.91	.95	.99	99.03	.07	.10	.14	.18	.22
740	99.25	.29	.33	.37	.41	.44	.48	.52	.55	.59
750	99.63	.67	.70	.74	.78	.82	.85	.89	.93	.96
760	100.00	.04	.07	.11	.15	.18	.22	.26	.29	.33

14. Universal constants.

Planck's Constant $h = 6.624 \times 10^{-34}$ Joule sec.

Charge on the Electron $e = 1.6008 \times 10^{-19}$ Coulomb

Avogadro's Number $N = 6.023 \times 10^{-23}$ per mole

Faraday = 96489 Coul/gm eq.

Millicurie $= 3.71 \times 10^{7}$ dis./sec.

Boetzmann's Constant $k = 1.38 \times 10^{-23}$

$$\frac{1}{4\pi e_0} = 8.988 \times 10^{9}\ N\ \ m^2\ C{-2}$$

$$\frac{\mu_0}{4\pi} = 10^{-7}\ \text{Wb/A} - \text{m}$$

$$1\ eV = 1.60 \times 10^{-19}\ \text{J}$$

LOGARITHMS TABLES

COMMON LOGARITHMS

	0	1	2	3	4	5	6	7	8	9	Mean Differences								
											1	2	3	4	5	6	7	8	9
10	0000	0043	0086	0128	0170	0212	0253	0294	0334	0374	4	8	12	17	21	25	29	33	37
11	0414	0453	0492	0531	0569	0607	0645	0682	0719	0755	4	8	11	15	19	23	26	30	34
12	0792	0828	0864	0899	0934	0969	1004	1038	1072	1106	3	7	10	14	17	21	24	28	31
13	1139	1173	1206	1239	1271	1303	1335	1367	1399	1430	3	6	10	13	16	19	23	26	29
14	1461	1492	1523	1553	1584	1614	1644	1673	1703	1732	3	6	9	12	15	18	21	24	27
15	1761	1790	1818	1847	1875	1903	1931	1959	1987	2014	3	6	8	11	14	17	20	22	25
16	2041	2068	2095	2122	2148	2175	2201	2227	2253	2279	3	5	8	11	13	16	18	21	24
17	2304	2330	2355	2380	2405	2430	2455	2480	2504	2529	2	5	7	10	12	15	17	20	22
18	2553	2577	2601	2625	2648	2672	2695	2718	2742	2765	2	5	7	9	12	14	16	19	21
19	2788	2810	2833	2856	2878	2900	2923	2945	2967	2989	2	4	7	9	11	13	16	18	20
20	3010	3032	3054	3075	3096	3118	3139	3160	3181	3201	2	4	6	8	11	13	15	17	19
21	3222	3243	3263	3284	3304	3324	3345	3365	3385	3404	2	4	6	8	10	12	14	16	18
22	3424	3444	3464	3483	3502	3522	3541	3560	3579	3598	2	4	6	8	10	12	14	15	17
23	3617	3636	3655	3674	3692	3711	3729	3747	3766	3784	2	4	6	7	9	11	13	15	17
24	3802	3820	3838	3856	3874	3892	3909	3927	3945	3962	2	4	5	7	9	11	12	14	16
25	3979	3997	4014	4031	4048	4065	4082	4099	4116	4133	2	3	5	7	9	10	12	14	15
26	4150	4166	4183	4200	4216	4232	4249	4265	4281	4298	2	3	5	7	8	10	11	13	15
27	4314	4330	4346	4362	4378	4393	4409	4425	4440	4456	2	3	5	6	8	9	11	13	14
28	4472	4487	4502	4518	4533	4548	4564	4579	4594	4609	2	3	5	6	8	9	11	12	14
29	4624	4639	4654	4669	4683	4698	4713	4728	4742	4757	1	3	4	6	7	9	10	12	13
30	4771	4786	4800	4814	4829	4843	4857	4871	4886	4900	1	3	4	6	7	9	10	11	13
31	4914	4928	4942	4955	4969	4983	4997	5011	5024	5038	1	3	4	6	7	8	10	11	12
32	5051	5065	5079	5092	5105	5119	5132	5145	5159	5172	1	3	4	5	7	8	9	11	12
33	5185	5198	5211	5224	5237	5250	5263	5276	5289	5302	1	3	4	5	7	8	9	10	12
34	5315	5328	5340	5353	5366	5378	5391	5403	5416	5428	1	3	4	5	6	8	9	10	11
35	5441	5453	5465	5478	5490	5502	5514	5527	5539	5551	1	2	4	5	6	7	9	10	11
36	5563	5575	5587	5599	5611	5623	5635	5647	5658	5670	1	2	4	5	6	7	8	10	11
37	5682	5694	5705	5717	5729	5740	5752	5763	5775	5786	1	2	3	5	6	7	8	9	10
38	5798	5809	5821	5832	5843	5855	5866	5877	5888	5899	1	2	3	5	6	7	8	9	10
39	5911	5922	5933	5944	5955	5966	5977	5988	5999	6010	1	2	3	4	5	7	8	9	10
40	6021	6031	6042	6053	6064	6075	6085	6096	6107	6117	1	2	3	4	5	6	8	9	10
41	6128	6138	6149	6160	6170	6180	6191	6201	6212	6222	1	2	3	4	5	6	7	8	9
42	6232	6243	6253	6263	6274	6284	6294	6304	6314	6325	1	2	3	4	5	6	7	8	9
43	6335	6345	6355	6365	6375	6385	6395	6405	6415	6425	1	2	3	4	5	6	7	8	9
44	6435	6444	6454	6464	6474	6484	6493	6503	6513	6522	1	2	3	4	5	6	7	8	9
45	6532	6542	6551	6561	6571	6580	6590	6599	6609	6618	1	2	3	4	5	6	7	8	9
46	6628	6637	6646	6656	6665	6675	6684	6693	6702	6712	1	2	3	4	5	6	7	7	8
47	6721	6730	6739	6749	6758	6767	6776	6785	6794	6803	1	2	3	4	5	5	6	7	8
48	6812	6821	6830	6839	6848	6857	6866	6875	6884	6893	1	2	3	4	4	5	6	7	8
49	6902	6911	6920	6928	6937	6946	6955	6964	6972	6981	1	2	3	4	4	5	6	7	8
50	6990	6998	7007	7016	7024	7033	7042	7050	7059	7067	1	2	3	3	4	5	6	7	8

LOGARITHMS (Contd.)

	0	1	2	3	4	5	6	7	8	9	Mean Differences								
											1	2	3	4	5	6	7	8	9
51	7076	7084	7093	7101	7110	7118	7126	7135	7143	7152	1	2	3	3	4	5	6	7	8
52	7160	7168	7177	7185	7193	7202	7210	7218	7226	7235	1	2	2	3	4	5	6	7	7
53	7243	7251	7259	7267	7275	7284	7292	7300	7308	7316	1	2	2	3	4	5	6	6	7
54	7324	7332	7340	7348	7356	7364	7372	7380	7388	7396	1	2	2	3	4	5	6	6	7
55	7404	7412	7419	7427	7435	7443	7451	7459	7466	7474	1	2	2	3	4	5	5	6	7
56	7482	7490	7497	7505	7513	7520	7528	7536	7543	7551	1	2	2	3	4	5	5	6	7
57	7559	7566	7574	7582	7589	7597	7604	7612	7619	7627	1	2	2	3	4	5	5	6	7
58	7634	7642	7649	7657	7664	7672	7679	7686	7694	7701	1	1	2	3	4	4	5	6	7
59	7709	7716	7723	7731	7738	7745	7752	7760	7767	7774	1	1	2	3	4	4	5	6	7
60	7782	7789	7796	7803	7810	7818	7825	7832	7839	7846	1	1	2	3	4	4	5	6	6
61	7853	7860	7868	7875	7882	7889	7896	7903	7910	7917	1	1	2	3	4	4	5	6	6
62	7924	7931	7938	7945	7952	7959	7966	7973	7980	7987	1	1	2	3	3	4	5	6	6
63	7993	8000	8007	8014	8021	8028	8035	8041	8048	8055	1	1	2	3	3	4	5	5	6
64	8062	8069	8075	8082	8089	8096	8102	8109	8116	8122	1	1	2	3	3	4	5	5	6
65	8129	8136	8142	8149	8156	8162	8169	8176	8182	8189	1	1	2	3	3	4	5	5	6
66	8195	8202	8209	8215	8222	8228	8235	8241	8248	8254	1	1	2	3	3	4	5	5	6
67	8261	8267	8274	8280	8287	8293	8299	8306	8312	8319	1	1	2	3	3	4	5	5	6
68	8325	8331	8338	8344	8351	8357	8363	8370	8376	8382	1	1	2	3	3	4	4	5	6
69	8388	8395	8401	8407	8414	8420	8426	8432	8439	8445	1	1	2	2	3	4	4	5	6
70	8451	8457	8463	8470	8476	8482	8488	8494	8500	8506	1	1	2	2	3	4	4	5	6
71	8513	8519	8525	8531	8537	8543	8549	8555	8561	8567	1	1	2	2	3	4	4	5	5
72	8573	8579	8585	8591	8597	8603	8609	8615	8621	8627	1	1	2	2	3	4	4	5	5
73	8633	8639	8645	8651	8657	8663	8669	8675	8681	8686	1	1	2	2	3	4	4	5	5
74	8692	8698	8704	8710	8716	8722	8727	8733	8739	8745	1	1	2	2	3	4	4	5	5
75	8751	8756	8762	8768	8774	8779	8785	8791	8797	8802	1	1	2	2	3	3	4	5	5
76	8808	8814	8820	8825	8831	8837	8842	8848	8854	8859	1	1	2	2	3	3	4	5	5
77	8865	8871	8876	8882	8887	8893	8899	8904	8910	8915	1	1	2	2	3	3	4	4	5
78	8921	8927	8932	8938	8943	8949	8954	8960	8965	8971	1	1	2	2	3	3	4	4	5
79	8976	8982	8987	8993	8998	9004	9009	9015	9020	9025	1	1	2	2	3	3	4	4	5
80	9031	9036	9042	9047	9053	9058	9063	9069	9074	9079	1	1	2	2	3	3	4	4	5
81	9085	9090	9096	9101	9106	9112	9117	9122	9128	9133	1	1	2	2	3	3	4	4	5
82	9138	9143	9149	9154	9159	9165	9170	9175	9180	9186	1	1	2	2	3	3	4	4	5
83	9191	9196	9201	9206	9217	9217	9222	9227	9232	9238	1	1	2	2	3	3	4	4	5
84	9243	9248	9253	9258	9263	9269	9274	9279	9284	9289	1	1	2	2	3	3	4	4	5
85	9294	9299	9304	9360	9315	9320	9325	9330	9335	9340	1	1	2	2	3	3	4	4	5
86	9345	9350	9355	9360	9365	9370	9375	9380	9385	9390	1	1	2	2	3	3	4	4	5
87	9395	9400	9405	9410	9415	9420	9425	9430	9435	9440	0	1	1	2	2	3	3	4	4
88	9445	9450	9455	9460	9465	9469	9474	9479	9484	9489	0	1	1	2	2	3	3	4	4
89	9494	9499	9504	9509	9513	9518	9523	9528	9533	9538	0	1	1	2	2	3	3	4	4
90	9542	9547	9552	9557	9562	9566	9571	9576	9581	9586	0	1	1	2	2	3	3	4	4
91	9590	9595	9600	9605	9609	9614	9619	9624	9628	9633	0	1	1	2	2	3	3	4	4
92	9638	9643	9647	9652	9657	9661	9666	9671	9675	9680	0	1	1	2	2	3	3	4	4
93	9685	9689	9694	9699	9703	9708	9713	9717	9722	9727	0	1	1	2	2	3	3	4	4
94	9731	9736	9741	9745	9750	9754	9759	9763	9768	9773	0	1	1	2	2	3	3	4	4
95	9777	9782	9786	9791	9795	9800	9805	9809	9814	9818	0	1	1	2	2	3	3	4	4
96	9823	9827	9832	9836	9841	9845	9850	9854	9859	9863	0	1	1	2	2	3	3	4	4
97	9868	9872	9877	9881	9886	9890	9894	9899	9903	9908	0	1	1	2	2	3	3	4	4
98	9912	9916	9921	9926	9930	9934	9939	9943	9948	9952	0	1	1	2	2	3	3	4	4
99	9956	9961	9965	9969	9974	9978	9983	9987	9991	9996	0	1	1	2	2	3	3	3	4

ANTILOGARITHMS

	0	1	2	3	4	5	6	7	8	9	Mean Differences 1	2	3	4	5	6	7	8	9
.00	1000	1002	1005	1007	1009	1012	1014	1016	1019	1021	0	0	1	1	1	1	2	2	2
.01	1023	1026	1028	1030	1033	1035	1038	1040	1042	1045	0	0	1	1	1	1	2	2	2
.02	1047	1050	1052	1054	1057	1059	1062	1064	1067	1069	0	0	1	1	1	1	2	2	2
.03	1072	1074	1076	1079	1081	1084	1086	1089	1091	1094	0	0	1	1	1	1	2	2	2
.04	1096	1099	1102	1104	1107	1109	1112	1114	1117	1119	0	1	1	1	1	2	2	2	2
.05	1122	1125	1127	1130	1132	1135	1138	1140	1143	1146	0	1	1	1	1	2	2	2	2
.06	1148	1151	1153	1156	1159	1161	1164	1167	1169	1172	0	1	1	1	1	2	2	2	2
.07	1175	1178	1180	1183	1186	1189	1191	1194	1197	1199	0	1	1	1	1	2	2	2	2
.08	1202	1205	1208	1211	1213	1216	1219	1222	1225	1227	0	1	1	1	1	2	2	2	3
.09	1230	1233	1236	1239	1242	1245	1247	1250	1253	1256	0	1	1	1	1	2	2	2	3
.10	1259	1262	1265	1268	1271	1274	1276	1279	1282	1285	0	1	1	1	1	2	2	2	3
.11	1288	1291	1294	1297	1300	1303	1306	1309	1312	1315	0	1	1	1	2	2	2	2	3
.12	1318	1321	1324	1327	1330	1334	1337	1340	1343	1346	0	1	1	1	2	2	2	2	3
.13	1349	1352	1355	1358	1361	1365	1368	1371	1374	1377	0	1	1	1	2	2	2	3	3
.14	1380	1384	1387	1390	1393	1396	1400	1403	1406	1409	0	1	1	1	2	2	2	3	3
.15	1413	1416	1419	1422	1426	1429	1432	1435	1439	1442	0	1	1	1	2	2	2	3	3
.16	1445	1449	1452	1455	1459	1462	1466	1469	1472	1476	0	1	1	1	2	2	2	3	3
.17	1479	1483	1486	1489	1493	1496	1500	1503	1507	1510	0	1	1	1	2	2	2	3	3
.18	1514	1517	1521	1524	1528	1531	1535	1538	1542	1545	0	1	1	1	2	2	2	3	3
.19	1549	1552	1556	1560	1563	1567	1570	1574	1578	1581	0	1	1	1	2	2	3	3	3
.20	1585	1589	1592	1596	1600	1603	1607	1611	1614	1618	0	1	1	1	2	2	3	3	3
.21	1622	1626	1629	1633	1637	1641	1644	1648	1652	1656	0	1	1	2	2	2	3	3	3
.22	1660	1663	1667	1671	1675	1679	1683	1687	1690	1694	0	1	1	2	2	2	3	3	3
.23	1698	1702	1706	1710	1714	1718	1722	1726	1730	1734	0	1	1	2	2	2	3	3	4
.24	1738	1742	1746	1750	1754	1758	1762	1766	1770	1774	0	1	1	2	2	2	3	3	4
.25	1778	1782	1786	1791	1795	1798	1803	1807	1811	1816	0	1	1	2	2	2	3	3	4
.26	1820	1824	1828	1832	1837	1841	1845	1849	1854	1858	0	1	1	2	2	3	3	3	4
.27	1862	1866	1871	1875	1879	1884	1888	1892	1897	1901	0	1	1	2	2	3	3	3	4
.28	1905	1910	1914	1919	1923	1928	1932	1936	1941	1945	0	1	1	2	2	3	3	4	4
.29	1950	1954	1959	1963	1968	1972	1977	1982	1986	1991	0	1	1	2	2	3	3	4	4
.30	1995	2000	2004	2009	2014	2018	2023	2028	2032	2037	0	1	1	2	2	3	3	4	4
.31	2042	2046	2051	2056	2061	2065	2070	2075	2080	2084	0	1	1	2	2	3	3	4	4
.32	2089	2094	2099	2104	2109	2113	2118	2123	2128	2133	0	1	1	2	2	3	3	4	4
.33	2138	2143	2148	2153	2158	2163	2168	2173	2178	2183	0	1	1	2	2	3	3	4	4
.34	2188	2193	2198	2203	2208	2213	2218	2223	2228	2234	1	1	2	2	3	3	4	4	5
.35	2239	2244	2249	2254	2259	2265	2270	2275	2280	2286	1	1	2	2	3	3	4	4	5
.36	2291	2296	2301	2307	2312	2317	2323	2328	2333	2339	1	1	2	2	3	3	4	4	5
.37	2344	2350	2355	2360	2366	2371	2377	2382	2388	2393	1	1	2	2	3	3	4	4	5
.38	2399	2404	2410	2415	2421	2427	2432	2438	2443	2449	1	1	2	2	3	3	4	4	5
.39	2455	2460	2466	2472	2477	2483	2489	2495	2500	2506	1	1	2	2	3	3	4	5	5
.40	2512	2518	2523	2529	2535	2541	2547	2553	2559	2564	1	1	2	2	3	4	4	5	5
.41	2570	2576	2582	2588	2594	2600	2606	2612	2618	2624	1	1	2	2	3	4	4	5	5
.42	2630	2636	2642	2649	2655	2661	2667	2673	2679	2685	1	1	2	2	3	4	4	5	6
.43	2692	2698	2704	2710	2716	2723	2729	2735	2742	2748	1	1	2	3	3	4	4	5	6
.44	2754	2761	2767	2773	2780	2786	2793	2799	2805	2812	1	1	2	3	3	4	4	5	6
.45	2818	2825	2831	2838	2844	2851	2858	2864	2871	2877	1	1	2	3	3	4	5	5	6
.46	2884	2891	2897	2904	2911	2917	2924	2931	2938	2944	1	1	2	3	3	4	5	5	6
.47	2951	2958	2965	2972	2979	2985	2992	2999	3006	3013	1	1	2	3	3	4	5	5	6
.48	3020	3027	3034	3041	3048	3055	3062	3069	3076	3083	1	1	2	3	4	4	5	6	6
.49	3090	3097	3105	3112	3119	3126	3133	3141	3148	3155	1	1	2	3	4	4	5	6	6
.50	3162	3170	3177	3184	3192	3199	3206	3214	3221	3228	1	1	2	3	4	4	5	6	7

ANTILOGARITHMS (Contd.)

—	0	1	2	3	4	5	6	7	8	9	Mean Differences								
											1	2	3	4	5	6	7	8	9
.51	3236	3243	3251	3258	3266	3273	3281	3289	3296	3304	1	2	2	3	4	5	5	6	7
.52	3311	3319	3327	3334	3342	3350	3357	3365	3373	3381	1	2	2	3	4	5	5	6	7
.53	3388	3396	3404	3412	3420	3428	3436	3443	3451	3459	1	2	2	3	4	5	6	6	7
.54	3467	3475	3483	3491	3499	3508	3516	3524	3532	3540	1	2	2	3	4	5	6	6	7
.55	3548	3556	3565	3573	3581	3589	3597	3606	3614	3622	1	2	2	3	4	5	6	7	7
.56	3631	3639	3648	3656	3664	3673	3681	3690	3698	3707	1	2	3	3	4	5	6	7	8
.57	3715	3724	3733	3741	3750	3758	3767	3776	3784	3793	1	2	3	3	4	5	6	7	8
.58	3802	3811	3819	3828	3837	3846	3855	3864	3873	3882	1	2	3	4	4	5	6	7	8
.59	3890	3899	3908	3917	3926	3936	3945	3954	3963	3972	1	2	3	4	5	5	6	7	8
.60	3981	3990	3999	4009	4018	4027	4036	4046	4055	4064	1	2	3	4	5	6	6	7	8
.61	4074	4083	4093	4102	4111	4121	4130	4140	4150	4159	1	2	3	4	5	6	7	8	9
.62	4169	4178	4188	4198	4207	4217	4227	4236	4246	4256	1	2	3	4	5	6	7	8	9
.63	4266	4276	4285	4295	4305	4315	4325	4335	4345	4355	1	2	3	4	5	6	7	8	9
.64	4365	4375	4385	4395	4406	4416	4426	4436	4446	4457	1	2	3	4	5	6	7	8	9
.65	4467	4477	4487	4498	4508	4519	4529	4539	4550	4560	1	2	3	4	5	6	7	8	9
.66	4571	4581	4592	4603	4613	4624	4634	4645	4656	4667	1	2	3	4	5	6	7	9	10
.67	4677	4688	4699	4710	4721	4732	4742	4753	4764	4775	1	2	3	4	5	7	8	9	10
.68	4786	4797	4808	4819	4831	4842	4853	4864	4875	4887	1	2	3	4	6	7	8	9	10
.69	4898	4909	4920	4932	4943	4955	4966	4977	4989	5000	1	2	3	5	6	7	8	9	10
.70	5012	5023	5035	5047	5058	5070	5082	5093	5105	5117	1	2	4	5	6	7	8	9	11
.71	5129	5140	5152	5164	5176	5188	5200	5212	5224	5236	1	2	4	5	6	7	8	10	11
.72	5248	5260	5272	5284	5297	5309	5321	5333	5346	5358	1	2	4	5	6	7	9	10	11
.73	5370	5383	5395	5408	5420	5433	5445	5458	5470	5483	1	3	4	5	6	8	9	10	11
.74	5495	5508	5521	5534	5546	5559	5572	5585	5598	5610	1	3	4	5	6	8	9	10	12
.75	5623	5636	5649	5662	5675	5689	5702	5715	5728	5741	1	3	4	5	7	8	9	10	12
.76	5754	5768	5781	5794	5808	5821	5834	5848	5861	5875	1	3	4	5	7	8	9	11	12
.77	5888	5902	5916	5929	5943	5957	5970	5984	5998	6012	1	3	4	5	7	8	10	11	12
.78	6026	6039	6053	6067	6081	6095	6109	6124	6138	6152	1	3	4	6	7	8	10	11	13
.79	6166	6180	6194	6209	6223	6237	6252	6266	6281	6295	1	3	4	6	7	9	10	11	13
.80	6310	6324	6339	6353	6368	6383	6397	6412	6427	6442	1	3	4	6	7	9	10	12	13
.81	6457	6471	6486	6501	6516	6531	6546	6561	6577	6592	2	3	5	6	8	9	11	12	14
.82	6607	6622	6637	6653	6668	6683	6699	6714	6730	6745	2	3	5	6	8	9	11	12	14
.83	6761	6776	6792	6808	6823	6839	6855	6871	6887	6902	2	3	5	6	8	9	11	13	14
.84	6918	6934	6950	6966	6982	6998	7015	7031	7047	7063	2	3	5	6	8	10	11	13	15
.85	7079	7096	7112	7129	7145	7161	7178	7194	7211	7228	2	3	5	7	8	10	12	13	15
.86	7244	7261	7278	7295	7311	7328	7345	7362	7379	7396	2	3	5	7	8	10	12	13	15
.87	7413	7430	7447	7464	7482	7499	7516	7534	7551	7568	2	3	5	7	9	10	12	14	16
.88	7586	7603	7621	7638	7656	7674	7691	7709	7727	7745	2	4	5	7	9	11	12	14	16
.89	7762	7780	7798	7816	7834	7852	7870	7889	7907	7925	2	4	5	7	9	11	13	14	16
.90	7943	7962	7980	7998	8017	8035	8054	8072	8091	8110	2	4	6	7	9	11	13	15	17
.91	8128	8147	8166	8185	8204	8222	8241	8260	8279	8299	2	4	6	8	9	11	13	15	17
.92	8318	8337	8356	8375	8395	8414	8433	8453	8472	8492	2	4	6	8	10	12	14	15	17
.93	8511	8531	8551	8570	8590	8616	8630	8650	8670	8690	2	4	6	8	10	12	14	16	18
.94	8710	8730	8750	8770	8790	8810	8831	8851	8872	8892	2	4	6	8	10	12	14	16	18
.95	8913	8933	8954	8974	8995	9016	9036	9057	9078	9099	2	4	6	8	10	12	15	17	19
.96	9120	9141	9162	9183	9204	9226	9247	9268	9290	9311	2	4	6	8	11	13	15	17	19
.97	9333	9354	9376	9397	9419	9441	9462	9484	9506	9528	2	4	7	9	11	13	15	17	20
.98	9550	9572	9594	9616	9638	9661	9683	9705	9727	9750	2	4	7	9	11	13	16	18	20
.99	9772	9793	9817	9840	9863	9886	9908	9931	9954	9977	2	5	7	9	11	14	16	18	20

NATURAL SINES

	0'	6'	12'	18'	24'	30'	36'	42'	48'	54'	Mean Difference				
	0°.0	0°.1	0°.2	0°.3	0°.4	0°.5	0°.6	0°.7	0°.8	0°.9	1	2	3	4	5
0	.00000	00175	00349	00524	00698	00873	01047	01222	01396	01571	29	58	87	117	145
1	.01745	01920	02094	02269	02443	02618	02792	02967	03141	03316	29	58	87	117	145
2	.03490	03664	03839	04013	04188	04362	04536	04711	04885	05059	29	58	87	116	145
3	.05234	05408	05582	05756	05931	06105	06279	06453	06627	06802	29	58	87	116	145
4	.06976	07150	07324	07498	07672	07846	08020	08194	08368	08542	29	58	87	116	145
5	.08716	08889	09063	09237	09411	09585	09758	09932	10106	10279	29	58	87	116	145
6	.10453	10626	10800	10973	11147	11320	11494	11667	11840	12014	29	58	87	116	145
7	.12187	12360	12533	12706	12880	13053	13226	13399	13572	13744	29	58	87	115	144
8	.13917	14090	14263	14436	14608	14781	14954	15126	15299	15471	29	58	86	115	144
9	.15643	15816	15988	16160	16333	16505	16677	16849	17021	17193	29	57	86	115	144
10	.17365	17537	17708	17880	18052	18224	18395	18567	18738	18910	29	57	86	114	143
11	.19081	19252	19423	19595	19766	19937	20108	20279	20450	20620	29	57	86	114	143
12	.20791	20962	21132	21303	21474	21644	21814	21985	22155	22325	28	57	85	114	142
13	.22495	22665	22835	23005	23175	23345	23514	23684	23853	24023	28	57	85	113	141
14	.24192	24362	24531	24700	24869	25038	25207	25376	25545	25713	28	56	85	113	141
15	.25882	26050	26219	26387	26556	26724	26892	27060	27228	27369	28	56	84	112	140
16	.27564	27731	27899	28067	28234	28402	28569	28736	28903	29070	28	56	84	112	139
17	.29237	29404	29571	29737	29904	30071	30237	30403	30570	30736	28	56	83	111	139
18	.30902	31068	31233	31399	31565	31730	31896	32061	32227	32392	28	55	83	110	138
19	.32557	32722	32887	33051	33216	33381	33545	33710	33874	34038	27	55	82	110	137
20	.34202	34366	34530	34694	34857	35021	35184	35347	35511	35674	27	55	82	109	136
21	.35837	36000	36162	36325	36488	36650	36812	36975	37137	37299	27	54	81	108	135
22	.37461	37622	37784	37946	38107	38268	38430	38591	38752	38912	27	54	81	107	134
23	.39073	39234	39394	39555	39715	39875	40035	40195	40355	40514	27	53	80	107	133
24	.40674	40833	40992	41151	41310	41469	41628	41787	41945	42104	27	53	79	106	132
25	.42262	42420	42578	42736	42894	43051	43209	43366	43523	43680	26	53	79	105	131
26	.43837	43994	44178	44307	44464	44620	44776	44932	45088	45243	26	52	78	104	130
27	.45399	45554	45710	45865	46020	46175	46330	46484	46639	46793	26	52	77	103	129
28	.46947	47101	47255	47409	47562	47716	47869	48022	48175	48328	26	51	77	102	128
29	.48481	48634	48786	48938	49090	49242	49394	49546	49697	49849	25	51	76	101	126
30	.50000	50151	50302	50453	50603	50754	50904	51054	51204	51354	25	50	75	100	125
31	.51504	51653	51803	51952	52101	52250	52399	52547	52696	52844	25	50	74	99	124
32	.52992	53140	53288	53435	53583	53730	53877	54024	54171	54317	25	49	74	98	123
33	.54464	54610	54756	54902	55048	55194	55339	55484	55630	55775	24	49	73	97	121
34	.55919	56064	56208	56353	56497	56641	56784	56928	57071	57215	24	48	72	96	120
35	.57358	57501	57643	57786	57928	58070	58212	58354	58496	58637	24	48	71	95	118
36	.58778	58920	59061	59201	59342	59482	59622	59763	59902	60042	23	47	70	94	117
37	.60181	60321	60460	60599	60738	60876	61015	61153	61291	61429	23	46	69	92	115
38	.61566	61704	61841	61978	62115	62251	62388	62524	62660	62796	23	46	68	91	114
39	.62932	63068	63203	63338	63473	63608	63742	63877	64011	64145	23	45	67	90	112
40	.64279	64412	64546	64679	64812	64945	65077	65210	65342	65474	22	44	66	88	111
41	.65606	65738	65869	66000	66131	66262	66393	66523	66653	66783	22	44	65	87	109
42	.66913	67043	67172	67301	67430	67559	67688	67816	67944	68072	22	43	64	86	107
43	.68200	68327	68455	68582	68709	68835	68962	69088	69214	69340	21	42	63	84	106
44	.69466	69591	69717	69842	69966	70091	70215	70339	70463	70587	21	41	62	83	104

NATURAL SINES (Contd.)

	0' 0°.0	6' 0°.1	12' 0°.2	18' 0°.3	24' 0°.4	30' 0°.5	36' 0°.6	42' 0°.7	48' 0°.8	54' 0°.9	Mean Difference 1	2	3	4	5
45	.70711	70834	70957	71080	71203	71325	71447	71569	71691	71813	20	41	61	82	102
46	.71934	72055	72176	72297	72417	72537	72657	72777	72897	73016	20	40	60	80	100
47	.73135	73254	73373	73491	73610	73728	73846	73963	74080	74198	20	39	59	79	98
48	.74314	74431	74548	74664	74780	74896	75011	75126	75241	75356	20	39	58	77	96
49	.75471	75585	75700	75813	.75927	76041	76154	76267	76380	76492	19	38	57	76	94
50	.76604	76717	76828	76940	77051	77162	77273	77384	77494	77605	19	37	56	74	93
51	.77715	77824	77934	78043	78152	78261	78369	78478	78586	78694	18	36	54	72	91
52	.78801	78908	79015	79122	79229	79335	79441	79547	79653	79758	18	35	53	71	89
53	.79864	79968	80073	80178	80282	80386	80489	80593	80696	80799	17	35	52	69	87
54	.80902	81004	81106	81208	81310	81412	81513	81614	81714	81815	17	34	51	68	84
55	.81915	82015	82115	82214	82314	82413	82511	82610	82708	82806	16	33	49	66	82
56	.82904	83001	83098	83195	83292	83389	83485	83581	83676	83772	16	32	48	64	80
57	.83867	83962	84057	84151	84245	84339	84433	84526	84619	84712	16	31	47	62	78
58	.84805	84897	84989	85081	85173	85264	85355	85446	85536	85627	15	31	46	61	76
59	.85717	85806	85896	85895	86074	86163	86251	86340	86427	86515	15	30	44	59	74
60	.86603	86690	86777	86863	86949	87036	87121	87207	87292	87377	14	29	43	57	72
61	.87462	87546	87631	87715	87798	87882	87965	88048	88130	88213	14	28	42	56	69
62	.88295.	88377	88458	88539	88620	88701	88782	88862	88942	89021	14	27	40	54	67
63	.89101	89180	89259	89337	89415	89493	89571	89649	89726	89803	13	26	39	52	65
64	.89879	89956	90032	90108	90183	90259	90334	90408	90483	90557	13	25	38	50	63
65	.90631	90704	90778	90851	90924	90996	91068	91140	91212	91283	12	24	36	48	60
66	.91355	91425	91496	91566	91636	91706	91775	91845	91914	91982	12	23	35	46	58
67	.92050	92119	92186	92254	92321	92388	92455	92521	92587	92653	11	22	33	45	56
68	.92718	92784	92849	92913	92978	93042	93106	93169	93232	.93295	11	21	32	43	54
69	.93358	93420	93483	93544	93606	93667	93728	93789	93849	93909	10	20	31	41	51
70	.93969	94029	94088	94147	94206	94264	94322	94380	94438	94495	10	19	29	39	49
71	.94552	94609	94665	94721	94777	94832	94888	94943	94997	95052	9	18	28	37	46
72	.95106	95159	95213	95266	95319	95372	95424	95476	95528	95579	9	17	26	35	44
73	.95630	95681	95732	95782	95832	95882	95931	95981	96029	96078	8	17	25	33	41
74	.96126	96174	96222	96269	96316	96363	96410	96456	96502	96547	8	16	23	31	39
75	.96593	96638	96682	96727	96771	96815	96958	96902	96945	96987	7	16	22	29	36
76	.97030	97072	97113	97155	97196	97237	97278	97318	97358	97398	7	14	20	27	34
77	.97437	97476	97515	97553	97592	97630	97667	97705	97742	97778	6	13	19	25	32
78	.97815	97851	97887	97922	97958	97992	98027	98061	98096	98129	6	12	17	23	29
79	.98163	98196	98229	98261	98294	98325	98357	98388	98420	98450	5	11	16	21	27
80	.98481	98511	98541	98570	98600	98629	98657	98686	98714	98741	5	10	14	19	24
81	.98769	98796	98823	98849	98876	98902	98927	98953	98978	99002	4	9	13	17	22
82	.99027	99051	99075	99098	99122	99144	99167	99189	99211	99233	4	8	11	15	19
83	.99255	99276	99297	99317	99337	99357	99377	99396	99415	99434	3	7	10	13	16
84	.99452	99470	99488	99506	99523	99540	99556	99572	99588	99604	3	6	8	11	14
85	.99619	99635	99649	99664	99678	99692	99705	99719	99731	99744	2	5	7	9	11
86	.99756	99768	99780	99792	99803	99813	99824	99834	99844	99854	2	4	5	7	9
87	.99863	99872	99881	99889	99897	99905	99912	99919	99926	99933	1	3	4	5	6
88	.99939	99945	99951	99956	99961	99966	99970	99974	99978	99982	1	2	2	3	4
89	.99985	99988	99990	99993	99995	99996	99998	99999	99999	1.0000	0	1	1	1	1
90	1.000														

NATURAL COSINES

Numbers in different columns to be subtracted, not added

	0'	6'	12'	18'	24'	30'	36'	42'	48'	54'	Mean Difference				
	0°.0	0°.1	0°.2	0°.3	0°.4	0°.5	0°.6	0°.7	0°.8	0°.9	1	2	3	4	5
0	1.000	99999	99999	99999	99998	99996	99995	99993	99990	99988	0	1	1	1	1
1	.99985	99982	99978	99974	99970	99966	99961	99956	99951	99945	1	2	2	3	4
2	.99939	99933	99926	99919	99912	99905	99897	99889	99881	99872	1	3	4	5	6
3	.99863	99854	99844	99834	99824	99813	99803	99792	99780	99768	2	4	5	7	9
4	.99756	99744	99731	99719	99705	99692	99678	99664	99649	99635	2	5	7	9	11
5	.99619	99604	99588	99572	99556	99540	99523	99506	99488	99470	3	6	8	11	14
6	.99452	99434	99415	99396	99377	99357	99337	99317	99297	99276	3	7	10	13	16
7	.99255	99233	99211	99189	99167	99144	99122	99098	99075	99051	4	8	11	15	19
8	.99027	99002	98978	98953	98927	98902	98876	98849	98823	98796	4	9	13	17	22
9	.98769	98741	98714	98686	98657	98629	98600	98570	98541	98511	5	10	14	19	24
10	.98481	98450	98420	98388	98357	98325	98294	98261	98229	98196	5	11	16	21	27
11	.98163	98129	98096	98061	98027	97992	97958	97922	97887	97851	6	12	17	23	29
12	.97815	97778	97742	97705	97667	97630	97592	97553	97515	97476	6	13	19	25	32
13	.97437	97398	97358	97318	97278	97237	97196	97155	97113	97072	7	14	20	27	34
14	.97030	96987	96945	96902	96858	96815	96771	96727	96682	96638	7	16	22	29	36
15	.96593	96547	96502	96456	96410	96363	96316	96269	96222	96174	8	16	23	31	39
16	.96126	96078	96029	95981	95931	95882	95832	95782	95732	95681	8	17	25	33	41
17	.95630	95579	95528	95476	95424	95372	95319	95266	95213	95159	9	17	26	35	44
18	.95106	95052	94997	94943	94888	94832	94777	94721	94665	94609	9	18	28	37	46
19	.94552	94495	94438	94380	94322	94264	94206	94147	94088	94029	10	19	29	39	49
20	.93969	93909	93849	93789	93728	93667	93606	93544	93483	93420	10	20	31	41	51
21	.93358	93295	93232	93169	93106	93042	92978	92913	92849	92784	10	21	32	43	54
22	.92718	92653	92587	92521	92455	92388	92321	92254	92186	92119	11	22	33	45	56
23	.92050	91982	91914	91845	91775	91706	91636	91566	91496	91425	11	23	35	46	58
24	.91355	91283	91212	91140	91068	90996	90924	90851	90778	90704	12	24	36	48	60
25	.90631	90557	90483	90408	90334	90259	90183	90108	90032	89956	13	25	38	50	63
26	.89879	89803	89726	89649	89571	89493	89415	89337	89259	89180	13	26	39	52	65
27	.89101	89021	88942	88862	88782	88701	88620	88539	88458	88377	13	27	40	54	67
28	.88295	88213	88130	88048	87965	87882	87798	87715	87631	87546	14	28	42	56	69
29	.87462	87377	87292	87207	87121	87036	86949	86863	86777	86690	14	29	43	57	72
30	.86603	86515	86427	86340	86251	86163	86074	85985	85896	85806	15	30	44	59	74
31	.85717	85627	85536	85446	85355	85264	85173	85081	84989	84897	16	30	46	61	76
32	.84805	84712	84619	84526	84433	84339	84245	84151	84057	83962	16	31	47	62	78
33	.83867	83772	83676	83581	83485	83389	83292	83195	83098	83001	16	32	48	64	80
34	.82904	82806	82708	82610	82511	82413	82314	82214	82115	82015	16	33	49	66	82
35	.81915	81815	81714	81614	81513	81412	81310	81208	81106	81004	17	34	51	68	84
36	.80902	80799	80696	80593	80489	80386	80282	80178	80073	79968	17	35	52	69	87
37	.79864	79758	79653	79547	79441	79335	79229	79122	79015	78908	18	35	53	71	89
38	.78801	78693	78586	78478	78369	78261	78152	78043	77934	77824	18	36	54	72	91
39	.77715	77605	77494	77384	77273	77162	77051	76940	76828	76717	18	37	56	74	93
40	.76604	76492	76380	76267	76154	76041	75927	75813	75700	75585	19	38	57	76	94
41	.75471	75356	75241	75126	75011	74896	74780	74664	74548	74431	19	39	58	77	96
42	.74314	74198	74080	73963	73846	73728	73610	73491	73373	73254	20	39	59	79	98
43	.73135	73016	72897	72777	72657	72537	72417	72297	72176	72055	20	40	60	80	100
44	.71934	71813	71691	71569	71447	71325	71203	71080	70957	70834	20	41	61	82	102

NATURAL COSINES (Contd.)

Numbers in different columns to be subtracted, not added

	0'	6'	12'	18'	24'	30'	36'	42'	48'	54'	Mean Difference				
	0°.0	0°.1	0°.2	0°.3	0°.4	0°.5	0°.6	0°.7	0°.8	0°.9	**1**	**2**	**3**	**4**	**5**
45	.70711	70587	70463	70339	70215	70091	69966	69842	69717	69591	21	41	62	83	104
46	.69466	69340	69214	69088	68962	68835	68709	68582	68455	68327	21	42	63	84	106
47	.68200	68072	67944	67816	67688	67559	67430	67301	67172	67043	21	43	64	86	107
48	.66913	66783	66653	66523	66393	66262	66131	66000	65869	65738	22	44	65	87	109
49	.65606	65474	65342	65210	65077	64945	64812	64679	64546	64412	22	44	66	88	110
50	.64279	64145	64011	63877	63742	63608	63473	63338	63203	63068	22	45	67	90	112
51	.62932	62796	62660	62524	62388	62251	62115	61978	61841	61704	23	45	68	91	114
52	.61566	61429	61291	61153	61015	60876	60738	60599	60460	60321	23	46	69	92	115
53	.60181	60042	59902	59763	59622	59482	59342	59201	59061	58920	23	47	70	94	117
54	.58779	58637	58496	58354	58212	58070	57928	57786	57643	57501	24	47	71	95	118
55	.57358	57215	57071	56928	56784	56641	56497	56353	56208	56064	24	48	72	96	120
56	.55919	55775	55630	55484	55339	55194	55048	54902	54756	54610	24	49	73	97	121
57	.54464	54317	54171	54024	53877	53730	53583	53435	53288	53140	24	49	74	98	123
58	.52992	52844	52696	52547	52399	52250	52101	51952	51803	51653	25	50	74	99	124
59	.51504	51354	51204	51054	50904	50754	50603	50453	50302	50151	25	50	75	100	125
60	.5000	49849	49697	49546	49394	49242	49090	48938	48786	48634	25	51	76	101	126
61	.48481	48328	48175	48022	47869	47716	47562	47409	47255	47101	26	51	77	102	128
62	.46947	46793	46639	46484	46330	46175	46020	45865	45710	45554	26	52	77	103	129
63	.45399	45243	45088	44932	44776	44620	44464	44307	44151	43994	26	52	78	104	130
64	.43837	43680	43523	43366	43209	43051	42894	42736	42578	42420	26	53	79	105	131
65	.42262	42104	41945	41787	41628	41469	41310	41151	40992	40833	26	53	79	106	132
66	.40674	40514	40355	40195	40035	39875	39715	39555	39394	39234	27	53	80	107	133
67	.39073	38912	38752	38591	38430	38268	38107	37946	37784	37622	27	54	81	107	134
68	.37461	37299	37137	36975	36812	36650	36488	36325	36162	36000	27	54	81	108	135
69	.35837	35674	35511	35347	35184	35021	34857	34694	34530	34366	27	54	82	109	136
70	.34202	34038	33874	33710	33545	33381	33216	33051	32887	32722	27	55	82	110	137
71	.32557	32392	32227	32061	31896	31730	31565	31399	31233	31068	28	55	83	110	138
72	.30902	30736	30570	30403	30237	30071	29904	29737	29571	29404	28	56	83	111	139
73	.29237	29070	28903	28736	28569	28402	28234	28067	27899	27731	28	56	84	112	139
74	.27564	27396	27228	27060	26892	26724	26556	26387	26219	26050	28	56	84	112	140
75	.25882	25713	25545	25376	25207	25038	24869	24700	24531	24362	28	56	85	113	141
76	.24192	24023	23853	23684	23514	23345	23175	23005	22835	22665	28	57	85	113	141
77	.22495	22325	22155	21985	21814	21644	21474	21303	21132	20962	28	57	85	114	142
78	.20791	20620	20450	20279	20108	19937	19766	19595	19423	19252	29	57	86	114	143
79	.19081	18910	18738	18567	18395	18224	18052	17880	17708	17537	29	57	86	114	143
80	.17365	17193	17021	16849	16677	16505	16333	16160	15988	15816	29	57	86	115	144
81	.15643	15471	15299	15126	14954	14781	14608	14436	14263	14090	29	58	86	115	144
82	.13917	13744	13572	13399	13226	13053	12880	12706	12533	12360	29	58	87	115	144
83	.12187	12014	11840	11667	11494	11320	11147	10973	10800	10626	29	58	87	116	145
84	.10453	10279	10106	09932	09758	09585	09411	09237	09063	08889	29	58	87	116	145
85	.08716	08542	08368	08194	08020	07846	07672	07498	07324	07150	29	58	87	116	145
86	.06976	06802	06627	06453	06279	06105	05931	05756	05582	05408	29	58	87	116	145
87	.05234	05059	04885	04711	04536	04362	04188	04013	03839	03664	29	58	87	116	145
88	.03490	03316	03141	02967	02792	02618	02443	02269	02094	01920	29	58	87	117	145
89	.01745	01571	01396	01222	01047	00873	00698	00524	00349	00175	29	58	87	117	145
90	.0000														

NATURAL TANGENTS

	0'	6'	12'	18'	24'	30'	36'	42'	48'	54'	Mean Difference				
	0°.0	0°.1	0°.2	0°.3	0°.4	0°.5	0°.6	0°.7	0°.8	0°.9	1	2	3	4	5
0	.00000	00175	00349	00524	00698	00873	01047	01222	01396	01571	29	58	87	116	146
1	.01746	01920	02095	02269	02444	02619	02793	02968	03143	03317	29	58	87	116	146
2	.03492	03667	03842	04016	04191	04366	04541	04716	04891	05066	29	58	87	117	146
3	.05241	05416	05591	05766	05941	06116	06291	06467	06642	06817	29	58	88	117	146
4	.06993	07168	07344	07519	07695	07870	08046	08221	08397	08573	29	59	88	117	146
5	.08749	08925	09101	09277	09453	09629	09805	09981	10158	10334	29	59	88	117	147
6	.10510	10687	10863	11040	11217	11394	11570	11747	11924	12101	29	59	88	118	147
7	.12278	12456	12633	12810	12988	13165	13343	13521	13698	13876	30	59	89	118	148
8	.14054	14232	14410	14588	14767	14945	15124	15302	15481	15660	30	59	89	119	149
9	.15838	16017	16196	16376	16555	16734	16914	17093	17273	17453	30	60	90	120	150
10	.17633	17813	17993	18173	18353	18534	18714	18895	19076	19257	30	60	90	120	150
11	.19438	19619	19801	19982	20164	20345	20527	20709	20891	21073	30	60	91	121	152
12	.21256	21438	21621	21804	21986	22169	22353	22536	22719	22903	30	61	92	122	153
13	.23087	23271	23455	23639	23823	24008	24193	24377	24562	24747	31	61	93	124	155
14	.24933	25118	25304	25490	25676	25862	26048	26235	26421	26608	31	62	93	124	155
15	.26795	26982	27169	27357	27545	27732	27920	28109	28297	28486	31	63	94	125	157
16	.28675	28864	29053	29242	29432	29621	29811	30001	30192	30382	32	63	95	127	158
17	.30573	30764	30955	31147	31338	31530	31722	31914	32106	32299	32	64	96	128	160
18	.32492	32685	32878	33072	33266	33460	33654	33848	34043	34238	32	65	97	129	162
19	.34433	34628	34824	35019	35216	35412	35608	35805	36002	36199	33	66	98	131	164
20	.36397	36595	36793	36991	37190	37388	37588	37787	37986	38186	33	66	99	133	166
21	.38386	38587	38787	38988	39190	39391	39593	39795	39997	40200	34	67	101	134	168
22	.40403	40606	40809	41013	41217	41421	41626	41831	42036	42242	34	68	102	136	170
23	.42447	42654	42860	43067	43274	43481	43689	43897	44105	44314	34	69	104	138	173
24	.44523	44732	44942	45152	45362	45573	45784	45995	46206	46418	35	70	105	141	176
25	.46631	46843	47056	47270	47483	47698	47912	48127	48342	48557	36	71	107	143	179
26	.48773	48989	49206	49423	49640	49858	50076	50295	50514	50733	36	73	109	145	182
27	.50953	51173	51393	51614	51835	52057	52279	52501	52724	52947	37	74	111	148	185
28	.53171	53395	53620	53844	54070	54296	54522	54748	54975	55203	38	75	113	151	188
29	.55431	55659	55888	56117	56347	56577	56808	57039	57271	57503	38	77	115	154	192
30	.57735	57968	58201	58435	58670	58905	59140	59376	59612	59849	39	78	118	157	196
31	.60086	60324	60562	60801	61040	61280	61520	61761	62003	62245	40	79	120	160	200
32	.62487	62730	62973	63217	63462	63707	63953	64199	64446	64693	41	82	123	164	205
33	.64941	65189	65438	65688	65938	66189	66440	66692	66944	67197	42	84	126	167	209
34	.67451	67705	67960	68215	68471	68728	68985	69243	69502	69761	43	86	129	171	214
35	.70021	70281	70542	70804	71066	71329	71593	71857	72122	72388	44	88	132	176	219
36	.72654	72921	73189	73457	73726	73996	74267	74538	74810	75082	45	90	135	180	225
37	.75355	75629	75904	76180	76456	76733	77010	77289	77568	77848	46	92	139	185	231
38	.78129	78410	78692	78975	79259	79544	79829	80115	80402	80690	47	95	142	190	237
39	.80978	81268	81558	81849	82141	82434	82727	83022	83317	83613	49	98	147	195	244
40	.83910	84208	84507	84806	85107	85408	85710	86014	86318	86623	50	100	151	201	252
41	.86929	87236	87543	87852	88162	88473	88784	89097	89410	89725	52	103	155	207	259
42	.90040	90357	90674	90993	91313	91633	91955	92277	92601	92926	53	107	160	214	268
43	.93252	93578	93906	94235	94565	94896	95229	95562	95897	96232	55	111	165	221	276
44	.96569	96907	97246	97586	97927	98270	98613	98958	99304	99652	57	114	171	229	286

NATURAL TANGENTS (Contd.)

	0'	6'	12'	18'	24'	30'	36'	42'	48'	54'	Mean Difference				
	0°.0	0°.1	0°.2	0°.3	0°.4	0°.5	0°.6	0°.7	0°.8	0°.9	1	2	3	4	5
45	1.00000	00350	00701	01053	01406	01761	02117	02474	02832	03192	58	118	177	237	296
46	1.03553	03915	04279	04644	05010	05378	05747	06117	06489	06862	61	123	184	245	307
47	1.07237	07613	07990	08369	08749	09131	09514	09899	10285	10672	63	127	191	255	319
48	1.11061	11452	11844	12238	12633	13029	13428	13828	14229	14632	66	132	199	265	331
49	1.15037	15443	15851	16261	16672	17085	17500	17916	18334	18754	69	138	207	276	344
50	1.19175	19599	20024	20451	20879	21310	21742	22176	22612	23050	72	143	216	288	359
51	1.23490	23931	24375	24820	25268	25717	26169	26622	27077	27535	75	150	225	300	375
52	1.27994	28456	28919	29385	29853	30323	30795	31269	31745	32224	78	157	235	314	392
53	1.32704	33187	33673	34160	34650	35142	35637	36134	36633	37134	82	164	247	329	411
54	1.37638	38145	38653	39165	39679	40195	40714	41235	41759	42286	86	172	259	345	431
55	1.42815	43347	43881	44418	44958	45501	46046	46595	47146	47700	91	181	272	362	453
56	1.48256	48816	49378	49944	50512	51084	51658	52235	52816	53400	95	191	286	382	477
57	1.53987	54576	55170	55767	56366	56969	57575	58184	58797	59414	100	201	302	403	504
58	1.60033	60657	61283	61914	62548	63185	63826	64471	65120	65772	106	213	319	426	533
59	1.66428	67088	67752	68419	69091	69766	70446	71129	71817	72509	113	226	339	452	564
60	1.73205	73905	74610	75319	76032	76749	77471	78198	78929	79665	120	240	360	481	600
61	1.80405	81150	81900	82654	83413	84177	84946	85720	86500	87283	128	255	383	511	639
62	1.88073	88867	89667	90472	91282	92098	92920	93746	94579	95417	136	273	409	546	683
63	1.96261	97111	97967	98828	99695	2.00569	2.01449	2.02335	2.03227	2.04125	146	292	438	584	731
64	2.05030	05942	06860	07785	08716	09654	10600	11552	12511	13477	157	314	471	629	786
65	2.14451	15432	16420	17416	18419	19430	20449	21475	22510	23553	169	338	508	677	846
66	2.24604	25663	26730	27806	28891	29984	31086	32197	33317	34447	183	366	549	732	915
67	2.35585	36733	37891	39058	40235	41421	42618	43825	45043	46270	199	397	596	795	994
68	2.47509	48758	50018	51289	52571	53865	55170	56487	57815	59156	Mean differences cease to be sufficiently accurate.				
69	2.60509	61874	63252	64642	66046	67462	68892	70335	71792	73263					
70	2.74748	76247	77761	79289	80833	82391	83965	85566	87161	88783					
71	2.90421	92076	93748	95437	97144	98868	3.00611	3.02372	3.04152	3.05950					
72	3.07768	09606	11464	13341	15240	17159	19100	21063	23048	25055					
73	3.27085	29139	31216	33317	35443	37594	39771	41973	44202	46458					
74	3.48741	51053	53393	55761	58160	60588	63048	65538	68061	70616					
75	3.73205	75828	78485	81177	83906	86671	89474	92316	95196	98117					
76	4.01078	04081	07127	10216	13350	16530	19756	23030	26352	29724					
77	4.33148	36623	40152	43735	47374	51071	54826	58641	62518	66458					
78	4.70463	74534	78673	82882	87162	91516	95945	5.00451	5.05037	5.09704					
79	5.14455	19293	24218	29235	34345	39552	44857	50264	55777	61397					
80	5.67128	72974	78938	85024	91236	97576	6.04051	6.10664	6.17419	6.24321					
81	6.31375	38587	45961	53503	61220	69116	77199	85475	93952	7.02637					
82	7.11537	20661	30018	39616	49465	59575	69957	80622	91582	8.02848					
83	8.14435	26356	38625	51259	64275	77689	91520	9.05789	9.20516	9.35724					
84	9.51436	9.6768	9.8448	10.019	10.199	10.385	10.5790	10.7800	10.9880	11.2050					
85	11.430	11.664	11.909	12.163	12.429	12.706	12.996	13.300	13.617	13.951					
86	14.301	14.669	15.056	15.464	15.895	16.350	16.832	17.343	17.886	18.464					
87	19.081	19.740	20.446	21.205	22.022	22.904	23.859	24.898	26.031	27.271					
88	28.636	30.145	31.821	33.694	35.801	38.188	40.917	44.066	47.740	52.081					
89	57.290	63.657	71.615	81.847	95.489	114.60	143.24	190.98	286.48	572.96					
90	∞														

IMPORTANT BOOKS ON PHYSICS

SOLID STATE PHYSICS AND ELECTRONICS

R.K. Puri & V.K. Babbar

CONTENTS : ● Crystal Structure ● X-Ray Diffraction and Reciprocal Lattice ● Bonding in Solids ● Lattice Vibrations ● Free Electron Theory of Metals ● Band Theory of Solids ● Semiconductors ● Magnetism in Solids ● Dielectric Properties of Solids ● Superconductivity ● Junction Diodes ● Rectifiers ● Transistors and Amplifiers ● Oscillators ● Modulation and Demodulation ● Cathode Ray Oscilloscope ● Radio Communication and Television ● Logic Gates ● **Appendix 1:** Voltage and Current Sources ● **Appendix II** : Thevenin's and Norton's Theorems ● **Appendix III:** Table of Physical Constants and Conversion Factors Index ● References ● Index

Code : 10 176 **2nd Edn. 2004** **ISBN:81-219-1475-2**

SOLID STATE PHYSICS

R.K. Puri & V.K. Babbar

CONTENTS: ● I-Crystal Structure ● II-X-Ray Diffraction and Reciprocal Lattice ● III Bonding in Solids ● IV Lattice Vibrations ● V Free Electron Theory of Metals ● VI Bond Theory of Solids ● VII Semiconductors ● VIII Magnetism in Solids ● IX Dielectric Properties of Solids ● X Superconductivity ● Appendix ● Index

Code : 10 181 **Rep. Edn. 2004** **ISBN: 81-219-1476-0**

ELEMENTS OF PROPERTIES OF MATTER

D.S. Mathur

CONTENTS: Units and Dimensions ● Motion along a Curve ●The Projectile Moment of Inertia ●Gyroscopic Motion ● Simple Harmonic Motion ● Measurement of Mass The Balance ● Acceleration due to Gravity ● Gravitation ● Elasticity ● Hydrostatics ● Jet Planes, Rockets and Satellites ● Friction and Lubrication-Principle of Virtual Work and its Simple Applications ● Flow of Liquids ● Viscosity ● Diffusion and Osmosis ● Surface Tension-Capillarity ● Gases-Kinetic Theory ● Special Theory of Relativity ● Appendices and Tables.

Code : 16 062 **11th Edn. 2005** **ISBN:81-219-0815-9**

MECHANICS

D.S. Mathur

CONTENTS: Vectors ● Reference Frames - Newton's Laws of Motion- Galilean Invariance ● Relativity ● Non-Relativistic Particle Dynamics ● Conservation Laws ● Law of Conservation of Energy ● Harmonic Oscillator ● Wave Motion ● Dynamics of Rigid Bodies ● Gravitation-Fields and Potentials ● Elasticity (Strength of Materials) ● Bending of Beams Columns ● Hydrodynamics–Viscosity ● Production and Measurement of Low Pressures (Pumps & Gauges).

Code : 16 063 **2nd Edn. 2004** **ISBN:81-219-0599-0**

PROPERTIES OF MATTER

Brij Lal & N. Subrahmanyam

CONTENTS: Units and Dimensions ● Non-Relativistic Motion ● Dynamics of Rigid Bodies ● Harmonic Oscillations and Superposition Principle ● Gravity and Gravitation ● Elasticity ● Fluid Motion, Viscosity and Low Pressure ● Surface Tension ● Universe ● Galilean Transformations and Special Theory of Relativity ● Appendices ● Index

Code : 16 082 **6th Edn. 2004** **ISBN:81-219-0280-0**

HEAT AND THERMODYNAMICS

Brij Lal & N. Subrahmanyam

CONTENTS : Thermometry ● Expansion ●Calorimetry ● Change of State ● Nature of Heat ●Thermodynamics ● Liquefaction of Gases ● Transmission of Heat ● Statistical Thermodynamics ● Appendices ● Index

Code : 16 060 **16th Edn. 2005** **ISBN-81-219-0417-X**

A TEXTBOOK OF OPTICS

N. Subrahmanyam & Brij Lal

CONTENTS: Part 1: Geometrical Optics: Refraction through Spherical Surfaces ●Refraction through Lenses ● Aberrations and Dispersion ● Optical Instruments ●Velocity of Light ● Photometry ● **Part II:** Physical Optics: Nature of Light ● Interference ● Diffraction ● Polarization ● Spectrum ● Crystal Structure and Atomic Physics.

Code : 16 056 **22nd Rev. Edn. 2005** **ISBN:81-219-0467-6**

ATOMIC AND NUCLEAR PHYSICS

N. Subrahmanyam & Brij Lal

CONTENTS: Electron • Positive Rays • X-Rays and Photo-electricity • Radioactivity • Particle Accelerators Atomic Structure • Nuclear Energy • Quantum Physics •Special Theory of Relativity • Statistical Mechanics • Crystal Structure • Exercises • Appendices.

Code : 16 132 **5th Edn. 2005** **ISBN : 81-219-0414-5**

ATOMIC PHYSICS (MODERN PHYSICS)

S.N. Ghoshal

CONTENTS: Atomic Structure of Matter • Cathode Rays and Positive Rays • Origin of the Quantum Theory: Plank's Theory of the Black Body Radiation • Structure of the Hydrogen like Atoms; Bohr-Sommerfeld Theory • Emission of Electrons from metal surfaces; Photoelectric and the thermionic effects • Alkalli Spectra, space quantization, electron spin; Periodic classification of elements • Laser • X-Rays • Wave Particle Duality; Heisenberg's Uncertainty Principle introduction to Wave Mechanics • Solutions of Schrodinger Equation in Some Simple Cases • Approximate Methods in Quantum Mechanics • Quantum Mechanical Theory of Atomic Structure • Atoms in Radiation Field • Special Theory of Relativity • Molecular Spectra •Structure and Properties of Solids • Metals and Semi-Conductors •Magnetic Properties of Solids •Statistical Mechanics • Appendices • Index **Volume II :** Rutherford Scattering of Alpha Particles and Nuclear Model of the Atom • Nuclear Structure and General Properties of Nuclei • Radioactivity • Alpha Particles and Alpha Radioactivity • Beta Particles and Beta Activity • Gamma Rays • Detection of Nuclear Radiations, and their Measurements • Determination of Some Nuclear Properties Nuclear Models • Nuclear Reactions • Accelerators of Charged Particles • Neutrons and Neutron Physics • Nuclear Fission and Nuclear Fusion • Peaceful use of Nuclear Energy • Transuranic and other Artificially Produced Elements • Nuclear Forces and Two Body Problem • Elementary Particles • Cosmic Rays • Appendices

Code : 16 173 **Rep. Edn. 2004** **ISBN:81-219-1095-1**

NUCLEAR PHYSICS

S.N. Ghoshal

CONTENTS:• Rutherford Scattering of Alpha Particles and Nuclear Model of the Atom • Nuclear Structure and General Properties of Nuclei • Radioactivity • Alpha Particles and Alpha Radioactivity • Beta Particles and Beta Activity • Gamma Rays • Detection of Nuclear Radiations and their Measurements • Determination of Some Nuclear Properties • Nuclear Models • Nuclear Reactions (I) • Nuclear Reactions (II) • Accelerators of Charged Particles • Neutrons and Neutron Physics • Nuclear Fission and Nuclear Fusion • Peaceful use of Nuclear Energy • Transuranic and other Artificially Produced Elements • Nuclear Forces and Two Body Problem • Elementary Particles • Cosmic Rays • Appendices • Index

Code : 16 187 **2nd Edn. 2004** **ISBN:81-219-0413-7**

QUANTUM MECHANICS

S.P. Singh, M.K. Bagde & Kamal Singh

CONTENTS: Development of Quantum Mechanics • Wave Properties of Matter • Heisenberg's Uncertainty Principle • Schrodinger's Wave Equation • Free States • Bound States • The Hydrogen Atom and the Rigid Rotator • Appendices

Code : 16 129 **2nd Edn. 2004** **ISBN: 81-219-0901-5**

OPTICS AND SPECTROSCOPY

R. Murugeshan

CONTENTS: • Geometrical Optics • Interference-I • Diffraction-I • Polarisation of Light-I Spectroscopy • Matrix Method in Paraxial Optics • Coherence • Fibre Optics • Holography • ElectromagneticWaves • Dispersion and Scatterin • Lasers • Interference-II • Fourier Transforms and Applications (Diffraction-II) • Diffraction of Light-III (Fresnel Class) • Polarisation-II (Matrix Method) • Fermat's Principle • Huygens' Principle and its Applications • Microwave Spectros-copy • Infra-red Spectroscopy • Doppler Effect in Light and its Applications • Cornu's Spiral • Lasers III • Rainbow • Magneto-Optics and Electro-Optics • Nonlinear Optics • Fibre Optics II • Holography-II • Interference III • Diffraction of Light–(IV) • Polarisation of Light–(III) • Twyman and Green Interferometer • Crystal Optics • Cauchy's Dispersion Formula • Optical Activity • Appendix • Examination Paper

Code : 16 206 **5th Rev. Edn. 2005** **ISBN:81-219-1441-8**

MODERN PHYSICS

R. Murugeshan, Kiruthiga Sivaprasath

CONTENTS: ● Part-I : Relativity ● Part-II: Atomic Physics ● Part-III: Quantum Mechanics ●Part-IV: Molecular Physics ● Part-V: Nuclear Physics ● Part-VI: Solid State Physics ● Part-VII: Solid State Devices ● Part-VIII: Electronics ● Part-IX: Statistical Mechanics ● Part-X : Classical Mechanics ● Part-XI: Astrophysics

Code : 16 089 **12th Rev. Edn. 2005** **ISBN:81-219-0210-X**

THERMAL PHYSICS

(For B.Sc., Major Students of Madras University)

R. Murugeshan & Kiruthiga Sivaprasath

CONTENTS: ● Thermometry and Calorimetry ● Thermodynamics ● Low Temperature Physics ● Conduction and Radiation ● Statistical Physics

Code : 16 290 **1st Edn. 2004** **ISBN:81-219-2391-3**

ELECTRICITY AND MAGNETISM

R. Murugeshan

CONTENTS: ● Charges and Fields ● Gauss's law and its Applications Electric Potential Capacitors ● Electrometers ● Current and Resistance ● Electrical Measurements ● Thermo-electricity ● Chemical effect of Electric Current ● Magnetic Effect of Electric Current ● Electromagnetic Induction ● Transient Currents ● Alternating Current ● Three phase A.C., Generators and Motors ● Magnetic Properties of Materials ● Maxwell's Equations and Electromagnetic Waves●Theory of Dielectrics ● Network Analysis and Network Theorems ● A.C. Bridges ●Motion of Charged Particles in Electric and Magnetic Fields ● Electromagnetic Induction II ● Magnetostatics ● Electric Multipoles ● The Electrostatic Field ● Work and Energy in Electrostatics onductors ● Electrostatic Field in Matter ● Maxwell's Equations and Electromagnetic Waves - II ● Magnatostatics - II ● Alternating Current Circuits ● Steady Current ● Thermo-electricity - II Electromagnetic Induction - II ● Chemical Effects of Currents ● Electric Field and Electric Potential ● Circuit Theory ● Dlectromagnetic Waves ● The Magnetic Dipole ● Moving Coil Galvanometer ● Coupled Circuits ● Magnetic Properties of Materials II ●Terrestrial Magnetism ●Vector Analysis ● *Index*

Code : 16 202 **5th Edn. 2004** **ISBN : 81-219-1705-0**

ALLIED PHYSICS

R. Murugeshan

CONTENTS : ● **Paper-I** ● Simple Harmonic Motion and Circular Motion ● Properties of Matters ● Heat, Thermodynamics and Sound ● Electricity and Magnetism ● Geometrical Optics ● Paper-II ● Physical Optics ● Atomic Physics ● Nuclear Physics ● Elements of Relativity and Quantum Mechanics ● Electronics ● Practical Physics ● Youngs modulus by Non-uniform Bending ● Youngs Modulus (E) Non-uniform Bending ● Rigidity Modulus (Static Toursion Method) ● Rigidity Modulus by Tosicenal Oscillations ● Surface Tension and Interfacial surface Tension Drop Weight Method ● Comparision of viscosities of two liquids—Burette Method Specific heat Capacity of a Liquid ● Sonometer— Frequency of A.C. Mains ● Determination of Radius of Curvature ● Air Wedge — Thickness of a wire ● Spectrumeter-Diffraction on Gravity- Wevelength of Hg Lines ● Potentiometer-Voltmeter Colibration ● Post office Box-Measure of resistance and Specific Resistance ● Ballistic Galvanometer Figure of Merit ● Logic Gates AND, OR, NOT ● Zener Diode Characteristics ● Nand Gate as a Universal Gate

Code : 16 294 **1st Edn. 2005** **ISBN:81-219-2444-8**

PROPERTIES OF MATTER AND ACOUSTICS

R. Murugeshan & Kiruthiga Sivaprasath

CONTENTS: ● Properties of Matter ● Unit 1 : Elasticity ● Unit 2: Bending of Beams ● Unit 3 : Fluids ● Acoustics ● Unit 4 : Waves and Oscillations ● Unit 5 : Ultrasonics and Acoustics of Buildings ● Madras University Question Paper

Code : 16 295 **1st Edition : 2005** **ISBN: 81-219-0272-X**

ELECTRICITY AND MAGNETISM

K.K. Tewari

CONTENTS: Units and Dimensions ● Vector Analysis (Algebra) ● Vector Differentiation and Integration ● **Electrostatics:** Electric Field ● Electrostatics-Electric Potential ● Capacitors and Dielectrics ● Electrometers and Electrostatics Machines ● Steady Current ● Magnetostatics ● The Magnetic Field due to Steady Currents ● Electromagnetic Induction ● Practical Applications of Electromagnetic Induction ● Dynamics of Charged Particles ● Magnetic Properties of Matter ● Maxwell's Equations and Electromagnetic Theory Alternating

Currents • Transformers and A.C. Bridges • Circuit Analysis • Appendices • Bibliography • Index

Code : 16 175 3rd Edn. 2004 ISBN:81-219-0667-9

ELECTRICITY AND MAGNETISM WITH ELECTRONICS

K.K. Tewari

CONTENTS : Units and Dimensions • Vector Analysis (Algebra) • Vector Differentiation and Integration • **Electrostatics :** Electric Field • Electrostatics • Electric Potential • Capacitors and Dielectrics • Electrometers and Electrostatics Machines • Steady Current • Magnetostatics • The Magnetic Field due to Steady Currents • Electromagnetic Induction • Practical Applications of Electromagnetic Induction • Dynamics of Charged Particles • Magnetic Properties of Matter • Maxwell's Equations and Electromagnetic Theory • Alternating Currents • Transformers and A.C.Bridges • Circuit Analysis • Electron Emission and Vaccum Tubes • Semi-Conductor Devices • Rectifiers • Amplifiers • Oscillators • Modulators and Detectors • Appendices • Source Books • Index

Code : 16 150 3rd Edn. 2004 ISBN:81-219-0855-8

BASIC ELECTRONICS (SOLID STATE)

B.L. Theraja

Code : 10 051 5th Edn. 2005 ISBN:81-219-0094-8

CONTENTS : Circuit Fundamentals • Resistive Circuits • Kirchhoffs Laws •Network Theorems • Passive Circuit Elements •Energy Sources • Magnetism & Electromagnetism • AC Fundamentals • Series AC Circuits • Time Constant Tuning Circuits and Filters • Solid State Physics • The P-N Junction • The P-N junction Diode • Special Diodes Optoelectronic Devices • DC Power Supplies • The Basic Transistor •Transistor Characteristics & Approximations • Load Lines and DC Bias Circuits • Transistor Equivalent Circuits & Models • Single-Stage Transistor Amplifiers • Multistage Amplifiers • Decibels & Frequency Response • Feedback Amplifiers •Field Effect Transistors • Breakdown Devices • Sinusoidal Oscillators • Nonsinusoidal Oscillators • Modulation & Demodulation • Integrated Circuits • Number Systems •Logic Gates • Boolean Algebra Logic Families • Transducers •Electronic Instruments • Fibre Optics • Index

PRINCIPLES OF ELECTRONICS

V.K. Mehta

CONTENTS: Introduction • Electron Emission • Vacuum Tubes • Vacuum Tube Rectifiers • Vacuum Tube Amplifiers • Gas-filled Tubes • Atomic Structure • Semiconductor Physics • Semiconductor Diode • Special Purpose Diodes •Transistors • Transistor Biasing • Single Stage Transistor Amplifiers • Multi-stage Transistor Amplifiers • Transistor Audio Power Amplifiers • Amplifiers with Negative Feedback • Sinusoidal Oscillators •Transistor Tuned Amplifiers • Modulation and Demodulation • Regulated D.C. Power Supply • Solidstate Switching Circuits • Field Effect Transistors • Silicon Controlled Rectifier • Power Electronics • Electronic Instruments • Integrated Circuits • Hybrid Parameters • Digital Electronics • Index

Code : 10 297 9th Rev. Edn. 2005 ISBN:81-219-2450-2

ELEMENTS OF ELECTRONICS

M.K. Bagde, S.P. Singh & Kamal Singh

CONTENTS: Basic Semiconductor Theory • Semiconductor Diodes • Bipolar Transistors • D.C. Bias and its Stabilization; Thermal Runaway • Hybrid Equivalent Circuits • Field Effect Transistors • Transistor Voltage Amplifiers • Feedback in Amplifiers • Power Amplifiers Transistor Oscillators • Logic Circuits • Cathode Ray Oscilloscope • Network Theorems.

Code : 16 133 5th Edn. 2002 ISBN:81-219-0294-0

MATHEMATICAL PHYSICS

H.K. Dass

Code : 16 224 4th Edn. 2004 ISBN:81-219-1469-8

CONTENTS: Partial Differentiation • Multiple Integrals • Differential Equations • Matrices • Vectors • Complex Number • Functions of a Complex Variable • Special Functions • Partial Differerential Equations • Statistics • Probability • Fourier Series • Laplace Transformation • Integral Transforms • Numerical Techniques • Numerical Method for Solution of Partial Differential Equations • Calculus of Variations • Tensor Analysis • Z-transforms • Infinite Series • Gamma, Beta Functions, Differentiation under the Integral Sign • Index

ATOMIC PHYSICS

J.B. Rajam

CONTENTS: Part I: The Beginnings of Atomic Physics • Discharge of electricity through Gases I The Electron • Positive Rays I X-Rays • Radioactivity • **Part: II** Fundamental Concepts of Atomic Physics • The

Theory of Relativity • The Quantum Theory of Radiation • The Wave Nature of Matter • Quantum Statistics • **Part III:** Physics of the Atom • The Atom Models • Molecular Spectra • Modern Problems of Radioactivity • Artificial Transmutation • Cosmic Rays • Structure and Properties of the Nucleus • Bibliography • Author Index & Subject • Index

Code : 16 241 **7th Edn. 2004** **ISN-81-219-1809-x**

FUNDAMENTALS OF MAGNETISM & ELECTRICITY

D.N. Vasudeva

CONTENTS: • **Part-I** Vector Analysis and its Applications • **Part-II** Electrosticts: Law of Inverse Squares • **Part-III** Magnetism: Magnetic Field • **Part-IV** Current Electricity: Steady Current (Cells) • **Part-V** Electromagnetism : Electromagnetic Induction • **Part-VI** Alternating Currents: A.C. Generators • **Part-VII** Electronics and Radio Communication • **Part-VIII** Atomic Physics: Conduction of Electricity: through Gases

Code : 16 058 **12th Edn. 2004** **ISBN:81-219-0955-4**

B.SC. PRACTICAL PHYSICS

C.L. Arora

CONTENTS: Section-I: General Properties of Matter • Section-II: Heat • Section-III: Sound •Section-IV: Light • Section V: Magnetism • Section VI: Electricity • Section VII: Electronics

Code : 16 070 **20th Edn. 2004** **ISBN: 81-219-0469-2**

QUESTION BANK FOR

B.SC. PART-I PHYSICS

S.P. Singh & P.S. Memne

CONTENTS: • Paper-I • Unit-I : Properties of Matter • Unit-II : Mathematical Background-I • Unit-III : Mathematical Background-II • Unit-IV : Motion of Charged Particles in Electric and Magnetic Fields • Question Paper • Paper-II • Unit-I • Unit-II • Unit-III • Unit-IV

Code : 16 291 **1st Edition : 2004**

BOOKS FOR OTHER COMPETITIVE EXAMINATIONS

OBJECTIVE GENERAL ENGLISH

R.S. Aggarwal
Monika Aggarwal

The book is meant for candidates appearing for various competitive examinations. The subject matter in this book is comprehensive, rigorous, yet very simple. Even an average student will find no difficulty in understanding the various concepts. The book has been divided into six units, namely Vocabulary Usage, Comprehension Ability, Selecting Phrases/Words, Error Detection, Rearrangement, General Usage. This book surely will lead the readers to success.

06 082 ISBN 81-219-1531-7 PP. 664

S. CHAND'S GENERAL STUDIES

C.S. Bedi & R.S. Aggarwal

This second updated edition of the book presents the subject of General Studies in a clear style that meets the most exacting standards of readability. One of the unique features of the book is that it **discusses** problems, issues, movements, etc., in such a way that it provide intellectual exercise to the reader. It is a valuable book for Central Services (Preliminary) Examination and all other High Grade Competitive examinations.

FEATURES: Part I - General Science • Part II — Indian History and Culture • Part III — Indian Polity and Economy • Part IV - World Geography • Part V - Miscellaneous Information • Part VI - Current Affairs.

06 081 ISBN:81-219-1496-5 pp. 1880

S. CHAND'S GENERAL ENGLISH FOR COMPETITIONS

A.N. Kapoor

This book is meant to equip students preparing for various competitive examinations. The aim of the book is to help students to master the correct use of English words and phrases on one hand and to develop in him/her a sense of belongingness for the English language.

06 093 ISBN:-81-219-1762-1 pp. 480

YOUR INTERVIEW

K.L. Kumar

The book will enable the young job seekers to discover his strengths and weaknesses and inspire him to prepare and face the interview board with courage and confidence. Sample interviews, caricatures and quick-review inset sheets have increased the utility of the book. It is a valuable book for the young job seekers.

06 091 ISBN:81-219-1618-6 pp. 144

ADVANCED OBJECTIVE GENERAL KNOWLEDGE

R.S. Aggarwal

The book comprises of subjective type questions based on Science, History, Indian Polity, Economics, Geography and General Awareness. The salient features of the book are that more than 10,000 multiple choice questions give a vast coverage of all the above subjects; references of the questions asked in various competitive examina-tions, given on memory basis, help to know the type of questions asked in a particular examination; and separate Model Set for rarely available Matching Type questions, the questions on Assertion - Reason and those based on diagrams.

06 068 ISBN:81-219-1227-X pp. 512

MANI RAM AGGARWAL'S

GENERAL KNOWLEDGE DIGEST AND GENERAL STUDIES

K. Mohan

This book is for the candidates who are appearing for Union and State Public Service Commissions' Competitive Examinations, L.I.C., G.I.C., A.A.O's, RBI Grade A and B, Bank PO's, CDS/NDA/OTA, Delhi Police S.I., Income Tax, Central Excise Inspectors, SSC's Asstt. Grade, L.D.C., Accounts & Auditors, M.BA, Hotel Management, etc., Entrance Examinations. The book is no doubt an appropriate and relevant guide.

06 001 ISBN:81-219-0798-5 pp. 788

R.S. Aggarwal

A MODERN APPROACH TO VERBAL AND NON-VERBAL REASONING

R.S. Aggarwal

This book contains all types of questions asked in various competitive examinations; fully solved examples with explanatory answers; and a huge collection of practisable questions. It is a unique book of Reasoning which is meant for competitive examinations like Bank Clerical, Bank P.O., LIC., GIC, M.B.A., Assistant Grade, Excise and Income Tax, IAS, IFS, AAO, Railways, Hotel Management and others. Question papers and references given on memory basis will lead the students to success.

06 055 ISBN:81-219-0551-6 pp. 912

ARITHMETIC SUBJECTIVE & OBJECTIVE FOR COMPETITIVE EXAMINATIONS

R.S. Aggarwal

This book is an asset to those who plan to appear in a competitive examination conducted by Banks, LIC, GIC, Excise & Income Tax, Railways, U.P.S.C and other departments for Clerical Grade, Assistant Grade, A.A.O., Inspectors, Probationary Officers, N.D.A. C.D.S., MBA, C.A.T. and other executive posts. The book carries both subjective and objective type of questions.

14 101 ISBN:81-219-0742-X pp. 610

QUANTITATIVE APTITUDE (Fully Solved)

R.S. Aggarwal

This revised edition of the book will be an asset to the candidates appearing for Bank P.O., Insurance A.A.O., Assistant Grade; Excise & Income Tax Inspectors; C.B.I. Sub-Inspectors; Teachers' Exam.; C.D.S; Railways; Hotel Management; M.B.A. and all other competitive examinations. It contains a huge accumulation of objective type questions with their solutions by short cut methods. Latest questions asked in various examinations have been added. It is hoped that the book will help the students to get through their competitive exams in flying colours.

06 008 ISBN:81-219-0632-6 pp. 592

A MODERN APPROACH TO VERBAL REASONING

R.S. Aggarwal

The book is unique for its coverage of all types of questions asked including those in logical deduction and all the study material available around. It contains a huge collection of practisable questions with fully solved examplyes and explanatory answers. This book is meant for competitive examinations like Bank Clerical, Bank P.O., LIC, GIC, M.B.A., Assistant Grade, Excise & Income Tax, IAS, IFS, AAO, Railways and others. It is assured that the book will do its deal in making students the masters in this competitive field.

06 053 ISBN:81-219-0552-4 pp. 496

A MODERN APPROACH TO NON-VERBAL REASONING

R.S. Aggarwal

The requirements of the candidates have been a major factor kept in mind during the compilation of this book and with this end in view all types of solved questions with a huge collection of practisable questions have been provided. This book also consists of well illustrated examples and fully solved exercises with explanatory answers and clues. This book is highly meant for the competitive examinations like Bank Clerical, Bank P.O., SBI P.O., R.B.I., L.I.C., G.I.C., A.A.O., M.B.A., I.A.S., I.F.S., Railways, V.D.C., Sub-inspectors of Police, Asstt. Grade, Income Tax and Central Excise, etc.

06 054 ISBN:81-219-0553-2 pp. 440

MATHEMATICS FOR C.D.S. (Fully Solved)

R.S. Aggarwal

Salient features of this book are: (a) Solutions by short-cut methods (b) Large number of challenging problems, fully solved (c) whole lot of solved examples to clarity the facts (d) Latest questions included on memory basis. This book on mathematics is really an asset to those who plan to appear in competitive examination for C.D.S.

06 066 ISBN:81-219-1019-6 pp. 700

MATHEMATICS FOR M.B.A. (Fully Solved)

R.S. Aggarwal

This book is an efficient Guide for the candidates preparing for the M.B.A. entrance examination. The book contains a huge accumulation of objective type questions fully solved by short-cut methods and it is hoped that the subject matter will lead candidates to success.

06 056 ISBN:81-219-0845-0 pp. 744

MATHEMATICS FOR N.D.A. (Fully Solved)

R.S. Aggarwal

This book provides a comprehensive study of all topics included in N.D.A. syllabus. It contains a large number of challenging problems with solutions; a lot of solved examples to clarify the facts; and latest questions included on memory basis. The subject matter of this book is according to the needs of the students and it is hoped that it will creat a confidence among students to successfully clear N.D.A. examination.

06 065 ISBN:81-219-1018-8 pp. 712

A MODERN APPROACH TO VERBAL AND NON-VERBAL REASONING

R.S. Aggarwal

This book contains all types of questions asked in various competitive examinations; fully solved examples with explanatory answers; and a huge collection of practisable questions. It is a unique book of Reasoning which is meant for competitive examinations like Bank Clerical, Bank P.O., LIC, GIC, M.B.A., Assistant Grade, Excise and Income Tax, IAS, IFS, AAO, Railways, Hotel Management and others. Question papers and references given on memory basis will lead the students to success.

06 055 ISBN:81-219-0551-6 pp. 912

MATHEMATICS FOR M.C.A.

R.S. Aggarwal & Deepak Aggarwal

The salient features of the book are: (i) Strictly according to the syllabi prescribed by various universities for M.C.A. entrance examinations (ii) Separate exercises for objective and descriptive type questions, fully solved (iii) previous year's questions included. The book is expected to prove a sune shot to success in Mathematics for all aspiring candidates for M.C.A. entrance examination.

06 084 ISBN:81-219-1482-5 pp. 320

अंकगणित (वस्तुनिष्ठ प्रश्नों सहित)

आर.एस. अग्रवाल

हिन्दी भाषी राज्यों की सिविल सेवा परीक्षा, रेलवे सेवा, जीवन बीमा निगम, साधारण बीमा निगम, पुलिस, आयकर, आबकारी, बैंकिंग सेवा, फारेस्ट रेंजर्स, क्लर्कस ग्रेड परीक्षा, असिस्टेंट ग्रेड तथा दूसरी प्रतियोगी परीक्षाओं के लिए यह सर्वोत्तम पुस्तक है। सभी परीक्षाओं की आपूर्ति हेतु विषयात्मक तथा वस्तुनिष्ठ प्रश्नों का अपार भंडार इस पुस्तक में निहित है। विभिन्न परीक्षाओं में पूछे गए प्रश्नों के विवरण परीक्षार्थियों से पूछ कर उनकी स्मृति के आधार पर एवं विभिन्न प्रकार की बाज़ार में उपलब्ध मैगजीन आदि से दिए गए हैं।

14 200 ISBN:81-219-0835-3 pp. 608

वस्तुनिष्ठ सामान्य हिन्दी
(Objective General Hindi)

आर.एस. अग्रवाल, मोनिका अग्रवाल

यह पुस्तक उ.प्र. सी.पी.एम.टी., बी.एड प्रवेश परीक्षा, यू.जी.सी. परीक्षा, असिस्टेंट ग्रेड स्टेनोग्राफर, लेखा-परीक्षक, हिन्दी अनुवादक, पुलिस सब-इन्सपैक्टर, डिप्टी जेलर, सी.बी.आई., बैंक पी.ओ., ग्रामीण बैंक, जीवन बीमा निगम, पी.सी. एस., रेलवे भर्ती बोर्ड परीक्षा, तथा अन्य प्रतियोगी परीक्षाओं के लिए अति-उपयोगी है। इस पुस्तक में विभिन्न प्रतियोगी परीक्षाओं में पूछे गए प्रश्नों के आधार पर पूर्ण सामाग्री समाहित की गई है। आशा है कि इस पुस्तक के अध्ययन से सभी प्रतियोगी परीक्षाओं में सामान्य हिन्दी में सफलता प्राप्त करने में प्रतियोगीयों को कोई कठिनाई नहीं होगी।

06 095 1st Edn. 1998 ISBN:81-219-1753-0 pp. 392

संख्यात्मक अभिरुची
(Quantitative Aptitude)

आर.एस. अग्रवाल

यह पुस्तक सिविल सेवा परीक्षा, रेलवे निगम, पुलिस, आयकर, वन्य सेवा, बैकिंग सेवा, क्लर्कस ग्रेड परीक्षा, असिस्टैंट ग्रेड तथा दूसरी प्रतियोगी परीक्षाओं के लिए यह पुस्तक अति-उपयोगी है। सभी परीक्षाओं की आपूर्ति हेतु वास्तुनिष्ठ प्रश्नों का अपार भंडार इस पुस्तक में निहित है। विभिन्न परीक्षाओं में पूछे गए प्रश्नों के विवरण परीक्षार्थियों से पूछ कर उनकी स्मृति के आधार पर एवं विभिन्न प्रकार की बाजार में उपलब्ध मैगजीन आदि से दिए गए हैं।

06 070 1st Edn. Rep. 1998 ISBN:81-219-1228-8 pp. 488

वस्तुनिष्ट अंकगणित
(पूर्ण हल सहित)

आर.एस. अग्रवाल

प्रस्तुत पुस्तक प्रतियोगी परीक्षाओं के लिए अंकगणित पर आधारित वस्तुनिष्ट प्रश्नों से विहित है। सभी प्रश्नों को संक्षिप्त विधि से हल किया गया है। विभिन्न परीक्षाओं में पूछे गए प्रश्नों का विवरण परीक्षार्थियों से पूछ कर उनकी स्मृति के आधार पर दिया गया है। बाजार में उपलब्ध विभिन्न मैगजीनों के प्रश्न भी दिए गए हैं

06 033 4th Edn. Rep. 1998 ISBN:81-219-0426-9 pp. 528

world. A few Indian scholars have also contributed a few topics on this much debated issue e.g. Ecologically Sustainable Institutions by Aseem Prakash & Anil K. Gupta and The Valuation and Pricing of Non-Timber Forest Products: Conceptual Iseus and a case study from India by Kauchan Chopra.

MERRIAM WEBSTER'S GEOGRAPHICAL DICTIONARY

The Merrian Webster's Geographical Dictionary is a reliable, easy to use reference. It gives vital information about key political and physical features around the globe. More than 48,000 entries provide census figures, size, Laudform deseriptions, political divisions, historced notes and more. Form the nations of the former union and Eastern Europe to the microstates of the Pacific, entries cover major countries, cities and physical features around the world. 252 clear, easy-to-rad maps prepared by the cartography department of Encyclopedia Britannica. Dozens of tables cover non-English geographical terms, U.S. and canadian national parks, administrative subdivisions of the government, and more. Entries include helpful pronumications and frastilerations of foreign characters including both Piryin and Wade-Giles transtiterations of chinese placenames. the book on the whole is very carefully researched and each and every information is accurate.

OXFORD CHILDREN'S ENCYCLOPEDIA

The Oxford Children's Encyclopedia is established as the essential reference work for children. Its wide-ranging selection of articles reflects the interests and needs of today's children, making it a must for enery home, classroom, and library shelf. Now completely revised and updated the nine volume set offers a solid foundation of knowledge in science, technology history geography, and the arts. It has 2083 articles, 130 Authors and 70 specialist consultants.

9 Volumes.

THE ENCYCLOPEDIA OF CAREERS & VOCATIONAL GUIDENCE

this book is the most comprehensive career publication available. The Encyclopedia of Careers and Vocational Guidance contains many major job articles. the four volume set is a must for career centres, guidance offices as well as high schools, public and college libraries. More than 500 articles detailing approximately 1,500 accupations, Volume I desirables 69 industries in detail, more than 100 new photographs, Indexes of federal government jobs classifiecation systems — DOT Index, GOE index, and the Canadian NOC index, Expanded Indexes of Resources and Association for the discabled are some of the features. More than 1,400 professional associations are listed for users to contact for more occupational information.

4 volumes

THE HANDBOOK FO REFERENCE METHODS FOR PLANT ANALYSIS

Edited by Yash P. Kalra

The book is an outstanding resource of plant analysis producers, outlined in easy to follow steps and laboratory-readyfor implementation. Plant laboratory preparation methods such as dry ashing and acid and microwave digestion are discussed in detail. Extraction techniques for analysis of readily soluble elements (petiole analysis) and quick test kits for field testing are also presented. this handbook consolidates proven, time-tested methods in one convenient source. Plant scientists in production agriculture, forestry, horticulture environmental sciences and other related disciplines will find the Handbook a standard laboratyr reference.